Industrial Combustion Pollution and Control

Environmental Science and Pollution Control Series

Additional Volumes in Preparation

Industrial Combustion Pollution and Control

Charles E. Baukal, Jr.

John Zink Company, LLC
Tulsa, Oklahoma, U.S.A.

CRC Press
Taylor & Francis Group
Boca Raton London New York

CRC Press is an imprint of the
Taylor & Francis Group, an **informa** business

Library of Congress Cataloging-in-Publication Data
A catalog record for this book is available from the Library of Congress.

First published 2004 by Marcel Dekker, Inc.

Published 2020 by CRC Press
Taylor & Francis Group
6000 Broken Sound Parkway NW, Suite 300
Boca Raton, FL 33487-2742

First issued in paperback 2020

© 2004 by Taylor & Francis Group, LLC
CRC Press is an imprint of Taylor & Francis Group, an Informa business

No claim to original U.S. Government works

ISBN 13: 978-0-367-57845-9 (pbk)
ISBN 13: 978-0-8247-4694-0 (hbk)

Visit the Taylor & Francis Web site at
http://www.taylorandfrancis.com

and the CRC Press Web site at
http://www.crcpress.com

Preface

This book, written primarily for the practicing engineer, is intended to fill a gap in the literature on pollution in industrial combustion. Many textbooks have been written on combustion, but these have included limited information about pollution from industrial applications. One of the purposes of this book is to synthesize, in a single, coherent reference source, a wealth of information from many relevant books, articles, and reports.

A distinguishing feature of this book is that examines each topic in a somewhat specialized context to show how it affects pollution from industrial combustion processes. In Chapter 2, for example, the basics of combustion are considered, but from the focused perspective of combustion's influence on pollution emissions. There is very little discussion of combustion kinetics, because this complicated subject is of more concern to researchers in the field than to practicing engineers. This book does not directly address subjects that have been more than adequately covered elsewhere, but, rather, examines those subjects in the context of how they influence the pollution generated by the combustion system.

This book is basically organized in three parts. The first part deals with the basics of combustion and pollution emissions common in industrial applications. It includes general chapters on combustion, systems, computer modeling, and experimental techniques. The middle part introduces general concepts of pollution from industrial combustion systems and includes chapters on NO_x, unburned hydrocarbons, SO_x, particulates, noise and vibration, and other pollutants such as CO_2, dioxins and furans, and thermal radiation. The last part describes pollution from specific applications of industrial combustion and includes chapters on production of metals (both ferrous and nonferrous), minerals (glass, cement, and bricks), chemicals (including petrochemical applications), waste incineration, and other industries (e.g., paper, printing, and foods).

This book basically concerns atmospheric combustion, which is the predominant type used in industry. Because the vast majority of industrial applications use gaseous fuels, that is the primary focus, with only a limited discussion of solid and liquid fuels. As with any book of this type, there are many topics that are not covered. The book does not address other aspects of pollutions in combustion, such as power generation (stationary turbines or boilers) and propulsion (internal combustion, gas turbine, or rocket engines). Nor does it treat, for example, packed bed combustion and material synthesis in flames. Other topics, such as pollution from paper and ink drying, are discussed only briefly. One reason for these limited discussions is that relatively little has been published on pollution from these applications.

The book focuses on topics relevant to the practicing engineer. It is not exhaustively comprehensive, but it does provide references for the interested reader seeking more information on a particular subject. In determining overall contents the guideline has been to minimize coverage of theory and maximize that of applications, while at least touching on the topics relevant to pollution in industrial combustion.

This is a rapidly changing field, especially as pollution regulations become increasingly stringent and new technology is developed to meet those requirements. The low pollutant emissions achieved today in many applications would, only a few years ago, have been thought impossible. Although rapid technological advances make it difficult to keep a book of this type current, the last chapter offers some examples of new developments that may become standards in the very near future. The dynamic nature of the field makes it exciting for researchers and technologists, but somewhat frustrating for those responsible for keeping up with the regulations and the new technologies. It will be interesting to see how regulations and technologies continue to change and how industry around the world adapts to those changes.

ACKNOWLEDGMENTS

This book is dedicated to my wife, Beth, to my children, Christine, Caitlyn, and Courtney, and to my mother, Elaine. It is also dedicated to the memories of my father Charles, Sr. and my brother Jim. I am thankful to Gordon Blizzard, formerly of Selas Corp (Dresher, PA), Tom Smith of Marsden, Inc. (Pennsauken, NJ), Buddy Eleazer of Air Products (Allentown, PA), and Dr. Roberto Ruiz of John Zink Company, LLC for the opportunity to learn firsthand about combustion and pollution in industrial applications. Many other colleagues, too numerous to mention, have also contributed to this book in a variety of ways. I also thank David Koch and Steve Pirnat of John Zink Company, LLC, for their support of the writing of this book.

Charles E. Baukal, Jr.

Contents

1

Introduction

The fields of combustion and pollution are very broad and touch directly or indirectly nearly all aspects of our lives. The electronic devices we use are generally powered by fossil-fuel-fired power plants. The cars we drive use internal combustion engines. The aircraft we fly in use jet-fuel-powered turbine engines. Most of the materials we use have been made through some type of heating or melting process. While this book is concerned specifically with industrial combustion processes, all of the above combustion processes share many features in common.

Industrial combustion pollution is complicated by many factors. First, the science of both combustion and pollution is still developing and has a long way to go until we have an adequate understanding of both so they can be better applied and controlled. While fire has been with us since the beginning of time, much remains to be learned about it. It combines heat transfer, thermodynamics, chemical kinetics, and multiphase turbulent fluid flow to name a few areas of physics. The study of pollution includes many of those same areas with the addition of others like environmental engineering. Therefore, the study of industrial combustion pollution is interdisciplinary by necessity.

1.1 INDUSTRIAL COMBUSTION

Combustion has been the foundation of worldwide industrial development for the past 200 years [1]. Industry relies heavily on the combustion process as shown in Table 1.1. The major uses for combustion in industry are shown in Table 1.2. Hewitt et al. [2] have listed some of the common heating applications used in industry, as shown in Table 1.3. Typical industrial combustion applications can also be characterized by their temperature ranges, as shown in Fig. 1.1. As can be seen in Fig. 1.2, the demand for energy is expected to continue to increase rapidly. Most of the energy (88%) is produced by the combustion of fossil fuels like oil, natural gas, and coal. According to the U.S. Department of Energy, the demand in the industrial sector is projected to increase by 0.8% per year to the year 2020 [3].

As shown in Fig. 1.3, three elements are required to sustain combustion processes: fuel, oxidizer, and an ignition source usually in the form of heat. Industrial combustion is defined here as the rapid oxidation of hydrocarbon fuels to generate large quantities of energy for use in industrial heating and

Table 1.1   The Importance of Combustion to Industry

	%Total energy from (at point of use)		
Industry	Steam	Heat	Combustion
Petroleum refining	29.6	62.6	92.2
Forest products	84.4	6.0	90.4
Steel	22.6	67.0	89.6
Chemicals	49.9	32.7	82.6
Glass	4.8	75.2	80.0
Metal casting	2.4	67.2	69.6
Aluminum	1.3	17.6	18.9

Source: U.S. Dept. of Energy, Energy Information Administration as quoted in the "Industrial Combustion Vision" prepared by the U.S. Dept. of Energy, May 1998.

Table 1.2   Major Process Heating Operations

Metal melting
- Steel making
- Iron and steel melting
- Nonferrous melting

Metal heating
- Steel soaking, reheat, ladle preheating
- Forging
- Nonferrous heating

Metal heat treating
- Annealing
- Stress relief
- Tempering
- Solution heat treating
- Aging
- Precipitation hardening

Curing and forming
- Glass annealing, tempering, forming
- Plastic fabrication
- Gypsum production

Fluid heating
- Oil and natural gas production
- Chemical/petroleum feedstock preheating
- Distillation, visbreaking, hydrotreating, hydrocracking, delayed coking

Bonding
- Sintering, brazing

Drying
- Surface film drying
- Rubber, plastic, wood, glass products drying
- Coal drying
- Food processing
- Animal food processing

Calcining
- Cement, lime, soda ash
- Alumina, gypsum

Clay firing
- Structural products
- Refractories

Agglomeration
- Iron, copper, lead

Nonmetallic materials melting
- Glass

Other heating
- Ore heating
- Textile manufacturing
- Food production
- Aluminum anode baking

Source: Industrial Combustion Vision, U.S. Dept. of Energy, May 1998.

Table 1.3 Examples of Processes in Process Industries Requiring Industrial Combustion

Process industry	Examples of processes using heat
Steel making	Smelting of ores, melting, annealing
Chemicals	Chemical reactions, pyrolysis, drying
Nonmetallic minerals (bricks, glass, cement and other refractories)	Firing, kilning, drying, calcining, melting, forming
Metal manufacture (iron and steel, nonferrous metals)	Blast furnaces and cupolas, soaking and heat treatment, melting, sintering, annealing
Paper and printing	Drying

Adapted from Ref. (2).

melting processes. Industrial fuels may be solids (e.g., coal), liquids (e.g., oil), or gases (e.g., natural gas). The fuels are commonly oxidized by atmospheric air (which is approximately 21% O_2 by volume) although it is possible in certain applications to have an oxidizer (sometime referred to as an "oxidant" or "comburent") containing less than 21% O_2 (e.g., turbine exhaust gas [4]) or more than 21% O_2 (e.g., oxy/fuel combustion [5]). The fuel and oxidizer are typically mixed in a device referred to as a burner, which is discussed in more detail below. An industrial heating process may have one or many burners depending on the specific application and heating requirements.

1.2 POLLUTION EFFECTS

The major concern for many years was improving the energy efficiency of industrial combustion processes. The concern has changed in recent years away from energy efficiency toward other issues including pollution emissions [1]. Environmental quality and greenhouse gases are seen as some of the key drivers shaping the future of industrial combustion. Interestingly, the burden of future improvements in industrial combustion processes is seen to fall primarily on academic researchers and equipment suppliers, rather than on end users. Golden [6] notes that the troposphere has historically been the "garbage dump" for pollution emissions.

Table 1.4 shows a list of typical air pollutants, their properties, and their significance [7]. Figure 1.4 shows a diagram of atmospheric pollution. As can be seen, in some cases there are multiple pollutants combining together and in some cases pollutants may participate in more than one process. Most air pollutants originate from combustion processes. This includes mobile sources such as automobiles and trucks and stationary sources such as power plants and industrial combustion processes. This book is not specifically concerned with what happens to pollutants once they are emitted into the atmosphere, but rather with how they are generated in industrial combustion processes. Therefore, this book does not consider topics like measuring ambient pollutant concentrations or plume dispersion modeling. However, the general effects of pollutants are discussed next to motivate the reader for the need to reduce or eliminate deleterious emissions. Table 1.5 shows the typical composition of unpolluted air.

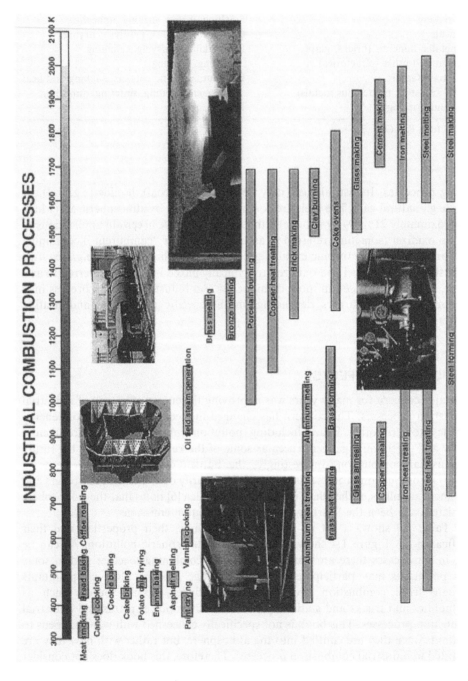

Figure 1.1   Temperature ranges of common industrial combustion applications. (Courtesy of Werner Dahm, University of Michigan, 1998.)

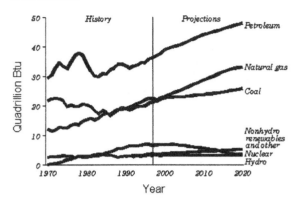

Figure 1.2   Historical and projected world energy consumption. (From Ref. 3.)

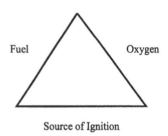

Figure 1.3   Combustion triangle.

Atmospheric ambient air quality trends show that regulatory efforts to control pollutants are working [8]. There have been dramatic reductions in five of the six criteria pollutants. From 1970 to 2000, CO, VOCs (volatile organic compounds) (indirect indicators of ozone, which is the actual criteria pollutant), SO_2, particulates, and lead emissions have declined by 25, 43, 44, 88, and 98%, respectively. NO_x was the only criteria pollutant to increase (by 20%) during that same period.

1.2.1   Global Warming

Global warming refers to what is believed to be a trend where the earth's temperature is increasing. This is believed to be primarily due to the increased emission of man-made pollutants that cause more of the sun's energy to be trapped in the atmosphere. These pollutants are referred to as so-called greenhouse gases. Most of these pollutants also occur naturally in the environment in some form or another and while man-made emissions are often relatively small compared to those naturally occurring, they may be affecting the natural balance in the environment. While it is estimated that it would take many years for there to be catastrophic consequences to the increase in greenhouse gas emissions, it could be too late to reverse the trend if something is not done now. Therefore, there has been a great deal of regulatory effort around the world to control the emission of a wide range of

Table 1.4 Gaseous Air Pollutants

Name	Formula	Properties of importance	Significance as air pollutants
Sulfur dioxide	SO_2	Colorless gas, intense acrid odor, forms H_2SO_3 in water	Damage to vegetation, building materials, respiratory system
Sulfur trioxide	SO_3	Soluble in water to form H_2SO_4	Highly corrosive
Hydrogen sulfide	H_2S	Rotten egg odor at low concentrations, odorless at high concentrations	Extremely toxic
Nitrous oxide	N_2O	Colorless; used as aerosol carrier gas	Relatively inert; not a combustion product
Nitric oxide	NO	Colorless; sometimes used as anesthetic	Produced during combustion and high-temperature oxidation; oxidizes in air to NO_2
Nitrogen dioxide	NO_2	Brown or orange gas	Component of photochemical smog formation; toxic at high concentration
Carbon monoxide	CO	Colorless and odorless	Product of incomplete combustion; toxic at high concentration
Carbon dioxide	CO_2	Colorless and odorless	Product of complete combustion of organic compounds; implicated in global climate change
Ozone	O_3	Very reactive	Damage to vegetation and materials; produced in photochemical smog
Hydrocarbons	C_xH_y	Many different compounds	Emitted from automobile crankcase and exhaust
Hydrogen fluoride	HF	Colorless, acrid, very reactive	Product of aluminium smelting; causes reactive flurosis in cattle; toxic

Source: Ref. 7.

pollutants. Common greenhouse gases of interest to industrial combustion processes include carbon dioxide, water vapor (H_2O), and nitrous oxide (N_2O) [9].

The issue of global warming is broad and diverse and covers a variety of topics including environmental, economics, political, regulatory/legal, technology, and a host of scientific areas. Geyer [10], a geophysicist, has edited a book that attempts to look at global warming from scientific, economic, and legal viewpoints. The U.S. Environmental Protection Agency (EPA) has a number of good resources concerning global warning [e.g., 11,12].

Global warming is caused by so-called greenhouse gases trapping the sun's energy within the Earth's atmosphere (see Fig. 1.5). This is caused when the sun's short wavelength ultraviolet rays penetrate through the earth's atmosphere and are absorbed by the earth. The energy is then reradiated from the earth back into space

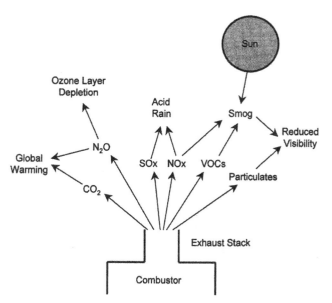

Figure 1.4 Diagram of atmospheric pollution.

Table 1.5 Gaseous Composition of Unpolluted Air (Dry and Wet Basis)

Gas	Ppmvd	ppmvw
Nitrogen	780,900	756,500
Oxygen	209,400	202,900
Water	—	31,200
Argon	9300	9000
Carbon dioxide	315	305
Neon	18	17.4
Helium	5.2	5.0
Methane	1.0–1.2	0.97–1.16
Krypton	1.0	0.97
Nitrous oxide	0.5	0.49
Hydrogen	0.5	0.49
Xenon	0.08	0.08
Organic vapors	~0.02	~0.02

but at much longer infrared wavelengths (see Fig. 1.6). Ideally, this process of the earth absorbing the sun's energy and reradiating back into space is an equilibrium process. If more energy is absorbed than reradiated, then the earth and the surrounding atmosphere will begin to heat up. If the earth were to reradiate more than it absorbed, then the earth and the surrounding atmosphere would begin to cool down.

Global warming occurs because the gases in the earth's atmosphere let proportionately more of the short wavelength ultraviolet rays from the sun pass through it than they let the longer wavelength infrared energy pass through them.

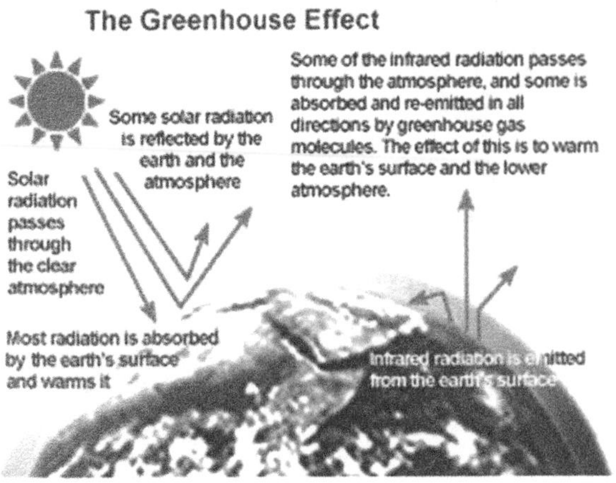

The Greenhouse Effect

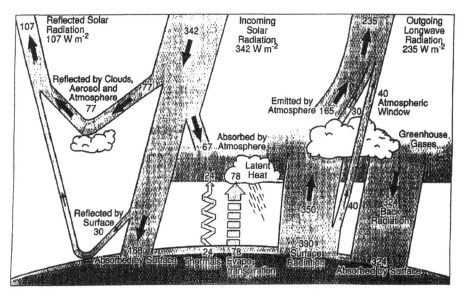

Figure 1.5 The greenhouse effect. (From Ref. 11.)

Figure 1.6 The greenhouse effect. (From Ref. 6.)

Some of the gases in the earth's atmosphere preferentially absorb the earth's long wavelength radiation, trapping the energy in the atmosphere and causing the earth and its global systems to increase in temperature. One way to illustrate global warming is what happens on a sunny day inside a greenhouse with walls of uninsulated windows. A very high proportion of the sun's ultraviolet rays pass through the windows of the greenhouse and are absorbed by the materials inside the greenhouse like the floor and the plants. Those materials inside the greenhouse then reradiate longer wavelength infrared energy in all directions. However, window glass transmits significantly less infrared energy compared to ultraviolet energy. The windows then absorb the infrared energy emitted from the materials inside the greenhouse and reradiate most of it back into the greenhouse. Thus, most of the sun's energy hitting the windows becomes trapped inside the greenhouse and the inside of the greenhouse heats up significantly. Even when it is very cold in the winter, a greenhouse with a direct view of the sun can heat up substantially inside due to this so-called "greenhouse effect."

There are certain gases that exhibit this selective transmissivity similar to window glass. Carbon dioxide (CO_2) is a gas in significant quantities in the earth's atmosphere and is a greenhouse gas. There is particular concern about CO_2 because it is produced in the burning of any fossil fuel. For example, the perfect combustion of coal used in power plants with air generates predominantly CO_2 and N_2, along with some O_2 and trace species, depending on the coal composition and the specific operation of the plant. The simplified process can be written as

$$C + 2(O_2 + 3.76N_2) \leftrightarrow CO_2 + 7.52N_2 \qquad (1.1)$$

where the constituents in parentheses are the main constituents of air which is approximately 21% O_2 and 79% N_2 by volume. As can be seen from the above chemical balance, the nitrogen goes through the process essentially unchanged (except for trace amounts of NO_x that may be produced, as discussed in Chap. 6). However, the carbon in the form of coal is converted nearly completely into CO_2. Plants and trees absorb and convert CO_2 back into carbon and oxygen (O_2). The carbon is retained by the plant while the O_2 is emitted back into the atmosphere. This process is sometimes referred to as forest sequestration. Under ideal conditions then, the CO_2 produced by burning fossil fuels will all be converted back into carbon and oxygen. The problem occurs because of the massive deforestation that continues to occur, particularly in third-world countries that are cutting down trees more rapidly than they can be replaced by new seedlings. The combination of increasing fossil-fuel consumption and the rapid depletion of plants and trees means that not all of the CO_2 is being converted back into C and O_2, resulting in a buildup of CO_2 in the atmosphere. Scientists generally agree that doubling the CO_2 concentration in the atmosphere increases the earth's surface temperature by 1.5°–4.5°C (3°–8°F). This has led many environmentalists to call for massive reductions in fossil-fuel consumption to stem this CO_2 buildup. The most common way this is being done is through what is referred to as a carbon tax, which is hoped to discourage fossil-fuel burning and encourage use of renewable resources like solar and wind energy. Nitrous oxide (N_2O) and methane (CH_4) are other important greenhouse gases, although they have a much lesser impact on global warming compared to CO_2.

The increase in temperature is very miniscule because the concentration of global warming gases in the atmosphere is small. Figure 1.7 shows the global temperature rise from 1880 to 2000, which shows that surface temperatures have risen 0.5°–1.0°F (0.3°–0.6°C) since the late 19th century. It is argued that if the concentration of these gases continues to increase, then global warming will accelerate. A primary concern is that this trend will not be reversible if it is not controlled now because it will be too late to do anything about the problem at some point in the foreseeable future. However, there are some conflicting satellite temperature data that do not indicate a warming trend although measurement

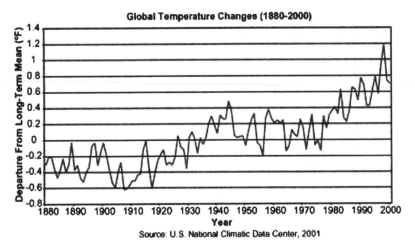

Figure 1.7   Global temperature changes from 1880–2000. (From Ref. 11.)

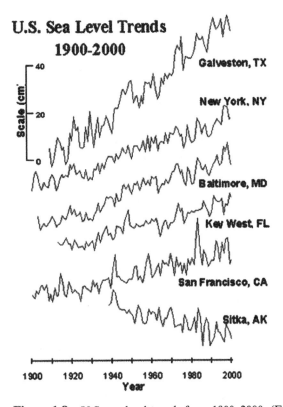

Figure 1.8   U.S. sea level trends from 1900–2000. (From Ref. 11.)

discrepancies and other factors may explain the discrepancy. One of the results of the global temperature rise has been large-scale melting of ice both in the oceans and in the mountains. As shown in Fig. 1.8, this has caused the sea level to rise 6–8 in. (15–20 cm) in the last century. If global warming continues, sea levels are expected to continue to rise as the polar ice caps continue to melt. A related trend indicating global warming is the apparent increase in global precipitation levels by about 1% over the past century.

There are other potential consequences of global warming including damaged crops, depleted water resources, increases in tropical diseases such as malaria, increased rain in some areas with drought in others, and flooding in coastal areas to name a few. In some cases, global warming may actually benefit certain geological areas where more rain and higher temperatures could increase agricultural production. However, most are agreed that the overall consequences of unabated global warming would be detrimental to most areas of the world.

However, the subject of global warming has become very controversial over the past few years as the data are far from conclusive about the causes and effects or that the environment is even in fact warming [e.g., 13,14]. The White House requested assistance from the National Academy of Science to study the question of climate change [15]. A group of eminent scientists noted that while there is a definite rise in global temperatures, which has likely been caused by human activities, they could not rule out that some of this rise may be due to natural variability. The report also noted that while some pollutants causing global warming are on the rise, others are actually decreasing for reasons that are not well understood. Further research was recommended in many areas to improve our understanding of climate changes.

This controversy over global warming is contrasted with the relatively recent ban on halogens and chlorofluorocarbons (CFCs) which are chemicals that have been universally accepted as damaging to the ozone layer. Part of the skepticism related to global warming is caused by the relatively recent cry by scientists in the 1970s about global cooling [16]. It is difficult to assess the veracity of the climatic data from many years ago as there is undoubtedly a large margin of error. An economist argues that global warming may actually be beneficial as cultures have tended to prosper during warmer times [17]. This same economist warns of economic disaster if all of the effort proposed by some is applied to reducing global warming. Another economist argues that some drastic actions need to be taken now to prevent global warming from spiraling out of control to the point that it will be too late to reverse [18]. He argues that a three-century horizon must be considered to make appropriate policy decisions today. Based on the estimated available fossil-fuel supply, it is argued that there is the long-term potential for much more than the 2°–3°C (4°–5°F) rise in temperature often used, up to a rise as much as 10°–18°C (18°–32°F). This economist tries to prove that a cost-benefit analysis argues for aggressive action to limit global carbon emissions. A Princeton University geoscientist points out the many problems with the scientific data that are used to argue for global warming [19]. It is argued that the earth is far more robust than most assume and that it is very unlikely we could do great harm to our planet. Williams [20] looks at the issue of global warming from a sociological perspective and considers how the public deals with the issue and proposed solutions. A proposed schematic for greenhouse gas sources and sinks is shown in Fig. 1.9.

The European Union (EU) agreed to a common greenhouse gas emission reduction of 8% (compared to 1990 levels) by the years 2008–2012 as part of the Kyoto protocol. This will translate into CO_2 reductions for the member countries. Tradable emissions permits and "green certificates" (credits for the use of renewable resources instead of fossil-fuel energy) are being considered as means to achieve these

Exhibit ES-2 Greenhouse Gas Sources and Sinks Associated with the Material Life Cycle

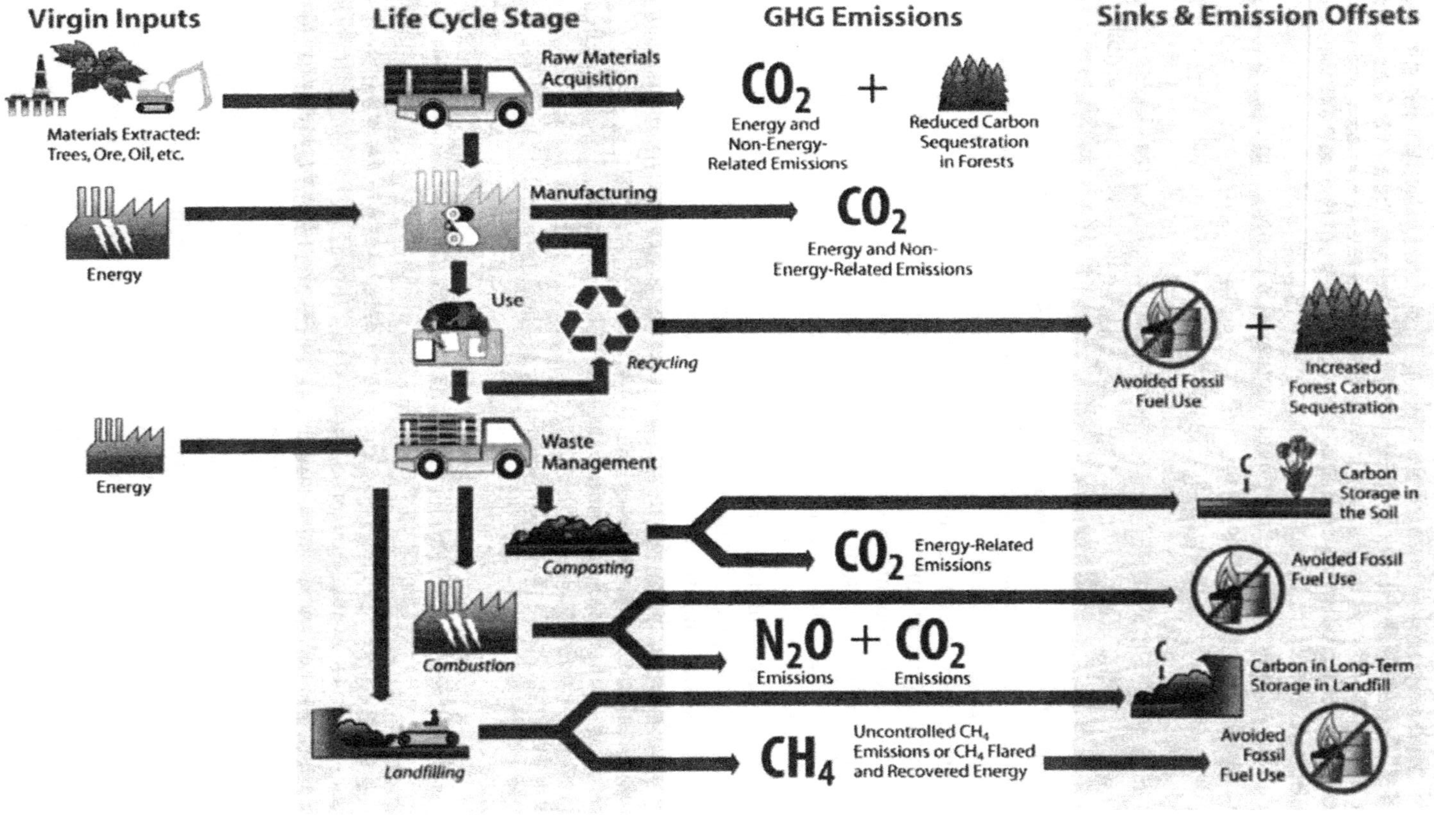

Figure 1.9 Greenhouse gas sources and sinks. (From Ref. 9.)

reductions [21]. The combination of permits and certificates is needed both to reduce emissions and encourage the use of renewable energy resources. However, the specific combination will significantly impact the price for energy to be paid by consumers [22]. Some countries have already begun to develop a program for tradable green certificates, however, there are concerns about the differences in the programs that need to be harmonized for an effective system for the entire EU [23]. The EU Commission would like to increase the use of renewable energy supplies to account for 12% of the energy consumption of the EU by 2010 [24]. Further information on European Union legislation, regulations, directives, and decisions can be found at the EU homepage [25].

It is not the purpose of this book to take a position relative to the validity of global warming. Rather, a pragmatic view is assumed where the existing regulations for pollution emissions are likely only to get more stringent with time. Therefore, until proven otherwise it is irrelevant to the industrial user what the long-term effects of certain pollutants like CO_2 have on the environment as it appears likely that these pollutants will be regulated, if not already, in the near future. Industrial combustion users will have to deal with these regulations whether they believe the affects on the environment are significant or not.

1.2.2 Acid Rain

Acid rain is sometimes referred to as acidic deposition. It is formed when air pollutants like NO_x and SO_x are transported over long distances and are carried to the earth after contacting some type of moisture in the atmosphere (e.g., rain) and form an acid like nitric acid or sulfuric acid (see Fig. 1.10). Note that although a natural background level of acid rain (e.g., from volcanic eruptions) has existed for ages, there has been a dramatic increase in acid rain levels in the last two centuries due to the proliferation of fossil-fuel combustion that accompanied the industrial

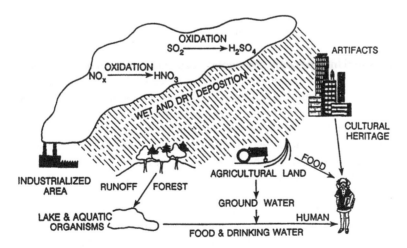

Figure 1.10 Acid rain deposition schematic. (From Ref. 26.)

revolution. The primary components causing the rain to be acidic are sulfuric acid (H_2SO_4) and hydrogen peroxide (H_2O_2), with nitric acid (HNO_3) also a factor.

Normal rain is slightly acidic (pH < 7) because of the presence of carbon dioxide in the atmosphere. In the absence of pollution, the typical pH of rain would be approximately 5.6. While not universally accepted, acid rain is sometimes defined as rain with a pH of less than 5.6. Acid rain was initially linked with burning coal in power plants but has since been expanded to include all types of fossil-fuel burning processes, which encompasses all of those considered here. The topic of acid rain not only pertains to science, but also to politics, which has often served only to cloud the real issues.

Forster [26] has written a fairly comprehensive book on the subject of acid rain that considers the sources, the impacts it has on a variety of aspects of the environment, control options, and the scientific validity of the arguments. Some of the impacts of acid rain include crop damage, soil degradation, forest decline, lake acidification affecting aquatic ecosystems, reduction in fish spawning resulting in the decline of fish populations, heavy metal accumulation, erosion and corrosion of structures, damage to buildings, and reduction in visibility. Acid rain is not known to be mutagenic or carcinogenic in humans directly, but is indirectly harmful because of the damage done to ecosystems and the food chain. Control strategies include increasing the use of renewable energy resources (e.g., solar, wind, hydro), using cleaner fuels (e.g., low sulfur coals), increased use of post-treatment equipment to remove NO_x and SO_x from exhaust stacks, and the use of new technologies that generate fewer pollutants.

Park [27] offers two general strategies for dealing with acid rain: prevention and cure. The latter consists, e.g., of neutralizing waters and soils that have been acidified. The former is one of the subjects of interest in this book, particularly for acid rain produced by industrial combustion exhaust products.

1.2.3 Ozone

Ozone (O_3) is an odorless, colorless gas that can be "good" or "bad" depending on where it is located. Ozone is present at all altitudes up to at least 100 km [28]. However most of the ozone resides in the upper atmosphere (about 10–30 miles above the earth's surface), in the stratosphere. Ozone there protects us from the harmful UV radiation from the sun. Unfortunately, the "good" ozone in the upper atmosphere is gradually being destroyed by man-made chemicals such as CFCs, which have been banned from use in the U.S., and by halons and other ozone-depleting substances. These substances break down in the presence of UV sunlight, releasing chlorine and bromine that react and scavenge "good" ozone from the atmosphere. These reactions are thinning the ozone layer in the upper atmosphere that protects us from the sun's harmful rays, which then increases health risks such as skin cancer. An area of ozone that has been significantly depleted is over the north and south poles. These are referred to as holes in the ozone layer.

Ozone in the troposphere (the atmospheric layer closest to the earth) where we live is "bad". Man-made chemicals are increasing the levels of ozone there. Nitrogen oxides and volatile organic compounds react in the presence of sunlight to form ground-level ozone. This is particularly a problem in the summer months when the sun's radiation is more intense. There are many potentially harmful effects of high levels of ground-level ozone, primarily related to respiratory disorders. For example, ozone can aggravate asthma, reduce lung function, and inflame and damage the cells lining the lungs. Ozone reacts chemically with internal body tissues, causing damage. Studies indicate that ozone may reduce the immune system's ability to fight off bacterial infections in the respiratory system.

Scientists have found that one out of every three people are at increased risk of experiencing ozone-related health effects. Children and the elderly are at increased risk. Athletes, workers who vigorously work outdoors, and people with pre-existing respiratory problems are particularly at risk. There are also some otherwise healthy individuals who are particularly susceptible to ozone health effects. As shown in Table 1.6, an air quality index (AQI) has been developed as a guide to ground-level ozone concentrations along with some cautionary guidelines for each index range [29]. Figure 1.11 shows some of the complex interactions between ozone depletion in the stratosphere and global warming in the troposphere [30]. It is estimated that between one-third and one-half of all Americans live in areas that exceed U.S. ozone standards at least once per year [31]. It is recommended that physical exertion outdoors be limited when the AQI is high and that sunscreen be used to prevent the detrimental effects of overexposure to the sun's UV rays. Ozone also damages crops and other plant life.

1.2.4 Visibility

Smog can be generated by air pollution and can significantly reduce visibility. The principal component of smog is ground-level ozone, which is produced by a combination of pollutants in the presence of sunlight. Smog formation is

Table 1.6 Air Quality Index

Air quality index	Air quality	Cautionary statement for ozone
0–50	Good	No health impacts are expected when air quality is in this range
51–100	Moderate	Unusually sensitive people should consider limiting prolonged outdoor exertion.
101–150	Unhealthy to sensitive groups	Active children and adults, and people with respiratory disease, such as asthma, should limit prolonged outdoor exertion
151–200	Unhealthy	Active children and adults, and people with respiratory disease, such as asthma, should limit prolonged outdoor exertion; everyone else, especially children, should limit outdoor exertion
201–300	Very unhealthy	Active children and adults, and people with respiratory disease, such as asthma, should avoid all outdoor exertion; everyone else, especially children, should limit out door exertion

Source: U.S. EPA, Air Quality Guide for Ozone, EPA Rep. #EPA-456/F-99-002, Washington, DC, 1999.

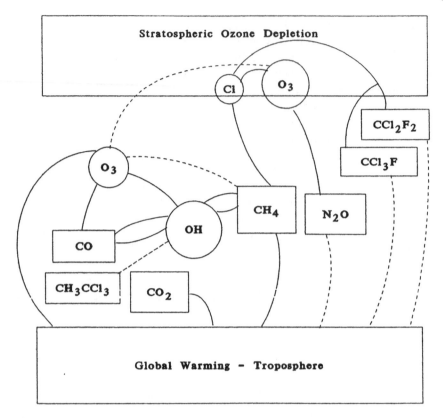

Figure 1.11 Atmospheric chemistry connecting ozone depletion in the stratosphere and global warming. (From Ref. 30.)

complicated by the weather, particularly wind patterns that carry pollutants away from their sources and temperature inversions that tend to keep pollutants near ground level. The 1990 Clean Air Act established five classes of nonattainment areas for smog ranging from marginal (relatively easy to clean up quickly) to extreme (difficult and lengthy to clean up).

Smoke generated by combustion processes can also deleteriously affect visibility. Smoke is not typically a problem in most industrial combustion processes that are run with excess O_2 in order to minimize CO emissions. Typically, CO would be produced before smoke unless the system was significantly deficient in O_2 in which case both CO and smoke may be formed. Smoke is an important issue for flares where performance is often measured by the "smokeless capacity" or how much fuel can be flared before visible smoke is produced (see Chap. 15). Smoke may also be emitted from an industrial combustion process where some of the raw materials are entrained in the exhaust gas flow and emitted into the atmosphere. For example, limestone dust in the cement process may be emitted from the exhaust stack (see Chap. 14). Fortunately, dustier processes typically have some type of particulate removal system, such as a baghouse, to remove particulates from the exhaust gases before they are emitted to the atmosphere (see Chap. 9).

Table 1.7 Possible Pollutant Emission of Some Important Industrial Processes.

Industrial process	Emission sources	Possible pollutants
Cement production	Rotary kiln	Particulates, NO_x, generally little SO_2, CO, H_2S
Glass production	Glass-melting furnaces or tanks	Particulate (saline and other), NO_x, SO_2 (oil-fired furnaces)
Casting	Cupola furnace	Particulates, CO
Pulp production	Furnaces for incineration of sulfite pulping agents	SO_2, particulates

Three different measurement scales are used to quantify visibility: extinction, deciviews, and visual range. There is not a linear correspondence between these scales. As an example, an extinction of 10 M/m is equivalent to 0 deciviews and a visual range of 400 km. The Ringleman scale is used to approximate smoke density (see Chap. 9).

1.2.5 Health Effects

Ozone, VOCs, nitrogen dioxide, carbon monoxide, particulate matter, sulfur dioxide, and lead are specified as criteria pollutants under the Clean Air Act. Each of these has some detrimental health effects on humans. Ozone, nitrogen dioxide, particulate matter, and sulfur dioxide can cause respiratory problems. These are particularly problematic for those who have lung diseases such as asthma and emphysema. VOCs can cause a variety of health problems including cancer. CO is a by-product of the incomplete combustion of a fossil fuel and can cause death in high enough concentrations due to asphyxiation. Fortunately, CO is normally easy to minimize or eliminate in most combustion processes. It is discussed in more detail in Chap. 7. Lead can cause damage to the brain and other parts of the nervous system. For many of the criteria pollutants, the ill-health affects are particularly damaging to the very young and to the very old. Table 1.7 shows a summary of the detrimental effects of the common pollutants from industrial combustion processes. Some of these affect the health of humans while others damage the environment.

There are many so-called "hazardous air pollutants" (HAPs) specified by the U.S. EPA that can cause cancer, other serious illnesses, or even death [32]. Some of these, for example benzene, may be present in the incoming feed materials to an industrial combustor. However, very few of these are emitted from properly operated combustion systems.

1.3 LITERATURE DISCUSSION

The subject of this book is pollution generated by industrial combustion systems. This section briefly considers some of the relevant literature on the subjects of pollution, combustion, and the combination of combustion with pollution. Many textbooks have been written on both pollution and combustion, but both types of book generally have only a limited amount of information concerning the combination of pollution and industrial combustion. Most of these books are written at a highly technical level for use in upper level undergraduate or graduate level courses. The books typically have a broad coverage with less emphasis on practical applications due to the nature of their target audience. This section briefly surveys books related to pollution, combustion, and pollution in combustion. A list of relevant journals and trade magazines on this subject is given in Appendix A.

1.3.1 Air Pollution Control

Crawford [33] has written an extensive work on the theory of the common techniques used to control air pollution [33]. Because of the rapid changes in this field, the book is somewhat out of date. The focus is more on controlling pollutants after they have been formed and not on minimizing their formation. Wark and Warner [34] have written a general-purpose book on air pollution that discusses the control of pollutants of interest here including particulates, carbon monoxide, SO_x, and NO_x. This book does not discuss industrial combustion applications. Stern et al. [35] have written a book on air pollution with six sections: (1) elements of air pollution, (2) effects of air pollution, (3) measurement and monitoring of air pollution, (4) meteorology of air pollution, (5) regulatory control of air pollution, and (6) engineering control of air pollution. While only the last section, which has some specific information on industrial combustion processes, is of direct interest here, the other sections provide good background on the general problem of air pollution.

Dullien [36] has written a more academic book on industrial gas cleaning, which includes details on the theories of the various air pollution-control techniques. Benítez [37] has written a textbook on the air pollution-control techniques [37]. One of the useful features of this book is the many worked examples, which include economic analyses, to demonstrate the use of the information, equations, and data. Cheremisinoff [38] has edited a book designed to aid in the practical understanding and control of air pollution problems. Cooper and Alley [39] provide much useful information on the equipment used in removing pollutants from dirty gas streams, including cost calculations. This book does not include any treatment of industrial combustion applications. Mycock et al. [40] have edited a general handbook on air pollution control engineering which is more focused on equipment and does not specifically consider the industries discussed here.

Baumbach [41] has written a book on air quality which has been translated into English, which has less emphasis on air pollution control equipment and more on what happens after pollutants have been emitted into the atmosphere. Although many of the references, regulations and data are German, the book is useful as it gives a more European flavor to air-pollution control. For example, there is more emphasis on pollutants like CO_2 emissions, dioxins, and furans that have generally received less attention in the U.S. The Environment Agency in the U.K. has prepared some useful guidelines on controlling pollution emissions from industrial combustion processes (e.g., [42]). Heumann (1997) has edited an extensive book on air-pollution control technologies [43]. While it does not specifically consider the industrial sources of interest here, it gives extensive descriptions of the theory, cost, and application of the common industrial pollutant-control technologies. The first third of the book is on theory and background of the control techniques, the second third on particulate-control technologies, and the last third on gaseous pollutant-control technologies. The book is pragmatic in that it attempts to view the control technologies from both regulatory and economic considerations.

Liu and Lipták [44] have edited an extensive handbook on all aspects of pollution including regulations, standards, control technologies, and prevention. It includes brief discussions of the more common combustion modification techniques for minimizing stack gas emissions. deNevers [45] has written a general textbook on air pollution control. The book contains a good general background on the subject, including effects on the environment and humans, and the theories behind the control techniques. While it is not specifically written for industrial combustion users, it does contain some practical information on pollution control equipment. Schnelle and Brown [46] have written a recent book on controlling air pollution emissions. The book has extensive sections on controlling VOCs, particulates, NO_x, and SO_x, a good historical background on regulations, and consideration of the economics of different control choices, but does not specifically discuss industrial combustion processes.

1.3.2 Industrial Combustion

Many theoretical books have been written on the subject of combustion but they have little if any discussion of industrial combustion processes [42–52]. Edwards [53] has a brief chapter on applications including both stationary (boilers and incinerators primarily) and mobile (primarily internal combustion engines) sources. Barnard and Bradley [54] have a brief chapter on industrial applications, but have little on pollution from those processes. A recent book by Turns [56], which is designed for undergraduate and graduate combustion courses, contains more discussions of practical combustion equipment than most similar books [55].

There have also been many books written on the more practical aspects of combustion. Griswold's book [56] has a substantial treatment of the theory of combustion, but is also very practically oriented and includes chapters on gas burners, oil burners, stokers and pulverized-coal burners, heat transfer, furnace refractories, tube heaters, process furnaces, and kilns. Stambuleanu's book [57] on industrial combustion has information on actual furnaces and on aerospace applications, particularly rockets. There are many data in the book on flame lengths, flame shapes, velocity profiles, species concentrations, and liquid and solid fuel combustion, with a limited amount of information on pollution. A book on industrial combustion has significant discussions on flame chemistry, but little on pollution from flames. Keating's book [59] on applied combustion is aimed at engines and has no treatment of industrial combustion processes. A recent book by Borman and Ragland [60] attempts to bridge the gap between the theoretical and practical books on combustion. However, the book has little discussion about the types of industrial applications considered here. Even handbooks on combustion applications have little if anything on industrial combustion systems [61–65].

1.3.3 Pollution and Industrial Combustion

The Los Angeles County Air Pollution Control District compiled a useful manual [66] on air pollution control for the industrial customers in their jurisdiction. While it is now considerably outdated, nevertheless it contains much useful information on a wide range of industrial combustion sources of air pollution including metallurgical processes, incineration, the minerals industry, petrochemical processes, and chemicals processing equipment. Unlike many other books on air-pollution control, it even has a chapter on combustion equipment. Interestingly, this book discusses many industrial processes no longer located in Los Angeles County due to pollution-related issues. The Air & Waste Management Association sponsored an extensive book, now in its second edition [67], designed to replace the book originally produced by the Los Angeles County Air Pollution Control District just discussed. Because of the many changes in technology and the Clean Air Act Amendments in 1990, there was a need for a new book of this type. As with most edited works, the book is not as integrated and homogeneous as those with only a handful of authors. However, it does benefit from the vast experience of numerous authors with specific expertise in the areas they write about. The book has specific chapters on waste incineration (Chap. 8), the chemical process industry (Chap. 12), the metallurgical industry (Chap. 14), the mineral products industry (Chap. 15), and the petroleum industry (Chap. 17).

Stern edited a five-volume series that reached a third edition on the subject of air pollution. Volume 4, entitled "Engineering Control of Air Pollution" [68], discusses most of the industrial combustion processes considered here. While some of the information has been obsoleted or outdated due to advances in technology, this book provides much useful information for the industrial user. Hesketh [69] has written a general textbook on air pollution control, but has more on combustion and industrial applications, particularly waste incineration, than most similar books. Baumbach [41] discusses some of the pollutants that may be emitted from important industrial processes. The U.S. EPA has prepared a manual [70] to aid in developing cost estimates for air pollution-control equipment. This manual is updated on an ongoing basis and is designed for "study" estimates with an accuracy on the order of ±30%. The manual is divided into seven sections: (1) introduction, (2) generic equipment and devices, (3) VOC controls, (4) NO_x controls, (5) SO_2 and acid gas controls, (6) particulate matter controls, and (7) mercury controls. Some of these are incomplete and in progress. The cost data include information about both new and retrofit applications.

1.4 COMBUSTION SYSTEM COMPONENTS

As shown in Fig.1.12, there are four components that are important in the transfer of thermal energy from a combustion process to some type of heat load: the burner which combusts the fuel with an oxidizer to release heat, the heat load, the combustor, and in some cases, a heat recovery device to increase the thermal efficiency of the overall combustion system [71]. Each of these components is briefly considered in this section and more fully discussed in Chap. 3.

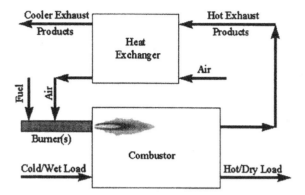

Figure 1.12 Schematic of the major components in a combustion system. (From Ref. 71.)

1.4.1 Burners

The burner is the device that is used to combust the fuel with an oxidizer to convert the chemical energy in the fuel into thermal energy. A given combustion system may have a single burner or many burners, depending on the size and type of the application. For example, in a rotary kiln, a single burner is located in the center of the wall on one end of a cylindrically shaped furnace (see Fig. 1.13). The heat from the burner radiates in all directions and is efficiently absorbed by the load. However, the cylindrical geometry has some limitations concerning size and load type, which make its use limited to certain applications such as melting scrap aluminum or producing cement clinker. A more common combustion system has multiple burners in a rectangular geometry (see Fig. 1.14).

There are many types of burner designs that exist due to the wide variety of fuels, oxidizers, combustion chamber geometries, environmental regulations, thermal input sizes, and heat transfer requirements, which includes things like flame temperature, flame momentum, and heat distribution.

1.4.2 Combustors

There are two predominant types of combustors used in the industries considered here: process heaters for lower temperature applications (less than about 2000°F) and furnaces for higher temperature applications (greater than about 2000°F). Process heaters include, e.g., ovens, heaters, reactors, and dryers. Furnaces include, e.g., kilns, incinerators, and thermal crackers. Combustors are used to transform incoming charge materials by, e.g., oxidation, reduction, melting, heat treating, curing, baking, and drying.

There are many important factors that must be considered when designing a combustor. A primary consideration for any combustor is the type of material that will be processed. One obvious factor of importance in handling the load and transporting it through the combustor is its physical state, whether it is a solid, liquid, or gas. The shape of the furnace will vary according to how the material will

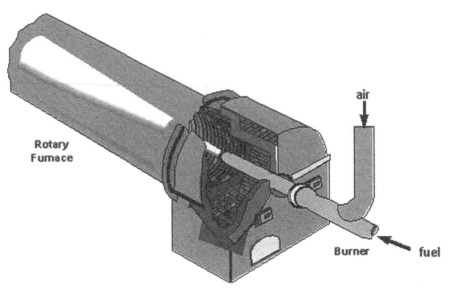

Figure 1.13 Single burner in a rotary kiln. (From Ref. 71.)

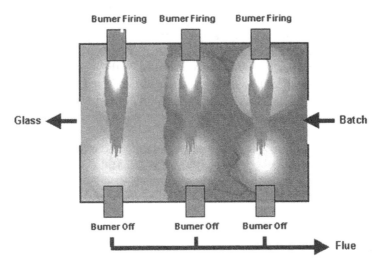

Figure 1.14 Multiple burners in a side-fired regenerative glass furnace. (From Ref. 71.)

be transported through it. When heat recovery is used in an industrial combustion process, it is an integral part of the system. The two most popular methods are regenerative and recuperative.

Combustors are often classified in a variety of ways. One is by the heating method as indirect or direct heating. In indirect heating, there is some type of intermediate heat-transfer medium between the flames and the load which keeps the combustion products separate from the load. As a result of the temperature limits of the heat-exchange materials, most higher temperature processes are of the direct heating type where the flames can directly radiate heat to the load. Another method of classifying combustors is by the processing system, which is either batch or continuous.

Two common furnace shapes are rectangular and cylindrical. The two most common orientations are horizontal and vertical, although inclined furnaces are used in certain applications such as cement production (see Chap. 14).

1.4.3 Heat Load

In some processes, fluids are transported through combustors in process tubes such as in refinery heaters. In some applications, heaters and burners are used to heat or dry moving substrates or webs such as in paper machines. Convection dryers are also used to heat and dry substrates such as textiles. Typically, high-velocity heated air is blown at the substrate from both sides so that the substrate is elevated between the nozzles. Some loads are opaque, which encompasses a wide range of materials including granular solids like limestone and liquids like molten metal. Another type of load is transparent, such as glass, which has selective radiant transmission properties.

1.4.4 Heat Recovery Devices

Heat recovery devices are often used to improve the efficiency of combustion systems. Some of these devices are incorporated into the burners, but more commonly they are another component in the combustion system, separate from the burners. These heat recovery devices incorporate some type of heat exchanger, depending on the application.

A recuperator is a low- to medium-temperature, continuous heat exchanger that uses the sensible energy from hot combustion products to preheat the incoming combustion air. Recuperators are typically used in lower temperature applications because of the limitations of the metals used to construct these heat exchangers. A regenerator is a higher temperature, transient heat exchanger that is used to improve the energy efficiency of high-temperature heating and melting processes, particularly in the high-temperature processing industries. In a regenerator, energy from the hot combustion products is temporarily stored in a unit constructed of firebricks.

1.5 COMBUSTION POLLUTANTS

In its broadest sense, an air pollutant is anything not normally present in the atmosphere or not present above a certain concentration. However, the definition is usually restricted to those contaminants that have a harmful effect on the environment. For example, if a process is emitting water in the form of steam into the atmosphere, it would usually not be considered a pollutant unless it was having some type of effect on the visibility at or near the discharge point. Table 1.8 lists some of the common gaseous airborne pollutants, common sources, and atmospheric removal reactions and sinks. The EPA quantifies the emissions from a wide range of sources and geographical locations in the U.S. to track how well regulations are controlling pollution [72]. One book offers four reasons for cleaning polluted gas streams [40]:

1. Concentration limits are harmful to humans, physical structures, and plant and animal life.

2. Legal limitations are imposed by the government for the protection of the public health and welfare.

3. Reducing pollution establishes civic goodwill.

4. It reduces or eliminates potential liability patterns.

Air pollutants have been categorized in a variety of ways. Some classify them into two broad categories: natural and man-made [40]. One organization defined three types of air pollutants: organic gases, inorganic gases, and aerosols [66]. Organic gases include hydrocarbons (e.g., paraffins, olefins, and aromatics), oxygenated hydrocarbons (e.g., aldehydes, ketones, alcohols, and acids), and halogenated hydrocarbons (e.g., carbon tetrachloride and trichloroethylene). These primarily cause eye irritation, oxidant formation, and visibility reduction. Inorganic gases include oxides of nitrogen (e.g., nitric oxide and nitrogen dioxide), oxides of sulfur (e.g., sulfur dioxide and sulfur trioxide), and carbon monoxide. These are of primary interest here and cause plant damage, eye irritation, oxidant formation, visibility reduction, and danger to human health. Aerosols include both solid (e.g., soot, metal oxides, and mineral dusts) and liquid (e.g., acid droplets and paints) particulates. These primarily cause visibility reduction. Solid aerosols are of specific interest here.

Table 1.8 Sources, Concentrations, and Scavenging Processes of Atmospheric Pollutants

Air Pollutant	Effects
Particulates	Speeds chemical reactions; obscures vision; corrodes metals; causes grime on belongings and buildings; aggravates lung illness
Sulfur oxides	Causes acute and chronic leaf injury; attacks a wide variety of trees; irritates upper respiratory tract; destroys paint pigments; erodes statuary; corrodes metals; ruins hosiery; harms textiles; disintegrates book pages and leather
Hydrocarbons (in solid and gaseous states)	May be cancer-producing (carcinogenic); retards plant growth; causes abnormal leaf and bud development
Carbon monoxide	Causes headaches, dizziness, and nausea; absorbs into blood; reduces oxygen content; impairs mental process
Nitrogen oxides	Causes visible leaf damage; irritates eyes and nose; stunts plant growth even when not causing visible damage; creates brown haze; corrodes metals
Oxidants: Ozone	Discolors the upper surface of leaves of many crops, trees, and shrubs; damages and fades textiles; reduces athletic performance; hastens cracking of rubber; disturbs lung function; irritates eyes, nose, and throat; induces coughing

Adapted from Ref. 44.

Dullien classified air pollutants into three categories: particulates, gases and vapors, and odorous substances [36]. The terminology in Table 1.9 shows how airborne particulates are classified. The gaseous and vapor-like pollutants were further subdivided into: sulfur-containing compounds such as SO_2 and H_2S that oxidize into compounds like SO_3, H_2SO_4, and MSO_4 (M = metal); nitrogen-containing compounds like NO and NH_3 that oxidize into NO_2 and MNO_3; organic compounds (C_1–C_5) that oxidize into ketones, aldehydes and acids; oxides of carbon (CO); and halogens and halides like Cl_2, HCl, HF, and F_2.

This book focuses on the common pollutants encountered in industrial combustion processes. Table 1.10 shows some of the possible pollutant emissions from industrial processes. However, there are numerous other pollutants that may be present in specific applications that are not considered here. For example, heavy metals, radioactive materials, and odors are not treated in this work.

1.5.1 Nitrogen Oxides

Nitric oxide is the primary NO_x compound emitted from industrial combustion sources. In the atmosphere, NO turns into NO_2, which can produce acid rain when it reacts with water in the atmosphere to form nitric acid; NO_2 also plays a role in ozone formation in the lower atmosphere. NO_2 can also be hazardous to the respiratory system, particularly for the very young, the elderly, and for those with existing respiratory problems. Ambient levels of NO_2 have increased slightly in the U.S. in the past couple of decades, which has prompted even tighter NO_x regulations for most industrial combustion processes. Over 55% of NO_2 emissions result from transportation (cars, trucks, etc.).

Table 1.11 shows a comparison of stationary nonutility industrial sources of NO_x emissions [73]. For the industrial sources considered here, cement kilns and indirect fired heaters represent two of the larger sources. However, these pale in comparison to industrial boilers and heaters. NO_x emissions are discussed in more detail in Chap. 6.

1.5.2 Unburned Combustibles

A principle unburned combustible is CO, which is formed by the incomplete combustion of carbon-containing fuels. CO is a colorless, odorless gas that can cause asphyxiation and death in high enough concentrations. Formation of CO is normally easily prevented by proper control of the combustion process and is not typically a problem for the vast majority of industrial combustion applications. Over three-quarters of the ambient CO in the atmosphere comes from automobile emissions and only a small portion comes from industrial combustion applications. The principle source of CO emissions from combustion processes, particularly internal combustion engines, is during cold start-up conditions where the lower temperatures are more favorable for CO generation even if there is sufficient oxygen present for complete combustion. The ambient levels of CO in the atmosphere have declined significantly in the past few decades in the U.S.

Table 1.9 Airborne Particulate Matter Industrial Terminology Classification

Contaminant	Major pollutant source	Estimated annual pollutant emissions (10^9kg/yr)	Estimated atmospheric residence time	Atmospheric removal reactions and sinks	Notes
SO_2	Fossil fuel	212	1–4 days	Oxidation to sulfate by photochemical reactions or in liquid droplets	High reaction rates in summer due to photochemical processes
CO	Auto exhaust, general combustion	700	1–3 months	Photochemical reactions with CH_4 and OH	No long-term changes in the atmosphere have been detected
NO, NO_2	Combustion	75 (as NO_2)	2–5 days	Oxidation to nitrate	Background concentrations in doubt but may be as low as 0.01 ppb
N_2O	Small amounts from combustion	3	20–100 yrs	Photochemical in stratosphere	
CH_4	Combustion, natural gas leakages	160	8 yrs	Reaction with OH to form CO	
Total non-CH_4 hydrocarbons	Combustion	40	Hours to a few days	Photochemical reactions with NO and O_3	Pollutant source includes biomass burning
CO_2	Combustion	22,000	2–4 yrs	Biogenic processes, photosynthesis, absorption in oceans	

Source: Ref. 36.

Table 1.10 Possible Pollutant Emissions from Some Important
Industrial Processes

Classification	Particulate diameter (μm)	Form
Coarse dust	> 100	Solid
Fine dust	1–100	Solid
Spray	> 10	Liquid
Mists	1–10	Liquid
Fumes and smoke	0.001–1	Solid or liquid

Adapted from Ref. 41.

Table 1.11 Comparison of Industrial Emission Sources

Industry	Main NO_x emission sources
Glass manufacturing	Glass-melting furnaces
Cement manufacturing	Cement kilns
Chemical industry	
ethylene/propylene production	Pyrolysis furnaces
ammonia and methanol synthesis	Steam reformers
nitric acid	Absorption towers
Petroleum and petrochemical industry	Various indirect fired heaters
Steel	Sinter plants, reheat, annealing and galvanizing furnaces
Pulp and paper	Lime kilns, recovery furnaces

Adapted from Ref. 73.

Volatile organic compounds or VOCs are a particular class of hydrocarbons that are normally only present in the exhaust gases of industrial combustion processes under certain conditions—primarily from the incomplete combustion of waste materials in incineration processes. VOCs are normally fairly easy to control and limit under proper combustion conditions. Only 5% of all VOC emissions into the atmosphere in the past two decades have come from fuel combustion processes. The vast majority come from transportation and industrial processes (e.g., paint production). Unburned combustibles are discussed in more detail in Chap. 7.

1.5.3 Sulfur Oxides

Sulfur dioxide is a pollutant emission formed in industrial combustion applications during the firing of fuels, like heavy oil, that contain sulfur. Under typical combustion conditions, essentially all of the sulfur in the fuel is converted into SO_2. The primary environmental concern with SO_x emissions is the generation of acid rain when SO_2 reacts with water in the atmosphere to form sulfuric acid. SO_x emissions also contribute to particulate matter formation and can cause respiratory problems for higher risk individuals when the concentrations are high enough. SO_x is not a concern when firing clean gaseous fuels like most typical natural gases that have little or no sulfur content. While not of interest here, SO_x emissions are of particular concern when firing coals, which often have significant sulfur contents and are used in very large quantities at many power plants. About 85% of the SO_x emissions in the atmosphere result from fuel combustion processes, with the vast majority coming from power plants burning coals containing sulfur. SO_x emissions are discussed in more detail in Chap. 8.

1.5.4 Particulate Matter

Particulate matter is a general term used for solid and/or liquid particles emitted into the atmosphere. If the concentration of the particulate emissions is dense enough then visibility can be impaired, which is the primary environmental concern for this pollutant. There are several possible sources of particulate emissions from industrial combustion. A primary source is from the combustion of liquid or solid fuels and may contain ash or char from the incomplete combustion of carbon-containing compounds. Another source of particulates is from carryover of solid particles in the production process. For example, fine sand particles are used in the production of glass. These particles can be carried out of the combustion chamber if the gas velocities are sufficient to cause entrainment. Larger particles are normally easily controlled as they can be dropped out of combustion exhaust gases in a variety of ways, including gravity in some cases. Fine particles are more of a problem because they can travel long distances in the atmosphere if they are not captured before exiting the combustion process. Fabric filters, electrostatic precipitators, and wet scrubbers are common post-treatment techniques for capturing fine particle emissions. $PM_{2.5}$ refers to particles that are less than or equal to 2.5 μm or microns (10^{-6} m) in diameter. PM_{10} refers to particles less than or equal to 10 μm in diameter. Current regulations are based on PM_{10} but it appears likely that this will change to $PM_{2.5}$ in the near future in an attempt to further reduce particulate emissions. About one-third of all particulate matter emissions into the atmosphere result from combustion processes of all types, with the largest coming from power generation. Particulates are discussed in more detail in Chap. 9.

1.5.5 Noise and Vibration

There are several possible sources of noise from combustion processes including high-speed flow through piping and piping components, fan noise from the rotation of the blades, and combustion roar from the combustion process to name a few. In some cases it is relatively easy to reduce the noise by modifying the process or configuration. In other cases, some type of noise suppression may be needed such as a muffle. Vibration can be caused by combustion instability. This can damage equipment and be harmful to people who are exposed to the vibration for long periods. In most cases, it is preferable to mitigate the source of the vibration especially if it is caused by some instability. In other cases, it may be desirable to merely dampen the vibration to tolerable levels. Noise and vibration are discussed in more detail in Chap. 10.

1.5.6 Other Pollutants

One important class of pollutants that may be emitted from industrial combustion processes is dioxins/furans (see Chaps. 11 and 16), which are referred to as HAPs. These are normally only a factor in a few limited applications such as waste incineration. However, they are of significant concern because of their high toxicity in relatively low concentrations. Ozone is an important atmospheric pollutant, but it is not produced directly by the combustion process. Rather, it is produced indirectly when some combustion products react in the atmosphere.

Carbon dioxide is a product of combustion from the burning of hydrocarbon fuels (see Chap. 11). It is considered by many to be a greenhouse gas responsible for global warming. However, this is a controversial subject that does not have widespread agreement among scientists. In Europe there are existing regulations and taxes to reduce CO_2 emissions compared to the U.S. where there are no such regulations and taxes. A primary concern is the increased burning of fuels over the past century which some believe has caused an imbalance in the ability of nature to absorb the additional CO_2 emissions. Because these are among the largest emissions from combustion processes, this has gained considerable attention in recent years.

There are some special pollutants related to the use of flares. These include smoke, thermal radiation, and noise. These are discussed in Chap. 7 under VOCs (Sec.7.4) and in Chap. 15 under flares (Sec.15.2.3). Figure 1.15 shows a flare used on an offshore oil platform for burning undesired flammable gases. Figure 1.16 shows a picture of the same flare with water injection used to minimize thermal radiation. Because of the physical limitations of the flare stack and platform, workers are relatively close to the flare. Water injection suppresses the radiation heat loading on the workers. The water consists of the salt water surrounding the platform, which is obviously in large supply. The equipment must be made of materials designed to tolerate the corrosive effects of the salt water.

There are a number of other pollutants that will not be considered in any great detail in this book because they are of lesser concern in industrial combustion applications. One example is lead emissions that are only present in some select applications such as lead production and waste incineration. These are normally easily controlled with the appropriate post-treatment equipment. Another pollutant that is not included here is odor which is not normally produced by the combustion process. Cheremisinoff et al. [74] give an extensive discussion of the various techniques available for handling odors from industrial processes. This book also does not consider wastewater effluents produced by some of the pollution-control technologies such as wet scrubbers. Chapter 11 discusses a variety of other pollutants of more minor consequence to most industrial combustion processes.

1.6 ENVIRONMENTAL REGULATIONS

There are many ways pollution emissions are regulated. This includes both the units used to quantify the emissions and the standards used to regulate the emissions. Both

Figure 1.15   Flare on an offshore oil platform without water injection. (Courtesy of John Zink Co., Tulsa, OK.)

Figure 1.16   Flare on an offshore oil platform with water injection for thermal radiation suppression. (Courtesy of John Zink Co., Tulsa, OK.)

are briefly discussed next. The U.S. EPA has a variety of web sites containing air pollution compliance information [e.g., 75].

The initial major regulation was the Air Pollution Control Act of 1955, followed by the Clean Air Act of 1963, the Air Quality Act of 1967, and the 1970 Clean Air Act, which established the national ambient air quality standards (NAAQS). The "criteria" pollutants that fell under this standard include SO_x, particulates, carbon monoxide, ozone, NO_x, and lead. Two types of standards were established. Primary standards were designed to protect public health and secondary standards were designed to protect public welfare. The next major legislation was the 1977 Clean Air Act Amendments, which established a prevention of significant deterioration (PSD) program, Non-Attainment Areas, and a controlled emissions trading program.

The Clean Air Act Amendments (CAAA) of 1990 were broken down into the following categories:

Title I: National Ambient Air Quality Standards
Title II: Mobile Sources
Title III: Air Toxics
Title IV: Acid Rain
Title V: Permits
Title VI: Stratospheric Ozone
Title VII: Enforcement
Title VIII: Miscellaneous
Title IX: Clean Air Research
Title X: Employment Transition

The 1990 Amendments listed 189 hazardous air pollutants, usually referred to as HAPs, that have to be addressed by the new air toxics program. These pollutants include organics, metals, and chemical compounds used in many industrial processes. Brownell [76] has edited a book, now in its third edition, designed to help in understanding the complex regulatory requirements of the Clean Air Act.

The Clear Skies Initiative was announced by the U.S. EPA in February 2002, which included a multipollutant cap and trade program [77]. Financial incentives have been established for companies to reduce emissions early and position themselves to be early sellers of credits. This means that even companies that today may not have pollution-control equipment may need to add them in the future. Those with existing control equipment may need to upgrade to get further reductions.

One of the problems with regulations is that inflexible limits are set that are technically, but not economically, feasible [78]. This added expenditure can inhibit an industry's ability to develop more advanced and cost-effective pollution-control technologies. Another problem is the lack of uniformity between the various levels of government and between different geographical locations. Regulations are also changing, sometimes rapidly, with time. All of these make it difficult to develop a cohesive pollution-control strategy and make compliance more difficult and costly for the end user.

The subject of permitting for modifications to existing sources and for construction of new sources is beyond the scope of the present work. Permitting is an entire topic unto itself as it varies dramatically based on geography, source type, and as a function of time as regulations continue to evolve. It is recommended that interested parties work with local environmental permitting authorities and with air-quality consultants familiar with local regulations. Bookchin and Farnsworth [79] provide good general discussions on environmental laws and regulations. Regulations are discussed in a general way throughout the book.

1.6.1 Units

The variety of emission units used to regulate pollutants is discussed in the sections pertaining to the individual pollutants. These units vary widely by geographic location and by application. For example, NO_x is commonly regulated on a volume basis, such as parts per million by volume (ppmv), in some places while in other places it may be regulated on a mass basis, such as mass of NO_x produced per some normalized unit. The normalized unit may be firing rate (e.g., lb $NO_x/10^6$ Btu fired) or may be the units produced (e.g., lb NO_x/ton of glass produced). Regulations on a volume basis can vary widely as they may be based on the exhaust gas volume with the water still in the sample such as parts per million by volume on a wet basis (ppmvw) or where the water has been removed from the gas sample such as parts per million by volume on a dry basis (ppmvd). There is even some variation of units for a given volume basis as the actual measured values are often corrected to some specific oxygen level in the gas. That level can vary from as low as 0% O_2 to as high as 15% O_2, depending on the application and source location. The methods used to make these corrections are discussed in more detail in the sections on the individual pollutants. The old adage "The solution to pollution is dilution" is no longer applicable because regulations are now written to correct for dilution.

U.S. EPA Method 19 describes data reduction procedures to calculate the emissions of nitrogen oxides (NO and NO_2), particulate matter (PM), and sulfur dioxide (SO_2) [80]. This includes correction factors for the O_2 level in the exhaust gases, whether the sample is on a dry or wet basis, and how to convert to various units. These are discussed in more detail below with some examples.

1.6.2 Standards

The units used vary not only by country, but also by geographical location within a country, which makes this a complicated issue requiring specific knowledge of the source location. For example, California and Texas generally have stricter regulations than other states in the U.S. because of the larger pollution problems there that require more aggressive action to resolve. Lawmakers in other states often look to California and Texas as leaders in this area. The leading industrialized nations often have much stricter pollution regulations than other nations, which may have no specific regulations in some cases. Zegel [81] has written a useful overview of environmental standards.

Pollution regulations also vary greatly by the specific application. For example, power generation from stationary gas turbines may require single digit NO_x emissions while there may be no specific NO_x emission limit for certain applications in certain locations. In general, the regulations are stricter for larger emission sources compared to those for smaller sources. For example, many states have tighter regulations for larger boilers compared to those for smaller boilers. This discrepancy is related to costs and benefits. It is proportionately less costly to reduce the emissions from a larger source than a smaller one. It is also easier to monitor and regulate a few larger sources compared to many smaller sources. Larger sources are often required to have continuous emission monitoring (CEM) systems that may even transmit emission data directly to the regulating agency. This is generally not viable for smaller sources where the equipment and administration for this type of monitoring would be prohibitively expensive at this time. These discrepancies are likely to change over time as technologies become more cost effective and regulations expand to cover ever smaller sources.

The U.S. EPA has compiled in AP-42 an extensive but not exhaustive list of air pollutant-emission factors for a wide range of applications and pollutants [82]. These are for both combustion and noncombustion processes such as waste incinerators and landfills. The general equation used for emission estimation is

$$E = A \times EF(1 - ER/100) \qquad (1.2)$$

where:

$E =$ emissions
$A =$ activity rate
$EF =$ emission factor
$ER =$ overall emission reduction efficiency (%)

ER is the product of the control device destruction/removal efficiency and the capture efficiency of the control system. The emission factor EF is intended to be a representative value for the specific application and conditions. Applications include residential, commercial, and industrial sources of air pollution. The actual units used in the factors depend on the application. For example, for waste incinerators the emission factors are given as the weight of pollutants emitted per weight of refuse combusted (e.g., pounds of particulate matter emitted per ton of refuse combusted). Both metric and English units are often given. The factors are also sometimes given for both uncontrolled and controlled emissions. For example, a factor may be given for combustion systems without and with an electrostatic precipitator. The factors are generally well-documented for the interested researcher.

These factors are compiled in AP-42 by application and are available both in printed and electronic format. They are not intended to represent a recommended emission limit or standard, nor the best available control technology. However, they are sometimes used to determine air permit levels where the actual pollution emissions have not been measured, e.g., in smaller sources where the emissions are insignificant compared to other larger sources in the plant. Use of these factors is much less expensive than certified source tests or continuous emission-monitoring systems. Because the factors are somewhat out of date by virtue of the lag between compiling the factors and publishing them, not to mention the added time between publication and when they are used, they are often conservative (overpredict) compared to the emissions from plants that have added new technology to reduce their pollution emissions. The factors are often based on actual emission test reports.

The factors are also given a rating from A to E related to their reliability or robustness. A rating of A, B, C, D, and E means the emission factor quality is excellent, above average, average, below average, and poor, respectively. The quality of the factor depends on the quality of the data collected. The highest quality data are collected with an accepted method and reported in enough detail for adequate validation. The lowest quality data are based on a generally unacceptable method but the data may provide an order of magnitude approximation of emissions.

Because of the dynamic nature of pollution regulations, only representative examples are given in this book to illustrate certain points. However, the trend is clear that emission regulations will continue to get stricter as new technologies become available to reduce pollution. In some cases, emissions regulations are not given as a specific value. These are briefly discussed next.

1.6.3   Variable Emission Regulations

Many approaches are being used to address the dynamic nature of pollution-emission regulations. The most common approach is simply to continue to rewrite the regulations as new technologies are developed to reduce emissions significantly compared to previous technologies. This process to rewrite the regulations can be expensive, tedious, lengthy, and fraught with problems. Regulators work in conjunction with users to determine what new levels make sense in the light of economics and technology. If the regulations are too strict, businesses may locate elsewhere which is normally undesirable as it reduces the tax base and transfers jobs out of the area, which hurts the economy of the original locale. If the regulations are not strict enough then the air quality of the area is adversely affected and the quality of life suffers. Regulators often introduce new regulations for a review period to get feedback from both citizens and from the companies affected by the regulations. There may be some negotiation to adjust the regulations according to the feedback received. Then, there is often a significant amount of time before the new regulations are enacted to allow companies to modify their equipment and processes to meet the new levels. Sometimes the new regulations only apply to new equipment that might be installed by a company, while excluding existing equipment. The exclusion of existing equipment is often referred to as "grandfathering." However, if there are any significant changes to the grandfathered equipment, then the new regulations are likely to apply. The definition of significant is often gray which further complicates the regulation process.

Another approach used to deal with changing regulations is not to have a specific value for a pollutant but to base the regulation on the technology available at the time of the permit application. There are a variety of acronyms used to refer to this approach: reasonably available control technology (RACT), best available control technology (BACT), maximum available control technology (MACT), lowest achievable emission rate (LAER), and best available retrofit technology (BART) are some examples. Government agencies like the U.S. EPA maintain databases that specify what technologies qualify under each of the above acronyms for each industry and process type. The RACT/BACT/LAER Clearinghouse (RBLC) maintains an extensive searchable list of emissions data based on permits that have been received from state and local air pollution control programs in the

U.S. [83]. The database lists data by process application (e.g., cement kilns), by company name (e.g., Continental Cement Company), and by permit number. The database includes information about the process such as the fuel and throughput rate, about the pollution emissions such as the mass of each pollutant per unit of throughput, and about the control equipment used to minimize pollutant emissions such as the use of an electrostatic precipitator.

By the very nature of this approach, the regulated limits are constantly changing with technological improvements. While this approach has the advantage of being more current than constantly modifying written regulations, there are other challenges including determining what limits should be associated with each new technology as applied to a variety of industrial processes. In some cases, the government agency may work with both the company that developed a new technology and with the first companies implementing the technology to determine fair and acceptable emission limits for future users. Bassett [84] lists the following factors to be considered when identifying BART:

- Technology available to control the pollutant(s) affecting visibility.
- Compliance cost.
- Energy and non-air quality environmental impacts of compliance.
- Existing air pollution-control equipment at the source.
- Remaining useful life of the source.
- How much visibility would be improved by implementing the technology.

Another approach to setting emission limits is to determine the overall allowable emission rate for a geographical location and then to let the companies operating in that region to determine the best way to meet that rate through a free-market emission-trading program. This gives each company more options in how they will meet their own emissions limit, which are typically determined on a plant-wide basis. A company may choose to install new technology that reduces pollution emissions. A company may choose to shut down some older, high-emitting processes where the cost to reduce emissions may not be economically viable. A company may purchase emission credits from other companies that are below their allowable limits. Conversely, a company may sell emission credits if they are below their limits. The cost of the credits is determined by free market economics. If more credits are available, then the cost will be lower than if there are fewer credits available. A company may also use a combination of strategies to meet its overall limit including shutting down high-emitting processes, installing new technologies in other processes, and buying or selling emission credits as needed. The right combination will vary for each company and will also vary with time for a given company. For example, if emission credits are relatively inexpensive in a given year, then it may be cheapest to simply buy credits. If emission credits are relatively expensive in another year, then it may make more sense to install new technology that either produces less pollution or that cleans up the pollution generated in a particular process. This overall approach has the advantages of giving the companies the option to determine the best strategy for their particular situation and simplifies the process of setting regulatory limits. The agency need only set the overall limit for the region and then the limits for the individual plants in that region. If a company wants to add a new process or expand its capacity in a given plant, then it must determine how to maintain its overall emission rate rather than have an agency determine what the rate should be for each new installation or significant modification within a given plant. This approach gives plants an incentive to evaluate new technologies continually to reduce pollution emissions, and the free market will eventually determine the winners and losers among those new technologies. Vatavuk [85] has developed a method to compare the costs of various air pollution control devices from one date to another to help in determining the economics of each alternative.

1.6.4 Sampling

Liu [85] has written some useful information on emission measurements. The U.S. EPA has prepared specific protocols, referred to as methods, for measuring a wide range of pollutants. For example, Method 5 discusses how to make proper particulate-emission measurements from a wide range of stationary sources. Figure 1.17 shows a typical set of equipment that can be purchased as a unit for making Method 5 measurements according to the EPA protocol. Figure 1.18 shows a commercially available sample probe specifically designed to make Method 5 measurements.

There are some specific requirements for how to sample the gas composition in exhaust gas stacks. Depending on the stack configuration, an array of sample points normally must be sampled across the cross-section of the stack. There are also some specific equipment requirements and sampling conditions for measuring certain types of pollutants. For example, NO in the presence of O_2 in a gas sample flow can convert to NO_2 in the sample lines if the temperature is not properly controlled. Specific guidelines for measuring various pollutants are discussed in more detail in their respective chapters.

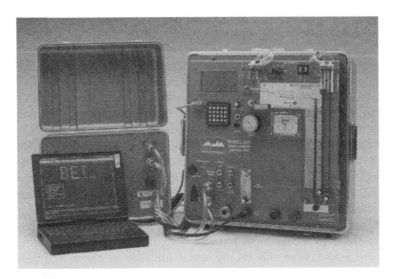

Figure 1.17   Commercial Method 5 sampling system. (Courtesy of Baldwin Environmental, Reno, NV.)

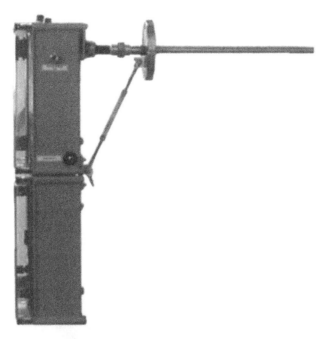

Figure 1.18 Sample probe for EPA Method 5 measurements. (Courtesy of PermaPure, Toms River, NJ.)

The gases should be sampled isokinetically, which means that the sample gases should be extracted from the exhaust stream at the same velocity as the gases going around the sample probe. If not, a nonrepresentative sample may result. If the gas sample velocity is too high, then more gas will be preferentially pulled into the probe while some particulates will continue flowing past the probe. This will give an inaccurately low particulate concentration. If the gas sample velocity is too low then the opposite may happen where particles will tend to flow into the probe while some of the gases may be diverted around the probe. This will tend to give an inaccurately high particulate concentration. One way to ensure isokinetic sampling is to measure the gas stream velocity where the sample probe will be inserted. However, the challenge is that the velocity varies significantly across the exhaust duct. The velocity distribution may be parabolically shaped if the flow is fully developed, or it could be much more complicated, depending on the upstream conditions such as being close to a damper or elbow.

A variety of equipment is available for sampling stack gases. Figure 1.19 shows a portable analyzer used for quick measurements of stack gases. These are not generally designed for continuous sampling operation and are not typically recognized by regulatory bodies for making certified emission measurements. However, they are very useful for spot checks and for research checks as their accuracy is comparable to full-blown continuous emission-monitoring systems (CEMS). Figure 1.20 shows a gas analysis system containing analyzers for CO, O_2, NO_x, and unburned hydrocarbon emissions. The system also includes a gas sampling

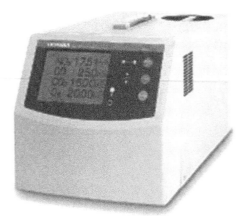

Figure 1.19   Typical portable multigas analyzer. (Courtesy of Horiba, Irvine, CA.)

Figure 1.20   Gas analysis system.

Figure 1.21   Unburned hydrocarbon analyzer.

and control unit and a computer for displaying and recording emissions. Figure 1.21 shows a typical single analyzer for measuring unburned hydrocarbon emissions (see Chap. 7).

1.7   INDUSTRIAL APPLICATIONS

As shown in Fig. 1.1 and Table 1.2, there are many industrial applications that use combustion to process some type of material. These range from lower temperature drying applications to higher temperature melting applications. A wide range of combustors are used depending on the production rate, the heat-transfer requirements, the material being processed, the fuel being combusted, the oxidizer being used, any heat recuperation technologies being employed, and in many cases the tradition of how the process has been done in the past. Each of the industries considered here has some unique aspect or aspects compared to the others, which is why they need to be considered separately as one solution does not fit all. Even within a given industry there are numerous permutations so that each individual case must be considered on its own. Space limitations preclude a detailed discussion of all possible industrial combustion applications. However, the larger sources of industrial combustion pollution are considered here. These are briefly discussed next and treated in more detail elsewhere in the book.

1.7.1   Metals Industries

Metals are used in nearly all aspects of our lives and play a very important role in society. The use of metals has been around for thousands of years. There are two predominant classifications of metals: ferrous (iron-bearing) and nonferrous (e.g., aluminum, copper, and lead). Ferrous metal production is often at high temperature because of the higher metal melting points compared to those of nonferrous metals. Many metals production processes are done in batch, compared to most other industrial combustion processes considered here, which are typically continuous. Another unusual aspect of metal production is the very high use of recycled materials. This often lends itself to batch production because of the somewhat unknown composition of the incoming scrap materials that may contain trace

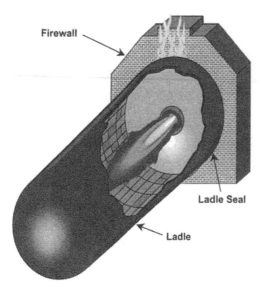

Firewall

Ladle Seal

Ladle

Figure 1.22   Schematic of a ladle preheater.

impurities that could be very detrimental to the final product if not removed. The metals are typically melted and then sampled to determine the chemistry so that the appropriate chemicals can be either added or removed to achieve the desired grade of material. Another unique aspect of the metals industry is that transfer vessels are preheated prior to the introduction of molten metals to minimize the thermal shock to the refractory. Figure 1.22 shows an example of preheating a transfer ladle.

Another somewhat unusual aspect of metals production is that supplemental heating may be required to reheat the metals for further processing. For example, ingots may be produced in one location and then transported to another location to be made into the desired shape (e.g., wheel castings are often made from remelting aluminum ingots or sows). While this process may be economically efficient, it is very energy and pollutant inefficient due to the additional heating. This is something that has begun to attract more attention in recent years where the entire life cycle of a product is considered rather than just its unit cost and initial energy requirements. For example, aluminum has a low life-cycle cost compared to that of many other metals because of its high recycle ratio. While the energy consumption to make aluminum from raw ore is fairly high, remelting scrap aluminum takes only a fraction of that energy, which also means less overall pollution. Pollution emissions and control in the metals industry is considered in more detail in Chap. 13.

1.7.2   Minerals Industries

Some common minerals processes include the production of glass, cement, and bricks. These are typically high-temperature heating and melting applications that require a significant amount of energy per unit of production. They also tend to have fairly high pollutant emissions as a result of the high temperatures and unit energy

Figure 1.23   Cement manufacturing plant.

requirements. Most of the minerals applications are continuous processes, but there is a wide range of combustors. Large glass furnaces are typically rectangularly shaped and have multiple burners. On the other hand, cement kilns are long refractory-lined rotating cylinders that are slightly inclined so that the materials flow gradually downhill. A typical cement plant is shown in Fig. 1.23.

Many of the minerals applications employ some type of heat recovery in the form of air preheating to improve energy efficiency. However, the heat recovery typically significantly increases NO_x emissions. Many of the minerals applications have dusty raw feed materials and require some type of particulate control system in the flue gas exhaust to minimize particulate emissions. Also, the incoming raw materials may contain contaminants such as niter (containing nitrogen), in glass production that can exacerbate NO_x emissions. While recycling of used glass (referred to as cullet) is practiced in some applications, there is generally must less recycling in the minerals industry than in the metals industry. However, both industries share the problem of high NO_x emissions due to the high processing temperatures. The minerals industry is considered in more detail in Chap. 14.

1.7.3   Chemicals Processing

This is a very broad classification that encompasses many different types of production processes that have been loosely subcategorized into chemicals (organic and inorganic) and petrochemicals (organic) applications. A typical refinery is shown in Fig. 1.24. There is some overlap in terms of the types of heating equipment used where many of the incoming feed materials are in liquid form (e.g., crude oil) that are processed in heaters with tubes running inside them. These are generally lower temperature applications ($<2300°F$ or $<1300°C$) that incorporate heat recovery to preheat the incoming feed materials. This is far more environmentally friendly compared to the combustion air preheating heat recovery employed in the metals

Figure 1.24 Refinery.

and minerals industries. Nearly all of the chemicals heating applications employ multiple burners but in a much more diverse configuration compared to that of other industries. Burners may be fired horizontally, vertically up, vertically down, or at some angles in between, depending on the specific process. There are numerous configurations for fired process heaters.

There are several aspects that make this industry unique compared to others. The first and one of the most important is the wide range of fuel compositions used to fire the heaters. These are mostly gaseous fuels that are by-products of the production process. These fuels often contain significant quantities of hydrogen and propane and may include large quantities of inert gases such as nitrogen and carbon dioxide. A given heater may need to be able to fire on multiple fuels that may be present during various times in the production process. Another unique aspect of this industry is that many of the heaters are fired with natural draft burners where no blower is used to supply the combustion air. These burners are designed differently from conventional forced draft burners and are more susceptible to variations in ambient conditions such as air temperature and humidity and wind speed. A third unique aspect of this industry is the widespread use of flares for burning very large quantities of fuel gases during upset or emergency conditions as well as burning smaller quantities during normal operation. These flares are not used to process any materials so all the energy is lost. In fact, the radiation from these flares is considered a pollutant because of its deleterious effect on surrounding equipment and nearby personnel. The chemicals industry is considered in more detail in Chap. 15.

1.7.4 Waste Incineration

The objective of waste incineration processes is to reduce or eliminate waste products which involves burning those materials. Not only is the incinerator (see Fig. 1.25) fired with burners, but the waste material itself is often part of the fuel that generates

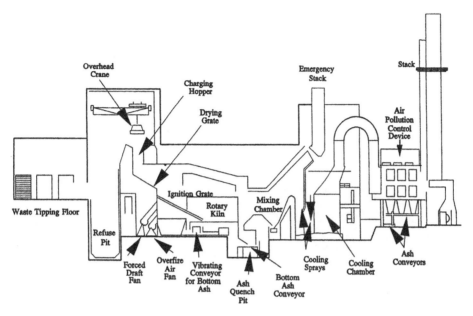

Figure 1.25   Schematic of a waste incinerator. (From Ref. 82.)

heat in the process. However, the waste usually has a very low heating value, hence the need for supplemental fuel. Incineration is a more complicated and dynamic process compared to other industrial combustion processes by nature of the variability of the feed material. The waste may be very wet after a rain storm which may put a huge extra heat load on the incinerator. In some locations where waste materials are separated for recycling, the waste actually fed into the incinerator may have a much higher heating value compared to that of other incinerators where there is no separation of the waste. Another complicating factor with incinerators is that the end product, e.g., the noncombustible waste, must also be disposed of which means that one of the goals of most incineration processes is to produce minimal waste output. Because of the waste material variability, other pollutants may be generated that are not normally associated with industrial combustion processes. An example is the burning of plastics that can produce dioxins and furans. The types of incinerators can vary greatly, depending on a variety of factors. Pollution emissions and control from incinerators is discussed in Chap. 16.

1.7.5   Other Industries

There are numerous other industries that could be considered in a book of this type. Here, only a few have been chosen as examples to show some of the unique challenges that must be addressed to minimize pollution emissions. The other industries considered here include paper (see a typical paper mill in Fig. 1.26), printing and publishing, textiles, and food processing. These are generally lower temperature applications where pollution emissions are often much lower than those of the higher temperature industries like metals and minerals production.

Figure 1.26   Paper mill.

Nevertheless, these "other" industries also produce pollution emissions that are being regulated and therefore must be monitored and controlled. In general, less information is available on these industries than on the larger pollution emission sources. These other industries are considered in Chap. 17.

1.8   POLLUTION CONTROL STRATEGIES

There are a number of issues to consider when designing a pollution-control system, which may include multiple technologies to handle multiple possible pollutants or even for handling the same pollutant in some cases. Some of these issues include flammability of the gases being treated and removed, applicable construction codes (e.g., National Electric Code), economics (see discussion below), corrosion, vent sizing, material handling, and chemical storage to name a few. One of the inherent difficulties in controlling pollutant emissions is that they are sometimes inextricably linked together. For example, some of the techniques for reducing NO_x emissions often increase CO emissions.

There are many factors that go into the selection of the appropriate control technologies to minimize pollution emissions from a process. Some broadly categorize these factors into: environmental, engineering, and economic [40]. Environmental includes things like equipment location, available space, ambient conditions, availability of utilities, and regulations. Engineering includes things like contaminant characteristics (physical and chemical properties, concentration, etc.), gas stream characteristics (flow rate, temperature, pressure, humidity, properties, etc.), and performance of the particular control system (power requirements, removal efficiency, temperature limitations, etc.). Economic considerations include the capital cost, ongoing operating costs, and the expected lifetime of the equipment. Spaite and Burckle [87] believe the two primary factors used to determine what control techniques to employ are technical feasibility and lowest cost.

The governing regulations dictate what emissions are permissible and therefore what reduction levels are required. Future regulations may also be taken into account if further reductions will be required in the near future. Economics play an important role in determining what technologies can even be considered based on a financial analysis of the specific industrial process in question. The economics include both the initial purchase price and the ongoing operating costs including maintenance. The economics also include the useful life of the equipment and the cost of capital for the plant.

Another economic factor to be considered in certain locations is emissions trading where emissions credits can be purchased from other companies to offset one's own emissions. The price is normally determined by free-market supply and demand, but generally increases over time as the regulations get more restrictive, which usually reduces the number of emissions credits available for purchase. There is a growing trend around the world to use emissions trading as part of the overall pollution-control strategy. As an example, the Texas Commission on Environmental Quality (TCEQ), formerly known as the Texas Natural Resource Conservation Commission (TNRCC), has an emissions banking and trading program designed to facilitate trading of emissions credits for VOCs, NO_x, and other criteria pollutants [88].

Removal efficiency is another important factor. While a given technology may be much less expensive than others, if it cannot reduce emissions enough to meet regulatory requirements, then that technology by itself is not a viable alternative. Space limitations at a given plant often play a role in deciding what technologies may be used. For example, a selective catalytic reduction process for removing NO_x from exhaust gas streams often requires a significant amount of real estate that may not be readily available in an older, land-locked facility. Ease of operation is often an important factor when considering emissions control technologies. For example, if a given technology is very sensitive to operating conditions and parameters, it may not be an effective choice for processes with less stability and control.

Because of the complexity of the problem, no solution will be applicable to every case even in the same industry and even for the same company at all locations. Each project should be evaluated on its own to determine the best solution for that particular set of conditions. The objective of this book is not to recommend any specific solutions, but rather to offer multiple alternatives for consideration when deciding how to control the pollution emissions from a given process in a given plant in a specific location. Because the regulations, technology, and economics are constantly changing, it would be futile to suggest one solution to fit any and all scenarios.

Davis et al. [89] list three categories of factors to be considered in selecting equipment for air pollution control:

Environmental

1. Equipment location
2. Available space
3. Ambient conditions
4. Utilities (e.g., power, water) availability
5. Regulatory requirements
6. Aesthetics (e.g., visible steam plume)
7. Wastes (e.g., solid or wastewater) produced by air pollution-control equipment
8. Contribution of air pollution-control system to plant noise levels

Engineering

1. Contaminant characteristics (e.g., shape, concentration, toxicity)
2. Gas stream characteristics (e.g., flow rate, humidity, temperature)
3. Equipment performance specifications (e.g., size, weight, efficiency, operating range)

Economic

1. Capital cost (e.g., engineering, equipment, installation)
2. Operating cost (e.g., utilities, maintenance)
3. Expected equipment lifetime and salvage value

There are often a myriad of possible choices for controlling pollution emissions from industrial combustion processes. Peirce et al. [7] list five possibilities for controlling air pollution emissions: (1) at the source, (2) where the pollutants are collected (e.g., exhaust hood), (3) cooling the gases to drop out any condensables, (4) post-treatment of the gases before they are emitted into the atmosphere, and (5) dispersion of the pollutants into the atmosphere to dilute their concentrations. The last possibility is not relevant here as it does not reduce emissions but merely redistributes them. One might say there are basically two strategies for reducing air pollution: prevention and remediation. Here, these are expanded into six general strategies that include:

1. Pretreating the incoming fuel, oxidizer, or feed materials.
2. Modifying the process in some way such as using alternative heating methods or raw materials, changing the product specification, or changing the product design.
3. Modifying the combustion process.
4. Removing the pollutants from the exhaust stream with some type of post-treatment equipment.
5. Process control to maintain design operating conditions that produce low emissions.
6. Other, which includes maintenance and operator training.

The first three strategies attempt to reduce or eliminate emissions by reducing or eliminating the production of the pollutants. The last strategy removes the pollutants after they have already been produced. Often, a combination of strategies may be most effective for a given situation. The general strategies presented here are discussed next. The alternatives for a specific pollutant are discussed under that pollutant (see Chaps. 6–11) and for a specific industry are discussed under that industry (see Chaps. 13–17).

Note that it is common to use more than one technique in a given combustion application which could combine pretreatment, process modification, combustion modification, and/or post-treatment techniques. This may be done for the removal of a single pollutant (e.g., NO_x) or for the removal of multiple pollutants (e.g., NO_x, SO_x, and particulates). For example, in the case of a single pollutant like NO_x, low NO_x burners might be used to reduce part of the emissions but may not be able to achieve reductions sufficient to meet regulatory requirements. A selective catalytic reduction system may be added to reduce the NO_x emissions further to meet permit limits. In the case of multiple pollutants, e.g., NO_x and particulates, water injection might be used to reduce NO_x emissions while an electrostatic precipitator might be used to remove particulates. Devices may be arranged in series to control pollutant emissions.

Note that it is beyond the scope of this book to discuss in detail some of the important factors concerning pollution-control equipment such as specific design, actual capital and operating costs, installation, maintenance, and troubleshooting. These are only discussed in a general way to compare the advantages and disadvantages of various control techniques. The interested reader should review the many available books discussed in Section 1.3.1 and equipment suppliers who can provide detailed information on their technologies.

1.8.1 Pretreatment

Pretreatment basically refers to modifying the incoming feed materials to either the process or the combustion system. This would include adding or removing elements from the fuel, oxidizer, or raw materials. The fuel could be modified by cleaning out some impurities, such as removing sulfur in a pretreatment process at the plant. The fuel could also be modified by putting in an additive designed to reduce pollutant emissions. This is commonly done in gasoline where a variety of additives are used to reduce pollution emissions from automobiles. The oxidizer could be pretreated by adding high-purity oxygen to increase the overall oxygen concentration to something above the approximately 21% O_2 by volume contained in normal air. The incoming raw materials could be cleaned to remove some potential pollutant-emitting chemicals. For example, discarded automobile engines contain parts made of a variety of materials which should be separated prior to recycling to minimize contamination and energy consumption to process unwanted materials.

1.8.2 Process Modification

Process modification refers to either replacing the incoming feed materials to either the process or the combustion system with some other alternative, or to modifying the end product in some way. Ramachandran [90] refers to the former as raw material substitution. An example of a fuel change would be to change from a high-sulfur coal to a low-sulfur coal to reduce SO_x emissions. In that example, the fuel type remains the same but a different grade or composition is being used. A more radical example would be to change from fossil-fuel heating to electrical heating. In that case, the emissions are moved from the industrial plant location to the power generation facility. Changing fuels is usually referred to as fuel switching. A variation of this technique is to switch part of the fuel, part or all of the time. For example, if the NO_x emissions from a combustion process are slightly over a new regulatory limit using oil fuel, then replacing some of that oil fuel with natural gas can reduce the emissions to below the new limit. This has the advantage of using as much of the cheaper fuel as possible. The potential disadvantage is that the combustion system must work on multiple fuels. However, this is not a significant problem today as there are standard burners available for multifuel firing.

In the case of the oxidizer, the combustion air can be replaced with high-purity oxygen.

Some even consider emissions trading and early equipment retirement under the category of operation changes to reduce pollutant emissions [91]. These are not considered in this book as they do not deal with technology but rather with management strategy and would normally only be viable options for a plant looking to reduce production. In the extreme case, it is possible to shut down an entire plant to sell the emissions credits if their value exceeds the profits that could be generated from selling the product.

1.8.3 Combustion Modification

Handy and Kelcher [92] showed that proper burner maintenance significantly reduces the fuel consumption by 7×10^6 Btu/hr for each ethylene-cracking furnace. Reducing fuel consumption indirectly reduces pollutant emissions because less flue gases are generated. Similarly, Stansifer [93] showed that cleaning up the fuel gas to an ethylene-cracking furnace reduces burner maintenance and improves thermal efficiency, which again reduces pollutant emissions. The fuel contains moisture and heavy hydrocarbons including tar-like material that causes coke buildup on the burners which detrimentally affects performance. A combined filter and coalescer system are used to remove the water and solids. McAdams et al. [94] list a number of potential maintenance issues that can adversely affect the performance of low NO_x burners. These issues include: fuel tip orifice plugging, tip coking, fuel line flow restrictions, air inlet blockage, air leakage into the furnace, tip overheating, refractory damage, and flame impingement on process tubes. Most of these can adversely affect burner performance and increase NO_x emissions. It is noted that the new ultra-low NO_x burners need more aggressive maintenance programs to keep them operating at optimal performance levels. Martinson [95] notes the importance of maintenance on the air pollution control equipment to achieve optimal performance and high pollutant removal efficiencies. Improper maintenance can cause downtime, which means lost production. It can also mean regulatory fines if emissions exceed permitted values to the failure of the equipment. Frequent, consistent, and regular inspections are recommended.

1.8.4 Post-Treatment

Post-treatment refers to removing the pollutants from the exhaust gas stream before they are emitted into the atmosphere. There may be many reasons for choosing this strategy. In some cases it may not be possible to reduce or eliminate the formation of one or more pollutants in the combustion process. It may be more economical to remove the pollutants rather than prevent them from forming in the first place. This may occur when pollutants can be easily recycled back into the process, e.g., in the case of particles carried out of the combustor. This could also occur if the changes that would need to be made to minimize pollutant formation have a negative impact on the processing of the product in the combustor. In some plants, it is more economical to duct multiple exhaust gas stacks together going to a single large post-treatment system, rather than modifying each of the individual combustion processes. It may also be necessary to use a post-treatment technique to supplement minimization strategies to be economically below regulated emission levels. Schifftner [96] has written a good general purpose book on air pollution-control equipment primarily focused on post-treatment.

In general, it is very desirable to concentrate the pollutants to be removed in order to make the post-treatment equipment more efficient. This may not always be economically feasible and there are often tradeoffs between the increasing the size and cost of the post-treatment equipment and using some technique to concentrate the pollutants by removing the benign chemicals. Nitrogen, carbon dioxide, and water vapor are usually in large quantities in typical industrial combustion processes. While it is often straightforward to condense the water vapor out of the exhaust gases, this may also cause some of the pollutants to be removed as well if this is not done properly. Membranes and adsorption processes are available to remove nitrogen, normally the largest constituent in a typical exhaust gas stream, but the high gas temperatures would present problems for the membranes and for the transport ducting system. This could mean that the flue gases would need to be cooled prior to removing the nitrogen. In that case, it may be preferred to make the post-treatment equipment larger to handle the larger gas flows and more dilute pollutant concentrations, compared to pretreating the gases prior to reaching the post-treatment system.

Many of the post-treatment systems for removing pollutants from exhaust streams involve catalysts. There are often difficulties in using catalysts because of the high gas temperatures and dirty exhaust streams from industrial combustion processes that can foul or damage catalytic removal systems. They can be thermally deactivated by sintering for example. However, they have been successfully used to remove a wide range of pollutants including NO_x, SO_x, CO, and VOCs. Heck and Farrauto [97] have written a good general-purpose book on the use of catalysts for air pollution control.

In general, it is usually desirable to concentrate the pollutants in the exhaust gas stream prior to treatment as it is often cheaper and easier to remove them [98]. In many cases it is relatively simple to remove the water vapor in the process, although this can be more complicated in particle-laden flows. The largest diluent in most industrial combustion exhaust streams is nitrogen which is not as simply removed. One of the benefits of oxygen-enhanced combustion is the removal of some or nearly all of the diluent nitrogen from the system [5].

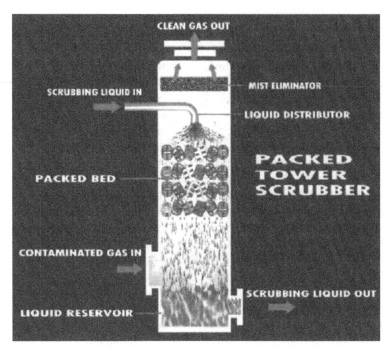

Figure 1.27   Packed tower scrubber. (Courtesy of Croll, Westfield, NJ.)

One of the potential issues with some of the post-treatment techniques that remove a pollutant from the exhaust gas stream is what to do with the collected pollutant. For example, particulates can be removed from an exhaust gas stream with a packed tower scrubber (see Fig. 1.27), a wet electrostatic precipitator (see Fig. 1.28), a dry electrostatic precipitator (see Fig. 1.29), or a wet scrubber (see Fig. 1.30). However, some of these techniques produce a sludge that must either be treated or disposed of as a result of the particulate removal. It would be desirable to recycle the collected materials but this may not be possible or economical in all cases. Table 1.12 shows the wide variety of techniques used either to prevent NO_x formation or to remove NO_x from exhaust gases [99]. These are discussed in more detail in Chap. 6. Table 1.13 shows the common techniques used to remove SO_x from exhaust gases [100]. These are discussed in more detail in Chap. 8.

1.8.5 Process Control

An important part of minimizing pollution emissions is to control all aspects of the process. This includes accurately metering the fuel and oxidizer flows and compositions to the burners, controlling the combustor itself (e.g., minimizing air infiltration into the process), and monitoring the incoming raw material feed rates into the combustor. If any of the system parameters are out of specification, then pollution emissions may be adversely affected. For example, if the fuel flow increases while the oxidizer flow rate remains the same, then carbon monoxide emissions will likely increase. If there are large air leaks into the furnace, then NO_x emissions are

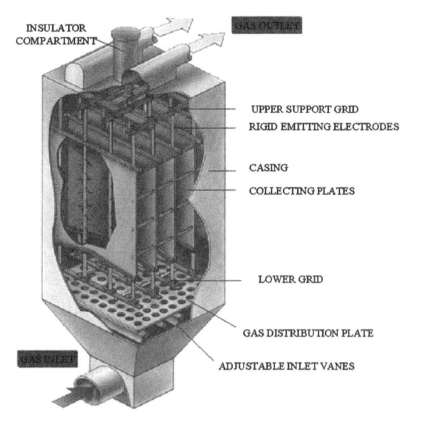

INSULATOR
COMPARTMENT

GAS OUTLET

UPPER SUPPORT GRID

RIGID EMITTING ELECTRODES

CASING

COLLECTING PLATES

LOWER GRID

GAS DISTRIBUTION PLATE

GAS INLET

ADJUSTABLE INLET VANES

Figure 1.28   Schematic of a wet electrostatic precipitator. (Courtesy of Belco, Parsippany, NJ.)

likely to increase. If one of the incoming raw materials contains more fine particles than normal, then particulate emissions may increase.

Another aspect of process control is for the proper operation of the pollution-control equipment. This includes operating the equipment within design specifications for the appropriate pollutants and also properly maintaining the equipment. For example, particulate emissions control devices are designed for a given maximum gas flow rate and particle loading density. If higher flow rates or particle loading densities are present, then the removal efficiency will likely suffer. If an electrostatic precipitator (ESP) is the only device used for controlling particulate emissions, then only particles that can be electrically charged can be removed (see Chap. 9). If particles that cannot be electrically charged are present in the system then the ESP will not be effective. If the pollution-control equipment is not properly maintained, then the removal efficiencies will decline. For example, if holes develop in the fabric filters in a baghouse then particles will pass through the system without being captured.

1.8.6   Other

Another technique that is often important in minimizing pollutant emissions is equipment maintenance. This includes the combustor, the burners, and the control

Figure 1.29 Photograph of a dry electrostatic precipitator. (Courtesy of United McGill, Columbus, OH.)

Figure 1.30 Wet scrubber. (Courtesy of Wheelabrator, Pittsburgh, PA.)

Table 1.12 NO$_x$ Control Techniques

Technique	Description	Advantages	Disadvantages	Impacts to consider	Applicability	NO$_x$ reduction
LEA	Reduces oxygen availability	Easy operational modification	Low NO$_x$ reduction potential	High carbon monoxide emissions, flame length, flame stability	All fuels	1–15%
OSC BOOS OFA	Staged combustion, creating fuel-rich and fuel-lean zones	Low operating cost, no capital requirement required for BOOS	(1) Typically requires higher air flow to control carbon monoxide; (2) relatively high capital cost; (3) moderate capital cost	Flame length, forced draft fan capacity, burner header pressure	All fuels; multiple-burner devices	30–60%
Air lances						
LNB	Provides internal staged cumbustion, thus reducing peak flame temperature and oxygen availability	Low operating cost, compatible with FGR as a combination technology to maximize NO$_x$ reduction	Moderately high capital cost; applicability depends on combustion device and fuels, design characteristics, and waste streams.	Forced-draft fan capacity, flame length, design compatibility, turndown flame stability	All fuels	30–50%
FGR	Up to 20–30% of the flue gas recirculated and mixed with the combustion air, thus decreasing peak flame temperatures.	High NO$_x$ reduction potential for natural gas and low-nitrogen fuels	Moderately high capital cost, moderately high operating cost, affects heat transfer and system pressures	Forced-draft fan capacity, furnace pressure, burner pressure drop, turndown flame stability	Gas fuels and low-nitrogen fuels	40–80%
W/SI	Injection of steam or water at the burner, which decreases flame temperature	Moderate capital cost, NO$_x$ reductions similar to FGR	Efficiency penalty due to additional water vapor loss and fan power requirements for increased mass flow	Flame stability, efficiency penalty	Gas fuels and low-nitrogen fuels	40–70%
RAPH	Air preheater modification to reduce preheat, thereby reducing flame temperature	High NO$_x$ reduction potential.	Significant efficiency loss (1% per 40°F)	Forced-draft fan capacity, efficiency penalty	Gas fuels and low-nitrogen fuels	25–65%

(continued)

Table 1.12 Continued

Technique	Description	Advantages	Disadvantages	Impacts to consider	Applicability	NO_x reduction
SCR	Catalysts located in flue gas stream (usually upstream of air heater) promotes reaction of ammonia with NO_x	High NO_x removal	Very high capital cost, high operating cost, extensive ductwork to and from reactor required; large volume reactor must be sited, increased pressure drop may require induced-draft fan or larger forced-draft fan, reduced efficiency, ammonium sulfate removal equipment for air heater required, water treatment of air heater wash required	Space requirements, ammonia slip, hazardous waste disposal	Gas fuels and low-sulfur liquid and solid fuels	70–90%
SNCR—urea injection	Injection of urea into furnace to react with NO_x to form nitrogen and water	Low capital cost, relatively simple system, moderate NO_x removal, non-toxic chemical, typically low energy injection sufficient	Temperature dependent, design must consider boiler operating conditions and design, NO_x reduction may decrease at lower loads	Furnace geometry and residence time, temperature profile	All fuels	25–50%
SNCR—ammonia injection	Injection of ammonia into furnace to react with NO_x to form nitrogen	Low operating cost, moderate NO_x removal	Moderately high, capital cost; ammonia handling storage vaporization	Furnace geometry and residence time, temperature profile	All fuels	25–50%

Source: Ref. 99.

Table 1.13 Flue Gas Desulfurization Process

Process genetics	Process operations	Active material	Key sulfur product
Throwaway processes			
1. Lime or limestone	Sturry scrubbling	$CaO, CaCO_3$	$CaSO_3/CaSO_4$
2. Sodium	Na_2SO_3 Solution	Na_2CO_3	Na_2SO_4
3. Dual alkali	Na_2SO_3 Solution, regenerated by CaO or $CaCO_3$	$CaCO_3/Na_2SO_3$ or $CaO/NaOH$	$CaSO_3/CaSO_4$
4. Magnesium promoted lime or limestone	$MgSO_3$ Solution, regenerated by CaO or $CaCO_3$	$MgO/MgSO_4$	$CaSO_3/CaSO_4$
Regenerative processes			
1. Magnesium oxide	$Mg(OH)_2$ slurry	MgO	15% SO_2
2. Sodium (Wellman–Lord)	Na_2SO_3 Solution	Na_2SO_3	90% SO_2
3. Citrate	Sodium citrate solution	H_2S	Sulfur
4. Ammonia	Ammonia solution, conversion to SO_2	NH_4OH	Sulfur (99.9%)
Dry processes			
1. Carbon adsorption	Adsorption at 400 K, reaction with H_2S to S, reaction with H_2 to H_2S	Activated carbon/H_2	Sulfur
2. Spray dryer	Absorption by sodium carbonate or slaked lime solutions	$Na_2CO_3/$ $Ca(OH)_2$	Na_2SO_3/Na_2SO_4 or $CaSO_3/CaSO_4$

Source: Ref. 100.

equipment. If combustors are not properly maintained, then emissions can be adversely affected. For example, if the combustor develops cracks that permit air infiltration, NO_x emissions will typically increase. If broken sight ports are not replaced, then large quantities of unwanted or 'tramp' air can infiltrate into the furnace. If the refractory becomes damaged and is not replaced, hot spots on the outer shell will develop and can injure personnel who may inadvertently come in contact with these hot surfaces.

It is very important to keep burners in good operating conditions to minimize pollutant emissions. If fuel-injection nozzles become plugged, the mixing patterns that produce the flame will be adversely affected, which normally would increase NO_x emissions, for example. This commonly occurs in the petrochemical industry when heavy hydrocarbons are improperly fired causing the injection nozzles to clog from carbon buildup due to coking. If the oil injectors become clogged then oil atomization can be adversely affected, which could increase particulate emissions and opacity. It is not uncommon for broken pieces of refractory to fall on to or into burners, significantly disturbing the designed burner performance. An important part of the burner that is sometimes overlooked is the burner block (also referred to as the burner tile or quarl). A cracked burner block can adversely affect the performance of the burner.

Another important component of the combustion system is the controls. These include things like valves, pressure and temperature measuring instruments, burner air control dampers, exhaust stack dampers, flow meters, controllers, fan and blower controls, and flame management systems to name a few. Proper operation of this equipment is integral to the optimal operation of the overall combustion system. For example, if the exhaust gas damper is not in good operating condition then excess air levels in the combustor may be adversely affected, causing the pollutant emissions to increase.

Related to the control system is proper training of the operators running the combustion systems. This ensures that the operators know how to operate the equipment as designed in order to minimize pollutant emissions. This also ensures that the operators will identify problems in a timely manner to minimize how long the equipment may be running out of specification. Proper training includes both preventative maintenance and troubleshooting.

REFERENCES

1. U.S. Department of Energy (DOE). *Industrial Combustion Vision: A Vision by and for the Industrial Combustion Community*. Washington, DC: U.S. DOE, 1998.
2. GF Hewitt, GL Shires, TR Bott. *Process Heat Transfer*. Boca Raton, FL: CRC Press, 1994.
3. U.S. Department. of Energy, Energy Information Administration. Annual Energy Outlook 1999. Rep. DOE/EIA-0383(99), Washington, DC.
4. P Barry, S Somers. Duct Burners. In: CE Baukal, ed. *The John Zink Combustion Handbook*. Boca Raton, FL: CRC Press, pp. 523–544, 2001.
5. CE Baukal, ed. *Oxygen-Enhanced Combustion*. Boca Raton, FL: CRC Press, 1998.
6. DM Golden. Interaction of combustion with the atmosphere. *Proc. Combustion Inst.*, Vol. 28, pp. 2383–2392, 2000.
7. JJ Peirce, RF Weiner, PA Vesilind. *Environmental Pollution and Control*. Boston, MA: Butterworth-Heinemann, 1998.
8. U.S. EPA. Latest Findings on National Air Quality: 2000 Status and Trends. Washington, DC: U.S. Environmental Protection Agency. Rep. EPA 454/K-01-002, 2001.
9. U.S. EPA. Solid Waste Management and Greenhouse Gases: A Life-Cycle Assessment of Emissions and Sinks. Rep. EPA-530-R-02-006. Washington, DC: Environmental Protection Agency, May 2002.
10. RA Geyer (ed.). *A Global Warming Forum: Scientific, Economic, and Legal Overview*. Boca Raton, FL: CRC Press, 1992.
11. www.epa.gov/globalwarming
12. U.S. EPA. Global Warming and Our Changing Climate. Washington, DC: U.S. Environmental Protection Agency, Rep. EPA 430-F-00-011, 2000.
13. PJ Michaels, RC Balling. Climate history during the recent greenhouse enhancement. In RA Geyer (ed.). *A Global Warming Forum: Scientific, Economic, and Legal Overview*. Boca Raton, FL: CRC Press, 1992.
14. Anon. Climate change belly-flop. *Oil & Gas J.*, Vol. 99(25): p. 19, 2001.
15. National Academy of Science. *Climate Change Science: An Analysis of Some Key Questions*. Washington, DC: National Academy Press, 2001.
16. SI Rasool, SH Schneider. Atmospheric carbon dioxide and aerosols: effects of large increases on global climate. *Science*, Vol. 173, pp. 138–141, 1971.
17. TG Moore. *Climate of Fear: Why We Shouldn't Worry about Global Warming*. Washington, DC: CATO Institute, 1998.

18. WR Cline. *The Economics of Global Warming*. Washington, DC: Institute for International Economics, 1992.

19. SG Philander. *Is the Temperature Rising?: The Uncertain Science of Global Warming*. Princeton, NJ: Princeton University Press, 1998.

20. JL Williams. *The Rise and Decline of Public Interest in Global Warming*: Toward a Pragmatic Conception of Environmental Problems. Huntington, NY: Nova Science, 2001.

21. PE Morthorst. National environmental targets and international emission reduction instruments. *Energy Policy*, Vol. 31, pp. 73–83, 2003.

22. SG Jensen, K Skytte. Simultaneous attainment of energy goals by means of green certificates and emission permits. *Energy Policy*, Vol. 31, pp. 63–71, 2003.

23. L Nielsen, T Jeppesen. Tradable green certificates in selected European countries – overview and assessment. *Energy Policy*, Vol. 31, pp. 3–14, 2003.

24. PE Morthorst, K Skytte, P Fristrup. Preface: Green certificates and emission trading. *Energy Policy*, Vol. 31, pp. 1–2, 2003.

25. http://europa.eu.int/eur-lex/en/index.html

26. BA Forster. The Acid Rain Debate: Science and Special Interests in Policy Formation. Ames, IA: Iowa State University, 1993.

27. CC Park. *Acid Rain: Rhetoric and Reality*. London: Methuen, 1987.

28. RA Cox. Stratospheric chemistry and the effect of pollutants on ozone. In C Vovelle (ed.). *Pollutants from Combustion: Formation and Impact on Atmospheric Chemistry*. Dordrecht, The Netherlands: Kluwer, 2000.

29. U.S. EPA. Air Quality Guide for Ozone. Rep. EPA-456/F-99-002. Washington, DC: U.S. Environmental Protection Agency, 1999.

30. RA Geyer. *A Global Warming Forum: Scientific, Economic, and Legal Overview*. Boca Raton, FL: CRC Press, 1993.

31. LW Canter. 5.1 Sources, Effects, and Fate of Pollutants. In DHF Liu, BG Lipták. *Environmental Engineers' Handbook*. Boca Raton, FL: Lewis Publishers, 1997.

32. CW Spicer, SM Gordon, MW Holdren, TJ Kelly, R Mukund. *Hazardous Air Pollutant Handbook: Measurements, Properties, and Fate in Ambient Air*. Boca Raton, FL: Lewis Publishers, 2002.

33. M Crawford. *Air Pollution Control Theory*. New York: McGraw-Hill, 1976.

34. K Wark, CF Warner. *Air Pollution: Its Origin and Control*, 2nd edn. New York: Harper and Row, 1981.

35. AC Stern, RW Boubel, DB Turner, DL Fox. *Fundamentals of Air Pollution*, 2nd edn. Orlando, FL: Academic Press, 1984.

36. FAL Dullien. *Introduction to Industrial Gas Cleaning*. San Diego, CA: Academic Press, 1989.

37. J Benítez. *Process Engineering and Design for Air Pollution Control*. Englewood Cliffs, NJ: Prentice Hall, 1993.

38. PN Cheremisinoff (ed.). *Air Pollution Control and Design for Industry*. New York: Marcel Dekker, 1993.

39. CD Cooper, FC Alley. *Air Pollution Control: A Design Approach*. Prospect Heights, IL: Waveland Press, 1994.

40. JC Mycock, JD McKenna, L Theodore. *Handbook of Air Pollution Control Engineering and Technology*. Boca Raton, FL: Lewis Publishers, 1995.

41. G Baumbach. *Air Quality Control*. Berlin: Springer, 1996.

42. Environment Agency. Combustion Processes Supplementary Guidance Note. Technical Guidance IPC S3 1.01. Bristol, U.K.: Environment Agency, 2000.

43. WL Heumann (ed.). *Industrial Air Pollution Control Systems*. New York: McGraw-Hill, 1997.

44. DHF Liu, BG Lipták (eds.). *Environmental Engineers' Handbook*, 2nd edn. Boca Raton, FL: Lewis Publishers, 1997.
45. N deNevers. *Air Pollution Control Engineering*, 2nd Edn. New York: McGraw-Hill, 2000.
46. KB Schnelle, CA Brown. *Air Pollution Control Technology*. Boca Raton, FL: CRC Press, 2002.
47. RA Strehlow. *Fundamentals of Combustion*, Scranton, PA: International Textbook Co., 1968.
48. FA Williams. *Combustion Theory*. Menlo Park, CA: Benjamin/Cummings, 1985.
49. B Lewis, G von Elbe. *Combustion, Flames and Explosions of Gases*; 3rd edn. New York: Academic Press, 1987.
50. W Bartok, AF Sarofim (eds.). *Fossil Fuel Combustion*. New York: John Wiley, 1991.
51. RM Fristrom. *Flame Structure and Processes*. New York: Oxford University Press, 1995.
52. I. Glassman. *Combustion*. 3rd edn. New York: Academic Press, 1996.
53. JB Edwards. *Combustion: The Formation and Emission of Trace Species*. Ann Arbor, MI: Ann Arbor Science Publishers, 1974.
54. JA Barnard, JN Bradley. *Flame and Combustion*. 2nd edn. London: Chapman and Hall, 1985.
55. SR Turns. *An Introduction to Combustion*. New York: McGraw-Hill, 1996.
56. J Griswold. *Fuels, Combustion and Furnaces*. New York: McGraw-Hill, 1946.
57. A Stambuleanu. *Flame Combustion Processes in Industry*. Tunbridge Wells, UK: Abacus Press, 1976.
58. E Perthuis. *La Combustion Industrielle*. Paris: Éditions Technip, 1983.
59. EL Keating. *Applied Combustion*. New York: Marcel Dekker, 1993.
60. G Borman, K Ragland. *Combustion Engineering*. New York: McGraw-Hill, 1998.
61. CG Segeler (ed.). *Gas Engineers Handbook*. New York: Industrial Press, 1965.
62. RD Reed. *Furnace Operations*, 3rd edn. Houston, TX: Gulf Publishing, 1981.
63. R Pritchard, JJ Guy, NE Connor. *Handbook of Industrial Gas Utilization*. New York: Van Nostrand Reinhold, 1977.
64. RJ. Reed. *North American Combustion Handbook*. Vol. I, 3rd edn. Cleveland, OH: North American Mfg. Co., 1986.
65. IHEA. *Combustion Technology Manual*. 5th edn. Arlington, VA Industrial Heating Equipment Assoc., 1994.
66. JA Danielson (ed.). *Air Pollution Engineering Manual*, 2nd edn. Research Triangle Park, NC: Environmental Protection Agency, 1973.
67. WT Davis (ed.). *Air Pollution Engineering Manual*, 2nd edn. New York: John Wiley, 2000.
68. AC Stern. *Air Pollution*, 3rd edn, Vol. 4: Engineering Control of Air Pollution. New York: Academic Press, 1977.
69. HE Hesketh. *Air Pollution Control: Traditional and Hazardous Pollutants*. Lancaster, PA: Technomic, 1991.
70. DC Mussatti (ed.). *EPA Air Pollution Control Cost Manual*, 6th edn. Rep. EPA/452/B-02-001. Washington, DC: U.S. Environmental Protection Agency, Jan. 2002.
71. CE Baukal. *Heat Transfer in Industrial Combustion*. Boca Raton, FL: CRC Press, 2000.
72. J Elkins, N Frank, J Hemby, D Mintz, J Szykman, A Rush, T Fitz-Simons, T Rao, R Thompson, E Wildermann, G Lear. National Air Quality and Emissions Trends Report, 1999. Washington, DC: U.S. Environmental Protection Agency, Rep. EPA 454/R-01-004, 2001.
73. G Braswell, Y Matros, G Bunimovich. NO_x in non-utility industries (Part 1). Environ. Protect. Vol. 12, No. 6, pp. 50–55, 2001.
74. PN Cheremisinoff, V Frega, TM Hellman. Industrial odor control. In PN Cheremisinoff (ed.). *Air Pollution Control and Design for Industry*. New York: Marcel Dekker, 1993.

75. S Bassett. Online Air Compliance Information Saves Time. *Pollut. Eng.* Vol. 33, No. 1, pp. 43–45, 2001.

76. FW Brownell. *Clean Air Handbook*, 3rd edn. Rockville, MD: Government Institutes, 1998.

77. P Zaborowsky. Getting a head start. *Environ. Protect.*, Vol. 13, No. 4, pp. 26–53, 2002.

78. Energetics, Inc. *Glass Technology Roadmap Workshop*. Columbia, MD, 1997.

79. D Bookchin, D Farnsworth. Environmental laws and regulations. In DHF Liu, BG Lipták (eds.). *Environmental Engineers' Handbook*, 2nd edn. Boca Raton, FL: Lewis Publishers, 1997.

80. U.S. EPA. Method 19: Determination of Sulfur Dioxide Removal Efficiency and Particulate Matter, Sulfur Dioxide, and Nitrogen Oxide Emission Rates. Code of Federal Regulations 40 Part 60, Appendix A-4. Washington, DC: U.S. Environmental Protection Agency, 2001.

81. WC Zegel. Standards. In DHF Liu, BG Lipták (eds.). *Environmental Engineers' Handbook*, 2nd edn. Boca Raton, FL: Lewis Publishers, 1997.

82. U.S. EPA. Compilation of Air Pollutant Emission Factors, Vol. I: Stationary Point and Area Sources, 5th edn. Washington, DC: U.S. Environmental Protection Agency Rep. AP-42, 1995.

83. U.S. EPA. RACT/BACT/LAER Clearinghouse, www.epa.gov/ttn/rblc/htm/bl02.cfm.

84. S Bassett. Avoiding best available retrofit technology. *Pollut. Eng.*, Vol. 33, No. 8, pp. 27–28, 2001.

85. WM Vatavuk. Air pollution control cost indexes: update #12. *Environ. Progr.*, Vol. 19, No. 3, F16–F18, 2000.

86. DHF Liu. 5.9 Emission Measurements; 5.11 Stack Sampling; 5.12 Continuous Emission Monitoring. In DHF Liu, BG Lipták. *Environmental Engineers' Handbook*. Boca Raton, FL: Lewis Publishers, 1997.

87. PW Spaite, JO Burckle. 2 Selection, Evaluation, and Application of Control Devices. In AC Stern (ed.). *Air Pollution*, Vol. 4: Engineering Control of Air Pollution, 3rd edn. New York: Academic Press, 1977.

88. www.tnrcc.state.tx.us/permitting/airperm/banking/index.htm

89. WT Davis, AJ Buonicore, L Theodore. Air Pollution Control Engineering. In WT Davis (ed.). *Air Pollution Engineering Manual*. New York: John Wiley, 2000.

90. G Ramachandran. Pollutants: minimization and control. In DHF Liu, BG Lipták (eds.). *Environmental Engineers' Handbook*, 2nd edn. Boca Raton, FL: Lewis Publishers, 1997.

91. K Yeager. Control alternatives and the acid rain issue. In TC Elliot, RG Schwieger (eds.). *The Acid Rain Sourcebook*. New York: McGraw-Hill, 1984.

92. T Handy, R Kelcher. Furnace burner maintenance and firing efficiency. *Proceedings of the 14th Ethylene Producers' Conference*. New York: American Institute of Chemical Engineers, Vol. 11, pp. 160–165, 2002.

93. MW Stansifer. Fuel gas clean up for low NO_x burners. *Proceedings of the 14th Ethylene Producers' Conference*. New York: American Institute of Chemical Engineers., Vol. 11, pp. 166–172, 2002.

94. J McAdams, J Karan, R Witte, M Claxton. Low NO_x Burner maintenance in high temperature furnaces. *Proceedings of the 14th Ethylene Producers' Conference*. New York: American Institute of Chemical Engineers, Vol. 11, pp. 351–370, 2002.

95. CM Martinson. Getting the most from air pollution control systems. *Chem. Process.*, Vol. 65, No. 2, pp. 25–28, 2002.

96. KC Schifftner. *Air Pollution Control Equipment Selection Guide*. Boca Raton, FL: Lewis Publishers, 2002.

97. RM Heck, RJ Farrauto. *Catalytic Air Pollution Control: Commercial Technology*. New York: Van Nostrand Reinhold, 1995.

98. MW First. Control of Systems, Processes, and Operations. In AC Stern (ed.). *Air Pollution*, Vol. 4: Engineering Control of Air Pollution, 3rd edn. New York: Academic Press, 1977.

99. DHF Liu. 5.19 Gaseous Emission Control. In DHF Liu, BG Lipták (eds.). Environmental Engineers' Handbook, 2nd edn. Boca Raton, FL: Lewis Publishers, 1997.

100. KT Chuang, AR Sanger. 5.20 Gaseous Emission Control: Physical and Chemical Separation. In DHF Liu, BG Lipták. *Environmental Engineers' Handbook*, 2nd edn. Boca Raton, FL: Lewis Publishers, 1997.

2

Some Combustion Fundamentals

2.1 INTRODUCTION

The purpose of this chapter is to present some of the basic fundamentals of combustion as a background for some of the pollution control strategies and techniques discussed in this book. Some of these fundamentals are also important for making pollutant emission calculations. The purpose is not to provide a detailed and theoretical treatment on combustion. Many good books are available to the interested reader on the subject of the theory of combustion [1–7].

There are three fundamental components in the combustion reaction [8] which will be briefly discussed next. The overall process can be written as

Fuel + Oxidizer + Diluent → Combustion Products

Each of these plays an important role in determining the pollutant emissions that are produced during combustion.

2.1.1 Fuel

A variety of fuels are used in industrial combustion processes. The fuel type varies by application and by geography. For example, the use of liquid fuels is much more common in South America compared to the predominant use of gaseous fuels in North America. Natural gas is commonly used in cement manufacturing compared to the use of refinery fuel gases containing hydrogen, methane, and propane in the petrochemical industry. The sections that follow are only intended to alert the reader to some of the problems and issues to be considered and is not intended to be comprehensive or exhaustive. The subject of fuel type is considered throughout the book as it pertains to a given industry and to pollution emissions. This section is merely an introduction to the topic.

2.1.1.1 Gaseous Fuels

Gaseous fuels are the predominant fuel source used in most of the applications considered here. In general, natural gas is the predominant gaseous fuel used because of its low cost and availability. However, a wide range of gaseous fuels are used in,

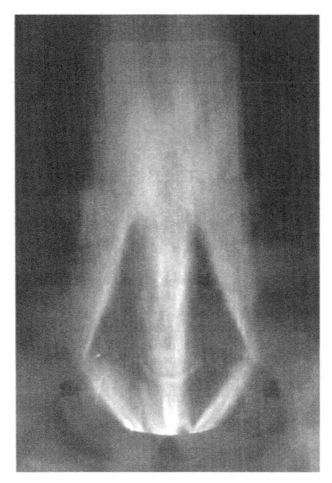

Figure 2.1   Typical gaseous fuel flame. (From Ref. 27. Courtesy of John Zink Co.)

for example, the chemicals industry [9]. These fuels contain multiple components such
as methane, hydrogen, propane, nitrogen and carbon dioxide and are sometimes
referred to as refinery fuel gases. Figure 2.1 shows an example of a typical
nonluminous gaseous flame from a burner used in the petrochemical industry.
Gaseous fuels are among the easiest to control because no vaporization is required as
with liquid and solid fuels. They are also often simpler to control to minimize
pollution emissions because they are more easily staged compared to liquid and solid
fuels. Table 2.1 shows typical data for the combustion of common hydrocarbons.

2.1.1.2   Liquid Fuels

Liquid fuels are used in some limited applications, but are more prevalent in certain
areas of the world such as South America. Number 2 and no. 6 oils are the most
commonly used liquid fuels. Waste liquid fuels are also used in incineration
processes. One of the specific challenges of using oils is vaporizing the liquid into
small enough droplets to burn completely. Improper atomization produces high

Table 2.1 Combustion Data for Hydrocarbons

Hydrocarbon	Formula	Higher healing value (vapor) (Btu/lb)	Theoretical airfuel ratio by mass	Max flame speed (ft/sec)	Adiabatic flame temp (in air) (°F)	Ignition temp (in air) (°F)	Flash point (°F)	Flammability limits (in air) (vol%)	
Paraffins or Alkanes									
Methane	CH_4	23,875	17.195	1.1	3484	130	Gas	5.0	15.0
Ethane	C_2H_6	22,323	15.899	1.3	3540	968–1166	Gas	3.0	12.5
Propane	C_3H_8	21,669	15.246	1.3	3573	871	Gas	2.1	10.1
n-Butane	C_4H_{10}	21,321	14.984	1.2	3583	761	−76	1.86	8.41
iseop-Butane	C_4H_{10}	21,271	14.984	1.2	3583	864	−117	1.80	8.44
n-Pemtame	C_5H_{12}	21,095	15.323	1.3	4050	588	< −40	1.40	7.80
iso-Pentane	C_5H_{12}	21,047	15.323	1.2	4055	788	< −60	1.32	9.16
Neopentane	C_5H_{12}	20,978	15.323	1.1	4060	842	Gas	1.38	7.22
n-Hexane	C_6H_{14}	20,966	15.238	1.3	4030	478	−7	1.25	7.0
Neohexane	C_6H_{14}	20,931	15.238	1.2	4055	797	−54	1.19	7.58
n-Heptane	C_7H_{16}	20,854	15.141	1.3	3985	433	25	1.00	6.00
Triptane	C_7H_{16}	20,824	15.141	1.2	4035	849	—	1.08	6.69
n-Octane	C_8H_{18}	20,796	15.093	—	—	428	56	0.95	3.20
iso-Octane	C_8H_{18}	20,770	15.093	1.1	—	837	10	0.79	5.94
Olefins or Alkeries									
Ethylene	C_2H_4	21,636	14.807	2.2	4250	914	Gas	2.75	28.6
Propylene	C_3H_6	21,048	14.807	1.4	4090	856	Gas	2.00	11.1
Butylene	C_4H_8	20,854	14.807	1.4	4030	829	Gas	1.98	9.65
iso-Butene	C_4H_8	20,737	14.807	1.2	—	869	Gas	1.80	9.0
n-Pentene	C_5H_{10}	20,720	14.807	1.4	4165	569	—	1.65	7.70
Aromatics									
Benzene	C_6H_6	18,184	13.297	1.3	4110	1044	12	1.35	6.65
Toluene	C_7H_8	18,501	13.503	1.2	4050	997	40	1.27	6.75
p-Xylene	C_8H_{10}	18,663	13.663	—	4010	867	63	1.00	6.00
Other Hydrocarbons									
Acetylene	C_2H_2	21,502	13.297	4.6	4770	763–824	Gas	2.50	81
Naphthalene	$C_{10}H_8$	17,303	12.923	—	4100	959	174	0.90	5.9

Source: Ref. 9. (Courtesy of CRC Press.)

unburned hydrocarbon emissions and reduces fuel efficiency. Steam and compressed air are commonly used to atomize liquid fuels. The atomization requirements often reduce the options for modifying the burner design to reduce pollutant emissions. Another challenge is that liquid fuel oils often contain impurities like nitrogen and sulfur that produce pollution emissions. In the case of fuel-bound nitrogen, so-called fuel NO_x emissions (see Chap. 6) increase. In the case of sulfur, essentially all of the sulfur in a liquid fuel converts to SO_x emissions (see Chap. 8).

Williams [10] has written a primer on the fundamentals of oil combustion that considers the effects of different types of oils, both light and heavy, on the combustion process [10]. The paper also includes a brief discussion of the primary pollutants from oil flames such as CO and NO$_x$. Waste oils represent a special class of liquid fuels. These may contain pollutants such as polychlorinated biphenyls (PCBs) that can be especially problematic because they may be carcinogenic. Table 2.2 shows the typical composition for waste oil [11]. Table 2.3 shows typical controlled and uncontrolled air emissions from the combustion of waste oil.

Table 2.2 Typical Composition of Waste Oil

Parameter	Average	Maximum
Sulfur	0.75% lb/lb oil	1.0% lb/lb oil
Chlorine	0.1% lb/lb oil	0.3% lb/lb oil
Fluorine	50 ppmw	80 ppmw
Lead	300 ppmw	350 ppmw
Cadmium	1 ppmw	15 ppmw
Nickel	10 ppmw	30 ppmw
Chromium	10 ppmw	50 ppmw
Copper	10 ppmw	50 ppmw
Vanadium	25 ppmw	50 ppmw
PCBs	Normally absent	10 ppmw
Ash	0.5% lb/lb oil	0.5% lb/lb oil
Suspended solids	0.5% lb/lb oil	0.5% lb/lb oil
Viscosity	39 centistokes at 40°C	49 centistokes at 40°C
Specific gravity	0.895 at 15°C	—
Flash point	85°C	—
Gross heating value	43.6 MJ/kg	—
Moisture content	2.5% lb/lb oil	3.0% lb.lb oil
Nitrogen	0.15% lb/lb oil	0.2% lb/lb oil

Source: Ref. 11.

Table 2.3 Typical Air Emissions from Combustion of Waste Oil

Substance	Typical uncontrolled air emissions (mg/m^3)	Typical controlled air emissions (mg/m^3 except as noted)
Hydrogen chloride	860 per % chlorine in fuel	30
Sulfur oxides	1700 per % sulfur in fuel	300
Nitrogen oxides	560 (without low NO$_x$ burners)	110–250 (with low NO$_x$ burners)
Particulates	50–125	25
Lead	28	Depends on fuel and process
Cadmium	1.2	Depends on fuel and process
Nickel	2.4	Depends on fuel and process
Copper, chromium, and vanadium	12	Depends on fuel and process
VOCs	Varies considerably by	
CO	process and	100
Dioxins/furans	combustor design	0.1–0.5 ng/m^3
PAH		0.1
PCBs		

Source: Ref. 11.

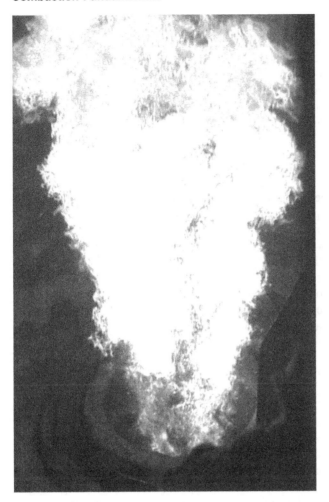

Figure 2.2 High-luminosity oil flame. (From Ref. 27. Courtesy of John Zink Co.)

One of the advantages of liquid fuel oils is that they often make the flames much more luminous. Figure 2.2 shows an example of a high-luminosity oil flame from a burner used in the petrochemical industry. The higher luminosity is caused by the high solid carbon content which produces infrared radiation when heated. This can significantly enhance the radiation heat transfer from the flame to the material being processed. This can indirectly reduce pollution emissions because the higher heat transfer can improve the thermal efficiency which means that less fuel needs to be burned for a given unit of production.

2.1.1.3 Solid Fuels

Solid fuels are not commonly used in most industrial combustion applications. The most common solid fuels are coal and coke. Coal is used in power generation and coke is used in some primary metals production processes. However, neither of these are considered traditional industrial combustion processes and therefore are not considered here. Another type of pseudosolid fuel is sludge that is processed in incinerators. Solid fuels also often contain impurities such as nitrogen and sulfur that can significantly increase pollutant emissions. Some solid fuels may also contain hazardous chemicals that can produce carcinogenic pollution emissions. Because solid fuels are not used frequently in the applications considered, they are only discussed in those specific cases.

2.1.2 Oxidizer

There is essentially only one oxidizer used in the industrial combustion processes considered here which is oxygen. Here, there is a distinction between the *oxidant* and the *oxidizer*. The oxidant in most industrial combustion processes is air, which contains approximately 79% nitrogen and 21% oxygen by volume. Only the oxygen in the air is the oxidizer, with nitrogen as a diluent.

Most industrial heating processes require substantial amounts of energy, which is commonly generated by combusting hydrocarbon fuels such as natural gas or oil. Most combustion processes use air as the oxidant. In many cases, these processes can be enhanced by using an oxidant that contains a higher proportion of O_2 than that in air. This is known as *oxygen-enhanced combustion* or OEC. Air consist of approximately 21% O_2 and 79% N_2 by volume. One example of OEC is using an oxidant consisting of air, blended with pure O_2. Another example is using high-purity O_2 as the oxidant, instead of air. This is usually referred to as *oxy/fuel* combustion.

New developments have made oxy/fuel combustion technology more amenable to a wide range of applications. In the past, the benefits of using oxygen could not always offset the added costs. New oxygen generation technologies, such as pressure and vacuum swing adsorption [12], have substantially reduced the cost of separating O_2 from air. This has increased the number of applications where using oxygen to enhance performance is cost justified. Another important development is the increased emphasis on the environment. In many cases, OEC can substantially reduce pollutant emissions [13]. This has also increased the number of cost-effective applications. The Gas Research Institute [14] and the U.S. Department of Energy [15] sponsored independent studies, which predict that OEC will be a critical combustion technology in the very near future.

Historically, air/fuel combustion has been the conventional technology used in nearly all industrial heating processes. Oxygen-enhanced combustion systems are becoming more common in a variety of industries. When traditional air/fuel combustion systems have been modified for OEC, many benefits have been demonstrated. Typical improvements include increased thermal efficiency, increased processing rates, reduced flue-gas volumes, and reduced pollutant emissions.

The use of oxygen in combustion has received relatively little attention from the academic combustion community. This may be for several reasons. Probably the most basic reason is the lack of research interest and funding to study OEC. The industrial gas companies that produce oxygen have been conducting research into OEC for many years which has been mostly applied R&D. Very little basic research has been done, compared to that on air/fuel combustion, to study the fundamental

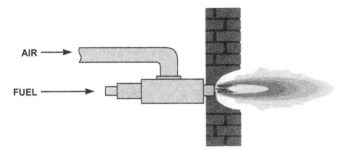

Figure 2.3 Schematic of an air/fuel burner.

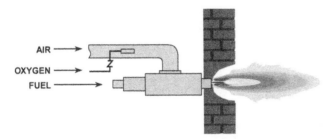

Figure 2.4 Burner with combustion air enriched with O_2. (From Ref. 23. Courtesy of CRC Press.)

processes in atmospheric flames utilizing OEC. The aerospace industry has done a considerable amount of work, e.g., to study the high-pressure combustion of liquid oxygen and liquid hydrogen used to propel space vehicles. However, that work has little relevance to the low-pressure combustion of fuels other than hydrogen in industrial furnace applications. Another reason why little research has been done may be due to concerns about the safety issues of using oxygen, as well as the very high-temperature flames that may be encountered using OEC. Another reason may be a cost issue since the small quantities of oxygen that might be used can be relatively expensive. Handling oxygen cylinders takes more effort than using either a houseline source of air or a small blower for the air used in small-scale flames.

2.1.2.1 Air/Fuel

Figure 2.3 shows a schematic of an air/fuel burner which is the most commonly used type in industrial combustion applications. In most cases the combustion is supplied by a fan or blower although there are many applications in the petrochemical industry where natural draft burners are commonly used. There are numerous variations of air/fuel burners, which are discussed throughout the book. Most of the research to date on reducing pollution emissions has been done on this class of burners.

2.1.2.2 Air Enriched With O_2

Figure 2.4 shows an air/fuel process where the air is enriched with O_2. This may be referred to as low-level O_2 enrichment or premix enrichment. Many conventional air/fuel burners can be adapted for this technology [16]. The O_2 is injected into the incoming combustion air supply, usually through a diffuser to ensure adequate mixing. This is usually an inexpensive retrofit that can provide substantial benefits. Typically, the added O_2 will shorten and intensify the flame. However, there may be some concerns if too much O_2 is added to a burner designed for air/fuel. The flame shape may become unacceptably short. The higher flame temperature may damage the burner or burner block. The air piping may need to be modified for safety reasons to handle higher levels of O_2.

2.1.2.3 O_2 Lancing

Figure 2.5 shows another method for enriching an air/fuel process with O_2. As in the first method, this is also generally used for lower levels of O_2 enrichment. However, oxygen lancing may have several advantages over air enrichment. No modifications to the existing air/fuel burner need to be made. Typically, the NO_x emissions are lower using O_2 lancing compared to premixing since this is a form of staging, which is a well-accepted technique for reducing NO_x [17]. Depending on the injection location, the flame shape may be lengthened by staging the combustion reactions. The flame heat release is generally more evenly distributed than with premix O_2 enrichment. Under certain conditions, O_2 lancing between the flame and the load causes the flame to be pulled toward the material. This improves the heat-transfer efficiency. Therefore, there is less likelihood of overheating the air/fuel burner, the burner block, and the refractory in the combustion chamber. Another variant of this staging method involves lancing O_2 not into the flame but somewhere else in the combustion chamber. One example of this technique is known as oxygen-enriched air staging or OEAS. That technology O_2 lancing is an inexpensive retrofit for existing processes. One potential disadvantage is the cost to add another hole in the combustion chamber for the lance. This includes both the installation costs and the lost productivity. However, the hole is typically very small.

One specific embodiment of O_2 lancing is known as undershot enrichment where O_2 is lanced into the flame from below. The lance is located between the burner and the material being heated. While air enrichment increases the flame temperature uniformly, the undershot technique selectively enriches the underside of

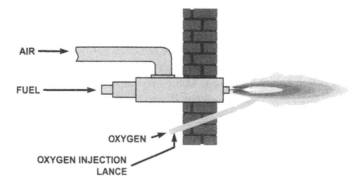

Figure 2.5   Burner with O_2 lanced into the flame. (From Ref. 23. Courtesy of CRC Press.)

the conventional flame, thereby concentrating extra heat downward toward the material being heated. While the mixing of oxygen and combustion air is not as complete with undershot oxygen as with premixing, this disadvantage is often outweighed by the more effective placement of the extra heat. Another benefit is that the refractory in the roof of the furnace generally receives less heat compared to air enrichment. This usually increases the life of the roof.

Jankes et al. [18] discuss the use of O_2 lancing in some industrial combustion applications including dolomite calcination, cement clinker production, glass melting, cast iron production, fritted glass tile production, and mineral wool production [18].

2.1.2.4 Oxy/Fuel

Figure 2.6 shows a third method of using OEC, commonly referred to as oxy/fuel combustion. In nearly all cases, the fuel and the oxygen remain separated inside the burner. They do not mix until reaching the outlet of the burner. This is commonly referred to as a nozzle-mix burner, which produces a diffusion flame. There is no premixing of the gases for safety reasons. Because of the extremely high reactivity of pure O_2, there is the potential for an explosion if the gases are premixed. In this method, high-purity oxygen ($>90\%$ O_2 by volume) is used to combust the fuel. As will be discussed later, there are several ways of generating the O_2. In an oxy/fuel system, the actual purity of the oxidizer will depend on which method has been chosen to generate the O_2. As will be shown later, oxy/fuel combustion has the greatest potential for improving a process, but it may also have the highest operating cost.

One specific variation of oxy/fuel combustion, known as "dilute oxygen combustion," is where fuel and oxygen are separately injected into the combustion chamber [19]. In order to ensure ignition, the chamber temperature must be above the autoignition temperature of the fuel. Depending on the exact geometry, this can produce an almost invisible flame, sometimes referred to as flameless oxidation. The advantage of this technique is very low NO_x emissions because hot spots in the "flame" are minimized, which generally reduces NO_x (see Chap. 6). A potential disadvantage, besides the safety concern, is a reduction in heat transfer as both the temperature and effective emissivity of the flame may be reduced.

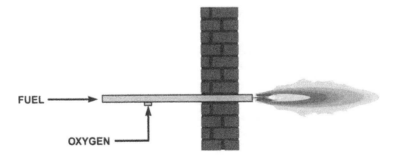

Figure 2.6 Burner with pure O_2. (From Ref. 23. Courtesy of CRC Press.)

OEC can dramatically reduce the carryover in a process originally designed for air/fuel combustion because of the reduction in the average gas velocity through the combustor which results from removing some or all of the diluent nitrogen from the system (see Fig. 2.7). For example, OEC was specifically used in a mobile waste incinerator to reduce the carryover of fine material from the primary to the secondary combustion chamber [20]. When the incinerator was fired on air/fuel, the material that was carried over into the secondary chamber adhered to the wall and eventually clogged the chamber, necessitating frequent shutdowns for maintenance. This problem was eliminated when the air/fuel system was replaced by an oxy/fuel system.

2.1.2.5 Air–Oxy/Fuel

Another method of using OEC involves separately injecting air and O_2 through a burner, as shown in Fig. 2.8. It is sometimes referred to as an air–oxy/fuel burner. This is a variation of the first three methods. In some cases, an existing air/fuel burner may be easily retrofitted by inserting an oxy/fuel burner through it [21]. In other cases, a specially designed burner may be used [22]. This method of OEC can have several advantages. It can typically use higher levels of O_2 than in methods one and two, which yields higher benefits. However, the operating costs are less than for oxy/fuel, which uses very high levels of O_2. The flame shape and heat release pattern may be adjusted by controlling the amount of O_2 used in the process. It is also a generally inexpensive retrofit. Many air/fuel burners are designed for dual fuels, usually a liquid fuel like oil, and a gaseous fuel like natural gas. The oil gun in the center of the dual fuel burner can usually be easily removed and replaced by either an O_2 lance or an oxy/fuel burner.

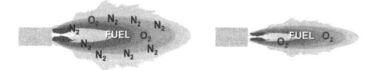

Figure 2.7   Enhanced chemical reactivity in oxy/fuel flames compared with air/fuel flames. (From Ref. 23. Courtesy of CRC Press.)

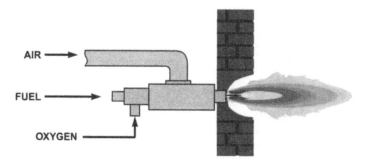

Figure 2.8   Air–oxy/fuel burner. (From Ref. 23. Courtesy of CRC Press.)

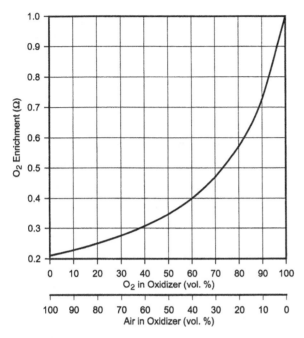

Figure 2.9 Oxidizer compositions for blends of air and pure O_2. (From Ref. 23. Courtesy of CRC Press.)

With this method of using OEC, the oxidizer composition may be specified in an alternative way. Instead of giving the overall O_2 concentration in the oxidizer, the oxidizer may be given as the fraction of the total oxidizer, which is air, and the fraction of the total oxidizer, which is pure O_2. The equivalent overall O_2 in the oxidizer can be calculated as follows:

$$\Omega = \frac{20.9}{0.209(\text{vol.\% } O_2) + (\text{vol.\% air})} \tag{2.1}$$

This conversion in Eq. (1.1) is shown graphically in Fig. 2.9. For example, the oxidizer may be specified as a blend of 60% O_2 and 40% air. That ratio of O_2 to air produces an equivalent of 39.8% overall O_2 in the oxidizer.

2.1.3 Diluent

The diluent refers to something other than the fuel or the oxidizer that is used to dilute either of them or the flame itself. The most common diluent is the products of combustion that are recycled back toward or into the burner. This is discussed in more detail next. Other diluents may also be used including water, steam, and gases like nitrogen or carbon dioxide. There may be a number of reasons for diluting the flame. An important reason is that it can reduce and moderate the flame temperatures that reduce NO_x emissions. Another reason is to change the heat-transfer distribution from the flame. The flame can be stretched to make the flame radiation more uniform by dilution. The convection heat transfer in the furnace may also be increased by adding diluent to the flame, which increases the mass flow rate and gas velocities in the furnace, although the gas temperature is usually less.

2.1.4 Recirculation

There are two types of gas recirculation utilized in industrial combustion processes: furnace gas recirculation (FuGR) and flue gas recirculation (FlGR). FuGR is where the combustion products are drawn back into the flame inside the furnace, while FlGR is where the combustion products are drawn back outside the furnace into the flame. The latter requires ductwork and some type of fan to move the hot gases from the exhaust end of the furnace to the burners. The generic term for either furnace or flue gas recirculation will be designated as FGR.

This gas recirculation is done for a variety of reasons, but primarily for improved thermal efficiency and reduced NO_x emissions. The thermal efficiency is improved because of the enhanced convective heat transfer inside the combustor due to the improved fluid flow. The heat transfer and thus the thermal efficiency are also enhanced because of the increased residence time of the hot gases in the combustor. NO_x emissions are often reduced using FGR because it reduces the peak flame temperatures in the combustion zone that are the primary source of thermal NO_x emissions (see Chap. 6).

There are a number of ways of quantifying the FGR process to determine the amount of gases recirculated back through the flame zone. The volume flow rate of flue gases (at standard temperature and pressure or STP) recirculated can be normalized to the volumetric flow rate of the incoming fuel (at STP):

$$R_1 = \frac{\text{Volume (STP) flue gases recycled}}{\text{Volume (STP) fuel}} \tag{2.2}$$

This is often used if the fuel is the primary motive force for inducing the FGR. An alternative ratio can be calculated using the volumetric flow of the incoming oxidant (at STP):

$$R_2 = \frac{\text{Volume (STP) flue gases recycled}}{\text{Volume (STP) oxidant}} \tag{2.3}$$

This is commonly used if the oxidant is the primary motive force for inducing the FGR. The FGR can be normalized by both the fuel and the oxidant:

$$R_3 = \frac{\text{Volume (STP) flue gases recycled}}{\text{Volume (STP) fuel} + \text{oxidant}} \tag{2.4}$$

The FGR can be normalized to the fuel flow rate, both on a mass basis:

$$R_4 = \frac{\text{Mass (STP) flue gases recycled}}{\text{Mass (STP) fuel}} \tag{2.5}$$

The FGR can be normalized to the oxidant flow rate on a mass basis:

$$R_5 = \frac{\text{Mass (STP) flue gases recycled}}{\text{Mass (STP) oxidant}} \tag{2.6}$$

The FGR can be normalized to the fuel and oxidant flow rates on a mass basis:

$$R_6 = \frac{\text{Mass (STP) flue gases recycled}}{\text{Mass (STP) fuel + oxidant}} \tag{2.7}$$

The actual volumetric flow of exhaust products can be normalized to the actual fuel flow rate:

$$R_7 = \frac{\text{Actual volume flue gases recycled}}{\text{Actual volume fuel}} \tag{2.8}$$

The actual volumetric flow of exhaust products can be normalized to the actual fuel flow rate:

$$R_8 = \frac{\text{Actual volume flue gases recycled}}{\text{Actual volume oxidant}} \tag{2.9}$$

The actual volumetric flow of exhaust products can be normalized to the actual fuel flow rate:

$$R_9 = \frac{\text{Actual volume flue gases recycled}}{\text{Actual volume fuel + oxidant}} \tag{2.10}$$

There are also numerous other possible combinations for calculating the recirculation ratio where the recycle gases and the fuel and/or oxidant are also included in the numerator. One example is:

$$R_{10} = \frac{\text{Volume (STP) fuel + flue gases recycled}}{\text{Volume (STP) fuel}} \tag{2.11}$$

As can be seen from the above, there are multiple possibilities for quantifying the exhaust gas recirculation rate. Figure 2.10 shows a comparison of the various methods of calculating the recirculation ratio. As can be seen, some of the methods are very similar and some are very different. Therefore, it is important to ensure what ratio is being used when comparing data.

The recirculation efficiency could also be written as a fraction of the total gases going through the burner or flame:

$$F_1 = \frac{\text{Volume (STP) flue gases recycled}}{\text{Volume (STP) fuel + oxidant + flue gas recycled}} \tag{2.12}$$

$$F_2 = \frac{\text{Mass flue gases recycled}}{\text{Mass fuel + oxidant + flue gas recycled}} \tag{2.13}$$

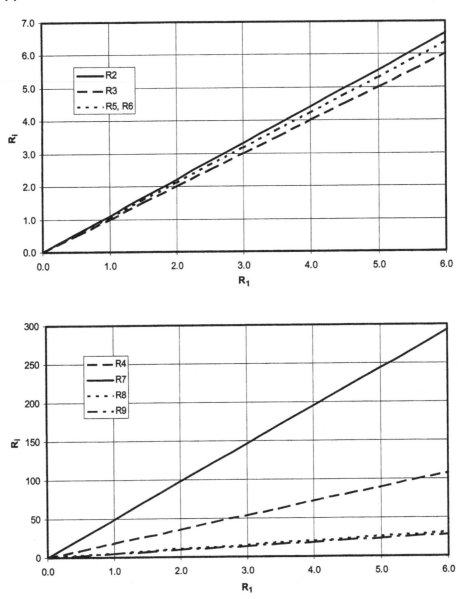

Figure 2.10 Comparison of exhaust gas recirculation ratios where R_1 has arbitrarily been chosen as the base case, for methane combusted stoichiometrically with air, with the recirculating gases at 2000°F (1366 K).

$$F_3 = \frac{\text{Volume flue gases recycled}}{\text{Volume fuel} + \text{oxidant} + \text{flue gas recycled}} \quad (2.14)$$

These fractions will range from 0 to something less than 1. Again, there are numerous other variations for these recycle fractions, although this is not commonly used for quantifying recirculation.

2.2 COMBUSTION CHEMISTRY

Combustion is usually considered to be the controlled release of heat and energy from the chemical reaction between a fuel and an oxidizer. This is in contrast to fires or explosions which are usually uncontrolled and undesirable. Virtually all of the combustion in industrial processes uses a hydrocarbon fuel. A generalized combustion reaction for a typical hydrocarbon fuel can be written as follows:

$$\text{Fuel} + \text{Oxidizer} \rightarrow CO_2 + H_2O + \text{Other species} \tag{2.15}$$

The "other species" depends on what oxidizer is used and what is the ratio of the fuel to oxidizer. The most commonly used oxidizer is air which consists of nearly 79% N_2 by volume. The N_2 is normally carried through in the combustion process. If the combustion is fuel-rich, meaning there is not enough oxygen to fully combust the fuel, then there will be unburned hydrocarbons in the exhaust products and little if any excess O_2. If the combustion is fuel lean, meaning there is more oxygen than required to fully combust the fuel, then there will be excess O_2 in the exhaust products. The exhaust gas composition is very important in determining the heat transfer in the system. Unburned hydrocarbons in the exhaust indicate that the fuel was not fully combusted and therefore not all of the available heat was released. High excess O_2 levels in the exhaust usually indicates that too much oxidizer was supplied. The excess oxidizer carries sensible energy out the exhaust which again means that some of the available heat of the fuel was not fully utilized to heat the load. If the oxidizer is air, then a large proportion of the available energy in the fuel will normally be carried out of the flue with the exhaust products.

2.2.1 Fuel Properties

Table F.2 gives some of the properties for gaseous fuels commonly used in industrial combustion systems (see App. F). As expected, the fuel has a significant influence on the heat transfer in a combustion system. One of the most important properties is the heating value of the fuel. This is used to determine how much fuel must be combusted to process the desired production rate of material that is being heated. The heating value is specified as either the higher heating value (HHV) or the lower heating value (LHV). The LHV excludes the heat of vaporization, which is the energy required to convert liquid water into steam. This means that the LHV assumes that all of the products of combustion are gaseous, which is generally the case for nearly all industrial combustion applications. If the combustion products were to exit the process at a temperature low enough that all of the water were converted from a gas into a liquid, then the heat of condensation would be released into the process as an additional source of energy. The HHV of a fuel includes that energy.

The composition of the fuel is important in determining the composition of the products of combustion and the amount of oxidizer that will be needed to combust the fuel, both of which are discussed below. It is also important for determining the soot-producing tendency of the fuel. The density of the fuel is needed to determine flow rates through the fuel-delivery system and the associated pipe sizes.

2.2.2 Oxidizer Composition

There are two common types of oxidizers used in industrial combustion processes. The majority of those processes use air as the oxidizer. However, many of the higher temperature processes use an oxidizer containing a higher concentration of oxygen than found in air (approximately 21% by volume). This type of combustion is referred to as oxygen-enhanced combustion and is discussed further in Ref. 23 In many cases the production rate in a heating process can be significantly increased even with only relatively small amounts of oxygen enrichment. In most cases, air/fuel burners can successfully operate with an oxidizer containing up to about 30% O_2 with little or no modifications [24]. At greater oxygen concentrations, the flame may become unstable or the flame temperature may become too high for a burner designed to operate under air/fuel conditions. In higher temperature applications where the benefits of higher-purity oxygen justify the added costs, higher-purity oxidizers may be used ($>90\%$ O_2). The heating process is greatly intensified by the high purity oxygen. It has only been in the last decade that a significant number of combustion systems have been operated in the intermediate oxygen purities (30–90% O_2), primarily for economic reasons as the cost of high purity oxidizers is not justified. The oxidizer purity has a great influence on the heat transfer in a combustion system as will be shown throughout this book.

2.2.3 Mixture Ratio

An important consideration when studying a combustion system is the ratio of the fuel to the oxidizer. There are many ways that this may be specified. These are briefly considered here.

A global combustion reaction using CH_4 as the fuel may be written as

$$CH_4 + (xO_2 + yN_2) \rightarrow CO, CO_2, H_2, H_2O, N_2, NO_x, O_2, \text{Trace species} \quad (2.16)$$

The stoichiometry of a reaction indicates the ratio of oxygen to fuel for a given system. One method of quantifying the stoichiometry is to consider only the O_2 in the oxidizer, since the inerts in the oxidizer are not needed for the reaction:

$$S_1 = \frac{\text{Volume flow rate of } O_2 \text{ in the oxidizer}}{\text{Volume flow rate of fuel}} \quad (2.17)$$

If CH_4 is again used as an example, a global simplified stoichiometric reaction with air can be written as

$$CH_4 + (2O_2 + 7.52N_2) \rightarrow CO_2 + 2H_2O + 7.52N_2 \quad (2.18)$$

where air is represented as $2O_2 + 7.52N_2$. In that case,

$$S_1 = \frac{2}{1} = 2$$

This method of specifying stoichiometry is more commonly used for combustion systems incorporating oxygen-enrichment. This is because the oxidizer composition can vary widely, but the amount of oxygen supplied to the combustion system is what is of importance.

The most common way of defining the stoichiometry or mixture ratio in industry in the U.S. is as follows:

$$S_2 = \frac{\text{Volume flow rate of oxidizer}}{\text{Volume flow rate of fuel}} \tag{2.19}$$

For the example reaction given above, this stoichiometry would be calculated as follows:

$$S_2 = \frac{2 + 7.52}{1} = 9.52$$

Example 2.1

Given: Hydrogen reacts completely with air in the theoretical amount required for complete combustion.

Find: Stoichiometric ratio S_2.

Solution: First balance the global reaction:

$$H_2 + x(O_2 + 3.76N_2) \rightarrow H_2O + 3.76xN_2$$

Solve for x using an O atom balance:
$$2x = 1 \quad \text{or} \quad x = 0.5$$

$$S_2 = \frac{0.5(1 + 3.76)}{1} = 2.38$$

Notice that the air requirements for H_2 are four times less than for CH_4.

Example 2.2

Given: Hydrogen reacts completely with pure oxygen in the theoretical amount required for complete combustion.

Find: Stoichiometric ratio S_1.

Solution: Use the global reaction balance from above but remove the N_2:

$$H_2 + 0.5O_2 \rightarrow H_2O$$

$$S_1 = \frac{0.5}{1} = 0.5$$

Notice that S_1 is considerably less than S_2 as noted above.

Table 2.4 shows the molar constants for the stoichiometric reactions of common fuels.

Table 2.4 Molar Constants for Stoichiometric Reactions of Common Fuels

Common name	Formula	Molecular weight	O_2	CO_2	H_2O
Hydrogen	H_2	2.02	0.5	0.0	1.0
Carbon monoxide	CO	28.01	0.5	1.0	0.0
Methane	CH_4	16.05	2.0	1.0	2.0
Ethane	C_2H_6	30.08	5.0	2.0	3.0
Ethene, ethylene	C_2H_4	28.06	4.0	2.0	2.0
Acetylene, ethyne	C_2H_2	26.04	3.0	2.0	1.0
Propane	C_3H_8	44.11	7.0	3.0	4.0
Propene, propylene	C_3H_6	42.09	6.0	3.0	3.0
Butane	C_4H_{10}	58.14	7.0	4.0	5.0
Butene, butylene	C_4H_8	56.12	8.0	4.0	4.0
Generic hydrocarbon	C_xH_y	$12.01x + 1.01y$	$x+y/2$	x	$y/2$

Source: Ref. 9. (Courtesy of CRC Press.)

The problem with the stoichiometry definition commonly used in industry (S_2) is that the stoichiometry must be recalculated whenever the oxidizer composition changes and stoichiometric conditions change for each oxidizer composition. This is not a concern if air is always used as the oxidizer, which is the case for the vast majority of combustion processes. For example, consider the reaction of methane and pure oxygen:

$$CH_4 + 2O_2 \rightarrow CO_2 + 2H_2O + 7.52N_2 \qquad (2.20)$$

The two stoichiometric ratios would be calculated as follows:

$$S_1 = \frac{2}{1} = 2, \qquad S_2 = \frac{2}{1} = 2$$

For this limited case where the oxidizer is pure O_2, $S_1 = S_2$, but in the example above using air as the oxidizer, $S_1 \neq S_2$. For both cases, S_1 is the same, while S_2 is different. The benefit of using S_1 is that the stoichiometry is independent of the oxidizer composition, so stoichiometric conditions are the same for any oxidizer composition. In Eq. (2.16), $S_1 = x/1 = x$. Theoretically, for the complete combustion of CH_4 with no excess O_2, $S_1 = 2.0$, no matter how much N_2 is in the oxidizer, while S_2 varies, depending on the actual N_2 content in the oxidizer, even though the same amount of O_2 may be present. We can define the stoichiometric ratio for theoretically perfect combustion as S_1^P, which is equal to 2.0 for methane, no matter what the quantity of other gases besides O_2 in the oxidizer. Then, the stoichiometric ratio for theoretically combustion for S_2 would be S_2^P which, unlike S_1^P, would vary depending on the actual oxidizer composition.

Actual flames generally require some excess O_2 for complete combustion of the fuel. This is due to incomplete mixing between the fuel and oxidant. For the fuel-rich combustion of CH_4, $S_1 < 2.0$. For the fuel lean combustion of CH_4, $S_1 > 2.0$. These generalizations are not possible for S_2 without specifying the oxidizer composition first.

Example 2.3

Given: Excess air level of 15%, methane flow rate of 1000 scfh.
Find: Required air flow rate.
Solution: As previously shown, S_2 for CH_4 and air at stoichiometric conditions
           is 9.52. If 15% excess air is required, then 115% of the stoichiometric
           amount of air is required $= 1.15(9.52$ scfh air/scfh $CH_4)(1000$ scfh
           $CH_4) = 10{,}948$ scfh air.

A common way of specifying the oxidizer composition is by calculating the O_2
mole fraction in the oxidizer, which may be defined as

$$W = \frac{\text{Volume flow rate of } O_2 \text{ in the oxidizer}}{\text{Total volume flow rate of oxidizer}} \tag{2.21}$$

Using Eq. (2.1), $\Omega = x/(x+y)$. If the oxidizer is air, which contains
approximately 21% O_2 by volume, $\Omega = 0.21$. If the oxidizer is pure O_2, $\Omega = 1.0$.
The O_2 enrichment level is sometimes used. This refers to the incremental O_2 volume
above that found in air. For example, if $\Omega = 0.35$ then the O_2 enrichment would be
14% (35% − 21% = 14%).

Example 2.4

Given: Oxidizer consists of 500 scfh of pure O_2 and 5000 scfh of air.
Find: O_2 enrichment level.
Solution: Compute total oxidizer flow rate $= 500 + 5000 = 5500$ scfh.
           Compute total O_2 flow rate $= 500 + 5000(0.21) = 1550$ scfh O_2.

$$\text{Compute } \Omega: \Omega = \frac{1550}{5500} = 0.282$$

The O_2 enrichment level then is 28.2% − 21% = 7.2%.

Another way of specifying the fuel mixture ratio is

$$\lambda_1 = \frac{\text{Actual volumetric ratio of oxygen:fuel}}{\text{Stoichiometric volumetric ratio of oxygen:fuel}} \tag{2.22}$$

The way it is commonly specified in industry in Europe is as follows:

$$\lambda_2 = \frac{\text{Actual volumetric ratio of oxygen:fuel}}{\text{Stoichiometric volumetric ratio of oxygen:fuel}} \tag{2.23}$$

or

$$\lambda_i = \frac{S_i}{S_i^P} \tag{2.24}$$

Using this definition for the mixture ratio, $\lambda_i < 1.0$ for fuel-rich flames and
$\lambda_i > 1.0$ for fuel-lean flames. Many industrial combustion processes run with
approximately 3% more O_2 than is theoretically needed for perfect combustion.

That is often the amount of excess O_2 which is required to minimize the emissions of unburned hydrocarbons and ensure the complete combustion of the fuel. This may be due to mixing limitations between the fuel and the oxidizer, especially in nonpremixed systems. Too much excess O_2 means that energy is being wasted heating up excess combustion air, instead of the load. Therefore, it is desirable to use only just enough excess O_2 to obtain low CO emissions. An example of a simplified global reaction for methane with 3% excess O_2 is the reaction below:

$$CH_4 + (2.06O_2 + 7.75N_2) \rightarrow CO_2 + 2H_2O + 0.06O_2 + 7.75N_2 \qquad (2.25)$$

For this reaction:

$$S_1 = \frac{2.06}{1} = 2.06, \qquad S_2 = \frac{2.06 + 7.75}{1} = 9.81$$

with $S_1^P = 2.0$ and $S_2^P = 9.52$. Then, for the reaction in Eq. (2.11):

$$\lambda_1 = \frac{S_1}{S_1^P} = \frac{2.06}{2.0} = 1.03, \qquad \lambda_2 = \frac{S_1}{S_1^P} = \frac{9.81}{9.52} = 1.03$$

As can be seen, these two ratios are the same, so a single ratio λ can be given, using either Eq. (2.8) or (2.9). This is an advantage of using this method of specifying the mixture ratio.

Another way of specifying the mixture ratio in industry is by the amount of either excess O_2 (XO2) or the amount of excess air (XA), on a volume basis. For the above example in reaction (2.11), XO2 = 3% and XA = 14.4%.

In the scientific community, it is common to use the equivalence ratio (ϕ) to specify the mixture ratio, which is the inverse of λ. Again, there are two possible ways this could be defined:

$$\phi_1 = \frac{\text{Stoichiometric volumetric ratio of oxygen:fuel}}{\text{Actual volumetric ratio of oxygen:fuel}} \qquad (2.26)$$

or

$$\phi_1 = \frac{\text{Stoichiometric volumetric ratio of oxygen:fuel}}{\text{Actual volumetric ratio of oxygen:fuel}} \qquad (2.27)$$

It can again be shown that these are the same so we have

$$\phi = \frac{S_i^P}{S_i} \qquad (2.28)$$

or

$$\phi = \frac{1}{\lambda} \qquad (2.29)$$

A comparison of these various methods of specifying the mixture ratio is shown in the appendices. Tables C1–C3 show various methods of specifying the mixture ratios for methane, propane, and hydrogen, respectively, as example fuels. As will be shown in the next section, this mixture is very important in determining the heat release and transport properties for the exhaust products.

2.2.4 Excess O₂

The amount of O_2 in the combustion exhaust products is often used to monitor and control the performance of combustion systems. In general, it is desirable to run a system at the minimum excess O_2 in the exhaust products without producing significant quantities of carbon monoxide. The actual O_2 level will depend on a number of variables. One variable is the amount of variation of the oxidizer used for combustion. In most cases the oxidizer is ambient air where the temperature and relative humidity can change significantly over the course of a day as well as from day to day. Another variable is the fuel composition, which can change significantly, e.g. in refineries where the fuel is usually an off-gas from the hydrocarbon production process that varies, depending on what is being processing. Another variable is how much air may leak into the combustor. This can especially be a problem in older furnaces that have not been well maintained and are run at negative pressures that induce air to leak into the furnace. Another variable is the combustion control system. An automatic control system that includes continuous monitoring of the exhaust gas O_2 content is normally run much differently from a manual control system that is infrequently checked and adjusted.

The global reaction of a hydrocarbon fuel with an oxidizer can be written as

$$C_xH_y + (1+XSO)(x+y/4)(O_2+aN_2) \rightarrow xCO_2+(y/2)H_2O + XSOO_2 \atop + (x+y/4)(1+XSO)aN_2 \quad (2.30)$$

where the trace species have been ignored; XSO is the percentage excess oxidizer, and a is the factor used to calculate the nitrogen in the oxidizer. An example will illustrate how to use the equation.

Example 2.5

Given: Methane combusted with 15% excess air.
Find: x, y, XSO, a for the above equation.
Solution: Methane is CH_4 which is understood to be C_1H_4: $x=1$, $y=4$.
            Excess air is given as 15% so $XSO=0.15$.
            Air is 20.9% O_2 by volume, the rest is usually assumed to be N_2 or 79.1%.
            Then, for each volume of O_2, there will be $0.791/0.209 = 3.78$ volumes of N_2.
            so $a=3.78$, and

$$CH_4 + (1 + 0.15)(1 + 4/4)(O_2 + 3.78\ N_2)$$
$$\rightarrow 1CO_2 + (4/2)H_2O + 0.15\ O_2 + (1 + 4/4)(1 + 0.15)3.78N_2 \quad \text{or}$$

$$CH_4 + 2.3O_2 + 8.69N_2 \rightarrow CO_2 + 2H_2O + 0.15O_2 + 8.69N_2.$$

The wet flue gas products are: $x\,CO_2 + (y/2)\,H_2O + XSO\,O_2 + (x+y/4)\times(1+XSO)aN_2$. The dry flue gas products are: $x\,CO_2 + XSO\,O_2 + (x+y/4)(1+XSO)\,a\,N_2$.

The volume percent O_2 in the dry flue products is often measured in combustion systems with an analyzer (see Chap. 5) and can be calculated from the above equation:

$$\text{Dry } O_2 = \frac{XSO(x + y/4)}{x + XSO(x + y/4) + (x + y/4)(1 + XSO)a}$$

$$\text{Dry } O_2 = \frac{XSO(X + y/4)}{x + (x + y/4)[a + (1 + a)XSO]} \qquad (2.31)$$

The use of this equation can be illustrated using the previous example.

Example 2.6

Given: Methane combusted with 15% excess air.
Find: The expected dry O_2 in the exhaust products.
Solution: From the previous example, $x = 1$, $y = 4$, $XSO = 0.15$, $a = 3.78$.

$$\text{Dry } O_2 = \frac{0.15(1 + 4/4)}{1 + (1 + 4/4)[3.78 + (1 + 3.78)0.15]} = 0.0300 = 3.00\%$$

This is what should be measured in the dry exhaust products assuming no air infiltration into the furnace.

A rule of thumb that is sometimes used is that the dry O_2 in the exhaust products is approximately equal to the incoming excess O_2 in the oxidizer. From the previous example, the excess combustion air was 15%, of which 20.9% of that air was excess O_2, or $(20.9\%)(15\%) = 3.14\%$ excess O_2 in the incoming combustion air. This is reasonably close to the calculated 3.00%. However, this rule of thumb is not very accurate as the excess air level increases, as shown in Figure 2.11. As the excess combustion air increases, the O_2 in the dry combustion products becomes significantly less than the excess O_2 in the combustion air. However, since most industrial combustion systems are operated at or near 10–15% excess air, the rule of thumb is reasonably accurate. Another problem with the rule of thumb is that it is only valid when the oxidizer is air or close to air, as shown in Figure 2.12. As the O_2 content in the oxidizer increases, the dry O_2 in the combustion products becomes significantly larger than the excess O_2 in the oxidizer, for a fixed level of excess oxidizer.

In most industrial combustion systems, the combustion air flow is not measured. Then, the excess oxidizer is instead calculated using the measured dry O_2 in the exhaust products, assuming negligible air infiltration into the furnace. The equation above for dry O_2 can then be rearranged to solve for XSO:

$$XSO = \frac{[x + a(x + y/4)](\text{dry } O_2)}{(x + y/4)[1 - (1 + a)(\text{dry } O_2)]} \qquad (2.32)$$

The use of this equation can be illustrated using the previous example.

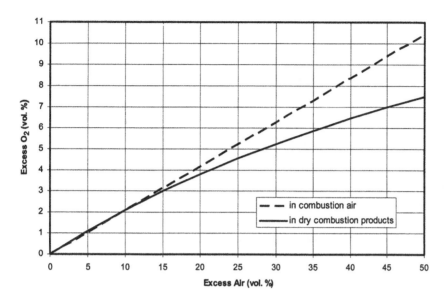

Figure 2.11 Calculated excess O_2 in the combustion air and in the dry combustion products as a function of the excess air for the combustion of methane.

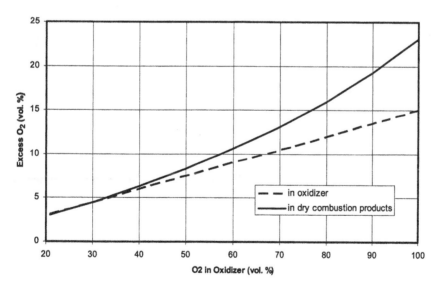

Figure 2.12 Calculated excess O_2 in the oxidizer (consisting of $O_2 + N_2$) and in the dry combustion products as a function of the O_2 in the oxidizer for the combustion of methane with 15% excess oxidizer.

Example 2.7

Given: Methane combusted with air, dry O_2 measured as 3.00%.
Find: The excess combustion air.
Solution: From the previous example, $x = 1$, $y = 4$, $a = 3.78$, and the dry O_2 is given as $3.00\% = 0.0300$.

$$XSO = \frac{[1 + 3.78(1 + 4/4)](0.0300)}{(1 + 4/4)[1 - (1 + 3.78)(0.0300)]} = 0.15 = 15\%$$

This checks with the previous calculation.

In the absence of an oxidizer flow meter, the oxidizer flow rate per unit flow rate of fuel can also be calculated using the dry O_2 measurement:

$$\frac{Q_{oxidizer}}{Q_{fuel}} = (1 + XSO)(x + y/4)(1 + a)$$

$$= \left\{ 1 + \frac{[x + a(x + y/4)](\text{dry } O_2)}{(x + y/4)[1 - (1 + a)(\text{dry } O_2)]} \right\} (x + y/4)(1 + a) \qquad (2.33)$$

The use of this equation can be illustrated using the previous example.

Example 2.8

Given: Methane combusted with air, dry O_2 measured as 3.00%.
Find: The combustion air flow rate.
Solution: From the previous examples, $x = 1$, $y = 4$, $a = 3.78$, and the dry O_2 is given as $3.00\% = 0.0300$.

$$\frac{Q_{air}}{Q_{methane}} = \left\{ 1 + \frac{[1 + 3.78(1 + 4/4)](0.0300)}{(1 + 4/4)[1 - (1 + 3.78)(0.0300)]} \right\} (1 + 4/4)(1 + 3.78) = 10.99$$

Therefore, 10.99 volumes of air are required for each volume of fuel for 15% excess air and 3.00% dry O_2 in the exhaust products.

2.2.5 Operating Regimes

Most industrial flames are turbulent. A turbulent Reynolds number can be defined as

$$Re_T = \frac{v' l_0}{v_0} \qquad (2.34)$$

One can define a turbulent characteristic length scale, commonly called the Kolmogorov length, l_K:

$$l_K = \frac{l_0}{(v l_0 / v_0)^{3/4}} \qquad (2.35)$$

where l_0 is a characteristic length scale usually associated with large eddies, v is a characteristic velocity, and v_0 is the characteristic kinematic viscosity. The Kolmogorov length is representative of the dimension where dissipation occurs.

The Taylor length scale can be defined as the ratio of the strain rate to the viscous forces:

$$l_T = \frac{l_0^2}{\mathrm{Re}_T} \tag{2.36}$$

where v' is a characteristic velocity, fluctuation from the mean velocity v, which is an indicator of the turbulence level. The flame thickness, l_L, is a characteristic length scale of the flame. The various lengths can be used to characterize the flame [25]:

$l_L < l_K$ Wrinkled flame
$l_K < l_L < l_T$ Severely wrinkled flame
$l_T < l_L < l_0$ Flamelets in eddies
$l_0 < l_L$ Distributed reaction front

A nondimensional Damköhler number can be defined that indicates the type of reaction time which is significant for the specific type of combustion reaction:

$$\mathrm{Da} = \frac{l_0 S_L}{v' l_L} \tag{2.37}$$

where S_L is the laminar flame speed. This number is the ratio of the reaction time to the flow rate. If this number is high then the assumption is often made in modeling that the reactions are infinite rate or equilibrium, which greatly simplifies modeling. If this ratio is low then finite-rate chemistry must be used, which increases the complexity and computation time of model simulations.

2.3 COMBUSTION PROPERTIES

This section briefly considers the combustion product composition, flame temperature, available heat, and flue-gas volume for combustion as commonly used in industrial applications. These are important in calculating the heat transfer from the flame, and exhaust gases to the furnace and to the load.

2.3.1 Combustion Products

There are a number of variables which can have a significant impact on the products of combustion. Some of the important variables include the oxidizer composition, mixture ratio, air and fuel preheat temperatures, and fuel composition. These are briefly discussed here.

2.3.1.1 Oxidizer Composition

The stoichiometric combustion of CH_4 with air may be represented by the following global equation:

$$CH_4 + 2O_2 + 7.52N_2 \rightarrow CO_2, 2H_2O, 7.52N_2, \text{Trace species} \tag{2.38}$$

It may be seen that over 70 vol.% of the exhaust gases are N_2. Similarly, a stoichiometric O_2/CH_4 combustion process may be represented by

$$CH_4 + 2O_2 \rightarrow CO_2, 2H_2O, \text{Trace species} \tag{2.39}$$

The volume of exhaust gases is significantly reduced by the elimination of N_2. In general, a stoichiometric oxygen-enhanced methane combustion process may be represented by

$$CH_4 + 2O_2 + xN_2 \rightarrow CO_2 + 2H_2O + xN_2 + \text{trace species} \qquad (2.40)$$

where $0 \le x \le 7.52$, depending on the oxidizer.

The actual composition of the exhaust products from the combustion reaction depends on several factors including the oxidizer composition, the temperature of the gases, and the equivalence ratio. A diagram showing an adiabatic equilibrium combustion reaction is presented in Fig. 2.13. An adiabatic process means that no heat is lost during the reaction, or that the reaction occurs in a perfectly insulated chamber. This is not the case in an actual combustion process where heat is lost from the flame by radiation. Figure 2.14 shows the predicted major species for the adiabatic equilibrium combustion of CH_4 as a function of the oxidizer composition. The calculations were made using a NASA computer program that minimizes the Gibbs free energy of a gaseous system [26]. An equilibrium process means that there is an infinite amount of time for the chemical reactions to take place, or the reaction products are not limited by chemical kinetics. In reality, the combustion reactions are completed in fractions of a second. As expected, Fig. 2.14 shows that as N_2 is removed from the oxidizer, the concentration of N_2 in the exhaust products decreases correspondingly. Likewise, there is an increase in the concentrations of CO, CO_2, and H_2O. For this adiabatic process, there is a significant amount of CO at higher levels of O_2 in the oxidizer.

Figure 2.15 shows the predicted minor species for the same conditions as Figure 2.14. Note that trace species have been excluded from this figure. The radical species H, O, and OH all increase with the O_2 in the oxidizer. NO initially increases and then decrease, after about 60% O_2 in the oxidizer as more N_2 is removed from the system. When the oxidizer is pure O_2, no NO is formed as no N_2 is available. Unburned fuel in the form of H_2 and unreacted oxidizer in the form of O_2 also increase with the O_2 concentration in the oxidizer. This increase in radical concentrations, unburned fuel in the form of CO and H_2, and unreacted O_2 are all due to chemical dissociation, which occurs at high temperatures.

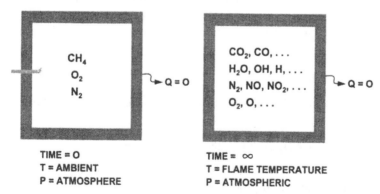

Figure 2.13 Adiabatic equilibrium reaction process. (From Ref. 23. Courtesy of CRC Press.)

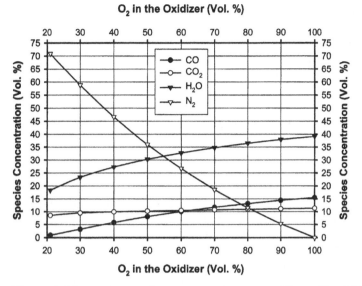

Figure 2.14 Major species concentrations vs. oxidant $(O_2 + N_2)$ composition for an adiabatic equilibrium stoichiometric CH_4 flame. (From Ref. 27. Courtesy of CRC Press.)

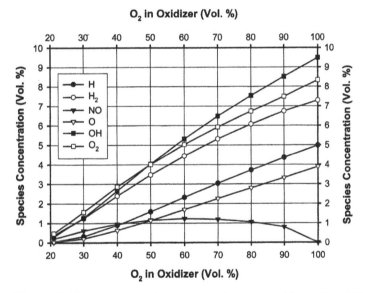

Figure 2.15 Minor species concentrations vs. oxidant $(O_2 + N_2)$ composition, for an adiabatic equilibrium stoichiometric CH_4 flame. (From Ref. 27. Courtesy of CRC Press.)

The actual flame temperature is lower than the adiabatic equilibrium flame temperature due to imperfect combustion and radiation from the flame. The actual flame temperature is determined by how well the flame radiates its heat and how well the combustion system, including the load and the refractory walls, absorbs that radiation. A highly luminous flame generally has a lower flame temperature than a highly nonluminous flame. The actual flame temperature will also be lower when the load and the walls are more radiatively absorptive. This occurs when the load and walls are at lower temperatures and have higher radiant absorptivities. These effects are discussed in more detail in Ref. 27. As the gaseous combustion products exit the flame, they typically lose more heat by convection and radiation as they travel through the combustion chamber. The objective of a combustion process is to transfer the chemical energy contained in the fuel to the load, or in some cases to the combustion chamber. The more thermally efficient the combustion process, the more heat that is transferred from the combustion products to the load and to the combustion chamber. Therefore, the gas temperature in the exhaust stack is desirably much lower than in the flame in a thermally efficient heating process. The composition of the combustion products then changes with gas temperature.

Figure 2.16 shows the predicted major species for the equilibrium combustion of CH_4 with "air" (21% O_2, 79% O_2) as a function of the gas temperature. The highest possible temperature for the air/CH_4 reaction is the adiabatic equilibrium temperature of 3537°F (2220 K). For the air/CH_4 reaction, there is very little change in the predicted gas composition as a function of temperature. Figure 2.17 shows the predicted minor species for the same conditions as in Figure 2.16. For the air/CH_4, none of the minor species exceeds 1% by volume. As the gas temperature increases, chemical dissociation increases.

2.3.1.2 Mixture Ratio

Figure 2.18 shows the predicted major gas species for the adiabatic equilibrium combustion of air/CH_4 as a function of the equivalence ratio. Figure 2.19 shows the predicted minor species as a function of the equivalence ratio. The O_2 and N_2

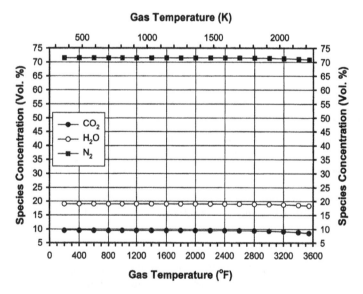

Figure 2.16 Equilibrium calculations for the predicted gas composition of the major species as a function of the combustion product temperature for stoichiometric air/CH_4 flames. (From Ref. 27. Courtesy of CRC Press.)

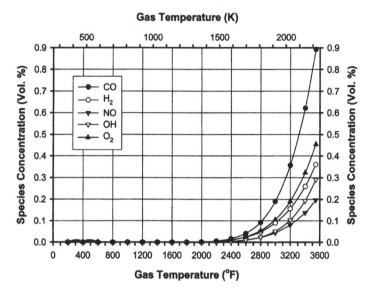

Figure 2.17 Equilibrium calculations for the predicted gas composition of the minor species as a function of the combustion product temperature for stoichiometric air/CH_4 flames (From Ref. 27. Courtesy of CRC Press.)

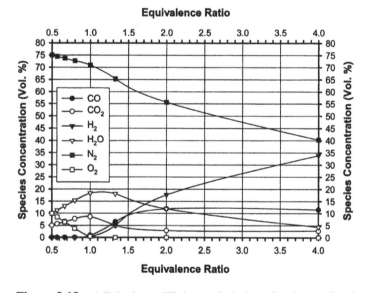

Figure 2.18 Adiabatic equilibrium calculations for the predicted gas composition of the major species as a function of the equivalence ratio for air/CH_4 flames. (From Ref. 27. Courtesy of CRC Press.)

concentrations in the exhaust gases strictly decreases with the equivalence ratio. The H_2O and the CO_2 concentrations peak at stoichiometric conditions ($\phi = 1.0$). This is important as both of these gases produce nonluminous gaseous radiation (see Chap. 3). As expected, the unburned fuels in the form of H_2 and CO both

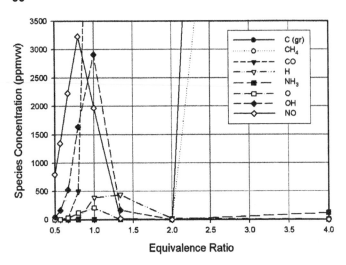

Figure 2.19 Adiabatic equilibrium calculations for the predicted gas composition of the minor species as a function of the equivalence ratio for air/CH_4 flames. (From Ref. 27. Courtesy of CRC Press.)

increase with equivalence ratio. This will be reflected in the available heat (discussed below) as not all of the fuel is fully combusted.

2.3.1.3 Oxidizer and Fuel Preheat Temperature

In many instances of industrial combustion, heat is recovered to improve the overall thermal efficiency of the process to reduce the operating costs. The recovered heat is most commonly used to preheat the incoming combustion air and is sometimes used to preheat the incoming fuel. Preheating either the air or the fuel affects the composition of the combustion products. Figure 2.20 shows the major species predicted for the combustion of air and CH_4 where the air is preheated up to as high as 2000°F (1366 K). CO_2, H_2O, and N_2 all decrease with air preheat, due to chemical dissociation. Figure 2.21 shows that the minor species increase with air preheat. Figure 2.22 and Figure 2.23 show the major and minor species, respectively, for the air/CH_4 flames as a function of the CH_4 preheat temperature up to temperatures as high as 1000°F (811 K). Due to safety considerations and the possibility of sooting up the fuel-supply piping, higher fuel preheat temperatures are not practical or recommended under most conditions. The figures show that there is only a slight decrease in the concentrations of the major species and a slight increase in the concentrations of the minor species. This is because the mass of fuel is relatively low compared to the mass of combustion air supplied to the system. This means that preheating the combustion air has a much more significant impact than preheating the fuel for a given preheat temperature.

2.3.1.4 Fuel Composition

Combustion products have been calculated here for the following four fuels: H_2, CH_4, C_3H_8, and blends of H_2 and CH_4. These are intended to be representative

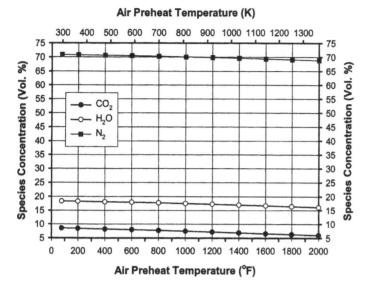

Figure 2.20 Adiabatic equilibrium stoichiometric calculations for the predicted gas composition of the major species as a function of the air preheat temperature for air/CH₄ flames. (From Ref. 27. Courtesy of CRC Press.)

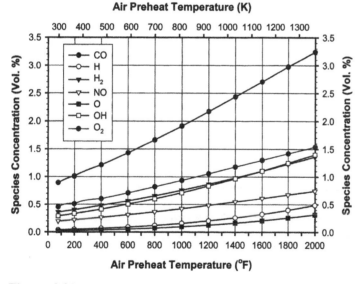

Figure 2.21 Adiabatic equilibrium stoichiometric calculations for the predicted gas composition of the minor species as a function of the air preheat temperature for air/CH₄ flames. (From Ref. 27. Courtesy of CRC Press.)

of fuels commonly used in industrial applications. In terms of luminosity, H_2 produces nonluminous flames, CH_4 produces low luminosity flames, and C_3H_8 produces higher luminosity flames. The predicted combustion product compositions for each fuel under a variety of operating conditions are given in Appendix D.

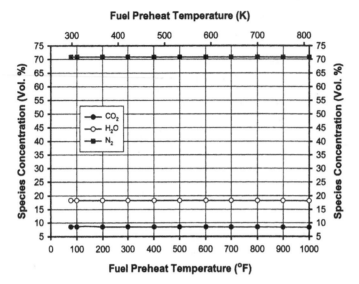

Figure 2.22 Adiabatic equilibrium stoichiometric calculations for the predicted gas composition of the major species as a function of the fuel preheat for air/CH$_4$ flames. (From Ref. 27. Courtesy of CRC Press.)

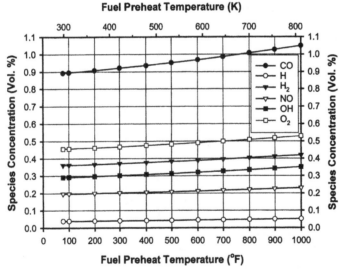

Figure 2.23 Adiabatic equilibrium stoichiometric calculations for the predicted gas composition of the minor species as a function of the fuel preheat for air/CH$_4$ flames. (From Ref. 27. Courtesy of CRC Press.)

2.3.2   Flame Temperature

The flame temperature is a critical variable in determining the heat transfer from the flame to the load as will be shown in Chapter 3. This section shows how the adiabatic flame temperature is affected by the oxidizer and fuel compositions, the mixture ratio, and the air and fuel preheat temperatures. As previously mentioned, real flame temperatures are not as high as the adiabatic flame temperature, but the trends are comparable and representative of actual conditions.

2.3.2.1 Oxidizer and Fuel Composition

The flame temperature increases significantly, when air is replaced with oxygen because N_2 acts as a diluent that reduces the flame temperature. Figure 2.24 is a plot of the adiabatic equilibrium flame temperature for CH_4 combustion, as a function of the oxidizer composition, for a stoichiometric methane combustion process. The flame temperature varies from 3600°F to 5000°F (2300–3000 K), for air and pure oxygen, respectively. The graph shows a rapid rise in the flame temperature from air up to about 60% O_2 in the oxidizer. The flame temperature increases at a slower rate at higher O_2 concentrations.

Table 2.5 lists the adiabatic flame temperatures for a number of fuels where the oxidizer is either air or pure O_2. From that table it can be seen that the fuel composition has a strong impact on the flame temperature. Figure 2.25 shows how the flame temperature varies for a fuel blend of H_2 and CH_4. The temperature increases as the H_2 content in the blend increases. It is important to note that the increase is not linear with a more rapid increase at higher levels of H_2. Because of the relatively high cost of H_2 compared to CH_4 and C_3H_8, it is not used in many industrial applications. However, high H_2 fuels are often used in many of the hydrocarbon and petrochemical applications for fluid heating. Those fuels are by-products of the chemical manufacturing process and, therefore, much less expensive than purchasing H_2 from an industrial gas supplier and are more cost effective than using other purchased fuels.

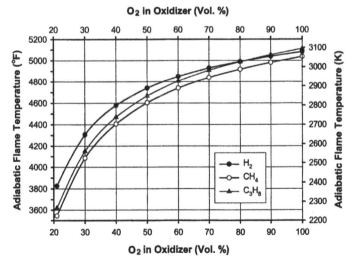

Figure 2.24 Adiabatic flame temperature vs. oxidizer composition for adiabatic equilibrium stoichiometric air/H_2, air/CH_4, and air/C_3H_8 flames. (From Ref. 27. Courtesy of CRC Press.)

Table 2.5   Adiabatic Flame Temperature for Some Common Gaseous Fuels

Fuel	Air		O₂	
	°F	K	°F	K
H_2	3807	2370	5082	3079
CH_4	3542	2223	5036	3053
C_2H_2	4104	2535	5556	3342
C_2H_4	3790	2361	5256	3175
C_2H_6	3607	2259	5095	3086
C_3H_6	4725	2334	5203	3138
C_3H_8	3610	2261	5112	3095
C_4H_{10}	3583	2246	5121	3100
CO	3826	2381	4901	2978

Source: Ref. 23. (Courtesy of CRC Press.)

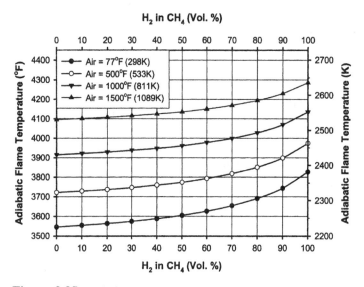

Figure 2.25   Adiabatic equilibrium flame temperature vs. fuel blend composition for stoichiometric air/fuel flames at different air preheat temperatures. (From Ref. 27. Courtesy of CRC Press.)

2.3.2.2 Mixture Ratio

Figure 2.26 is a similar plot of the adiabatic equilibrium flame temperature for CH_4 flames as a function of the stoichiometry for four different oxidizer compositions ranging from air to pure O_2. The peak flame temperatures occur under stoichiometric conditions. The lower the O_2 concentration in the oxidizer, the more the flame temperature is reduced by operating under nonstoichiometric conditions (either fuel-rich or fuel-lean). This is due to the higher concentration of N_2 which absorbs heat and lowers the overall temperature. Figure 2.27 shows the adiabatic flame temperature as a function of the equivalence ratio for three fuels: H_2, CH_4, and

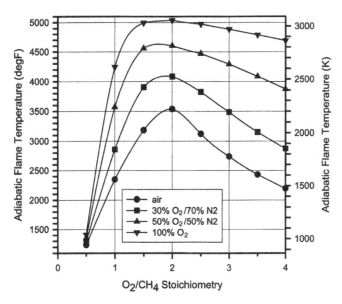

Figure 2.26 Adiabatic equilibrium flame temperature vs. stoichiometry for a CH₄ flame and various oxidizers. (From Ref. 23. Courtesy of CRC Press.)

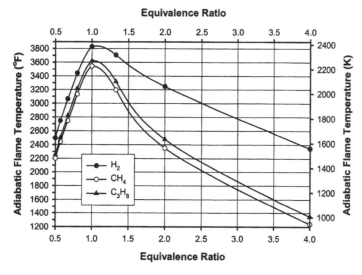

Figure 2.27 Adiabatic equilibrium flame temperature vs. equivalence ratio for air/H₂, air/CH₄ and air/C₃H₈ flames. (From Ref. 27. Courtesy of CRC Press.)

C₃H₈. The peak temperature occurs in stoichiometric conditions (φ = 1.0). In that case there is just enough oxidizer to combust fully all of the fuel. Any additional oxidizer absorbs sensible energy from the flame and reduces the flame temperature. In most real flames, the peak flame temperature often occurs at slight fuel-lean conditions (φ < 1.0). This is due to imperfect mixing where slightly more O₂ is needed to combust fully all of the fuel. Nearly all industrial combustion applications are run under fuel-lean conditions to ensure that the CO emissions are low. Therefore, depending on the actual burner design, the flame temperature may be close to its peak, which is often desirable for maximizing heat transfer. One problem often encountered by maximizing the flame temperature is that the NO_x emissions are also maximized since NO_x increases approximately exponentially with gas temperature. This has led to many design concepts for reducing the peak flame temperature in the flame to minimize NO_x emissions [28].

Figure 2.28 shows a contour plot of the adiabatic flame temperature (AFT) for the combustion of methane as a function of both the oxidant composition and the $O_2:CH_4$ stoichiometry. The AFT is highest for stoichiometries near 2.0 (which is stoichiometric for methane) and increases with the O_2 in the oxidizer.

2.3.2.3 Oxidizer and Fuel Preheat Temperature

Figure 2.29 shows how the AFT varies as a function of the oxidizer preheat temperature for air/CH_4 and O_2/CH_4 flames. The increase in flame temperature is relatively small for the O_2/CH_4 flame because the increased sensible heat of the O_2 is only a fraction of the chemical energy contained in the fuel. For the air/CH_4

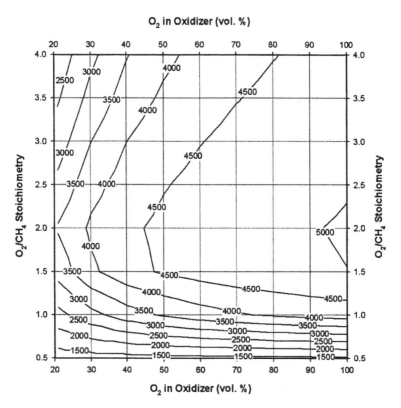

Figure 2.28a Contour plot of the adiabatic equilibrium flame temperature (°F) vs. oxidizer composition and O_2:NG stoichiometry for the combustion of ambient temperature methane.

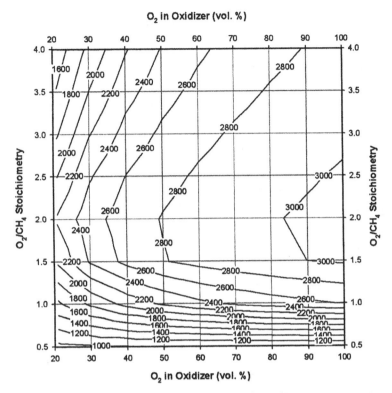

Figure 2.28b Contour plot of the adiabatic equilibrium flame temperature (K) vs. oxidizer composition and O_2:NG stoichiometry for the combustion of ambient temperature methane.

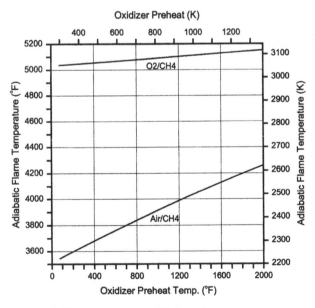

Figure 2.29 Adiabatic equilibrium flame temperature vs. oxidant preheat temperature for stoichiometric air/CH_4 and O_2/CH_4 flames. (From Ref. 23. Courtesy of CRC Press.)

flames, preheating the air has a more dramatic impact because the increase in sensible heat is very significant due to the large mass of air in the combustion reaction. Figure 2.30 shows that the AFT increases rapidly for air/fuel flames for each of the three fuels shown. Figure 2.25 also shows how preheating the air in the combustion of a blended fuel affects the flame temperature. Again, the higher the air preheat, the higher the temperature of the combustion products. Figure 2.31

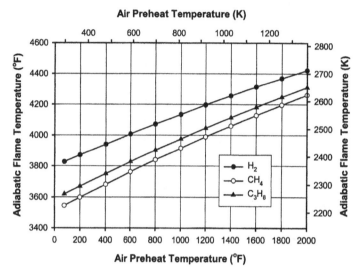

Figure 2.30 Adiabatic equilibrium flame temperature vs. air preheat temperature for stoichiometric air/H_2, air/CH_4, and air/C_3H_8 flames. (From Ref. 27. Courtesy of CRC Press.)

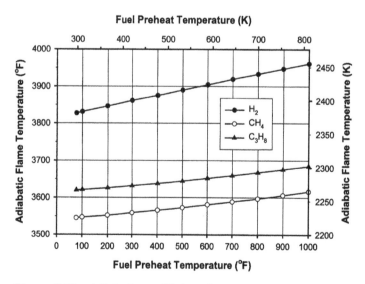

Figure 2.31 Adiabatic equilibrium flame temperature vs. fuel preheat temperature for stoichiometric air/H_2, air/CH_4, and air/C_3H_8 flames. (From Ref. 27. Courtesy of CRC Press.)

shows that preheating the fuel has less impact on the flame temperature than preheating the oxidizer because of the mass flow rate differences as previously discussed.

2.3.2.4 Flue Gas Recirculation

Figure 2.32 shows the calculated AFT as a function of the flue gas recirculation ratio for the combustion of hydrogen, methane, and propane, where the recycled gases are assumed to be at 2000°F. The curves show that flue gas recirculation significantly reduces the adiabatic flame temperature. This has an important effect on reducing NO_x emissions (see Chap. 6). Figure 2.33 shows the AFT as a function of the recirculation ratio and the air preheat temperature for methane. Again, the predicted temperature reduces dramatically with flue gas recirculation. Figure 2.34 is a similar plot, but as a function of the recirculation ratio and methane preheat temperature.

2.3.3 Available Heat

The available heat in a combustion system is important in determining the overall thermal efficiency and is, therefore, a factor when calculating the heat transfer in the process. It would be less effective to try to maximize the heat transfer in a system that inherently has a low available heat. Available heat is defined as the gross heating value of the fuel, less the energy carried out of the combustion process by the hot exhaust gases. The heat lost from a process through openings in the furnace, through the furnace walls, or by air infiltration is not considered in calculating the theoretical available heat as those are dependent on the process. A Sankey diagram is often used

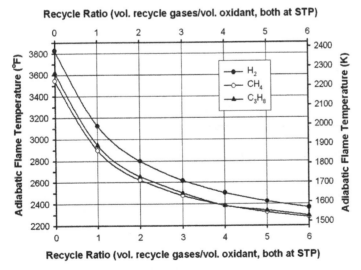

Figure 2.32 Adiabatic equilibrium flame temperature as a function of the flue gas recirculation ratio (vol. of recycle gases/vol. of oxidant, both at STP) for the combustion of hydrogen, methane, and propane where the recycled gases are assumed to be at 2000°F (1366 K).

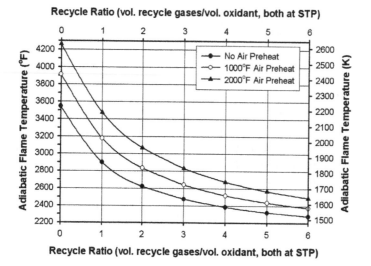

Figure 2.33 Adiabatic equilibrium flame temperature as a function of the flue gas recirculation ratio and air preheat temperature (vol. of recycle gases/vol. of oxidant, both at STP) for the combustion of methane where the recycled gases are assumed to be at 2000°F (1366 K).

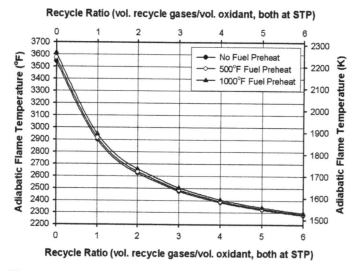

Figure 2.34 Adiabatic equilibrium flame temperature as a function of the flue gas recirculation ratio and fuel preheat temperature (vol. of recycle gases/vol. of oxidant, both at STP) for the combustion of methane where the recycled gases are assumed to be at 2000°F (1366 K).

to show where all the energy goes in a combustion system. A simplified example is shown in Figure 2.35. There can be as much or as little detail as needed. For example, there can be finer detail in the losses to include air infiltration, conduction losses through the wall, and radiation losses through openings. The theoretical available heat should be proportional to the amount of energy actually absorbed by

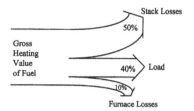

Figure 2.35 Example of a Sankey diagram. (From Ref. 27. Courtesy of CRC Press.)

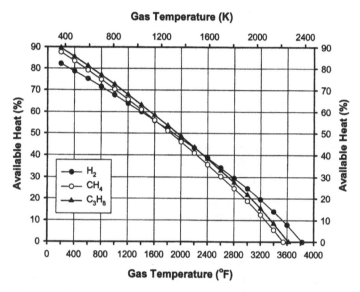

Figure 2.36 Available heat vs. gas temperature, for stoichiometric air/H_2, air/CH_4, and air/C_3H_8 flames. (From Ref. 27. Courtesy of CRC Press.)

the load in an actual process, which is directly related to the thermal efficiency of the system. Therefore, the theoretical available heat is used here to show the thermal efficiency trends as functions of exhaust gas temperature, oxidizer and fuel compositions, mixture ratio, and air and fuel preheat temperatures.

Figure 2.36 shows how the available heat decreases rapidly with the exhaust gas temperature and is relatively independent of the fuel composition for the three fuels shown. Then, to maximize the thermal efficiency of a process, it is desirable to minimize the exhaust gas temperature. This is usually done by maximizing the heat transfer from the exhaust gases to the load (and furnace walls) and by recovering some of the heat in the exhaust gases by preheating the fuel and/or the oxidizer. Figure 2.37 shows how the available heat, for stoichiometric air/CH_4 and O_2/CH_4 flames, varies as a function of the exhaust gas temperature. As the exhaust temperature increases, more energy is carried out of the combustion system and less remains in the system. The available heat decreases to zero at the adiabatic equilibrium flame temperature where no heat is lost from the gases. The figure shows that even at gas temperatures as high as 3500°F (2200 K), the available heat of an

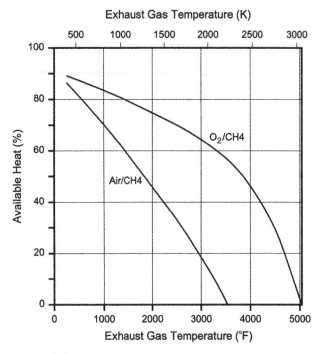

Figure 2.37 Available heat vs. exhaust gas temperature, for stoichiometric air/CH$_4$ and O$_2$/CH$_4$ flames. (From Ref. 23. Courtesy of CRC Press.)

O$_2$/CH$_4$ system is still as high as 57%. The figure also shows that it is usually not very economical to use air/CH$_4$ systems for high-temperature heating and melting processes. At an exhaust temperature of 2500°F (1600 K), the available heat for the air/CH$_4$ system is only a little over 30%. Heat recovery in the form of preheated air is commonly used for higher temperature heating processes to increase the thermal efficiencies.

Figure 2.38 is a graph of the available heat for the combustion of CH$_4$ as a function of the O$_2$ concentration in the oxidizer, for three different exhaust gas temperatures. As the exhaust gas temperature increases, the available heat decreases because more energy is carried out of the exhaust stack. There is an initial rapid increase in available heat as the O$_2$ concentration in the oxidizer increases from the 21% found in air. That is one reason why O$_2$ enrichment has been a popular technique for using OEC because the incremental increase in efficiency is very significant.

Figure 2.39 shows how the available heat increases with the oxidizer preheat temperature. The thermal efficiency of the air/CH$_4$ doubles by preheating the air to 2000°F (1400 K). For the O$_2$/CH$_4$ flames, the increase in efficiency is much less dramatic by preheating the O$_2$. This is because the initial efficiency with no preheat is already 70% and because the mass of the O$_2$ is not nearly as significant in the combustion reaction as compared to the mass of air in an air/fuel flame. There are also safety concerns when flowing hot O$_2$ through piping, heat recuperation equipment, and a burner. Figure 2.40 shows that the available heat increases rapidly for any of the three fuels shown as the combustion air preheat temperature increases.

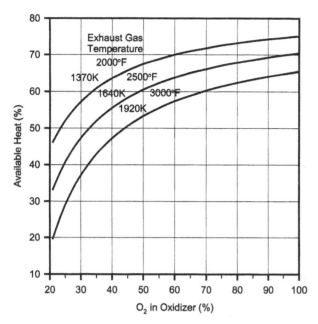

Figure 2.38 Available heat vs. oxidant composition, for a stoichiometric CH₄ flame, at exhaust temperatures of 2000°, 2500°, and 3000°F. (From Ref. 23. Courtesy of CRC Press.)

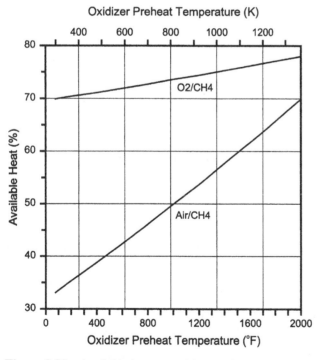

Figure 2.39 Available heat vs. oxidant preheat temperature for equilibrium stoichiometric air/CH₄ and O₂/CH₄ flames at an exhaust gas temperature of 2500°F (1644 K). (From Ref. 23. Courtesy of CRC Press.)

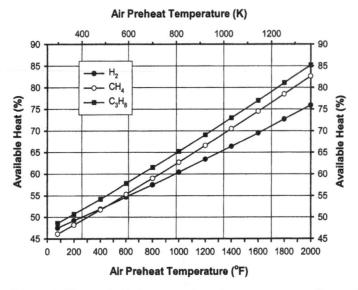

Figure 2.40 Available heat vs. air preheat temperature, for equilibrium stoichiometric air/H$_2$, air/CH$_4$, and air/C$_3$H$_8$ flames at an exhaust gas temperature of 2000°F (1366 K). (From Ref. 27. Courtesy of CRC Press.)

The fuel savings for a given technology can be calculated using the available heat curves:

$$\text{Fuel savings (\%)} = \left(1 - \frac{AH_2}{AH_1}\right) \times 100 \qquad (2.41)$$

where AH$_1$ is the available heat of the base case process and AH$_2$ is the available heat using a new technology. For example, if the base case process has an available heat of 30% and the available heat using the new technology is 45%, then the fuel savings $= (1 - 45/30) \times 100 = -50\%$, which means that 50% less fuel is needed for process 2 compared to process 1.

2.3.4 Flue Gas Volume

The flow rate of gases through a combustion chamber is proportional to the convective heat transfer to the load. There are several factors which influence that flow rate. One is the gas temperature since higher temperature gases have higher actual flow rates (e.g., actual cubic feet per hour or ACFH) due to the thermal expansion of the gases. This means that preheating the fuel or the oxidizer, which both normally increase the flame temperature, would produce higher actual flow rates. However, the flow rate of the gases is the same when corrected to standard temperature and pressure conditions (STP). Another factor that has a very strong influence on the gas flow rate through the combustion system is the oxidizer composition. Oxygen-enhanced combustion basically involves removing N$_2$ from

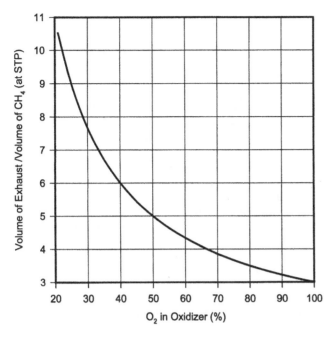

Figure 2.41 Normalized flue gas volume vs. oxidizer composition, for a stoichiometric CH_4 flame. (From Ref. 23. Courtesy of CRC Press.)

the oxidizer. A major change compared to air/fuel combustion is the reduction in the flue gas volume. Figure 2.41 shows the exhaust gas flow rate, normalized to the fuel flow rate at standard temperature and pressure conditions (e.g., standard cubic feet per hour or SCFH), for the stoichiometric combustion of CH_4 where it has been assumed that all the combustion products are CO_2, H_2O, and N_2 (except when the oxidizer is pure O_2 when there is no N_2). This means that for each unit volume of fuel, three normalized volumes of gas are produced for oxy/fuel compared to 10.5 volumes for air/fuel. This reduction can have both positive and negative effects, but the effect on convective heat transfer is a reduction in the average gas velocity through a given combustor and a resulting reduction in convection to the load.

REFERENCES

1. JA Barnard, JN Bradley. *Flame and Combustion*, 2nd edn. London: Chapman and Hall, 1985.
2. FA Williams. *Combustion Theory*. Menlo Park, CA: Benjamin/Cummings, 1985.
3. B Lewis, G von Elbe. *Combustion, Flames and Explosions of Gases*, 3rd edn. New York: Academic Press, 1987.
4. W Bartok, AF Sarofim (eds.). *Fossil Fuel Combustion*. New York: John Wiley, 1991.
5. RM Fristrom. *Flame Structure and Processes*. New York: Oxford University Press, 1995.
6. SR Turns. *An Introduction to Combustion*. McGraw-Hill, New York, 1996.
7. G Borman, K Ragland. *Combustion Engineering*. New York: McGraw-Hill, 1998.
8. JB Edwards. *Combustion: The Formation and Emission of Trace Species*. Ann Arbor, MI: Ann Arbor Science Publishers, 1974.

9. CE Baukal (ed.). *The John Zink Combustion Handbook*. Boca Raton, FL: CRC Press, 2001.

10. A Williams. Fundamentals of oil combustion. *Prog. Energy Combust. Sci.*, Vol. 2, pp. 167–179, 1976.

11. U.K. Environment Agency. Combustion Processes: Waste and Recovered Oil Burners 3 MW(th) and Over. Chief Inspector's Guidance Note S2 1.04. London: HMSO, 1995.

12. RM McGuinness, WT Kleinberg. Oxygen production. In CE Baukal (ed.). *Oxygen-Enhanced Combustion*. Boca Raton, FL: CRC Press, 1998.

13. CE Baukal. Pollutant emissions. In CE Baukal (ed.). *Oxygen-Enhanced Combustion*. Boca Raton, FL: CRC Press, 1998.

14. SJ Williams, LA Cuervo, and MA Chapman. High-Temperature Industrial Process Heating: Oxygen–Gas Combustion and Plasma Heating Systems. Gas Research Institute rep. GRI-89/0256, Chicago, IL, July 1989.

15. AS Chace, HR Hazard, A Levy, AC Thekdi and EW Ungar. Combustion Research Opportunities for Industrial Applications—Phase II. U.S. Dept. of Energy rep. DOE/ID-10204-2, Washington, DC, 1989.

16. SV Joshi, JS Becker, and GC Lytle. Effects of oxygen enrichment on the performance of air–fuel burners. In MA Lukasiewicz (ed.). *Industrial Combustion Technologies*. Amer. Society of Metals, Materials Park, OH, pp. 165–170, 1986.

17. U.S. Environmental Protection Agency. Alternative Control Techniques Document—NO_x Emissions from Utility Boilers. EPA rep. EPA-453/R-94-023, Research Triangle Park, NC, 1994.

18. G Jankes, M Stanjevic, M Karan, M Kuburovic, M Adzic. The use of technical oxygen for combustion processes in industrial furnaces. *Proceedings of the 5th European Conference on Industrial Furnaces and Boilers*, Lisbon, Portugal, Vol. 1, pp. 649–658, 2000.

19. H Kobayashi. Segregated Zoning Combustion. U.S. Patent 5 076 779, 31 Dec. 1991.

20. CR Griffith. PCB and PCP destruction using oxygen in mobile incinerators. *Proceedings of the 1990 Incineration Conference*, San Diego, CA 14–18 May 1990.

21. ER Bazarian, JF Heffron, and CE Baukal. Method for Reducing NO_x Production During Air–Fuel Combustion Processes. U.S. Patent 5 308 239, 1994.

22. GM Gitman. Method and Apparatus for Generating Highly Luminous Flame. U.S. Patent 4797087, 1989.

23. CE Baukal (ed.), *Oxygen-Enhanced Combustion*. Boca Raton, FL: CRC Press, 1998.

24. AI Dalton, DW Tyndall. Oxygen Enriched Air/Natural Gas Burner System Development. NTIS Rep. PB91-167510, Springfield, VA, 1989.

25. I Glassman. *Combustion*, 3rd edn. New York: Academic Press, 1996.

26. S Gordon, BJ McBride. Computer Program for Calculation of Complex Chemical Equilibrium Compositions, Rocket Performance, Incident and Reflected Shocks, and Chapman-Jouguet Detonations. NASA Rep. SP-273, 1971.

27. CE Baukal. *Heat Transfer in Industrial Combustion*. Boca Raton, FL: CRC Press, 2001.

28. JL Reese, GL Moilanen, R Borkowicz, C Baukal, D Czerniak, R Batten. State-of-the-Art of NO_x Emission Control Technology, ASME. Paper 94-JPGC-EC-15, *Proceedings of International Joint Power Generation Conference*, Phoenix, AZ, 3–5 Oct 1994.

3

Combustion Systems

3.1 INTRODUCTION

There are four components that are important in the transfer of thermal energy from a combustion process to some type of heat load (see Fig. 3.1) [1]. One component is the burner, which combusts the fuel with an oxidizer to release heat. Another component is the load itself, which can greatly affect how the heat is transferred from the flame. In most cases, the flame and the load are located inside of a combustor, which may be a furnace, heater, or dryer, which is the third component in the system. In some cases, there may be some type of heat recovery device to increase the thermal efficiency of the overall combustion system, which is the fourth component of the system. Each of these components is briefly considered in this chapter. Various aspects of these components are discussed in more detail in other chapters of the book.

Although there are other important components in a combustion system, such as the flow control system, they do not normally have a significant impact on the pollution from the flame. An exception would be the flow controls for a pulsed combustion system where the cycling of either the fuel or oxidizer supply valves can cause the pulsing which can significantly reduce the pollution from the flame to the load. In general, however, the other components in a combustion system do not usually influence the pollution, which is the subject of this book.

3.2 BURNERS

The burner is the device that is used to combust the fuel with an oxidizer to convert the chemical energy in the fuel into thermal energy. This device is specifically designed to properly mix the fuel in the air to produce controlled burning to achieve the desired heat release distribution. A given combustion system may have a single burner or many burners, depending on the size and type of the application. For example, in a vertical cylindrical furnace, one configuration has a single burner in the floor firing vertically upward (see Fig. 3.2). The heat from the burner radiates in all directions and is efficiently absorbed by the process tubes at the perimeter. However, the cylindrical geometry has some limitations concerning size and load type, which make its use limited to certain applications such as melting scrap aluminum or

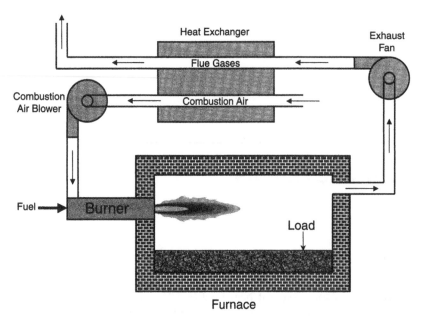

Figure 3.1   Schematic of major components in a combustion system.

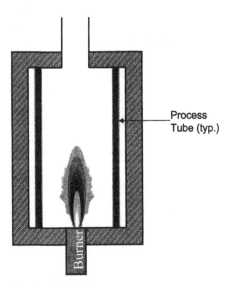

Figure 3.2   Elevation view of a single burner in a vertical cylindrical furnace.

producing cement clinker. A more common combustion system has multiple burners in a rectangular geometry (see Fig. 3.3). This type of system is generally more difficult to analyze because of the multiplicity of heat sources and because of the interactions between the flames and their associated products of combustion.

There are many factors that go into the design of a burner. This section will briefly consider some of the important factors that are taken into account for a particular type of burner, with specific emphasis on how those factors impact

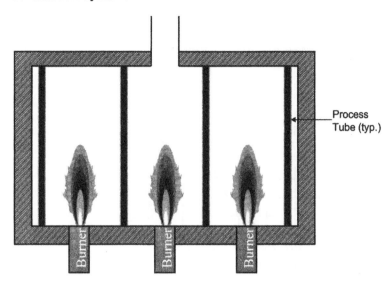

Figure 3.3 Elevation view of multiple burners in a cabin heater.

pollution. These factors also affect other things like heat transfer from the flame to the load, which will only briefly be discussed since they normally only influence the pollution characteristics for a given burner design under fairly limited and special conditions.

3.2.1 Competing Priorities

There have been many changes in the traditional designs that have been used in burners, primarily because of the recent interest in reducing pollutant emissions. In the past, the burner designer was primarily concerned with efficiently combusting the fuel and transferring the energy to a heat load. New and increasingly more stringent environmental regulations have added the need to consider the pollutant emissions produced by the burner. In many cases, reducing pollutant emissions and maximizing combustion efficiency are at odds with each other. For example, a well-accepted technique for reducing NO_x emissions is known as staging, where the primary flame zone is deficient in either fuel or oxidizer [2]. The balance of the fuel or oxidizer may be injected into the burner in a secondary flame zone or, in a more extreme case, may be injected somewhere else in the combustion chamber. Staging reduces the peak temperatures in the primary flame zone and also alters the chemistry in a way which reduces NO_x emissions because fuel-rich or fuel-lean zones are less conducive to NO_x formation than near-stoichiometric zones.

Figure 3.4 shows how the NO_x emissions are affected by the exhaust product temperature. Since thermal NO_x is exponentially dependent on the gas temperature, even small reductions in the peak flame temperature can dramatically reduce NO_x emissions. However, lower flame temperatures often reduce the radiant heat transfer from the flame since radiation is dependent on the fourth power of the absolute temperature of the gases. Another potential problem with staging is that it may increase CO emissions, which is an indication of incomplete combustion and reduced combustion efficiency. However, it is also possible that staged combustion may

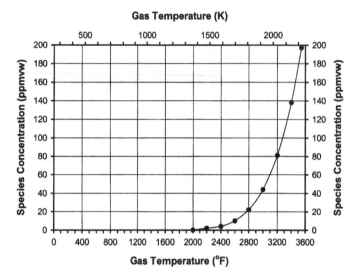

Figure 3.4 Dependence of NO on gas temperature. (From Ref. 1. Courtesy of CRC Press.)

produce soot in the flame, which can increase flame radiation. The actual impact of staging on the heat transfer from the flame is highly dependent on the actual burner design.

In the past, the challenge for the burner designer was often to maximize the mixing between the fuel and the oxidizer to ensure complete combustion, especially if the fuel was difficult to burn, as in the case of low heating value fuels such as waste liquid fuels or process gases from chemicals production. Now, the burner designer must balance the mixing of the fuel and the oxidizer to maximize combustion efficiency while simultaneously minimizing all types of pollutant emissions. This is no easy task as, e.g., NO_x and CO emissions often go in opposite directions as shown in Fig. 3.5. When CO is low, NO_x may be high and vice versa. Modern burners must be environmentally friendly, while simultaneously efficiently transferring heat to the load.

3.2.2 Design Factors

There are many types of burner designs that exist due to the wide variety of fuels, oxidizers, combustion chamber geometries, environmental regulations, thermal input sizes, and heat transfer requirements, which includes things like flame temperature, flame momentum, and heat distribution. Some of these design factors are briefly considered here.

Another important consideration for the pollution emission performance of a given burner design is how well the burner is maintained so that it operates as designed. McAdams et al. [3] note the importance of burner maintenance for maintaining the emission performance of low NO_x burners. Some potential problems include fuel tip orifice plugging, tip coking, fuel line flow restrictions, air inlet blockage, air leakage into the furnace, tip overheating, and refractory damage. These

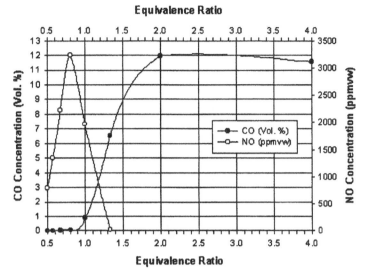

Figure 3.5   Dependence of NO and CO on equivalence ratio. (From Ref. 1. Courtesy of CRC Press.)

problems can be mitigated by proper burner specification, appropriate materials of construction selection, rigorous preventative maintenance practices, proper operating procedures, and fuel conditioning where needed. Burners must operate as designed in order to optimize performance and minimize pollution emissions

3.2.2.1   Fuel

Depending on many factors, certain types of fuels are preferred for certain geographic locations due to cost and availability considerations. Gaseous fuels, particularly natural gas, are commonly used in most industrial heating applications in the United States. In Europe, natural gas is also commonly used along with light fuel oil. In Asia and South America, heavy fuel oils are generally preferred although the use of gaseous fuels is on the rise. Fuels also vary, depending on the application. For example, in incineration processes, waste fuels are commonly used either by themselves or with other fuels like natural gas. In the petrochemical industry, fuel gases often consist of a blend of several fuels, including gases like hydrogen, methane, propane, butane, and propylene.

The fuel choice has an important influence on the pollution from a flame [4]. In general, solid fuels, such as coal and liquid fuels, like oil produce very luminous flames (see Fig. 3.6) which contain soot particles that radiate like black bodies to the heat load. Gaseous fuels like natural gas often produce nonluminous flames (see Fig. 3.7) because they burn so cleanly and completely without producing soot particles. A fuel like hydrogen is completely nonluminous as there is no carbon available to produce soot. Heavier hydrocarbon gaseous fuels like propane generally produce more luminous flames (see Fig. 3.8) than those of lighter hydrocarbon fuels like methane. In cases where highly radiant flames are required, a luminous flame is preferred. In cases where convection heat transfer is required, a nonluminous flame may be preferred in order to minimize the possibility of contaminating the heat load

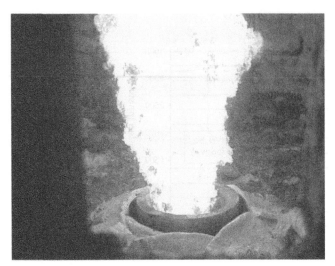

Figure 3.6 Example of a highly luminous heavy oil flame. (From Ref. 4. Courtesy of John Zink Co.)

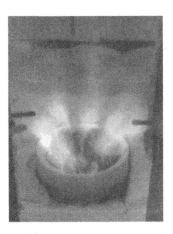

Figure 3.7 Example of a nonluminous natural gas flame. (From Ref. 4. Courtesy of John Zink Co.)

with soot particles from a luminous flame. Where natural gas is the preferred fuel and highly radiant flames are desired, new technologies are being developed to produce more luminous flames. These include things like pyrolyzing the fuel in a partial oxidation process [5], using a plasma to produce soot in the fuel [6], and generally controlling the mixing of the fuel and oxidizer to produce fuel-rich flame zones that generate soot particles [7]. Therefore, the fuel itself has a significant impact on the heat transfer mechanisms between the flame and the load. In most cases, the fuel choice is dictated by the customer as part of the specifications for the system and is not chosen by the burner designer. The designer must make the best of whatever fuel has been selected. In most cases, the burner design is optimized based on the choice for the fuel.

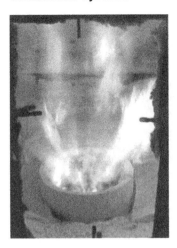

Figure 3.8   Example of a luminous propane flame. (From Ref. 4. Courtesy of John Zink Co.)

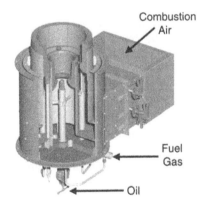

Figure 3.9   Typical combination oil and gas burner. (From Ref. 8. Courtesy of CRC Press.)

In some cases, the burner may have more than one type of fuel. An example is shown in Fig. 3.9 [8]. Dual-fuel burners are designed to operate typically on either gaseous or liquid fuels. These burners are used where the customer may need to switch between a gaseous fuel like natural gas and a liquid fuel like oil, usually for economic reasons. These burners normally operate on one fuel or the other, and occasionally on both fuels. Another application where multiple fuels may be used is in waste incineration. One method of disposing of waste liquids contaminated with hydrocarbons is to combust them by direct injection through a burner (see Fig. 3.10). The waste liquids are fed through the burner, which is powered by a traditional fuel such as natural gas or oil. The waste liquids often have very low heating values and are difficult to combust without auxiliary fuel. This further complicates the burner design where the waste liquid must be vaporized and combusted concurrently with the normal fuel used in the burner.

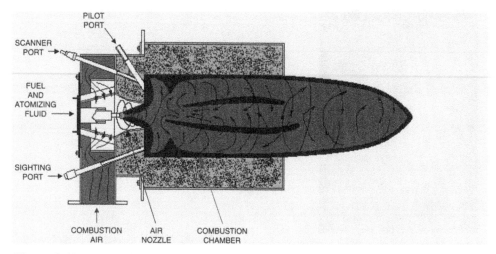

Figure 3.10 Example of a waste liquid fuel burner. (Courtesy of the American Society of Metals, Warren, PA.)

Stansifer [9] notes the importance of the condition of the fuel in NO_x emissions. In refineries and chemical plants, the fuel used in the burners often contains by-products from the chemical manufacturing process. In addition to fuels like methane, propane, and hydrogen, these fuels may also contain moisture and heavy hydrocarbons (sometimes tar-like) that can cause carbon buildup on the burner injector nozzles. This buildup is referred to as coking and can significantly degrade the performance of the burners, including increasing NO_x emissions, as nozzles get plugged off and flame patterns become irregular. Fuel gas-treatment systems (e.g., filters, coalescers, separators) upstream of the burners to remove moisture and heavy hydrocarbons can greatly improve overall system performance and reduce burner maintenance by reducing or eliminating coking of the burner nozzles. Clean fuels are especially important with newer generations of low NO_x burners that are more sensitive to system changes by virtue of their designs that are needed to achieve low NO_x.

3.2.2.2 Oxidizer

The predominant oxidizer used in most industrial heating processes is atmospheric air. This can present challenges in some applications where highly accurate control is required due to the daily variations in the barometric pressure and humidity of ambient air. The combustion air is sometimes preheated and sometimes blended with some of the products of combustion, which is usually referred to as flue gas recirculation (FlGR). In certain cases, preheated air is used to increase the overall thermal efficiency of a process. FlGR is often used both to increase thermal efficiency and to reduce NO_x emissions. The thermal efficiency is increased by capturing some of the energy in the exhaust gases, which are used to preheat the incoming combustion oxidizer.

NO_x emissions may also be reduced because the peak flame temperatures are reduced which can reduce the NO_x emissions, which are highly temperature dependent. There are also many high temperature combustion processes that use an oxidizer, which contains a higher proportion of oxygen than the 21% (by volume) that is found in normal atmospheric air. This is referred to as oxygen-enhanced combustion (OEC) and has many benefits that include increased productivity and thermal efficiency while reducing the exhaust gas volume and pollutant emissions [10]. A simplified global chemical reaction for the stoichiometric combustion of methane with air is given as follows:

$$CH_4 + 2O_2 + 7.52N_2 \rightarrow CO_2 + 2H_2O + 7.52N_2 + \text{trace species} \quad (3.1)$$

This compares to the same reaction where the oxidizer is pure O_2 instead of air:

$$CH_4 + 2O_2 \rightarrow CO_2 + 2H_2O + \text{trace species} \quad (3.2)$$

The volume of exhaust gases is significantly reduced by the elimination of N_2. In general, a stoichiometric oxygen-enhanced methane combustion process may be represented by

$$CH_4 + 2O_2 + xN_2 \rightarrow CO_2 + 2H_2O + xN_2 + \text{trace species} \quad (3.3)$$

where $0 \leq x \leq 7.52$, depending on the oxidizer. The N_2 contained in air acts as a ballast, which may inhibit the combustion process and have negative consequences. The benefits of using oxygen-enhanced combustion must be weighed against the added cost of the oxidizer, which in the case of air is essentially free except for the minor cost of the air-handling equipment and power for the blower. The use of a higher purity oxidizer has many consequences with regard to pollution generated by the flame, which are considered elsewhere in this book.

3.2.2.3 Gas Recirculation

Figure 3.11 shows a schematic of flue gas recirculation. Recirculating flue gases back through the flame is often used to mitigate pollution emissions by reducing the gas

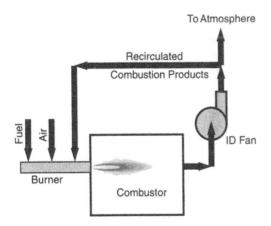

Figure 3.11 Schematic of flue gas recirculation. (From Ref. 1. Courtesy of CRC Press.)

temperatures in the flame. It may be done through the aerodynamic design of the gas injectors in the burner to induce product gases in the furnaces back through the flame. It may also be done by using an external fan or eductor to take gases at or near the stack exit and send them back to the burners. While this was traditionally done for increasing the thermal efficiency of the system, it is now more commonly done to reduce NO_x emissions (see Chap. 6).

A common technique used in combustion systems is to design the burner to induce furnace gases to be drawn into the burner to dilute the flame, usually referred to as furnace gas recirculation (FuGR). Figure 3.12 shows a schematic of flue gas recirculation. Even though the furnace gases are hot, they are still much cooler than the flame itself. This dilution may accomplish several purposes. One is to minimize NO_x emissions by reducing the peak temperatures in the flame, as in FlGR. However, furnace gas recirculation may be preferred to FlGR because no external high temperature ductwork or fans are needed to bring the product gases into the flame zone. Another reason to use furnace gas recirculation may be to increase the convective heating from the flame because of the added gas volume and momentum. An example of furnace gas recirculation into a burner is shown in Fig. 3.13 [11,12].

3.2.3 General Burner Types

There are numerous ways that burners can be classified. Some of the common ones are discussed in this section. One classification type is by the burner outlet shape, which is typically round (see Fig. 3.6) or rectangular (see Fig. 3.14). The burner shape is not considered here as a separate type.

3.2.3.1 Mixing Type

One common method for classifying burners is according to how the fuel and the oxidizer are mixed. In premixed burners, shown in a diagram in Fig. 3.15 and schematically in Fig. 3.16, the fuel and the oxidizer are completely mixed before

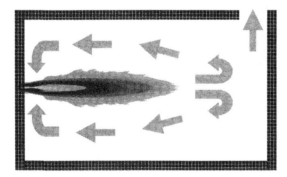

Figure 3.12   Schematic of furnace gas recirculation. (From Ref. 4. Courtesy of John Zink Co.)

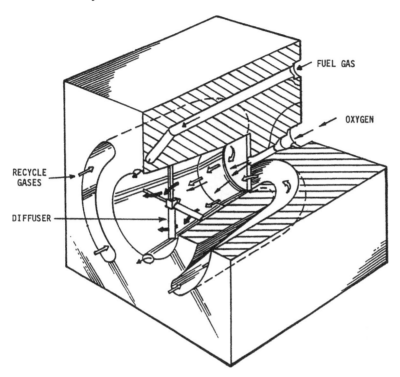

Figure 3.13 Burner incorporating furnace gas recirculation. (From Ref. 1. Courtesy of CRC Press.)

combustion begins. Porous radiant burners usually are of the premixed type. Premixed burners often produce shorter and more intense flames, compared to diffusion flames. This can produce high-temperature regions in the flame leading to nonuniform heating of the load and higher NO_x emissions. However, in flame impingement heating, premixed burners are useful because the higher temperatures and shorter flames can enhance the heating rates.

In diffusion-mixed burners, shown schematically in Fig. 3.17, the fuel and the oxidizer are separated and unmixed prior to combustion, which begins where the oxidizer/fuel mixture is within the flammability range. Early "burners" consisted of injectors where the fuel and oxidizer were separately injected into a hot furnace where the streams mixed [13]. This has some obvious safety problems but was used for some time until improvements such as the Bunsen burner, were developed. Oxygen/fuel burners are usually diffusion burners, primarily for safety reasons, to prevent flashback and explosion in a potentially dangerous system. Diffusion gas burners are sometimes referred to as "raw gas" burners as the fuel gas exits the burner essentially intact with no air mixed in. Diffusion burners typically have longer flames than premixed burners, do not have as high a temperature hot spot, and usually have a more uniform temperature and heat flux distribution.

It is also possible to have partially premixed burners, shown schematically in Figs. 3.16 and 3.18, where a portion of the fuel is mixed with the oxidizer. This is often done for stability and safety reasons where the partial premixing helps anchor

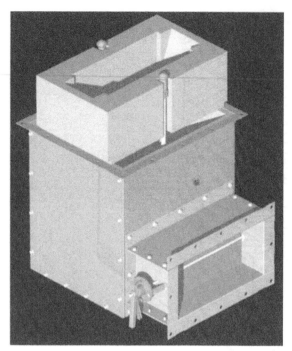

Figure 3.14   Rectangular "flat" flame shape burner. (Courtesy of John Zink Co.)

Figure 3.15   Premix burner. (From Ref. 1. Courtesy of CRC Press.)

the flame, but not fully premixing lessens the chance for flashback. This type of burner often has a flame length and temperature and heat flux distribution that is in between the fully premixed and diffusion flames.

Another burner classification based on mixing is known as staging: staged air and staged fuel. A staged-air burner is shown in a diagram in Fig. 3.19, schematically in Fig. 3.20, and pictorially in Fig. 3.21. A staged-fuel burner is shown in a diagram in Fig. 3.22, schematically in Fig. 3.23, and pictorially in Fig. 3.24. Secondary and sometimes tertiary injectors in the burner are used to inject a portion of the fuel and/or the oxidizer into the flame, downstream of the root of the flame. Staging is often done to reduce NO_x emissions and to produce longer flames. These longer flames typically have a lower peak temperature and more uniform heat-flux distribution than those of nonstaged flames.

3.2.3.2   Oxidizer Type

Burners and flames are often classified according to the type of oxidizer that is used. The majority of industrial burners use air for combustion. In many of the higher

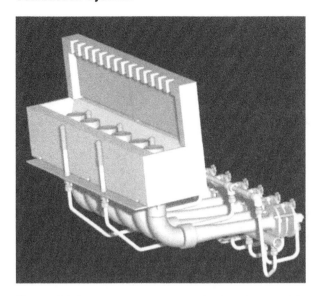

Figure 3.16 Typical partially premixed gas burner. (Courtesy of John Zink Co.)

Figure 3.17 Diffusion burner. (From Ref. 1. Courtesy of CRC Press.)

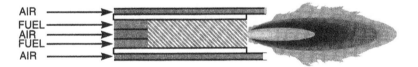

Figure 3.18 Partially premixed burner. (From Ref. 1. Courtesy of CRC Press.)

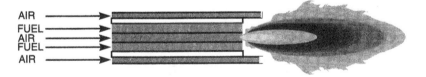

Figure 3.19 Schematic of a staged-air burner. (From Ref. 1. Courtesy of CRC Press.)

temperature heating and melting applications, such as glass production, the oxidizer is pure oxygen. In other applications, the oxidizer is a combination of air and oxygen, often referred to as oxygen-enriched air combustion.

Another way to classify the oxidizer is by its temperature. It is common in many industrial applications to recover heat from the exhaust gases by preheating the incoming combustion air, either with a recuperator or a regenerator (discussed below). Such a burner is often referred to as a preheated air burner.

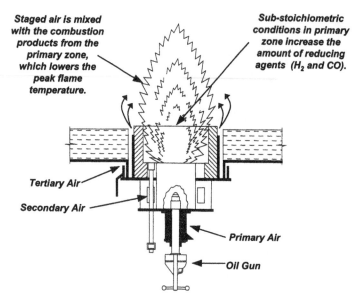

Figure 3.20   Staged-air oil burner. (From Ref. 8. Courtesy of John Zink Co.)

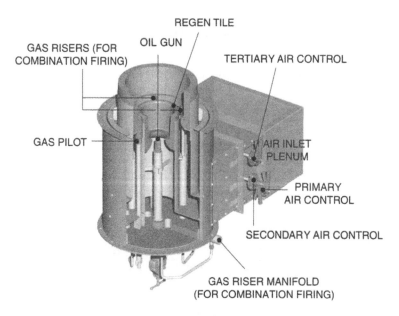

Figure 3.21   Typical staged-air combination oil and gas burner. (From Ref. 4. Courtesy of John Zink Co.)

3.2.3.3   Draft Type

Most industrial burners are known as forced-draft burners. This means that the oxidizer is supplied to the burner under pressure. For example, in a forced-draft air burner, the air used for combustion is supplied to the burner by a blower.

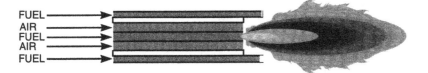

FUEL
AIR
FUEL
AIR
FUEL

Figure 3.22 Schematic of a staged-fuel burner. (From Ref. 1. Courtesy of CRC Press.)

In natural-draft burners, the air used for combustion is induced into the burner by the negative draft produced in the combustor. A schematic is shown in Fig. 3.25 and an example is shown in Fig. 3.26. In this type of burner, the pressure drop and combustor stack height are critical in producing enough suction to induce sufficient combustion air into the burners. This type of burner is commonly used in the chemical and petrochemical industries in fluid heaters. The main consequence of the draft type on heat transfer is that the natural-draft flames are usually longer than the forced-draft flames so that the heat flux from the flame is distributed over a longer distance and the peak temperature in the flame is often lower. Forced-draft burner systems often produce more noise (see Chap. 10) compared to natural draft systems because of the noise produced by the fans and blowers.

3.2.3.4 Heating Type

Burners are often classified as to whether they operate by direct or indirect heating. In direct heating, there is no intermediate heat-exchange surface between the flame and the load (see Fig. 3.27). In indirect heating, such as radiant tube burners, there is an intermediate surface between the flame and the load. This is usually done because the combustion products cannot come into contact with the load because of possible contamination.

In an indirect-fired furnace, there is some intermediate surface between the combustion products and the load (see Fig. 3.28). The surface is commonly some type of ceramic due to the high temperatures, although metals are used in some cases. This surface is designed to prevent the combustion products from contacting the load and reducing the quality of the finished product. That quality can be reduced in two ways: (1) by chemically altering the product and (2) by physically changing the surface. An example of the first is in many metallurgical processes, especially heat treating, where the metal product must be heated in a protective atmosphere containing H_2, N_2, and CO, with only negligible quantities of O_2 and H_2O, which are detrimental to the quality of the metal during heat treating (see Chap. 13) [14]. An example of the second type of product quality reduction is in drying a coating or an ink on a web surface in a dryer where the combustion gases could disturb the surface before the coating is dry or the ink is set. Another type of indirect-fired furnace is where the flame gases are separated from the load for either transport or safety reasons. For example, in a process fluid heater, the fluid is transported through metal tubes located inside a furnace. The metal tubes separate the combustion products from the fluid. One reason for the tubes is to transport the fluids at an elevated pressure through the heaters, which are fired at approximately atmospheric pressure. Another reason is related to safety where many of the fluids being heated are hydrocarbons that are potentially explosive if overheated and exposed to sufficient oxygen.

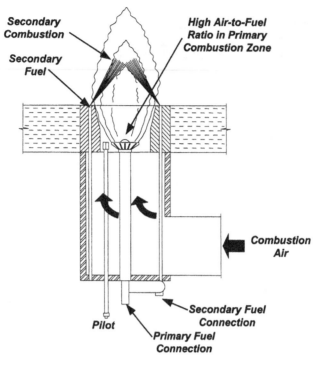

Figure 3.23 Typical staged-fuel gas burner. (From Ref. 1. Courtesy of John Zink Co.)

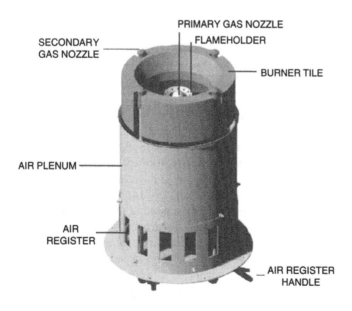

Figure 3.24 Typical staged-fuel diffusion burner. (From Ref. 4. Courtesy of John Zink Co.)

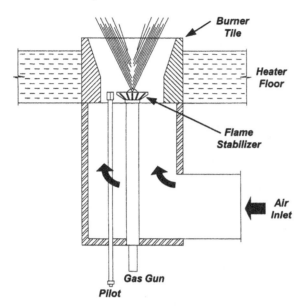

Figure 3.25 Typical natural draft gas burner. (From Ref. 8. Courtesy of John Zink Co.)

Figure 3.26 Natural draft burner. (Ref. 1. Courtesy of John Zink Co.)

Two methods are commonly used to separate the combustion products from the load. One is to use open-flame burners but to have a separator, sometimes referred to as a muffle, across the entire combustion space between the flames and the load. Some of the challenges of this method include supporting the separator because of the high temperatures, maximizing the heat transfer from the flames to

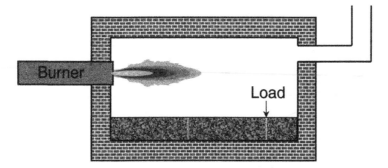

Figure 3.27   Elevation view of a direct-fired furnace.

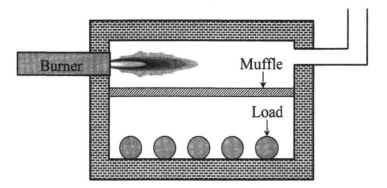

Figure 3.28   Elevation view of an indirect-fired furnace.

the separator to optimize the thermal efficiency, and getting a good gas seal around the perimeter of the separator due to thermal expansion. The second method commonly used in industrial combustion applications to separate the exhaust products from the load is to use radiant tube burners. In that method, the flame from each individual burner is contained in a ceramic tube. This often improves the overall heat transfer to the separator because of the improved forced convection inside the tube. As with the muffle, supporting long radiant tubes as well as the seal between the typically metal burner and ceramic tube can be problems. A different challenge from that with the muffle is obtaining uniform heat flux from the radiant tubes as the hottest gases are produced at the burner end with the coldest gases exiting from the tube. This can be mitigated through proper design of the radiant tube burner. Perforated ceramic or wire-mesh radiant burners are used in certain types of industrial heating applications. Examples of these types of burners are shown in Fig. 3.29 [15].

3.3   COMBUSTORS

This section briefly introduces the combustors that are commonly used in industrial heating and melting applications. Dryers (see Fig. 3.30 for examples of continuous dryers) are commonly used in lower temperature applications. Heaters are used in

Figure 3.29 Examples of porous ceramic and wire-mesh radiant burners. (From Ref. 15. Courtesy of Solaronics, Rochester, MI.)

moderate temperature applications. Furnaces are used in high-temperature applications. Trinks and Mawhinney have written a very useful two-volume guide on industrial furnaces for the reader interested in more details on combusters [25,26].

3.3.1 Design Considerations

There are many important factors that must be considered when designing a combustor. This section only briefly considers a few of those factors and how they may influence the pollution from the system.

3.3.1.1 Load Handling

A primary consideration for any combustor is the type of material that will be processed. One obvious factor of importance in handling the load and transporting it through the combustor is its physical state, whether it is a solid, liquid, or gas. Another factor is the transport properties of the load. For example, the solid may be granular or it might be in the form of a sheet (web) as shown in Fig. 3.31. Related to that is how the solid will be fed into the combustor. A granular solid could be fed into a combustor continuously with a screw conveyor or it could be fed in with discrete charges from a front-end loader. The shape of the furnace will vary according to how the material will be transported through it. For example, limestone is fed continuously into a rotating and slightly downwardly inclined cylinder.

3.3.1.2 Temperature

In this book, industrial heating applications have been divided into two categories: higher and lower temperatures. The division between the two is somewhat arbitrary but mainly concerns the different types of applications used in each. For example, most of the metal and glass melting applications fall into the higher temperature categories as the furnace temperatures are often well over 2000°F (1400 K). They use technologies like air preheating and oxygen enrichment to achieve those higher temperatures. Lower temperature applications include dryers, process heaters, and heat treating and are typically below about 2000°F (1400 K). Although many of these processes may use air preheating, it is primarily to improve the thermal efficiency and not to achieve higher flame temperatures. Obviously the combustors are designed differently for higher and lower temperature processes.

3.3.1.3 Heat Recovery

When heat recovery is used in an industrial combustion process, it is an integral part of the system. The two most popular methods are regenerative and recuperative, which are discussed briefly below. The heat-recovery system is important in the design of the combustor as it determines the thermal efficiency of the process and the flame temperatures in the system. It also influences the heat-transfer modes as it may increase the both the radiation and convection because of higher flame temperatures. Another type of heat recovery that is used in some processes is furnace or flue gas recirculation where the exhaust products are recirculated back through the flame. This also influences the heat transfer and furnace design as it can moderate the flame temperature but increase the volume flow of gases through the combustion chamber. Heat-exchanger classifications are shown in Fig. 3.32.

3.3.2 General Classifications

There are several ways that a combustor can be classified, which are briefly discussed in this section. Each type has an impact on the heat-transfer mechanisms in the furnace.

3.3.2.1 Load Processing Method

Furnaces are often classified as to whether they are batch or continuous. In a batch furnace, the load is charged into the furnace at discrete intervals where it is heated. There may be multiple load charges, depending on the application. Normally, the firing rate of the burners is reduced or turned off during the charging cycle. On some furnaces, a door may also need to be opened during charging. These significantly impact the heat transfer in the system as the heat losses during the charge cycle are very large. The radiation losses through open doors are high and the reduced firing rate may not be enough to maintain the furnace temperature. In some cases, the temperature on the inside of the refractory wall, closest to the load, may actually be lower than the temperature of the refractory at some distance from the inside, due to the heat losses during charging. The heating process and heat transfer are dynamic and constantly changing as a result of the cyclical nature of the load charging. This makes analysis of these systems more complicated because of the need to include time in the computations.

In a continuous furnace, the load is fed into and out of the combustor constantly. The feed rate may change sometimes due to conditions upstream or downstream of the combustor or due to the production needs of the plant, but the process is nearly steady state. This makes continuous processes simpler to analyze as there is no need to include time in the computations. It is often easier to make meaningful measurements in continuous processes due to their steady-state nature. Figure 3.30 shows some examples of continuous dryers [16].

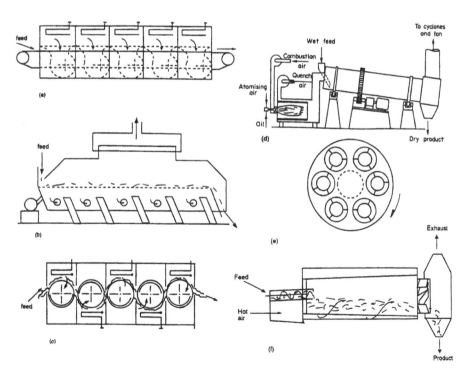

Figure 3.30 Examples of continuous dryers. (From Ref. 16. Courtesy of CRC Press.)

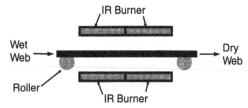

Figure 3.31 Gas-fired infrared burners heating a moving substrate. (From Ref. 1. Courtesy of CRC Press.)

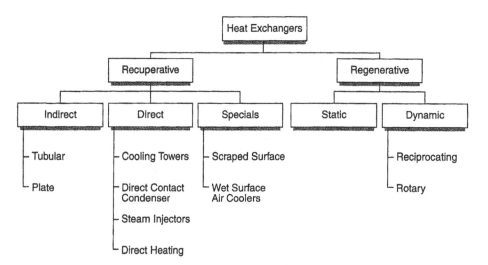

Figure 3.32 Heat-exchanger classifications. (From Ref. 22. Courtesy of CRC Press.)

There are some furnaces that are semicontinuous where the load may be charged in a nearly continuous fashion, but the finished product may be removed from the furnace at discrete intervals. An example is an aluminum reverberatory furnace, which is charged using an automatic side-well feed mechanism. In that process, shredded scrap is continuously added to a circulating bath of molten aluminum. When the correct alloy composition has been reached and the furnace has a full load, some or all of that load is then tapped out of the furnace. The effect on heat transfer is somewhere between that for batch and continuous furnaces.

3.3.2.2 Heating Type

As described above for burners, combustors are often classified as operating by indirect or direct heating. An important consideration for the choice of the type of oven or dryer for the specific application depends on whether direct or indirect heating is needed [17]. In indirect heating, there is some type of intermediate heat-transfer medium between the flames and the load which keeps the combustion products separate from the load (see Fig. 3.33). One example is a muffle furnace where there is a high-temperature ceramic muffle between the flames and the load. The flames transfer their heat to the muffle, which then radiates to the load which is usually some type of metal. The limitation of indirect heating processes is the

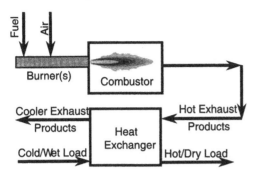

Figure 3.33 Schematic of an indirect heating system. (From Ref. 1. Courtesy of CRC Press.)

temperature limit of the intermediate material. Although ceramic materials have fairly high temperature limits, other issues such as structural integrity over long distance spans and thermal cycling can still reduce the recommended operating temperatures. Another example of indirect heating is in process heaters where fluids are transported through metal tubes that are heated by flames. Indirect heating processes often have fairly uniform heat-flux distributions because the heat-exchange medium tends to homogenize the energy distribution from the flames to the load. The heat transfer from the heat exchange surface to the load is often fairly simple and straightforward to compute because of the absence of chemical reactions in between. However, the heat transfer from the flames to the heat-exchange surface and the subsequent thermal conduction through that surface are as complicated as if the flame was radiating directly to the load.

In direct heating, there is no heat exchange surface between the flames and the load, as shown in Fig. 3.34. As a result of the temperature limits of the heat-exchange materials, most higher temperature processes are of the direct heating type where the flames can directly radiate heat to the load.

3.3.2.3 Geometry

Another common way of classifying combustors is according to their geometry, which includes their shape and orientation. The two most common shapes are rectangular and cylindrical. The two most common orientations are horizontal and vertical, although inclined furnaces are often used in certain applications such as rotary cement furnaces. An example of using the shape and orientation of the furnace as a means of classification would be a vertical cylindrical heater (sometimes referred to as a VC) used to heat fluids in the petrochemical industry. Both the furnace shape and orientation have important effects on the heat transfer in the system. They also determine the type of analysis that will be used. For example, in a VC heater it is often possible to model only a slice of the heater due to its angular symmetry, in which case cylindrical coordinates would be used. On the other hand, it is usually not reasonable to model a horizontal rectangular furnace using cylindrical coordinates, especially if buoyancy effects are important. Some furnaces are classified by what they look like. One example is a shaft furnace used to make

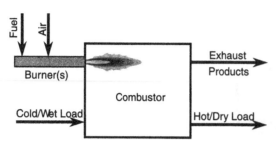

Figure 3.34   Schematic of a direct heating system. (From Ref. 1. Courtesy of CRC Press.)

iron. The raw materials are loaded into the top of a tall thin vertically oriented cylinder. Hot combustion gases generated at the bottom through the combustion of coke flow up through the raw materials which are then heated. The melted final product is tapped out of the bottom. The furnace looks and acts almost like a shaft because of the way the raw materials are fed in through the top and exit at the bottom. A transfer chamber used to move molten metal around in a steel mill is often referred to as a ladle because of its function and appearance. These ladles are preheated using burners before the molten metal is poured into them to prevent the refractory-lined vessels from thermally shocking.

Another aspect of the geometry that is important in some applications is whether the furnace is moving or not. For example, in a rotary furnace for melting scrap aluminum, the furnace rotates to enhance mixing and heat-transfer distribution. This again affects the type of analysis that would be appropriate for that system and can add some complexity to the computations. The burner orientation with respect to the combustor is also sometimes used to classify the combustor. For example, a wall-fired furnace has burners located in and firing along the wall.

3.3.2.4   Heat Recuperation

In many heat-processing systems, energy recuperation is an integral part of the combustion system. Often the heat recuperation equipment is a separate component of the system and not part of the burners themselves. Depending on the method used to recover the energy, the combustors are commonly referred to as either recuperative or regenerative (see discussion below). The heat transfer in these systems is a function of the energy recovery system. For example, the higher the combustion air-preheat temperature, the hotter the flame and the more radiant heat that can be produced by that flame. The convective heat transfer may also be increased due to the higher gas temperature and also due to the higher thermal expansion of the gases, which increases the average gas velocity through the combustor.

3.4   HEAT LOAD

This section is a brief introduction to some of the important issues concerning the heat load in a furnace or combustor.

3.4.1 Process Tubes

In petrochemical production processes, process heaters are used to heat petroleum products up to operating temperatures. The fluids are transported through the process heaters in process tubes. These heaters often have a radiant section and a convection section. In the radiant section, radiation from burners heats the process tubes. In the convection section, the combustion products heat the tubes by flowing over the tubes. The design of the radiant section is especially important as flame impingement on the tubes can cause premature failure of the tubes or cause the hydrocarbon fluids to coke inside the tubes, which reduces the heat transfer to the fluids.

3.4.2 Moving Substrate

In some applications, heaters and burners are used to heat or dry moving substrates or webs. An example is shown in Fig. 3.31. One common application is the use of gas-fired infrared (IR) burners to remove moisture from paper during the forming process [18]. These paper webs can travel at speeds over 300 m/sec (1000 ft/sec) and are normally dried by traveling over and contacting steam-heated cylinders. IR heaters are often used selectively to dry certain portions of the web that may be wetter than others. For example, if the target moisture content for the paper is 5%, then the entire width of the paper must have no more than 5% moisture. Streaks of higher moisture areas often occur in sections along the width of the paper. Without selectively drying these areas, those streaks would be dried to the target moisture level, which means that the rest of the sheet would be dried to even lower moisture levels. This creates at least two important problems. The first is lost revenue because paper is usually sold on a weight basis. Any water unnecessarily removed from the paper decreases its weight and therefore results in lost income. Another problem is a reduction in the quality of the paper. If areas of the paper are too dry, they do not handle as well in devices like copiers and printers and are not nearly as desirable as paper of uniform moisture content. Therefore, selective drying of the paper only removes the minimum amount of water from the substrate. The challenge of this application is to measure the moisture content profile across the width of a sheet that may be several meters wide and moving at hundreds of meters per second. That information must then be fed to the control system for the IR heaters, which must then be able to react almost instantaneously. This is possible today because of advances in measurement and controls systems.

Another example of a moving substrate application is using IR burners to remove water during the production of fabrics in textile manufacturing [19]. Moving substrates present unique challenges for burners. Often the material being heated can easily be set on fire if there is a line stoppage and the burner is not turned off quickly enough. This means that the burner control system must be interlocked with the web-handling equipment so that the burners can be turned off immediately in the event of a line stoppage. If the burners have substantial thermal mass, then the burners may need to be retracted away from the substrate during a stoppage or heat shields may need to be inserted between the burners and the substrate to prevent overheating.

Convection dryers are also used to heat and dry substrates. Typically, high-velocity heated air is blown at the substrate from both sides so that the substrate is elevated between the nozzles. In many cases, the heated air is used for both heat and mass transfer, to volatilize any liquids on or in the substrate such as water, and then carry the vapor away from the substrate.An important aspect of heating webs is how the energy is transferred into the material. For example, dry paper is known to be a good insulator. When steam cylinders are used to heat and dry paper, they become less and less effective as the paper becomes drier because the heat from the cylinder cannot conduct through the paper as well as when it is moist since the thermal conductivity of the paper increases with moisture content. IR burners are effective for drying paper because the radiant energy transfers into the paper and is absorbed by the water. The radiant penetration into the paper actually increases as the paper becomes drier, unlike with steam cylinders which become less effective.

3.4.3 Opaque Materials

This type of load encompasses a wide range of materials including granular solids like limestone and liquids like molten metal. For this type of load, the heat transfers to the surface of the load and must conduct down into the material. This process can be enhanced by proper mixing of the materials so that new material is constantly exposed to the surface as in rotary kilns or in aluminum reverberatory furnaces, which have molten-metal pumps to recirculate the metal continuously through the heating zone. The potential problems with this method include overheating the surface materials or having lower thermal efficiencies by limiting the heat transfer to the surface to prevent overheating.

3.4.4 Transparent Materials

The primary example of this type of load is glass, which has selective radiant transmission properties. In glass-melting processes (see Chap. 11), the primary mode of heat transfer is by radiation. Flames have specific types of radiant outputs that vary as a function of wavelength. If the flame is nonluminous, the flame usually has higher radiant outputs in the preferred wavelengths for water and carbon dioxide bands. If the flame is luminous, it has a broader, more graybody-type spectral radiant profile. Chapter 4 shows that luminous flames are preferred in melting glass because of the selective transmission properties of molten glass. This allows a significant portion of the radiation received at the surface of the glass to penetrate into the glass, which enhances heat-transfer rates and reduces the chances of overheating the surface which would reduce product quality.

3.5 HEAT RECOVERY DEVICES

Heat recovery devices are often used to improve the efficiency of combustion systems and to achieve higher flame temperatures for processes like glass production that have high melting temperatures. Some of these devices are incorporated into the burners, but more commonly they are another component in the combustion system, separate from the burners. These heat recovery devices incorporate some type of heat exchanger, depending on the application. The two most common types have been recuperators and regenerators, which are briefly discussed next. Reed [20] predicts an increasing importance for heat recovery devices in industrial combustion systems for increasing heat transfer and thermal efficiencies. Katsuki and Hasegawa [21] have written a review article on the used of highly preheated air (above 1300 K or 1900°F) in industrial combustion applications including an extensive discussion of NO emissions. The two most common types of heat exchanger used in industrial combustion applications are regenerators and recuperators. There are various types of each as shown in Fig. 3.32 [22]. These are discussed next. The primary pollution problem associated with using heat recovery devices is higher NO_x emissions. This is discussed in other chapters in the book.

3.5.1 Recuperators

A recuperator is a low- to medium-temperature (up to about 1300°F or 700°C), continuous heat exchanger that uses the sensible energy from hot combustion products to preheat the incoming combustion air. These heat exchangers are commonly counterflow where the highest temperatures for both the combustion products and the combustion air are at one end of the exchanger with the coldest temperatures at the other end. Lower temperature recuperators are normally made of metal, while higher temperature recuperators may be made of ceramics. Recuperators are typically used in lower temperature applications because of the limitations of the metals used to construct these heat exchangers. Figure 3.35 shows an example of a recuperative burner.

3.5.2 Regenerators

A regenerator is a higher-temperature, transient heat exchanger that is used to improve the energy efficiency of high-temperature heating and melting processes,

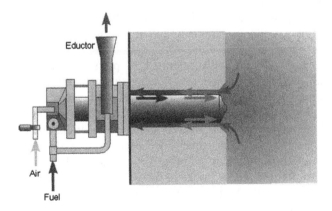

Figure 3.35 Example of a recuperative burner. (Courtesy of WS Thermal Process Technology, Inc., Elyria, OH.)

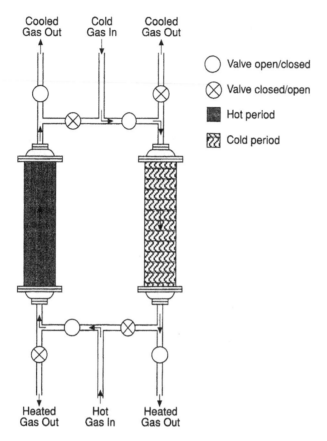

Figure 3.36 Schematic of a fixed-bed regenerator. (From Ref. 23. Courtesy of CRC Press.)

particularly in the high-temperature processing industries like glass production. In a regenerator, energy from the hot combustion products are temporarily stored in a unit constructed of firebricks. This energy is then used to heat the incoming combustion air during a given part of the firing cycle up to temperatures in excess of 2000°F (1000°C).

Regenerators (see Fig. 3.36) are normally operated in pairs [23]. During one part of the cycle, the hot combustion gases are flowing through one of the regenerators and heating up the refractory bricks, while the combustion air is flowing through and cooling down the refractory bricks in the second regenerator. Both the exhaust gases and the combustion air directly contact the bricks in the regenerators, although not both at the same time since each is in a different regenerator at any given time. After a sufficient amount of time (usually from 5 to 30 min), the cycle is reversed so that the cooler bricks in the second regenerator are then reheated while the hotter bricks in the first regenerator exchange their heat with the incoming combustion air. A reversing valve is used to change the flow from one gas to another in each regenerator. Figure 3.37 shows an example of a regenerative burner.

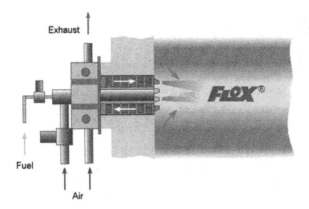

Figure 3.37 Example of a regenerative burner. (Courtesy of WS Thermal Process Technology, Inc., Elyria, OH.)

Davies [24] notes that some of the following questions should be considered concerning regenerative burners:

- What will the fuel savings be (compared to no heat recuperation)?
- What is the maximum allowable flue gas temperature?
- How large is the heat exchanger?
- What is the air pressure drop through the exchanger?
- How long will the exchanger run without plugging?
- Is the flue gas path through the furnace altered?

REFERENCES

1. CE Baukal. *Heat Transfer in Industrial Combustion*. Boca Raton, FL: CRC Press, 2000.
2. JL Reese, GL Moilanen, R Borkowicz, C Baukal, D Czerniak, R Batten. State-of-the-art of NO_x emission control technology. ASME paper 94-JPGC-EC-15, *Proceedings of International Joint Power Generation Conference*, Phoenix, AZ, 3–5 Oct. 1994.
3. J McAdams, J Karan, R Witte, M Claxton. Low NO_x burner maintenance in high temperature furnaces. *Proceedings of the 14th Ethylene Producers' Conference*, New York: American Institute of Chemical Engineers, Vol. 11, pp. 351–370, 2002.
4. CE Baukal (ed.). *The John Zink Combustion Handbook*. Boca Raton, FL: CRC Press, 2001.
5. ML Joshi, ME Tester, GC Neff, SK Panahi. Flame particle seeding with oxygen enrichment for NO_x reduction and increased efficiency. *Glass*, Vol. 68, No. 6, pp. 212–213, 1990.
6. R Ruiz, JC Hilliard. Luminosity enhancement of natural gas flames. *Proceedings of 1989 International Gas Research Conf.*, ed. TL Cramer, Government Institutes, Rockville, MD, pp. 1345–1353, 1990.
7. AG Slavejkov, TM Gosling, RE Knorr. Low-NO_x staged combustion device for controlled radiative heating in high temperature furnaces. U.S. Patent 5611682, 18 March 1997.
8. CE Baukal (ed.). *Computational Fluid Dynamics in Industrial Combustion*. Boca Raton, FL: CRC Press, 2001.

9. MW Stansifer. Fuel gas clean up for low NO$_x$ burners. *Proceedings of the 14th Ethylene Producers' Conference*. New York: American Institute of Chemical Engineers, Vol. 11, pp. 166–172, 2002.

10. CE Baukal (ed.). *Oxygen-Enhanced Combustion*. Boca Raton, FL: CRC Press, 1998.

11. KJ Fioravanti, LS Zelson, CE Baukal. Flame stabilized oxy–fuel recirculating burner. U.S. Patent 4954076, 4 Sept. 1990.

12. CE Baukal, KJ FIoravanti, L Vazquez del Mercado. The REFLEX burner. *Proceedings of the 3rd Fossil Fuel Combustion Symposium*, Houston, TX ASME PD-Vol. 33, pp. 61–67, 1991.

13. D Bennett. Burner History 101, Part 1. *Process Heat.*, Vol. 9, No. 10, pp. 20–21, 2002.

14. HS Kurek, M Khinkis, W Kunc, A Touzet, A de La Faire, T Landais, A Yerinov, O Semernin. Flat radiant panels for improving temperature uniformity and product quality in indirect-fired furnaces. *Proceedings of 1998 International Gas Research Conference*, Vol. V: *Industrial Utilization*, ed. DA Dolenc, Gas Research Institute, Chicago, pp. 15–23, 1998.

15. F. Ahmady. *Process Heat.*, Vol. 1, No. 2, pp. 38–43, 1994.

16. RB Keey. Dryers. In GF Hewitt, GL Shires, YV Polezhaev (eds). *International Encyclopedia of Heat & Mass Transfer* pp. 337–342, Boca Raton, FL, CRC Press 1997.

17. D Traub. Indirect vs. direct: the heat transfer method does affect the process *Process Heat.*, Vol. 6, No. 2, pp. 26–29, 1999.

18. S Longacre. Using infrared to dry paper and its coatings, *Process Heat.*, Vol. 4, No. 2, pp. 45–49, 1997.

19. TM Smith, CE Baukal. Space-age refractory fibres improves gas-fired infrared generators for heat processing textile webs. *J. Coated Fabrics*, Vol. 12, No. 3, pp. 160–173, January 1983.

20. RJ Reed, Future consequences of compact highly effective heat recovery devices. In C Presser, DG Lilley (eds.). *Heat Transfer in Furnaces*, ASME HTD-Vol. 74, pp. 23–28, New York, 1987.

21. M Kastuki, T Hasegawa. The science and technology of combustion in highly preheated air. *Twenty-Seventh Symposium (International) on Combustion*. Pittsburgh, PA: The Combustion Institute, pp. 3135–3146, 1998.

22. RJ Brogan. Heat exchangers. In GF Hewitt, GL Shires, YV Polezhaev (eds.). *International Encyclopedia of Heat & Mass Transfer*, Boca Raton, FL: CRC Press, 1997.

23. AJ Willmont. Regenerative heat exchangers. In GF Hewitt, GL Shires, YV Polezhaev (eds.). *International Encyclopedia of Heat & Mass Transfer*, Boca Raton, FL: CRC Press 1997.

24. T Davies. Regenerative burners for radiant tubes–field testing experience. In MA Lukasiewicz (eds.). *Industrial Combustion Technologies*, Warren, PA: American Society of Metals, pp. 65–70, 1986.

25. W Trinks, MH Mawhinney. Industrial Furnaces, Vol. 1, Ed. 5. New York: John Wiley & Sons, 1961.

26. W. Trinks, MH Mawhinney. Industrial Furnaces, Vol. 2, Ed. 4. New York: John Wiley & Sons, 1967.

4

Computer Modeling

4.1 COMBUSTION MODELING

There are many reasons to model a combustion system. The most obvious is to gain insight into a particular configuration in order to optimize it. Optimization means different things to different people. It may mean maximizing thermal efficiency, minimizing pollutant emissions, maximizing throughput, minimizing operating costs, or some combination of these. Another reason to model is in the development of new technologies. New geometries can be tested relatively quickly compared to building an entire combustion system. Ideally, modeling is done in conjunction with experimentation to validate a particular new design. Doing modeling first can save considerably on prototype development time and costs by eliminating particular designs without having actually to test them. However, in most cases it is not possible to eliminate prototype testing completely because of the uncertainty and limitations of combustion modeling, especially when it comes to new configurations that may never have been tried before.

Another reason for modeling is to aid in scaling systems to either larger or smaller throughputs. Simple velocity or residence time-scaling laws often do not apply to complicated combustion problems [1]. Modeling can be used for predictive purposes to test different scenarios that may be too risky or expensive to try in an existing operational industrial combustion system. For example, a glass producer may want to evaluate the impact of replacing an existing air preheat system with pure oxygen.

Another reason to do combustion modeling is to help determine the location for instrumentation. For example, models can be used to help decide where to locate thermocouples in a furnace wall at potential hot spots in order to prevent refractory damage. Although experiments are normally used to validate modeling results, the opposite may also be true. In industrial combustion systems, large-scale probes may be necessary due to water-cooling requirements for survivability. These large probes can cause significant disturbances in the process, which can be simulated with models. The model results can then be used to determine the relevance of experimental measurements. Models can also be used to simulate potentially dangerous conditions to assess the consequences in order to design the proper safety equipment and procedures.

A more recent use of computer modeling is for control of processes where the models are used to predict the results under the given conditions and then adjust the operating parameters to produce the desired results. This includes the use of artificial intelligence where the control system has a large database of past operating conditions and the associated results so that the system can then predict and adjust itself to meet new operating conditions. Examples include making adjustments as equipment ages and deteriorates as well as for new materials being processed. In the past, these adjustments would have been based on the knowledge and experience of the operators and were often trial-and-error. Newer control systems promise more sophisticated and systematic evaluation of the given operating conditions and desired results.

One of the risks of computer modeling is that too much faith may be placed in the results. Some tend to believe anything generated by a computer. However, if the computer models have not been properly validated, then the results may be highly suspect. For the foreseeable future, it is likely that models will continue to use various approximations (e.g., turbulence) in order to obtain solutions in a reasonable amount of time. Therefore, the user must exercise good judgment and not try to overextend the results beyond what is warranted. For example, in many cases models are very useful in predicting pollutant emission trends but are often very inaccurate in predicting the actual emissions. Knowledge of the model's capabilities helps one understand which results are more reliable and which ones are less reliable. The bulk fluid flow and heat transfer in a combustion system can usually be predicted with a high degree of accuracy, while the small-scale turbulence and trace species predictions may be less reliable. Therefore, it is recommended that computer modeling of combustion systems only be done by those who have been properly trained in that area.

Patankar and Spalding [2] note some of the important aspects of the problem statement for industrial combustion modeling problems:

- Geometry of the combustion chamber
- Fuel and air input conditions
- Thermal boundary conditions
- Thermodynamic, transport, radiative, and chemical–kinetic properties
- The desired outputs of models:
    velocity, temperature, composition, etc. throughout the chamber
    heat flux and temperature at the wall

Figure 4.1 shows a schematic of the elements of computational fluid dynamics (CFD) modeling [3].

A number of books have been written on the subject of modeling combustion processes. However, very few have specifically concerned large-scale industrial combustion systems. Khalil [4] presented modeling results for six large-scale industrial furnaces with published experimental data for comparison. These six studies involved burners with and without quarls (burner tiles); methane, natural gas, and propane fuels; firing rates ranging from 0.74 to 13 MW (2.5–44×10^6 Btu/hr); furnace lengths ranging from 4.5 to 11 m (15–36 ft); and swirl numbers ranging from 0 to 5.0. The modeling results using the k–ε turbulence model were in good agreement with the published experimental data. Oran and Boris [5] have edited a large book on combustion modeling. Part 1 of the book concerns modeling the chemistry of

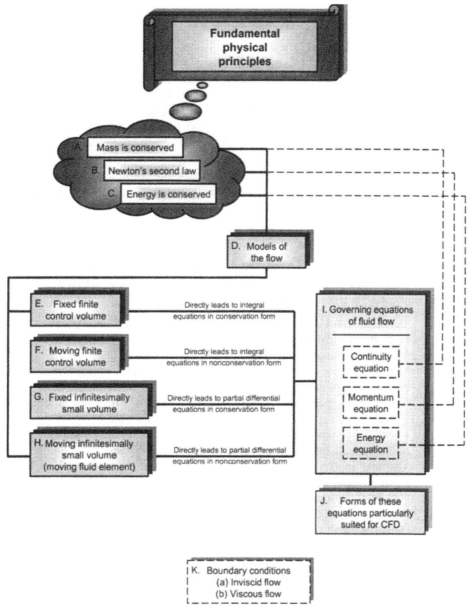

Figure 4.1 Elements of CFD modeling. (From Ref. 3. Courtesy of John Zink.)

combustion. Part 2 contains information on flames and flame structure. Part 3 is on high-speed reacting flows. Part 4 is humorously entitled "(Even More) Complex Combustion Systems" and has chapters on liquid and solid fuel combustion, as well as on pulse combustion. This book is more theoretical in nature and is intended for aerospace combustion. However, it does have some useful information pertinent to industrial combustion, which is referred to later in this chapter.

A book edited by Larrouturou [6] looks at modeling of some fundamental processes in combustion science, but does not specifically consider large-scale industrial flames. The papers have significant discussions of flame chemistry and fluid dynamics, but very little on heat transfer from the flames. Chung [7] has edited a book that contains chapters on the various techniques used to model combustion processes, but without any specific applications to industrial combustion problems. In a handbook on fluid dynamics, Lilley [8] has a brief treatment of combustion modeling with only very brief discussions of industrial applications. Baukal et al. [9] have edited a book specifically on modeling industrial combustion processes. Poinsot and Veynante [10] discuss the theory and fundamentals of computational combustion modeling, but have no discussions of industrial combustion applications. Some books have been written on the subject of environmental computer modeling [11–15].

4.2 MODELING APPROACHES

A complete combustion system may be extremely complex and can include a wide range of physical processes that are often highly interactive and interdependent. A given combustion system may include:

- Turbulent fluid dynamics in the flame with laminar fluid dynamics in the bulk of the combustor.
- Multidimensional flows which could include swirl.
- Multiple phases that could include gases, liquids, and solids, depending on the fuel composition.
- Very high temperature, velocity, and species gradients in the flame region with much lower gradients in the bulk of the combustor.
- Large material property variations caused by the wide range of temperatures, species, and solids present in the system.
- Multiple modes of heat transfer, especially radiation which is highly nonlinear and may include wavelength dependence.
- Complex chemistry involving numerous reactions and many species, most of which are in trace amounts.
- Porous media.
- Catalytic chemical reactions in some limited applications.
- Complex, nonsymmetrical furnace geometries.
- Multiple flame zones produced by burners that may be operated under different conditions and whose flames interact with each other.
- A heat load that may be moving and interacting with the combustion space above it in a nonlinear manner.
- A heat load that may produce volatile species during the heating process.
- A heat load whose properties may vary greatly with temperature, physical state, and even wavelength for radiation.
- A transient heating and melting process that may include discrete material additions and withdrawals.

There are many challenges caused by this complexity, which include inadequate physics to model the problem properly, large numbers of grid points requiring large amounts of computer memory, and long computation times.

The simulation results may be difficult to validate as many of the experimental measurements are difficult, time consuming, and costly to make in industrial combustors. Therefore, in most combustion simulations, simplifying assumptions must be made to obtain cost-effective solutions in the amount of time available for a given problem. The actual simplifications depend on many factors including the level of accuracy required, the available amount of computing power, the skill and knowledge of the modeler, the experience with the given system being simulated, and the time available to attain a solution. These simplifying approaches are briefly discussed here. More detailed information on each aspect of the modeling is given later in this chapter. Spalding [16] discussed simplifying approaches to combustion modeling and noted that the main concern is which modeling rules can be ignored to simplify the problem and then to estimate the errors in the resulting predictions. He also noted the difficulty in matching all the dimensionless groups in a large-scale problem with small-scale experiments. Weber et al. [17] classified models for designing industrial burners into three categories. First-order methods give rough qualitative estimates of heat fluxes and flame shapes. Second-order methods give results of higher accuracy than those of first-order methods for temperature, oxygen concentration, and heat flux. Third-order methods further improve accuracy over second-order methods and give detailed species predictions in the flame, which are useful for pollutant formation rates. The order used will in large part depend on the information and accuracy that are needed.

4.2.1 Fluid Dynamics

There are a variety of methods available to simulate the fluid flow in a combustion system. The Navier–Stokes equations are generally accepted as providing an "exact" model for turbulent fluid-flow systems [18]. Unfortunately, these equations for systems of practical interest are too complicated to solve exactly either analytically or numerically. Therefore, different types of approximations have been suggested for solving these equations. These are very briefly discussed next with appropriate references for the reader interested in more detail.

4.2.1.1 Moment Averaging

This has been by far the most popular method used in simulating large-scale industrial combustion problems, primarily because of the ready availability of commercial software programs like PHOENICS, FLUENT, FLOW-3D, TEACH, PCGC-3 [19], Harwell-3D, GENMIX, and others to solve these problems. Nikolova [20] discusses the use of a code called MOSCOW 78 to study pollution emissions from industrial combustion processes. Figure 4.2 shows the outline of a glass furnace modeled with PCGC-3. In this method, the turbulent velocity components are decomposed into average and fluctuating terms and solved using the famous k–ε closure equations [21]. Despite the well-known limitations of this approach, it remains the most popular choice for solving practical combustion problems. This may be because it has been around for decades and therefore the software has been highly developed. Finite difference [22,23], finite element [24,25], finite volume [26,27], and spectral element [28] techniques have been used to simulate fluid flows,

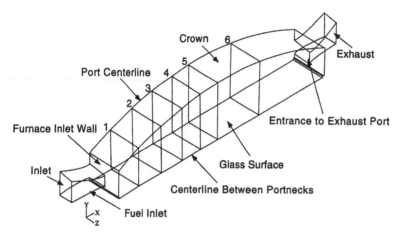

Figure 4.2 Schematic representation of a single portneck in an industrial glass furnace modeled using PCGC-3. (From Ref. 9. Courtesy of CRC Press.)

some including turbulent combustion. The commercial codes today are very user-friendly and have excellent pre- and post-processing packages to make setting up the problem and viewing the results relatively simple and straightforward. Because of the popularity and widespread use of this method, it is discussed in more detail later in this chapter.

4.2.1.2 Vortex Methods

Most numerical approaches for solving fluid-flow problems use a Eulerian scheme with a fixed coordinate system that is discretized into small parts. One problem with this approach is that there may often be areas in the flow where the gradients are very high and require very fine discretization, while in nearby areas the gradients may be much lower and need much less discretization. To complicate this disparity further, these areas may be moving. Finite difference solution convergence problems result from having fine cells next to coarse cells. Therefore, the choice is to use either finer or coarser cells for both areas. If finer cells are used, then accuracy is improved, but with a significant penalty in solution times. If coarser cells are used, then solution times are improved, but accuracy is sacrificed. An alternative approach is to use a Lagrangian system with a moving coordinate system that can keep track of the finer details of high gradient areas, without the burden of unnecessary detail in areas that do not require it. Some Lagrangian methods use grid points that are transported along flow trajectories while other Lagrangian methods are grid free [29]. The Navier–Stokes equations are set up and solved in terms of vorticity:

$$\frac{\partial \vec{\omega}}{\partial t} + \vec{u} \cdot \nabla \vec{\omega} = \vec{\omega} \cdot \nabla \vec{u} \tag{4.1}$$

where $\omega = \nabla \times \vec{u}$, $\nabla = (\partial/\partial x, \partial/\partial y, \partial/\partial z)$, $\vec{u} = (u,v,w)$ and $\vec{x} = (x,y,z)$. Velocities are then calculated from the vorticity solutions. This method has been applied to industrial combustion simulations [30]. Variations of this method have also been referred to as large-eddy simulations (LES) [31,32]. Dahm and co-workers have developed a method known as the local integral moment (LIM), which is based on large-eddy simulation concepts [32–37].

4.2.1.3 Spectral Methods

This is an approximation method where the solutions for the scalar variables in the partial differential equations are simulated as a truncated series expansion [38]:

$$C(x,t) = \sum_{k=0}^{N} c_k(t)\phi(x) \tag{4.2}$$

where $C(x, t)$ is a scalar variable like temperature, N is the finite wave-number truncation cutoff, $c_k(t)$ are the expansion coefficients, and $\phi_k(x)$ are the basis functions that are chosen to best represent the flow. This solution approach is more global than finite difference discretization approaches, which tend to be more local. Therefore, spectral methods can provide more accurate approximations of the solution compared to moment methods, although this is not always the case. Solution times may be longer and the selection of the proper basis functions is critical to the success of this approach, which has been used in combustion problems [39,40] but has not been a popular method for solving industrial combustion problems. This method could become more popular if the appropriate user-friendly software were developed and commercialized.

4.2.1.4 Direct Numerical Simulation

In this method, usually referred to as DNS [41–45], no assumptions are made regarding the turbulent behavior of the flow. The exact Navier–Stokes equations are solved at small enough length and time scales that the complete physics of the problem can be captured. This approach obviously requires tremendous computing power and is not currently used for solving industrial combustion problems. However, as rapid advance in computers continue including parallel processing, large memories, and fast computing speeds, this method may become more prevalent in the future. The method is currently being used to solve fundamental combustion [46] and aerospace propulsion problems using supercomputers where the simulation costs are not a significant portion of the overall cost of new developments. At present, the economics of DNS are not justified for most industrial combustion equipment manufacturers and end-users where the cost of these calculations could dwarf the actual cost of the combustion system itself.

4.2.2 Geometry

There are several different levels of complexity concerning the geometry of a given combustion system, ranging from zero-dimensional up to fully three-dimensional. These levels are briefly discussed here and have been discussed in more detail by Khalil [4].

4.2.2.1 Zero-Dimensional

Numerous modeling approaches to handling the complexity of large-scale combustors are possible and have been used. Before the advent of CFD codes, a common modeling approach was to do an overall heat and material balance on the system. This is often referred to as zero-dimensional modeling because it does not give any spatial resolution. This type of zero-dimensional modeling does not involve any analysis of the fluid dynamics. It can, however, include detailed analysis of the chemical reactions and is often referred to as a stirred reactor or stirred vessel. This type of modeling was made easier with the advent of electronic spreadsheets, but still requires numerous assumptions and simplifications. A more recent type of zero-dimensional model may include very detailed chemistry but still no fluid flow. In that type, the reactor is assumed to be typically either constant pressure or constant volume. The main variable then becomes time, which may be finite or infinite (equilibrium). Zero-dimensional models give a reasonable approximation of the overall performance of the system, but give very little information on the detailed performance, such as, for example, where potential hot spots in the furnace wall might be. Despite the obvious disadvantages, there are some advantages of zero-dimensional modeling. One is that solutions can be obtained very quickly. This is important in parameter studies where a large number of variables are to be investigated and where fast results are needed. Another advantage is that these models can be very helpful in developing an understanding of the system performance, which can sometimes be lost when detailed analyses are done. One can see the forest, before looking at the individual trees in the forest. Another advantage is that this type of modeling does not require the same level of training as with complicated modeling, so it can be done by a wider group of personnel. One example is the zero-dimensional model of furnaces for rapidly heating cylindrical metal billets [47]. Another example of a zero-dimensional model is given by Kuo [48] to simulate a batch-fed solid-waste incineration process.

4.2.2.2 One-Dimensional

The next level of complexity involves one-dimensional modeling. This is where only one spatial dimension is considered. Although this greatly simplifies the number of equations, these models may still be fairly complicated and provide many details on the spatial changes of a given parameter. One-dimensional modeling is often used to examine the detailed chemistry in a combustion process, which may be simulated as a plug-flow reactor.

Despite the limitations, there are advantages in using this type of geometrical simplification. In certain applications, these models are particularly relevant, with little or no sacrifice in accuracy and resolution. An example is in porous radiant burners and flat flames, which are both essentially one-dimensional in nature. Another obvious advantage is that faster results are possible, compared to multidimensional modeling. One-dimensional models also greatly simplify the task of radiation modeling, which can become very complicated in multidimensional geometries. However, it should be noted that one-dimensional models may still be fairly complicated and may include very detailed chemistry, multiple phases, porous media, and radiation. As an example, Singh et al. [49] report on a one-dimensional model used to simulate ceramic radiant burners. For that type of burner, the one-dimensional model is generally very adequate.

4.2.2.3 Multidimensional

The highest level of geometrical complexity involves multidimensional modeling, both two- and three-dimensional geometries. Geometry simplifications are often used to reduce the computing requirements for simulating combustion systems. Wherever possible, three-dimensional (3D) problems are simulated by two-dimensional (2D) models or by axisymmetric geometries, which are 3D problems that can be solved in two spatial variables. In the early days of CFD, it was not uncommon to simulate a rectangular furnace as a cylindrical axisymmetric geometry to reduce the problem from 3D to 2D. Smooke and Bennett [50] discuss 2D axisymmetric modeling of laminar flames, as shown in Fig. 4.3. A related simplification is modeling certain types of cylindrical problems as angular slices, instead of modeling the entire cylinder. For example, if a burner has four injectors equally spaced angularly and radially from the centerpoint, then this can be modeled as a 90° slice of a cylinder by using symmetric boundary conditions.

Another type of geometric simplification in multidimensional modeling involves limiting the number of grid points due to the limitations of the computer speed and memory. It may not always be possible to model the entire combustion system, so an approach that has often been taken is to simulate separately the flame region where small-scale effects are important and the combustor where large-scale effects are predominant. The results of the flame simulation may then be used as inputs to the large-scale modeling of the combustor itself. For example, a single flame can be more accurately input as a heat source using the detailed modeling results for that flame. Another common method for minimizing the number of grid points is to model only a small portion or section of a combustor. For example, most

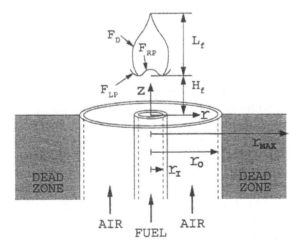

Figure 4.3 Schematic of an axisymmetric coflow burner. (From Ref. 50. Courtesy of CRC Press.)

glass furnaces have multiple burners symmetrically firing parallel to the molten-glass bath. Often, only a single slice of the furnace containing one burner is modeled. Although this precludes simulating the flame-to-flame interactions, it is a reasonable assumption to make in order to achieve timely and cost-effective solutions of acceptable accuracy.

The obvious advantage of multidimensional modeling is that much higher spatial resolution is possible. This can provide important insight into the problem, which is not possible with simpler geometrical models. This resolution is particularly important in simulating burner performance since the burner geometry is normally too complicated to model as a one-dimensional problem. However, there are some obvious disadvantages to multidimensional modeling including longer computational times, difficulties in visualizing and interpreting the results, and more difficulty in separating the effects of individual parameters. Gillis and Smith [51] evaluated a 3D model for industrial furnaces and compared modeling results against experimental data for two pilot-scale furnaces.

4.2.3 Reaction Chemistry

The reaction chemistry is the second important aspect of most industrial combustion problems. Modeling approaches for this chemistry ranges from nonreacting up to multiple reactions with multiple species, and finite-rate kinetics. The different approaches commonly used in modeling combustion problems are briefly discussed next. Note that Gardiner [52] has edited a book on combustion chemistry specifically designed for those who are relatively new to the field of combustion modeling.

4.2.3.1 Nonreacting Flows

When CFD codes first became commercially available, the chemistry submodels were very primitive and greatly increased the computation time, often beyond the capability of the available hardware. Therefore, a common approach to simulating combustion problems was to model them as nonreacting flows. This has sometimes been referred to as "cold-flow" modeling, which is really a misnomer as the flame was often simulated as a flow input of hot inert gases to the combustor. A variation of this approach is to use a nonreacting gas which has the thermophysical properties, like viscosity, thermal conductivity, and specific heat, of the combustion products as a function of temperature. Those properties are separately calculated, typically using some type of equilibrium chemistry calculation. The properties are then curve fit with temperature and included in the CFD codes. In many cases, this type of nonreacting flow model can give fairly accurate predictions for the overall energy transfer in a large-scale combustor.

Nonreacting flow modeling may grossly oversimplify a problem, but it can give considerable insight into the flow patterns inside the combustor. The flame can also be simplified to be a heat source, in order to avoid modeling the chemical reactions in the flame zone. The difficulty is how to specify the heat-release profile of the flame, especially since that is something usually desired of the modeling itself.

Although there are some advantages in using nonreacting chemistry, such as simplicity and speed, this approach is rarely used in most types of combustion modeling today because it is too limited and unrealistic.

4.2.3.2 Simplified Chemistry

The term "simplified" chemistry is a somewhat relative term, but generally refers to reducing the number of chemical equations used to represent a system, reducing the complexity of the reaction mechanism, or a combination of both. In the first approach, a very limited number of reactions and species are used to represent the actual combustion reaction system, which may involve hundreds of reactions and dozens of species. In this approach, a greatly reduced set of reactions are used [53]. Often the goal of this approach is to predict flow and heat-transfer information, but not detailed species such as pollutant emissions like NO_x. Simplified chemistry was often used in the early days of CFD modeling because of the limitations of the submodels and computer memory and because the main interest of the modeler were things like the heat transfer to the load and the walls and the bulk gas flow in the system. For example, the earliest models for simulating the combustion of methane used a single-step reaction such as the following:

$$CH_4 + O_2 + N_2 \rightarrow CO_2 + H_2O + O_2 + N_2 \tag{4.3}$$

Infinite-rate kinetics were used and no minor species were included. This simplified chemistry could obviously not be used to predict pollutant emissions like NO_x, but was useful for simulating the flow patterns and heat transfer in the combustor. An example of a slightly more complicated reaction set is given by Westbrook and Dryer [54]:

$$CH_4 + 1.5O_2 \rightarrow CO + 2H_2O \tag{4.4a}$$

$$CO + 0.5O_2 \rightarrow CO_2 \tag{4.4b}$$

A more popular approach in recent years is to use slightly more complicated reduced sets. An example of a four-step reduced mechanism set for methane flames is given by [55]

$$CH_4 + 2H + H_2O \rightarrow CO + 4H_2 \tag{4.5a}$$

$$CO + H_2O \rightarrow CO_2 + H_2 \tag{4.5b}$$

$$H + H + M \rightarrow H_2 + M \tag{4.5c}$$

$$O_2 + 3H_2 \rightarrow +2H + 2H_2O \tag{4.5d}$$

Another aspect to simplified chemistry models involves not only the number of equations used, but also the type of chemical kinetics that are being simulated. This second approach to simplifying the chemistry is sometimes referred to as reaction mechanism simplification, or mechanism reduction. Infinite-rate kinetics, or equilibrium chemistry, is an example of this type of approach, which is often used in combustion modeling. This means that the chemical reactions are assumed to be infinitely fast and therefore independent of time. This is often a reasonable assumption to make but is again dependent on the specific problem and required level of accuracy. Another variation of this approach is an empirical correlation for the chemistry of a given system.

Brouwer et al. [56] developed a seven-step reduced reaction mechanism for the selective catalytic reduction process to reduce NO_x emissions, compared to the complete chemical mechanism of 327 reactions. Good agreement for the conditions studied was found for the reduced reaction set compared to the full set. Hewson and Bollig [57] developed a five-step reduced reaction mechanism set for methane flames and a six-step reduced reaction set for nitrogen chemistry, both to be used together to predict NO_x emissions from nonpremixed hydrocarbon diffusion flames. Xiao et al. [58] discuss the use of an automatic method for reducing chemical kinetics in combustion modeling using a technique of intrinsic low-dimensional manifolds or ILDM. Sung et al. [59] developed a reduced mechanism for methane oxidation consisting of 16 species and 12 lumped reaction steps, which yielded better results than those of typical four- or five-step reduced mechanisms. Lovas et al. [60] discuss a reduced reaction set for the combustion of fuels containing nitrogen. Cremer et al. [61] proposed an automated method for determining a reduced set of reactions with only 10 species for selective noncatalytic reduction of NO_x for use in CFD modeling. Linan et al. [62] developed a reduced kinetic mechanism for simulating lean premixed combustion.

Brink et al. [63] describe the use of the eddy break-up model for turbulence and a four-step reaction to model gas-fired combustion processes. The numerical temperature predictions were in good agreement with published experimental data. Zhou et al. [64] have developed a unified second-order moment (USM) turbulence-chemistry model for simulating NO_x formation under turbulent combustion conditions. Simulations of a methane–air jet flame compared favorably with a pure probability density function (PDF) model. This simplified model converged in less than 1 h on a Pentium 2 computer.

Giral and Alzueta [65] discuss a reduced chemical mechanism for simulating the reburning process, which is an established technique for reducing NO_x emissions in industrial combustion processes. The full chemistry consists of 438 reactions and 65 chemical species compared to the reduced mechanism consisting of 15 reactions and 19 species. Results for the reduced mechanism model compared favorably to those for the full reaction chemistry set.

4.2.3.3 Complex Chemistry

Another approach to modeling of combustion systems is to use very detailed chemistry. This approach is commonly used if detailed information on gas species is required, such as when, e.g., NO_x emissions need to be predicted. Again, "complex" chemistry is a somewhat vague and relative term, but here refers to multistep reactions with multiple species. The actual numbers of reactions and species depends on a given problem and the level of detail required. Complex chemistry also concerns finite-rate kinetics where the reaction rates are time dependent. Gardiner et al. [66] describe the use of GRI-Mech^{TM} in detailed chemical modeling of the formation and destruction of NO in natural gas flames. Han et al. [67] discuss the use of the two-stage Lagrangian model to predict NO_x control in gas reburning using chemistry consisting of 50 species and 315 reaction mechanisms. Hill and Smoot [68] have written a detailed article on modeling the chemistry of the formation and destruction of NO_x in combustion systems. Schwer et al. [69] discuss an automated technique for upgrading the numerical solving routines in older codes where complex chemistry is included with numerous species and reaction mechanisms.

4.2.4 Radiation

Kocaefe et al. [70] give a brief review of some of the methods used for radiation modeling. They conclude that the imaginary planes and discrete transfer methods have good accuracy and low computation times while the zone method has the lowest computation time if the interchange factors are known or only calculated once. Some of the approaches used to handle radiation are discussed in this section. Malalasekera et al. [71] discuss the various methods to model radiative heat transfer in combustion systems.

4.2.4.1 Nonradiating

Another type of simplification involves using known empirical relationships for the problem at hand. These empirical correlations normally only apply to a specific set of conditions and problems, but can be very useful for reducing the size and complexity of the problem. For example, it may be possible to simulate the nonlinear radiation from the flame to the load and combustor walls as a type of radiation heat-transfer coefficient in order to make the radiation linear with temperature and therefore much easier to solve:

$$h_{\mathrm{rad}} = f\left(T_{\mathrm{source}}^4 - T_{\mathrm{sink}}^4\right) \approx f(T_{\mathrm{source}} - T_{\mathrm{sink}}) \tag{4.6}$$

This approach should be used with caution only after careful examination and understanding of the system under investigation. Although this may limit the generality of the problem, this type of simplification may greatly reduce the time to obtain solutions. This makes it possible to do more simulations of the problem and may be especially useful for finding optimized conditions.

The key to using any simplifications is to understand the resulting inaccuracies they introduce. Therefore, it is usually prudent to have experimental data to compare against any simplified numerical simulations. It is also advisable to use the most complicated possible model for at least a base case problem, which can then be used to compare against the simplified results. If the simplified results compare favorably to the full-blown simulation, there is some justification for using the simplifications. However, if the simplifications do not compare favorably with comprehensive model results, then further analysis is warranted to understand the discrepancies.

As computer power continues to improve, fewer and fewer simplifications will be necessary. Eventually it will be possible to do direct numerical simulations (DNS) so that even the turbulent fluid flow will not need to be approximated because it will be possible and practical to model the small length scales present in such flows.

4.2.4.2 Participating Media

Participating media includes nonluminous gaseous radiation and luminous radiation from particle-laden flows. Bhattacharjee and Grosshandler [72] note three factors that complicate gaseous radiation modeling: (1) the spectral variation of the properties requires calculations over the spectrum, (2) the gaseous composition is not homogeneous over the entire space, which means that integrations must be done over every line of sight, and (3) the asymmetry of most real problems means integration over all solid angles. Sivathanu et al. [73] noted that there are accurate methods for calculating weakly radiating turbulent diffusion flames, but that it is much more difficult to model the strongly radiating turbulent diffusion flames that are used in many industrial combustion systems. Turbulence can significantly increase the mean radiation levels from diffusion flames [74–79].

Hoogendoorn et al. [80] used a 15-band gas radiation model in a well-stirred furnace zone method to simulate the nonluminous radiation in a natural gas fired regenerative glass melter. The results showed that as much as 99% of the heat transfer to the melt was by radiation. They compared the axial heat-flux distribution using both a simplified plug-flow model and a more complete model. The plug-flow model gave both unrealistically high fluxes in the middle of the melter and low fluxes near the ends of the melter. The addition of additives to the combustion products was shown to increase slightly the heat flux to the glass. However, this would lead to lower flue gas outlet temperatures, which reduces the performance of the regenerative air preheater so that the overall effect of the additive in this process would be minimal. Increasing the roof emissivity from 0.4 to 1 was shown to increase the heat transfer to the glass melt by 8%.

Zhenghua and Holmstedt [81] present a fast narrow-band computer model (FASTNB) for predicting the radiation intensity in a general nonisothermal and nonhomogeneous combustion environment. The model is used to calculate the spectral absorption coefficients for CO_2, H_2O, and soot. It is claimed to be as much as 20 times faster than a benchmark model called RADCAL [82–84] with only a 1% deviation from that model. Further development was proposed for inclusion of other gases like carbon monoxide, methane, propylene, and acetylene. Liu et al. [85] present a new approximate method for nongray gas radiative heat transfer using a statistical narrow-band model that utilizes a local absorption coefficient, which is calculated using local properties rather than global properties. The main advantage of the proposed approximate method is considerable savings in computational time, up to two orders of magnitude reduction. The method also improves the accuracy of the calculations compared to methods using global properties.

Gritzo and Strickland [86] present a gridless, integral method for solving the radiative transport equations for use in combustion calculations using Lagrangian techniques to solve the fluid dynamics. Their approach is particularly compatible with parallel computing. It is shown that this method compares favorably with other popular methods used in grid-based solution techniques, which can have significant errors when adapted to gridless solution schemes. Previous methods to solve for radiation have relied on grid-based calculations and are not optimal for transport element methods.

For some combustion processes, primarily those involving solid and liquid fuels, spectral radiation from particulates may be significant. Ahluwalia and Im [87] present a three-dimensional spectral radiation model that they used to model the burning of deeply cleaned coals in a pulverized coal furnace. Spectroscopic data were used to calculate the absorption coefficients of the gases. The extinction and scattering efficiencies of the particulates were calculated using Mie theory. The optical properties of the char, ash, and soot were determined from reflectivity, transmissivity, and extinction measurements. The radiation from the char was as much as 30% of the nonluminous gaseous radiation. The heat transfer in the furnace ranged from 168 to 221 MW (5.73–7.54×10^8 Btu/hr) depending on the specific fuel used. It is noted that ashes rich in iron enhance radiative heat transfer and fine grinding of the coal improves furnace heat absorption. In a later paper, Ahluwalia and Im [88] used a hybrid technique to solve spectral radiation involving gases and particulates in coal furnaces. To optimize computational speed and accuracy, the discrete ordinate method (S_4), modified differential approximation (MDA), and P_1 approximation were combined and used in different ranges of optical thicknesses. The MDA method has been shown to be sufficiently accurate for all optical thicknesses but computationally slow for the optically thin and thick limits [89]. There were significant discrepancies between the predicted and calculated heat fluxes. This was explained by the difficulty in making heat flux measurements in industrial furnaces. The soot, char, and ash contributions to heat transfer were approximately 15%, 3%, and up to 14%, respectively.

4.2.5 Time Dependence

Another critical aspect of combustion modeling is whether or not the solution is time dependent. Nearly all industrial combustion processes are time dependent at small length scales due to turbulence. However, these processes are normally modeled as steady-state systems because of the limitations of the turbulent submodels, the large increase in computer time required to simulate transient combustion, and the lack of need for such detailed information in most industrial combustion systems.

4.2.5.1 Steady State

In steady-state simulations, there is no time dependence of the solution. The problem with turbulent combustion problems is that the time scale, especially in the near-flame region, is very small. To simulate accurately an entire system using such a small time scale is normally computationally prohibitive. However, this is not usually a problem in most cases because that type of detail is not required. The standard approach has been to simulate the average properties. There is considerable debate about averaging turbulent properties, but the reality is that virtually all commercial codes have some type of turbulence averaging (discussed in more detail below). For many continuous industrial applications, the heating process is essentially steady state with fixed and usually know combustion chamber wall temperatures, fuel firing rates, and material feed rates. Processes that are truly varying with time, such as batch heating processes, may be simulated using average conditions over the entire cycle or by making a series of steady-state calculations to simulate various steps in the process.

4.2.5.2 Transient

Transient or time-varying calculations are rarely made for large-scale industrial combustion processes due to the large computational time required, which usually exceeds the amount of time allowed for the needed simulations. As computer speeds continue to increase, this type of computation will increase in popularity for those applications that have significant variations during a given cycle. A good example of such an application is scrap-metal melting. Initially, a charge of cold solid scrap metal is charged into a colder furnace. Then, the burners begin to heat up and melt down the metal, as well as heat up the furnace. At any given time there may be a mixture of solid and liquid metal. When the charge is at or nearly fully molten, a second charge of cold scrap may be added to the first melted charge. Several more charges are possible, depending on the application. An accurate simulation of this process should include a fully transient computation.

4.3 SIMPLIFIED MODELS

Wilson [90] presents a new and more efficient technique for simulating the first-order dependence of a flame in a furnace. The technique is called the moving-boundary flame model. Dynamic state variables are created to keep track of the size of the flame in the furnace. The technique is claimed to be a considerable improvement over point reactor models, but is not intended to replace detailed multidimensional CFD models. Its accuracy depends on data from either experimental results or from more detailed combustion simulations. This reduced model for the flame makes it possible to do on-line dynamic simulations that may be used for burner diagnostics and controls.

Gray et al. [91] developed a simplified model called COMBUST for use on a personal computer to solve the radiant heat transfer in gas-fired furnaces. The model was capable of solving for either direct- or indirect-fired furnaces and used the Hottel zone model for solving for the radiation heat transfer. Rumminger et al. [92] used a one-dimensional model to simulate a porous radiant burner. The bilayered reticulated ceramic burner consisted of an outer layer, known as the flame support, with large pores ($4\,pores/cm^2$) that extracts heat from the postflame gases and an inner layer, known as the diffuser, with small pores ($25\,pores/cm^2$), that prevents the flame from flashing back. The flame was assumed to be one-dimensional, laminar, and steady-state. The gas was assumed to be optically thin and the solid was taken to be spectrally gray. Detailed chemical kinetics were used with the assumption that the gases were adiabatic at the gas outlet. Radiation was simulated as a gray gas. The surface temperature of the porous radiant burner was computed where the radiant losses from the burner were equivalent to the convective heat loss from the gases. The following properties were used for the outer-layer porous medium: radiation extinction coefficient of $115\,m^{-1}$ ($35\,ft^{-1}$), scattering albedo of 0.72, pore diameter of

0.22 cm (0.087 in.), porosity of 0.8, and bulk thermal conductivity of 1.0 W/m-K (0.58 Btu/hr-ft-°F). For the inner-layer porous medium the following properties were used: radiation extinction coefficient of $1000\,m^{-1}$ ($300\,ft^{-1}$), scattering albedo of 0.77, pore diameter of 0.022 cm (0.0087 in.), porosity of 0.65, and bulk thermal conductivity of 1.0 W/m-K (0.58 Btu/hr-ft-°F). The convection correlation for gas flow in reticulated ceramics was taken from Younis and Viskanta [93]. Kendall and Sullivan [94] presented a one-dimensional model, similar to that of Rumminger et al. [92], where the flow was simplified but the chemistry was detailed, using a code called PROF (premixed one-dimensional flame). A sample result is given in Fig. 4.4. The figure shows that the calculated radiant output has little dependence on emissivity in the range 0.4–1.0.

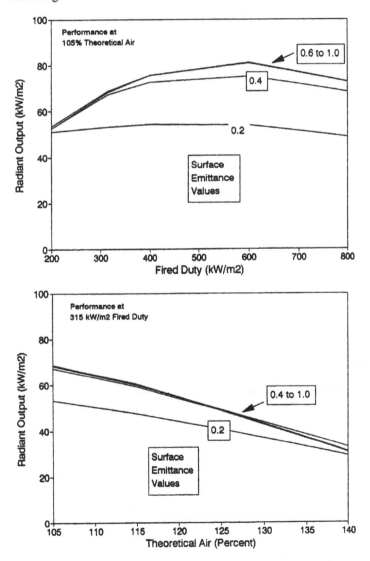

Figure 4.4 Calculated radiant output for a porous radiant burner using the PROF code. (From Ref. 94. Courtesy of GTI.)

4.4 COMPUTATIONAL FLUID DYNAMIC MODELING

Although the term computational fluid dynamics (CFD) may be applied to a wide range of simulations; here, CFD represents multidimensional (2D and 3D) modeling of the fluid flow including chemistry and heat transfer. Any of these three components (fluid flow, chemistry, and heat transfer) may range from simplified to highly complex. The simulations may be computed by any number of different types of schemes (e.g., finite difference, finite volume).

4.4.1 Increasing Popularity of CFD

Computational fluid dynamics is a numerical tool for simulating the complicated fluid flow, heat transfer, and chemical reactions in a combustor. This tool has been gaining in popularity in recent years because of a number of factors. One obvious change is the dramatic increase in computer power that is available at a cost that is affordable for virtually all businesses. Each new generation of computer hardware continues to have more speed and more memory at a lower cost. The personal computers of today are more powerful than the workstations of only a few years ago. Another reason for the growing popularity of CFD codes is that they are now available at a reasonable cost, which usually includes some type of support to aid the user both in the general use of the code as well as the application of the code to the specific needs of the user. Another change is that the CFD computer codes now have very easy-to-use front ends for setting up problems and viewing the results. Before the days of the commercial CFD codes, data had to be input in a certain format, which generally required detailed knowledge of both the code and of the computer. The results were generally only available in tabular form, which made it difficult and unwieldy to visualize the predictions. Now the data can be input without any knowledge of the details inside the code and without regard for the specific computer operating system. The graphical user interfaces are much more powerful and can be used to set up quickly very complicated flow geometries including smoothly contoured walls, which had to modeled using a stairstep type of approach in the past. The results of the modeling can be quickly and easily viewed in visually appealing and useful formats. One researcher has renamed "CFD" to be "colorized fluid dynamics" because of the explosion of color and pretty pictures that can be generated with the codes that are available today [95].

Another important factor in the adoption of CFD is the improvements in the physics and number of submodels that are available to the user. It is now possible to model compressible and incompressible, viscous and nonviscous, laminar and turbulent, high and low pressure, reacting and nonreacting, multiphase, and many other types of flows. The user often has a choice of several submodels for thermal radiation, turbulence, soot, and pollution chemistry, for example. Another important factor that has accelerated the use of CFD is experimental validation where experimental data have been compared against numerical predictions to show the validity of the predictions. Where the correlation between the predictions and the experimental data was poor, changes were made to the code to improve its weaknesses. As with most tools of this type, increased adoption of the codes has led to greater acceptance by the engineering community. This growth in popularity is expected to continue to increase as more and more engineers are trained in its use.

4.4.2 Potential Problems of CFD

As with most things, CFD modeling is not a panacea which can be used to solve all problems. As the saying goes, "garbage in = garbage out," or the results are only as good as the input data. It is incumbent on the user to input the relevant geometry and boundary conditions into the simulation and to select the appropriate submodels for a given problem. Because of the ease of use of the codes available today, it is possible for someone with little or no training in fluid dynamics, heat transfer, and chemical reactions to make predictions that may or may not be credible. It is often observed that anything predicted by a computer must by definition be correct. Those who are skilled in this art know that this is far from the truth and tend to use the codes to predict trends, rather than to guarantee absolute numbers, unless they have a great deal of experience with a particular type of problem and have experimental data to validate the code under those conditions. It is very easy for the CFD codes to be misapplied to problems that are beyond the range of their validity. Therefore, it is appropriate to use the caveat "let the user beware."

There are still many limitations of the physical models in the codes. One example, which is of considerable importance in most combustion problems, is turbulence. The empirical k–ε model has been around for many years and has been widely used despite the many known limitations. Two other submodels of great importance in combustion modeling are radiation and chemical reactions. These are all discussed in some detail later. Suffice it to say that further research is required and is ongoing to improve those submodels.

Combustion problems are among the most complex that CFD codes are used to solve because they usually involve complicated geometries and fluid dynamics, heat transfer including nonlinear thermal radiation, and chemically reacting flows, which may include many species and literally hundreds of chemicals reactions. Many industrial combustors have two very different length scales because the combustor itself may be relatively large, while the length scale required to simulate properly the individual flames in the combustor may be several orders of magnitude smaller. Therefore, large-scale problems may require hundreds of thousands, if not millions, of grid points. As the complexity of the problem increases, so does the number of iterations required to reach a converged solution. The large number of grid points and complicated physics often equate to long computation times, depending on the available hardware. A calculation for a large industrial furnace with multiple burners and the full set of physics, modeled on a typical workstation, can take literally weeks to get a converged solution. Normally, multiple simulations of a given problem are required to find the optimum set of operating conditions. However, it is rare that weeks are available to obtain those solutions. This means that some simplification of the problem is required. This usually involves fairly intimate knowledge of the physics of the problem and preferably some prior knowledge of typical results. It is usually desirable to have some base case against which the model results can be compared to determine the validity of the numerical predictions. Sometimes, the time, experience with related problems, and base case data are lacking so that the modeler is left to use his or her best judgment in how to simplify a given problem. Experienced users know the inherent dangers in blindly simplifying a problem in order to get a "solution" within the time available.

It is tempting to use only CFD to design new combustion equipment because it is often much cheaper and faster than building prototypes, which are usually tested first under controlled laboratory conditions before trying them out in actual field installations. Further, it is also tempting to use CFD to guarantee the performance of combustion equipment because it may be difficult, if not impossible, to test the equipment in every conceivable type of application. Too much confidence in CFD codes is potentially dangerous without proper experimental validation. A more logical approach is to use a combination of numerical modeling in conjunction with experimental measurements. CFD modeling can be used to dramatically reduce the cycle times for developing new products by rapidly simulating and wide range of configurations which would be both time-consuming and expensive to do with prototypes. CFD modeling can also be used to scale-up laboratory or field results from one specific application to another type of application. Only proper experimental validation can assure the user of the usefulness of the modeling results. Unfortunately, this step is often overlooked or ignored and can result in spurious predictions.

4.4.3 Equations

In this section, the equations are given without derivation for flows in rectangular coordinates. Equations for cylindrical and spherical coordinate systems are given in Appendix E. A more complete discussion of these equations and there derivation is given in many other places (e.g., [96]) and has not been repeated here for the sake of brevity.

4.4.3.1 Fluid Dynamics

The unsteady equations of motion for an incompressible Newtonian fluid with constant viscosity in rectangular coordinates (x, y, z) are given as follows:

$$\frac{\partial u}{\partial \tau} + u\frac{\partial u}{\partial x} + v\frac{\partial u}{\partial y} + w\frac{\partial u}{\partial z} = f_x - \frac{1}{\rho}\frac{\partial p}{\partial x} + \nu\left(\frac{\partial^2 u}{\partial x^2} + \frac{\partial^2 u}{\partial y^2} + \frac{\partial^2 u}{\partial z^2}\right) \tag{4.7}$$

$$\frac{\partial v}{\partial \tau} + u\frac{\partial v}{\partial x} + v\frac{\partial v}{\partial y} + w\frac{\partial v}{\partial z} = f_y - \frac{1}{\rho}\frac{\partial p}{\partial y} + \nu\left(\frac{\partial^2 v}{\partial x^2} + \frac{\partial^2 v}{\partial y^2} + \frac{\partial^2 v}{\partial z^2}\right) \tag{4.8}$$

$$\frac{\partial w}{\partial \tau} + u\frac{\partial w}{\partial x} + v\frac{\partial w}{\partial y} + w\frac{\partial w}{\partial z} = f_z - \frac{1}{\rho}\frac{\partial p}{\partial z} + \nu\left(\frac{\partial^2 w}{\partial x^2} + \frac{\partial^2 w}{\partial y^2} + \frac{\partial^2 w}{\partial z^2}\right) \tag{4.9}$$

where f_i is some type of body force such as buoyancy.

In industrial combustion, it is normally assumed that the flows are of low Mach number, which simplifies the fluid dynamics where the flow is incompressible [97]. One of the earliest and still widely used algorithms was developed by Patankar and Spalding [98,99]. The algorithm is known by the acronym SIMPLE, which stands for semi-implicit pressure-linked equation.

A discussion of turbulence modeling has been adequately treated in many other places and is only briefly considered here [100–124]. Spalding [125] discussed the eddy break-up model for turbulent combustion. Prudnikov [126], Spalding [127],

Bray [128,129], Pope [130], and Ashurst [131] have provided reviews of combustion in turbulent flames. Jones and Whitelaw [132] compared some of the available turbulence models against experimental data. Faeth [133] reviewed the interactions between turbulence and heat and mass transfer processes in flames. Arpaci [134] discussed a method for including the interaction between turbulence and radiation. Yoshimoto et al. [135] gave a typical example of modeling a furnace using the k–ε turbulence model. A comparison of the numerical results against experimental measurements is shown in Fig. 4.5. Lindstedt and Váos [136] gave a good discussion of the closure problem using the Reynolds stress equations to solve turbulent flame models. Swaminathan and Bilger [137] discussed the stationary laminar flamelet and conditional moment closure submodels used for simulating turbulent combustion. Bilger [138] modeled turbulent diffusion flames using cylindrical coordinates and the following generalized governing equation:

$$\frac{\partial}{\partial x}(\bar{\rho}\tilde{u}\phi) + \frac{1}{r}\frac{\partial}{\partial r}(r\bar{\rho}\tilde{v}\phi) = \frac{1}{r}\frac{\partial}{\partial r}\left(r\mu_{\text{eff},\phi}\frac{\partial\phi}{\partial r}\right) + S_\phi \tag{4.10}$$

where x is the streamwise direction, u is the streamwise velocity, r is the radial direction, v is the radial velocity, ρ is the density, $\mu_{\text{eff},\phi}$ and S_ϕ are given in Table 4.1, and ϕ is the variable under consideration: 1, $\tilde{u}$, $\tilde{f}$ (mixture fraction), k (turbulent kinetic energy), ε (eddy dissipation rate), or g (mixture fraction fluctuations). The formulation is based on Favre (mass-weighted) averaged quantities:

$$\tilde{\phi} = \frac{\overline{\rho\phi}}{\bar{\rho}} \tag{4.11}$$

where the overbar represents a conventional time average. The constants in Table 4.1 are empirical and were determined by matching predictions and measurements for constant-density round jets.

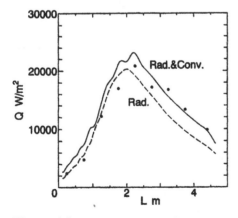

Figure 4.5 Calculated heat flux rates to a furnace wall compared with experimental measurements. (From Ref. 135. Courtesy of Taylor & Francis.)

Table 4.1 Turbulence Model Parameters

ϕ	$\mu_{\text{eff},\phi}$	S_ϕ
1	–	0
$\tilde{u}$	$\mu + \mu_t$	$a(\bar{\rho} - \rho_\infty)$
$\tilde{f}$	$(\mu/Sc) + (\mu_t/\sigma_f)$	0
k	$\mu + (\mu_t/\sigma_k)$	$\mu_t(\partial\tilde{u}/\partial r)^2 - \rho\varepsilon$
ε	$\mu + (\mu_t/\sigma_\varepsilon)$	$[C_{\varepsilon 1}\mu_t(\partial\tilde{u}/\partial r)^2 - C_{\varepsilon 2}\bar{\rho}\varepsilon](\varepsilon/k)$
g	$(\mu/Sc) + (\mu_t/\sigma_g)$	$C_{g1}\mu_t(\partial\tilde{f}/\partial r)^2 - C_{g2}\bar{\rho}g\varepsilon/k$
C_μ $C_{\varepsilon 1}$ C_{g1}	$C_{\varepsilon 2} = C_{g2}$ σ_k σ_ε	$\sigma_f = \sigma_g$ Sc
0.09 1.44 2.8	1.87 1.0 1.3	0.7 0.7
$\mu_t = \bar{\rho}C_\mu k^2/\varepsilon$		

Source: Ref. 138. Courtesy of CRC Press.

Turbulence is very important in most industrial combustion applications that involve high speed flows. Therefore, this phenomenon must be included in most types of models if representative results are expected. Some examples of modeling of swirling flows, which may be important in certain types of combustors, are given in Sec. 4.5.1.

4.4.3.2 Heat Transfer

The energy equation for an incompressible fluid can be written as

$$\rho c_p \left(\frac{\partial t}{\partial \tau} + u\frac{\partial t}{\partial x} + v\frac{\partial t}{\partial y} + w\frac{\partial t}{\partial z} \right) = \frac{\partial}{\partial x}\left(k\frac{\partial t}{\partial x} \right) + \frac{\partial}{\partial y}\left(k\frac{\partial t}{\partial y} \right) + \frac{\partial}{\partial z}\left(k\frac{\partial t}{\partial z} \right) + \dot{q} + \Phi$$

$$(4.12)$$

where

$$\Phi = 2\mu\left[\left(\frac{\partial u}{\partial x}\right)^2 + \left(\frac{\partial v}{\partial y}\right)^2 + \left(\frac{\partial w}{\partial z}\right)^2 + \frac{1}{2}\left(\frac{\partial v}{\partial x} + \frac{\partial u}{\partial y}\right)^2 + \frac{1}{2}\left(\frac{\partial w}{\partial y} + \frac{\partial v}{\partial z}\right)^2 + \frac{1}{2}\left(\frac{\partial u}{\partial z} + \frac{\partial w}{\partial x}\right)^2 \right]$$

$$(4.13)$$

The source term $\dot{q}$ contains the terms for calculating radiation which are dependent on the absolute temperatures raised to the fourth power, which makes the system highly nonlinear and much more difficult to solve. One technique that is often employed in numerical solution schemes is to "turn off" the radiation terms for some of the calculations to allow the solution to stabilize and for the iterations to be completed more quickly. For example, a particular solution scheme may only solve for the radiation source term every 10 iterations.

The neglect of radiation cannot be justified in combustion system modeling [78]. Nonluminous and luminous radiation can greatly complicate a problem because of the spectral dependence of the solution. Cess and Tiwari [139] and Ludwig et al. [140] gave a very extensive treatment of gaseous radiation, including the methods available at that time for analyzing those types of problems. They give different techniques used for computing nonluminous radiation and both experimental and computational data on a wide variety of gases, including CO, CO_2, H_2O, HCl, HF, NO, and OH. Beér and coworkers [141,142] presented a discussion of some of the early methods used for radiation analysis including the zone method and some of the flux methods. Lowes et al. [143] reviewed some of the methods used to analyze radiation in furnaces, including two flux and multi-flux models. Buckius and Tien [144] showed that computations using nongray homogeneous and nonhomogeneous radiation models for infrared flame radiation compared favorably with experimental measurements. Crosbie and Dougherty [145] gave an extensive review of exact methods for solving the radiative transport equation. However, it is noted that exact solutions are not practical for engineering problems [78].

Wall et al. [146] used a simple zoned model using the Monte Carlo technique for the radiative heat transfer in a pilot-scale furnace for oil and gas flames. They used a convective coefficient of $5.8\,W/m^2$-K ($1.0\,Btu/hr$-ft^2-°F) for the transfer from the gas to the furnace walls. Predictions showed good agreement with experimental measurements. Hayasaka [147] described a method called radiative heat ray (RHR) which is intended to model the actual radiation phenomenon from an atom, which is claimed to be more computationally efficient than either the Hottel zone or Monte Carlo methods. Bhattacharjee and Grosshandler [72,148] developed a model termed the effective angle model (EAM), which promises computer storage and computational time savings compared to other models. The EAM should be effective for calculating radiation in two-dimensional or cylindrical combustors with black walls that have either a specified temperature or heat-flux condition. Viskanta and Mengüç [149] gave an extensive review of radiative heat transfer in combustion systems, including some simple examples. They note that the following characteristics are needed from a radiation model:

- Capability of handling inhomogeneous and spectrally dependent properties.
- Capability of handling highly anisotropic radiation fields due to large temperature gradients and anisotropically scattering particles present in the medium.
- Compatibility with finite difference/finite element algorithms for solving transport equations.

They also compared the different techniques for modeling radiative heat transfer as shown in Table 4.2. Komornicki and Tomeczek [150] developed a modification of the wide-band gas model for use in calculating flame radiation. The model compared well against experimental data and both a narrow-band model and an unmodified wide-band model. Soufiani and Djavdan [151] compared the weighted sum of gray gases (WSGG) and the statistical narrow-band (SNB) radiation models. The WSGG model is much less computationally intensive than the SNB model. They found that the WSGG only introduced small errors when the gas mixture was nearly isothermal and surrounded by cold walls. However, significant inaccuracies were found when using the WSGG where large temperature gradients existed. Lallemant et al. [152] compared nine popular total emissivity models used in CFD modeling for H_2O–CO_2 homogenous mixtures with the exponential wide-band model (EWBM) [153,154]. They recommended the use of the EWBM in conjunction

Table 4.2 Comparison of Techniques for Modeling Radiative Heat Transfer

Method	Remarks	Advantages	Disadvantages
Mean beam length	Approximation of radiation heat flux using concept of gas emissivity	Simple, possible to include detailed spectral information	Isothermal system; uncertain accuracy; insufficient detail; difficult to generalize
Zone	Approximation of system by finite-size zones containing uniform temperature and composition gases	Nonhomogeneities in temperature and concentration of gases can be accounted for	Cumbersome; restricted to relatively simple geometries; difficult to account for scattering and spectral information of gases; not compatible with numerical algorithms for solving transport equations
Differential spherical harmonics, moment	Approximation of RTE in terms of the moments of intensity	RTE is recast into a system of differential equations; absorption and scattering can be accounted for; compatible with numerical algorithms for solving transport equation	Unknown accuracy as the relationship between RTE and the flux equations is not always explicit
Flux and discrete ordinates	Approximation of angular intensity distribution along discrete directions and solutions of these equations numerically	Flexible; higher order approximations are accurate; can account for spectral absorption by gases and scattering by particles; compatible with numerical algorithms for solving transport equations	Time consuming; requires iterative solution of finite-difference equations; simple flux approximations are not accurate

Method	Description	Advantages	Limitations
Discrete transfer, ray tracing, numerical	Solves RTE approximately along a long-of-sight	Can use spectral information; flexible; compatible with numerical algorithms	Time consuming if scattering by particles is to be accounted for; accuracy is poor if few rays are considered in scattering media
Monte Carlo	Simulation of physical process using purely statistical techniques and following individual photons	Flexibility for application to complex geometries; absorption and scattering by particles can be accounted for	Can be time consuming; not compatible with numerical algorithms for solving transport equations
Hybrid	New procedures, relatively untested, which use a combination of two or more methods	Different methods can be developed to account for geometric effects; flexible; may be compatible with the numerical algorithm	Relatively untested; cannot be generalized to all system

RTE = radiation transport equation.
Source: Ref. 149. Courtesy of Gulf Publishing.

with WSGG models. A more recent review by Carvalho and Farias [155] presents the various models that have been used to simulate radiation in combustion systems. These methods include:

- The zone method, usually referred to as Hottel's zonal method [156].
- The Monte Carlo method, which is a statistical method [157].
- Schuster–Hamaker-type flux models [2,158,159].
- Schuster–Schwarzschild-type flux models [160–163].
- Spherical harmonic flux models (P–N approximations) [164].
- Discrete ordinates approximations [165–169].
- Finite volume method [170,171].
- Discrete transfer method [172–174].

A schematic of some of the popular radiation models is shown in Fig. 4.6 [77]. The discrete exchange factor (DEF) method has been used by Naraghi and coworkers [175–177]. Denison and Webb [178] presented a spectral radiation approach for generating WSGG models. Unlike some other methods, the absorption coefficient is the modeled radiative property, which permits arbitrary solution of the radiative transfer equation.

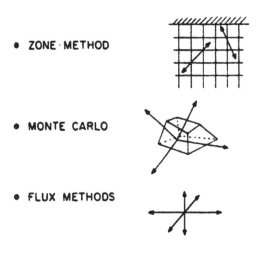

Figure 4.6 Common approaches to radiation modeling. (From Ref. 77. Courtesy of The Combustion Institute.)

Some examples of the application of various radiation models to industrial combustion processes are given next. Siddall and Selcuk [179] describe the application of the two-flux method to a process gas heater. Docherty and Fairweather [174] showed that their predictions using the discrete-transfer method for radiation from nonhomogeneous combustion products compared favorably with narrow-band calculations, as shown in Fig. 4.7. Cloutman and Brookshaw [180] described a numerical algorithm for solving radiative heat losses from an experimental burner. Abdullin and Vafin [181] modeled the radiative properties of a waterwall combustor to determine their effects on the heat transfer in a tube furnace. They modeled downfired burners bounded by rows of vertical tubes with the exhaust at the bottom of the furnace. The results showed peak heat fluxes at about 20% of the distance from the ceiling and the floor, with radiation far exceeding convection. As expected the tube emissivity was an important parameter in the heat flux in the combustor. The partial pressure of the combustion products also had an interesting effect because it affected the gas radiation and absorptivity. The peak radiation to the waterwall was predicted for a gas partial pressure of the combination of CO_2 and H_2O of 0.27 atm (0.27 barg).

Ahluwalia and Im [182] presented an improved technique for modeling the radiative heat transfer in coal furnaces. Coal furnaces differ from gas-fired combustion processes because of the presence of char and ash, which produce significant quantities of luminous radiation. This improved technique was developed to help solve three-dimensional spectral radiation transport equations for the case of absorbing, emitting, and anisotropically scattering media, which are present in coal systems. The incorporation of spectral radiation can significantly increase the computational time and complexity, depending on how the spectra are discretized.

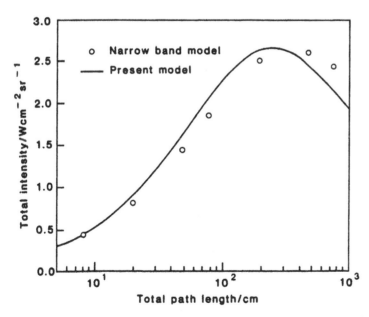

Figure 4.7 Predicted radiation intensity as a function of path length for narrow-band and discrete transfer radiation models. (From Ref. 174. Courtesy of The Combustion Institute.)

The technique is a hybrid combination of the discrete ordinate method (S_4) [183], modified differential approximation (MDA) [184], and P_1 approximation for use in different ranges of optical thicknesses. It combines a char burnout model and spectroscopic data for H_2O, CO_2, CO, char, soot, and ash. It is used to determine the influence of ash composition, ash content, and coal preparation on heat absorption by the furnace. In the simulation of an $80\,MW_e$ corner-fired coal boiler, predicted wall heat fluxes ranging from approximately 100 to $600\,kW/m^2$ (32,000–190,000 Btu/ hr-ft^2) compared favorably with experimental measurements.

Song and Viskanta [185] discuss the modeling of radiation and turbulence as applied to combustion. Figure 4.8 shows the predicted heat flux for models with and without turbulence/radiation interactions. Köylü and Faeth [186] discuss modeling the properties of flame-generated soot. They evaluated approximate methods for calculating the following properties for both individual aggregates and polydisperse aggregate populations: the Rayleigh scattering approximation [187], Mie scattering for an equivalent sphere [188], and Rayleigh–Debye–Gans (R-D-G) scattering [189] for both given and fractal aggregates. Available measurements and computer simulations were not adequate to evaluate the approximate prediction methods properly. Given those limitations, Rayleigh scattering generally underestimated scattering, Mie scattering for an equivalent sphere was unreliable, and R-D-G scattering gave the most reliable results. Bressloff et al. [190] presented a coupled strategy for predicting soot and gas species concentrations, and radiative exchange in turbulent combustion. Good agreement was found with experimental data on temperature, mixture fraction, and soot volume fraction. Bai et al. [191] discuss soot modeling in turbulent jet diffusion flames. Brookes and Moss [192] showed the intimate connection between soot production and flame radiation, which must be accurately accounted for in modeling.

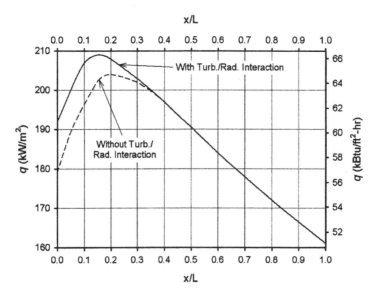

Figure 4.8 Heat flux distributions along the sink: comparison of the total (convective + radiative) with and without turbulence/radiation interactions. (From Ref. 23. Courtesy of CRC Press.)

Numerous methods exist for modeling the nonluminous spectral radiation from combustion products like H_2O, CO_2, and CO. Wide-band models were the first to be used because they are the simplest to implement [193,194]. More advanced models incorporated narrow-band approximations. Taine [195] and Hartmann et al. [196] have computed line-by-line calculations for single absorption bands of CO_2 and H_2O, respectively. Goody [197] developed a statistical narrow-band model. Properties used in the band models are often taken from Ludwig et al. [140]. Song et al. [198] used a statistical narrow-band hybrid model to calculate the gaseous radiation heat transfer in a side-port-fired glass furnace firing on natural gas. This study showed the need to include spectral calculations because a gray-medium assumption overestimates the heat transfer and produces inconsistent results [199].

4.4.3.3 Chemistry

Many schemes have been proposed for the number of equations and reaction rates that may be used to simulate combustion reactions. One that has commonly been used is known as CHEMKIN[TM], which was developed at Sandia National Labs [200]. Another source of chemical kinetic data is a database, formed through funding by the Gas Research Institute (Chicago, IL), known as GRI-Mech [201]. The National Institute of Standards and Technology or NIST (Washington, DC) has also assembled and maintains a very extensive database (over 37,000 separate reactions for over 11,400 distinct reactant pairs) of chemical kinetic data taken from over 11,000 papers [202]. The CEC Group on Evaluation of Kinetic Data for Combustion Modeling was established by the European Energy Research and Development Programme to compile a database of critically evaluated chemical kinetic data [203,204]. Gaz de France has sponsored research towards improving the chemical kinetic modeling of natural gas combustion [205]. Gardiner [206], Sloane [207], and Libby and Williams [208] have edited books concerning combustion chemistry. Golden [209] reviewed the rate parameters used in combustion chemistry modeling. A number of papers have been written which discuss chemistry in combustion processes [210–219]. Some papers specifically consider the interaction of turbulence with chemical reactions [220], and some other papers specifically consider the interaction between radiation and chemistry [221].

One approach that has been used to simplify the chemistry in combustion modeling is to use a statistical approach referred to as the probability density function (PDF) approach [222,223]. This is coupled with the solution of the energy and momentum equations along with the species equations. The PDF approach is most suited to turbulent, reactive flows since the complex chemical reactions can be treated without modeling assumptions [224]. However, some simplifications are usually required because of the excess computer time requirement for complete PDF modeling.

Modeling soot formation in flames is also a challenging aspect of the simulation. Coelho and Carvalho [225] compared different soot formation models for turbulent diffusion propane flames with 500°C (930°F) air preheat. They used the soot formation models given by Khan and Greeves [226] and Stewart et al. [227]. They used soot oxidation models given by Magnussen and Hjertager [228], Lee et al.

[229], and Nagle and Strickland-Constable [230]. By comparing with available soot data, it was found that the Stewart et al. model gave better predictions than the Khan and Greeves model, once the constants were properly tuned. There were not enough data to determine which soot oxidation model gave the best predictions. Delichatsios and Orloff [231] studied the interaction between luminous flame radiation and turbulence. They concluded that soot formation was determined by the straining rate of the small (Kolmogorov) scales. Boerstoel et al. [232] found that experimental data compared favorably with several different soot formation and oxidation models for a high-temperature furnace. Said et al. [233] proposed a simple two-equation model for soot formation and oxidation in turbulent flames. Xu et al. [234] studied the soot produced by fuel-rich, oxygen/methane, atmospheric pressure laminar premixed flames. Their measurements showed good agreement with the soot computational models proposed by Frenklach and Wang [235] and Leung and Lindstedt [236].

In certain industrial heating processes, there may be additional chemical reactions besides those involved in the combustion. The additional reactions may come from the material processing in the combustor or they may also come from downstream processing of the exhaust gases, especially in the case of many of the post-treatment pollutant-reduction technologies such as selective noncatalytic reduction (SNCR) or methane reburn for NO_x reduction. In the latter case, it may be argued that the downstream treatment technologies involving combustion are a part of the overall combustion system. Several examples will suffice to illustrate these "other" chemical reactions. In the glass-melting process, there are many chemical reactions produced during the melt-in of the incoming batch materials including the production of CO_2 and some corrosive species. In the flash smelting process used for the processing of copper, there are many chemical reactions involving copper, sulfur, and iron, which are separate from the combustion reactions. In the methane reburn NO_x reduction technology, methane is injected downstream of the main combustion zone where it chemically reduces much of the NO generated in the flame region back to N_2 and O_2. It is not the purpose here to detail the noncombustion reactions, but merely to point out that they may need to be included in the model as they are directly or indirectly coupled with the combustion system.

4.4.3.4 Multiple Phases

In some combustion systems, multiple phases are present. The most notable involve the combustion of liquid and solid fuels. In the case of a liquid fuel, the fuel is atomized or vaporized into a fine spray which can then be combusted. In the case of a solid fuel, the fuel normally must be finely ground so that complete combustion can be achieved. In both cases, the modeling effort is significantly complicated. In the United States, the vast majority of industrial combustion processes use gaseous fuels. Therefore, modeling the combustion of liquid or solid fuels will not be treated here. It is recommended that the interested reader should consult some of the numerous references that are available for liquid and solid fuel combustion modelling, which are given throughout this chapter and in the general references cited in Chap. 1.

4.4.4 Boundary and Initial Conditions

There are several types of boundary condition that may be applicable to a particular problem, depending on the specific conditions. For the sake of argument, a two-dimensional rectangular coordinate system (x, y) will be used for illustration purposes. A boundary condition of the first kind is where a variable is specified as a function along a given boundary:

$$\phi = f(x, y, \tau) \tag{4.14}$$

where ϕ is the variable and τ is time. Examples include a constant-temperature wall $t(x, y = 0, \tau) = 2000°\text{F}$ or a constant inlet velocity $u(x = 0, y, \tau) = 100\,\text{ft/sec}$. If the scalar is zero everywhere on a given boundary, this is referred to as a homogeneous boundary condition of the first kind.

A boundary condition of the second kind is when the gradient of a variable is a function of the position and time:

$$\frac{\partial \phi}{\partial n} = f(x, y, \tau) \tag{4.15}$$

where n is the normal to a surface. An example would be a constant-heat flux at a horizontal wall: $\partial t/\partial y(y = 0) = 100\,\text{Btu/ft}^2\text{-hr}$. If the gradient is equal to zero everywhere on a boundary, this is known as a homogeneous boundary condition of the second kind.

A boundary condition of the third kind is a linear combination of the first and second kinds:

$$a\phi + b\frac{\partial \phi}{\partial n} = f(x, y, \tau) \tag{4.16}$$

where a and b are functions not dependent on ϕ, and n is the normal to a surface. An example would be where the convection to a surface is equal to the conduction into the surface: $k\partial t/\partial y(y = 0) + ht_1 = ht_0$, where k is the thermal conductivity of the solid and h is the convection coefficient. If the function is equal to zero everywhere on a boundary, this is known as a homogeneous boundary condition of the third kind.

These three types cover many common boundary conditions. There are also other types of boundary conditions that do not fit those types. For example, a radiative boundary condition is a function of a fourth-power temperature law, which makes it highly nonlinear. These types of boundary condition are sometimes approximated by boundary conditions of the third kind where it is possible to simplify the solution to the problem.

Each type of equation (conservation of mass, momentum, species, etc.) must have a sufficient number of boundary conditions to yield a unique solution. A simple example will illustrate this. The steady-state, one-dimensional heat conduction through an infinitely long slab of finite width can be written as

$$q = -k\frac{\partial t}{\partial x} \tag{4.17}$$

This equation can be integrated to find the heat flux through the slab:

$$\int_{x=a}^{x=b} q\,dx = -\int_{x=a}^{x=b} \frac{\partial t}{\partial x}\,dx = -k[t(x=b) - t(x=a)] \tag{4.18}$$

where it has been assumed that the thermal conductivity k is a constant. Boundary conditions of the first kind are needed at each side of the slab to obtain a unique solution.

Initial conditions refer to the conditions which exist at time = zero. If the problem is steady-state, then no initial conditions are required. However, for transient problems, the initial conditions must be specified for all variables, prior to initiating the computations for obtaining unique solution.

4.4.4.1 Inlets and Outlets

The simplest type of inlet condition is to have uniform scalar and vector properties, such as a fluid flowing into a chamber at constant temperature, pressure, and velocity, with a single species. This applies to many problems, but there are many more that are not quite so simple. Many problems have some type of velocity profile at the inlet, for example, gases flowing through a pipe entering a burner. A blended fuel with multiple constituents needs the capability of having multiple species in a single inlet stream. These nonuniform inlet conditions are available in commercial codes.

In older commercial software packages, outlet cells were specified as such but no other conditions could be assigned to them. Newer packages now have the capability of specifying outlet conditions. For many problems where the fluids flow strictly out of the system, this is not necessary. However, for problems with recirculating flows, it is often possible that gases may be drawn into the system through some of the "outlet" cells. This can be useful for including, for example, ambient air infiltration into a furnace.

4.4.4.2 Surfaces

The primary variables used most often for surfaces are constant temperature or constant heat flux. It is also possible to put in temperature and heat flux profiles. Newer codes also have provisions to put in the thermal conductivity as a function of temperature for the wall to calculate conduction through the walls. Radiation characteristics such as emissivity can also be specified for surfaces. This is important for virtually all industrial combustion problems. Many codes also have the provision to put in porous surfaces for gases to flow through. These are especially relevant for simulating porous refractory burners.

4.4.4.3 Symmetry

There are many problems having some type of symmetry which can be used to reduce the number of computations needed to simulate a combustion system. One type of symmetry is where, for example, the left half of a furnace is a mirror image of the right half. In that case, only one of the halves needs to be modeled, with a symmetry boundary condition along the axis of symmetry in the middle of the furnace. Another type of symmetry is slice symmetry where only a single slice of a furnace is modeled to reduce the computation time. For example, a furnace with multiple burners symmetrically located on both sides would be time consuming to model accurately. A single burner in the middle of a side could be modeled with symmetry boundary conditions on both sides and on one end (the middle of the furnace). Although the symmetry is not strictly correct as the exhaust is typically at one of the furnace ends, the symmetry model may have sufficient accuracy for the problem under consideration, for example, the heat flux rates to the wall or to the load. Another example of where a symmetry boundary condition can be used is in a cylindrical furnace that has burners located symmetrically located on one end of the furnace. For example, assume that the furnace has six burners located on one end of the cylinder and firing into the cylinder. Assuming that these burners are located symmetrically every 60° around the wall, only one-sixth of the furnace needs to be simulated where periodic symmetric boundary conditions are used to simulate a single slice. The beauty of the codes available today is that the results for the entire furnace can be displayed if desired even though only a fraction was actually modeled.

4.4.5 Discretization

As CFD initially developed, the physical space being modeled was discretized into many smaller spaces in order to convert the partial differential equations into algebraic equations. In general, the more grids used the higher the accuracy. Unfortunately, this often meant that solutions were grid dependent. That is, different solutions were computed for different size grids. Therefore, it was often necessary to run many simulations with different size grids in order to find the relationship between the grid and the solution so that a truly grid-independent solution could be found. This was very computationally intensive. Early codes only had the capability of a uniform grid everywhere. Later codes had provisions for varying grid sizes so that only fine gridding was used in areas of high gradients while a coarser grid could be used where variables were only changing slowly. The newest codes have adaptive grid capability where the grid is automatically adjusted by the code to achieve the optimal mesh for a given problem under the constraints given by the user. Another grid improvement over early codes is known as boundary-fitted coordinates (BFC), where smooth curving surfaces can be simulated as such without having to do a stair-step approximation that was previously necessary for orthogonal gridding methods. This section briefly discusses the various approaches that are commonly used for discretizing CFD problems.

4.4.5.1 Finite Difference

Özisik [237] discusses the general use of the finite difference technique in heat transfer and fluid flow problems, but without reactions. Figure 4.9 shows a typical grid for the finite difference method. The finite difference technique is one of the oldest methods for numerically solving fluid-flow problems and dates back to at least 1910 [23]. In the finite difference method, the differential fluid-flow equations are replaced by discrete difference approximations. The gridwork of points used to compute these

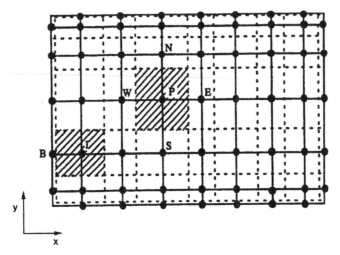

Figure 4.9 Finite difference grid with boundary condition. (From Ref. 23. Courtesy of CRC Press.)

approximations is known as the finite difference grid, which is regularly spaced for optimal problem convergence. The regular spacing is one of the largest constraints for using this method. It is possible for some problems to convert irregularly spaced grid points into regular spacing by means of a transformation, but it is often difficult to determine the transformation for complicated problems. Forward, centered, and backward differencing schemes may be used. To illustrate this, assume that there is a two-dimensional curve with the function $y = f(x)$. Take three points on the curve: $[x_0, f(x_0)]$, $[x_1, f(x_1)]$, and $[x_2, f(x_2)]$, where x_1 is spaced at Δx from x_0 and x_2 is spaced Δx from x_1. Then the slope (m) of the line $f(x)$ at x_1 may be approximated with difference equations using a forward difference technique:

$$m_{forward} = \frac{f(x_2) - f(x_1)}{x_2 - x_1} = \frac{f(x_1 + \Delta x) - f(x_1)}{\Delta x} \tag{4.19}$$

The slope of the line may be estimated using a backward difference technique:

$$m_{backward} = \frac{f(x_1) - f(x_0)}{x_1 - x_0} = \frac{f(x_1) - f(x_1 - \Delta x)}{\Delta x} \tag{4.20}$$

Or the slope of the line may be estimated using a centered difference technique:

$$m_{centered} = \frac{f(x_2) - f(x_0)}{x_2 - x_0} = \frac{f(x_1 + \Delta x) - f(x_1 - \Delta x)}{2\Delta x} \tag{4.21}$$

This same type of discretization is used to solve the fluid-flow equations in difference form. The centered difference technique often produces the fastest rate of convergence in most problems.

4.4.5.2 Finite Volume

The finite volume technique is probably the most popular in commercial codes, possibly because it has been used since the earliest days of commercial CFD codes. Figure 4.10 shows a typical two-dimensional grid for the finite volume method. This technique developed from a modification of the finite difference technique, in order to solve more complicated flow geometries [23]. In this method, finite volumes are used around each grid point to calculate the discretized fluid-flow equations, along with neighboring grid points. In the figure, the point of interest is (i, j), which communicates with its neighboring grid points $(i + 1, j)$, $(i, j + 1)$, $(i-1, j)$, and $(i, j-1)$ through the four faces of the hatched volume around (i, j). In this method, the spacing between points does not need to be equal, nor does the size of the control volumes. The differential equations are then discretized in terms of control volumes and fluxes through the faces (see Ref. 99 for the discretized finite volume equations). Murthy and Mathur [238] discuss the use of unstructured mesh (see Fig. 4.11) finite volume methods for combustion problems.

4.4.5.3 Finite Element

Reddy and Gartling [239] present the general use of the finite difference technique in heat transfer and fluid-flow problems, but without reactions. This technique is well known for solving solid modeling problems to predict, for example, stresses in a deformed solid body. It has only relatively recently been applied to fluid-flow problems. The fundamental difference between the finite element, finite difference, and finite volume techniques is that the finite element technique uses the integral equations for fluid flow, rather than the differential equations. The fluid-flow equations are discretized in terms of flux vectors. Some functional form is assumed for the approximation that will be used to solve the integral equations, where the time and space independent variables are separated. Various solution techniques are then used to solve the resulting integral equations with the assumed approximation

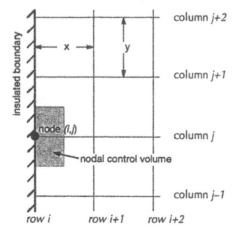

Figure 4.10 Finite volume grid and control volumes. (From Ref. 27. Courtesy of CRC Press.)

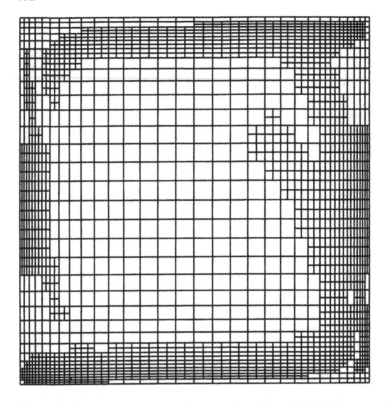

Figure 4.11 Nonconformal mesh adaptation. (From Ref. 9. Courtesy of CRC Press.)

form. A common method is to assume that the solution is in the form of a polynomial, where the coefficients in the polynomial are then solved numerically to obtain the best approximation for the assumed form. The reader should consult the relevant references for more details on this technique [23,24,238].

4.4.5.4 Mixed

Pember et al. [97] used a hybrid discretization scheme to simulate an industrial burner. Their grid consisted of uniform rectangular cells everywhere except at the furnace walls for which they used finite elements to solve what they called mixed cells where the cells contained some fluid and some solid boundary. This method was termed an "embedded boundary method" and was offered as an alternative to using body-fitted coordinates.

4.4.5.5 None

As discussed for the vortex methods of solution (see Sec. 4.2.1.2), no discretization of the space is used to simulate the fluid flow. Instead, the properties of the vortices are tracked as they evolve and dissipate, without regard to any grid system. This is a more accurate way of simulating the space as it eliminates the problems of how to discretize the space. There are also no problems with the numerical errors associated with discretization.

4.4.6 Solution Methods

Shyy and Mittal [240] give a number of common methods used to solve the Navier–Stokes equations. In most cases, the objective is to reduce integro-differential equations into algebraic equations, usually by discretization, that can then be solved in a variety of ways. The choice for the solution method depends on the type of problem, especially as to whether the problem can be linearized or not. Linearized equations are commonly put into matrix form and then solved using a variety of available matrix reduction techniques, such as the Gauss elimination method, the Thomas algorithm, Gauss–Seidel iteration, successive over-relaxation, red and black ordering scheme, or LU (lower/upper triangular) decomposition. Nonlinear systems may be solved with other techniques such as Newton–Raphson Iteration.

Bilger [241] describes the use of the conditional moment closure (CMC) method for modeling nonpremixed turbulent combustion, including pollution emissions. This is a relatively new technique, which has shown promising results compared to experimental measurements and direct numerical simulations. One advantage is that it is not as computationally intensive as some other methods, which makes it attractive for use in combustion pollution modeling. Larroya et al. [242] describe a new Monte Carlo method for solving the PDF equations in turbulent combustion modeling. The CPU time and memory requirements are too high to use the method for simulating actual combustors at this time.

4.4.7 Model Validation

Roache [243] has written an entire book about verification and validation of computation fluid dynamics modeling. He notes the distinction between verification ("solving the equations right") and validation ("solving the right equations"). Here, no discussion will be given on verification as this is available in any good CFD book (including the book by Roache), which as Roache notes is a more mathematical issue. For the purposes of this section, it will be assumed that the mathematical models in a given code have been properly debugged and produce reliable results within a given accuracy range. Of more concern in combustion modeling is validation, to make sure the appropriate physics are being used for the problem under consideration and that those physics are properly simulated. Validation is a much more difficult problem than verification, especially in combustion simulations, due to the difficulties of making relevant and accurate measurements in harsh environments.

One of the seductive aspects of computer modeling is that virtually any type of problem can be simulated. How a problem is modeled depends on many things, but if enough assumptions are made it is possible to generate computational "results." For the naïve and inexperienced, the tendency may be to believe anything that is generated on a computer, because how can a computer be wrong. The caveat "garbage in, garbage out" definitely applies to computer modeling of complex industrial combustion problems. Any given problem may have many assumptions that need to be made, so the results are only as good as the model and the accompanying assumptions. Paraphrasing an anonymous researcher: "Everyone believes a computer analysis except the one who did it, and no one believes experimental results except the one who made them." This is to say that most people inherently realize the difficulties of making experimental measurements in complex geometries, but most naturally believe the results generated by a computer.

As any good modeler knows, a model is only as good as its validation. Models must be constantly be tested against experimental measurements when they are applied to new problems. Model validation is particularly difficult for industrial combustion problems because of the difficulty in making measurements in harsh environments and because of the cost involved in making those measurements. Most measurements made in industrial combustors are with intrusive water-cooled probes because many of the nonintrusive laser-based techniques have not been developed yet for large scales, are not rugged enough for the environments, or are too costly to use outside the laboratory. These intrusive probes are often larger than those used in laboratories because more water cooling is required in high-temperature combustors. Therefore, the flow is disturbed by the probes, which makes it more difficult to compare the measurements with the modeling results. In general, there are relatively few experimental data available for industrial combustors with sufficient information to do a complete model validation. This is an important research need for the future to generate comprehensive data sets in a wide range of industrial combustion systems which can be used for model validation. Some typical model validation cases are given next that are representative of those available to date.

Fiveland et al. [244] presented four validation cases for comparison against codes developed by Babcock and Wilcox (Alliance, OH). Although the codes are primarily directed at large-scale boilers, the validations were done for a broader range of cases including flow in a curved duct, nonreacting flow and natural gas combustion in a swirl-stabilized flame, and swirling-flow coal combustion in a one-sixth scale model of a utility boiler. Kaufman and Fiveland [245] generated a large set of experimental data for the swirl-stabilized natural gas flame case in the Burner Engineering Research Laboratory at Sandia National Laboratories (Livermore, CA) as part of a program partially funded by the Gas Research Institute (Chicago, IL). These data were used for the model validation. The model results were generally very good, except in the recirculation zones. Further work was recommended to improve the chemistry and turbulence models.

4.5 INDUSTRIAL COMBUSTION EXAMPLES

The two major parts of industrial combustion problems include the burners and the combustors. Often these are modeled separately for a variety of reasons as previously discussed. Examples of modeling different types of burners and combustors are given next.

4.5.1 Burner Modeling

There have been numerous papers on modeling industrial burners. A sampling of references are given for modeling such burners:

- Radiant tube burners [246–248]
- Tubular burners [249]

- Swirl burners [250,251]
- Pulse combustion burner [252]
- Porous radiant burners [253–258]
- Industrial hydrogen sulfide burner [259]

Butler et al. [260] gave a general discussion of modeling burners using the finite volume techniques. Schmücker and Leyens [261] described the use of CFD to design a new nozzle-mix burner referred to as the Delta Burner. Schmidt et al. [262] described the use of CFD to redesign a burner, originally firing on coal, to fire on natural gas for use in a rotary kiln. Gershtein and Baukal [263] discuss the use of CFD to develop new burner designs. Figure 4.12 shows an oxy/fuel burner designed to produce flat-shaped flames for melting glass. Figure 4.13 shows the predicted flow field inside the burner using FLUENT.

Modeling radiant burners poses the additional challenge of simulating a porous medium [264]. Perrin et al. [265] discuss the use of a numerical model for the design of a single-ended radiant tube for immersion in and heating of a bath of molten zinc. A sample result of the temperature and heat flux distribution from the tube is shown in Fig. 4.14. As can be seen, most of the heat transfer is by radiation. Hackert et al. [266] simulated the combustion and heat transfer in two-dimensional porous burners. Two different porous geometries were simulated: a honeycomb consisting of parallel nonconnecting passages and a separated plate geometry consisting of parallel but broken walls where there was no continuous solid path, which minimized the importance of solid conduction. Spatial calculated temperatures compared favorably with measured values. The radiant efficiency in the stable region of the flame ranged from 15 to 17 and 22% for the two geometries,

Figure 4.12 Photo of Cleanfire®; flat flame burner body without the burner block. (From Ref. 9. Courtesy of CRC Press.)

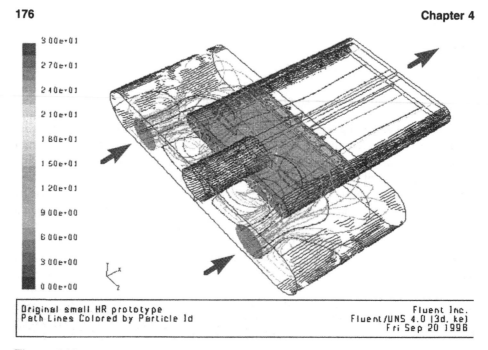

Figure 4.13 Predicted flow field inside a burner used to produce a flat flame for glass melting. (From Ref. 263. Courtesy of CRC Press.)

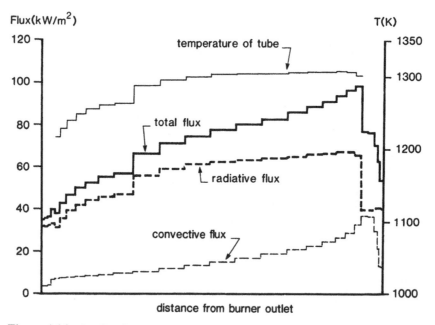

Figure 4.14 Predicted temperature and heat flux on the shell of a single-ended radiant tube immersed in molten zinc. (From Ref. 265. Courtesy of the Amer. Society of Metals.)

respectively. They determined a volumetric Nu of 5.4 ± 0.3, regardless of burning rate or pore size, for the separated plates geometry, which compares favorably with other values reported in the literature. Fu et al. [267] used a one-dimensional model to simulate the performance of a porous radiant burner. The model accounted for the interaction of convection, conduction, radiation and chemical reaction in the burner, which consisted of two layers of reticulated ceramics having different porosities. The model showed that the radiant efficiency increased with the volumetric heat-transfer coefficient and the effective thermal conductivity of the solid matrix and decreased when the firing rate increased and the equivalence ratio decreased.

4.5.2 Combustor Modeling

Numerous papers have been presented over the past few decades on modeling of industrial combustors. Some of the earliest work was done at Imperial College (U.K.) [2,268,269]. A sampling of references are given for modeling industrial combustors:

- General combustors and boilers [270–285]
- Pulse combustion system [286–289]
- Glass furnaces [290–296]
- Aluminum reverberatory furnaces (see Fig. 4.15) [263,297]
- Metal reheat furnaces [47,298–302]
- Radiant tube batch furnace [303]
- Flash smelting furnaces [304,305]
- Cokeless natural gas-fired cupola metal-melting furnace [306]
- Industrial coal combustors [307–311]
- Steam reformers for liquefied petroleum gas conversion [312]
- Fluid catalytic crackers [313,314]
- Furnace for carbon black production [315]
- Rotary hearth calciners [316]
- Rotary kilns [319]
- Cement rotary kiln [319]
- Vertical lime kilns [320]
- Roller kiln for ceramics production [321]
- Generic oxy/fuel-fired furnace [322]
- Rotary kiln incinerator [323–325]
- Municipal solid-waste incinerator [326,327]
- Hazardous waste incineration furnace [328]
- Even baking ovens [329].

Carvalho and Nogueira [330] modeled glass-melting furnaces, cement kilns, and baking ovens. Song and Viskanta [331] did a parametric study of the thermal performance of a generic natural gas-fired furnace. Some papers have specifically focused on radiation modeling in combustors [332], and some other papers have considered swirling flows in furnaces [333,334].

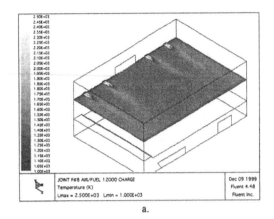

a.

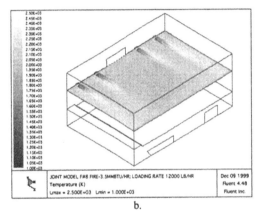

b.

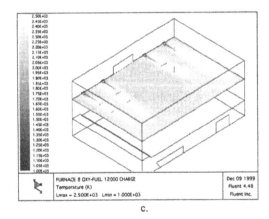

c.

Figure 4.15 Horizontal temperature distribution through the burner level in an aluminum furnace: (a) air/fuel case; (b) air–oxy/fuel case; and (c) oxy/fuel case. (From Ref. 263. Courtesy of CRC Press.)

Carvalho and Nogueira [335] reviewed the modeling of industrial glass-melting processes. They noted that modeling glass melting is especially important because of the difficulty in making measurements due to the very high temperatures and corrosive environment. The models are useful for improving glass quality, increasing thermal efficiency, reducing pollution emissions, and improving equipment reliability. They recommended further research in simulating flow modeling, batch melting, fining and refining, foam formation/elimination, homogenizing, refractories, radiative transfer, and the mass transfer between the glass melt and the combustion chamber. Glass modeling is particularly difficult if the combustion space and molten glass are coupled together because of the large disparity in flow types. The combustion space may have turbulent gas flow while the molten glass is very low-speed flow of a highly viscous material. The system is further complicated by the transparent radiative characteristic of the glass and by the chemical reactions occurring in both the gas space and liquid glass phase. In many cases, electrodes are located in the liquid glass and gases may be bubbled through the glass to stimulate stirring and circulation patterns in the glass. The load consists of fine solid materials at the inlet of the tank and liquid glass at the outlet of the tank. Despite this complexity, models have been successfully used to further the understanding of and improve the glass production process. Paarhuis et al. [336] modeled the coupling of the molten glass and the combustion space above the molten glass in a typical glass furnace. Numerical predictions were in good agreement with experimental data. The primary emphases of the study were temperatures and heat fluxes in the furnace.

Viskanta and coworkers did parametric computational studies of both direct- and indirect-fired furnaces [337–340]. Chapman et al. [337] presented the results of parametric studies of a direct-fired continuous reheating furnace. They developed a simplified mathematical model that accurately calculated the heat balance throughout the furnace. The combustion space was divided into zones that were considered to be well-stirred reactors. The load in each zone was further subdivided into smaller control volumes. Radiation was modeled using Hottel's zone method. A primary objective of the study was to compute the furnace thermal efficiency as a function of a variety of parameters including the load velocity, load emissivity, furnace combustion space height, and refractory emissivity. The furnace efficiency increased rapidly with the initial increase in load velocity and then leveled off with further increases in velocity. The load heat flux was relatively insensitive to the load velocity, except for the lowest speed where the flux was considerably lower. The heat flux to the load was most sensitive to the load emissivity, but was relatively insensitive to the height of the combustion space. Although the model used in the study was fairly simple, with no detailed fluid-flow calculations, it was useful for studying a wide range of values for different parameters, which makes it a valuable design tool for studying other configurations. In a companion study, Chapman et al. [338] also did a parametric modeling study of direct-fired batch reheating furnaces. Again it was shown that the heat flux to the load was very sensitive to the emissivity of the load. Chapman et al. [340] modeled a direct-fired metal reheat furnace with impinging flame jets. The model simulated an actual steel reheat furnace at an Inland Steel Co. plant in East Chicago, IN.

Henneke et al. [3] modeled an ethylene cracking furnace containing radiant wall burners and tubes to carry the hydrocarbons being processed in the furnace. Fig. 4.16 shows a view inside the furnace including gas flow patterns near the burners. Fig. 4.17 shows the geometry of half a xylene reboiler furnace being modeled. Fig. 4.18 shows predicted oxygen concentrations in a thermal oxidizer designed to destroy chlorinated hydrocarbons.

Figure 4.16   Flow patterns inside a simulated ethylene cracking furnace containing radiant wall burners. (From Ref. 3. Courtesy of John Zink Co.)

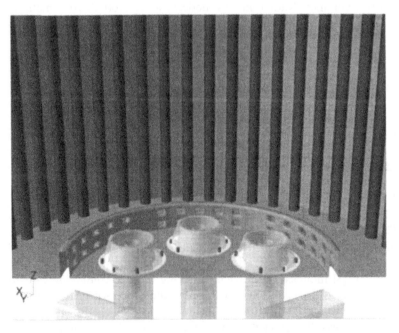

Figure 4.17   Geometry of a xylene reboiler showing half of the furnace and three of the six burners in the bottom of the furnace. (From Ref. 3. Courtesy of John Zink Co.)

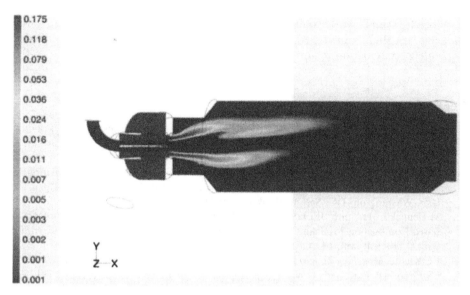

Figure 4.18 Oxygen mass fraction contours in a thermal oxidizer for destroying chlorinated hydrocarbons. (From Ref. 3. Courtesy of John Zink Co.)

4.5.3 Combustion Pollution Modeling

Some glass furnace combustion models have included only the thermal NO_x mechanism [341], while others (typically based on FLUENT [342]) have included all of the mechanisms [343–345]. May et al. [346] used CFD to model high-temperature processes in glass furnaces to study the effect of furnace design on NO_x formation. The model included turbulent-chemistry interactions and spectral radiation effects. It also included air preheating, which is commonly used in glass furnaces. Both thermal and prompt NO_x reactions were included. A parametric study of the geometry showed that increasing the inside height of the furnace reduced NO_x emissions. Further parametric studies were recommended to optimize the furnace design for both low emissions and high thermal efficiency.

Droesbeke et al. [347] modeled the NO_x emissions from an industrial burner; NO_x reductions using flue gas recirculation were numerically shown. Yamashita et al. [348] numerically modeled steam injection in counterflow diffusion flames as a technique for reducing NO_x emissions. Holmes [349] used CFD to model a newly developed burner designed for use in a variety of industrial applications including continuous reheating and aluminum melting that incorporated dilution, staging, and flameless combustion to minimize NO_x emissions. Zakaria et al. [350] have modeled NO_x emissions from a municipal solid-waste incinerator. This is a particularly challenging modeling problem because of the added complexities of volatilization, pyrolysis, and gasification of the solid-waste materials.

Johnson et al. [351] described the use of optimization software used in conjunction with CFD to find the best design according to preset criteria. The optimization was successfully demonstrated to find the optimum shape of a combustor to minimize CO emissions in the exhaust. Only a relatively minor change in geometry significantly reduced the CO emissions. The combustor internal diameter was slightly increased and the geometry near the burner was modified to eliminate a recirculation region.

REFERENCES

1. R Weber. Scaling Characteristics of Aerodynamics, Heat Transfer, and Pollutant Emissions in Industrial Flames. *Twenty-Sixth Symposium (International) on Combustion*, The Combustion Institute, Pittsburgh, PA, pp. 3343–3354, 1996.

2. S Patankar, B Spalding. Simultaneous Predictions of Flow Patterns and Radiation for Three-Dimensional Flames. pp. 73–94, in *Heat Transfer in Flames*, ed. NH Afgan, JM Beer. Washington, DC: Scripta Book Co., 1974.

3. M Henneke, JD Smith, JD Jayakaran, M Lorra. Computational Fluid Dynamics (CFD) Based Combustion Modeling. Chap. 9 in the *John Zink Combustion Handbook*, ed. CE Baukal. Boca Raton, FL: CRC Press, 2001.

4. EE Khalil. *Modelling of Furnaces and Combustors*. Kent, UK: Abacus Press, 1982.

5. ES Oran, JP Boris (eds.). *Numerical Approaches to Combustion Modeling*, Vol. 135, *Progress in Astronautics and Aeronautics*, Washington, DC: American Institute of Aeronautics and Astronautics, 1991.

6. B Larrouturou. *Recent Advances in Combustion Modelling*. Singapore: World Scientific, 1991.

7. TJ Chung. *Numerical Modeling in Combustion*. Washington, DC: Taylor & Francis, 1993.

8. DG Lilley. Chemically Reacting Flows (Combustion). Chap. 16 in *The Handbook of Fluid Dynamics*, ed. RW Johnson. Boca Raton, FL: CRC Press, 1998.

9. CE Baukal, VY Gershtein, X Li (eds.). *Computational Fluid Dynamics in Industrial Combustion*. Boca Raton, FL: CRC Press, 2001.

10. T Poinsot, D Veynante. *Theoretical and Numerical Combustion*. Philadelphia, PA: R.T. Edwards, 2001.

11. DF Elger, K Horii (eds.). *Industrial and Environmental Applications of Fluid Mechanics — 1995*. New York: ASME Press, 1995.

12. DL Boyer, JS Fernando. *Advances in Environmental Fluid Mechanics*. Billerica, MA: Computational Mechanics Inc., 2003.

13. H Rubin, JF Atkinson. *Environmental Fluid Mechanics*. New York: Marcel Dekker, 2001.

14. O Kolditz. *Computational Methods in Environmental Fluid Mechanics*. 2nd Edn. New York: Springer-Verlag, 2002.

15. HH Shen, AH Cheng, K-H Wang, MH Teng (eds.). *Environmental Fluid Mechanics*. Reston, VA: American Society of Civil Engineers, 2002.

16. DB Spalding. The Art of Partial Modeling. *Ninth Symposium (International) on Combustion*. New York: Academic Press, pp. 833–843, 1963.

17. R Weber, AA Peters, PP Breithaupt, BMV Visser. Mathematical modeling of swirling pulverized coal flames: what can combustion engineers expect from modeling? *Am. Soc. Mech. Eng. (ASME) FACT*, Vol. 17, pp. 71–86, 1993.

18. ES Oran, JP Boris. Detailed modeling of combustion systems. *Prog. Energy Comb. Sci.*, Vol. 7, No. 1, pp. 1–72, 1981.

19. SC Hill, AM Eaton, LD Smoot. PCGC-3. In *Computational Fluid Dynamics in Industrial Combustion*, ed. CE Baukal, VY Gershtein, X Li. Boca Raton, FL: CRC Press, 2001.

20. I Nikolova. Possibility to Model Undesirable Combustion Emissions. *Fifth International Conference on Technologies and Combustion for a Clean Environment*, Lisbon, Portugal, Vol. 2, The Combustion Institute — Portuguese Section, pp. 839–846, 1999.

21. BE Launder, DB Spalding. The Numerical Computation of Turbulent Flows. *Lectures in Mathematical Modeling of Turbulence.* London: Academic Press, 1972.

22. JI Ramos. Finite-Difference Methods in Turbulent Combustion. In *Numerical Modeling in Combustion*, ed. TJ Chung. Washington, DC: Taylor & Francis, pp. 281–373, 1993.

23. KD Mish. Finite Difference Method. Chap. 26 in *The Handbook of Fluid Dynamics*, ed. RW Johnson. Boca Raton, FL: CRC Press, 1998.

24. TJ Chung. Finite Element Methods in Turbulent Combustion. In *Numerical Modeling in Combustion*, ed. TJ Chung. Washington, DC: Taylor & Francis, pp. 375–397, 1993.

25. AJ Baker. Finite Element Method. Chap. 28 in *The Handbook of Fluid Dynamics*, ed. RW Johnson. Boca Raton, FL: CRC Press, 1998.

26. HA Dwyer. Finite-Volume Methods in Turbulent Combustion. In *Numerical Modeling in Combustion*, ed. TJ Chung. Washington, DC: Taylor & Francis, pp. 399–408, 1993.

27. SV Patankar, KC Karki, KM Keldar. Finite Volume Method. Chap. 27 in *The Handbook of Fluid Dynamics*, ed. RW Johnson. Boca Raton, FL: CRC Press, 1998.

28. GE Karniadakis, RD Henderson. Spectral Element Methods for Incompressible Flows. Chap. 29 in *The Handbook of Fluid Dynamics*, ed. RW Johnson. Boca Raton, FL: CRC Press, 1998.

29. AF Ghoniem. Vortex Simulation of Reacting Shear Flow. In *Numerical Approaches to Combustion Modeling*, ed. ES Oran, JP Boris. Vol. 135, *Progress in Astronautics and Aeronautics.* Washington, DC: American Institute of Aeronautics and Astronautics, pp. 305–348, 1991.

30. L-F Martins, AF Ghonien. Simulation of the Nonreacting Flow in a Bluff-Body Burner – Effect of the Diameter Ratio. In *Heat and Mass Transfer in Fires and Combustion Systems*, ed. WL Grosshandler, HG Semerjian. ASME HTD-Vol. 148, pp. 33–44, New York, 1990.

31. C Fureby, E Lundgren, S-I Möller. Large Eddy Simulation of Combustion. In *Tranport Phenomena in Combustion*, Vol. 2, ed. SH Chan. Washington, DC: Taylor & Francis, pp. 1083–1094, 1996.

32. HG Weller, G Tabor, AD Gosman, C Fureby. Application of a Flame-Wrinkling LES Combustion Model to a Turbulent Mixing Layer. *Twenty-Seventh Symposium (International) on Combustion*, The Combustion Institute, Pittsburgh, PA, pp. 899–907, 1998.

33. G Tryggvason, WJA Dahm. An integral method for mixing, chemical reactions, and extinction in unsteady strained diffusion layers. *Comb. Flame*, Vol. 83, Nos. 3–4, pp. 207–220, 1990.

34. CHH Chang, WJA Dahm, G Tryggvason. Lagrangian model simulations of molecular mixing, including finite rate chemical reactions, in temporally developing shear layer. *Phys. Fluids A*, Vol. 3, No. 5, pp. 1300–1311, 1991.

35. WJA Dahm, G Tryggvason, M Zhuang. Integral method solution of time-dependent strained diffusion-reaction layers with multistep kinetics. *SIAM J. Appl. Math.*, Vol. 56, No. 4, pp. 1039–1059, 1996.

36. WJA Dahm, G Tryggvason, JA Kezerle, RV Serauskas. Simulation of Turbulent Flow and Complex Chemistry by Local Integral Moment (LIM) Modeling, proceedings of *1995 International Gas Research Conference*, ed. DA Dolenc. Rockville, MD: Government Institutes, pp. 2169–2178, 1996.

37. WJA Dahm, G Tryggvason, RD Frederiksen, MJ Stock. Local Integral Moment (LIM) Simulations. In *Computational Fluid Dynamics in Industrial Combustion.*, ed. CE Baukal, VY Gershtein, X Li. Boca Raton, FL: CRC Press, 2001.

38. MY Hussaini, TA Zang. Spectral methods in fluid dynamics, *Ann. Rev. Fluid Mech.*, Vol. 19, pp. 339–367, 1987.

39. PA McMurtry, P Givi. Spectral Simulations of Reacting Turbulent Flows. In *Numerical Approaches to Combustion Modeling*, ed. ES Oran, JP Boris. Vol. 135, *Progress in*

Astronautics and Aeronautics, Washington, DC: American Institute of Aeronautics and Astronautics, pp. 257–303, 1991.

40. P Givi, CK Madnia. Spectral Methods in Combustion. In *Numerical Modeling in Combustion*, ed. TJ Chung. Washington, DC: Taylor & Francis, pp. 409–452, 1993.

41. V Eswaran, SB Pope. Direct numerical simulations of the turbulent mixing of a passive scalar. *Phys. Fluids*, Vol. 31, No. 3, pp. 506–520, 1988.

42. JH Chen, JM Card, M Day, S Mahalingam. Direct Numerical Simulation of Turbulent Non-Premixed Methane-Air Flames. In *Tranport Phenomena in Combustion*, Vol. 2, ed. SH Chan. Washington, DC: Taylor & Francis, pp. 1049–1060, 1996.

43. T Poinsot. Using Direct Numerical Simulations to Understand Premixed Turbulent Combustion. *Twenty-Sixth Symposium (International) on Combustion*. Pittsburgh, PA: The Combustion Institute, pp. 219–232, 1996.

44. T Baritaud, T Poinsot, M. Baum (eds.). *Direct Numerical Simulation for Turbulent Reacting Flows*, Éditions Technip, Paris, 1996.

45. M Boger, D Veynante, H Boughanem, A Trouvé. Direct Numerical Simulation Analysis of Flame Surface Density Concept for Large Eddy Simulation of Turbulent Premixed Combustion. *Twenty-Seventh Symposium (International) on Combustion*. Pittsburgh, PA: The Combustion Institute, pp. 917–925, 1998.

46. PA McMurtry, P Givi. Direct numerical simulations of mixing and reaction in non-premixed homogeneous turbulent flows. *Comb. Flame*, Vol. 77, pp. 171–185, 1989.

47. RM Davies, DM Lucas, BE Moppett, RA Galsworthy. Isothermal model studies of rapid heating furnaces. *J. Inst. Fuel*, Vol. 44, pp. 453–461, 1971.

48. JT Kuo. System Simulation and Control of Batch-Fed Solid Waste Incinerators. In *Heat Transfer in Fire and Combustion Systems*, ed. WW Yuen, KS Ball. ASME HTD-Vol. 272, pp. 55–62, New York, 1994.

49. S Singh, M Ziolkowski, J Sultzbaugh, R Viskanta. Mathematical Model of a Ceramic Burner Radiant Heater. In *Fossil Fuel Combustion 1991*, ed. R. Ruiz. ASME PD-Vol. 33, pp. 111–116, New York, 1991.

50. MD Smooke, BAV Bennett. Numerical Modeling of Multidimensional Laminar Flames. In *Computational Fluid Dynamics in Industrial Combustion*, ed. CE Baukal, VY Gershtein, X Li. Boca Raton, FL: CRC Press, 2001.

51. PA Gillis, PJ Smith. An Evaluation of Three-Dimensional Computational Combustion and Fluid Dynamics for Industrial Furnace Geometries. *Twenty-Third Symposium (International) on Combustion*. Pittsburgh, PA: The Combustion Institute, pp. 981–991, 1990.

52. WC Gardiner. *Gas-Phase Combustion Chemistry*. New York: Springer, 2000.

53. M Frenklach. Reduction of Chemical Reaction Models. In *Numerical Approaches to Combustion Modeling*, ed. ES Oran, JP Boris. Vol. 135, *Progress in Astronautics and Aeronautics*. Washington, DC: American Institute of Aeronautics and Astronautics, pp. 129–154, 1991.

54. CK Westbrook, FL Dryer. Simplified reaction mechanisms for the oxidation of hydrocarbon fuels in flames. *Comb. Sci. Technol.*, Vol. 27, pp. 31–43, 1981.

55. N Peters. Systematic Reduction of Flame Kinetics: Principles and Details. In *Dynamics of Reactive Systems*, ed. AL Kuhl, JR Bowen, J-C Leyer, A Borisov. Vol. 113, *Progress in Astronautics and Aeronautics*, Washington, DC: American Institute of Aeronautics and Astronautics, pp. 67–86, 1988.

56. J Brouwer, MP Heap, DW Pershing, PJ Smith. A Model for Prediction of Selective Noncatalytic Reduction of Nitrogen Oxides by Ammonia, Urea, and Cyanuric Acid with Mixing Limitations in the Presence of CO. *Twenty-Sixth Symposium (International) on Combustion*. Pittsburgh, PA: The Combustion Institute, pp. 2117–2124, 1996.

57. JC Hewson, M Bollig. Reduced Mechanisms for NO$_x$ Emissions from Hydrocarbon Diffusion Flames. *Twenty-Sixth Symposium (International) on Combustion*. Pittsburgh, PA: The Combustion Institute, pp. 2171–2179, 1996.

58. K Xiao, D Schmidt, U Maas. PDF Simulation of Turbulent Non-Premixed CH$_4$/H$_2$–Air Flames Using Automatically Reduced Chemical Kinetics. *Twenty-Seventh Symposium (International) on Combustion*. Pittsburgh, PA: The Combustion Institute, pp. 1073–1080, 1998.

59. CJ Sung, CK Law, J-Y Chen. An Augmented Reduced Mechanism for Methane Oxidation with Comprehensive Global Parametric Validation. *Twenty-Seventh Symposium (International) on Combustion*. Pittsburgh, PA: The Combustion Institute, pp. 295–304, 1998.

60. T Lovas, D Nilsson, F Mauss. Development of Reduced Chemical Mechanisms for Nitrogen Containing Fuels. *Fifth International Conference on Technologies and Combustion for a Clean Environment*, Lisbon, Portugal, Vol. 1, The Combustion Institute — Portuguese Section, pp. 139–143, 1999.

61. MA Cremer, CJ Montgomery, DH Wang, MP Heap, J-Y Chen. Development and Implementation of Reduced Chemistry for Computational Fluid Dynamics Modeling of Selective Non-Catalytic Reduction. *Proceedings of the Combustion Institute*, Pittsburgh, PA, Vol. 28, pp. 2427–2434, 2000.

62. A Linan, AL Sanchez, A Lepinette, M Bollig, B Lazaro. The Reduced Kinetic Description of Lean Premixed Combustion. *Fifth International Conference on Technologies and Combustion for a Clean Environment*, Lisbon, Portugal, Vol. 1, The Combustion Institute — Portuguese Section, pp. 313–322, 1999.

63. A Brink, C Mueller, P Kilpinen, M Hupa. Modeling of Gaseous Fuel Combustion Using the Eddy Break-Up Model Combined with Fast Chemistry or Finite Rate Chemistry. *Proceedings of Fifth European Conference on Industrial Furnaces and Boilers*, Portugal, Vol. II, pp. 335–343, 2000.

64. LX Zhou, L Qiao, XL Chen, J Zhang. A USM turbulence-chemistry model for simulating NO$_x$ formation in turbulent combustion. *Fuel*, Vol. 81, pp. 1703–1709, 2002.

65. I Giral, MU Alzueta. An augmented reduced mechanism for the reburning process. *Fuel*, Vol. 81, pp. 2263–2275, 2002.

66. WC Gardiner, VV Lissianski, Z Qin, GP Smith, DM Golden, M Frenklach, B Eiteneer, M Goldenberg, NW Moriarty, CT Bowman, RK Hanson, S Song, CC Schmidt, RV Serauskas. The GRI-Mech^{TM} Model for Natural Gas Combustion and NO Formation and Removal Chemistry. *Fifth International Conference on Technologies and Combustion for a Clean Environment*, Lisbon, Portugal, Vol. 1, The Combustion Institute — Portuguese Section, pp. 153–155, 1999.

67. D Han, MG Mungal, VM Zamansky, TJ Tyson. Prediction of NO$_x$ Control by basic and advanced gas reburning using the two-stage Lagrangian model. *Comb. Flame*, Vol. 119, pp. 483–493, 1999.

68. SC Hill, LD Smoot. Modeling of nitrogen oxides formation and destruction in combustion systems. *Prog. Energy Combust. Sci.*, Vol. 26, Nos. 4–6, pp. 417–458, 2000.

69. DA Schwer, JE Tolsma, WH Green, PI Barton. On upgrading the numerics in combustion chemistry codes. *Combust. Flame*, Vol. 128, pp. 270–291, 2002.

70. YS Kocaefe, A Charette, M Munger. Comparison of the Various Methods for Analysing the Radiative Heat Transfer in Furnaces. *Proceedings of the Combustion Institute Canadian Section Spring Technical Meeting*, Vancouver, Canada, May, pp. 15–17, 1987.

71. W Malalasekera, HK Versteeg, JC Henson, JC Jones. Calculation of Radiative Heat Transfer in Combustion Systems. *Fifth International Conference on Technologies and Combustion for a Clean Environment*, Lisbon, Portugal, Vol. 2, The Combustion Institute — Portuguese Section, pp. 1001–1015, 1999.

72. S Bhattacharjee, WL Grosshandler. Effect of radiative heat transfer on combustion chamber flows, *Combust. Flame*, Vol. 77, pp. 347–357, 1989.

73. YR Sivathanu, JP Gore, J Dolinar. Transient Scalar Properties of Strongly Radiating Flames. In *Heat and Mass Transfer in Fires and Combustion Systems*, ed. WL Grosshandler, HG Semerjian. ASME HTD-Vol. 148, pp. 45–56, New York, 1990.

74. G. Cox. On radiant heat transfer in diffusion flames. *Combust. Sci. Technol.*, Vol. 17, pp. 75–78, 1977.

75. VP Kabashnikov, GI Kmit. Influence of turbulent fluctuations on thermal radiation. *Appl. Spectrosc.*, Vol. 31, pp. 963–967, 1979.

76. WL Grosshandler, P Joulain. The effect of large scale fluctuations on flame radiation. *Prog. Astro. and Aero.*, Vol. 105, Part II, Washington, DC: AIAA, pp. 123–152, 1986.

77. AF Sarofim. Radiative Heat Transfer in Combustion: Friend or Foe, Hoyt C. Hottel Plenary Lecture. *Twenty-First Symposium (International) on Combustion*, Pittsburgh, PA: The Combustion Institute, pp. 1–23, 1986.

78. R Viskanta, MP Menguc. Radiation heat transfer in combustion systems. *Prog. Energy Combust. Sci.*, Vol. 8, pp. 97–160, 1987.

79. GM Faeth, JP Gore, SG Chuech, SM Jeng. Radiation from Turbulent Diffusion Flames, *Ann. Rev. Numerical Fluid Mech. & Heat Trans.*, ed. CL Tien, TC Chawla. New York: Hemisphere, Vol. 2, pp. 1–38, 1989.

80. CJ Hoogendoorn, L Post, JA Wieringa. Modelling of combustion and heat transfer in glass furnaces. *Glastech. Ber.*, Vol. 63, No. 1, pp. 7–12, 1990.

81. Y Zhenghua, G Holmstedt. Fast, narrow-band computer model for radiation calculations. *Num. Heat Transfer*, Part B, Vol. 31, pp. 61–71, 1997.

82. WL Grosshandler. Radiation from Nonhomogeneous Fires, Tech. Rep. FMRC, Sept. 1979.

83. WL Grosshandler. Radiative heat transfer in nonhomogeneous gases: a simplified approach. *Int. J. Heat Mass Transfer*, Vol. 23, pp. 1447–1459, 1980.

84. WL Grosshandler. RADCAL: A Narrow-Band Model for Radiation Calculation in a Combustion Environment. NIST Tech. Note 1402, Apr 1993.

85. F Liu, ÖL Gülder, GJ Smallwood. Non-grey gas radiative transfer analyses using the statistical narrow-band model. *Int. J. Heat Mass Transfer*, Vol. 41, No. 14, pp. 2227–2236, 1998.

86. LA Gritzo, JH Strickland. A gridless solution of the radiative transfer equation for fire and combustion calculations. *Combust. Theory Model.*, Vol. 3, pp. 159–175, 1999.

87. RK Ahluwalia, KH Im. Radiative Heat Transfer in PC Furnaces Burning Deeply Cleaned Coals. U.S. Dept. of Energy Rep. DE91 006022, Argonne National Laboratory, Argonne, IL, 1990.

88. RK Ahluwalia, KH Im. Spectral radiative heat-transfer in coal furnaces using a hybrid technique. *J. Inst. Energy*, Vol. 67, pp. 23–29, 1994.

89. HM Park, RK Ahluwalia, KH Im. Three-dimensional radiation in absorbing–emitting–scattering media using the modified differential approximation. *Int. J. Heat Mass Transfer*, Vol. 36, No. 5, pp. 1181–1189, 1993.

90. TL Wilson. A Three-Region, Moving Boundary Model of a Furnace Flame. *Proceedings of the Fifth Symposium on High Performance Computing*, Atlanta, GA, Apr 1997, U.S. Dept. of Energy Rep. DE97003334, Oak Ridge National Laboratory, TN, 1997.

91. WA Gray, E Hampartsoumian, JM Taylor, M Gover, J Sykes. Modelling of Radiant Heat Transfer in Gas-Fired Furnaces Using a Practical Personal Computer Package. *Proceedings of 1992 International Gas Research Conference*, ed. HA Thompson, Rockville, MD, Government Institutes, pp. 2184–2194, 1993.

92. MD Rumminger, RW Dibble, NH Heberle, DR Crosley. Gas Temperature Above a Porous Radiant Burner: Comparison of Measurements and Model Predictions.

Twenty-Sixth Symposium (International) on Combustion. Pittsburgh, PA: The Combustion Institute, pp. 1755–1762, 1996.

93. LB Younis, R. Viskanta. Experimental determination of the volumetric heat transfer coefficient. *Int. J. Heat Mass Transfer*, Vol. 36, No. 6, pp. 1425–1434, 1993.

94. RM Kendall, JD Sullivan. Selective and Enhanced Radiation from Porous Surface Radiant Burners. Gas Research Institute Rep. GRI-93/0160, Chicago, 1993.

95. R Dibble. University of California at Berkeley, pers. comm., 1997.

96. RB Bird, MD Graham. General Equations of Newtonian Fluid Dynamics. Chap. 3, in *The Handbook of Fluid Dynamics*, ed. RW Johnson. Boca Raton, FL: CRC Press, 1998.

97. RB Pember, AS Almgren, WY Crutchfield, LH Howell, JB Bell, P Colella, VE Beckner. An Embedded Boundary Method for the Modeling of Unsteady Combustion in an Industrial Gas-Fired Furnace. U.S. Dept. of Commerce Rep. no. DE96004155, Springfield, VA, 1995.

98. SV Patankar, DB Spalding. *Heat and Mass Transfer in Boundary Layers: A General Calculation Procedure*, 2nd edn. London: Intertext Books, 1970.

99. SV Patankar. *Numerical Heat Transfer and Fluid Flow*. New York: McGraw-Hill, 1980.

100. GK Batchelor. *The Theory of Homogeneous Turbulence*. New York: Cambridge University Press, 1953.

101. JT Davies. *Turbulence Phenomena*. New York: Academic Press, 1972.

102. BE Launder, DB Spalding. *Mathematical Models of Turbulence*. New York: Academic Press, 1972.

103. JO Hinze. *Turbulence*, 2nd edn. New York: McGraw-Hill, 1975.

104. P Bradshaw (ed.). *Turbulence* New York: Springer-Verlag, 1978.

105. W Kollmann. *Prediction Methods for Turbulent Flows*. Washington, DC: Hemisphere, 1980.

106. P Bradshaw. *Engineering Calculation Methods for Turbulent Flow*. New York: Academic Press, 1981.

107. T. Tatsumi (ed.). *Turbulence and Chaotic Phenomena in Fluids*. New York: North Holland, 1984.

108. MM Stanisic. *The Mathematical Theory of Turbulence*. New York: Springer-Verlag, 1985.

109. M Lesieur. *Turbulence in Fluids*, Boston, MA: Dordrecht, 1987.

110. HC Mongia, RMC So, JH Whitelaw (eds.). *Turbulent Reactive Flow Calculations*. New York: Gordon and Breach, 1988.

111. BS Petukhov, AF Polyakov. *Heat Transfer in Turbulent Mixed Convection*. New York: Hemisphere, 1988.

112. VR Kuznetsov. *Turbulence and Combustion*. New York: Hemisphere, 1990.

113. WD McComb. *The Physics of Fluid Turbulence*. Oxford, UK: Oxford University Press, 1990.

114. W Rodi, EN Ganic (eds.). *Engineering Turbulence Modeling and Experiments*. Amsterdam, The Netherlands: Elsevier Science, 1990.

115. M Landahl. *Turbulence and Random Processes in Fluid Mechanics*. Cambridge, UK: Cambridge University Press, 1992.

116. SW Churchill. *Turbulent Flows*. Boston, MA: Butterworth-Heinemann, 1993.

117. DC Wilcox. *Turbulence Modeling for CFD*. La Cãnada, CA: DCW Industries, 1993.

118. AJ Chorin. *Vorticity and Turbulence*. New York: Springer-Verlag, 1994.

119. RJ Garde. *Turbulent Flow*. New York: John Wiley, 1994.

120. U Frisch. *Turbulence*. Cambridge, UK: Cambridge University Press, 1995.

121. K Hanjalic, JCF Pereira (eds.). *Turbulence, Heat, and Mass Transfer*. New York: Begell House, 1995.

122. CJ Chen. *Fundamentals of Turbulence Modeling*. Washington, DC: Taylor & Francis, 1998.

123. CG Speziale, RMC So. Turbulence Modeling and Simulation. Chap. 14 In *The Handbook of Fluid Dynamics*, ed. RW Johnson. Boca Raton, FL: CRC Press, 1998.

124. J Baldyga. *Turbulent Mixing and Chemical Reactions*. New York: John Wiley, 1999.

125. DB Spalding. Development of the Eddy-Break-Up Model of Turbulent Combustion. *Twentieth Symposium (International) on Combustion*. The Combustion Institute, Pittsburgh, PA, pp. 1657–1663, 1976.

126. AG Prudnikov. Flame Turbulence. *Seventh Symposium (International) on Combustion*, London: Butterworths Scientific, pp. 575–582, 1959.

127. DB Spalding. Mathematical models of turbulent flames; a review. *Combust. Sci. Technol.*, Vol. 13, Nos. 1–6, pp. 3–25, 1976.

128. KNC Bray. The Interaction Between Turbulence and Combustion. *Seventeenth Symposium (International) on Combustion*. The Combustion Institute, Pittsburgh, PA, pp. 223–233, 1978.

129. KNC Bray. The Challenge of Turbulent Combustion. *Twenty-Sixth Symposium (International) on Combustion*. The Combustion Institute, Pittsburgh, PA, pp. 1–26, 1996.

130. SB Pope. Computations of Turbulent Combustion: Progress and Challenges. *Twenty-Third Symposium (International) on Combustion*, The Combustion Institute, Pittsburgh, PA, pp. 591–612, 1990.

131. WT Ashurst. Modeling Turbulent Flame Propagation. *Twenty-Fifth Symposium (International) on Combustion*, The Combustion Institute, Pittsburgh, PA, pp. 1075–1089, 1994.

132. WP Jones, JH Whitelaw. Modelling and Measurements in Turbulent Combustion. *Twentieth Symposium (International) on Combustion*. The Combustion Institute, Pittsburgh, PA, pp. 233–249, 1984.

133. GM Faeth, Heat and Mass Transfer in Flames. In *Heat Transfer 1986*, Vol. 1, ed. CL Tien, VP Carey, JK Ferrell. *Proceedings of the Eighth International Heat Transfer Conference*, San Francisco, CA, pp. 151–160, 1986.

134. VS Arpaci. Radiative Turbulence: Radiation Affected Turbulent Forced Convection. In *Heat Transfer in Fire and Combustion Systems – 1993*, ed. B Farouk, MP Menguc, R Viskanta, C Presser, S Chellaiah. ASME HTD-Vol. 250, pp. 155–160, New York, 1993.

135. T Yoshimoto, T Okamoto, T Takagi. Numerical Simulation of Combustion and Heat Transfer in a Furnace and Its Comparison with Experiments. In *Tranport Phenomena in Combustion*, Vol. 2, ed. SH Chan. Washington, DC: Taylor & Francis, pp. 1153–1164, 1996.

136. RP Lindstedt, EM Váos. Modeling of premixed turbulent flames with second moment methods. *Combust. Flame*, Vol. 116, pp. 461–485, 1999.

137. N Swaminathan. RW Bilger. Assessment of combustion submodels for turbulent nonpremixed hydrocarbon flames. *Combust. Flame*, Vol. 116, pp. 519–545, 1999.

138. RW Bilger. Turbulent jet diffusion flames. *Prog. Energy Combust. Sci.*, Vol. 1, pp. 87–109, 1976.

139. RD Cess, SN Tiwari. Infrared Radiative Energy Transfer in Gases. In *Advances in Heat Transfer*, Vol. 8, ed. JP Hartnett, TF Irvine. New York: Academic Press, pp. 229–283, 1972.

140. CB Ludwig, W Malkmus, JE Reardon, JAL Thomson. *Handbook of Infrared Radiation*, NASA Rep. SP-3080, Washington, DC, 1973.

141. TR Johnson, JM Beér. Radiative Heat Transfer in Furnaces: Further Development of the Zone Method of Analysis. *Fourteenth Symposium (International) on Combustion*, The Combustion Institute, Pittsburgh, PA, pp. 639–649, 1972.

142. JM Beér. Methods for Calculating Radiative Heat Transfer from Flames in Combustors and Furnaces. In *Heat Transfer in Flames*, ed. NH Afgan, JM Beer. Washington, DC: Scripta Book Co., pp. 29–45, 1974.

143. TM Lowes, H Bartelds, MP Heap, S Michelfelder, BR Pai. Prediction of Radiant Heat Flux Distribution. In *Heat Transfer in Flames*, ed. NH Afgan, JM Beer. Washington, DC: Scripta Book Company, Chap. 10, pp. 179–190, 1974.

144. RO Buckius, CL Tien. Infrared flame radiation. *Int. J. Heat Mass Transfer*, Vol. 20, pp. 93–106, 1977.

145. AL Crosbie, RL Dougherty. Two-dimensional radiative transfer in cylindrical geometry with anisotropic scattering, *J. Quant. Spectrosc. Radiat. Transfer*, Vol. 25, No. 6, pp. 551–569, 1981.

146. TF Wall, HT Duong, IM Stewart JS Truelove. Radiative Heat Transfer in Furnaces: Flame and Furnace Models of the IFRF M1- and M2-Trials. *Nineteenth Symposium (International) on Combustion*. The Combustion Institute, Pittsburgh, PA, pp. 537–547, 1982.

147. H Hayasaka. A Direct Simulation Method for the Analysis of Radiative Heat Transfer in Furnaces. In *Heat Transfer in Furnaces*, ed. C Presser, DG Lilley. ASME HTD-Vol. 74, pp. 59–63, New York, 1987.

148. S Bhattacharjee, WL Grosshandler. A simplified model for radiative source term in combusting flows. *Int. J. Heat Mass Transfer*, Vol. 33, No. 3, pp. 507–516, 1990.

149. R Viskanta, MP Mengüç. Principles of Radiative Heat Transfer in Combustion Systems. In *Handbook of Heat and Mass Transfer*, ed. N Cheremisinoff. Vol. 4, Houston, TX: Gulf Publishing, Chap. 22, pp. 925–978, 1990.

150. W Komornicki, J Tomeczek. Modification of the wide-band gas radiation model for flame calculation. *Int. J. Heat Mass Transfer*, Vol. 35, No. 7, pp. 1667–1672, 1992.

151. A Soufiani, E Djavdan. A comparison between weighted sum of gray gases and statistical narrow-band radiation models for combustion applications. *Combust. Flame*, Vol. 97, pp. 240–250, 1994.

152. N Lallemant, A Sayre, R Weber. Evaluation of emissivity correlations for H_2O–CO_2–N_2/air mixtures and coupling with solution methods of the radiative transfer equation. *Prog. Energy Combust. Sci.*, Vol. 22, pp. 543–574, 1996.

153. DK Edwards. Molecular Gas Band Radiation. In *Advances in Heat Transfer*, TF Irvine. JP Hartnett, Vol. 12. New York: Academic Press, pp. 115–193, 1976.

154. AT Modak. Exponential wide band parameters for the pure rotational band of water vapor. *J. Quant. Spectosc. Radiat. Transfer*, Vol. 21, No. 2, pp. 131–142, 1979.

155. MG Carvalho, TL Farias. Modelling of heat transfer in radiating and combusting systems. *Trans. IChemE*, Vol. 76, Part A, pp. 175–184, 1998.

156. HC Hottel, AF Sarofim. *Radiative Transfer*. New York: McGraw-Hill, 1967.

157. JR Howell, M. Perlmutter. Monte Carlo solution of thermal transfer through radiant media between gray walls. *J. Heat Transfer*, Vol. 86, No. 1, pp. 116–122, 1964.

158. HC Hamaker. Philips Research Reports 3, 103, 112, and 142, 1947.

159. AD Gosman, FC Lockwood. Incorporation of a Flux Model for Radiation into a Finite Difference Procedure for Furnace Calculations. *14th Symposium (International) on Combustion*. The Combustion Institute, Pittsburgh, PA, pp. 661–671, 1973.

160. TM Lowes, H Bartelds, MP Heap, S Michelfelder, BR Pai. Prediction of Rheat Flux Distributions, International Flame Research Foundation. Rep. GO2/A/26, Ijmuiden, The Netherlands, 1973.

161. W Richter, R Quack. A mathematical Model of a Low-Volatile Pulverised Fuel Flame. In *Heat Transfer in Flames*, ed. NH Afgan, JM Beer. Washington, DC: Scripta Book Co., pp. 95–110, 1974.

162. RG Siddall, N Selçuk. Two-flux modelling of two-dimensional radiative transfer in axi-symmetrical furnaces. *J. Inst. Fuel*, Vol. 49, pp. 10–20, 1976.

163. RG Siddall, N Selçuk. Evaluation of a new six-flux model for radiative transfer in rectangular enclosures. *Trans. IChem*, Vol. 57, pp. 163–169, 1979.

164. R Viskanta. Radiative Transfer and Interaction of Convection with Radiation Heat Transfer. In *Advances in Heat Transfer*, ed. TF Irvine, JP Hartnett. Vol. 3, New York: Academic Press, pp. 175–252, 1966.

165. S Chandrasekhar. *Radiative Transfer*. New York: Dover Publications, 1960.

166. WA Fiveland. Discrete-ordinates solutions of the radiative transport equation for rectangular enclosures. *J. Heat Transfer*, Vol. 106, pp. 699–706, 1984.

167. AS Jamaluddin, PJ Smith. Predicting radiative transfer in rectangular enclosures using the discrete ordinates method. *Combust. Sci. Technol.*, Vol. 59, Nos. 4–6, pp. 321–340, 1988.

168. AS Jamaluddin, PJ Smith. Predicting radiative transfer in axisymmetric cylindrical enclosures using the discrete ordinates method. *Combust. Sci. Technol.*, Vol. 62, Nos. 4–6, pp. 173–186, 1988.

169. WA Fiveland, AS Jamaluddin. Three-Dimensional Spectral Radiative Heat Transfer Solutions by the Discrete-Ordinates Method. In *Heat Transfer Phenomena in Radiation, Combustion, and Fires*, ed. RK Shah. ASME HTD-Vol. 106, pp. 43–48, New York, 1989.

170. GD Raithby, EH Chui. A finite-volume method for predicting radiant heat transfer in enclosures with participating media. *J. Heat Transfer*, Vol. 112, pp. 414–423, 1990.

171. JC Chai, HS Lee, SV Patankar. Finite volume method for radiation heat transfer. *J. Thermophys. Heat Transfer*, Vol. 8, No.3, pp. 419–425, 1994.

172. NG Shah. New method of computation of radiative heat transfer in combustion chambers. Ph.D thesis, Imperial College, London, 1979.

173. FC Lockwood, NG Shah. A New Radiation Solution Method for Incorporation in General Combustion Prediction Procedures. *18th Symposium (International) on Combustion*, The Combustion Institute, Pittsburgh, PA, pp. 1405–1414, 1981.

174. P Docherty, M Fairweather. Predictions of radiative transfer from nonhomogeneous combustion products using the discrete transfer method. *Combust. Flame*, Vol. 71, pp. 79–87, 1988.

175. MHN Naraghi, M Kassemi. Radiative Transfer in Rectangular Enclosures: A Discretized Exchange Factor Solution. *ASME Proceedings of the 1988 National Heat Transfer Conference*. ed. HR Jacobs. Vol. 1, pp. 259–267, New York, 1988.

176. MHN Naraghi. Radiative Heat Transfer in Non-Rectangular Enclosures. In *Heat Transfer Phenomena in Radiation, Combustion, and Fires*, ed. RK Shah ASME HTD-Vol. 106, pp. 17–25, New York, 1989.

177. MHN Naraghi, B Litkouhi. Discrete Exchange Factor Solution of Radiative Heat Transfer in Three-Dimensional Enclosures. In *Heat Transfer Phenomena in Radiation, Combustion, and Fires*, ed. RK Shah. ASME HTD-Vol. 106, pp. 221–229, New York, 1989.

178. MK Denison, BW Webb. A spectral line-based weighted-sum-of-gray-gases model for arbitrary RTE solvers. *J. Heat Transfer*, Vol. 115, pp. 1004–1012, 1993.

179. RG Siddall, N Selçuk. The Application of Flux Methods to Prediction of the Behavior of a Process Gas Heater. In *Heat Transfer in Flames*, ed. NH Afgan, JM Beer. Washington, DC: Scripta Book Company, Chap. 11, pp. 191–200, 1974.

180. LD Cloutman, L Brookshaw. Numerical Simulation of Radiative Heat Loss in an Experimental Burner. Lawrence Livermore National Laboratory, U.S. Dept. of Energy Rep. UCRL-JC-115048, 1993.

181. AM Abdullin, DV Vafin. Numerical investigation of the effect of the radiative properties of a tube waterwall and combustion products on heat transfer in tube furnaces. *J. Eng. Phys. Thermophys.*, Vol. 65, No. 2, pp. 752–757, 1994.

182. RK Ahluwalia, KH Im. Radiative Heat Transfer in Coal Furnaces. U.S. Dept. of Energy Rep. DE92018770, Argonne National Laboratory, IL, 1992.

183. WA Fiveland, AS Jamaluddin. Three-dimensional spectral radiative heat transfer solutions by the discrete-ordinate method, *J. Thermophys.*, Vol. 5, No. 3, pp. 335–339, 1991.

184. MF Modest. Modified differential approximation for radiative transfer in general three-dimensional media. *J. Thermophys.*, Vol. 3, No. 3, pp. 283–288, 1989.

185. TH Song, R Viskanta. Interaction of radiation with turbulence: application to a combustion system. *J. Thermophys.*, Vol. 1, No. 1, pp. 56–62, 1987.

186. ÜÖ Köylü, GM Faeth. Radiative properties of flame-generated soot. *J. Heat Transfer*, Vol. 115, pp. 409–417, 1993.

187. CL Tien, SC Lee. Flame radiation. *Prog. Energy Combust. Sci.*, Vol. 8, pp. 41–59, 1982.

188. WH Dalzell, GC Williams, HC Hottel. A light scattering method for soot concentration Measurements. *Combust. Flame*, Vol. 14, pp. 161–170, 1970.

189. JE Martin, AJ Hurd. Scattering from fractals. *J. Appl. Cryst.*, Vol. 20, pp. 61–78, 1987.

190. NW Bressloff, JB Moss, PA Rubini. CFD Prediction of Couple Radiation Heat Transfer and Soot Production in Turbulent Flames. *Twenty-Sixth Symposium (International) on Combustion.* The Combustion Institute, Pittsburgh, PA, pp. 2379–2386, 1996.

191. XS Bai, M Balthasar, F Mauss, L Fuchs. Detailed Soot Modeling in Turbulent Jet Diffusion Flames. *Twenty-Seventh Symposium (International) on Combustion.* The Combustion Institute, Pittsburgh, PA, pp. 1623–1630, 1998.

192. SJ Brookes, JB Moss. Predictions of soot and thermal radiation properties in confined turbulent jet diffusion flames. *Combust. Flame*, Vol. 116, pp. 486–503, 1999.

193. DK Edwards, LK Glassen, WS Hauser, JS Tuchscher. Radiation heat transfer in nonisothermal nongray gases. *J. Heat Transfer*, Vol. 86C, pp. 219–229, 1967.

194. B Leckner. Spectral and total emissivity of water vapor and carbon dioxide. *Combust. Flame*, Vol. 19, pp. 33–48, 1972.

195. J Taine. A line-by-line calculation of low-resolution radiative properties of CO_2–CO transparent nonisothermal gaseous mixtures up to 3000 K. *J. Quant. Spectrosc. Radiat. Transfer*, Vol. 30, pp. 371–379, 1983.

196. JM Hartmann, L Leon, J Taine. Line-by-line and narrow-band statistical model calculations for H_2O. *J. Quant. Spectrosc. Radiat. Transfer*, Vol. 32, No. 2, pp. 119–127, 1984.

197. RM Goody. *Atmospheric Radiation*, Vol. I, Oxford, UK: Oxford University Press, 1964.

198. G Song, T Bjørge, J Holen, BF Magnussen. Simulation of fluid flow and gaseous radiation heat transfer in a natural gas-fired furnace. *Int. J. Num. Meth. Heat Fluid Flow*, Vol. 7, No. 2/3, pp. 169–180, 1997.

199. RK Ahluwalia, KH Im. Spectral radiative heat-transfer in coal furnaces using a hybrid technique. *J. Inst. Energy*, Vol. 67, pp. 23–29, 1994.

200. RJ Kee, RM Rupley, JA Miller. CHEMKIN-II: A Fortran Chemical Kinetics Package for the Analysis of Gas Phase Chemical Kinetics, Sandia National Laboratory Rep. SAND89–8009B, Livermore, CA, 1989.

201. CT Bowman, RK Hanson, DF Davidson, WC Gardiner, V Lissianski, GP Smith, DM Golden, M Frenklach, M Goldberg. http://www.me.berkeley.edu/gri_mech/, 2001.

202. WG Mallard, F Westley, JT Herron, RF Hampson, DH Frizzell. *NIST Chemical Kinetics Database User's Guide – Windows Version 2Q98*, Washington, DC: National Institute of Standards and Technology, 1998.

203. DL Baulch, CJ Cobos, RA Cox, C Esser, P Frank et al. Evaluated kinetic data for combustion modelling, *J. Phys. Chem. Ref. Data*, Vol. 21, No. 3, pp. 411–734, 1992.

204. DL Baulch, CJ Cobos, RA Cox, P Frank, G Hayman, Th Just, JA Kerr, T Murrells, MJ Pilling, J Troe, RW Walker, J Warnatz. Summary table of evaluated kinetic data for combustion modeling: Supplement 1. *Combust. Flame*, Vol. 98, pp. 59–79, 1994.

205. A Turbiez, P Desgroux, JF Pauwels, LR Sochet, S Poitou, M Perrin, GDF.kin®: A New Step Towards a Detailed Kinetic Mechanism for Natural Gas Combustion Modeling. Proceedings of the *1998 International Gas Research Conference*, Vol. V: *Industrial Utilization*, ed. DA Dolenc. Gas Research Institute, Chicago, pp. 210–221, 1998.

206. WC Gardiner (ed.). *Combustion Chemistry*. New York: Springer-Verlag, 1984.

207. TM Sloane. *The Chemistry of Combustion Processes*. Washington, DC: American Chemical Society 1984.

208. PA Libby, FA Williams (eds). *Turbulent Reacting Flows*. London: Academic Press, 1994.

209. DM Golden. Evaluation of Chemical Thermodynamics and Rate Parameters for Use in Combustion Modeling. In *Fossil Fuel Combustion*, ed. W Bartok, AF Sarofim. New York: John Wiley, Chap. 2, 1991.

210. CK Westbrook, FL Dryer. Chemical Kinetics and Modeling of Combustion Processes. *Eighteenth Symposium (International) on Combustion*. The Combustion Institute, Pittsburgh, PA, pp. 749–767, 1980.

211. F Kaufman. Chemical Kinetics and Combustion: Intricate Paths and Simple Steps. *Nineteenth Symposium (International) on Combustion*. The Combustion Institute, Pittsburgh, PA, pp. 1–10, 1982.

212. J Wofrum. Chemical Kinetics in Combustion Systems: The Specific Effect of Energy, Collisions, and Transport Processes. *Twentieth Symposium (International) on Combustion*. The Combustion Institute, Pittsburgh, PA, pp. 559–573, 1984.

213. SW Benson. Combustion, A Chemical and Kinetic View. *Twenty-First Symposium (International) on Combustion*. The Combustion Institute, Pittsburgh, PA, pp. 703–711, 1986.

214. VY Basevich. Chemical Kinetics in the Combustion Process. In *Handbook of Heat and Mass Transfer*, ed. N Cheremisinoff, Vol. 4, Chap. 18, Houston, TX: Gulf Publishing, 1990.

215. P Gray. Chemistry and Combustion. *Twenty-Third Symposium (International) on Combustion*, The Combustion Institute, Pittsburgh, PA, pp. 1–19, 1990.

216. FL Dryer. The Phenomenology of Modeling Combustion Chemistry. In *Fossil Fuel Combustion*, ed. W Bartok, AF Sarofim. John Wiley, New York: Chap. 3, 1991.

217. E Ranzi, A Sogaro, P Gaffuri, G Pennati, CK Westbrook, WJ Pitz. A new comprehensive reaction mechanism for combustion of hydrocarbon fuels. *Combust. Flame*, Vol. 99, pp. 201–211, 1994.

218. HC Magel, U Schnell, KRG Hein. Simulation of Detailed Chemistry in a Turbulent Combustor Flow. *Twenty-Sixth Symposium (International) on Combustion*, The Combustion Institute, Pittsburgh, PA, pp. 67–74, 1996.

219. JA Miller. Theory and Modeling in Combustion Chemistry. *Twenty-Sixth Symposium (International) on Combustion*. The Combustion Institute, Pittsburgh, PA, pp. 461–480, 1996.

220. AY Federov, VA Frost, VA Kaminsky. Turbulent Transfer Modeling in Flows with Chemical Reactions. In *Tranport Phenomena in Combustion*, Vol. 2, ed. SH Chan. Washington, DC: Taylor & Francis, pp. 933–944, 1996.

221. A de Lataillade, M El Hafi, R Fournier, JL Dufresne. Radiative Transfer Modeling for Radiation-Chemistry Coupling Analysis. *Fifth International Conference on Technologies and Combustion for a Clean Environment*. Lisbon, Portugal, Vol. 1, The Combustion Institute – Portuguese Section, pp. 185–188, 1999.

222. SB Pope. PDF methods for turbulent reactive flows, *Prog. Energy Combust. Sci.*, Vol. 11, No. 2, pp. 119–192, 1985.

223. WP Jones, M Kakhi. Pdf modeling of finite-rate chemistry effects in turbulent nonpremixed jet flames. *Comb. Flame*, Vol. 115, pp. 210–229, 1998.

224. V Saxena, SB Pope. PDF simulations of turbulent combustion incorporating detailed chemistry. *Combust Flame*, Vol. 117, pp. 340–350, 1999.

225. PJ Coelho, MG Carvalho. Modelling of Soot Formation in Turbulent Diffusion Flames. In *Heat Transfer in Fire and Combustion Systems*, ed. WW Yuen, KS Ball. ASME HTD-Vol. 272, pp. 29–39, New York, 1994.

226. IM Khan, G Greeves. A Method for Calculating the Formation and Combustion of Soot in Diesel Engines. In *Heat Transfer in Flames*, ed. NH Afgan and JM Beer. Washington, DC: Scripta Book Co., Chapter 25, 1974.

227. CD Stewart, KJ Syed, JB Moss. Modelling soot formation in non-premixed kerosene-air flames. *Combust. Sci. Tech.*, Vol. 75, pp. 211–266, 1991.

228. BF Magnussen, BH Hjertager. On Mathematical Modelling of Turbulent Combustion with Special Emphasis on Soot Formation and Combustion. *16th Symposium (International) on Combustion*, The Combustion Institute, Pittsburgh, PA, pp. 719–728, 1977.

229. KB Lee, MW Thring, JM Beer. On the rate of combustion of soot in a laminar soot flame. *Combust. Flame*, Vol. 6, pp. 137–145, 1962.

230. J Nagle, RF Strickland-Constable. Oxidation of Carbon Between 1000–2000°C, Proceedings of 5th Conference on Carbon, New York: Pergamon Press, Vol. 1, pp. 154-, 1961.

231. MA Delichatsios, L Orloff. Effects of Turbulence on Flame Radiation from Diffusion Flames, *Twenty-Second Symposium (International) on Combustion*, The Combustion Institute, Pittsburgh, PA, pp. 1271–1279, 1988.

232. P Boerstoel, TH van der Meer, CJ Hoogendoorn. Numerical Simulation of Soot-Formation and -Oxidation in High Temperature Furnaces. In *Tranport Phenomena in Combustion*, Vol. 2, ed. SH Chan. Washington, DC: Taylor & Francis, pp. 1025–1036, 1996.

233. R Said, A Garo, R. Borghi. Soot formation modeling for turbulent flames. *Combust. Flame*, Vol. 108, pp. 71–86, 1997.

234. F Xu, K-C Lin, GM Faeth. Soot formation in laminar premixed methane/oxygen flames at atmospheric pressure. *Combust. Flame*, Vol. 115, pp. 195–209, 1998.

235. M Frenklach, H Wang. Detailed Modeling of Soot Particle Nucleation and Growth. *Twenty-Third Symposium (International) on Combustion*. The Combustion Institute, Pittsburgh, PA, pp. 1559–1566, 1990.

236. KM Leung, RP Lindstedt. Detailed kinetic modeling of C_1C_3 alkane diffusion flames. *Combust. Flame*, Vol. 102, Nos. 1–2, pp. 129–160, 1995.

237. MN Özisik. *Finite Difference Methods in Heat Transfer*, Boca Raton, FL: CRC Press, 1994.

238. JY Murthy, SR Mathur. Unstructured Mesh Methods for Combustion Problems. In *Computational Fluid Dynamics in Industrial Combustion*, ed. CE Baukal, VY Gershtein, X Li. Boca Raton, FL: CRC Press, 2001.

239. JN Reddy, DK Gartling. *The Finite Element Methods in Heat Transfer and Fluid Dynamics*. Boca Raton, FL: CRC Press, 1994.

240. W Shyy, R. Mittal. Solution Methods for the Incompressible Navier-Stokes Equations. Cha. 31 in *The Handbook of Fluid Dynamics*, ed. RW Johnson. Boca Raton, FL: CRC Press, 1998.

241. RW Bilger. Recent Progress in CMC Methods of the Prediction of Pollutant Species in Nonpremixed Turbulent Combustion. *Fifth International Conference on Technologies and Combustion for a Clean Environment*, Lisbon, Portugal, Vol. 1, The Combustion Institute – Portuguese Section, pp. 167–176, 1999.

242. JC Larroya, C Francois, M Cazalens, L Vervisch. Testing a New Monte Carlo Method for Solving PDF Equation in Turbulent Combustion Aeronautical Applications. Fifth International Conference on Technologies and Combustion for a Clean Environment,

Lisbon, Portugal, Vol. 1, The Combustion Institute – Portuguese Section, pp. 177–183, 1999.

243. PJ Roache. *Verification and Validation in Computational Science and Engineering.* Albuquerque, NM: Hermosa Publishers, 1998.

244. WA Fiveland, KC Kaufman, JP Jessee. Validation of an Industrial Flow and Combustion Model. In *Computational Heat Transfer in Combustion Systems*, ed. MQ McQuay, W Schreiber, E Bigzadeh, K Annamalai, D Choudhury, A Runchal. *ASME Proceedings of the 31st Annual National Heat Transfer Conference*, Vol. 6, ASME HTD-Vol. 328, pp. 147–157, New York, 1996.

245. KC Kaufman, WA Fiveland. Pilot Scale Data Collection and Burner Model Numerical Code Validation, Topical report for Gas Research Institute contract 5093–260–2729, 1995.

246. AM Lankhorst, JFM Velthuis. Ceramic Recuperative Radiant Tube Burners: Simulations and Experiments. In *Tranport Phenomena in Combustion*, Vol. 2, ed. SH Chan. Washington, DC: Taylor & Francis, pp. 1330–1341, 1996.

247. F Mei, H Meunier. Numerical and Experimental Investigation of a Single Ended Radiant Tube. In *ASME Proceedings of the 32nd National Heat Transfer Conf.*, Vol. 3: *Fire and Combustion*, ed. L Gritzo, J-P Delplanque, pp. 109–118, New York: ASME, 1997.

248. H Ramamurthy, S Ramadhyani, R. Viskanta. Development of fuel burn-up and wall heat transfer correlations for flows in radiant tubes. *Num. Heat Transfer, Part A*, Vol. 31, pp. 563–584, 1997.

249. B Zamuner, R Borghi. Influence of Physical Phenomena on the Formation of Pollutants in Combustion. In C Vovelle (ed.). *Pollutants from Combustion: Formation and Impact on Atmospheric Chemistry.* Dordrecht, The Netherlands: Kluwer, 2000.

250. S Bortz, A. Hagiwara. Inviscid Model for the Prediction of the Near Field Region of Swirl Burners. In *Industrial Combustion Technologies*, ed. MA Lukasiewicz. Warren, PA: American Society of Metals, pp. 89–97, 1986.

251. C Mueller, H Kremer, A Brink, P Kilpinen, M Hupa. Modeling of an Industrial Scale Gas-Burner Using Finite-Rate Chemistry – Options and Restrictions Resulting from Eddy-Break Up Models. Fifth International Conference on Technologies and Combustion for a Clean Environment, Lisbon, Portugal, Vol. 1, The Combustion Institute – Portuguese Section, pp. 157–165, 1999.

252. Y Tsujimoto, N Machii. Numerical Analysis of Pulse Combustion Burner, *Twenty-First Symposium (International) on Combustion*, The Combustion Institute, Pittsburgh, PA, pp. 539–546, 1986.

253. TW Tong, SB Sathe, RE Peck. Improving the Performance of Porous Radiant Burners Through Use of Sub-Micron Size Fibers. In *Heat Transfer Phenomena in Radiation, Combustion, and Fires*, ed. RK Shah. ASME HTD-Vol. 106, pp. 257–264, New York, 1989.

254. SB Sathe, RE Peck, TW Tong. A numerical analysis of combustion and heat transfer in porous radiant burners. *Int. J. Heat Mass Transfer*, Vol. 33, No. 6, pp. 1331–1338, 1990.

255. SH Chan, K Kumar. Analytical Investigation of SER Recuperator Performance. In *Fossil Fuel Combustion Symposium 1990*, ed. S Singh. ASME PD-Vol. 30, pp. 161–168, New York, 1990.

256. P-F Hsu, JR Howell, RD Matthews. A numerical investigation of premixed combustion within porous inert media. *J. Heat Transfer*, Vol. 115, No. 3, pp. 744–750, 1993.

257. R Mital, JP Gore, R Viskanta, S Singh. Radiation Efficiency and Structure of Flames Stabilized Inside Radiant Porous Ceramic Burners. In *Combustion and Fire*, ed. MQ McQuay, W Schreiber, E Bigzadeh, K Annamalai, D Choudhury, A. Runchal. *ASME*

Proceedings of the 31st National Heat Transfer Conference, Vol. 6, ASME HTD-Vol. 328, pp. 131–137, New York, 1996.

258. CL Hackert, JL Ellzey, OA Ezekoye. Numerical Simulation of a Porous Honeycomb Burner. In *ASME Proceedings of the 32nd National Heat Transfer Conference*. Vol. 3: *Fire and Combustion*, ed. L Gritzo, J-P Delplanque. pp. 147–153, ASME, New York, 1997.

259. MM Sidawi, B Farouk, U Parekh. A Numerical Study of an Industrial Hydrogen Sulfide Burner with Air- and Oxygen-Based Operations. In *Heat Transfer in Fire and Combustion Systems – 1993*, ed. B Farouk, MP Menguc, R Viskanta, C Presser, S Chellaiah, ASME HTD-Vol. 250, pp. 227–234, New York, 1993.

260. GW Butler, J Lee, K Ushimaru, S Bernstein, AD Gosman. A Numerical Simulation Methodology and its Application in Natural Gas Burner Design. In *Industrial Combustion Technologies*, ed. MA Lukasiewicz. Warren, PA: American Society of Metals, pp. 109–116, 1986.

261. A. Schmücker, RE Leyens. Development of the Delta Burner Using Computational Fluid Dynamics. Proceedings of *1998 International Gas Research Conference*, Vol. V: *Industrial Utilization*, ed. DA Dolenc. Gas Research Institute, Chicago, pp. 516–526, 1998.

262. B Schmidt, B Spiegelhauer, NB Kampp Rasmussen, F Giversen. Development of a Process Adapted Gas Burner Through Mathematical Modelling and Practical Experience. Proceedings of *1998 International Gas Research Conference*, Vol. V: *Industrial Utilization*, ed. DA Dolenc. Gas Research Institute, Chicago, pp. 578–584, 1998.

263. VY Gershtein, CE Baukal. CFD in Burner Development. In *Computational Fluid Dynamics in Industrial Combustion*. ed. CE Baukal, VY Gershtein, X Li. Boca Raton, FL: CRC Press, 2001.

264. JR Howell, MJ Hall JL Ellzey. Combustion Within Porous Media. In *Heat Transfer in Porous Media*, ed. Y Bayazitoglu, U.B. Sathuvalli. ASME, HTD-vol. 302, pp. 1–27, New York, 1995.

265. M Perrin, P Lievoux, R Borghi, M Gonzalez. Utilization of a Numerical Model for the Design of a Gas Immersion Tube. In *Industrial Combustion Technologies*, ed. MA Lukasiewicz. Warren, PA: American Society of Metals, pp. 127–134, 1986.

266. CL Hackert, JL Ellzey, OA Ezekoye. Combustion and heat transfer in model two-dimensional porous burners. *Combust. Flame*, Vol. 116, pp. 177–191, 1999.

267. X Fu, R Viskanta, JP Gore. Modeling of Thermal Performance of a Porous Radiant Burner. In *Combustion and Radiation Heat Transfer*, ed. RA Nelson, KS Ball, ZM Zhang. *Proceedings of the ASME Heat Transfer Division – 1998*, Vol. 2, ASME HTD-Vol. 361–2, pp. 11–19, New York, 1998.

268. SV Patankar, DB Spalding. A Computer Model for Three-Dimensional Flows in Furnaces. *Fourteenth Symposium (International) on Combustion*. The Combustion Institute, Pittsburgh, PA, pp. 605–614, 1973.

269. LS Caretto, AD Gosman, SV Patankar, DB Spalding. Two calculation procedures for Steady, Three-Dimensional Flows with Recirculation. *Proceedings of 3rd International Conference on Numerical Methods in Fluid Dynamics*, Springer, Berlin, pp. 60–68, 1972.

270. SV Patankar, DB Spalding. A Computer Model for Three-Dimensional Flow in Furnaces, *Fourteenth Symposium (International) on Combustion*, The Combustion Institute, Pittsburgh, PA, pp. 605–614, 1972.

271. TM Lowes, MP Heap, S Michelfelder, BR Pai. Paper 5. Mathematical modelling of combustion chamber performance. *J. Inst. Fuel*, Vol. 46, No. 38, pp. 343–351, 1973.

272. W Richter. Prediction of heat and mass transfer in a pulverised fuel furnace, *Letters in Heat & Mass Transfer*, Vol. 1, pp. 83–94, 1978.

273.   MMM Abou Ellail, AD Gosman, FC Lockwood, IEA Megahed. Description and validation of a three-dimensional procedure for combustion chamber flows. *J Energy,* Vol. 2, No. 2, pp. 71–80, 1978.

274.   AD Gosman, FC Lockwood, AP Salooja. The Prediction of Cylindrical Furnaces Gaseous Fueled with Premixed and Diffusion Burners. *Seventeenth Symposium (International) on Combustion*, The Combustion Institute, Pittsburgh, PA, pp. 747–760, 1978.

275.   EE Khalil, P Hutchinson, JH Whitelaw. The Calculation of the Flow and Heat-Transfer Characteristics of Gas-Fired Furnaces. *Eighteenth Symposium (International) on Combustion.* The Combustion Institute, Pittsburgh, PA, pp. 1927–1938, 1980.

276.   K Görner. Prediction of the Turbulent Flow, Heat Release and Heat Transfer in Utility Boiler Furnaces. In *Coal Combustion*, ed. J Feng. New York: Hemisphere Publishing, pp. 273–282, 1988.

277.   K Görner, W Zinser. Prediction of three-dimensional flows in utility boiler furnaces and comparison with experiments. *Combust. Sci. Technol.*, Vol. 58, pp. 43–58, 1988.

278.   PA Gillis, PJ Smith. An Evaluation of Three-Dimensional Computational Combustion and Fluid Dynamics in Industrial Furnace Geometries. In *23rd Symposium (International) on Combustion.* The Combustion Institute, Pittsburgh, PA, pp. 981–991, 1990.

279.   MG Carvalho, JB Lopes, M Nogueira. A Three-Dimensional Procedure for Combustion and Heat Transfer in Industrial Furnaces. in *Advanced Computational Methods in Heat Transfer*, Vol. 3: *Phase Change and Combustion Simulation*, ed. LC Wrobel, CA Brebbia, AJ Nowak. Berlin: Springer-Verlag, pp. 171–183, 1990.

280.   H Meunier. Modelling of Industrial Furnaces. *Proceedings of 2nd European Conference on Industrial Furnaces and Boilers*. Portugal, ed. R Collin, W Leuckel, A Reis, J. Ward. pp. 1–21, Apr. 1991.

281.   JM Rhine, RJ Tucker. *Modelling of Gas-fired Furnaces and Boilers and Other Industrial Heating Processes*. British Gas and McGraw-Hill, New York, 1991.

282.   V Sidlauskas, M Tamonis. Mathematical modeling of the thermal process in industrial combustion chambers. *Heat Transfer – Soviet Res.*, Vol. 23, No. 7, pp. 897–914, 1991.

283.   M Matsumura, S Ito, Y Ichiraku, T Saeki. Heat Transfer Simulation in Industrial Gas Furnaces. *Proceedings of 1992 International Gas Research Conference*, ed. HA Thompson. Goverenment Institutes, Rockville, MD, pp. 2195–2204, 1993.

284.   CJ Hoogendoorn. Full Modelling of Industrial Furnaces and Boilers. In *Tranport Phenomena in Combustion*, Vol. 2, ed. SH Chan. Washington, DC: Taylor & Francis, 1177–1188, 1996.

285.   A Scherello, U Konold, H Kremer, M Lorra. Mathematical Modelling of Industrial Furnaces. *Proceedings of 5th European Conference on Industrial Furnaces and Boilers*, Portugal, Vol. II, pp. 427–433, 2000.

286.   PK Barr, JO Keller, JA Kezerle. SPCDC: A User-Friendly Computation Tool for the Design and Refinement of Practical Pulse Combustion Systems. *Proceedings of 1995 International Gas Research Conference*, ed. DA Dolenc. Government Institutes, Rockville, MD, pp. 2150–2159, 1996.

287.   B Ponizy, S Wojcicki. On Modeling of Pulse Combustors, *Twentieth Symposium (International) on Combustion.* The Combustion Institute, Pittsburgh, PA, pp. 2019–2024, 1984.

288.   PK Barr, HA Dwyer, Pulse Combustor Dynamics: A Numerical Study. In *Numerical Approaches to Combustion Modeling*, ed. ES Oran, JP Boris. Vol. 135, *Progress in Astronautics and Aeronautics*, Washington, DC: American Institute of Aeronautics and Astronautics, 1991, pp. 673–710.

289.   E Lundgren, U Marksten, S-I Möller. The Enhancement of Heat Transfer in the Tail Pipe of a Pulse Combustor. *Twenty-Seventh Symposium (International) on Combustion.* The Combustion Institute, Pittsburgh, PA, pp. 3215–3220, 1998.

290. AD Gosman, FC Lockwood, IEA Megahed, NG Shah. The Prediction of the Flow, Reaction and Heat Transfer in the Combustion Chamber of a Glass Furnace. In *AIAA 18th Aerospace Sciences Meeting*, Pasadena, CA, pp. 14–46, Jan. 1980.

291. L Post, CJ Hoogendoorn. Heat transfer in gas-fired glass furnaces. *VDI, Berichte*, No. 645, pp. 457–466, 1987.

292. CJ Hoogendoorn, L Post, JA Wieringa. Modelling of combustion and heat transfer in glass furnaces, *Glastech. Ber.* Vol. 63, No. 1, pp. 7–12, 1990.

293. VB Kut'in, SN Gushchin, VG Lisienko. Heat transfer in the cross-fired glass furnace. *Glass Ceramics*, Vol. 54, Nos. 5–6, pp. 135–138, 1997.

294. MG Carvalho, M Nogueira, J Wang. Mathematical Modelling of the Glass Melting Industrial Process. In *Proceedings of XVII International Congress on Glass*. Beijing: International Academic Publishers, Vol. 6, pp. 69–74, 1995.

295. VB Kut'in, SN Gushchin, VG Lisienko. Heat exchange in the cross-fired glass furnace. *Glass Ceramics*, Vol. 54, Nos. 5–6, pp. 172–174, 1997.

296. BC Hoke, P Schill. CFD Modeling for the Glass Industry. In *Computational Fluid Dynamics in Industrial Combustion*, ed. CE Baukal, VY Gershtein, X Li. Boca Raton, FL: CRC Press, 2001.

297. VY Gershtein, CE Baukal. Model Prediction Comparison for Aluminum Reverberatory Furnace Firing on Air-, Air-Oxy-, and Oxy/Fuel. Presented at the 1999 Minerals, Metals & Materials Society Annual Meeting & Exhibition, San Diego, CA, Feb. 28–Mar. 4, 1999.

298. YK Lee, HS Park, KW Cho. Effect of Fuel Gas Preheating on Combustion and Heat Transfer in Reheating Furnace. *Proceedings of the 13th Energy Engineering World Congress; Energy & Environmental Strategies for the 1990s*, Atlanta, GA, Chap. 78, pp. 461–466, Oct. 1990, 1991.

299. R Klima. Improved knowledge of gas flow and heat transfer in reheating furnaces. *Scandi. J. Metall. (Suppl.)*, Vol. 26, pp. 25–32, 1997.

300. JM Blanco, JM Sala. Improvement of the efficiency and working conditions for reheating furnaces through computational fluid dynamics. *Indust. Heat.*, Vol. LXVI, No. 5, pp. 63–67, 1999.

301. CE Baukal. Modeling Impinging Flame Jets. In *Computational Fluid Dynamics in Industrial Combustion*. ed. CE Baukal, VY Gershtein, X Li. Boca Raton, FL: CRC Press, 2001.

302. P Marino. Numerical Modeling of Steel Tube Reheating in Walking Beam Furnaces. *Proceedings of 5th European Conference on Industrial Furnaces and Boilers*, Portugal, Vol. II, pp. 237–247, 2000.

303. H Ramamurthy, S Ramadhyani, R Viskanta. Thermal System Model for a Radiant Tube Batch Reheating Furnace. *Proceedings of 1992 International Gas Research Conference*, ed. HA Thompson. Goverenment Institutes, Rockville, MD, pp. 2205–2216, 1993.

304. NDH Munroe. Experimental and Numerical Modeling of Transport Phenomena in a Particulate Reacting System. In *Heat Transfer in Fire and Combustion Systems – 1993*, ed. by B Farouk, MP Menguc, R Viskanta, C Presser, S Chellaiah. ASME HTD-Vol. 250, pp. 69–78, New York, 1993.

305. T Ahokainen, A Jokilaakso, O Teppo, Y Yang. Flow and Heat Transfer Simulation in a Flash Smelting Furnace. NTIS Rep. DE95779247, U.S. Dept. of Commerce, Springfield, VA, 1994.

306. M Davies, C Weichert, R Scholz. Energetic Process Model for the Description of the Melting Behaviour [sic] of a Natural Gas Fired Cokeless Cupola Furnace (CLCF). *Proceedings of 5th European Conference on Industrial Furnaces and Boilers*, Portugal, Vol. II, pp. 415–426, 2000.

307. DL Smoot. Pulverized Coal Diffusion Flames: a Perspective Through Modelling. In *18th Symposium (International) on Combustion*. The Combustion Institute, Pittsburgh, PA, pp. 1185–1202, 1981.

308. WA Fiveland, RA Wessel. Numerical model for predicting performance of three-dimensional pulverized-fuel fired furnaces, *ASME J. Eng. Gas Turbines Power*, Vol. 110, pp. 117–126, 1988.

309. S Li, B Yu, W Yao, W Song. Numerical Modelling for Pulverised Coal Combustion in Large Furnace. *Proceedings of 2nd International Symposium on Coal Combustion*. Beijing, Report ed. X Xu, L Zhou, W. Fu. Beijing, China: China Machine Press, pp. 167–173, 1991.

310. R Boyd, A Lowe. Three-Dimensional Modelling of a Pulverised Coal Fired Utility Furnace. In *Coal Combustion*. New York: Hemisphere, pp. 165–172, 1988.

311. BS Brewster, SC Hill, PT Radulovic, LD Smoot. Comprehensive Modeling. In *Fundamentals of Coal Combustion for Clean and Efficient Use*, Chap. 8, ed. LD Smoot. Amsterdam: Elsevier, 1993.

312. K Kudo, H Taniguchi, K Guo. Heat-transfer simulation in a furnace for steam reformer. *Heat Transfer – Japanese Research*, Vol. 20, No. 8, pp. 750–764, 1992.

313. KN Theologos, NC Markatos. Advanced modeling of fluid catalytic cracking riser-type reactors. *AIChE J.*, Vol. 36, No. 6, pp. 1007–1017, 1993.

314. SL Chang, CQ Zhou, SA Lottes, B Golchert, M Petrick. A Numerical Investigation of the Scaled-Up Effects on Flow, Heat Transfer, and Kinetics Processes of FCC Units. In *Combustion and Radiation Heat Transfer*, ed. RA Nelson, KS Ball, ZM Zhang. *Proceedings of the ASME Heat Transfer Division – 1998*, Vol. 2, ASME HTD-Vol. 361–2, pp. 73–81, New York, 1998.

315. TM Gruenberger, M Moghiman, PJ Bowen, N Syred. Improving Mixing Behaviour [sic] in New Design of Carbon Black Gas Furnace Using 3D CFD Modelling. *Proceedings of 5th European Conference on Industrial Furnaces and Boilers*, Portugal, Vol. II, pp. 465–473, 2000.

316. HC Meisingset, JG Balchen, R Fernandez. Mathematical Modelling of a Rotary Hearth Calciner, *Light Metals 1996*, ed. W Hale. Warren, PA: The Minerals, Metals & Materials Society, pp. 491–497, 1996.

317. JR Ferron, DK Singh. Rotary kiln transport processes. *AIChE J.*, Vol. 37, No. 5, pp. 747–758, 1991.

318. AA Boateng. On Flow Induced Kinetic Diffusion and Rotary Kiln Bed Burden Heat Transport. *ASME Proceedings of the 32nd National Heat Transfer Conference*, Vol. 3: *Fire and Combustion*, ed. L Gritzo, J-P Delplanque. ASME, New York, pp. 183–191, 1997.

319. PS Ghoshdastidar, VK Anandan Unni. Heat Transfer in the Non-Reacting Zone of a Cement Rotary Kiln. In *Heat Transfer Phenomena in Radiation, Combustion, and Fires*, ed. RK Shah. ASME HTD-Vol. 106, pp. 113–122, New York, 1989.

320. EE Khalil. Flow and Combustion Modeling of Vertical Lime Kiln Chambers. In *Industrial Combustion Technologies*, ed. MA Lukasiewicz. Warren, PA: American Society of Metals, pp. 99–107, 1986.

321. J Uche, JM Marin. Simulation and Optimization of a Single-Deck Roller Kiln for Ceramic Products. *Proceedings of 5th European Conference on Industrial Furnaces and Boilers*, Portugal, Vol. II, pp. 445–453, 2000.

322. B Farouk, MM Sidawi. Effects of Nitrogen Removal in a Natural Gas Fired Industrial Furnace: A Three Dimensional Study. in *Heat Transfer in Fire and Combustion Systems – 1993*, ed. B Farouk, MP Menguc, R Viskanta, C Presser, S Chellaiah. ASME HTD-Vol. 250, pp. 173–183, New York, 1993.

323. WD Owens, GD Silcox, JS Lighty, XX Deng, DW Pershing, VA Cundy, CB Leger, AL Jakway. Thermal analysis of rotary kiln incineration: comparison of theory and experiment. *Combust. Flame*, Vol. 86, pp. 101–114, 1991.

324. D Pal, JA Khan, JS Morse. Computational Modelling of an Industrial Rotary Kiln Incinerator. In *Heat Transfer in Fire and Combustion Systems*, ed. AM Kanury, MQ Brewster. ASME HTD-Vol. 199, pp. 167–173, New York, 1992.

325. FC Chang, CA Rhodes. Computer Modeling of Radiation and Combustion in a Rotary Solid-Waste Incinerator, Argonne National Laboratory, U.S. Dept. of Energy Rep. ANL/ET/CP-85778, 1995.

326. N Machii, K Nishimura, D Liu, K Shibata. Development of CAE for MSWI Ash Melting Furnace. *Proceedings of the 1998 International Gas Research Conference*, Vol. V: *Industrial Utilization*, ed. DA Dolenc. Gas Research Institute, Chicago, pp. 696–704, 1998.

327. T Klasen, K Gorner. The Use of CFD for the Prediction of Problem Areas Inside a Waste Incinerator with Regard to Slagging, Fouling and Corrosion. *Proceedings of 5th European Conference on Industrial Furnaces and Boilers*, Portugal, Vol. II, pp. 393–402, 2000.

328. SE Bayley, RT Bailey, DC Smith. Heat Transfer Analysis of Hazardous Waste Containers Within a Furnace. In *Combustion and Fire*, ed. MQ McQuay, W Schreiber, E Bigzadeh, K Annamalai, D Choudhury, A Runchal. *ASME Proceedings of the 31st National Heat Transfer Conference*, Vol. 6, ASME HTD-Vol. 328, pp. 61–69, New York, 1996.

329. MG Carvalho, N Martins. Mathematical Modelling of Heat and Mass Transfer Phenomena in Baking Ovens. *5th International Conference On Computational Methods and Experimental Measurements*, in *Computational Methods and Experimental Measurements V* ed. A Sousa, CA Brebbia, GM Carlomagno. The Netherlands: Elsevier, pp. 359–370, 1991.

330. M Carvalho, M Nogueira. Improvement of energy efficiency in glass-melting furnaces, cement kilns and baking ovens. *Appl. Therm. Eng.*, Vol. 17, Nos. 8–10, pp. 921–933, 1997.

331. TH Song, R Viskanta. Parametric Study of the Thermal Performance of a Natural Gas-Fired Furnace. In *Industrial Combustion Technologies*, ed. MA Lukasiewicz. Warren, PA: Amer. Society of Metals, 1989.

332. JA Wieringa, JJ Elich, CJ Hoogendoorn. Spectral Radiation Modelling of Gas-Fired Furnaces. In *Proceedings of 2nd European Conference on Industrial Furnaces and Boilers*, Portugal, ed. R Collin, W Leuckel, A Reis, J Ward. pp. 36–53, Apr. 1991.

333. M.J.S. de Lemos. Computation of Heated Swirling Flows with a Fully-Coupled Numerical Scheme. In *Combustion and Fire*, ed. MQ McQuay, W Schreiber, E Bigzadeh, K Annamalai, D Choudhury, A Runchal. *ASME Proceedings of the 31st National Heat Transfer Conference*, Vol. 6, ASME HTD-Vol. 328, pp. 139–145, New York, 1996.

334. M.J.S. de Lemos. Simulation of Vertical Swirling Flows in a Model Furnace with a High Performance Numerical Method. In *Combustion and Radiation Heat Transfer*, ed. RA Nelson, KS Ball, ZM Zhang. *Proceedings of the ASME Heat Transfer Division – 1998*, Vol. 2, ASME HTD-Vol. 361–2, pp. 21–28, New York, 1998.

335. MG Carvalho, M Nogueira. Modelling of glass melting industrial process, *J. Phys.*, Vol. 3, Part 7.2, pp. 1357–1366, 1993.

336. BD Paarhuis, AM Lankhorst, M Riepen, JFM Velthuis. A Rapid Simulation Model of the Combustion Space of Glass Melting Furnaces. *Fifth International Conference on Technologies and Combustion for a Clean Environment*, Lisbon, Portugal, Vol. 2, The Combustion Institute – Portuguese Section, pp. 697–703, 1999.

337. KS Chapman, S Ramadhyani, R Viskanta. Modeling and Analysis of Heat Transfer in a Direct-Fired Continuous Reheating Furnace. In *Heat Transfer in Combustion Systems*, ed. N Ashgriz, JG Quintiere, HG Semerjian, SE Slezak. ASME HTD-Vol. 122, pp. 35–43, New York, 1989.

338. KS Chapman, S Ramadhyani, R Viskanta. Modeling and Analysis of Heat Transfer in a Direct-Fired Batch Reheating Furnace. In *Heat Transfer Phenomena in Radiation, Combustion, and Fires*, ed. RK Shah, ASME HTD-Vol. 106, pp. 265–274, New York, 1989.

339. R Viskanta, KS Chapman, S Ramadhyani. Mathematical Modeling of Heat Transfer in High-Temperature Industrial Furnaces. In *Advanced Computational Methods in Heat Transfer*, Vol. 3: *Phase Change and Combustion Simulation*, ed. LC Wrobel, CA Brebbia, AJ Nowak, Berlin: Springer-Verlag, pp. 117–131, 1990.

340. KS Chapman, S Ramadhyani, R Viskanta. Two-dimensional modeling and parametric studies in a direct-fired furnace with impinging jets. *Combust. Sci. Technol.*, Vol. 97, pp. 99–120, 1994.

341. MG Carvalho, N Speranskaia, J Wang, M Nogueira. Modeling of Glass Melting Furnaces: Applications to Control, Design and Operation Optimization. *Advances in Fusion and Processing of Glass II*, Westerville, OH: The American Ceramic Society, pp. 109–135, 1998.

342. *FLUENT User's Guide*, Fluent, Inc., Lebanon, NH, July, 1998.

343. MK Choudhary. A modelling study of flows and heat transfer in an electric melter. *J. Non-Cryst. Solids*, Vol. 101, pp. 41–53, 1988.

344. X Li. Effect of Oxygen Enriched Air Staging on CO Emission for an Owens-Brockway Side-Port Regenerative Container Glass Furnace. In *Final Report – Demonstration of Oxygen-Enriched Air Staging at Owens-Brockway Glass Containers*, GRI-97/0292, Gas Research Institute, Chicago, Oct. 1997.

345. J Wang, BW Webb, MQ McQuay, K Bhatia, Numerical Simulation of an Oxy-Fuel-Fired Float Glass Furnace by Means of a Model Coupling the Combustion Space and the Glass Tank. In *Collected Papers from the 60th Conference on Glass Problems*. The American Ceramic Society, Westerville, OH, 1999.

346. F May, S Linka, H Kremer, S Wirtz. The Influence of Furnace Design on the NO-Formation in High Temperature Processes. *Fifth Inernational Conference on Technologies and Combustion for a Clean Environment*, Lisbon, Portugal, Vol. 2, The Combustion Institute – Portuguese Section, pp. 691–696, 1999.

347. A Droesbeke, F Missaire, O Berten, B Leduc, M Vansnick. Numerical Simulation of NO_x Emissions Reductions of an Industrial Burner. *Proceedings of 5th European Conference on Industrial Furnaces and Boilers*, Portugal, Vol. 1, pp. 235–244, 2000.

348. H Yamashita, D Zhao, SN Danov, T Furuhata, N Arai. A numerical study on NO_x Reduction by Steam Addition in Counterflow Diffusion Flame Using Detailed Chemical Kinetics. In MA Abraham, RP Hesketh (eds.). *Reaction Engineering for Pollution Prevention*. Amsterdam: Elsevier, 2000.

349. M Holmes. Process-Friendly NO_x Reduction Techniques. *Proceedings of 5th European Conference on Industrial Furnaces and Boilers*, Portugal, Vol. 1, pp. 447–453, 2000.

350. R Zakaria, YR Goh, YB Yang, CN Lim, J Goodfellow, KH Chan, G Reynolds, D Ward, RG Siddall, V Nasserzadeh, J Swithenbank. Fundamentals Aspects of Emissions from the Burning Bed in a Municipal Solid Waste Incinerator. *Proceedings of 5th European Conference on Industrial Furnaces and Boilers*, Portugal, Vol. II, pp. 287–300, 2000.

351. RW Johnson, MD Landon, EC Perry. Design Optimization. In *Computational Fluid Dynamics in Industrial Combustion*, ed. CE Baukal, VY Gershtein, X Li. Boca Raton, FL: CRC Press, 2001.

5

Source Testing and Monitoring

5.1 INTRODUCTION

There are a number of reasons for sampling the exhaust gases from an industrial combustion process [1]. One is to determine whether the emissions are within regulatory limits. Another is to determine the efficiency and effectiveness of the pollution-control technology being used in a given process. Stricter limits or ineffective pollution-control technologies may lead to design changes in the equipment. Sampling is also useful for predicting any potential problems in the development of a new process or modification of the existing process. The U.S. Environmental Protection Agency has established sampling methods for different pollutants in 40 CFR 60 [2]. A list of all the methods is given in Appendix G. These are available electronically over the Internet [3]. There are many useful general references on experimental measurements [4–8] as well as specific references on pollution-measurement techniques [9].

Crawford (1976) lists the following steps in the process of measuring pollutants in an air stream [10]:

1. Collection of the air sample containing the pollutant.
2. Conditioning of the sample.
3. Separation of the pollutants from the air in the sample.
4. Measurement of the desired properties of the pollutants.
5. Measuring the sample flow rate.
6. Disposal of the sample gas after completing the measurements.

It is noted that not all of these steps are required in particular circumstances. The last step is normally not a factor for most industrial combustion processes where the gas sample is vented to the atmosphere. This is only a factor where the pollutants in the sample may be poisonous, radioactive, or toxic in the concentrations present.

The objective of this chapter is to discuss the general principles in measuring pollutants in exhaust gas streams. The specific techniques and equipment for a particular pollutant are discussed in the chapter devoted to that pollutant.

5.2 SOME STATISTICS

There are some limited statistics that may be needed for measuring and reporting pollution emissions. These are briefly discussed here without derivation or proof. The reader is referred to environmental statistics books for those details [11–22]. The first concept is the mean or average value of a set of numbers:

$$\bar{x} = \frac{1}{n} \sum_{i=1}^{n} x_i \tag{5.1}$$

where $\bar{x}$ is the average value, x_i is the ith value, i is the index going from 1 to n, and n is the total number of observations.

Example 5.1

Given: 10 NO_x measurements (all in ppmvd at 3% O_2): 41, 36, 43, 40, 37, 36, 42, 35, 38, and 39.

Find: Find the average NO_x value.

Solution: $n = 10$, $x_1 = 41$, $x_2 = 36$, $x_3 = 43$, $x_4 = 40$, $x_5 = 37$, $x_6 = 36$, $x_7 = 42$, $x_8 = 35$, $x_9 = 38$, $x_{10} = 39$.

$$\bar{x} = \frac{1}{n} \sum_{i=1}^{n} x_i = \frac{1}{10} \sum_{i=1}^{10} x_i = \frac{1}{10} \sum (x_1 + x_2 + x_3 + x_4 + x_5 + x_6 + x_7 + x_8 + x_9 + x_{10}).$$

$$\bar{x} = \frac{1}{10} \sum (41 + 36 + 43 + 40 + 37 + 36 + 42 + 35 + 38 + 39) = \frac{1}{10}(387) = 38.7.$$

Then, the average of the 10 NO_x measurements is 38.7 ppmvd at 3% O_2. Many calculators and spreadsheet software packages have averaging functions to simplify this calculation. Note the importance of correcting all measurements to the same basis (3% O_2 in the above example) so that they can be properly compared.

From the above data set, the minimum NO_x reading is 35 ppmvd and the maximum reading is 43 ppmvd, both at 3% O_2. Therefore, the range of readings is

$$\text{Range} = x_{\min} \to x_{\max} \tag{5.2}$$

which in this case is from 35 to 43 ppmvd. Some operating permits may be written on a not-to-exceed basis, which means that a given pollutant concentration cannot exceed a given value. In that case, only the maximum value over a given time period may be of interest. For the above data set, that would be 43 ppmvd at 3% O_2. If the permitted NO_x emission limit is not-to-exceed 40 ppmvd at any time, then the above process would have exceeded that limit even though the average of 38.7 ppmvd is below the limit. Some permits are written so that the average concentration cannot exceed a specific value, so it is important to know how the permit is written.

Another statistical concept that may be used in reporting pollutant emissions is referred to as a moving average. Some regulations are written based on the average of a set of readings taken over a specific time period that is moving. For example, a pollutant may be limited to a certain maximum concentration over an 8-hr moving average. The following example will illustrate the concept.

Example 5.2

Given: NO_x measurements (all in ppmvd) taken once each hour over a 10 hr period: 41, 36, 43, 40, 37, 36, 42, 35, 38, and 39.

Find: Find the 8-hr moving average NOx at the 8th, 9th, and 10th hours.

Solution:

$$\bar{x}_{8th} = \frac{1}{8}\sum_{i=1}^{8} x_i = \frac{1}{8}\sum (x_1 + x_2 + x_3 + x_4 + x_5 + x_6 + x_7 + x_8)$$

$$\bar{x}_{8th} = \frac{1}{8}\sum (41 + 36 + 43 + 40 + 37 + 36 + 42 + 35) = \frac{1}{8}(310) = 38.8$$

$$\bar{x}_{9th} = \frac{1}{8}\sum_{i=1}^{8} x_i = \frac{1}{8}\sum (x_2 + x_3 + x_4 + x_5 + x_6 + x_7 + x_8 + x_9)$$

$$\bar{x}_{9th} = \frac{1}{8}\sum (36 + 43 + 40 + 37 + 36 + 42 + 35 + 38) = \frac{1}{8}(307) = 38.4$$

$$\bar{x}_{10th} = \frac{1}{8}\sum_{i=1}^{8} x_i = \frac{1}{8}\sum (x_3 + x_4 + x_5 + x_6 + x_7 + x_8 + x_9 + x_{10})$$

$$\bar{x}_{8th} = \frac{1}{8}\sum (43 + 40 + 37 + 36 + 42 + 35 + 38 + 39) = \frac{1}{8}(310) = 38.8.$$

The moving average takes the most recent data for the specified period. In the above case, the data were taken from the most recent 8 hr. Figure 5.1 shows a set of daily averages for a 10 day span with a line through the data, which is a linear regression fit. Note that the vertical axis has been exaggerated to show the relatively small shift in the averages during that time span. The graph shows that average daily NO_x appears to be increasing. This may indicate some type of problem that needs to be corrected before regulatory limits are exceeded. An example might be that the removal efficiency of a downstream NOx post-treatment system may be declining and the equipment is in need of service.

Another statistical measure that is useful for reporting data is referred to as the standard deviation. This is a measure of the dispersion of the data from the average value. Figure 5.2 shows two sets of data taken on two different days over the same hours of the day. The data have been artificially chosen so that the average over the given period is the same for both days, but there is considerably more "scatter" on Day 2, which means that the data are farther from the average value compared to Day 1. The Day 2 data then have a higher standard deviation than that of the Day 1 data.

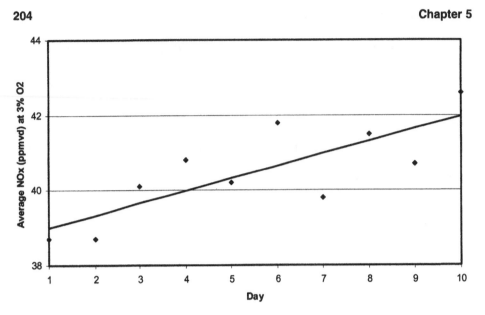

Figure 5.1 Average emissions over a 10-day time span with a linear regression fit.

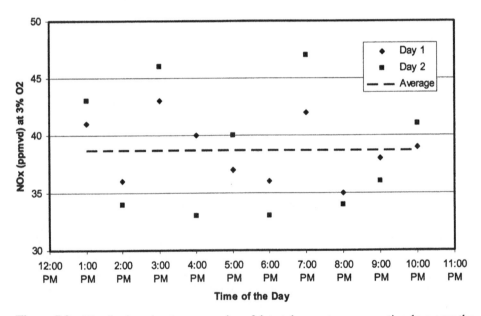

Figure 5.2 Distribution about a mean value of data taken on two consecutive days over the same time span.

The unbiased or "$n-1$" standard deviation is defined as

$$\sigma_u = \sqrt{\frac{\sum_{i=1}^{n}(x_i = \bar{x})^2}{n-1}} \qquad (5.3)$$

The biased or "n" standard deviation is defined as

$$\sigma_b = \sqrt{\frac{\sum_{i=1}^{n}(x_i = \bar{x})^2}{n}} \tag{5.4}$$

where the difference between the two is the divisor. For a large sample size where n is large, these two will be essentially the same.

Example 5.3

Given: 10 NO_x measurements (all in ppmvd at 3% O_2): 41, 36, 43, 40, 37, 36, 42, 35, 38, and 39.

Find: Find the biased and unbiased standard deviations.

Solution: $n = 10$, $x_1 = 41$, $x_2 = 36$, $x_3 = 43$, $x_4 = 40$, $x_5 = 37$, $x_6 = 36$, $x_7 = 42$, $x_8 = 35$, $x_9 = 38$, $x_{10} = 39$, $\bar{x} = 38.7$ as calculated in a previous example above

$$\sigma_u = \sqrt{\frac{\sum_{i=1}^{n}(x_i - \bar{x})^2}{n-1}} = \sqrt{\frac{\sum_{i=1}^{10}(x_i - \bar{x})^2}{10-1}}$$

$$\sigma_u = \sqrt{\frac{(x_1 - \bar{x})^2 + (x_2 - \bar{x})^2 + (x_3 - \bar{x})^2 + (x_3 - \bar{x})^2 + \cdots + (x_{10} - \bar{x})^2}{9}}$$

$$\sigma_u = \sqrt{\frac{(41 - 38.7)^2 + (36 - 38.7)^2 + (43 - 38.7)^2 + (40 - 38.7)^2 + \cdots + (39 - 38.7)^2}{9}}$$

$$\sigma_u = 2.75\,\text{ppmvd}$$

$$\sigma_b = \sqrt{\frac{(41 - 38.7)^2 + (36 - 38.7)^2 + (43 - 38.7)^2 + (40 - 38.7)^2 + \cdots + (39 - 38.7)^2}{10}}$$

$$\sigma_b = 2.61\,\text{ppmvd}.$$

As expected, the biased standard deviation is slightly less than the unbiased because of the larger denominator. Many calculators and spreadsheet software packages have standard deviation functions, although one must be careful to know which one is being used or how to get the desired standard deviation if both forms are available. For example, in Microsoft EXCEL®, both functions are available. The $n-1$ or unbiased function is STDEV and the n or biased function is STDEVP.

The unbiased standard deviation is typically used for emissions reporting and will be used here for illustration purposes. The Day 1 data are assumed to be normally distributed about the mean value as shown in Fig. 5.3. This means that if enough data were collected and the data were truly normally distributed about the mean, they would fit the curve shown in the figure. The graph also shows the limits of one standard deviation from the mean both to the left and to the right. As can be seen, the majority of the data would fall in the region bounded by one standard

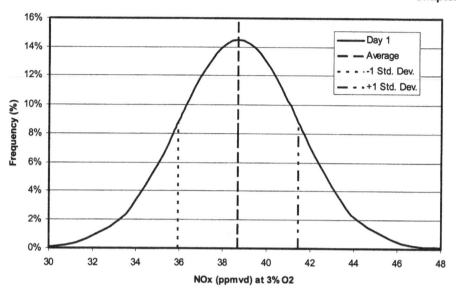

Figure 5.3 Normal distribution of Day 1 data about the mean of 38.7 ppmvd NO_x at 3% O_2.

deviation to the left and to the right of the mean. Therefore, it is common to report data in the following form:

$$x = \bar{x} \pm \sigma \tag{5.5}$$

In the example above for the Day 1 data, this would be written as

$$NO_x = 38.7 \pm 2.8 \text{ ppmvd at 3\% } O_2$$

The standard deviation shows how tightly distributed the data are about the mean. The smaller the standard deviation, the more closely the data are to the mean value. Figure 5.2 showed that the Day 2 data were more scattered compared to the Day 1 data. Figure 5.4 shows a comparison of the normal distribution curves for the Day 1 and Day 2 data, both of which have the same mean value. The curve for Day 1 is taller and thinner, which shows that most of the data are closer to the mean. The curve for Day 2 is shorter and fatter, which shows that most of the data are farther from the mean.

A high relative standard deviation could indicate that there may be a problem with the measurement system where there is not high repeatability if the data are not truly changing. A high standard deviation could also indicate that there are wider swings in the data that could be caused by changing conditions. For example, if the fuel composition is changing with time, which is sometimes the case in petrochemical applications, this could cause the NO_x to change significantly. Another example is a changing ambient environment where there may be wide swings in the temperature and relative humidity of the incoming combustion air that can affect NO_x emissions. A high relative standard deviation is not generally desirable because more cushion will be needed between the permitted or regulated emission limits and the measured

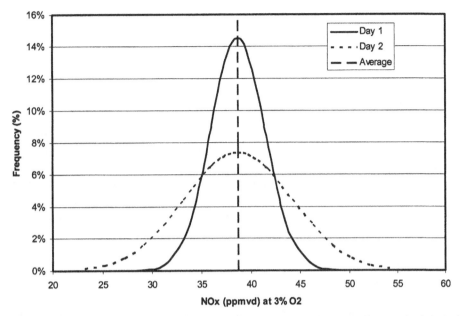

Figure 5.4 Comparison of normal distributions for Day 1 with a smaller standard deviation and Day 2 with a larger standard deviation, both with the same mean value.

emissions because of the higher possibility of significant excursions above the mean value.

5.3 SAMPLING

An intersociety committee put together a guidebook for air sampling, primarily in the ambient atmosphere [23]. The following general physical precautions relevant to extractive gas sampling were noted: ensure that the sample is homogeneous, be aware of any absorption and diffusion effects in the sampling system, check for mechanical defects in the sampling system, and calibrate at the design sampling flow rate. It is also noted that any interferences between chemicals in a sampling system must also be known and corrected for in order to ensure accurate and reliable measurements. The U.S. EPA has developed an extensive manual to aid in estimating the costs of pollution-control equipment [24]. A chapter in the manual is devoted to emissions monitoring [25].

It is important that the position of the sample probe is known accurately if there is a considerable variation in the gas composition profile across the duct. It is preferred that the measurements be made sufficiently downstream of flow disturbances like elbows or dampers to minimize the variation across the duct. If this is not possible as is the case when physical restrictions limit how the exhaust ducting can be run and where the measurements can be made, then more traverses are recommended.

EPA Method 1 recommends that both sample and velocity measurements be made at least eight internal diameters downstream of any flow disturbance (e.g., elbow, duct contraction, or expansion) and at least two internal diameters upstream from any flow disturbance. If this is not possible, then measurements may be made as close as two internal diameters downstream and 0.5 internal diameters upstream of disturbances, but more measurements across the duct cross-section are needed. Figure 5.5 shows the recommended minimum number of traverse (sampling) points for measuring either the gas sample composition or the gas velocity where no particulates are in the gas stream. The figure shows that at least eight or nine measurement points are recommended for internal stack diameters between 12 and 24 in. where the measurement location is at least eight internal diameters downstream and two internal diameters upstream from any upstream disturbances. For internal stack diameters greater than 24 in. in internal diameter, at least 12 sampling points are recommended if the measurement location is at least seven internal diameters downstream and 1.75 internal diameters upstream of any flow disturbance. A minimum of 16 sample points are recommended for internal stack diameters 12 in. in internal diameter or greater where the sampling location is between two and seven (eight for internal stack diameters between 12 and 24 in.) internal diameters downstream and between 0.5 and 1.75 (two for internal stack diameters between 12 and 24 in.) internal diameters upstream from a flow disturbance. Figure 5.6 shows the recommended minimum number of traverse points if there are particulates in the flow. As can be seen, more traverses are generally needed when particulates are present.

The minimum distance to a disturbance determines the minimum number of recommended traverse points. For example, if the stack diameter is 18 in. in internal diameter, the gas stream contains no particulates, and the sample location is 10

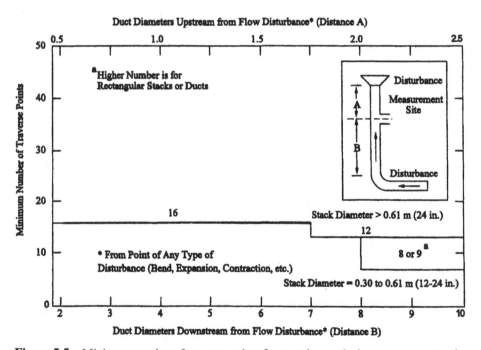

Figure 5.5 Minimum number of traverse points for sample or velocity measurements when particulates are not present. (From Ref. 2.)

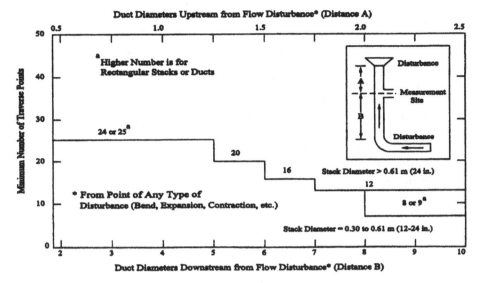

Figure 5.6 Minimum number of traverse points for sample or velocity measurements when particulates are present. (From Ref. 2.)

internal diameters downstream from the closest downstream disturbance but only one internal diameter upstream from the closest upstream disturbance, then the distance to the upstream disturbance determines the minimum number of traverses. According to Fig. 5.5, only eight or nine traverses would be needed based on the downstream distance (10 diameters) to a disturbance, but at least 16 traverses are recommended because the sample location is only one internal diameter upstream from the location.

For rectangular ducts, an equivalent diameter is calculated using the following:

$$D_e = \frac{2LW}{L + W} \tag{5.6}$$

where L is the length of the rectangular duct cross-section, W is the width of the rectangular duct cross-section, and D_e is the equivalent diameter. Note that $L = W$ for square ducts. Note that all of these dimensions are internal.

Example 5.4

Given: Duct with rectangular cross-sectional area with internal dimensions of 1 ft by 2 ft.

Find: Find the equivalent internal diameter.

Solution: $L = 2$ ft, $W = 1$ ft

$$D_e = \frac{2(2\,\text{ft})(1\,\text{ft})}{(2\,\text{ft}) + (1\,\text{ft})} = 1.33\,\text{ft}.$$

This value is then used with the figures to determine the number of traverse points

Table 5.1 shows the recommended layout for rectangular ducts based on the number of traverse points from the above figures. For example, if 12 traverse points are recommended, then the table suggests a 4×3 layout. Figure 5.7 shows an example of a rectangular duct divided into 12 equal subareas in a 4×3 matrix. The samples are then measured in the centroid of each subarea as shown in the figure.

Table 5.2 shows the recommended layout for circular ducts based on the number of traverse points from the above figures. For example, if 12 traverse points are recommended, then the table suggests the location of the sample points. Figure 5.8 shows an example of a circular duct divided into 12 equal subareas in a symmetric matrix. The samples are then measured in the centroid of each subarea as shown in the figure. The distances to each centroid are given as a percentage of the total diameter.

Example 5.5

Given: Circular 5 ft external diameter duct with 6 in. of insulation inside the duct, located 30 ft downstream of the nearest disturbance and 10 ft upstream of the nearest disturbance; no particulates in the gas stream.

Table 5.1   Recommended Measurement Layout for Rectangular Cross-Sectional Area Ducts

# Traverse points in layout	Recommended matrix
9	3×3
12	4×3
16	4×4
20	5×4
25	5×5
30	6×5
36	6×6
42	7×6
49	7×7

Source: Ref. 2.

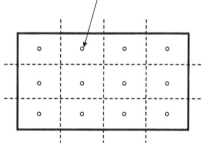

Figure 5.7   Rectangular cross-sectional duct divided into 12 equal subareas.

Table 5.2 Recommended Measurement Layout for Circular Cross-Sectional Area Ducts

Traverse point on a diameter	Number of traverse points on a diameter											
	2	4	6	8	10	12	14	16	18	20	22	24
1	14.6	6.7	4.4	3.2	2.6	2.1	1.8	1.6	1.4	1.3	1.1	1.1
2	85.4	25.0	14.6	10.5	8.2	6.7	5.7	4.9	4.4	3.9	3.5	3.2
3		75.0	29.6	19.4	14.6	11.8	9.9	8.5	7.5	6.7	6.0	5.5
4		93.3	70.4	32.3	22.6	17.7	14.6	12.5	10.9	9.7	8.7	7.9
5			85.4	67.7	34.2	25.0	20.1	16.9	14.6	12.9	11.6	10.5
6			95.6	80.6	65.8	35.6	26.9	22.0	18.8	16.5	14.6	13.2
7				89.5	77.4	64.4	36.6	28.3	23.6	20.4	18.0	16.1
8				96.8	85.4	75.0	63.4	37.5	29.6	25.0	21.8	19.4
9					91.8	82.3	73.1	62.5	38.2	30.6	26.2	23.0
10					97.4	88.2	79.9	71.7	61.8	38.8	31.5	27.2
11						93.3	85.4	78.0	70.4	61.2	39.3	32.3
12						97.9	90.1	83.1	76.4	69.4	60.7	39.8
13							94.3	87.5	81.2	75.0	68.5	60.2
14							98.2	91.5	85.4	79.6	73.8	67.7
15								95.1	89.1	83.5	78.2	72.8
16								98.4	92.5	87.1	82.0	77.0
17									95.6	90.3	85.4	80.6
18									98.6	93.3	88.4	83.9
19										96.1	91.3	86.8
20										98.7	94.0	89.5
21											96.5	92.1
22											98.9	94.5
23												96.8
24												99.9

Source: Ref. 2.

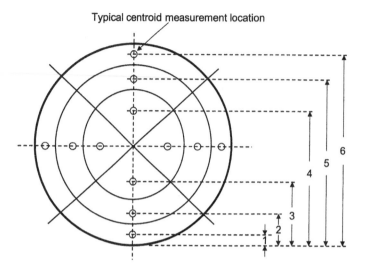

Figure 5.8   Circular cross-sectional duct divided into 12 equal subareas.

Find: Minimum number and location of sample points.
Solution: Internal duct diameter = 5 ft − (2) × 6 in. = 4 ft
              Distance to nearest downstream disturbance = 30/4 = 7.5 internal
              diameters
              Distance to nearest upstream disturbance = 10/4 = 2.5 internal
              diameters
              From Fig. 5.5, a minimum of 12 traverses are recommended
              From Table 5.2, the sample points should be located at 4.4, 14.6, 29.6,
              70.4, 85.4, and 95.6% of the distance across the internal diameter of
              4 ft or 2.1, 7.0, 14.2, 33.8, 41.0, and 45.9 in. from the edge of the duct
              (see Fig. 5.9).

EPA Method 1A applies to ducts with equivalent internal diameters of less than 12 in. but greater than or equal to 4 in. The method does not apply to swirling or cyclonic flows. If particulates are present, one or two traverse points should be located at least eight internal diameters downstream and 10 internal diameters upstream from any flow disturbances. Note that velocity measurements should be made eight internal diameters downstream from the gas-sampling location for this case. Figure 5.10 shows the recommended sampling arrangement for small ducts. Again, Fig. 5.5 should be used to determine the number of sample points for flows without particulates and Fig. 5.6 should be used for flows with particulates. In either case, the minimum number of sample points is eight for circular ducts and nine for rectangular ducts. For circular ducts, the number of sample points should be divisible by four. The main difference is that the distances from the gas-sampling location and from the velocity measurement location to the nearest upstream and downstream disturbances should be calculated (four total equivalent internal diameters). The distance requiring the most number of traverse points should be used as the minimum number of traverse points.

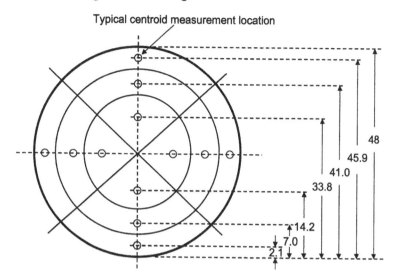

Typical centroid measurement location

Figure 5.9 Solution to example problem for location of 12 traverse points in a 4 ft diameter stack.

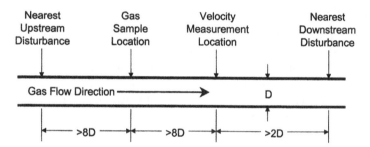

Figure 5.10 Recommended sampling arrangement for small ducts.

Example 5.6

Given: Circular duct with 6 in. internal diameter, gas sample location is 3 ft downstream from the nearest downstream disturbance, the velocity measurement location is 2 ft downstream of the gas sample location, and the velocity measurement location is 1 ft downstream of the nearest upstream disturbance (see Fig. 5.11).

Find: Minimum number of sample points.

Solution: First calculate four distances:

$D_1 =$ distance from gas sample probe to nearest upstream flow disturbance $= 3\,\text{ft}/6\,\text{in.} = $ six equivalent internal diameters;

$D_2 =$ distance from gas sample probe to nearest downstream flow disturbance (velocity probe) $= 2\,\text{ft}/6\,\text{in.} = $ four equivalent internal diameters; $D_3 =$ distance from velocity probe to nearest upstream flow disturbance (gas sample probe) $= 2\,\text{ft}/6\,\text{in.} = $ four equivalent internal diameters;

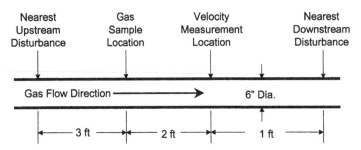

Figure 5.11 Probe locations for sample problem.

$D_4 =$ distance from velocity probe to nearest downstream flow disturbance $= 1\,\text{ft}/6\,\text{in}. =$ two equivalent internal diameters;
Note that D_1 and D_3 are less than the recommended eight diameters, but this may have been all the available space in the system. The worst case is D_3, so from Fig. 5.5 at least 16 traverse points are recommended. Note that this is evenly divisible by four as recommended.

5.3.1 Extractive Sampling

Isokinetic sampling pertains to extractive gas sampling where the sample is withdrawn out of the gas stream at the same velocity as the exhaust gas stream velocity. The EPA defines isokinetic sampling as "sampling in which the linear velocity of the gas entering the sampling nozzle is equal to that of the undisturbed gas stream sample point" [26]. This is particularly important when measuring larger particulates, especially those greater than about 5 μm. Figure 5.12 shows the three possibilities for withdrawing a sample from the exhaust gas stream. When the sample velocity is less than the gas stream velocity, some particles in the gas stream flow around the probe and the measured particulate concentration is lower than the actual particulate concentration. When the sample velocity is greater than the gas stream velocity, then some particles in the gas stream are preferentially drawn into the sample probe and the measured particulate concentration is greater than the actual particulate concentration. The sample flow rate is normally easily varied by either controlling the sample pump suction rate or more typically by adjusting a control valve in the sampling system.

The challenge in isokinetic sampling is measuring both the sample gas velocity and the exhaust gas stream velocity. In most cases, the exhaust gas stream velocity varies across the exhaust duct and also varies as a function of time as changes in the combustion system cause the exhaust gas volume to vary due to changing process requirements. It is not sufficient to measure the exhaust gas velocity at the location where the sample probe will be located with, for example, a pitot tube and assume that the exhaust gas velocity will be constant at all times and under all conditions. Isokinetic conditions are approximated when the static pressure is equal inside and outside of a properly designed sample probe (see Fig. 5.13). The combined pitot–sample probe is used to measure the gas velocity in the vicinity of the probe and then to adjust the sample flow rate so that the average velocity at the sample probe inlet is

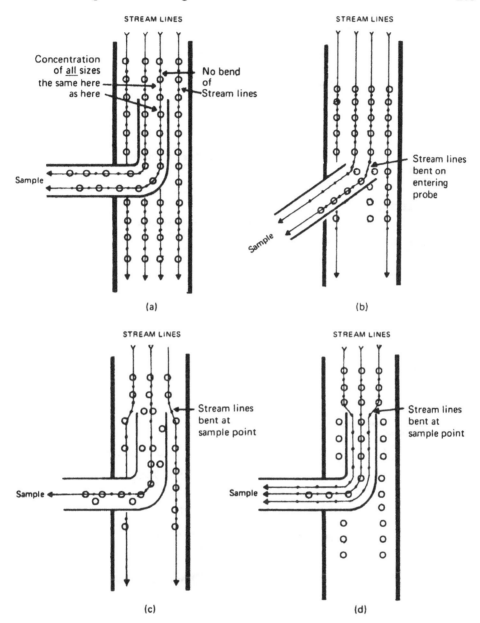

Figure 5.12 Schematics of isokinetic and nonisokinetic sampling. (From Ref. 23. Courtesy of CRC Press.)

equal to the gas velocity as measured by the pitot probe. The gas velocity is measured using the pitot probe by determining the pressure difference across the probe as shown in Fig. 5.14. This can be done by simply using a differential pressure gauge (Fig. 5.15), manometer (Fig. 5.16), or differential pressure transmitter (Fig. 5.17) and controlling the sample flow rate until the measured differential pressure is zero.

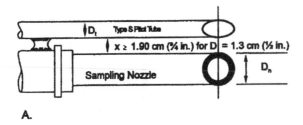

A.

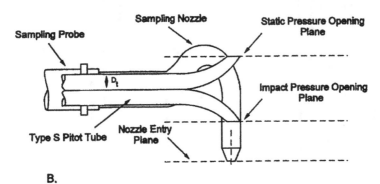

B.

Figure 5.13   Isokinetic sample probe. (From Ref. 2.)

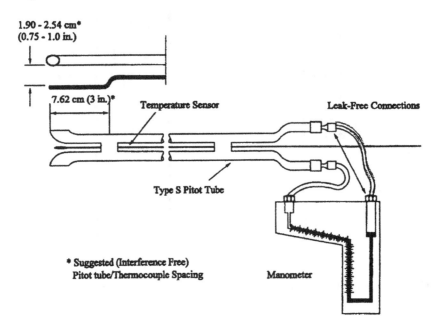

Figure 5.14   Type S pitot tube manometer assembly for measuring gas velocity. (From Ref. 2.)

Another type of isokinetic sampling probe is shown in Fig. 5.18. While this probe is more difficult to build because of the small passage sizes, it is more easy to operate than the probe recommended by the U.S. EPA. Two static pressure measurements are made: one for the outer stream and one for the inner stream.

Figure 5.15 Differential pressure gauge.

Figure 5.16 Manometers for measuring differential pressure.

Figure 5.17 Differential pressure transmitter.

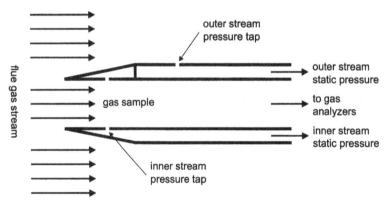

Figure 5.18 Nozzle for isokinetic sampling probe (not to scale).

These are connected to a differential pressure measuring device (see discussion above) and the gas sample flow is adjusted until the static pressure differential is zero, which means that the static pressure is the same inside and outside the probe. The static pressure holes on both the inside and outside of the probe go around the entire circumference at equally spaced distances apart. These holes are connected to the appropriate manifold inside the probe so that the static pressure measurements are actually averaged values around the circumference of the probe in case there are any significant flow deviations. The probe tip is aerodynamically designed to minimize disturbances to the out exhaust gas stream flow.

Another important aspect of extractive sampling is proper conditioning of the sample stream. If the exhaust gas stream contains particulates, these can clog up the sample lines. Some type of high-temperature filter (see Fig. 5.19) or shield (see Fig. 5.20) is often attached to the sample probe to prevent particulates from entering the sampling system. Figure 5.21 shows a schematic of a typical sampling system. In order to prevent further reactions in the sampling system, the gas sample is typically quenched as quickly as possible to "freeze" the gas chemistry or the sample is maintained above the dew point to prevent water from condensing out until it reaches the water knockout unit. This is sometimes referred to as hot–wet sampling where the sample is maintained above the dew point until the sample reaches the condenser. Figure 5.22 shows an example of a heated sample probe to prevent catalytic reactions inside the probe itself. The sample is not diluted after it has been withdrawn from the exhaust gas stream.

Figure 5.19 Sample frit filter. (Courtesy of Millennium Instruments, Spring Grove, IL.)

Figure 5.20 Sample shield. (Courtesy of Millennium Instruments, Spring Grove, IL.)

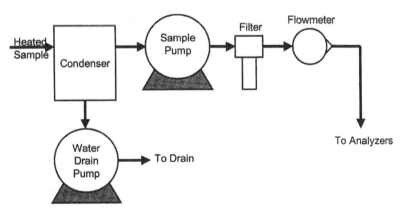

Figure 5.21 Gas sample flow schematic.

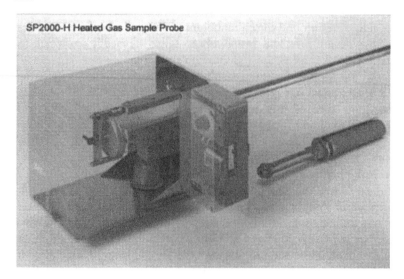

Figure 5.22   Heated gas sample probe. (Courtesy of M&C, Ratingen, Germany.)

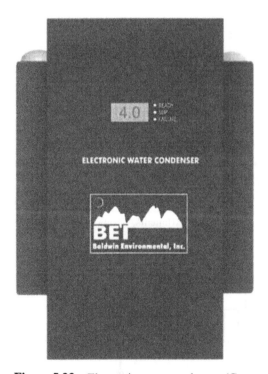

Figure 5.23   Electronic water condenser. (Courtesy of Baldwin Environmental, Reno, NV.)

One common type of water knockout unit is an electronic water condenser (see Fig. 5.23). Another type uses a permeation dryer where nitrogen on one side of a membrane removes water from the gas sample on the other side of the membrane (see Fig. 5.24). The gas sample conditioning system can be purchased prepackaged as

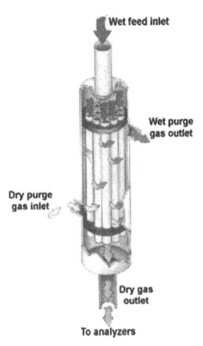

Figure 5.24   Permeation dryer. (Courtesy of Perma Pure, Toms River, NJ.)

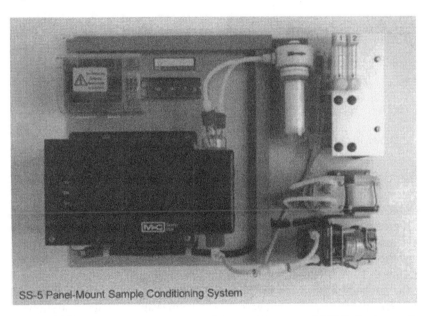

Figure 5.25   Prepackaged sample conditioner. (Courtesy of M&C, Ratingen, Germany.)

shown in Fig. 5.25. It is also necessary to measure the sample flow rate to ensure that the proper flows are going to the analyzers. Figure 5.26 shows a sampling system that includes flow meters for measuring and controlling the sample flow rates to the analyzers.

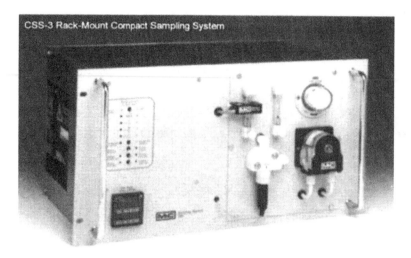

Figure 5.26   Sampling system. (Courtesy of M&C, Ratingen, Germany.)

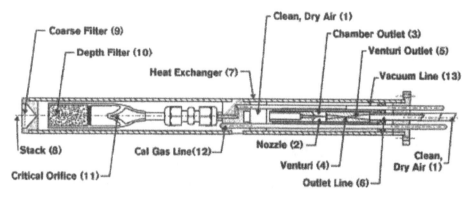

Figure 5.27   Schematic of a stack dilution probe. (Courtesy of EPM, Mt. Prospect, IL.)

Another alternative to removing water from the gas sample prior to sending it to the analyzers is to dilute the sample so that the effective moisture content is not a problem for the analyzers. This is sometimes referred to as cool–dry sampling where the sample is greatly diluted immediately after extraction from the exhaust gas stack. Dilution systems can be built into the probe as shown schematically in Fig. 5.27 and pictorially in Fig. 5.28. Specially designed probes are used to sample particle-laden flows according to EPA Method 5, as shown in Fig. 5.29.

Once the gas sample has been properly conditioned, it can be analyzed in one or more analyzers, depending on what gases are being measured. Figure 5.30 shows a single analyzer used to measure the oxygen concentration in the sample. Figure 5.31 shows a multigas analyzer used to measure gases such as nitrogen oxides, sulfur oxides, oxygen, carbon dioxide, and carbon monoxide.

There are a number of potential sampling problems [25]:

- Probes and lines clogging with contamination.
- Heated lines failing in cold climates causing water to freeze and block lines.

Figure 5.28   Stack dilution probe. (Courtesy of EPM, Mt. Prospect, IL.)

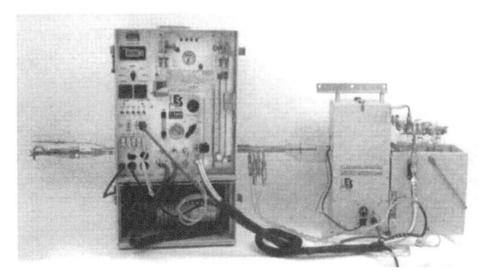

Figure 5.29   Method 5 sampling probe. (Courtesy of Environmental Supply, Durham, NC.)

- Probe filter causing loss of pollutant as it passes through the probe media (scrubbing).
- Dilution probe causing temperature, pressure, gas density effects, and water droplet evaporation when dilution air is added to the sample gas.
- Water entrainment.
- Leaks in the tubing or elsewhere in the system.
- Adsorption of pollutant to the wall, filter, tubing, or other components.
- Adsorption of pollutant to the water, which is removed by a conditioning system.

5.3.2 In-Situ Sampling

In-situ sampling involves analyzing the exhaust gases in the exhaust stack without extracting them. In-situ analyzers are available for SO_2, CO, O_2, NO_x, O_2, and CO_2.

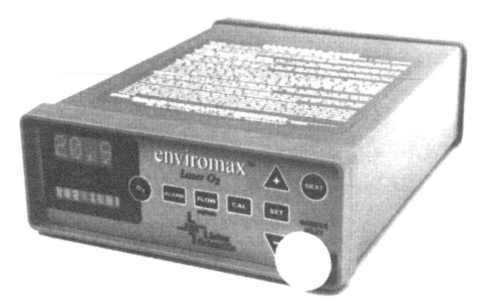

Figure 5.30   Oxygen analyzer. (Courtesy of Liston, Irvine, CA.)

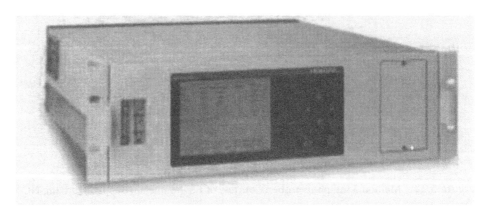

Figure 5.31   Multigas analyzer. (Courtesy of Horiba, Irvine, CA.)

They are also available for measuring opacity, particulates, and gas flow rate. In-situ sampling eliminates the need for sample transport and conditioning that are required in extractive sampling. There are two types of in-situ sampling: point and path. Point sampling measures at a specific point or location in the exhaust stack while path measures across the stack, giving an averaged value. Another important advantage of in-situ sampling is rapid response since the transport time in extractive sampling can be significant.

However, there are some additional challenges of using in-situ sensors due to the harsh conditions that usually include high temperatures and may include particulates and high gas flow rates. Another potential problem is that the analytical equipment may be located in inconvenient locations, possibly at elevated locations on tall exhaust stacks, which can make maintenance and troubleshooting more difficult compared to extractive systems that are typically located in conditioned enclosures or rooms at ground level. In-situ sampling systems can also become plugged or clogged in exhaust gas streams with high particulate levels.

One common type of in-situ sensor is made of zirconia (ZrO_2) which is used for wet O_2 measurements. Light-absorption techniques involving UV or IR spectrometers are also used to measure gases like CO, CO_2, SO_2, and NO.

5.4 TESTING EQUIPMENT

This section briefly considers the equipment commonly used to measure variables of importance when quantifying pollution emissions. These variables include the exhaust gas flow rate, the gas composition, and particulate emissions (solids in the gas stream). Note that there are many other variations that may be used in a given installation and new types of equipment are constantly being developed to make these measurements easier, more repeatable, and more accurate.

5.4.1 Exhaust Gas Flow

The exhaust gas flow is usually not directly measured but is often determined through a combination of measurements, calculations, and assumptions. For example, the gas velocity is measured at one or more points across the exhaust duct. Since it is not known a priori if the gas sample is homogeneous, it must be assumed that there is some variation across the exhaust gas duct until proven otherwise. This is done by making multiple traverses across the duct. The cross-sectional area of the duct is subdivided into smaller areas for the purposes of making measurements. The gas velocity in each subarea is assumed to be constant. Then the gas flow rate through that subarea is simply

$$Q_i = \frac{v_i}{A_i} \tag{5.7}$$

where Q_i is the gas flow rate through sub area i, v_i is the gas velocity through the subarea, and A_i is the area. The sum of all the subareas equals the total cross-sectional area of the duct:

$$A = \sum_{i=1}^{n} A_i \tag{5.8}$$

where A is the total duct cross-sectional area and n is the total number of subareas. Then, the total exhaust gas flow rate equals the sum of the individual gas flow rates:

$$Q = \sum_{i=1}^{n} Q_i \tag{5.9}$$

5.4.1.1 Pitot Tube

EPA Method 2 discusses how to determine the volumetric stack gas flow rate using an S-Type Pitot tube that includes a temperature sensor (see Fig. 5.14). The method only applies to flows that are not swirling or cyclonic. This method combines measurements and calculations to determine the gas flow rate. Specific guidelines are given for what materials should be used and what dimensions are critical. Each pitot probe must be calibrated with a known coefficient that should be engraved on the probe. The static and dynamic measuring holes on the probe must not be plugged, which can be a problem in particulate-laden flows. The pressure "head" or difference between the static and dynamic measurements is typically very small and can be measured with an inclined manometer (see Fig. 5.16) or a similar device like a magnehelic gauge. Both of these need to be calibrated to ensure accuracy.

The equation governing the response of a Pitot probe is commonly given as [27]:

$$v = \sqrt{\frac{2(p_t - p_{st})}{C_p \rho}} \tag{5.10}$$

where p_t is the total or impact pressure, p_{st} is the static pressure, and C_p is the probe calibration constant. According to EPA Method 2, the average stack gas velocity can be calculated for an S-Type Pitot tube using

$$v_s = K_p C_p \sqrt{\frac{(p_t - p_{st}) T_s}{P_s M_s}} \tag{5.11}$$

where

$\quad\quad\quad\quad v_s = $ average stack gas velocity (ft/sec.)
$\quad\quad\quad\quad K_p = $ velocity equation constant
$\quad\quad\quad\quad C_p = $ Pitot tube probe constant
$\quad\quad p_t - p_{st} = $ difference between total and static pressures measured with Pitot tube (in. of H_2O)
$\quad\quad\quad\quad T_s = $ absolute stack temperature (°R)
$\quad\quad\quad\quad P_s = $ absolute stack pressure (in. of Hg)

$$M_s = M_d(1 - B_{ws}) + 18 B_{ws} \tag{5.12}$$

where

$\quad\quad\quad\quad M_s = $ molecular weight of wet stack gas (lb/lb-mole)
$\quad\quad\quad\quad M_d = $ molecular weight of dry stack gas (lb/lb-mole)
$\quad\quad\quad\quad B_{ws} = $ fraction by volume of water in the gas stream (determined by Methods 4 or 5)

EPA Method 2C concerns determination of the exhaust gas flow rate using a standard hemispherically nosed Pitot probe as shown in Fig. 5.32. The previous equations for the S-Type Pitot probe apply to the standard Pitot tube, although the probe constant will be different. Becker and Brown [27] studied the response of Pitot

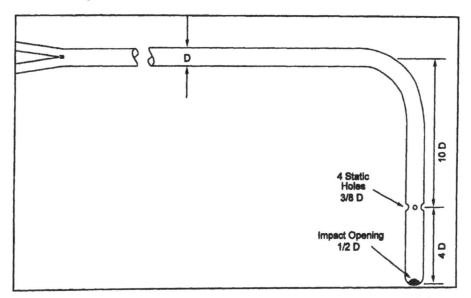

Figure 5.32   Standard hemispherically nosed Pitot tube from 40 CFR 60 Method 2C. (From Ref. 2.)

probes in turbulent nonreacting gas streams. The following terminology is used to specify the recommendations of their study:

d_1 = total pressure hole i.d.
d_2 = probe o.d.
d_3 = static pressure hole i.d.
l_1 = internal length before any change in diameter or direction of total impact tube
l_2 = external length before any change in diameter or direction of probe
l_3 = length from probe tip to centerline of static pressure hole

In that study, the following conditions were assumed:

1.   The probe is long enough ($l_2/d_2 > 6$) so that the downstream geometry has a negligible effect on the response.
2.   The internal geometry ($l_1/d_1 > 3$) has a negligible effect on the response.
3.   Compressibility effects are small (Ma < 0.3).

Cho and Becker recommend a single static pressure inlet [28].

A thermocouple or similar device can be used to measure the gas temperature, which is needed to compute the gas density. A barometer is needed to measure the atmospheric pressure, also needed to compute the gas density. In lieu of a barometer, the barometric pressure can be obtained from the National Weather Service.

According to EPA Method 2, the average dry stack gas volumetric flow rate is calculated as follows:

$$Q = 3600(1 - B_{ws})v_s A\left[\frac{T_{std}P_s}{T_s P_{std}}\right] \tag{5.13}$$

where

  $Q =$ Volumetric flow rate in dry standard cubic feet per hour (dscf/hr)

  $B_{ws} =$ Fraction by volume of water in the gas stream (determined by Methods
      4 or 5)

  $v_s =$ Average stack gas velocity (ft/sec)

  $A =$ Cross-sectional area of the stack (ft^2)

  $T_{std} =$ Standard absolute temperature $= 528°R$

  $P_s =$ Absolute stack pressure (in. of Hg)

  $T_s =$ Absolute stack temperature (°R)

  $P_{std} =$ Standard absolute pressure (29.92 in. of Hg)

EPA Methods 2G and 2F discuss the use of two- and three-dimensional Pitot probes, respectively, which are useful for swirling or cyclonic flows, but are used infrequently in industrial combustion applications and are therefore not discussed further here. The calibration procedures are considerably more difficult than those for one-dimensional Pitot probes.

EPA Method 2H discusses how to determine a correction for the velocity decay near the wall in stacks greater than 1.0 m (3.3 ft) in diameter. Two procedures are discussed: one uses velocity measurements and the other uses a generic adjustment factor.

5.4.1.2 Direct Measurement

EPA Method 2A discusses the direct measurement of the gas volume flow rate through pipes and small ducts. These devices include, for example, positive displacement meters and turbine meters. This is a much simpler and faster technique than measuring multiple traverse points in the exhaust gas stack and then computing the flow rate based on several different types of measurements (pressure difference, gas temperature, barometric conditions, and moisture in the stack gases). Direct measuring devices also need to be properly calibrated to ensure accuracy. These usually cost much more than using Pitot tubes and manometers but require far less labor to operate. They can also mitigate the problem of varying conditions across the exhaust duct. However, they are typically only used on smaller diameter ducts and they may introduce a significant pressure drop in the exhaust system that may require either additional fan power or a reduction in the exhaust gas volume flow rate, which is usually not desirable.

EPA Method 2D discusses the measurement of the gas volume flow rates in small pipes and ducts using rotameters, orifice plates, or other similar devices to measure flow rate or pressure drop. A rotameter consists of a glass vertical cylinder with a slight taper inside and some type of float. There are graduated markings along the length of the cylinder. The gas flow comes in from the bottom and causes the float to rise proportionally to the flow rate. Appropriate curves are then used for the particular gas flowing through the rotameters to determine the flow rate. Orifice plates (see Fig. 5.33) measure the pressure drop across a known and precise hole or orifice. Using the upstream pressure and the gas temperature and properties, the gas flow rate can then be calculated.

Figure 5.33 Orifice plate flow measurement system from 40 CFR 60 Method 2D. (From Ref. 2.)

5.4.1.3 Calculation

This is a simple and straightforward method for calculating the average exhaust gas flow rate, which is often of sufficient accuracy for many situations. As will be shown, this method uses a combination of measurements, assumptions, and calculations to determine the gas flow rate. There are also some variations of the technique as will be shown. However, in general, the technique involves calculating the gas flow rate based on either the measured input flow rates, on the measured stack gas composition, or a combination of the two as a check to make sure that the results make sense.

In nearly all industrial combustion systems, the fuel flow rate is measured. In many cases it must be measured in order to calculate emissions according to regulatory requirements. For example, some air permits are based on a given maximum fuel flow rate. For some applications that do not have continuous emissions monitoring systems, an emission factor may be assumed using EPA AP-42 [29]. These factors are usually calculated based on an emission rate per unit fuel input (e.g., lb $NO_2/10^6$ Btu-input). If there is a total emission limit on the plant, then the maximum total fuel flow rate can be calculated. This type of permit often requires regular reports on the fuel consumption that is used to calculate the estimated emissions.

The combustion air flow rate is also sometimes measured. Together with the fuel flow rate, the estimated exhaust gas flow rate can be calculated.

Example 5.7

Given:	Measured methane gas flow rate of 10,000 scfh, air flow rate of 115,000 scfh, and exhaust gas temperature of 2000°F (1366 K).
Find:	The estimated exhaust gas flow rate in both scfh and acfh.
Solution:	Assume that there is no air leakage into the furnace and no exhaust gases leaking out of the furnace.

$$10,000CH_4 + 115,000(0.21O_2 + 0.79N_2) = 10,000CO_2 + 20,000H_2O + 4150O_2 + 90,850N_2$$

Then the total scfh $= 10,000 + 20,000 + 4150 + 90,850 = 125,000$ scfh

The total acfh $= 125,000 \times [(2000 + 460)/(70 + 460)] = 580,189$ acfh.

The above example shows how assumptions, measurements, and calculations can be used to calculate the exhaust gas flow rate.

The above assumption can be eliminated by using an O_2 measurement in the exhaust stack to estimate air infiltration into the furnace (assuming that the measured wet O_2 is higher than the calculated wet O_2).

Example 5.8

Given:	In addition to the measurements from the previous example, the measured wet O_2 in the stack is 4%.
Find:	The estimated exhaust gas flow rate in both scfh and acfh.
Solution:	Based on the measured fuel and air measurements and assuming no air leakage into the furnace, then the calculated wet O_2 would be:

4150 scfh/125,000 scfh $= 3.32\%$ (since the measured wet O_2 is 4% then there must be air infiltration)

$$10,000CH_4 + x(0.21O_2 + 0.79N_2) = 10,000CO_2 + 20,000H_2O + (0.21x - 20,000)O_2 + 0.79xN_2$$

$4\% = 0.04 = (0.21x - 20,000)/[30,000 + (0.21x - 20,000) + 0.79x]$

Solving for x: $x = 120,000$ scfh

Then the total scfh $= 10,000 + 20,000 + 5200 + 94,800 = 130,000$ scfh

The total acfh $= 130,000 \times [(2000 + 460)/(70 + 460)] = 603,396$ acfh.

Note that the previous example also applies when the combustion air flow rate is not measured, but the wet O_2 is measured. The same technique also applies if the dry O_2 is measured instead of the wet O_2. The only difference is that the water is deleted from the combustion products as illustrated in the next example.

Example 5.9

Given:	Same as the previous example, except the measured *dry* O_2 in the stack is 3.5%, and the combustion air flow is not measured.
Find:	The estimated exhaust gas flow rate in both scfh and acfh.
Solution:	Based on the measured fuel and air measurements and assuming no air leakage into the furnace, then the calculated wet O_2 would be:

$$10,000CH_4 + x(0.21O_2 + 0.79N_2) = 10,000CO_2 + 20,000H_2O + (0.21x - 20,000)O_2 + 0.79xN_2$$

$3.5\% = 0.035 = (0.21x - 20,000)/[10,000 + (0.21x - 20,000) + 0.79x]$

Solving for x: $x = 112,286$ scfh

Then the total scfh $= 10,000 + 20,000 + 3580 + 88,706 = 122,286$ scfh
The total acfh $= 122,286 \times [(2000 + 460)/(70 + 460)] = 567,592$ acfh.

While measuring the combustion air flow may not be necessary to calculate the exhaust gas flow, measuring only the O_2 in the stack (either wet or dry) only tells us how much air has gotten into the furnace. It does not tell where it came from so it is possible that the burners are running fuel rich with a large amount of air infiltration. Therefore, it is usually desirable to measure the combustion air flow to ensure the operating conditions of the burners.

5.4.1.4 Optical

Optical sensing systems are available for measuring the exhaust gas flow rate. These are similar to direct measurement techniques except that those are normally thermal or mechanical, as opposed to optical. A schematic of an optical flow measurement system is shown in Fig. 5.34. A photograph of an optical sensor is shown in Fig. 5.35. Another optical sensor is shown in Fig. 5.36. At this time, the method is in its infancy. It can be somewhat expensive and may involve high maintenance in a dirty environment.

5.4.2 Gas Composition

EPA Method 3 is used to determine the dry molecular weight of an exhaust gas stream. This is needed, for example, to calculate the exhaust gas flow rate used in

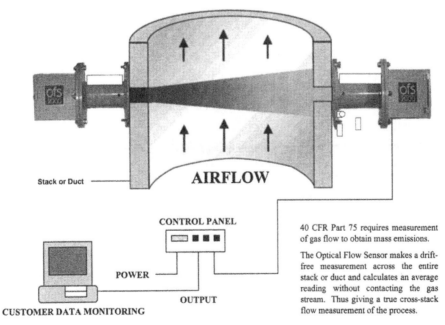

Stack or Duct

AIRFLOW

CONTROL PANEL

40 CFR Part 75 requires measurement of gas flow to obtain mass emissions.

The Optical Flow Sensor makes a drift-free measurement across the entire stack or duct and calculates an average reading without contacting the gas stream. Thus giving a true cross-stack flow measurement of the process.

POWER

OUTPUT

CUSTOMER DATA MONITORING

Figure 5.34 Schematic of an optical exhaust gas-flow sensing system. (Courtesy of OSI, Gaithersburg, MD.)

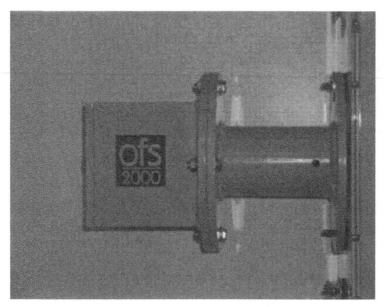

Figure 5.35 Optical flow sensor for measuring stack gas flow rates. (Courtesy of OSI, Gaithersburg, MD.)

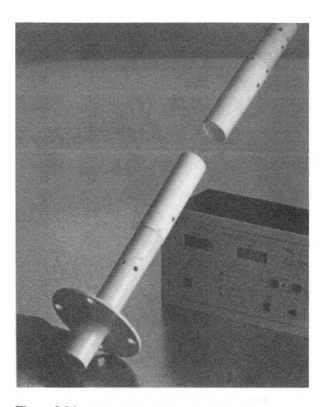

Figure 5.36 In-situ exhaust gas flow meter. (Courtesy of Monitor Labs, Denver, CO.)

calculating the pollutant emission rates. The method specifically refers to determining the CO_2 and O_2 concentrations in a gas stream. The technique uses single-point grab sampling (batch extractive sample), single-point integrated sampling, or multipoint integrated sampling. The sample is typically analyzed using an Orsat analyzer. This method is not commonly used in industrial combustion applications where instrument analyzers are typically used.

Method 3A concerns the determination of O_2 and CO_2 using instrument analyzers. A typical O_2 analyzer is shown in Fig. 5.37. This method discusses test and calibration procedures. Method 3B is specific to gas streams that contain significant quantities of gases other than O_2, CO_2, CO, and N_2 (excluding water, which can be easily removed from the stream) that could affect the gas analysis results. For example, significant quantities of SO_2 or HCl can interfere with CO_2 measurements using an Orsat analysis. Method 4 is used to determine the moisture content in an exhaust gas stream. This is important in determining the total exhaust gas volume flow rate. A gas sample is removed from the exhaust gas stream at a constant flow rate. Moisture is removed from the sample and determined either volumetrically or gravimetrically. The sample probe and transport lines are heat traced to prevent water from condensing out of the sample until it reaches the condenser, which might be an electronic device (see Fig. 5.23) or an ice bath for example. Figure 5.38 shows a schematic of the recommended sampling train used to measure the moisture content in an exhaust gas stream.

EPA Method 6 provides procedures for measuring sulfur dioxide emissions from stationary sources where the gas sample is extracted from the exhaust stack. Ammonia, water-soluble cations, and fluorides cause interferences with SO_x measurements. Method 6A concerns sulfur dioxide, moisture, and carbon dioxide measurements from fossil-fuel combustion sources by chemically separating the SO_2 and CO_2 components, where different reagent chemicals are used. Method 6C discusses the use of instrument analyzers to measure sulfur dioxide emissions from stationary sources. This is the most commonly used method in industrial combustion processes. These analyzers typically use ultraviolet (UV), nondispersive infrared (NDIR), or fluorescence techniques. A schematic of an accepted sampling system is shown in Fig. 5.39. Method 8 discusses determining sulfuric acid and sulfur dioxide emissions from stationary sources.

EPA Method 7 concerns the determination of nitrogen oxide emissions from stationary sources. This method is based on chemical separation using a reagent from a grab sample taken from the exhaust stack. Methods 7A and 7B give

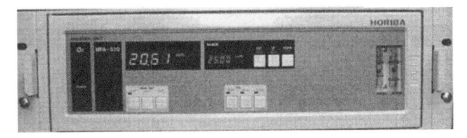

Figure 5.37 O_2 instrument analyzer.

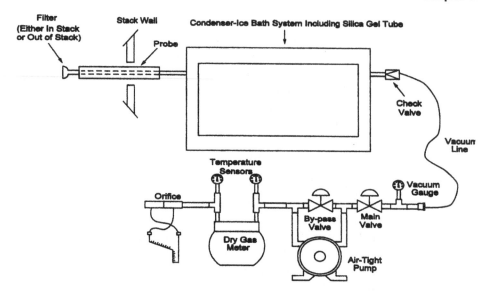

Figure 5.38 EPA Method 4 moisture sampling train—reference method. (From Ref. 2.)

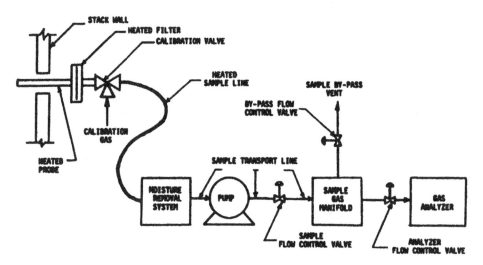

Figure 5.39 Method 6C sampling system schematic for SO_2 measurements. (From Ref. 2.)

procedures for the use of ion chromatography and UV spectrophotometry, respectively, for measuring nitrogen oxide emissions, also from a grab sample. Methods 7C and 7D discuss the use of the alkaline permanganate/calorimetric and alkaline permanganate/ion chromatographic methods, respectively, with extractive samples from the exhaust stack. Method 7E discusses the use of instrument analyzers (see Fig. 5.40) to measure NO_x emissions from gas samples continuously extracted from the exhaust stack. This is the most commonly used technique in industrial combustion applications.

EPA Method 10 discusses measuring carbon monoxide emissions from stationary sources from continuous samples extracted from an exhaust stack

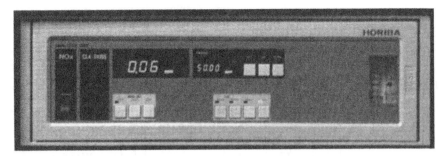

Figure 5.40 Typical NO$_x$ analyzer.

where the sample is measured with an NDIR analyzer. Possible interferences include water, carbon dioxide, and carbon monoxide. Method 10A tells us how to make certified carbon monoxide measurements from continuous emission monitoring systems at petroleum refineries.

EPA Method 18 discusses measurement techniques for determining gaseous organic compound emissions from exhaust stacks. The technique involves separating the major organic compounds from the sample and then individually quantifying the constituents using flame ionization, photoionization, electron capture, or other appropriate techniques. Diluting the sample is also used in certain cases to mitigate the effects of moisture. Method 25 is a similar process for determining the total gaseous nonmethane organic [volatile organic compounds (VOCs)] emissions from a stack. These measurements are made on a continuous gas sample with a flame ionization detector (FID). Carbon dioxide and water vapor can potentially interfere with the measurements. Figure 5.41 shows a schematic of the recommended sampling train to determining VOC emissions. Method 25A concerns measuring VOCs with a flame ionization analyzer (FIA) for determining the concentrations of samples containing primarily alkanes, alkenes, and/or arenes. Method 25B concerns the use of an NDIR analyzer to measure VOCs consisting primarily of alkanes.

EPA Method 23 discusses how to measure the emissions of dioxins and furans from stationary sources. A sample is withdrawn isokinetically and collected on a glass-fiber filter. The sample is then separated using high-resolution gas chromatography and measured by high-resolution mass spectrometry. A schematic of an approved gas sampling train is shown in Fig. 5.42. The condenser and adsorbent trap must be specially designed for this application. Specific filters, reagents, adsorbents, and sample recovery are required. The sample analysis is also extensive. These types of measurements should be done by qualified professionals to ensure high accuracy, especially because of the toxic nature of the constituents and the strict regulatory requirements.

5.4.3 Particulates

EPA Method 5 is used to determine particulate emissions from stationary sources. Method 5B refers to the determination of nonsulfuric acid particulate matter emissions from stationary sources. Method 5D concerns determining particulate matter emissions from positive pressure fabric filters, also known as baghouses, where this method has particular emphasis on the sample location. Method 5E

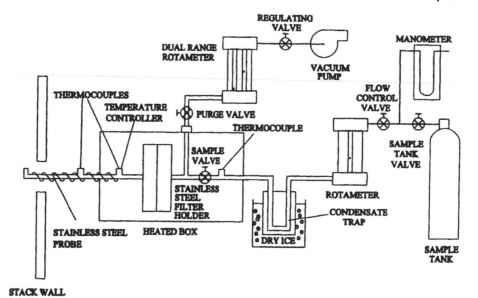

Figure 5.41 EPA Method 25 sampling train schematic for measuring volatile organic compounds. (From Ref. 2.)

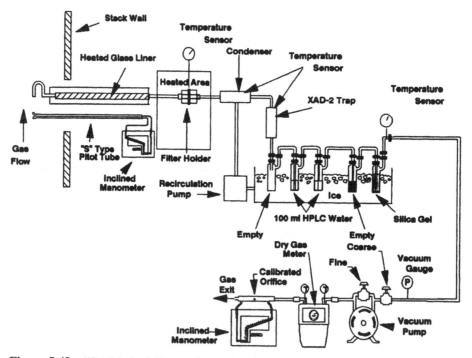

Figure 5.42 EPA Method 23 sampling train schematic for measuring dioxins and furans. (From Ref. 2.)

describes the procedure for determining particulate emissions from fiberglass wool manufacturing. Method 5F concerns nonsulfate particulate emissions from stationary sources, which includes a modified method for separating the sample from the filter. Method 5I concerns the determination of low concentrations of particulate emissions from stationary sources. This method differs from Method 5 with the following modifications: improved sample-handling procedures, use of a lightweight sample filter assembly, and the use of low-residue grade acetone to recover the particulate matter collected on the filter.

For particulate measurements, the gas sample must be withdrawn isokinetically (see previous discussion) to ensure that a representative sample of particles is withdrawn from the gas stream. If the extraction rate is too high, then the measured particulate concentration will be too high. If the extraction rate is too low, then the measured particulate concentration will be too low. A specially designed probe to sample gases from particle-laden flows is shown in Fig. 5.43. The construction of the sampling system is also important to ensure water and particulates do not come out of the extracted sample until it reaches the measurement location in the system. For example, the probe should be lined with borosilicate or quartz glass tubing and heated to prevent condensation and to prevent the particulates from coating the walls.

The extracted sample containing particulates passes through a glass-fiber filter maintained at a temperature of $120° \pm 14°C$ ($248 \pm 25°F$). A schematic of a typical particulate sampling train is shown in Fig. 5.44. The sample is extracted for a known period of time, long enough to collect a sample of sufficient mass. Moisture is removed from the particulate sample captured on the filter. The dried sample and

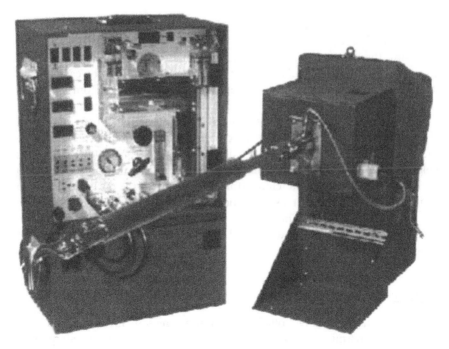

Figure 5.43 Method 5 sampling probe. (Courtesy of Millennium Instruments, Spring Grove, IL.)

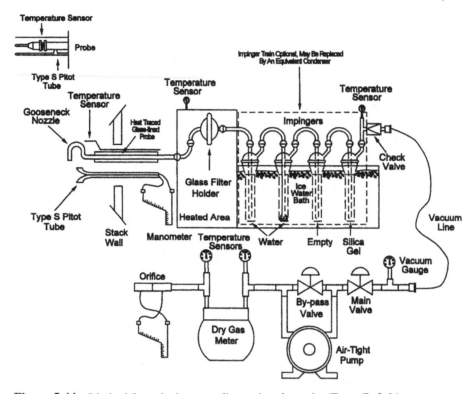

Figure 5.44 Method 5 particulate sampling train schematic. (From Ref. 2.)

filter are then weighed. This is compared to the weight of the clean filter prior to sampling. The difference in the weights then gives the weight of particulates captured. Using the measured sample extraction time, sample flow rate, and total exhaust gas flow rate, the particulate matter emission rate can then be calculated.

EPA Method 9 concerns the visual determination of opacity from emissions from stationary sources. The opacity may be caused by particulates and/or by steam. This method requires a certified observer trained in the measurement technique. An alternative technique known as LIDAR based on light backscattering can be used to determine opacity remotely as described in Alternate Method 1. Method 18 discusses a particulate matter measurement technique that can be used in lieu of Method 5 when it is known that there are no temperature effects so that the measurements can be made directly in the exhaust stack. Method 22 discusses the visual determination of smoke emissions from flares. This method does not require a certified observer as in Method 9 as the opacity is not quantified. It is merely a visual determination if smoke is present from flaring.

5.5 SOURCE TESTING

Source testing involves making "official" pollution emission measurements used either to determine permit levels or to determine compliance with an air permit. Often an outside third-party company specializing in making these measurements is hired. These companies usually must be certified by a recognized agency to make these measurements. The company will usually bring a truck to the site equipped with the appropriate measurement and calibration equipment. This source testing must be done in applications where there is no certified continuous emission monitoring system (CEMS). It may also be done even where there is a CEMS as an independent verification. It is also used by equipment vendors trying to certify new technology. For example, a burner manufacturer may hire a firm to make official measurements to demonstrate reduced pollution emissions compared to other technologies. Figure 5.45 shows an example of measurements being made from an exhaust stack.

The frequency of source testing depends on a number of factors. Measurements are usually made when an air permit is being established for the first time and whenever significant modifications are made to an existing permitted system. Annual verification tests may also be required, especially where no emissions monitoring system exists. These verification tests may not require "official" third-party measurements, depending on the rules of the given governing agency. An inspector may make unannounced spot checks with a portable analyzer to verify compliance, which is discussed in Sec. 5.7.

An alternative to source testing is to measure the emissions of the burner in a test facility (see Fig. 5.46). Figure 5.47 shows a pilot-scale test furnace that could be used for regulatory source monitoring. While this is not always permissible, depending on the rules of the local regulatory agency, it may be used to establish permit limits. It may also be used in lieu of expensive field source tests that may not be economically practical for smaller emission sources.

Figure 5.45 Measuring emissions from an exhaust stack.

Figure 5.46   John Zink R&D Test Center. (Courtesy of John Zink Co.)

Figure 5.47   Test furnace to simulate a section of an ethylene cracking furnace. (Courtesy of John Zink Co.)

5.6 CONTINUOUS MONITORING

Depending on regulations and the size of the emission source, some industrial combustion processes may be required to have a continuous emissions monitoring system. For example, some air permits may be written in such a way as to limit emissions over some time interval, such as a 24-hr rolling average. Another common limit is that emissions of a particular pollutant may not exceed a given concentration at any time. There are many other variations as well. In order to determine whether a plant is in compliance, a certified continuous emission monitoring system may be required. These are not only fairly expensive to purchase, but also have rigorous and ongoing maintenance and calibration requirements to ensure high accuracy and reliability. This is compared to other pollutant emission sources that may only require an annual test. In some cases, only an initial test may be required after the equipment is first started or after any significant modifications have been made.

The U.S. EPA has developed a series of performance specifications (PS) for continuous emissions monitoring systems (see Fig. 5.48) for measuring a wide range of pollutant emissions [30]. These specifications are listed in Table 5.3. Note that two of the specifications (4 and 4A) have identical titles. PS-4 is for the general measurement of CO emissions while PS-4A is for measuring lower concentrations

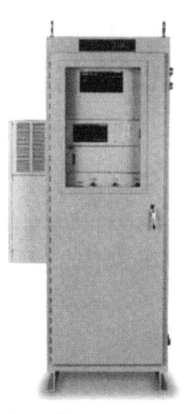

Figure 5.48 Continuous emission monitoring system. (Courtesy of Horiba, Irvine, CA.)

Table 5.3 EPA Performance Specifications for Continuous Emission Monitoring Systems

Specification #	Specification title
1	Specifications and test procedures for continuous opacity monitoring systems in stationary sources
2	Specifications and test procedures for SO_2 and NO_x continuous emission monitoring systems for stationary sources
3	Specifications and test procedures for O_2 and CO_2 continuous emission monitoring systems for stationary sources
4	Specifications and test procedures for carbon monoxide continuous emission monitoring systems for stationary sources
4A	Specifications and test procedures for carbon monoxide continuous emission monitoring systems for stationary sources
4B	Specifications and test procedures for carbon monoxide and oxygen continuous emission monitoring systems for stationary sources
5	Specifications and test procedures for TRS (total reduced sulfur) continuous emission monitoring systems for stationary sources
6	Specifications and test procedures for continuous emission rate monitoring systems for stationary sources
7	Specifications and test procedures for hydrogen sulfide continuous emission monitoring systems for stationary sources
8	Specifications and test procedures for volatile organic compound continuous emission monitoring systems for stationary sources
8A	Specifications and test procedures for total hydrocarbon continuous emission monitoring systems for stationary sources
9	Specifications and test procedures for gas-chromatographic continuous emission monitoring systems for stationary sources
15	Specifications and test procedures for extractive FTIR (Fourier transform infrared) continuous emission monitoring systems for stationary sources

Source: Ref. 2.

(< 200 ppmv) of CO emissions. These are to be used in conjunction with the EPA methods discussed above. The performance specifications include discussions of calibration procedures and relative accuracy requirements. The EPA has also developed quality assurance procedures to be used for compliance determination [31].

Parker lists the following factors as important in a continuous monitoring program [32]:

1. Knowledge of the process to be monitored.
2. Knowledge of the relevant regulatory requirements.
3. Using the correct monitors.
4. Using appropriately trained personnel with the proper calibration, quality control, and quality assurance procedures.
5. Relevant data collection, review, and reporting.
6. Ongoing maintenance program.
7. Comprehensive cost tracking, control, and budgeting.

The three major components of a continuous emissions monitoring system are: (1) the sampling and conditioning system, (2) the gas analyzers and monitors, and (3) the data-acquisition system (DAS). The first two have been discussed above. The DAS may be as simple as manually recording emissions data at scheduled and regular time intervals. It may be as sophisticated as a completely automated computer-recorded data management system. It may also be some combination of the two where some of the data are manually recorded while other data are automatically recorded by a computer or some similar type of electronic device. For example, some of the emissions such as particulates may not be easily collected on a continuous basis while other emissions such as NO_x are easily collected through continuous monitoring.

5.7 SPOT CHECKING

Spot checks may be made by the firm owning the industrial combustion equipment or by inspectors checking compliance with air permits. Portable analyzers are usually used to make spot check measurements of exhaust gas emissions. Figure 5.49 shows an example of a portable kit used for making field measurements. The kit includes the analyzer, the sample probe, a sample conditioner, and the appropriate tubing and connectors for the system. This can

Figure 5.49 Field kit for making gas emission measurements. (Courtesy of ETA, Garner, NC.)

be easily transported to a site and quickly assembled for portable measurements. Figure 5.50 shows an example of a similar portable measurement analysis system. Figure 5.51 shows a more sophisticated "portable" analyzer, which will generally have higher accuracy and repeatability compared to other smaller analyzer systems. However, it is not nearly as portable and is less convenient, for example, for carrying on airplanes to remote sites.

Figure 5.50 Portable gas analyzer. (Courtesy of ECOM, Gainsville, GA.)

Figure 5.51 Portable gas emission measurement system. (Courtesy of APEX, Holly Springs, NC.)

REFERENCES

1. JT Yeh. Modeling Atmospheric Dispersion of Pollutants. In PN Cheremisinoff (ed.). *Air Pollution Control and Design for Industry*. New York: Marcel Dekker, 1993.
2. EPA. 40 CFR 60—Standards of Performance for New Stationary Sources. Washington, DC: U.S. Environmental Protection Agency, 2001.
3. http://www.access.gpo.gov/nara/cfr/cfrhtml_00/Title_40/40cfr60a_main_00.html
4. AJ Wheeler, AR Ganji. *Introduction to Engineering Experimentation*. New York: Prentice Hall, 1995.
5. GW Cobb. *Introduction to Design and Analysis of Experiments*. New York: Springer-Verlag, 1998.
6. JP Holman. *Experimental Methods for Engineers*, 7th edn. New York: McGraw-Hill, 2000.
7. DC Montgomery. *Design and Analysis of Experiments*, 5th edn. New York: John Wiley, 2000.
8. WJ Diamond. *Practical Experiment Designs: For Engineers and Scientists*, 3rd edn. New York: John Wiley, 2001.
9. BJ Finlayson-Pitts, JN Pitts. *Atmospheric Chemistry: Fundamentals and Experimental Techniques*. New York: John Wiley, 1986.
10. M Crawford. *Air Pollution Control Theory*. New York: McGraw-Hill, 1976.
11. United Nations (ed.). *Concepts and Methods of Environment Statistics: Statistics of the Natural Environment*. New York: United Nations Publications, 1992.
12. O Newman, A Foster (eds). *Environmental Statistics Handbook: Europe*. Farmington Hills, MI: Gale Group, 1993.
13. GP Patil, CR Rao (eds.). *Multivariate Environmental Statistics*. New York: Elsevier Science, 1994.
14. CR Rao, GP Patil (eds.). *Environmental Statistics*. New York: Elsevier Science, 1994.
15. WR Ott. *Environmental Statistics and Data Analysis*. Boca Raton, FL: CRC Press, 1995.
16. AR Hoshmand. *Statistical Methods for Environmental and Agricultural Sciences*, 2nd edn. Boca Raton, FL: CRC Press, 1997.
17. NT Kottegoda, R Russo. *Statistics, Probability and Reliability Methods for Civil and Environmental Engineers*. New York: McGraw-Hill, 1997.
18. WW Piegorsch, LH Cox, D Nychka (eds.). *Case Studies in Environmental Statistics*. New York: Springer-Verlag, 1998.
19. Novartis Foundation. *Environmental Statistics: Analysing [sic] Data for Environmental Policy*. New York: John Wiley, 1999.
20. BF Manly. *Statistics for Environmental Science and Management*. Boca Raton, FL: CRC Press, 2000.
21. PM Berthouex, LC Brown. *Statistics for Environmental Engineers*, 2nd edn. Boca Raton, FL: Lewis Publishers, 2002.
22. J Townsend. *Practical Statistics for Environmental and Biological Scientists*. New York: John Wiley, 2002.
23. JP Lodge (ed.). *Methods of Air Sampling and Analysis*, 3rd edn. Chelsea, MI: Lewis Publishers, 1977.
24. DC Mussatti (ed.). *EPA Air Pollution Control Cost Manual*, 6th edn. Rep. EPA/452/B-02-001. Washington, DC: U.S. Environmental Protection Agency, Jan. 2002.
25. DC Mussatti, M Groeber, D Maloney, W Koucky, PM Hemmer. Section 2: Generic Equipment and Devices, Chap. 4: Monitors. In DC Mussatti (ed.). *Air Pollution Control Cost Manual*, 6th edn. Rep. EPA/452/B-02-001. Washington, DC: U.S. Environmental Protection Agency, Jan. 2002.
26. Office of the Federal Register. 60.2 Definitions. U.S. Code of Federal Regulations Title 40, Part 60. Washington, DC: U.S. Government Printing Office, 2001.

27. HA Becker, APG Brown. Response of Pitot probes in turbulent streams. *J. Fluid Mech.*, Vol. 62, pp. 85–114, 1974.

28. HS Cho, HA Becker. Response of static pressure probes in turbulent streams. *Exp. Fluids*, Vol. 3, pp. 93–102, 1985.

29. EPA. *Compilation of Air Pollutant Emission Factors*, Vol. I: *Stationary Point and Area Sources*, 5th edn. Washington, DC: U.S. Environmental Protection Agency Rep. AP-42, 1995.

30. EPA. Appendix B: Performance Specifications. Code of Federal Regulations 40, Part 60. Washington, DC: U.S. Environmental Protection Agency, 2001.

31. EPA. Appendix F: Quality Assurance Procedures. Code of Federal Regulations 40, Part 60. Washington, DC: U.S. Environmental Protection Agency, 2001.

32. JM Parker. Continuous Emission Monitoring (CEM). In JC Mycock, JD McKenna, L Theodore (eds.). *Handbook of Air Pollution Control and Engineering*. Boca Raton, FL: Lewis Publishers, 1995.

6

Nitrogen Oxides (NO$_x$)

6.1 INTRODUCTION

Most of the world's nitrogen occurs naturally in the atmosphere as an inert gas contained in air, which consists of approximately 78% N$_2$ by volume. NO$_x$ refers to oxides of nitrogen. These generally include nitrogen monoxide also known as nitric oxide (NO) and nitrogen dioxide (NO$_2$). They may also include nitrous oxide (N$_2$O), also known as laughing gas, as well as other less common combinations of nitrogen and oxygen such as nitrogen tetroxide (N$_2$O$_4$) and nitrogen pentoxide (N$_2$O$_5$). The EPA defines nitrogen oxides as "all oxides of nitrogen except nitrous oxide" [1].

In most high-temperature heating applications, the majority of the NO$_x$ exiting the exhaust stack is in the form of nitric oxide (NO) [2]. NO is a colorless gas that rapidly combines with O$_2$ in the atmosphere to form NO$_2$ (see Fig. 6.1 [3]). Nitric oxide is poisonous to humans and can cause irritation of the eyes and throat, tightness of the chest, nausea, headache, and gradual loss of strength. Prolonged exposure to NO can cause violent coughing, difficulty in breathing and cyanosis and could be fatal. It is interesting to note that *Science* magazine named NO as its 1993 Molecule of the Year. The reason is that NO is absolutely essential in human physiology. A growing body of research indicates its importance in everything from aiding digestion and regulating blood pressure to acting as a messenger in the nervous system. It is also a promising drug in the treatment of persistent pulmonary hypertension, which is a life-threatening lung condition affecting about 4000 babies each year. Louis Ignarro, Ferid Murad, and Robert Furchgott won the 1998 Nobel Prize in Physiology or Medicine for their discovery that NO is a signaling molecule in the cardiovascular system [4]. Nitric oxide is a precursor to the formation of NO$_2$ and an active compound in ozone formation. It is the predominant NO$_x$ compound formed in combustion processes and then reacts in the atmosphere to form NO$_2$. Hori et al. [5] have shown experimentally in a flow reactor that ethylene and propane are more effective at oxidizing NO to NO$_2$ while methane is less effective.

Nitrogen dioxide (NO$_2$) is a reddish-brown, highly reactive gas that has a suffocating odor and is a strong oxidizing agent. It is highly toxic and hazardous because of its ability to cause delayed chemical pneumonitis and pulmonary edema. Nitrogen dioxide vapors are a strong irritant to the pulmonary tract. Inhalation may also cause irritation of the eyes and throat, tightness of the chest, headache, nausea, and gradual loss of strength. Severe symptoms may be delayed and include cyanosis,

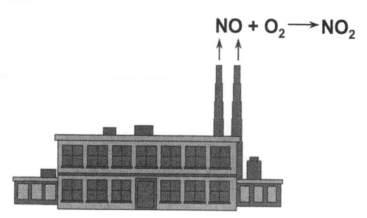

Figure 6.1 Diagram of NO exiting a stack and combining with O_2 to form NO_2. (From Ref. 3. Courtesy of John Zink Co, Tulsa, OK.)

increased difficulty in breathing, irregular respiration, lassitude, and possible death due to pulmonary edema. Chronic or repeated exposure to NO_2 can cause a permanent decrease in pulmonary function. Intermittent low-level NO_2 exposure may also induce kidney, liver, spleen, and red blood cell and immune system alterations [6]. Exposure to typical NO_2 concentrations found in the atmosphere has not proven to cause lung problems. Nitrogen dioxide is present in urban air and reacts with air to form corrosive nitric acid and toxic organic nitrates. It also plays a major role in the atmospheric reactions that cause ground-level ozone.

Nitrous oxide (N_2O) contributes to ozone destruction in the stratosphere and is a relatively strong greenhouse gas [7]. Although it is not an important factor in most of the high-temperature industrial combustion sources considered here, it is important at lower temperatures (705–950°C), especially when solid fuels are used. Tomeczek and Gradon [8] discuss the contribution of N_2O reactions to the formation of NO in flames.

Many combustion processes are operated at elevated temperatures and high excess air levels. The combustion products may have long residence times in the combustion chamber. These conditions produce high thermal efficiencies and product throughput rates. Unfortunately, those conditions also favor the formation of NO_x. NO_x emissions are among the primary air pollutants because of their contribution to smog formation, acid rain, and ozone depletion in the upper atmosphere. It is interesting to note that only about 5% of typical NO_x sources in an industrial region of the United States come from industrial sources, compared to 44% from highway and off-road vehicles [9]. Figure 6.2 shows typical stationary sources of NO_x emissions including the applications of specific interest here [10].

6.2 FORMATION MECHANISMS

There are three generally accepted mechanisms for NO_x production: thermal, prompt, and fuel. These are discussed next. Hill and Smoot [11] have written a detailed article on modeling the chemistry of the formation and destruction of NO_x

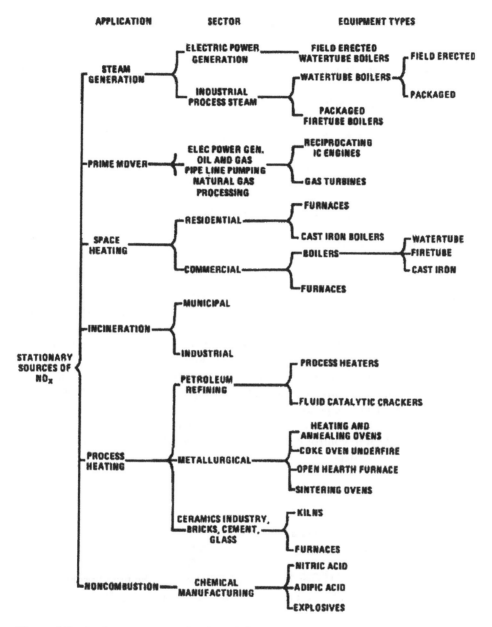

Figure 6.2 Stationary sources of NO$_x$ emissions. (From Ref. 10.)

in combustion systems, which includes information on formation mechanisms and simplifications used in modeling.

6.2.1 Thermal NO$_x$

Thermal NO$_x$ is formed by the high-temperature reaction (hence the name thermal NO$_x$) of nitrogen with oxygen, by the well known Zeldovich mechanism

[12]. It is sometimes referred to as Zeldovich NO_x [13]. It is given by the simplified reaction:

$$N_2 + O_2 \rightarrow NO, NO_2 \tag{6.1}$$

The two predominant reactions are

$$O + N_2 \leftrightarrow NO + N \quad k_f = 2 \times 10^{14} \exp[(-76{,}500 \text{ kcal/mol})/RT)] \tag{6.2}$$

$$N + O_2 \leftrightarrow NO + O \quad k_f = 6.3 \times 10^9 \exp[(-6300 \text{kcal/mol})/RT)] \tag{6.3}$$

The extended Zeldovich mechanism includes a third equation:

$$N + OH \leftrightarrow NO + H \tag{6.4}$$

Thermal NO_x increases exponentially with temperature. Above about 2000°F (1100°C), it is generally the predominant mechanism in combustion processes, making it important in most high-temperature heating applications. This means that this mechanism becomes more important when air preheating or oxygen enrichment [14] of the combustion air are used, which normally increases the flame temperature. The strategies for minimizing thermal NO_x emissions include limiting local oxygen concentrations in the vicinity of the peak flame temperatures, limiting local nitrogen concentrations in the vicinity of the peak flame temperatures, reducing the residence time at the peak flame temperatures, and/or reducing the peak flame temperatures. As can be seen, the peak flame temperature is prominent in all of these strategies. It is usually not practical to control nitrogen levels so control techniques typically focus on the other three strategies.

6.2.2 Prompt NO_x

Prompt NO_x is formed by the relatively fast reaction (hence the name prompt NO_x) between nitrogen, oxygen, and hydrocarbon radicals. It is sometimes referred to as Fenimore NO_x after C.P. Fenimore from the General Electric Research and Development Center in Schenectady, NY, who first coined the term "prompt NO_x" and showed experimentally the presence of this mechanism [15]. It is given by the overall reaction:

$$CH_4 + O_2 + N_2 \rightarrow NO, NO_2, CO_2, H_2O, \text{ species} \tag{6.5}$$

In reality, this very complicated process consists of hundreds of reactions. The hydrocarbon radicals are intermediate species formed during the combustion process. Prompt NO_x is generally an important mechanism in lower temperature combustion processes, but is generally much less important compared to thermal NO_x formation at the higher temperatures found in most industrial combustion processes. It is also generally important in very fuel-rich conditions, which are not normally encountered except under certain circumstances.

6.2.3 Fuel NO$_x$

Fuel NO$_x$ is formed by the direct oxidation of organonitrogen compounds contained in the fuel (hence the name fuel NO$_x$). It is given by the overall reaction:

$$R_xN + O_2 \rightarrow NO, NO_2, CO_2, H_2O, \text{trace species} \tag{6.6}$$

In reality, there are many intermediate reactions for this formation mechanism as indicated in Fig. 6.3. Fuel NO$_x$ is not a concern for high-quality gaseous fuels like natural gas or propane, which normally have no organically bound nitrogen. However, fuel NO$_x$ may be important when oil (e.g., residual fuel oil), coal, or waste fuels are used, which may contain significant amounts of organically bound nitrogen. Table 6.1 shows typical thermal and fuel NO$_x$ emissions for process heaters [16]. The conversion of fuel-bound nitrogen to NO$_x$ ranges from 15 to 100%. The conversion efficiency is generally higher the lower the nitrogen content in the fuel. Controlling excess oxygen is an important strategy for minimizing fuel NO$_x$ emissions.

6.3 SOURCES FOR NO$_x$

6.3.1 Oxidizer

The oxidizer is usually the largest source of both oxygen and nitrogen that are needed to form NO$_x$. Air, which is approximately 21% oxygen and 79% nitrogen by volume, is the predominant oxidizer used in industry. Most of the oxygen combines with the fuel although there is normally some excess O$_2$ to ensure complete combustion. The nitrogen does not participate in the combustion reaction and is a diluent that both absorbs heat and contributes to NO$_x$ formation. For many common fuels like natural gas, the combustion air flow rate is approximately 10 times the fuel flow rate. Therefore, there is a large volume of nitrogen available to

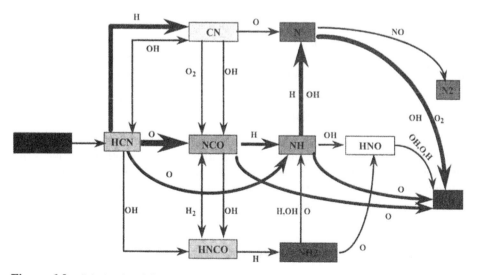

Figure 6.3 Schematic of fuel NO$_x$ formation pathways. (From Ref. 3. Courtesy of John Zink Co.)

Table 6.1 Uncontrolled NO_x Emission Factors for Typical Process Heaters

Model heater type	Uncontrolled emission factor (lb/10⁶ Btu)		
	Thermal NO_x	Fuel NO_x	Total NO_x[a]
ND, natural gas-fired[b]	0.098	N/A	0.098
MD, natural gas-fired[b]	0.197	N/A	0.197
ND, distillate oil-fired	0.140	0.060	0.200
ND, residual oil-fired	0.140	0.280	0.420
MD, distillate oil-fired	0.260	0.060	0.320
ND, residual oil-fired	0.260	0.280	0.540
ND, pyrolysis, natural gas-fired	0.104	N/A	0.104
ND, pyrolysis, high-hydrogen fuel gas-fired[c]	0.140[d]	N/A	0.140

[a]Total NO_x = Thermal NO_x + Fuel NO_x.
[b]Heaters firing refinery fuel gas with up to 50 mol% hydrogen can have upto 20% higher NO_x emissions than similar heaters firing natural gas.
[c]High-hydrogen fuel gas is fuel gas with 50 mol% or greater hydrogen content.
[d]Calculated assuming approximately 50 mol hydrogen.
N/A = Not applicable.
ND = Natural draft.
MD = Mechanical draft.
Source: Ref. 16.

form NO_x, which means that nitrogen availability is not the limiting factor. Excess oxygen is a limiting factor as will be discussed in more detail later. When the oxidizer is pure oxygen, then nitrogen availability is the limiting factor as will also be discussed later in more detail.

6.3.2 Fuel

Certain fuels contain bound nitrogen that can lead to fuel NO_x. Heavy fuel oils, coal, coke, and waste oils are examples of fuels that typically contain bound nitrogen. Light oils and natural gas may contain some bound nitrogen but it is often a very small constituent. Heavy fuel oil is the most important fuel containing bound nitrogen in industrial combustion applications.

6.3.3 Incoming Raw Materials

Bound nitrogen in the incoming raw materials can also increase NO_x emissions. One example is the niter used in the glass-manufacturing process. Another example is the waste materials fed into incinerators that may contain used oils or sewage sludge containing bound nitrogen. The incoming raw materials to a cement kiln may have up to 1000 ppm bound nitrogen [17].

6.4 IMPORTANT FACTORS AFFECTING NO_x

There are many factors that have an impact on NO_x formation. These include the oxidizer and fuel compositions and temperatures, the ratio of the fuel to the oxidizer,

Table 6.2 Hierarchy of Parameters Affecting Thermal NO_x Emissions

Primary equipment and fuel parameters	Secondary combustion parameters	Fundamental parameters
Inlet temperature, velocity	Combustion intensity	Oxygen level
Firebox design	Heat removal rate	Peak temperature
Fuel composition	Mixing of combustion	Exposure time at
Injection pattern of fuel and air	products into flame	peak temperature
Size of droplets or particles	Local fuel/air ratio	
Burner swirl	Turbulent distortion	
External mass addition	of flame zone	

Source: Ref. 18.

the burner and heater designs, the furnace and flue gas temperatures, and the operational parameters of the combustion system. Table 6.2 shows some of these parameters and their importance in the production of thermal NO_x emissions [18]. Some of these are considered next.

6.4.1 Oxidizer Composition

Figure 6.4 shows the results of those calculations for NO as a function of the oxygen concentration in the oxidizer. Nitric oxide is given in both mass (lb NO/MMBtu) (pounds per million Btu) and volume units (ppmvd), for comparison. The mass unit (lb/MMBtu) has been normalized to a unit flow rate of fuel, which also equates to a given unit of energy based on the higher heating value of the fuel. The volume unit (ppmvd) has not been corrected to any specific O_2 level. There are two competing effects that produce the parabolic shape of the NO curves. As the oxygen concentration increases, the flame temperature increases. This accelerates NO_x formation, because of the exponential dependence of the thermal NO_x reactions on temperature. However, as the O_2 concentration increases, the N_2 concentration simultaneously decreases. This lowers NO_x because less N_2 is available to make NO_x. The simultaneous effects of increasing flame temperature and reduction in the amount of N_2 produce peak NO_x values in middle ranges of oxygen enrichment.

Figure 6.5 shows NO and CO as functions of the O_2/CH_4 stoichiometry for CH_4 combusted with an oxidizer consisting of 95% O_2 and 5% N_2. This represents one type of oxygen-enhanced system. The graph shows the strong dependence of both pollutants on the stoichiometry. Under fuel rich conditions, NO decreases, while CO increases. Also, the fuel efficiency is reduced since the fuel is not fully combusted. Under fuel-lean conditions, NO increases, while CO decreases. Again, the fuel efficiency is reduced because the excess oxidizer carries heat out of the process. In order to minimize both CO and NO_x, the combustion system should be operated near stoichiometrically, which also maximizes the fuel efficiency.

Figure 6.6 shows a plot of the adiabatic equilibrium flame temperature for an air/CH_4 flame and an O_2/CH_4 flame, as functions of the flame stoichiometry.

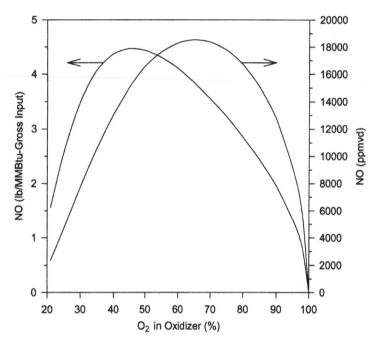

Figure 6.4 Adiabatic equilibrium NO as a function of oxidant (O_2/N_2) composition for a stoichiometric CH_4 flame. (From Ref. 14. Courtesy of CRC Press.)

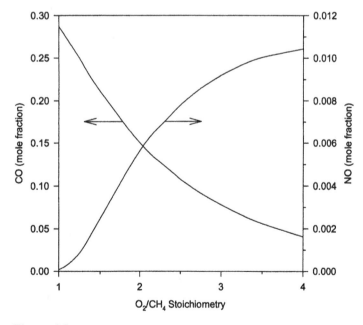

Figure 6.5 Adiabatic equilibrium NO and CO as functions of O_2/CH_4 stoichiometry for an oxidant of 95% O_2–5% N_2 and a fuel of pure CH_4. (From Ref. 14. Courtesy of CRC Press.)

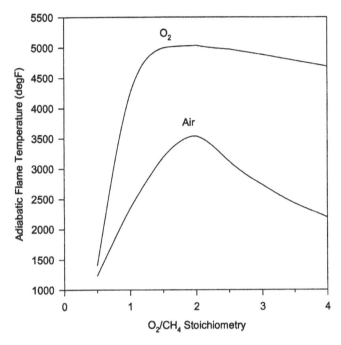

Figure 6.6 Adiabatic equilibrium flame temperature as a function of O_2/CH_4 stoichiometry for CH_4 flames and oxidants of air and pure O_2. (From Ref. 14. Courtesy of CRC Press.)

There are several things to notice. The flame temperature for the air/CH_4 flame is very dependent on the stoichiometry. For the O_2/CH_4 flame, the temperature is very dependent on the stoichiometry only under fuel-rich conditions. The temperature is not very dependent on the stoichiometry when the O_2/CH_4 flame is fuel lean.

Figure 6.6 helps to explain why, for example, NO_x is reduced dramatically under fuel-rich conditions. One reason is the dramatic reduction in the flame temperature. Another reason concerns the chemistry. In a reducing atmosphere, CO is formed preferentially to NO. This is exploited in some of the NO_x reduction techniques. An example is methane reburn [19]. The exhaust gases from the combustion process flow through a reduction zone that is under reducing conditions; NO_x is reduced back to N_2. Any CO that may have formed in the reduction zone and other unburned fuels are then combusted downstream of the reduction zone. However, they are combusted at temperatures well below those found in the main combustion process. These lower temperatures are not favorable to NO_x formation. Figure 6.7 shows the importance of the gas temperature on thermal NO_x formation. Many combustion modification strategies for reducing NO_x involve reducing the flame temperature. Bilger [20] describes the use of pure oxygen to produce zero pollutant emissions as the "oxygen economy". Howard et al. [21] did a fundamental experimental study of oxygen-enriched turbulent diffusion flames, which included measurements of NO_x emissions.

Sung and Law [22] have shown computationally that air leakage into an oxy-fuel system can significantly increase NO_x emissions. This is important to note as most industrial combustion processes are not well sealed and have the potential for

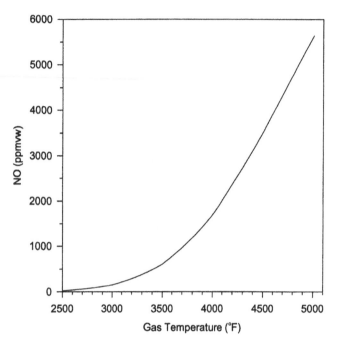

Figure 6.7 Adiabatic equilibrium NO as a function of gas temperature for an oxidant of 95% O_2–5% N_2 and a fuel of pure CH_4. (From Ref. 14. Courtesy of CRC Press.)

air leakage into the process. Some processes are run at positive pressure so that furnace gases leak out rather than ambient air leaking in. Other processes have huge air leakage as they are batch processes where large doors are open on a regular basis. For example, in aluminum melting furnaces, there may be five or more times during the process when a door is opened to charge more material into the furnace. This means that large quantities of ambient air enter the furnace.

One of the benefits of using oxygen-enhanced combustion (OEC) is that the lean flammability limit can be significantly extended, which can be used to reduce NO_x emissions, as shown by Qin et al. [23]. This was shown both experimentally and computationally for laminar premixed methane flames. The NO_x reduction is caused both by the reduced N_2 concentration as more O_2 is added to the oxidizer, and by the temperature reduction by using a more fuel-lean mixture.

6.4.2 Fuel Composition

Certain components in a fuel gas mixture may increase NO_x while others may decrease NO_x. Ren et al. [24] showed that adding hydrogen or carbon monoxide to methane increased NO_x, primarily due to an increase in the adiabatic flame temperature. They also showed that adding diluents like CO_2, N_2, or H_2O to the fuel decreased NO_x on a volumetric basis.

Figure 6.8 shows how the fuel composition affects NO for a blend of CH_4 and H_2. First, it is important to note that NO increases as the H_2 content in the blend

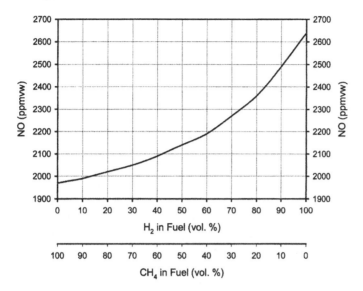

Figure 6.8 Adiabatic equilibrium NO as a function of fuel composition (CH$_4$–H$_2$) for a stoichiometric air/fuel flame. (From Ref. 3. Courtesy of John Zink Co.)

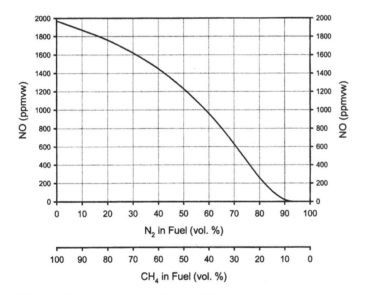

Figure 6.9 Adiabatic equilibrium NO as a function of fuel composition (CH$_4$–N$_2$) for a stoichiometric air/fuel flame. (From Ref. 3. Courtesy of John Zink Co.)

increases. This is similar to the effect on the adiabatic flame temperature, as shown in Fig. 2.25. The second thing to note is that the effect is not linear between pure CH$_4$ and pure H$_2$. The NO$_x$ increases more rapidly as the H$_2$ content increases. The third thing to notice is that there is a significant difference between the two extremes as the NO$_x$ ranges from a little less than 2000 ppmvw to a little more than 2600 ppmvw.

Figure 6.9 shows how the fuel composition affects NO for a blend of CH$_4$ and N$_2$; NO (ppmvw) drops off rapidly as the N$_2$ in the fuel blend increases.

At 100% N_2, the "fuel" produces no NO. The additional quantity of N_2 in the fuel does not increase NO_x since there is already plenty of N_2 available from the combustion air to make NO_x.

In the case where high-purity oxygen is used as the oxidizer (oxy–fuel combustion), then nitrogen in the fuel significantly impacts the NO_x emissions. Sung and Law [22] showed computationally that NO increased linearly for the combustion of pure O_2 with a fuel containing methane and nitrogen where the nitrogen ranged from 0 to 1% per cent by weight of the fuel.

The composition of the fuel supplied to a combustion system has a significant impact on the NO_x emissions. In the petrochemical and chemical process industries, there is a very wide range of fuel blends used for process heating. These fuels are often by-products from a refining process. They typically contain hydrocarbons ranging from C1 to C4, hydrogen, and inert gases like N_2 and CO_2. In a given plant or refinery, burners used in process heaters may need to be capable of firing on multiple fuels that are present at different times (e.g., start-up, normal operation, upset conditions). In many cases, the NO_x emissions from the heaters may not exceed a given value regardless of what fuel composition is being fired. Therefore, it is critical that the effects of the fuel composition on NO_x emissions be understood and quantified to ensure that permitted values are not exceeded.

This section shows the results of an extensive series of tests to study the effects of fuel composition on NO_x emissions from an industrial-scale natural draft diffusion flame burner shown in Fig. 6.10 [25]. A close up of the front of the burner

Figure 6.10   Diffusion flame John Zink Co. model VYD burner on a stand. (Courtesy of John Zink Co.)

is shown in Fig. 6.11. Figure 6.12 shows the hemispherical tip for the fuel delivery. The tip is surrounded by a cone diffuser that acts as a flame stabilizer and homogenizes the air flow. The data provide additional insight into the effects on NO_x over the entire range of fuel compositions consisting of various fractions of three primary components: H_2, C_3H_8, and CH_4. Figures 6.13, 6.14, and 6.15 show how NO_x theoretically varies for two-component fuel mixtures of CH_4–C_3H_8, CH_4–H_2, and C_3H_8–H_2, respectively. These figures show the predicted adiabatic equilibrium NO concentrations for flames with 15% excess air. Figure 6.16 shows a ternary diagram of the calculated adiabatic flame temperatures (figure on left) over the range of three-component fuel blends tested and another ternary diagram showing the predicted adiabatic equilibrium NO (figure on right) for three-component fuel blends containing CH_4, C_3H_8, and H_2 combusted with 15% excess air.

Testing was conducted using a conventional-type burner (see Fig. 6.17) with a single fuel gas tip and flameholder. The burner was fired vertically upward in a rectangular furnace (see Fig. 6.18). The test furnace was a rectangular heater 8 ft (2.4 m) wide, 12 ft long (3.7 m), and 15 ft (4.6 m) tall. The furnace was cooled by a water jacket on all four walls. The interior of the water-cooled walls was covered with varying layers of refractory lining to achieve the desired furnace temperature. The burner was tested at a nominal heat release rate of 7.5×10^6 Btu/hr (2.2 MW).

A velocity thermocouple (also known as a suction thermocouple or suction pyrometer) was used to measure the furnace and stack gas temperatures. The furnace

Figure 6.11 Close-up of John Zink Co. model VYD burner. (Courtesy of John Zink Co.)

Figure 6.12 Close-up of tip and cone diffuser for John Zink Co. VYD burner. (Courtesy of John Zink Co.)

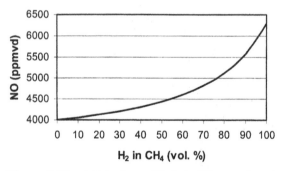

Figure 6.13 Adiabatic equilibrium NO as a function of the fuel-blend composition for H_2–CH_4 blends combusted with 15% excess air where both the fuel and the air are at ambient temperature and pressure. (From Ref. 3. Courtesy of John Zink Co.)

draft was measured with an automatic, temperature-compensated, pressure transducer as well as an inclined manometer connected to a pressure tap in the furnace floor. Fuel flow rates were measured using calibrated orifice meters, fully corrected for temperature and pressure. Emission levels were measured using state-of-the-art continuous emissions monitors (CEMs) to measure emissions species concentrations of NO_x, CO, and O_2.

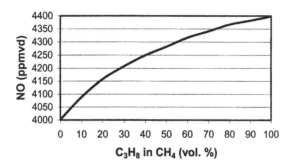

Figure 6.14 Adiabatic equilibrium NO as a function of the fuel-blend composition for C_3H_8-CH_4 blends combusted with 15% excess air where both the fuel and the air are at ambient temperature and pressure. (From Ref. 3. Courtesy of John Zink Co.)

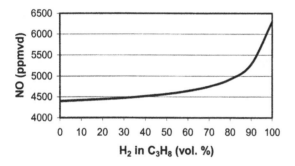

Figure 6.15 Adiabatic equilibrium NO as a function of the fuel-blend composition for H_2–C_3H_8 blends combusted with 15% excess air where both the fuel and the air are at ambient temperature and pressure. (From Ref. 3. Courtesy of John Zink Co.)

The experimental matrix consisted of firing the burner at a constant heat release (7.5×10^6 Btu/hr or 2.2 MW) and excess air level (15%) with 15 different fuel blends consisting of varying amounts of H_2, C_3H_8, and Tulsa Natural Gas (TNG).* For testing and analysis purposes, TNG was treated as a single fuel component for convenience. TNG, which consists of approximately 93% CH_4, is a more economical choice than pure CH_4 for experimental work and the analysis is simplified by treating it as a single component. All 15 fuel compositions were tested on each of six different fuel gas tips, which differed in port diameter sizes, to enable the acquisition of additional information regarding effects resulting from differing fuel pressures.

Figure 6.19 shows the variation in relative measured NO_x emissions resulting from different concentrations (volume basis) of H_2 in a fuel blend composed with a balance of TNG for each of the six different fuel gas tips tested. The plot, which

---

*The nominal composition by volume of TNG is: 93.4% CH_4, 2.7% C_2H_6, 0.60% C_3H_8, 0.20% C_4H_{10}, 0.70% CO_2, and 2.4% N_2.

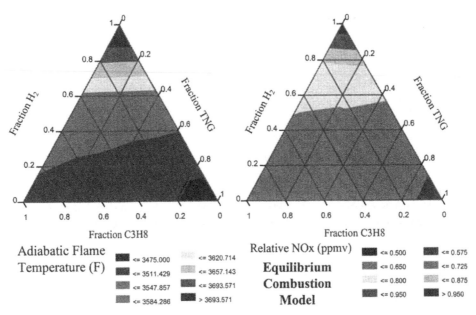

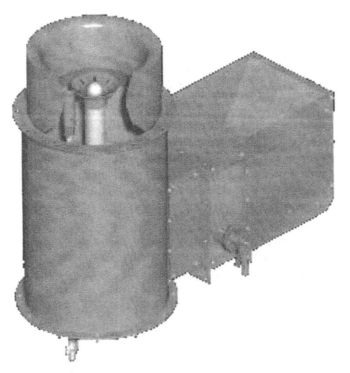

Figure 6.16 Ternary plot of adiabatic equilibrium NO (fraction of the maximum value) as a function of the fuel-blend composition for H_2–CH_4–C_3H_8 blends combusted with 15% excess air where both the fuel and the air are at ambient temperature and pressure. (From Ref. 3. Courtesy of John Zink Co.)

Figure 6.17 Raw gas (VYD) burner. (From Ref. 3. Courtesy of John Zink Co.)

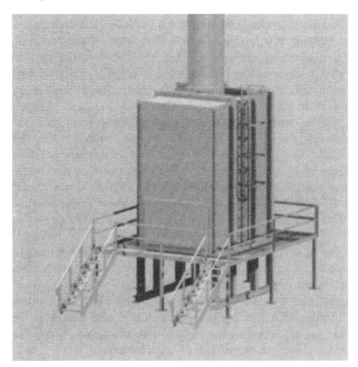

Figure 6.18 Test furnace. (From Ref. 3. Courtesy of John Zink Co.)

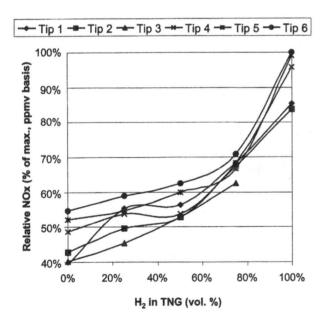

Figure 6.19 Measured NO$_x$ (percentage of the maximum ppmv value) as a function of the fuel-blend composition for H$_2$–TNG blends combusted with 15% excess air where both the fuel and the air were at ambient temperature and pressure. (From Ref. 3. Courtesy of John Zink Co.)

illustrates NO$_x$ levels on a concentration basis, clearly shows the correlation between increased H$_2$ content and higher NO$_x$ emission levels. The slope of the profile is exponentially increasing, qualitatively similar to that predicted by the plotted theoretical calculations shown previously in Fig. 6.13. The effect of H$_2$ is significant, with the sharpest increase in NO$_x$ levels taking place as concentration levels of H$_2$ in the fuel mixture rise from 75 to 100%.

The variation in relative measured NO$_x$ emissions resulting from different concentrations (volume basis) of C$_3$H$_8$ in a fuel blend composed with a balance of TNG is shown in Fig. 6.20. The slope of the increase in NO$_x$ levels corresponding to increased concentrations of C$_3$H$_8$ is shown to be relatively constant or slightly declining over the gradient in C$_3$H$_8$ concentration, in contrast with the exponentially increasing profile of the H$_2$–TNG plot in Fig. 6.19. The profile showing the effect of C$_3$H$_8$ content is also seen to be similar to the corresponding calculated trends shown previously in Fig. 6.14.

Figure 6.21 shows the final two-component fuel blend results being examined, which describe the variation in relative measured NO$_x$ emissions resulting from different concentrations (volume basis) of H$_2$ in a fuel blend composed with a balance of C$_3$H$_8$. The upper plot, which shows measured relative NO$_x$ on a volume concentration basis, illustrates that for a given tip geometry and port size the measured NO$_x$ concentrations actually decrease slightly with increasing H$_2$ content up to 75% H$_2$ content, then sharply increase with H$_2$ concentration.

Due to the decrease in total dry products of combustion from the burning of H$_2$, expressing NO$_x$ in terms of concentration (ppmv) does not fully represent the

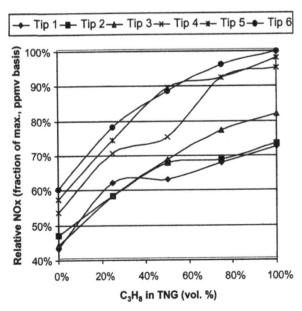

Figure 6.20   Measured NO$_x$ (percentage of the maximum ppmv value) as a function of the fuel-blend composition for C$_3$H$_8$–TNG blends combusted with 15% excess air where both the fuel and the air were at ambient temperature and pressure. (From Ref. 3. Courtesy of John Zink Co.)

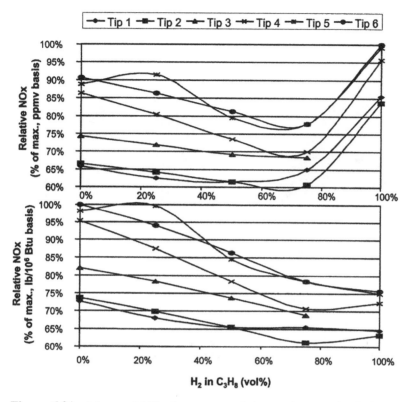

Figure 6.21 Measured NO$_x$ (percentage of the maximum value in both ppmv and lb/MMBtu) as a function of the fuel-blend composition for H$_2$–C$_3$H$_8$ blends combusted with 15% excess air where both the fuel and the air were at ambient temperature and pressure. (From Ref. 3. Courtesy of John Zink Co.)

actual mass rate of NO$_x$ emissions produced. The lower plot, which shows the variation in measured NO$_x$ levels on a mass per unit heat release basis, illustrates that the overall emissions of NO$_x$ on a mass basis decrease with increasing fuel hydrogen content and continue to decrease or remain relatively flat even in the high-hydrogen content region, which produced a sharp increase in NO$_x$ levels on a volume concentration basis.

Three-component interaction results were also examined by considering results from several of the tested fuel gas tip designs. Figures 6.22, 6.23, and 6.24 show contoured ternary plots for tip designs 2, 4, and 6, respectively, of variation in relative measured NO$_x$ levels corresponding to different fractions of H$_2$, C$_3$H$_8$, and TNG in the fuel blend. Plots for three tip designs are shown, the tips differing only in fuel port area size, which results in different fuel pressures for a given heat release on each tip. The results are shown for tips in order of increasing port area size, or in other words, decreasing fuel pressure levels for the design heat release. Two plots are shown for each of three tips, with one illustrating NO$_x$ levels on a volume concentration basis and the other illustrating NO$_x$ levels on a mass per unit heat-release basis.

For each given tip, the highest NO$_x$ emissions on a concentration basis occur in the high-hydrogen content region, while the highest NO$_x$ emissions on a mass

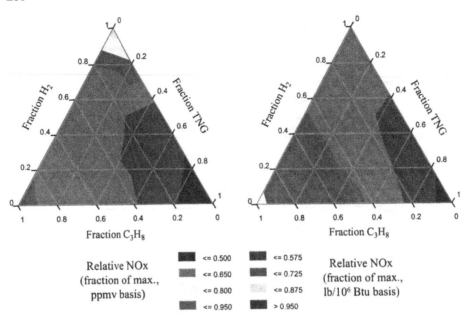

Figure 6.22 Measured NO$_x$ (fraction of the maximum value in both ppmv and lb/MMBtu) as a function of the fuel-blend composition for TNG–H$_2$–C$_3$H$_8$ blends combusted with 15% excess air where both the fuel and the air were at ambient temperature and pressure for gas tip #2. (From Ref. 3. Courtesy of John Zink Co.)

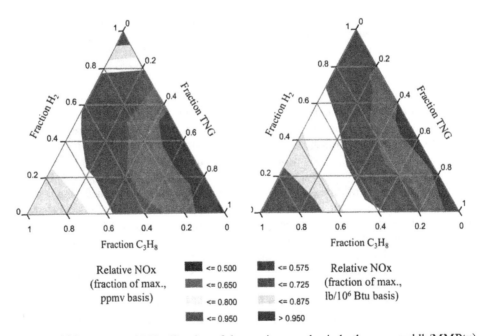

Figure 6.23 Measured NO$_x$ (fraction of the maximum value in both ppmv and lb/MMBtu) as a function of the fuel-blend composition for TNG–H$_2$–C$_3$H$_8$ blends combusted with 15% excess air where both the fuel and the air were at ambient temperature and pressure for gas tip #4. (From Ref. 3. Courtesy of John Zink Co.)

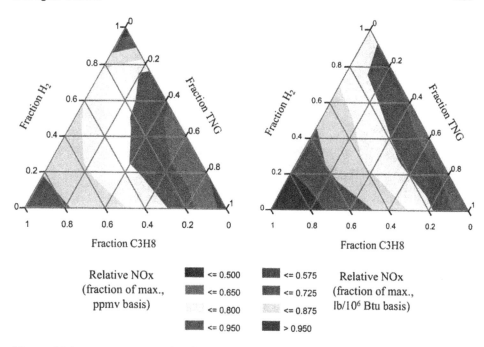

Figure 6.24 Measured NO_x (fraction of the maximum value in both ppmv and lb/MMBtu) as a function of the fuel-blend composition for TNG–H_2–C_3H_8 blends combusted with 15% excess air where both the fuel and the air were at ambient temperature and pressure for gas tip #6. (From Ref. 3. Courtesy of John Zink Co.)

per unit heat-release basis occur in the high-propane region. The contoured gradients illustrate the interaction of the three fuel components and how each of the components affects NO_x emission in different regions of the fuel mixture, such as the steep NO_x concentration gradients in the high-hydrogen content regions. The effect of C_3H_8 content can be seen to dominate the NO_x level gradients on a mass per unit heat-release basis with a relatively constant slope. It is also interesting to note that NO_x levels overall appear to increase as fuel gas tips change from having a less open fuel port area (higher fuel pressures for a given heat release) to having a greater open fuel port area (lower fuel pressure for a given heat release).

Figure 6.25 shows ternary plots of fuel composition effects on NO_x at a nominal constant fuel pressure of 21 psig (145 kPa g). This analysis, made possible by testing a range of fuel gas tips, enables the examination of fuel composition effects on NO_x emissions relatively independently from fuel pressure variations. A qualitative comparison of the plot on the left with the theoretical plots previously shown in Fig. 6.16 reveals that, on a volume concentration basis, the change in NO_x level as a function of fuel composition, for a relatively constant pressure and constant heat release, varies similarly to the trends predicted by the adiabatic flame temperature variation and predicted relative NO_x concentrations from the equilibrium combustion model over the same regions. This result is expected due to the well-established correlation of the dependence of thermal NO_x formation on flame temperature. The mass basis plot on the right in Fig. 6.25 shows that variation in NO_x levels with fuel composition, from a constant fuel pressure perspective, are

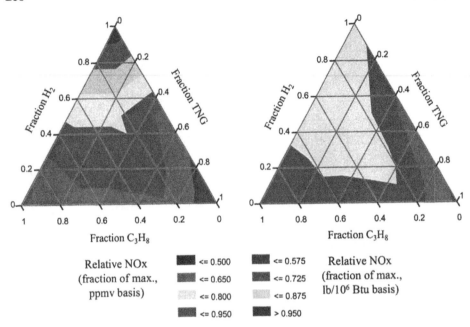

Figure 6.25 Measured NO_x (fraction of the maximum value in both ppmv and lb/MMBtu) as a function of the fuel-blend composition for $TNG–H_2–C_3H_8$ blends combusted with 15% excess air where both the fuel and the air were at ambient temperature and pressure for a constant fuel gas pressure of 21 psig. (From Ref. 3. Courtesy of John Zink Co.)

less severe than seen in the analysis of a single fuel gas tip with fixed port sizes, for which fuel pressures may vary greatly to maintain a given heat release with fuel composition variation.

From both the two- and three-component analyses it is evident that fuel pressure has a significant affect on NO_x emission levels. Figure 6.26 shows a plot of relative NO_x levels versus fuel pressure for each of the 15 different fuels tested. This plot shows a consistent decrease in NO_x levels correlated with an increase in fuel pressure. This phenomenon is explained by the burner configuration, which allows significant amounts of inert flue gas to be entrained into the flame zone with increasing fuel jet momentum, thus decreasing thermal NO_x formation.

Figure 6.27 shows an overall view of the data collected from all six tips with each of the 15 different fuel compositions (90 data points in total) from both a NO_x volume concentration basis and mass per unit heat-release viewpoint. The plots use fuel pressure and adiabatic flame temperatures as the primary axes to illustrate usefully some overall trends. The plot of relative NO_x concentration levels shows the minimum NO_x levels occur in the region with the lowest adiabatic flame temperature and highest fuel pressures. Inversely, the highest NO_x concentration levels are found in the region of high adiabatic flame temperatures and low fuel pressures, when high concentrations of hydrogen are present. The mass per unit heat-release NO_x levels are also at a minimum in the same region as the concentration-based profiles; however, the maximum NO_x levels, when measured on a mass basis, are not found in the same region, but occur in areas of lowest fuel pressures, with a mildly elevated adiabatic flame temperature, which correspond to high C_3H_8 concentration regions.

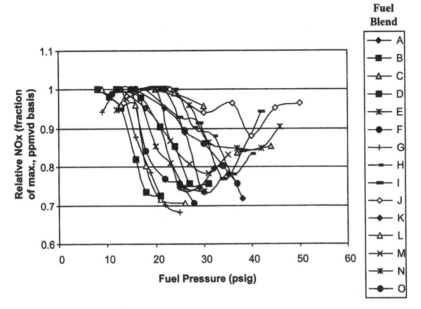

Figure 6.26 Measured NO$_x$ (fraction of the maximum value in ppmvd) as a function of the fuel pressure for all 15 different TNG–H$_2$–C$_3$H$_8$ blends (A–O) combusted with 15% excess air where both the fuel and the air were at ambient temperature and pressure. (From Ref. 3. Courtesy of John Zink Co.)

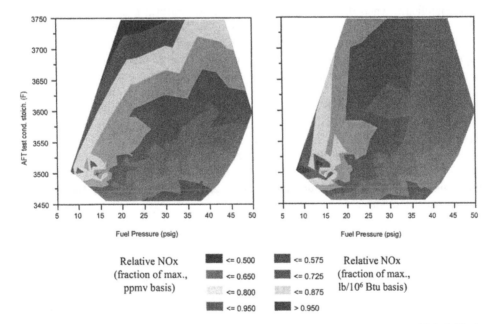

Figure 6.27 Measured NO$_x$ (fraction of the maximum value in both ppmv and lb/MMBtu) as a function of the fuel blend composition, fuel gas pressure and calculated adiabatic flame temperature for TNG–H$_2$–C$_3$H$_8$ blends combusted with 15% excess air where both the fuel and the air were at ambient temperature and pressure. (From Ref. 3. Courtesy of John Zink Co.)

These overall trends concur with the previously discussed results and agree with the correlations shown by the three- and two-component interaction analyses.

Adiabatic flame temperature and fuel pressure are both identified as significant fundamental parameters affecting NO_x emission levels when considering the effect of fuel composition on NO_x levels. For a conventional burner, with NO_x on a concentration basis, the adiabatic flame temperature is dominant, with fuel pressure remaining significant in affecting NO_x emission levels. The highest NO_x levels on a volume concentration basis occurred at the highest hydrogen content fuel compositions at lower fuel pressures. On a mass per heat-release basis, however, the highest relative NO_x levels were achieved for fuel compositions containing large fractions of C_3H_8. This appears to result from some combined characteristics of a high-propane mixture including: very low fuel pressure for a given heat release in comparison with the other fuels; somewhat higher adiabatic flame temperature than that of CH_4; and a substantially larger amount of total dry products of combustion produced for a given heat release when compared with H_2. In summary, the results of this work provide both quantitative and qualitative information to improve emission performance prediction and design of burners with application to a wide variation of fuel compositions.

Yuan et al. [26] modeled NO_x emissions from bark boilers used in the manufacture of paper products. Because the temperatures were relatively low, the contribution from fuel NO_x (from the nitrogen content in the wood chips) was dominant compared to that of thermal NO_x when the boiler was fired mainly on bark. NO_x increased significantly with fuel nitrogen content. Thermal NO_x became more significant when the boiler was fired mainly on natural gas.

Sorum et al. [27] discuss the formation of NO from the combustion of volatiles from municipal solid wastes. This was studied numerically and experimentally with a pilot-scale fixed-bed reactor. One of the parameters studied was the oxygen content in the oxidizer which was varied from 12 to 100% by volume. Another parameter studied was the conversion of fuel-bound nitrogen into NO for a variety of single-component wastes including newspaper, cardboard, glossy paper, low-density polyethylene (LDPE), and polyvinylchloride (PVC), as well as blends of these components. The conversion efficiency for paper and cardboard ranged from 26 to 99%. Efficiencies above 100% were calculated for LDPE and PVC, which suggest that NO was formed by thermal and/or prompt mechanisms in addition to fuel NO_x. Nitric oxide increased with the oxygen content in the oxidizer.

6.4.3 Mixture Ratio

Figure 6.28 shows the predicted NO as a function of the flame stoichiometry for an air/CH_4 flame. Nitric oxide increases under fuel-lean conditions and decreases at fuel rich conditions. Fig. 2.27 shows a plot of the adiabatic equilibrium flame temperature for an air/CH_4 flame as a function of the flame equivalence ratio. There are several things to notice. The flame temperature for the air/CH_4 flame is very dependent on the stoichiometry. The figure helps to explain why, for example, NO_x is reduced dramatically under fuel-rich conditions. One reason is the dramatic reduction in the flame temperature. Another reason concerns the chemistry. In a

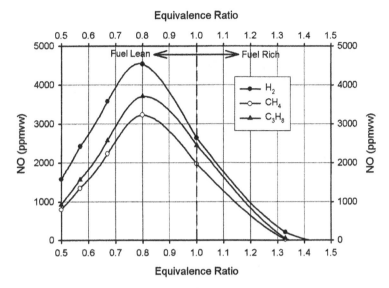

Figure 6.28 Adiabatic equilibrium NO as a function of equivalence ratio for air/fuel flames.

reducing atmosphere, CO is formed preferentially to NO. This is exploited in some of the NO_x reduction techniques. An example is methane reburn [19]. The exhaust gases from the combustion process flow through a reduction zone that is under reducing conditions; NO_x is reduced back to N_2. Any CO that may have formed in the reduction zone and other unburned fuels are then combusted downstream of the reduction zone. However, they are combusted at temperatures well below those found in the main combustion process. These lower temperatures are not favorable to NO_x formation. Figure 6.29 shows how NO_x increases with the O_2 content in the flue gas, which is equivalent to an increase in excess air. Therefore, controlling excess oxygen is an important strategy for minimizing NO_x emissions.

Figure 6.30 shows the effect of air infiltration into a furnace. At positive furnace pressures, combustion products leak out of the furnace, which precludes ambient air from leaking into the furnace. Negative furnace pressures allow ambient air to leak into the furnace. This additional air increases NO_x emissions as it indirectly increases the mixture ratio. While very few studies have examined the effect of air leakage (e.g., leakage rate and location) in industrial furnaces, it is expected that the farther the leaks from the flame zones, the less the impact on NO_x emissions. This is because most or all of the extra O_2 does not enter the highest temperature zones that cause most of the thermal NO_x emissions.

Figure 6.31 shows the combined effects of the oxidizer composition and the oxidizer:fuel mixture ratio. As previously shown, the highest NO_x occurs for oxidant compositions in the middle range (approximately 40–60% O_2) and for higher O_2:fuel mixture ratios. Figure 6.31a shows the results in ppmvd, compared to Fig. 6.31b in $lb/10^6$ Btu. The results are similar but slightly different for the reasons previously discussed due to the difference in the flue gas volume as the oxidant composition changes. Figures 6.31b and 6.31c are similar except for the difference in units.

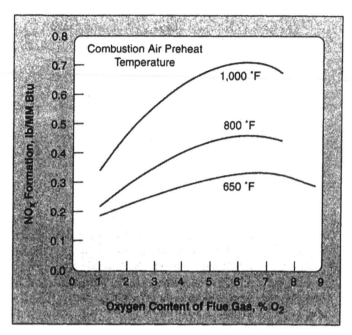

Figure 6.29 NO_x as a function of O_2 content in the flue gas. (From Ref. 50.)

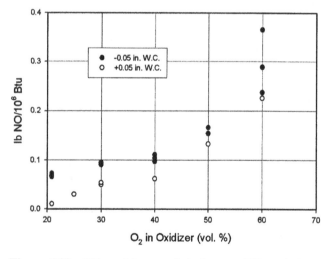

Figure 6.30   Effect of furnace air leakage on NO_x emissions.

6.4.4   Combustion Product Temperature

Figure 6.32 shows the importance of the gas temperature on thermal NO_x formation. The NO_x rises rapidly at temperatures above 2000°F (1100°C) for all three fuels shown. This is a demonstration of the increase in thermal NO_x as a function of temperature. Many combustion modification strategies for reducing NO_x involve reducing the flame temperature because it has such a large impact on NO_x. For example, one strategy is to inject water into the flame to reduce NO_x by cooling down the flame to a lower temperature where NO_x formation is less favorable.

6.4.5 Oxidizer Temperature

Figure 6.33 shows how NO_x increases when the combustion air is preheated. Air preheating is commonly done to increase the overall thermal efficiency of the heating process. However, it can dramatically increase NO_x emissions because of the strong temperature dependence of NO formation. Fig. 2.30 shows how the adiabatic flame temperature increases with air preheating. The increase in NO emissions mimics the increase in flame temperature. Figure 6.34 shows how NO_x increases with air preheat for both oil-fired and gas-fired conventional burners.

Katsuki and Hasegawa [28] have written a review article on the use of highly preheated air (above 1300 K or 1900°F) in industrial combustion applications

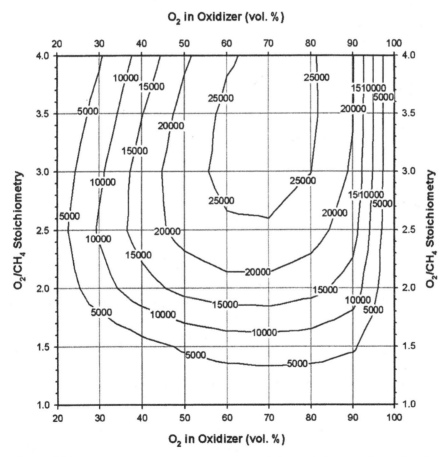

Figure 6.31a Contour plot of the adiabatic equilibrium NO (in ppmvd) as a function of the oxygen concentration in the oxidizer and of the oxygen:CH$_4$ mixture ratio where both the fuel and the oxidizer are at ambient temperatures.

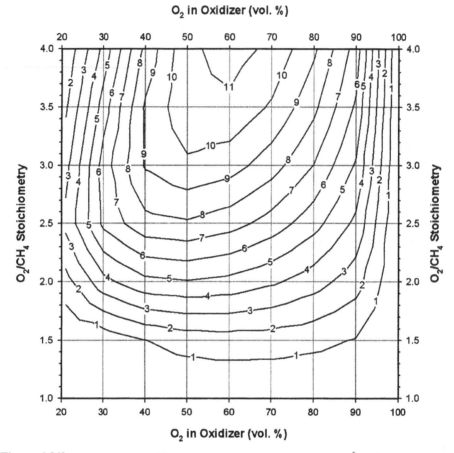

Figure 6.31b   Contour plot of the adiabatic equilibrium NO (in lb/10^6 Btu) as a function of the oxygen concentration in the oxidizer and of the oxygen:CH$_4$ mixture ratio where both the fuel and the oxidizer are at ambient temperatures.

including an extensive discussion of NO emissions. They note that this is potentially the most significant disadvantage of using preheated air compared to the main advantage of substantial energy savings from using heat recovery. The primary ways to minimize NO$_x$ in systems with high levels of air preheat is to dilute the gases with combustion products to lower the O$_2$ concentration and to control the mixing between the preheated air and the incoming fuel. Weber et al. [29] experimentally demonstrated a furnace design methodology for using air preheating by external flue gas recirculation to increase thermal efficiency without high NO$_x$ emissions. The burner used fuel staging and the furnace is well mixed to prevent high peak gas temperatures that increase NO$_x$ emissions. Scherello et al. [30] experimentally studied the effect of regenerative air preheating, commonly used in glass-melting furnaces, on NO$_x$ emissions. Increasing the air preheat temperature increased NO$_x$. The burner angle and gas momentum were less important factors concerning NO$_x$. Kapros and Solyom [31] developed a burner using air preheated up to 900°C (1700°F) by a rotary regenerator. NO$_x$ emissions were within European Community

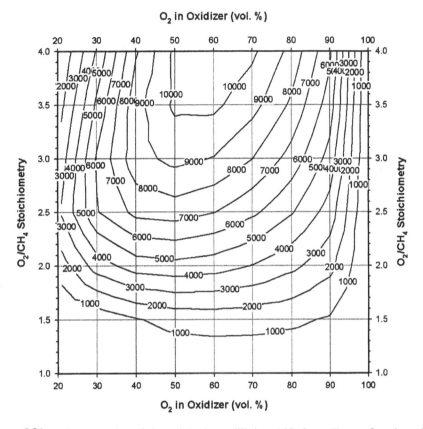

Figure 6.31c Contour plot of the adiabatic equilibrium NO (in ng/J) as a function of the oxygen concentration in the oxidizer and of the oxygen:CH₄ mixture ratio where both the fuel and the oxidizer are at ambient temperatures.

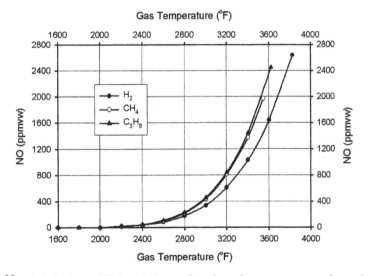

Figure 6.32 Adiabatic equilibrium NO as a function of gas temperature for stoichiometric air/fuel flames. (From Ref. 3. Courtesy of John Zink Co.)

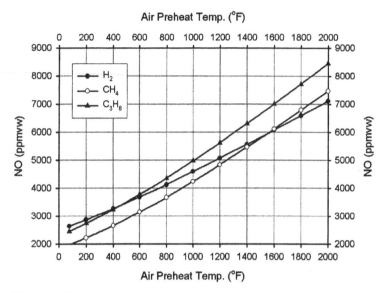

Figure 6.33 Adiabatic equilibrium NO as a function of air preheat temperature for stoichiometric air/fuel flames. (From Ref. 3. Courtesy of John Zink Co.)

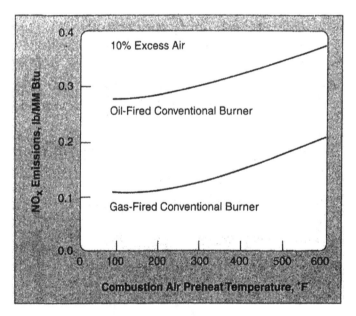

Figure 6.34 Impact of air preheat on NO_x generation. (From Ref. 50.)

standards. Giammartini et al. [32] experimentally demonstrated significant NO_x reductions in combustion systems with high levels of air preheat by diluting the flames with furnace gases to reduce the peak flame temperatures. Gaba [33] developed a recuperative burner for preheating the combustion air up to 450°C (840°F) to improve thermal efficiency and increase throughput in industrial forge furnaces while minimizing NO_x emissions.

6.4.6 Fuel Temperature

Figure 6.35 shows how NO_x increases with the fuel preheat temperature. Fuel preheating is another method used to improve the overall thermal efficiency of a heating process. Fig. 2.31 shows how the adiabatic flame temperature increases due to fuel preheating. The increase in NO_x emissions follows the same pattern as the increase in flame temperature.

A study by Yap et al. [34] indicates that fuel preheating may actually reduce NO_x emissions for oxygen-enriched flames. While there was not a broad enough set of conditions tested to determine this conclusively, experimental and modeling results for oxygen enrichment up to 11% and fuel preheat up to 600 K (600°F) showed that fuel preheating could reduce NO_x. As previously shown, small amounts of oxygen enrichment can dramatically increase NO_x emissions. The reduction using fuel preheat was explained by the increased radiant cooling of the flame, which reduces the gas temperatures in the flame and hence reduces NO_x emissions that are highly dependent on gas temperature.

6.4.7 Gas Recirculation

Figure 6.36 shows that recirculating flue gases at 2000°F can dramatically reduce NO_x emissions for a variety of fuels. Although the flue gases are hot, they are still considerably cooler than the adiabatic flame temperatures of each fuel. The recycle gases quench the flame and reduce thermal NO_x emissions. Even a recycle ratio as small as one volume of recycle gases per volume of incoming combustion air (both at standard temperature and pressure or STP) can reduce the NO_x emissions by about 80%. Since this plot is theoretical, it does not specify how the recycle gases got back into the flame. This could have been by external or internal flue gas recirculation.

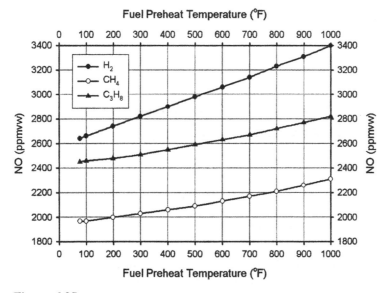

Figure 6.35 Adiabatic equilibrium NO as a function of fuel preheat temperature for a stoichiometric air/CH$_4$ flame. (From Ref. 3. Courtesy of John Zink Co.)

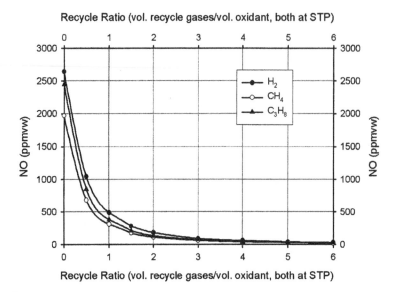

Figure 6.36 Adiabatic equilibrium NO as a function of the recycle ratio (moles recycle gas/ mole combustion air) for the stoichiometric combustion of hydrogen, methane, and propane with the recycle gases at 2000°F (1366 K).

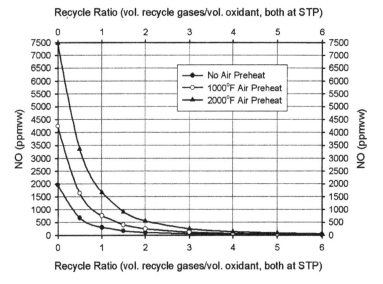

Figure 6.37 Adiabatic equilibrium NO as a function of the recycle ratio (moles recycle gas/ mole combustion air) for the stoichiometric combustion of methane at different air preheat temperatures with the recycle gases at 2000°F (1366 K).

Figure 6.37 shows a similar phenomenon for the effect of gas recirculation on reducing NO_x emissions for systems with air preheat. As previously shown in Fig. 6.33, preheating the incoming combustion air significantly increases NO_x emissions because of the increased flame temperatures. In fact, air preheating is typically done both for increasing fuel efficiency and for increasing flame temperatures for high-temperature applications such as glass melting (see Chap. 14). Figure 6.37 shows that only a relatively small amount of gas recirculation can dramatically reduce NO_x emissions in systems with air preheating.

Figure 6.38 again shows that gas recirculation can dramatically reduce NO_x emissions, in this case for fuel preheating where methane has been used as a typical fuel. Fuel preheating is commonly done to increase fuel efficiency and flame temperatures, as described above for the case of air preheating. In some cases, fuel preheating may also be used to increase the soot generation in the flame to increase the radiation from the flame to the load. As the mass of fuel is much less than the mass of incoming oxidant, preheating the fuel has a much smaller effect on NO_x as shown by how close together the three different curves are for various preheat levels. In some applications in the petrochemical and refining industries, the fuel is available at relatively high pressure and is used to inspirate flue gases into the flame for NO_x reduction (see Chap. 15).

6.4.8 Flame Radiation

It has been shown by many researchers that flame radiation plays an important role in NO formation in flames [35]. Hydrocarbon fuels with higher carbon-to-hydrogen ratios tend to produce flames with higher radiant fractions. Higher flame radiation releases more energy from the flame and thus lowers the gas temperatures in the flames. Since thermal NO_x is exponentially dependent on gas temperature, more radiant flames tend to produce less NO_x. Since it is not always possible to change the fuel composition, other schemes are available for increasing the radiation from the

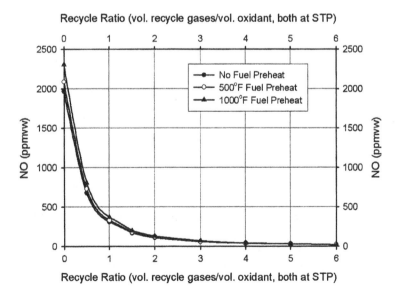

Figure 6.38 Adiabatic equilibrium NO as a function of the recycle ratio (moles recycle gas/mole combustion air) for the stoichiometric combustion of methane at different fuel preheat temperatures with the recycle gases at 2000°F (1366 K).

flame. These include air staging where the initial part of the flame is fuel rich to produce soot particles that are good radiators, especially compared to gases. Particulates like carbon may be deliberately added to the fuel to radiate to enhance flame radiation. Fujimori et al. [36] studied the effects of radiation and air preheat on NO_x emissions from turbulent ethylene jet flames.

6.5 ENVIRONMENTAL AND HEALTH CONCERNS

In addition to the poisoning effect that NO_x has on humans, there are also other problems associated with these chemicals. NO_x contributes to ozone depletion, reacts with water in the atmosphere to form acid rain, and participates in the formation of smog. These are discussed next.

Neither NO or NO_2 are known to cause any direct damage to materials. At sufficient concentrations, NO_2 absorbs visible light and can cause a noticeable reduction in visibility. At sufficient concentrations and exposure times, NO_2 has suppressed the growth of certain fruits and vegetables [37]. Nitrogen dioxide can damage leaves by causing irregular, white or brown collapsed lesions on the leaf tissue [38].

6.5.1 Ozone Generation

Ozone is present in the upper atmosphere and shields the earth from high-intensity ultraviolet rays from the sun. One of the recent environmental threats that has received much media attention is the hole in the ozone layer close to Antarctica due to chlorofluorocarbons, also known as freons. However, ozone in the lower atmosphere is undesirable. There, NO reacts with oxygen to form ozone, in addition to NO_2:

$$NO + HC + O_2 + Sunlight \rightarrow NO_2 + O_3$$

Ozone (O_3) is also a health hazard that can cause respiratory problems in humans. It is also an irritant to the eye, nose, and throat. Ozone can cause damage to plants including crops, and deterioration of textiles and other materials. Nitrogen dioxide is extremely reactive and is a strong oxidizing agent. It explodes on contact with alcohols, hydrocarbons, organic materials, and fuels.

6.5.2 Acid Rain

Rain is effective at removing NO_2 from the atmosphere. However, NO_2 decomposes on contact with water to produce nitrous acid (HNO_2) and nitric acid (HNO_3), which are highly corrosive (see Fig. 6.39). Nitric oxide is the predominant form of NO_x produced in industrial combustion processes. It reacts with O_2 in the atmosphere to form NO_2. When NO_2 forms in the atmosphere and comes into contact with water in the form of rain, nitric acid is formed. The two step reaction to form the acid is

$$NO + 0.5O_2 \leftrightarrow NO_2 \tag{6.7}$$

$$3NO_2 + H_2O \rightarrow 2HNO_3 + NO \tag{6.8}$$

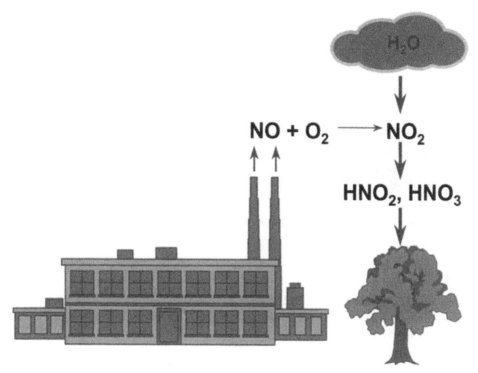

Figure 6.39   Diagram of acid rain. (From Ref. 3. Courtesy of John Zink Co.)

The first reaction is relatively slow but has sufficient time to occur in the atmosphere. The nitric acid formed in the second reaction is the acid rain that damages the environment. Acid rain is destructive to anything it contacts, including plants, trees, and man-made structures like buildings, bridges, and the like.

6.5.3 Smog

Besides acid rain, another problem with NO_2 is its contribution to smog. Smog is the combination of smoke and fog and occurs where there are high concentrations of pollutants combined with fog. Smog impairs visibility through the atmosphere. When sunlight contacts a mixture of NO_2 and unburned hydrocarbons in the atmosphere, photochemical smog is produced (see Fig. 6.40). The qualitative relationship between the major chemical and atmospheric variables in the smog formation process can be summarized as follows [38]:

$$PPL = \frac{(ROG)(NO_x)(\text{Light intensity}) (\text{Temperature})}{(\text{Wind velocity}) (\text{Inversion height})} \tag{6.9}$$

where PPL = photochemical pollution level, ROG = reactive organic gas concentration, and NO_x = nitrogen oxides concentration.

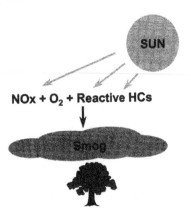

Figure 6.40 Diagram of smog formation. (From Ref. 3. Courtesy of John Zink Co.)

6.5.4 Health Effects

Smog causes eye and respiratory irritation and reduced visibility. Nitrogen dioxide can irritate the lungs and lowers resistance to respiratory infections such as influenza. It is particularly troublesome for the very young and the very old.

6.6 REGULATIONS

Regulations for NO_x vary by country and region. The United States, Japan, and Germany have the strictest regulations. Perhaps the most stringent standards in the world are those enforced by the South Coast Air Quality Management District (SCAQMD). SCAQMD governs the greater Los Angeles area and has proposed rules restricting NO_x from burners to less than 5 ppmvd corrected to 3% O_2 for new sources. Currently, there are no burners that can meet these emissions without postcombustion controls.

The U.S. Environmental Protection Agency (EPA) tracks the emissions of NO_x by the source type and location [39]. Figure 6.41 shows how NO_x emissions in the United States have varied since 1970 for some of the industrial combustion applications considered in this book. The EPA established regional guidelines for reducing NO_x emissions through State Implementation Plans (SIPs). Table 6.3 shows how the federal legislation compares with that introduced by California and Texas, which have some of the more stringent NO_x regulations in the United States [40]. Their regulations are much lower than the federal guidelines because of the more urgent need to reduce smog in those states. Table 6.4 shows emission factors developed for process heaters [41].

6.6.1 Unit Conversions

It is often necessary to convert pollutant measurements (e.g., of NO_x and CO) into a standard basis for both regulatory and comparison purposes. One conversion that is often necessary is from the measured O_2 level in the exhaust gases to a standard

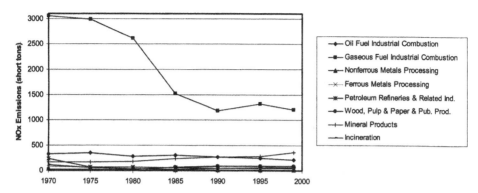

Figure 6.41 NO$_x$ emissions in the United States since 1970 based on the application. (From Ref. 39.)

Table 6.3 U.S. Legislation Targeted to Reducing NO$_x$ Emissions from Nonutility Stationary Sources

Legislation	U.S. states affected	Approximate reduction required (%)
NO$_x$ State Implementation Plan Call (EPA)	AL, CT, DE, GA, IL, IN, KY, MD, MA, MI, MO, NJ, NY, NC, OH, PA, RI, SC, TN, VA, WV, Washington DC	30–70
South Coast Air Quality Management District (SCAQMD) rule 1146	CA	75
Texas Natural Resource Conservation Commission (TNRCC)	TX	90 for large units, 70 for small units

Source: Ref. 40.

Table 6.4 Combustion Emission Factors (lb/10^6 Btu) by Fuel Type

Fuel type	SO$_x$	NO$_x$	CO	Particulates	VOCs
Distillate fuel	0.160	0.140	0.0361	0.010	0.002
Residual fuel	1.700	0.370	0.0334	0.080	0.009
Other oils	1.700	0.370	0.0334	0.080	0.009
Natural gas	0.000	0.140	0.0351	0.003	0.006
Refinery gas	0.000	0.140	0.0340	0.003	0.006
LPG	0.000	0.208	0.0351	0.007	0.006
Propane	0.000	0.208	0.0351	0.003	0.006
Steam coal	2.500	0.950	0.3044	0.720	0.005
Petroleum coke	2.500	0.950	0.3044	0.720	0.005
Electricity	1.450	0.550	0.1760	0.400	0.004

Source: Ref. 41.

basis O_2 level. The method for converting measurements to a standard basis is given by [42]

$$\text{ppm}_{\text{corr}} = \text{ppm}_{\text{meas}} \left(\frac{20.9 - O_{2_{\text{ref}}}}{20.9 - O_{2_{\text{meas}}}} \right) \tag{6.10}$$

where

$\qquad \text{ppm}_{\text{meas}} = $ Measured pollutant concentration in flue gases (ppmvd)
$\qquad \text{ppm}_{\text{corr}} = $ Pollutant concentration corrected to a reference O_2 basis (ppmvd)
$\qquad O_{2_{\text{meas}}} = $ Measured O_2 concentration in flue gases (vol%, dry basis)
$\qquad O_{2_{\text{ref}}} = $ Reference O_2 basis (vol%, dry basis)

Example 6.1

Given: Measured $NO_x = 20$ ppmvd, measured $O_2 = 2\%$ on a dry basis.
Find: NO_x at 3% O_2 on a dry basis.
Solution: $\text{ppm}_{\text{meas}} = 20$, $O_{2\text{meas}} = 2$, $O_{2\text{ref}} = 3$

$$\text{ppm}_{\text{corr}} = (20) \left(\frac{20.9 - 3}{20.9 - 2} \right) = 18.9 \text{ ppmvd}.$$

This example shows that NO_x values will be lower when the basis O_2 is higher than the measured O_2 because higher O_2 levels mean more air dilution and, therefore, lower NO_x concentrations. The reverse is true when the basis O_2 is lower than the measured O_2 level.

     Another correction that may be required is to convert the measured pollutants from a measured furnace temperature to a different reference temperature. This may be required when a burner is tested at one furnace temperature and needs to be modified to find out the equivalent at the another furnace temperature. The correction for temperature is

$$\text{ppm}_{\text{corr}} = \text{ppm}_{\text{meas}} \left(\frac{T_{\text{ref}} - T_{\text{basis}}}{T_{\text{meas}} - T_{\text{basis}}} \right) \tag{6.11}$$

where

$\qquad \text{ppm}_{\text{meas}} = $ Measured pollutant concentration in flue gases (ppmvd)
$\qquad \text{ppm}_{\text{corr}} = $ Pollutant concentration corrected to a reference temperature basis (ppmvd)
$\qquad T_{\text{ref}} = $ Reference furnace temperature (°F)
$\qquad T_{\text{meas}} = $ Measured furnace temperature (°F)
$\qquad T_{\text{basis}} = $ Basis furnace temperature (°F)

Example 6.2

Given: Measured $NO_x = 20$ ppmvd, measured furnace temp. $= 1800°$F.
Find: NO_x at a reference temperature of 2000°F.
Solution: $\text{ppm}_{\text{meas}} = 20$, $T_{\text{meas}} = 1800°$F, assume $T_{\text{ref}} = 400°$F

$$\text{ppm}_{\text{corr}} = (20) \left(\frac{2000 - 400}{1800 - 400} \right) = 22.9 \text{ ppmvd}.$$

There are two things to notice in the above example. The first is that the basis temperature was chosen as 400°F, which is an empirically determined value that applies to many burners commonly used in the hydrocarbon and petrochemical industries. However, this equation should be used with care for more unusual burner designs and when there is a very large difference between the measured and the reference furnace temperatures. The second thing to notice is that the NO_x increases when the reference temperature is higher than the measured temperature and vice versa. As will be shown later, NO_x generally increases with the furnace temperature.

These two corrections can also be combined into a single correction when both the measured O_2 level and furnace temperature are different from the reference O_2 level and furnace temperature:

$$ppm_{corr} = ppm_{meas}\left(\frac{20.9 - O_{2_{ref}}}{20.9 - O_{2_{meas}}}\right)\left(\frac{T_{ref} - T_{basis}}{T_{meas} - T_{basis}}\right) \tag{6.12}$$

where the variables are as defined above.

Example 6.3

Given: Measured $NO_x = 20$ ppmvd, measured $O_2 = 2\%$ on a dry basis, measured furnace temp. $= 1800°F$.

Find: NO_x at 3% O_2 on a dry basis at a reference temperature of 2000°F.

Solution: $ppm_{meas} = 20$, $O_{2_{meas}} = 2$, $O_{2_{ref}} = 3$, $T_{meas} = 1800°F$, assume $T_{ref} = 400°F$

$$ppm_{corr} = (20)\left(\frac{20.9 - 3}{20.9 - 2}\right)\left(\frac{2000 - 400}{1800 - 400}\right) = 21.6$$

In this case, the increase in NO_x due to the temperature correction is greater than the reduction in NO_x due to the higher O_2 reference.

The correction in Eq. (6.10) assumes that the excess O_2 comes from air. This may or may not be the case with OEC. If the excess O_2 measured in the flue gases came from an oxygen-enriched oxidant, then Eq. (6.10) should be modified as follows:

$$ppm_{CORR} = ppm_{MEAS}\left[\frac{O_{2_{OXID}} - O_{2_{BASIS}}}{O_{2_{OXID}} - O_{2_{MEAS}}}\right] \tag{6.13}$$

where

   $ppm_{MEAS} =$ Measured pollutant concentration in flue gases (ppmvd)
   $ppm_{CORR} =$ Pollutant concentration corrected to a standard O_2 basis (ppmvd)
   $O_{2_{OXID}} = O_2$ concentration in oxidant (vol%)
   $O_{2_{MEAS}} =$ Measured O_2 concentration in flue gases (vol%, dry basis)
   $O_{2_{BASIS}} =$ Standard O_2 basis (vol%, dry basis)

The conversion from ppmvd to mg/Nm³ at a base oxygen concentration and normal conditions of 0°C (32°F) and atmospheric pressure is

$$NO_x(\text{mg/Nm}^3) = 2.05 NO_x(\text{ppmvd}) \tag{6.14}$$

Example 6.4

Given: Measured $NO_x = 20$ ppmvd.
Find: NO_x in mg/Nm^3.
Solution: NO_x (mg/Nm^3) = 2.05 (20 ppmvd) = 41.0 mg/Nm^3.

Most industrial heating systems are operated at negative pressures to prevent process gases from leaking into the work environment. Therefore, some amount of air generally leaks into the combustion chamber. In an OEC system, the "oxidant" is then a combination of the air leaking into the process and the oxidant supplied through the burner(s). Air leakage into an oxygen-enhanced combustion process further complicates the problem of determining what O_2 concentration should be used for the oxidant.

From Fig. 2.41, it may be seen that the exhaust volume per unit of fuel input is dramatically reduced as the O_2 in the oxidizer increases. It may also be seen that the available heat increases as well (see Fig. 2.38). These both have a dramatic impact on the relevance of some of the NO_x units. Because the flue gas volume may be reduced by $>90\%$ when replacing air with pure oxygen, comparing NO_x from an air/fuel system to the NO_x from an oxy/fuel system on a ppmv basis does not make sense. For example, 200 ppmvd NO_x in an oxy/fuel system is actually less NO_x by mass than 100 ppmvd NO_x in an air/fuel system. This is because of the vast differences in flue gas volumes. Figure 6.42 shows a comparison of NO_x in ppmvd at 0% O_2 for systems using air and pure oxygen.

As can be seen, care must be exercised when comparing NO_x on a volume basis for systems using different oxidizers. Hence, assuming that air/fuel and oxy/fuel systems are equally efficient, it makes more sense to compare the NO_x on a mass basis. Figure 6.4 shows NO_x in both mass and volume units, as a function of the oxidizer composition. However, as shown in Fig. 6.43, oxy/fuel systems are significantly more efficient than air/fuel systems. Less fuel is required for a given unit

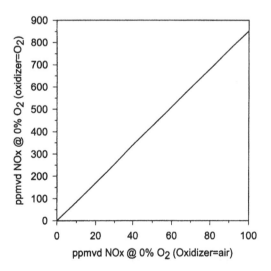

Figure 6.42 NO_x (ppmvd at 0% O_2) when oxidant is air compared to pure O_2. (From Ref. 14. Courtesy of CRC Press.)

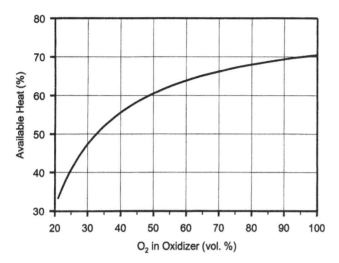

Figure 6.43 Available heat as a function of oxidant (O_2 in N_2) composition for a stoichiometric CH_4 at an exhaust gas temperature of 2500°F (1640 K).

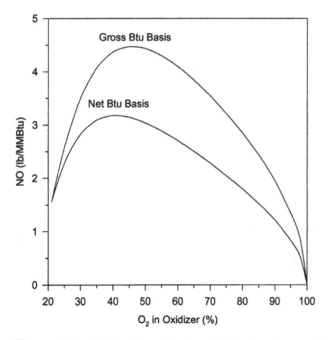

Figure 6.44 Adiabatic equilibrium NO (net and gross bases) as a function of oxidant composition for a CH_4 flame. (From Ref. 14. Courtesy of CRC Press.)

of production. Therefore, a better NO_x unit would be the mass of NO_x produced per unit mass of material processed.

This concept can be illustrated with a unit like lb/MMBtu on a net basis. This unit is derived by combining the NO_x/MMBtu (on a gross basis) curve in Fig. 6.43 and the available heat curve in Fig. 6.44. It is assumed that the base case is air/fuel combustion. Then, as the O_2 in the oxidizer increases, the thermal efficiency increases. The gross and net Btu curves are shown in Fig. 6.44. The net Btu curve is equivalent to a throughput basis. It is assumed that the heat losses from the system and the heat transfer to the product would not vary significantly for different oxidizers. This figure shows that it may be very deceiving to compare NO_x based only on the gross firing rate when the oxidizer compositions are different.

6.6.2 Reporting Units

Baukal and Eleazer [43] have discussed potential sources of confusion in the existing NO_x regulations. These sources of confusion may be classified as either general or specific. General sources of confusion include, for example, the wide variety of units that have been used, reporting on either a dry or wet sample basis, measuring NO but reporting NO_2, and reporting on a volume versus a mass basis.

Historically, governing bodies have sprung up regionally in order to regulate particular sources. The governing bodies have generally adopted units related to a traditional industry metric. This has led to a wide variety of NO_x units. For example, internal combustion (IC) engines are generally regulated on a gram per brake horsepower (g/bhp) basis—a mass-based unit normalized by the output power of the engine. Gas turbines on the other hand are generally regulated on a part per million (ppm) basis. Because this unit is volume based, it must be referenced to a standard condition. Gas turbines usually operate near 15% excess oxygen, and traditionally NO_x measurement required removal of water before analysis. So, gas turbines often use a ppm measurement referenced on a dry volume basis (ppmdv) to 15% oxygen.

On the other hand, one typically operates industrial boilers and process heaters nearer to 3% excess oxygen. So, NO_x emissions from those units are generally referenced as ppmvd corrected to 3% oxygen. However, these units may also be regulated on a mass basis normalized by the heat release of the burner; for example pounds per million Btu (lb/MMBtu). Large electrical utilities operate their boilers under very tight oxygen limits. Therefore, some U.S. agencies regulate utility boilers on a pound per megawatt basis (lb/MW). A further complication is whether to normalize the unit by gross output power (gross MW) or to subtract parasitic power losses (net MW). Foreign regulatory agencies use SI units such as grams per normal cubic meter (mg/Nm^3).

It is important to be able to convert field measurements into specific units in order to determine whether the emissions from a specific burner or heater are below their allowable limits. In nearly all cases, NO_x is measured on a ppmvd basis. The following example will show how to convert these units to a specific basis.

Example 6.5

Given:	Fuel = methane with a gross or higher heating value of 1012 Btu/ft^3, NO = 20 ppmvd, measured O_2 = 2% on a dry basis.
Find:	NO_x as NO_2 in lb/10^6 Btu (gross).
Solution:	First calculate the dry flue gas products. The global chemical reaction

is as follows:

$$CH_4 + x(O_2 + 3.76N_2) = CO_2 + 2H_2O + yO_2 + 3.76xN_2$$

where $O_2 + 3.76N_2$ is the composition of air (79% N_2, 21% O_2).

(1) Given 2% O_2 in dry flue gases: $\dfrac{y}{1 + y + 3.76x} = 0.02$

(2) O atom balance: $2x = 2 + 2 + 2y = 4 + 2y$ or, $x = 2 + y$

Solving (1) and (2) simultaneously:

$$CH_4 + 2.188(O_2 + 3.76N_2) = CO_2 + 2H_2O + 0.188O_2 + 8.23N_2$$

This shows the moles of products for each mole of CH_4. Note that NO in the flue products has been ignored because it is only in trace amounts. Assume that all NO is converted into NO_2 in the atmosphere.

(988 ft³ of CH_4)(1012 Btu/ft³ CH_4) $= 1 \times 10^6$ Btu (gross)

(988 ft³ of CH_4)(1 + 0.188 + 8.23) = 9305 ft³ of dry combustion products at STP per 10^6 Btu

Given 20 ppmvd NO_2 = (20 ft³ $NO_2/10^6$ ft³ dry products)(9305 ft³ dry products/ 10^6 Btu) = 0.186 ft³ $NO_2/10^6$ Btu

Density of NO_2 = 0.111 lb/ft³

Mass of NO_2 in exhaust products = (0.186 ft³ $NO_2/10^6$ Btu)(0.111 lb NO_2/ft³ NO_2) = 0.021 lb $NO_2/10^6$ Btu (gross)

6.7 MEASUREMENT TECHNIQUES

Accurate measurements of pollutants, such as NO and CO, from industrial sources are of increasing importance in view of strict air-quality regulations. Based on such measurements, companies may have to pay significant fines, stop production, install expensive flue-gas treatment systems, buy NO_x credits in certain nonattainment areas, or change the production process to a less polluting technology. If compliance is achieved, however, the company may continue their processes without interruption and, sometimes, sell their NO_x credits. Mandel [44] notes that the equipment cost for the gas analysis system is relatively small compared to the maintenance and repair costs.

Numerous studies have been done and recommendations made on the best ways to sample hot gases from high-temperature furnaces. For example, EPA Method 7E [45] applies to gas samples extracted from an exhaust stack that are analyzed with a chemiluminescent analyzer (see Fig. 6.45). A typical sampling system is shown in Fig. 6.46. The major components are: a heated sampling probe, heated filter, heated sample line, moisture removal system, pump, flow control valve, and then the analyzer. The EPA method states that the sample probe may be made of glass, stainless steel, or other equivalent materials. The probe should be heated to prevent water in the combustion products from condensing inside the probe. While usually not approved for certification testing, portable gas analyzers (see Fig. 6.47) are often used for spot checks for those installations that do not have continuous

Figure 6.45   Chemiluminescent NO$_x$ analyzer. (Courtesy of California Analytical, Orange, CA.)

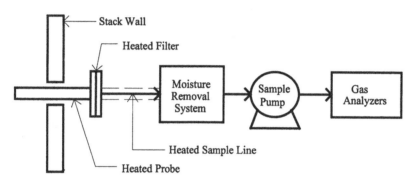

Figure 6.46   Sampling system schematic as recommended by the EPA. (From Ref. 14. Courtesy of CRC Press.)

emission monitoring systems. Figure 6.48 shows that portable analyzers can provide accurate data compared to permanent gas analysis systems.

The EPA method is appropriate for a lower temperature, nonreactive gas sample obtained, for example, from a utility boiler. However, this method should not be used to obtain samples from higher temperature industrial furnaces used in glass or metals production. Flue gas temperatures from such furnaces, as well as from some incinerators, can be as high as 2400°F (1300°C). This would cause the probe to overheat and affect the measurements because of high-temperature surface reactions inside the probe.

The effects of probe materials, such as metal and quartz, as well as the probe cooling requirements, have been investigated for sampling gases in combustion systems [46]. Several studies have found that both metal and quartz probe materials can significantly affect NO measurements in air/fuel combustion systems, especially under fuel-rich conditions with high CO concentrations [47,48]. However, the NO readings were not affected under fuel-lean conditions.

The probe materials and cooling requirements are of even greater importance in oxy/fuel combustion systems, where pollutant concentrations are much higher, because of the lack of diluent N$_2$. It has been shown that the use of nonwater-cooled probes can lead to false readings of both NO and CO, when sampling high-temperature flue gases [49]. A series of tests was done in a large-scale research furnace. Both quartz and inconel sampling probes were used. The results showed

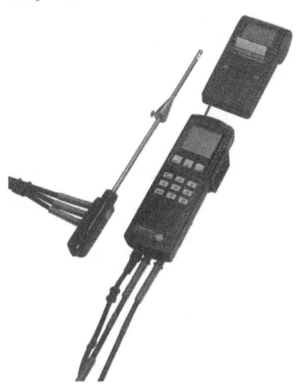

Figure 6.47 Portable gas analyzer. (Courtesy of Testo, Flanders, NJ).

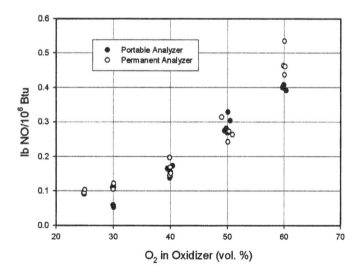

Figure 6.48 NO$_x$ emissions taken from a pilot-scale combustion test furnace using a portable in-situ analyzer compared to an extractive permanent gas analysis system.

a discrepancy in the NO_x measurements, between these two probes. This discrepancy, as well as a temperature dependence, indicated that the probe material played a key role. Inconel is a good catalyst of the following reactions at temperatures above 1300°F (980 K):

$$CO + NO \rightarrow CO_2 + N \tag{6.15}$$

$$CO + O_2 \rightarrow CO_2 + O \tag{6.16}$$

The quartz tube also catalyzed these reactions, but with a much lower efficiency.

Excess quantities of oxidizers in the above reactions resulted in complete consumption of CO. In a fuel-rich case, a high NO concentration was almost completely eliminated by sampling through a hot inconel tube. However, the NO started increasing after the inconel tube temperature was lowered by pulling the tube out of the furnace. This temperature dependence is shown in Fig. 6.49.

The higher NO readings at lower temperatures, especially below 1500°F (1100 K), indicate a decrease in the inconel catalytic efficiency for the above reactions. Other test data showed that the reactions above were weakly catalyzed in the quartz tube too, which lowered the NO readings.

These experiments demonstrated some of the problems that may be encountered in using a hot probe for sampling gases in high-temperature processes. For reducing conditions inside a combustion chamber, the measured NO may be much lower than the actual NO. This should not be the case for most combustors, which are normally operated under at least slightly oxidizing conditions. Under

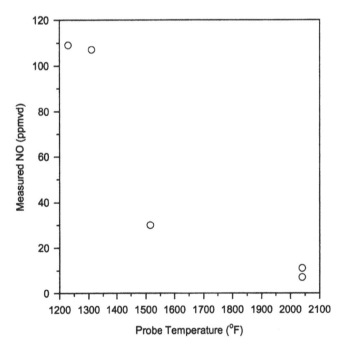

Figure 6.49 NO vs. probe temperature for an inconel probe with high CO concentrations in the exhaust. (From Ref. 14. Courtesy of CRC Press.)

oxidizing conditions, the exhaust gas measurements may show much lower CO readings than are actually present. The experiments have shown that both metal and quartz uncooled probes can affect the readings provided that the probe surface temperature exceeds 1300°F (980 K).

To avoid the surface catalytic effects, a water-cooled probe should be used when sampling high-temperature combustion products. This is of particular importance to oxy/fuel combustion, where measured pollutant species concentrations are much higher, due to the elimination of N_2. Recommended probe materials and cooling requirements, to avoid reaction of the various gases inside the probe, are given elsewhere [42]. It should be noted that the exhaust gas sample should not be cooled below the condensation point of water. This could condense the water in the gas sample inside the sampling probe or sample lines. Since NO_2 is water soluble, false low readings of NO_2 could result.

6.8 ABATEMENT STRATEGIES

Before air-quality regulations, the flue gases from combustion processes were vented directly to the atmosphere. As air-quality laws tightened and the public's awareness increased, industry began looking for new strategies to curb NO_x emissions. The strategies for reducing NO_x are discussed next. Table 6.5 shows a summary of NO_x control techniques developed for the U.S. EPA [10]. Table 6.6 shows a summary of common NO_x control technologies including combustion modifications, process modifications, and post-treatment [50]. Reese et al. [51] presented a useful review of the common NO_x control technologies. Table 6.7 shows typical NO_x reduction efficiencies as functions of the burner draft type (natural or forced), fuel (distillate or residual oil), and reduction technique [16]. The NO_x emissions from gas-fired process heating equipment are highly variable (see Table 6.8) [52]. Therefore, the technique or techniques chosen to reduce NO_x emissions are very site and equipment dependent. This section is not intended to be exhaustive, but is extensive and includes many of the commonly used techniques for minimizing NO_x emissions.

6.8.1 Pretreatment

The first NO_x reduction strategy can be referred to as pretreatment. Pretreatment is a preventive technique to minimize NO_x generation. In pretreatment, the incoming feed materials (fuel, oxidizer, and/or the material being heated) are treated in such a way as to reduce NO_x. Some of these treatments include fuel switching, using additives, fuel treatment, and oxidizer switching.

6.8.1.1 Fuel Switching

Fuel switching is simply replacing a more polluting fuel with a less polluting fuel. For example, fuel oils generally contain some organically bound nitrogen that produces fuel NO_x. Natural gas does not normally contain any organically bound nitrogen and usually has only low levels of molecular nitrogen (N_2). Partial or complete substitution of natural gas for fuel oil can significantly reduce NO_x emissions by

Table 6.5 Summary of NO_x Control Techniques

Technique	Principle of Operation	Status of Development	Limitations	Applications	
				Near-Term	Long-Term
Combustion modification	Suppress thermal NO_x through reduced flame temperature, reduced O_2 level; suppress fuel NO_x through delaying fuel/air mixing or reduced O_2 level in primary flame	Operational for point sources; pilot-scale and full scale studies on combined modifications, operational problems, and advanced design concepts for area sources	Degree of control limited by operational problems	Retrofit utility, industrial boilers, gas turbines; improved designs; new utility boilers	Optimized design area, point sources
Flue gas—noncombustion tail gas treatment	Additional absorption of NO_x to HNO_3; conversion of NO_x to NH_4NO_3; reduction of NO_x to N_2 by catalytic treatment	Operational for existing and new nitric acid plants meeting NSPS; pilot scale feasibility studies for conventional combustion systems	New wet processes developing experience in applications; old catalytic processes have high costs, interference by fuel sulfur or metallic compounds	Noncombustion sources (nitric acid plants)	Possible supplement to combustion modifications; simultaneous SO_x/NO_x removal

Fuel switching	Simultaneous SO_x and NO_x control by conversion to clean fuels; synthetic gas or oil from coal; SNC; methanol; hydrogen	Synthetic fuel plants in pilot-scale stage; commercial plants due by mid 1980s	Fuel cost differential may exceed NO_x, SO_x, control costs with coal	Negligible use	New point sources, (combined cycle); convert area sources (residential)
Fuel additives	Reduce or suppress NO by catalytic action of fuel additives	Inactive; preliminary screening studies indicated poor effectiveness	Large make-up rate of additive for significant effects; presence of additives as pollutant	Negligible use	Not promising
Fuel denitrification	Removal of fuel nitrogen compounds by pretreatment	Oil desulfurization yields partial denitrification	Effectiveness for coal doubtful; no effect on thermal NO_x	Negligible use	Supplement to combustion modification
Catalytic combustion	Heterogeneously catalyzed reactions yield low combustion temperature, low thermal NO_x	Pilot-scale test beds for catalyst screening, feasibility studies	Limited retrofit applications; requires clean fuels	Small space heaters	Possible use for residential heating, small boilers, gas turbines
Fluidized bed combustion	Coal combustion in solid bed yields low temperature, low NO_x	Pilot-scale study of atmospheric and pressurized systems; focus on sulfur retention devices	Fuel nitrogen conversion may require control (staging); may require large make-up of limestone sulfur absorbent	Negligible use	Utility, industrial boilers beginning 1980s; possible combined cycle, waste fuel application

Source: EPA Report 450/1–78-001. Washington, DC: U.S. Environmental Protection Agency, 1978.

Table 6.6 NO$_x$ Reduction Technologies

Technology	Approximate reduction (%)	Approximate emissions (lb/MMBtu)
Standard burners	Base case	0.14
Low-NO$_x$ burners (LNB)	60	0.06
Ultralow-NO$_x$ burners (ULNB)	80–95	0.007–0.03
Flue gas recirculation	55	0.025
Selective noncatalytic reduction (SNCR)	40	0.33–0.085
Selective catalytic reduction (SCR)	90–97	0.006–0.015

Source: Ref. 50.

reducing the amount of nitrogen in the fuel. Figure 6.8 shows that CH$_4$ produces less NO$_x$ than H$_2$. Fuels composed entirely of hydrogen can produce twice as much NO$_x$ as fuels with no hydrogen [53]. Fuel switching may or may not be an option, depending on the availability of fuels and on the economics of switching to a different fuel.

6.8.1.2 Additives

Another type of pretreatment involves adding a chemical to the incoming feed materials (raw materials, fuel, or oxidizer) to reduce emissions by changing the chemistry of the combustion process. One example would be injecting ammonia into the combustion air stream as a type of in-situ de-NO$_x$ process. Several factors must be considered to determine the viability of this option. These include economics, the effects on the process, and the ease of blending chemicals into the process.

6.8.1.3 Fuel Pretreatment

A third type of pretreatment involves treating the incoming fuel prior to its use in the combustion process. An example would be removing fuel-bound nitrogen from fuel oil or removing molecular nitrogen from natural gas. This is sometimes referred to as fuel denitrification. This is normally an expensive process, depending on how much treatment must be done and how the fuel is treated. For example, it is generally more difficult to remove nitrogen from fuel oil than from natural gas. In Europe, some natural gas supplies have as much as 15% N$_2$ by volume. If only a few per cent of N$_2$ needs to be removed from that type of natural gas, this can be done relatively easily and inexpensively with adsorption or membrane separation techniques.

MacKenzie et al. [54] discuss the four common techniques for separating nitrogen from producer gas, particularly for improving the quality of fuel gases removed from naturally occurring natural gas reserves. These techniques include: single high-pressure column process, double column (high and low pressure) process, three column (high, intermediate, and low pressure) process, and two column (high and low pressures) process. Simulations were done to show the differences in power requirements among the four techniques as a function of the inlet nitrogen concentration.

Table 6.7 Reduction Efficiencies for NO$_x$ Control Techniques

Draft and fuel type	Control technique	Total effective NO$_x$ reduction (%)
ND, distillate	(ND) LNB	40
	(MD) LNB	43
	(ND) ULNB	76
	(MD) ULNB	74
	SNCR[a]	60
	(MD) SCR	75
	(MD) LNB + FGR	43
	(ND) LNB + SNCR	76
	(MD) LNB + SNCR	77
	(MD) LNB + SCR	86
ND, residual	(ND) LNB	27
	(MD) LNB	33
	(ND) ULNB	77
	(MD) ULNB	73
	SNCR	60
	(MD) SCR	75
	(MD) LNB + FGR	28
	(ND) LNB + SNCR	71
	(MD) LNB + SNCR	73
	(MD) LNB + SCR	83
MD, distillate	(MD) LNB	45
	(ND) ULNB	74
	(MD) SNCR	60
	(MD) SCR	75
	(MD) LNB + FGR	48
	(MD) LNB + SNCR	78
	(MD) LNB + SCR	92
MD, residual	(MD) LNB	37
	(ND) ULNB	73
	(MD) SNCR	60
	(MD) SCR	75
	(MD) LNB + FGR	34
	(MD) LNB + SNCR	75
	(MD) LNB + SCR	91

[a]Reduction efficiencies for ND or MD SNCR are equal.
MD = mechanical draft, ND = natural draft, LNB = low NO$_x$ burner, ULNB = ultralow NO$_x$ burner, SNCR = selective noncatalytic reduction, SCR = selective catalytic reduction, FCR = flue gas recirculation.
Source: Ref. 16.

6.8.1.4 Oxidizer Switching

The fourth type of pretreatment is oxidizer switching, where a different oxidizer is used. Air is the most commonly used oxidizer. It can be shown that a substantial NO$_x$ reduction can be achieved by using pure oxygen, instead of air, for combustion [55].

Table 6.8 NO_x Control Technologies in Process Heaters

Control Technology	Controlled emissions	Percent reduction
Low-NO_x burners	0.1–0.3 lb/MMBtu	25–65
Staged-air lances	Not available	35–51
Fiber-burner	10–20 ppm	
Ammonia injection	Not available	43–70
Urea injection + low NO_x burner	Not available	55–70
Selective catalytic reduction	20–40 ppm	65–90
Selective catalytic reduction + low NO_x burner	25–40 ppm	70–90

Note: uncontrolled emissions are in the range 0.1–0.53 lb/MMBtu.
Source: Ref. 52.

For example, in the extreme case of combusting a fuel like CH_4 with pure O_2, instead of air that contains 79% N_2 by volume, it is possible to eliminate NO_x completely as no N_2 is present to produce NO_x. For example, if H_2 is combusted with pure O_2, the global reaction can be represented by

$$H_2 + O_2 \rightarrow H_2O \tag{6.17}$$

By drastically reducing the N_2 content in the system, NO_x is minimized.

To assess the validity of the theoretical NO_x reduction using oxygen enrichment as previously discussed, a comprehensive R&D program was conducted, with partial funding from the Gas Research Institute [56]. The two regimes of low- and high-level oxygen enrichment were studied. Low-level enrichment typically involves adding pure oxygen to air to increase the total O_2 concentration from 21% to as high as 35%. In high-level enrichment, air is replaced with oxygen of varying purity, depending on the oxygen production method. These two regimes are important because they encompass most industrial applications.

Experiments in a pilot-scale furnace were conducted to determine which parameters impact NO_x emissions. The normalized NO_x emissions were independent of the firing rate for the relatively narrow range of firing rates that were tested. NO_x emissions were much higher under fuel-lean conditions compared to fuel-rich conditions. This is as predicted by the theoretical calculations (see Fig. 6.6). For low-level O_2 enrichment, NO_x increased rapidly as the oxygen concentration increased. For high level O_2 enrichment, NO_x increased rapidly as the oxidizer purity decreased. Both of these trends validated the theoretical calculations (see Fig. 6.4). Those results are important because some oxygen-generation methods produce O_2 that may contain several per cent of nitrogen and because some natural gas sources can have as high as 15% nitrogen by volume.

There were several important results from those experiments. The first is that the NO_x emissions for low-level oxygen enrichment were nearly an order of magnitude higher than for high-level enrichment. The second is that the experimental NO_x trends were the same as those predicted by theory. However, the experimental measurements were about an order of magnitude lower than the theoretical predictions. This is because actual flames are not adiabatic processes, since a large amount of heat is radiated from the flames. The actual flame temperature is usually much lower than the adiabatic equilibrium flame temperature.

However, there are significant challenges to using high-purity oxygen, instead of air, for combustion [14]. This technique has not been used widely in some applications such as in the hydrocarbon and petrochemical industries, but could become more popular in the future as the cost of oxygen continues to decline as less expensive methods for separating it from air are developed.

6.8.2 Process Modification

There are a number of techniques that may be employed to change the existing process in such a way as to reduce NO_x emissions. These methods are often more radical and expensive and are not often used except under somewhat unusual circumstances. These must be analyzed on a case-by-case method to see if they are viable.

6.8.2.1 Reduced Production

If the mass of NO_x emitted from a plant is too high, an alternative is to reduce the firing rate, which means a corresponding reduction in production. The reduction in NO_x is proportional to the reduction in firing rate as less fuel is burned and therefore less NO_x is formed. However, this is generally not a preferred alternative for obvious reasons as less of the product being made is available to sell. Depending on the costs to reduce NO_x, this may be the most economic alternative in some cases.

In boilers, reducing the firing rate reduces the overall temperature inside the boiler, which reduces the thermal NO_x formation [57]. This technique is known as *derating* and is not desirable if the boiler is capacity limited, but in certain limited applications it may be a viable alternative.

6.8.2.2 Electrical Heating

One process modification that is sometimes used to minimize or eliminate NO_x emissions is to replace some or all of the fossil-fuel-fired energy with electrical energy. The electrical energy produces no NO_x emissions at the point of use and moves the emissions to the power plant. In general, the resulting NO_x emissions at the power plant are often lower than at an industrial site because of the strict limits imposed on the plant and the various methods employed to minimize NO_x that are often more cost effective on a unit mass basis because of the economies of scale.

There are a number of potential problems with this method. The first is that the economics are usually very unfavorable when replacing fossil fuels with electrical energy. In most hydrocarbon and petrochemical processes, the fuel used in the heaters is a by-product that is available at little or no cost. On the other hand, electrical energy is often much more expensive than even purchased fossil fuels like natural gas or oil. Besides the higher operating costs, there would be substantial capital costs involved in converting some or all of the existing fossil-energy heating to electricity. Besides the removal of the existing burners, there would be the cost of the new electrical heaters and often large costs of installing electrical substations that would be required for all of the additional power. In many parts of the country, large additional sources of electricity are not readily available so a new source of electricity may need to be built at the plant, such as a cogeneration facility. However, although the electrical costs may be reduced in that scenario because the transmission losses are much lower, the NO_x emissions are now at different locations at the site and little may then be gained in reducing overall NO_x emissions for the plant. It is likely in the future that regulations will consider the net NO_x generated during the production of a product and would include the NO_x formed in the generation of electricity. This will make replacement of fossil energy with electricity less attractive as most of the power generated in the United States is by fossil-fuel-fired power plants.

6.8.2.3 Improved Thermal Efficiency

By making a heating process more efficient, less fuel needs to be burned for a given unit of production. Since the firing rate is directly proportional to the NO_x emissions, less fuel used equals less NO_x produced. There are many ways to improve the efficiency of a process. A few representative examples will be given. One is to repair the refractory and air-infiltration leaks on an existing heater. This is often relatively inexpensive and saves fuel while reducing NO_x. Another is to add heat recovery to a heating process that does not have it currently. The heat recovery could be in several forms. One method is to preheat the incoming combustion air. As previously discussed, this can increase NO_x emissions due to the higher flame temperatures if it is not done properly. Another method is to add a convection section on to a heater that does not presently have it. This has other operational benefits as well and is often a good choice. A more drastic method of increasing the thermal efficiency of a heating process is to replace an old existing heater with a new, more modern design. This may make sense if the existing heater is very old, requires high maintenance, and is not easily repairable or upgradable.

6.8.2.4 Product Switching

Another radical process modification that can reduce NO_x is to switch the product being produced to one that requires less energy to process. In a process heater, this would involve replacing the existing process fluid with one that requires less energy to heat. For example, the heavier crude oils require more energy to process than do lighter, purer crudes so less energy would be needed to process the latter. Less energy consumption means less NO_x generated. However, this is obviously not an option in most cases and is only considered under extreme circumstances. In the above example, purer or "sweeter" crudes are much more expensive raw materials than less pure or more "sour" crudes. Therefore, the savings in energy may be more than offset by the higher raw material costs.

6.8.3 Combustion Modification

The second strategy for reducing NO_x is known as combustion modification. Combustion modification prevents NO_x from forming by changing the combustion process. There are numerous methods that have been used to modify the combustion process for low NO_x. A popular method is low NO_x burner design where specially designed burners generate less NO_x than generated by previous burner technologies. Low NO_x burners may incorporate a number of techniques for minimizing NO_x including flue gas recirculation, staging, pulse combustion, and advanced mixing. Common combustion modification techniques are discussed next. Rhine [58] has written a useful review of many of the techniques used to reduce NO_x emissions in industrial burners used in higher temperature applications.

6.8.3.1 Air Preheat Reduction

One combustion modification technique is reducing the combustion air-preheat temperature. As shown in Fig. 6.33, reducing the level of air preheat can significantly reduce NO_x emissions. Air preheat greatly increases NO_x for processes that use heat recuperation. However, reduction of air preheat also reduces the overall system efficiency as shown in Fig. 2.40. The loss of efficiency can be somewhat mitigated if the heater is equipped with a convection section. This is a fairly easy technique to implement and may be cost effective if the lost efficiency is more than offset by alternative NO_x reduction techniques.

6.8.3.2 Low Excess Air

As shown in Fig. 6.29, excess air increases NO_x emissions. The excess air generally comes from two sources: the combustion air supplied to the burner and air infiltration into the heater. Excess air produced by either source is detrimental to NO_x emissions. Excess air increases NO_x formation by providing additional N_2 and O_2 that can combine in a high-temperature reaction zone to form NO. In many cases, NO_x can be reduced by simply reducing the excess air through the burners.

Air infiltration, sometimes referred to as tramp air, into a combustion system affects the excess air in the combustor and can affect NO_x emissions. The quantity and location of the leakage are important. Small leaks far from the burners are not nearly as deleterious as large leaks near the flames. By reducing air infiltration (leakage) into the furnace, NO_x can be reduced because excess O_2 generally increases NO_x.

There is also an added benefit of reducing excess air. Reducing the excess O_2 in a combustion system is also useful for maximizing thermal efficiency because any unnecessary air absorbs heat that is then carried out of the stack with the exhaust products. However, there is a practical limit to how low the excess O_2 can be. Since the mixing of the fuel and air in a diffusion flame burner is not perfect, some excess air is necessary to ensure both complete combustion of the fuel and minimization of CO emissions. The limit on reducing the excess air is CO emissions. If the excess O_2 is reduced too much, then CO emissions will increase. Carbon monoxide is not only a pollutant, but it is an indication that the fuel is not being fully combusted, resulting in lower system efficiencies.

There are some special techniques that control the O_2 in the flame to minimize NO_x. One example is pulse combustion, which has been shown to reduce NO_x because the alternating very fuel-rich and very fuel-lean combustion zones (see Fig. 6.50 [59]) minimize NO_x formation. The overall stoichiometry of the oxidizer and fuel is maintained by controlling the pulsations. Pulse combustion has not been

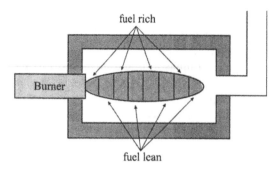

Figure 6.50 Alternating fuel-rich and fuel-lean zones in pulse combustion. (From Ref. 59. Courtesy of CRC Press.)

used in many industrial combustion processes at this time due to some operational problems, especially the high-frequency cycling of the switching valves, that have not yet been satisfactorily resolved.

6.8.3.3 Staging

Staged combustion, sometimes referred to as off-stoichiometric combustion, is an effective technique for lowering NO_x. Staging means that some of the fuel or oxidizer or both is added downstream of the main combustion zone. The fuel, oxidizer, or both may be staged into the flame. For example, there may be primary and secondary fuel inlets where a portion of the fuel is injected into the main flame zone and the balance of the fuel is injected downstream of that main flame zone. In fuel staging, some of the fuel is directed into the primary combustion zone while the balance is directed into secondary and even tertiary zones in some cases (see Fig. 3.24). This makes the primary zone fuel lean, which is less conducive to NO_x formation when compared to stoichiometric conditions. The excess O_2 from the primary zone is then used to combust the fuel added in the secondary and tertiary zones. While the overall stoichiometry may be the same as in a conventional burner, the peak flame temperature is much lower in the staged fuel case because the combustion process is staged over some distance while heat is simultaneously being released from the flame. The lower temperatures in the staged fuel flame help to reduce the NO_x emissions. Then, fuel staging is effective for two reasons: (1) the peak flame temperatures are reduced which reduces NO_x, and, (2) the fuel-rich chemistry in the primary flame zone also reduces NO_x. Waibel et al. [60] have shown that fuel staging is one of the most cost-effective methods for reducing NO_x in process heaters Buinevicius et al. [61] experimentally studied natural gas fuel staging to reduce NO_x emissions [61]. NO_x reductions up to 80% were demonstrated by providing 35–40% of the fuel through a secondary injector located 2.8–3 burner block diameters from the burner centerline.

In air staging, some of the combustion air is directed into the primary combustion zone while the balance is directed into secondary and even tertiary zones in some cases (see Fig. 3.20). This makes the primary zone fuel rich, which is less conducive to NO_x formation when compared to stoichiometric conditions. The unburned combustibles from the primary zone are then combusted in secondary and tertiary zones. While the overall stoichiometry may be the same as in a conventional burner, the peak flame temperature is much lower in the staged-air case because the combustion process is staged over some distance while heat is simultaneously being released from the flame. The lower temperatures in the staged air flame help to reduce the NO_x emissions. Giese et al. [62] demonstrated the use of air staging to reduce NO_x emissions in a new burner development project.

Baukal and Dalton [63] demonstrated that lancing oxygen into a flame (see Fig. 6.51) at a particular location can reduce NO_x compared to other locations. Pure oxygen was added to an air/fuel flame to bring the total oxygen in the combined oxidant up to 22.8%. The O_2:CH_4 stoichiometry without the pure O_2 was 1.8 and with the O_2 it was 2.0 (stoichiometric). The pure oxygen was added at different locations in the flame to determine the effects on NO_x. As shown in Fig. 6.52, lancing O_2 into the flame downstream from the base produced lower NO_x than by premixing the O_2 with the combustion air. It is interesting to note that it was even possible to reduce NO_x below the levels for air/fuel-only combustion.

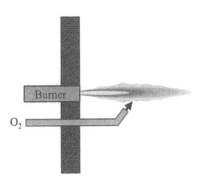

Figure 6.51 O_2 lancing into a flame. (From Ref. 63.)

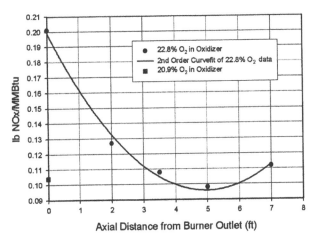

Figure 6.52 Oxygen enrichment of air to 22.8% overall O_2 in the oxidizer as a function of the location of the oxygen addition (From Ref. 63.)

Dearden et al. [64] experimentally and numerically demonstrated that oxidizer staging in an oxy/fuel flame can reduce NO_x emissions. Jager et al. [65] discuss the use of air staging to reduce NO_x emissions in a prototype burner. For the flat-shaped flame configuration tested, staging 60% of the oxygen produced the lowest NO_x emissions. NO_x reductions up to 71% were demonstrated for the patented burner designed for use in high-temperature glass furnaces. Fleck et al. [66] experimentally demonstrated a burner using staging of both the fuel and the oxidizer to produce low NO_x emissions (6 ppmv at 3% O_2) in a pilot-scale furnace including some air preheat. Pickenacker et al. [67] discuss the use of staging either fuel or air to reduce NO_x emissions from a porous inert-media burner. An oxy/fuel version of the burner was also shown to have very low NO_x.

Some of the practical techniques for staging include biased firing of the burners. An example would be firing some burners fuel rich and other burners fuel lean so that the overall mixture ratio is close to stoichiometric. Another example is referred to as burners-out-of-service, which is discussed in Sec. 6.8.3.8.

6.8.3.4   Gas Recirculation

There are two common ways to recirculate combustion exhaust products through a flame—by furnace gas recirculation (FuGR) and flue gas recirculation (FlGR). In flue gas recirculation, exhaust gases are recirculated from the exhaust stack or flue back through the burner (see Fig. 6.53). This requires some type of fan to circulate the gases external to the furnace and back through the burner. The burner must be designed to handle both the added volume and different temperature of the recirculated gases that are often partially or fully blended with the combustion air.

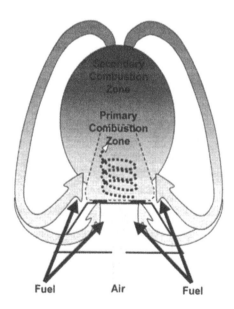

Figure 6.53   Schematic of a burner incorporating furnace gas recirculation. (From Ref. 3. Courtesy of John Zink Co.)

Flue gas recirculation causes the exhaust gases in the flue to be recirculated back through the burner into the flame via ductwork external to the furnace. Although the furnace or flue gases are hot, they are considerably cooler than the flame itself. The cooler furnace or flue gases act as a diluent, reducing the flame temperature, which in turn reduces NO_x (see Fig. 6.36). Advanced mixing techniques use carefully designed burner aerodynamics to control the mixing of the fuel and the oxidizer. The goals of this technique are to avoid hot spots and make the flame temperature uniform, to increase the heat release from the flame which lowers the flame temperature, and to control the chemistry in the flame zone to minimize NO_x formation. Garg [68] estimated NO_x reductions of up to 50% using flue gas recirculation. Brune et al. [69] showed that NO_x emissions increased with the temperature of the recirculating flue gas for a self-recuperative burner.

Furnace gas recirculation (FuGR) is a process that causes the products of combustion inside the combustion chamber to be recirculated back into the flame. Figure 6.54 shows a schematic of a burner that recirculates furnace gases through the flame zone to minimize NO_x emissions.

In FuGR, the combustion products are recirculated inside the furnace back to the burner and are inspirated into the flame to moderate its temperature (see Fig. 6.55). Even though the recirculated combustion products are hot, they are still considerably cooler than the flame. These recirculated gases cool the flame and lower NO_x emissions. Droesbeke et al. [70] demonstrated significant reductions in NO_x from an industrial burner using flue gas recirculation. Bob et al. [71] describe a burner design incorporating FuGR and preheated air staging to minimize NO_x emissions while still achieving high thermal efficiencies. Smirnov [72] describes a new high-momentum radiant tube burner for reducing NO_x emissions.

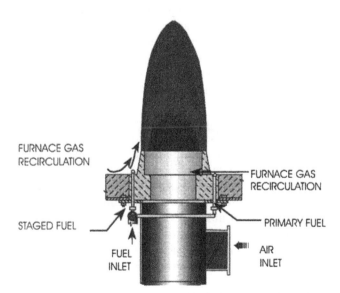

Figure 6.54 Schematic of a burner incorporating furnace gas recirculation. (From Ref. 3. Courtesy of John Zink Co.)

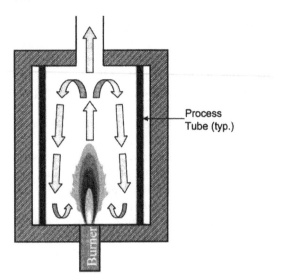

Process
Tube (typ.)

Figure 6.55 Schematic of furnace gas recirculation. (From Ref. 3. Courtesy of John Zink Co.)

The high-momentum gases induce increased FuGR through the burner, which reduces NO_x generation.

6.8.3.5 Water/Steam Injection

Many of the combustion modification methods attempt to reduce the temperature of the flame to lower NO_x emissions. In many cases, this may result in a reduction of the combustion efficiency [73]. For example, if water is injected into the flame to lower NO_x, the water absorbs heat from the flame and carries most of that energy out with the exhaust gases, preventing the transfer of much of that energy to the load. The water acts as a thermal ballast to reduce the peak flame temperatures. Combustion modification methods are usually less capital intensive than most post-treatment methods. In many cases, there is a limit to how much NO_x reduction can be achieved using these methods.

Another form of water injection is to inject water in the form of steam. There are several reasons for this. One is that steam is much hotter than liquid water and already includes the latent heat of vaporization needed to change the liquid water to a vapor. When liquid water is injected into a combustion process, it can put a large heat load on the process because liquid water can absorb a large amount of energy before becoming a vapor due to its high latent heat of vaporization. Steam puts a much smaller load on the process because it absorbs less energy than liquid water. Another reason for using steam instead of liquid water is that steam is already in vapor form and mixes readily with the combustion gases. Liquid water must be injected through nozzles to form a fine mist to disperse it uniformly with the combustion gases. Therefore, it is often easier to blend steam than liquid water into the combustion products. Steam is commonly used to vaporize liquid fuels like oil, which can reduce NO_x emissions compared to vaporizing the fuel with compressed air as is sometimes done.

6.8.3.6 Reburning

Reburning is a technique similar to fuel staging but uses a different strategy. An example is methane reburn where some methane is injected into the exhaust gases, usually well after the primary combustion zone, where the gases are at a lower temperature. As previously shown in Fig. 6.28, fuel-rich conditions are not favorable to NO_x. As the exhaust gases from the combustion process flow through this fuel-rich reducing zone, NO_x is reduced back to N_2. Any CO and other unburned fuels in the exhaust gases are then combusted downstream of the reduction zone at temperatures well below those found in the main combustion process. These lower temperature reactions are not favorable to NO_x formation so the net effect is that NO_x is reduced. This is a type of nitrogen reformer [74].

There are some challenges with this technique. One is to get proper injection of the reburning gas and the exhaust products. Another is that the reburn zone must be capable of sustaining combustion. Because it needs to be done in a lower temperature part of the process to minimize the subsequent formation of NO_x, it may need to be done, for example, in a previously uninsulated portion of ductwork that may have to be replaced with higher temperature materials and it may need to be insulated. A third challenge is trying to take advantage of some of the energy produced during the reburning. A heat-recovery system may need to be added.

Dagaut et al. [74] have shown that the reburning fuel has some effect on the NO_x reduction efficiency at a given temperature [74]. For the fuels considered, the ranking from lowest to highest NO_x reduction efficiency was as follows: methane, natural gas, ethane, ethylene, and acetylene. Zamansky et al. [75] showed experimentally that adding nitrogen agents like ammonia and urea and sodium salts into the flue gases with a delay time of 0.1–0.5 sec after injection of the reburning fuel can enhance the reburning process. At that point in the process, the oxygen content in the flue gases has been depleted to the point that the injection of nitrogen agents does not promote NO generation. The enhanced NO reduction can be explained by the presence of additional active radicals that promote the reduction.

Miller et al. [76] experimentally demonstrated that fuel-lean reburning can reduce NO_x emissions in exhaust gas streams. The application of reburning in field applications has often been restricted to overall fuel-rich or reducing conditions in the reburning zone. This research shows it is technically possible to utilize reburning for NO_x reduction of up to 50% in fuel-lean (up to 6% O_2 in the flue gases) environments. Watts et al. [77] have written a brief review of the use of reburn technology to reduce NO_x emissions in coal-fired boilers. Braun-Unkhoff et al. [78] compared experimental data on reburning technology applied on a large scale with numerical predictions. There was a good correlation between the measurements and predictions. Wendt [79] has shown that nonideal mixing can enhance the effectiveness of reburning for diffusion flames.

6.8.3.7 Low NO_x Burners

Garg [80] discusses the use of various low NO_x burners to achieve emissions reductions compared to standard gas burners. Table 6.9 shows typical NO_x reductions using various low NO_x burner techniques. An EPA study found that ultra-low NO_x burners were the most cost-effective means to reduce NO_x [16].

Table 6.9 NO_x Reductions for Different Low-NO_x Burner Types

Burner type	Typical NO_x reduction (%)
Staged-air burner	25–35
Staged-fuel burner	40–50
Low-excess-air burner	20–25
Burner with external flue gas recirculation (FGR)	50–60
Burner with internal flue gas recirculation	40–50
Air or fuel-gas staging with internal FGR	55–75
Air or fuel-gas staging with external FGR	60–80

Source: Ref. 80.

A relatively new development in industrial burner technology for reducing NO_x is called lean premix [81]. Lean premix involves running the primary part of the flame with as much air as possible and as little fuel as possible, that is, to make the air/fuel mixture as fuel lean as possible. This is not conducive to NO_x emissions as previously shown in Fig. 6.28. The challenge running near the flammability limit is to maintain flame stability. Ren et al. [24] discuss the use of lean premix combined with methane reforming as an option either to preheating the fuel or adding expensive components like hydrogen or carbon monoxide to the fuel to increase stability. Both of those techniques add cost and complexity to the combustion system. As previously shown, the addition of hydrogen also increases flame temperatures and usually NO_x emissions.

Okajima et al. [82] studied a noncircular burner shaped like a star with six points to determine the impact of burner shape on NO_x. Experimental results showed the star burner produces less NO_x than a comparable round burner due in part to the change in blow-off characteristics and flame stabilization.

6.8.3.8 Burners-Out-Of-Service (BOOS)

This is a technique primarily used in boilers where the fuel is turned off to the upper burners, while maintaining the air flow [83]. The fuel removed from the upper burners is then redirected to the lower burners, while maintaining the same air flow to the lower burners. Therefore, the overall fuel and air flow to the boiler remains the same, but is redistributed. This makes the lower burners fuel rich, which is less conducive to NO_x formation due to the lower flame temperatures and fuel-rich chemistry. The upper burners running on air only provide the rest of the air needed to fully combust the fuel. Rather than air staging in individual burners, the BOOS technique stages air over the entire boiler. This technique is relatively inexpensive to implement. Ensuring proper heat distribution is important to prevent overheating the tubes or derating the firing capacity.

6.8.3.9 Burner Spacing

The interaction between burners, especially low NO_x burners, can increase NO_x emissions. These burners are designed to have carefully controlled mixing, for example, fuel staging, to minimize the peak temperatures in the flame, which reduces NO. However, if the burners are spaced too closely together, the designed mixing is disturbed, which normally increases NO_x emissions. This is a relatively newer problem because of the advent of low NO_x burners that are inherently less intense than many older-style burners. The reduced intensity makes them more susceptible to the gas-flow patterns in a combustor, including the gas flows coming from adjacent burners.

There are two common options to mitigate this problem: modify the burner design to accommodate the tight spacing or modify the spacing to accommodate the low NO_x burners. Except in new furnace construction or in a major furnace rebuild, modifying the burner spacing is usually not preferred because of the high cost of modifying an existing furnace. However, modifying the burner design is also not necessarily an easy proposition either. If the flames are shortened, for example, by intensifying the mixing between the fuel and the combustion air, this normally increases NO_x emissions. In some cases it may be possible to modify the burners only where they come in close contact with adjacent burners. For example, one technique would be to reduce or eliminate fuel staging in the region between adjacent burners. If the burner has fuel nozzles around the perimeter, those closest to adjacent burners can be eliminated or have smaller outlet ports to reduce the fuel flow.

6.8.3.10 Pulsed Combustion

Pulsed combustion is a proven technique for reducing NO_x emissions but it has only been used on a limited number of industrial combustion applications. Martins et al. [84] experimentally studied NO_x emissions from a Rijke-type pulse combustor. While the system efficiency increased with pulse combustion compared to nonpulse combustion, NO_x emissions also increased. Kegasa [85] experimentally showed that forced oscillating combustion can reduce NO_x emissions under certain circumstances of a particular oscillation frequency, phase difference between the fuel and the air, and duty ratio. Barham et al. [86] studied the effect of the flapper-valve thickness on the operation of a pulsed combustor including pollutant emissions. The experimental results indicated that increasing the air/fuel ratio decreased NO_x emissions and did not have a significant impact on the system thermal efficiency. Carbon monoxide emissions were minimized at a specific air/fuel ratio.

6.8.4 Post-Treatment

The fourth strategy for minimizing NO_x is known as post-treatment. Post-treatment removes NO_x from the exhaust gases after the NO_x has already been formed in the combustion chamber. SO_x emissions are typically removed by reaction with a reagent like lime (CaO) to form particles that can be scrubbed out of the exhaust gas stream (see Chap. 8). However, there is no comparable inexpensive reagent and scrubbing system for NO_x. Another problem is that the reaction of NO_2 with water to form nitric acid is much slower than the comparable reactions for SO_x. The reaction is too slow to go to completion under normal conditions in the short time NO_x is in the treatment equipment. Therefore, a variety of other techniques are used to remove NO_x from exhaust gas streams. The general strategy is to use a reducing agent, such as CO, CH_4, other hydrocarbons, or ammonia, to remove the oxygen from the NO and convert it into N_2 and O_2. Often some type of catalyst is required for the reactions.

Two of the most common methods of post-treatment are selective catalytic reduction (SCR) and selective noncatalytic reduction (SNCR) [52]. SCR is generally used instead of SNCR when higher NO_x reduction is required. A catalyst is a substance that causes or speeds up a chemical reaction without undergoing a chemical change itself. Wet techniques for post-treatment include oxidation–absorption, oxidation–absorption–reduction, absorption–oxidation, and absorption–reduction. Dry techniques for post-treatment, besides SCR and SNCR, include activated carbon beds, electron beam radiation, and reaction with hydrocarbons. One of the advantages of post-treatment methods is that multiple exhaust streams can be treated simultaneously, thus achieving economies of scale. Most of the post-treatment methods are relatively simple to retrofit to existing processes.

Many of these techniques are fairly sophisticated and are not trivial to operate and maintain in industrial furnace environments. For example, the catalytic reduction techniques require a catalyst that may become plugged or poisoned fairly quickly by dirty flue gases. Post-treatment methods are often capital intensive. They usually require halting production if there is a malfunction of the treatment equipment. Also, post-treatment does not normally benefit the combustion process in any way. For example, it does not increase production or energy efficiency. It is strictly an add-on cost. Some trade organizations like the American Petroleum Institute have issued recommended practices for post-combustion NO_x control for fired equipment [87].

6.8.4.1 Selective Catalytic Reduction

Selective catalytic reduction (SCR) involves injecting a NO_x-reducing chemical into an exhaust stream in the presence of a catalyst within a specific temperature window. This process applied to a boiler is shown in Fig. 6.56 [88]. The chemical is typically ammonia and the temperature window is approximately 500° to 1100°F (230°–600°C). Figure 6.57 shows how the gas temperature affects the NO_x removal efficiency for an SCR process. The NO_x and NH_3 react on the catalyst surface to form N_2 and H_2O. The major reactions are

$$6NO + 4NH_3 \rightarrow 5N_2 + 6H_2O$$
$$2NO + 4NH_3 + 2O_2 \rightarrow 3N_2 + 6H_2O$$

The ammonia may be in the form of anhydrous ammonia, which is nearly 100% pure ammonia or mixed with water where the ammonia concentration is typically in the range 20–30%. SCR using ammonia as the reductant or reagent typically uses one of three types of catalysts: noble metal, base metal, and zeolites [89]. A common catalyst configuration used in selective catalytic reduction processes is shown in Fig. 6.58. The noble metals are normally washcoated on to an inert ceramic or metal monolith. These are used to treat particle-free, low-sulfur exhausts. Base metal catalysts are either washcoated or extruded on to honeycombs and are only used in particle-free exhausts. Zeolites may be washcoated or extruded into

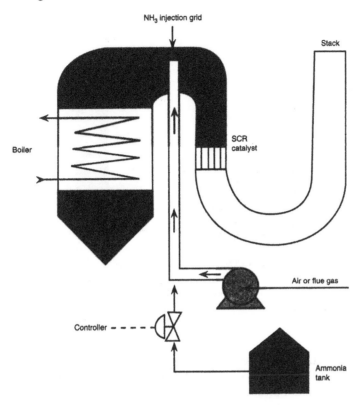

Figure 6.56   Schematic of the selective catalytic reduction process. (From Ref. 91.)

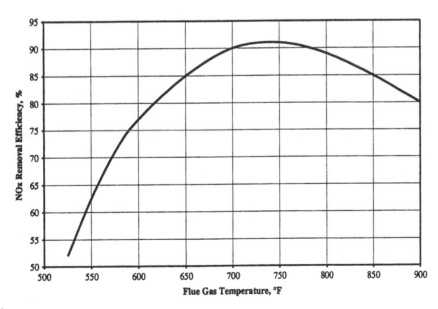

Figure 6.57   NO$_x$ removal efficiency vs. temperature for selective catalytic reduction. (From Ref. 91.)

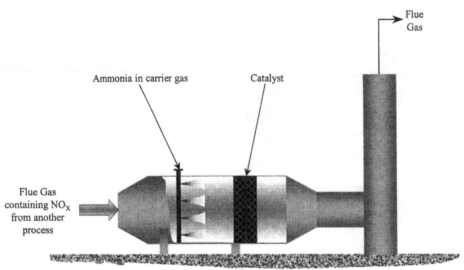

Figure 6.58 Common catalyst configuration used in selective catalytic reduction systems. (From Ref. 3. Courtesy of John Zink Co.)

honeycombs and can function at higher temperatures (650°–940°F). The most commonly used commercial SCR catalyst is vanadia supported on a high surface area anatase titania. Heck and Farrauto [90] recommend the following catalysts depending upon the temperature range:

- Low temperature (175°–250°C or 350°–480°F): platinum
- Medium temperature (300°–450°C or 570°–840°F): vanadium
- Low temperature (350°–600°C or 660°–1100°F): zeolite

An overall schematic of selective catalytic reduction is shown in Fig. 6.59. The U.S. EPA has prepared a helpful manual for estimating the cost of air pollution control using various techniques, which includes a chapter on SCR for NO_x control [91].

There are a number of potential problems and challenges with SCR techniques. The catalyst introduces a pressure drop into the system, which often increases the power requirements for the gas-handling equipment. The catalyst may become plugged or fouled in dirty exhaust streams, which is especially a challenge when firing liquid fuels like residual oil. The catalyst may also become poisoned or deactivated under certain conditions [92]. The ammonia must be properly injected into the flue gases to achieve proper mixing, must be injected at the right location to be in the proper temperature window, and must be injected in the proper amount to obtain adequate NO_x reduction without allowing ammonia to slip through unreacted. The presence of certain chemicals in certain concentrations can significantly affect the performance of an SCR [93]. SCR systems are not very tolerant of constantly changing conditions as a stable window of operation is required for optimum efficiency. Another problem is handling the spent catalyst. Figure 6.60 shows how the activity of the catalyst declines over time. Regeneration is often the most attractive but may be more expensive than buying new catalyst. Disposal of the spent catalyst may be expensive as it may be classified as a hazardous

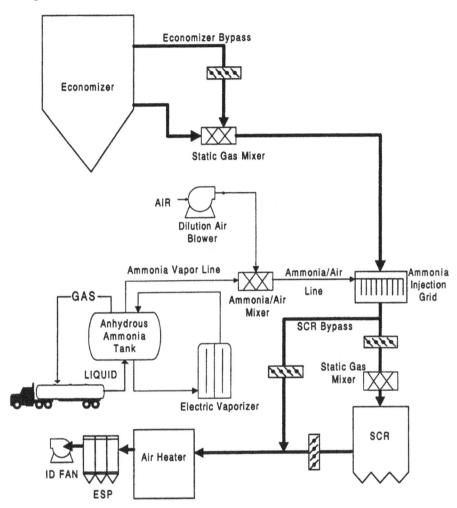

Figure 6.59 Selective catalytic reduction process flow diagram. (From Ref. 91.)

waste, especially if the catalyst contains vanadium as is commonly the case. An EPA study found that SCR was the most expensive means to reduce NO_x [16]. SCR systems are capital intensive, have high operating costs, and require significant amounts of real estate in the plant that is often difficult to find.

Research continues on finding new combinations of chemicals, catalysts, and catalyst supports for economically reducing NO_x. An example is research looking at NO_x reduction using ammonia with both supported and unsupported vanadia catalysts [94]. Ozkan et al. [95] have edited a book that includes a wide range of research in the general area of SCR. Aerogels of titania, silica, and vanadia are being investigated for use in SCR processes [96].

Laplante and Lindenhoff [97] discuss the use of SCRs in refineries to meet the increasingly more stringent NO_x regulations. Reductions of up to 95% are possible. According to the authors, this may be the only currently available technology to achieve single-digit NO_x performance. The NO_x catalyst is made of fiber-reinforced

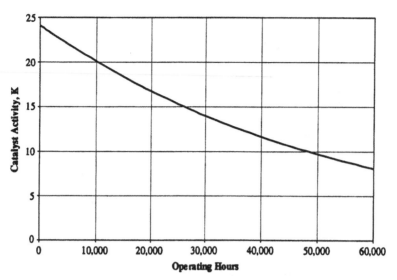

Figure 6.60 Typical catalyst deactivation for an SCR as a function of operating time. (From Ref. 91.)

titanium oxide carrier that is impregnated with vanadium oxide and tungsten oxide on the catalyst surface. The catalyst is assembled into modules for ease of handling and installation.

6.8.4.2   Selective Noncatalytic Reduction

Selective noncatalytic reduction (SNCR) involves injecting NO_x-reducing chemicals, such as ammonia, into the exhaust products from a combustion process within a specific temperature window (see Fig. 6.61). No catalyst is involved in the process, which is one advantage over SCR. The most commonly used chemicals are ammonia and urea. Other chemicals like hydrogen, hydrogen peroxide, and methanol may be added to improve the performance and lower the minimum threshold temperature.

The Exxon thermal de-NO_x process is one common SNCR technique using ammonia that is employed in a wide variety of industrial applications. A typical global reaction for this technique may be written as

$$2NO + 2NH_3 + 2O_2 \rightarrow 2N_2 + 3H_2O$$

The optimum temperature window, without the addition of other chemicals to increase the temperature window, is 1600°–2200°F (870°–1200°C). The effect of temperature on NO_x reduction performance is shown in Fig. 6.62 [98].

The Nalco Fuel Tech NO_xOUT® is a common SNCR technique employing urea:

$$CO(NH_2)_2 + 2NO + 1/2O_2 \rightarrow 2N_2 + CO_2 + 2H_2O$$

The optimum temperature window, without the addition of other chemicals to increase the temperature window, is 1600°–2000°F (870°–1100°C). At higher

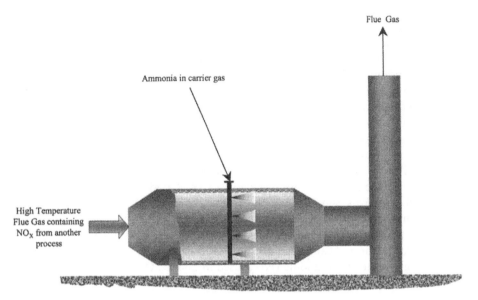

Figure 6.61 Selective noncatalytic reduction system. (From Ref. 3. Courtesy of John Zink Co.)

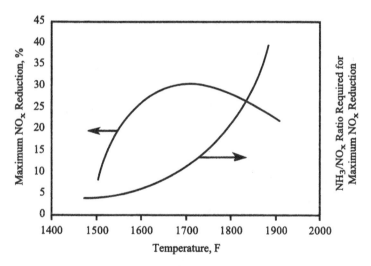

Figure 6.62 SNCR temperature window. (From Ref. 98. Courtesy of CRC Press.)

temperatures, the reagent (e.g., ammonia or urea) can actually oxidize and make NO, which is both counterproductive and wastes reagent. Rota et al. [99] experimentally and numerically studied the $NO_x OUT$ process in the temperature range 950–1450 K (1250°–2150°F) using a laboratory reactor. The experiments showed that the process is very effective in a narrow temperature window of 1250–1300 K (1800°–1900°F). The nitrogen in the reducing agent and the O_2 concentration in the gas being treated are also important parameters in the removal process. For the conditions studied, the CO/NO ratio did not significantly affect the removal efficiency.

Besides the temperature operating window, residence time is another important factor for optimum performance. Sufficient reaction time can effectively broaden the operating window [100]. Figure 6.63 shows how increasing the residence time from 100 to 500 msec greatly expands the operating window and improves the NO_x reduction efficiency. Mixing is another important factor for maximizing NO_x removal efficiencies [101].

There are many potential problems with SNCR. The first is the initial and operating costs, which are usually significantly more than nonpost-treatment techniques like low NO_x burners. Although the use of SNCR decreases NO_x, it may increase other undesirable emissions such as CO, N_2O, and NH_3 (which can occur if the injected chemicals slip through the exhaust without reacting, referred to as ammonia slip) [16]. However, as shown in Fig. 6.64, the NO_x removal efficiency increases as the ammonia slip increases so that acceptable limits must be determined either by design or regulatory constraints. There are also safety concerns with transporting and storing ammonia (NH_3) used in SNCR. Other major challenges of this technology include:

- Finding the proper location in the process to inject the chemicals (the chemicals must be injected where the flue gases are within a relatively narrow temperature window for optimum efficiency).
- Injecting the proper amount of chemicals (too much will cause some chemicals to slip through unreacted and too little will not get sufficient NO_x reductions).
- Getting proper mixing of the chemicals with the flue gas products (there must be both adequate mixing and residence time for the reactions to go to completion).

Both physical modeling and computer modeling are often used to determine the optimal place, amount, and method of injection.

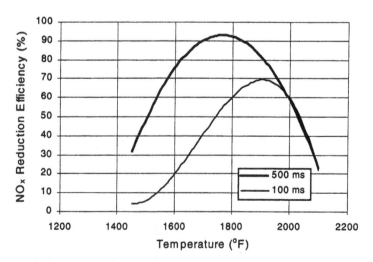

Figure 6.63   Effect of residence time on SNCR NO_x reduction efficiency. (From Ref. 103.)

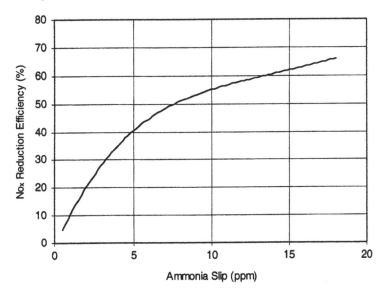

Figure 6.64 Effect of ammonia slip on SNCR NO$_x$ reduction efficiency. (From Ref. 103.)

Under ideal laboratory conditions, SCNRs can be shown to achieve significant NO$_x$ reductions. However, in practical applications the reductions are usually much less due to the nonuniformity of the temperature profile in the combustor, difficulties with completely mixing the nitrogen agent (ammonia, urea, cyanuric acid, etc.) into the exhaust stream, limited residence times, and ammonia slip. These can reduce the effectiveness by up to 50%. Zamansky et al. [102] experimentally and computationally demonstrated that it is possible to alleviate these practical problems with the injection of inexpensive and nontoxic inorganic salts such as sodium carbonate with the nitrogen agent. The results showed that the inorganic salts can significantly broaden the effective temperature window, which can be a significant obstacle in industrial processes due to the nonhomogeneity of the gas temperatures. The U.S. EPA has prepared a helpful manual for estimating the cost of air pollution control using various techniques, which includes a chapter on SNCR for NO$_x$ control [103].

6.8.4.3 Catalytic Reduction

One method of reducing NO$_x$ is referred to as catalytic cleaning [104]. In this process, NO formed during combustion is converted into N$_2$ in the presence of a catalyst according to the basic reaction:

$$NO + CO \leftrightarrow N_2 + CO_2$$

Figure 6.65 shows a schematic of a typical catalytic reduction system.

In the application of this technology to recuperative radiant tube burners, the catalyst is placed in the recuperator section of the burner in a position where the gases will be in the proper temperature range to optimize the catalytic reactions. A second catalyst is located further downstream in the burner to oxidize the CO. The flue gas temperatures and excess air ratios in the vicinity of the catalysts affect the

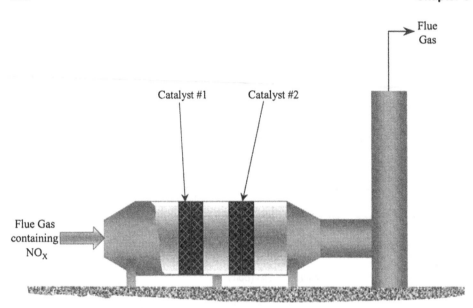

Figure 6.65 Catalytic cleaning NO_x reduction system.

efficiency of the reduction processes. NO_x reductions of up to 87% are projected for this newly developed process. The advantage of this technique compared to SCR and SNCR is that no reagent is needed. However, further work is required to make catalytic cleaning effective in industrial combustion applications.

6.8.4.4 Other

Thomas and Vanderschuren [105] describe the use of hydrogen peroxide (H_2O_2) to reduce NO_x emissions in a wet-absorption process. Their work shows that there is an optimum concentration of H_2O_2 beyond which there is no increase in NO_x removal efficiency. Figure 6.66 shows a patented packed tower scrubber used to remove NO_x from gases.

One technique currently under development is termed nonthermal plasma [106]. A pulsed corona discharge generates short high-frequency pulses that increase radical production in an exhaust stream. These radicals then make it easier to transform the toxic pollutants into benign compounds. A key to the use of these plasmas is that they have little effect on the bulk carrier gas and preferentially react with the trace constituents. A catalyst is used to convert the radical species produced by the plasma into benign species. In one form of this technology, NO and NO_2 are oxidized to N_2O_5, which can be removed with a wet scrubber.

Another new development is a device called a fungal vapor-phase bioreactor that is designed to remove NO and VOCs from waste gas streams [107]. This technology, generally referred to as biofiltration, shows a resistance to adverse operating conditions and is able to maintain high removal efficiencies over an extended period of time, which are problems with some other post-treatment technologies. Removal efficiencies exceeding 90% for both NO_x and VOCs have been demonstrated on a bench-scale apparatus.

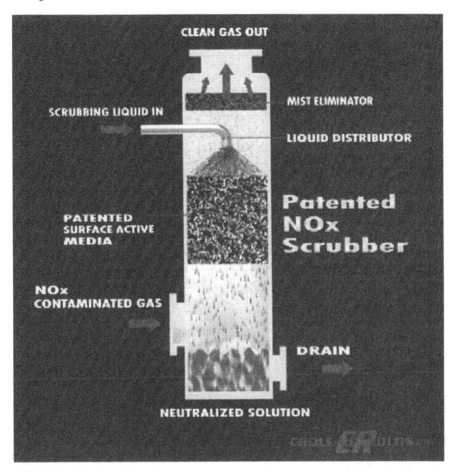

Figure 6.66 Packed tower scrubber used for NO_x removal. (Courtesy of Croll, Westfield, NJ.)

A relatively new technique referred to as low-temperature oxidation with absorption converts NO and NO_2 into another form of NO_x (N_2O_5) that is very soluble and is easily removed with a wet scrubber [98]:

$$NO + O_3 \rightarrow NO_2 + O_2$$

$$NO_2 + O_3 \rightarrow N_2O_5 + O_2$$

$$N_2O_5 + H_2O \rightarrow HNO_3$$

In this technique, the oxidizer is ozone (O_3), which can be generated using either air or pure oxygen. The process takes place at about 300°F, which makes it impractical for many industrial combustion processes unless the flue gases are cooled prior to reaching the ozone. At temperatures above about 500°F, ozone decomposes very rapidly. The process can have NO_x removal efficiencies as high as 99%. Other reactions with CO and SO_2 are slow compared to the NO_x reactions and do not compete for the ozone.

REFERENCES

1. Office of the Federal Register. 60.2 Definitions. U.S. Code of Federal Regulations, Title 40, Part 60. Washington, DC: U.S. Government Printing Office, 2001.
2. U.S. Environmental Protection Agency. Nitrogen Oxide Control for Stationary Combustion Sources. U.S. EPA Rep. EPA/625/5-86/020, 1986.
3. CE Baukal (ed.). *The John Zink Combustion Handbook.* Boca Raton, FL: CRC Press, 2001.
4. www.nobel.se/announcement-98/medicine98.html
5. M Hori, N Matsunaga, N Marinov, W Pitz, C Westbrook. An Experimental and Kinetic Calculation of the Promotion Effect of Hydrocarbons on the NO–NO$_2$ Conversion in a Flow Reactor. *Twenty-Sixth Symposium (International) on Combustion.* Pittsburgh, PA: The Combustion Institute, pp. 389–396, 1996.
6. M Sandell. Putting NO$_x$ in a box. *Pollut. Eng.,* Vol. 30, No. 3, pp. 56–58, 1998.
7. K Svoboda, J Cermák, V Vesely. NO$_x$ Chemistry and Emissions—II: Heterogeneous Reactions (N$_2$O). In C Vovelle (ed.). *Pollutants from Combustion: Formation and Impact on Atmospheric Chemistry.* Dordrecht, The Netherlands: Kluwer, 2000.
8. J Tomeczek, B Gradon. Contribution of Nitrous Oxide Reactions to the NO Formation in Flames. *Fifth International Conference on Technologies and Combustion for a Clean Environment.* Lisbon, Portugal, Vol. 1, The Combustion Institute—Portuguese Section, pp. 129–137, 1999.
9. M Moreton, S Beal. Controlling NO$_x$ emissions. *Pollut. Eng. Int.,* pp. 14–16, Winter, 1998.
10. U.S. EPA. Control Techniques for Nitrogen Oxides Emissions from Stationary Sources. EPA Rep. 450/1-78-001. Washington, DC: U.S. Environmental Protection Agency, 1978.
11. SC Hill, LD Smoot. Modeling of nitrogen oxides formation and destruction in combustion systems. *Prog. Energy Combust. Sci.,* Vol. 26, Nos. 4–6, pp. 417–458, 2000.
12. YB Zeldovich. *Acta Physicochem. (USSR),* Vol. 21, p. 557, 1946.
13. AM Dean, JW Bozzelli. Chap. 2: Combustion Chemistry of Nitrogen. In WC Gardiner (ed.). *Gas-Phase Combustion Chemistry.* New York: Springer, 2000.
14. CE Baukal (ed.). *Oxygen-Enhanced Combustion.* Boca Raton, FL: CRC Press, 1998.
15. CP Fenimore. Formation of Nitric Oxide in Premixed Hydrocarbon Flames. *Thirteenth Symposium (International) on Combustion.* Pittsburgh, PA: The Combustion Institute, pp. 373–380, 1971.
16. EB Sanderford. Alternative Control Techniques Document—NO$_x$ Emissions from Process Heaters. U.S. Environ. Protection Agency Rep. EPA-453/R-93-015, Feb. 1993.
17. R Battye, S Walsh, J Lee-Greco. NO$_x$ Control Technologies for the Cement Industry. EPA Rep., Washington, DC: Environmental Protection Agency, September 2000.
18. U.S. EPA. Control Techniques for Nitrogen Oxide Emissions from Stationary Sources. Rep. EPA-450/3-83-002. Washington, DC: Environmental Protection Agency, Jan. 1983.
19. U.S. EPA. Alternative Control Techniques—NO$_x$ Emissions from Utility Boilers. Rep. EPA-453/R-94-023, Washington, DC: Environmental Protection Agency, 1994.
20. RW Bilger. Zero Release Combustion Technologies and the Oxygen Economy. *Fifth International Conference on Technologies and Combustion for a Clean Environment.* Lisbon, Portugal, Vol. 2. The Combustion Institute—Portuguese Section, pp. 1039–1046, 1999.
21. LM Howard, PM Patterson, M Pourkashanian, A Williams, CW Wilson, LT Yap. Experimental Investigation of Oxygen-Enriched Lifted and Attached Turbulent Diffusion Flames. *Fifth International Conference on Technologies and Combustion for a Clean Environment.* Lisbon, Portugal, Vol. 2. The Combustion Institute—Portuguese Section, pp. 1133–1142, 1999.

22. CJ Sung, CK Law. Dominant Chemistry and Physical Factors Affecting NO Formation and Control in Oxy–Fuel Burning. *Twenty-Seventh Symposium* (*International*) *on Combustion*. Pittsburgh, PA: The Combustion Institute, pp. 1411–1418, 1998.

23. W Qin, J-Y Ren, FN Egolfopoulos, S Wu, H Zhang, TT Tsotsis. Oxygen Composition Modulation Effects on Flame Propagation and NO_x Formation in Methane/Air Premixed Flames. *Proceedings of the Combustion Institute*, Pittsburgh, PA, Vol. 28, pp. 1825–1831, 2000.

24. J-Y Ren, F-N Egolfopoulos, TT Tsotsis. NO_x emission control of lean methane–air combustion with addition of methane reforming products. *Combust. Sci. Technol.*, Vol. 174, pp. 181–205, 2002.

25. RR Hayes, CE Baukal, P Singh, D Wright. Fuel Composition Effects on NO_x. *Proceedings of the Air & Waste Management Association's 94th Annual Conference & Exhibition*, Paper 434, Orlando, FL, 2001.

26. J Yuan, Z Xiao, I Garshore, M Salcudean. NO_x Emission Modeling and Control in Bark Boilers. *Proceedings of 5th European Conference on Industrial Furnaces and Boilers*. Lisbon, Portugal, Vol. 1, pp. 659–667, 2000.

27. L Sorum, O Skreiberg, P Glarborg, A Jensen, K Dam-Johansen. Formation of NO from combustion of volatiles from municipal solid wastes. *Combust. Flame*, Vol. 123, pp. 195–212, 2001.

28. M Katsuki, T Hasegawa. The Science and Technology of Combustion in Highly Preheated Air. *Twenty-Seventh Symposium* (*Intenational*) *on Combustion*. Pittsburgh, PA: The Combustion Institute, pp. 3135–3146, 1998.

29. R Weber, AL Verlaan, S Orsino, N Lallemant. On Emerging Furnace Design Methodology that Provides Substantial Energy Savings and Drastic Reductions in CO_2, CO and NO_x Emissions. *Proceedings of 5th European Conference on Industrial Furnaces and Boilers*. Lisbon, Portugal, Vol. 1, pp. 43–57, 2000.

30. A Scherello, M Flamme, H Kremer. Optimization of Burner Systems for Glass Melting Furnaces with Regenerative Air Preheating. *Proceedings of 5th European Conference on Industrial Furnaces and Boilers*. Lisbon, Portugal, Vol. 1, pp. 323–382, 2000.

31. T Kapros, J Solyom. Development of Gas Firing System Using Rotary Regenerator. *Proceedings of 5th European Conference on Industrial Furnaces and Boilers*. Lisbon, Portugal, Vol. 1, pp. 363–369, 2000.

32. S Giammartini, G Girardi, R Cipriani, F Cuoco, M Sica. Diluted Combustion with High Air-Preheating: Experimental Characterization of Laboratory Furnace by Means of Advanced Diagnostics. *Proceedings of 5th European Conference on Industrial Furnaces and Boilers*. Lisbon, Portugal, Vol. 1, pp. 371–381, 2000.

33. A Gaba. Low NO_x Recuperative High Velocity Recirculating Burner. *Proceedings of 5th European Conference on Industrial Furnaces and Boilers*. Lisbon, Portugal, Vol. 1, pp. 397–405, 2000.

34. LT Yap, M Pourkashanian, L Howard, A Williams, RA Yetter. Nitric-Oxide Emissions Scaling of Buoyancy-Dominated Oxygen-Enriched and Preheated Methane Turbulent-Jet Diffusion Flames. *Twenty-Seventh Symposium* (*International*) *on Combustion*. Pittsburgh, PA: The Combustion Institute, pp. 1451–1460, 1998.

35. JH Frank, RS Barlow, C Lundquist. Radiation and Nitric Oxide Formation in Turbulent Non-Premixed Flames. *Proceedings of the Combustion Institute*. Pittsburgh, PA, Vol. 28, pp. 447–454, 2000.

36. T Fujimori, Y Hamano, J Sato. Radiative Heat Loss and NO_x Emission of Turbulent Jet Flames in Preheated Air up to 1230 K. *Proceedings of the Combustion Institute*. Pittsburgh, PA, Vol. 28, pp. 455–461, 2000.

37. K Wark, CF Warner. *Air Pollution: Its Origin and Control*, 2nd edn. New York: Harper & Row, 1981.

38. AC Stern, RW Boubel, DB Turner, DL Fox. *Fundamentals of Air Pollution*, 2nd edn. Orlando, FL: Academic Press, 1984.

39. J Elkins, N Frank, J Hemby, D Mintz, J Szykman, A Rush, T Fitz-Simons, T Rao, R Thompson, E Wildermann, G Lear. National Air Quality and Emissions Trends Report, 1999. Washington, DC: U.S. Environmental Protection Agency, Rep. EPA 454/R-01-004, 2001.

40. G Braswell, Y Matros, G Bunimovich. NO_x in non-utility industries (Part 1). *Environ. Protect.*, Vol. 12, No. 6, pp. 50–55, 2001.

41. U.S. Dept. of Energy Office of Industrial Technology. Petroleum—Industry of the Future: Energy and Environmental Profile of the U.S. Petroleum Refining Industry. U.S. DOE, Washington, DC, Dec. 1998.

42. American National Standards Institute/American Society Mechanical Engineers. Performance Test Code PTC 19.10, Part 10: Flue and Exhaust Gas Analyses. New York: ASME, 1981.

43. CE Baukal, PB Eleazer. Quantifying NO_x for industrial combustion processes. *J. Air Waste Mgmt. Assoc.*, Vol. 48, pp. 52–58, 1997.

44. SB Mandel. What is the total cost for emissions monitoring? *Hydrocarbon Process.* Vol. 76, No. 1, pp. 99–102, 1997.

45. U.S. Government. Code of Federal Regulations 40, Part 60, Revised July 1, 1994.

46. MC Drake. Kinetics of Nitric Oxide Formation in Laminar and Turbulent Methane Combustion, Gas Research Institute (Chicago, IL) Rep. No. GRI-85/0271, 1985.

47. MF Zabielski, LG Dodge, MB Colket, DJ Seery. The Optical and Probe Measurement of NO: A Comparative Study. *Eighteenth Symposium (International) on Combustion.* Pittsburgh, PA, The Combustion Institute, p. 1591, 1981.

48. A Berger, G Rotzoll. Kinetics of NO reduction by CO on quartz glass surfaces. *Fuel*, Vol. 74, p. 452, 1995.

49. AG Slavejkov, CE Baukal. Flue Gas Sampling Challenges in Oxygen–Fuel Combustion Processes. In Vol. 2. SH Chan (ed.). *Tranport Phenomena in Combustion*. Washington, DC: Taylor & Francis, 1230, 1996.

50. M Bradford, R Grover, P Paul. Controlling NO_x Emissions—Part 1. *Chem. Eng. Prog. Mag.*, Vol. 98, No. 3, pp. 42–46, 2002.

51. JL Reese, R Batten, GL Molianen, CE Baukal, R Borkowicz, DO Czerniak. State of the Art of NO_x Emission Control Technologies. ASME Paper 94-JPGC-EC-15, 1994.

52. J Bluestein. NO_x Controls for Gas-Fired Industrial Boilers and Combustion Equipment: A Survey of Current Practices. Gas Research Institute (Chicago, IL) Rep. GRI-92/0374, 1992.

53. HM Gomaa, LG Hackemesser, DT Cindric. NO_x/CO Emissions and control in ethylene plants. *Environ. Prog.* Vol. 10, No. 4, pp. 267–272, 1991.

54. D MacKenzie, I Cheta, D Burns. Removing nitrogen. *Hydrocarbon Eng.*, Vol. 7, No. 11, pp. 57–63, 2002.

55. CE Baukal, AI Dalton. Nitrogen Oxide Measurements in Oxygen Enriched Air-Natural Gas Combustion Systems. *Proceedings of 2nd Fossil Fuel Combustion Symposium*, ASME PD-Vol. 30, pp. 75–79, New Orleans, LA, Jan. 15, 1990.

56. AI Dalton, DW Tyndall. Oxygen Enriched Air/Natural Gas Burner System Development. Final Report, July 1984 – Sep. 1989, Gas Research Institute Rep. GRI-90/0140, Chicago, IL, 1989.

57. J Colannino. NO_x Reduction for Stationary Sources. *AIPE Facilities*, Vol. 23, No. 1, pp. 63–66, 1996.

58. J Rhine. Developments in Gas-Fired Industrial Heating Processes for the New Millenium. *Proceedings of 5th European Conference on Industrial Furnaces and Boilers.* Lisbon, Portugal, Vol. 1, pp. 203–219, 2000.

59. CE Baukal. *Heat Transfer in Industrial Combustion.* Boca Raton, FL: CRC Press, 2000.

60. R Waibel, D Nickeson, L Radak, W Boyd. Fuel Staging for NO_x Control. In MA Lukasiewicz (ed.). *Industrial Combustion Technologies.* Warren, PA: American Society of Metals, pp. 345–350, 1986.

61. K Buinevicius, E Puida, L Narbutas. Reduction of Nitrogen Oxides Formation During Gas Burning. *Proceedings of 5th European Conference on Industrial Furnaces and Boilers.* Lisbon, Portugal, Vol. II, pp. 617–622, 2000.

62. R Giese, A Al-Halbouni, R Sontag, P Kaferstein. The Influence of Air and Fuel Impulses on the Emission Behavior of the COSTAIR Gas Burner. *Proceedings of 5th European Conference on Industrial Furnaces and Boilers.* Lisbon, Portugal, Vol. II, pp. 525–532, 2000.

63. CE Baukal, AI Dalton. Reduction of Nitrogen Oxides in Oxygen-Enriched Combustion Processes. U.S. Patent 5 413 476 9 May 1995.

64. LM Dearden, M Pourkashanian, GT Spence, A Williams, J Connors, D Satchell, LT Yap. Development of a Staged Oxy–Fuel Flat-Jet Burner—Experimental Study on Turbulent Diffusion Oxygen Gas Flames. *Fifth International Conference on Technologies and Combustion for a Clean Environment*, Lisbon, Portugal, Vol. 2, The Combustion Institute—Portuguese Section, pp. 1127–1132, 1999.

65. FK Jager, H Kohne, K Luck, D Goebel. Scaling-Method for the Dimensioning of Low NO_x-Burners in the Capacity Range of 15 kW to 350 kW and 2 MW to 10 MW. *Proceedings of 5th European Conference on Industrial Furnaces and Boilers.* Lisbon, Portugal, Vol. 1, pp. 273–282, 2000.

66. B Fleck, HA Becker, A Sobiesiak, AD Lawrence. Ultra-Low-NO_x Natural Gas Burner: Development and Performance. *Proceedings of 5th European Conference on Industrial Furnaces and Boilers.* Lisbon, Portugal, Vol. 1, pp. 171–180, 2000.

67. O Pickenacker, A Kesting, D Trimis. Novel Low NO_x Burner Designs for Boilers and Furnaces by Using Staged Combustion in Inert Porous Media. *Proceedings of 5th European Conference on Industrial Furnaces and Boilers.* Lisbon, Portugal, Vol. 1, pp. 303–314, 2000.

68. A Garg. Trimming NO_x. *Chem. Eng.*, Vol. 99, No. 11, pp. 122–124, 1992.

69. M Brune, M Bob, M Flamme, A Lynen, JA Wunning, JG Wunning, HJ Dittmann. Optimisation [sic] of Ceramic Self-Recuperative Burners by Mathematical Modelling. *Proceedings of 5th European Conference on Industrial Furnaces and Boilers.* Lisbon, Portugal, Vol. 1, pp. 191–201, 2000.

70. A Droesbeke, F Missaire, O Berten, B Leduc, M Vansnick. Numerical Simulation of NO_x Emissions Reductions of an Industrial Burner. *Proceedings of 5th European Conference on Industrial Furnaces and Boilers.* Lisbon, Portugal, Vol. 1, pp. 235–244, 2000.

71. M Bob, M Brune, M Flamme. Low NO_x Combustion Systems for Intensified High Temperature Processes. *Proceedings of 5th European Conference on Industrial Furnaces and Boilers.* Lisbon, Portugal, Vol. II, pp. 533–542, 2000.

72. V Smirnov. High-Momentum Flame technology for low NO_x in SER tube burners. *Indust. Heat.*, Vol. LXVIII, No. 11, pp. 49–52, 2001.

73. HL Shelton. Find the right low-NO_x solution. *Environ. Eng. World*, Vol. 2, No. 6, pp. 24–27, 1996.

74. P Dagaut, J Luche, F Lecomte, M Cathonnet. The Kinetics of C_1 to C_4 Hydrocarbons–NO Reactions in Relation with Reburning. *Proceedings of 5th European Conference on Industrial Furnaces and Boilers.* Lisbon, Portugal, Vol. II, pp. 499–509, 2000.

75. VM Zamansky, L Ho, PM Maly, WR Seeker. Reburning Promoted by Nitrogen- and Sodium-Containing Compounds. *Twenty-Sixth Symposium (International) on Combustion.* Pittsburgh, PA: The Combustion Institute, pp. 2075–2082, 1996.

76. CA Miller, AD Touati, J Becker, JOL Wendt. NO_x Abatement by Fuel-Lean Reburning: Laboratory Combustor and Pilot-Scale Package Boiler Results. *Twenty-Seventh*

Symposium (International) on Combustion. Pittsburgh, PA: The Combustion Institute, pp. 3189–3195, 1998.

77. JU Watts, A Mann, J Harvilla, D Engelhardt. NO$_x$ Control by Utilization of Reburn Technologies in the United States. *Fifth International Conference on Technologies and Combustion for a Clean Environment*, Lisbon, Portugal, Vol. 2, The Combustion Institute—Portuguese Section, pp. 1017–1021, 1999.

78. M Braun-Unkhoff, P Frank, S Koger, W Leuckel. Evaluation of NO$_x$ Reburning Models Under Large Scale Conditions. *Fifth International Conference on Technologies and Combustion for a Clean Environment*, Lisbon, Portugal, Vol. 2, The Combustion Institute—Portuguese Section, pp. 1023–1032, 1999.

79. JOL Wendt. Pollutant Formation in Furnaces: NO$_x$ and Fine Particulates. *Proceedings of 5th European Conference on Industrial Furnaces and Boilers*. Lisbon, Portugal, Vol. 1, pp. 129–156, 2000.

80. A Garg. Specify better low-NO$_x$ burners for furnaces. *Chem. Eng. Prog.*, Vol. 90, No. 1, pp. 46–49, 1994.

81. JG Seebold, RT Waibel, TL Webster. Control refinery NO$_x$ emissions cost-effectively. *Hydrocarbon Process*. Vol. 80, No. 11, pp. 55–59, 2001.

82. S Okajima, T Watanabe, MLV Lu. Combustion Characteristics of Premixed Flames on Non-Circular Burner. *Proceedings of 5th European Conference on Industrial Furnaces and Boilers*. Lisbon, Portugal, Vol. 1, pp. 383–390, 2000.

83. J Colannino. Low-cost techniques reduce boiler NO$_x$. *Chem. Eng.*, pp. 100–106, Feb. 1993.

84. CA Martins, JA Carvalho, MA Ferreira, CAG Veras. An Experimental Investigation of NO$_x$ Formation in a Rijke Type Pulse Combustor. *Fifth International Conference on Technologies and Combustion for a Clean Environment*. Lisbon, Portugal, Vol. 1, The Combustion Institute—Portuguese Section, pp. 489–494, 1999.

85. A Kegasa. NO$_x$ Reduction Effect of Forced Oscillating Combustion. *Fifth International Conference on Technologies and Combustion for a Clean Environment*. Lisbon, Portugal, Vol. 1, The Combustion Institute—Portuguese Section, pp. 495–500, 1999.

86. P Barham, KJA Hargreaves, H Ipakchi, WC Maskell. Emission Characteristics and Influence of Flapper Valve Thickness upon Operation of a Premixed Pulsed Combustor. *Fifth International Conference on Technologies and Combustion for a Clean Environment*. Lisbon, Portugal, Vol. 1, The Combustion Institute—Portuguese Section, pp. 481–488, 1999.

87. API Recommended Practice 536: Post-Combustion NO$_x$ Control for Fired Equipment in General Refinery Serices, 1st Edn. Washington, DC: American Petroleum Institute, Mar. 1998.

88. KT Chuang, AR Sanger. 5.20 Gaseous Emission Control: Physical and Chemical Separation. In DHF Liu, BG Lipták (eds.). *Environmental Engineers' Handbook*, 2nd edn. Boca Raton, FL: Lewis Publishers, 1997.

89. CJ Pereira, MD Amiridis. NO$_x$ Control from Stationary Sources. In: US Ozkan, SK Agarwal, G Marcelin (eds.). *Reduction of Nitrogen Oxide Emissions*. Washington, DC: American Chemical Society, 1995.

90. RM Heck, RJ Farrauto. *Catalytic Air Pollution Control: Commercial Technology*. New York: Van Nostrand Reinhold, 1995.

91. DC Mussatti, R Srivastava, PM Hemmer, R Strait. Section 4.2: NO$_x$ Post-Combustion, Chap. 2: Selective Catalytic Reduction. In DC Mussatti (ed.). *Air Pollution Control Cost Manual*, 6th edn. Rep. EPA/452/B-02-001. Washington, DC: U.S. Environmental Protection Agency, Jan. 2002.

92. E Hums, GW Spitznagel. Deactivation Behavior of Selective Catalytic Reduction of DeNO$_x$ Catalysts. In US Ozkan, SK Agarwal, G Marcelin (eds.). *Reduction of Nitrogen Oxide Emissions*. Washington, DC: American Chemical Society, 1995.

93. BW-L Jang, JJ Spivey, MC Kung, B Yang, HH Kung. Low-Temperature Selective Catalytic Reduction of NO by Hydrocarbons in the Presence of O_2 and H_2O. In US Ozkan, SK Agarwal, G Marcelin (eds.). *Reduction of Nitrogen Oxide Emissions*. Washington, DC: American Chemical Society, 1995.

94. US Ozkan, Y Cai, MW Kumthekar. Selective Catalytic Reduction of Nitric Oxide with Ammonia over Supported and Unsupported Vanadia Catalysts. In RG Silver, JE Sawyer, JC Summers (eds.). *Catalytic Control of Air Pollution*, Washington, DC: American Chemical Society, 1992.

95. US Ozkan, SK Agarwal, G Marcelin (eds.). *Reduction of Nitrogen Oxide Emissions*. Washington, DC: American Chemical Society, 1995.

96. MD Amiridis, BK Na, EI Ko. Selective Catalytic Reduction of NO by NH3 over Aerogels of Titania, Silica, and Vanadia. In US Ozkan, SK Agarwal, G Marcelin (eds.). *Reduction of Nitrogen Oxide Emissions*, Washington, DC: American Chemical Society, 1995.

97. MP Laplante, P Lindenhoff. How low can you go?—Catalytic NO_x reduction in refineries. *World Refining*, Vol. 13, No. 2, pp. 46–51, 2002.

98. KB Schnelle, CA Brown. *Air Pollution Control Technology Handbook*. Boca Raton, FL: CRC Press, 2002.

99. R Rota, D Antos, EF Zanoelo, M Morbidelli. Experimental and modeling analysis of the NO_xOUT process. *Chem. Eng. Sci.*, Vol. 57, pp. 27–38, 2002.

100. ML Lin, JR Comparato, WH Sun. Applications of Urea-Based Selective Noncatalytic Reduction in Hydrocarbon Processing Industry. In US Ozkan, SK Agarwal, G Marcelin (eds.). *Reduction of Nitrogen Oxide Emissions*, Washington, DC: American Chemical Society, 1995.

101. R Rota, EF Zanoela, M Morbidelli, S Carra. Effect of Mixing on Selective Non-Catalytic Nitric Oxide Reduction. *Fifth International Conference on Technologies and Combustion for a Clean Environment*, Lisbon, Portugal, Vol. 1, The Combustion Institute—Portuguese Section, pp. 119-128, 1999.

102. VM Zamansky, PM Maly, L Ho, VV Lissianski, D Rusli, WC Gardiner. Promotion of Selective Non-Catalytic Reduction of NO by Sodium Carbonate. *Twenty-Seventh Symposium (International) on Combustion*. Pittsburgh, PA: The Combustion Institute, pp. 1443–1449, 1998.

103. DC Mussatti, R Srivastava, PM Hemmer, R Strait. Section 4.2: NO_x Post-Combustion, Chap. 1: Selective Noncatalytic Reduction. In DC Mussatti (ed.). *Air Pollution Control Cost Manual*, 6th edn. Rep. EPA/452/B-02-001. Washington, DC: U.S. Environmental Protection Agency, Jan. 2002.

104. M Naslund, N Lindskog. Catalytic cleaning in radiant tube burners reduces NO_x emissions. *Indust. Heat.*, Vol. LXVIII, No. 8, pp. 41–46, 2001.

105. D Thomas, J Vanderschuren. The Use of Hydrogen Peroxide for the NO_x Reduction in Wet Processes. In N Piccinini, R Delorenzo (eds.). *Chemical Industry and Environment II*, Vol. 2. Turin, Italy: Politecnico di Torino, 1996.

106. I Orlandini, U Riedel. Modelling of NO and HC removal by non-thermal plasmas. *Combust. Theory Model.* Vol. 5, pp. 447–462, 2001.

107. JR Woertz, KA Kinney, PJ Szaniszlo. A fungal vapor-phase bioreactor for the removal of nitric oxide from waste gas streams. *J. Air Waste Mgmt Assoc.*, Vol. 51, pp. 895–902, 2001.

7

Unburned Combustibles

7.1 INTRODUCTION

This section has been broken into two types of combustibles. The first involves the incomplete combustion of the fuel, which usually produces carbon monoxide and, in some limited cases, not all of the hydrocarbon fuel is consumed and passes through the combustor unreacted. The second type of combustible is volatile organic compounds (VOCs), which are generally only important in a limited number of processes, typically involving contaminated or otherwise hazardous waste streams.

7.2 CARBON MONOXIDE

While carbon monoxide (CO) is produced naturally by processes like agricultural fires, methane oxidation, plant growth and decay, and other natural processes, the largest source is by the incomplete burning of fossil fuels. It is generally produced in trace quantities in many combustion processes as a product of incomplete combustion. The largest source is from automobile emissions, which tend to peak in a given location during the morning and evening rush hours where city traffic is heaviest.

Carbon monoxide is a flammable gas, which is nonirritating, colorless, odorless, tasteless, and normally noncorrosive. It is highly toxic and acts as a chemical asphyxiant by combining with hemoglobin in the blood that normally transports oxygen inside the body. The affinity of CO for hemoglobin is approximately 300 times greater than that of oxygen for hemoglobin [1]. CO preferentially combines with hemoglobin to the exclusion of oxygen so that the body becomes starved of oxygen, which can eventually lead to asphyxiation. CO weakens the contractions of the heart, which then decreases the blood flow to the muscles and organs. It is particularly dangerous for those with respiratory problems. Even at low concentrations, CO can affect mental function, vision, and alertness. It is insidious in that the victim may not even realize what is happening because the effects are gradual. Depending on the CO level and exposure time, effects on humans can range from mild headaches to death. The very young, the very old, and those with respiratory problems are at the greatest risk. There is no evidence that CO has any deleterious effects on material surfaces or plant life [2]. Because of its effects on human health, CO is a regulated pollutant with specific emissions guidelines, depending on the application and the geographical location.

7.2.1 Formation Mechanisms

Carbon monoxide is generally produced by the incomplete combustion of a carbon-containing fuel. While CO is generally very easy to convert into CO_2, it may be formed if there is insufficient oxygen for complete combustion, if there is insufficient temperature to react the CO fully, and if there is insufficient residence time in a combustion zone with adequate temperature and oxygen for complete combustion.

Before the advent of some of the new NO_x reduction technologies like air staging (see Chap. 6), there was generally no concern for CO emissions from industrial combustion processes, which had more than sufficient oxygen, temperature, and residence time to burn CO completely. However, industrial combustion systems have been significantly modified to control other pollutants so that it may now be possible in some systems to have CO emissions in the exhaust stream.

Normally, a combustion system is operated slightly fuel lean (excess O_2) to ensure complete combustion and to minimize CO emissions. Figure 7.1 shows the calculated CO as a function of the equivalence ratio (ratio of 1 is stoichiometric, >1 is fuel rich, and <1 is fuel lean). Because these are adiabatic calculations with very high flame temperatures, the dissociation in the flame produces high quantities of CO even under fuel-lean conditions. This is graphically shown in Fig. 7.2 where much more CO is produced at higher gas temperatures, all other variables remaining

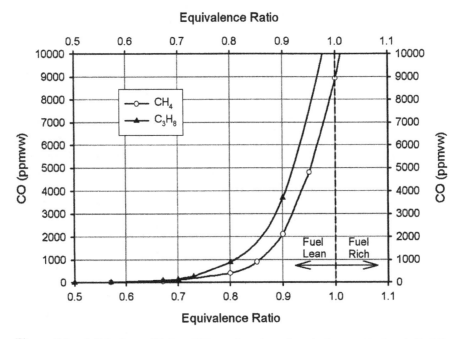

Figure 7.1 Adiabatic equilibrium CO as a function of equivalence ratio for air/fuel flames.

Gas Temperature (°F)

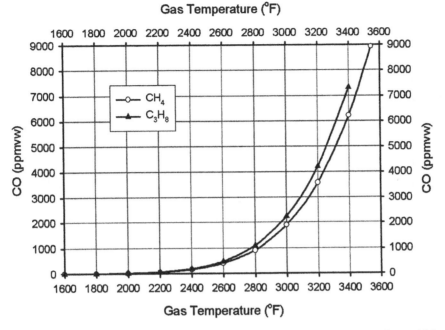

Figure 7.2 Adiabatic equilibrium CO as a function of gas temperature for stoichiometric air/fuel flames.

the same. The lifetime of CO in the atmosphere is fairly long, on the order of 2 to 4 months, before it is oxidized to CO_2 [2].

Figures 7.3 and 7.4 show the effects on CO production of air and fuel preheating, respectively. In both cases, the higher flame temperatures produced by preheating cause more CO formation as the preheat temperature increases. Figure 7.5 shows the effect of fuel composition for H_2/CH_4 blends. As expected, higher concentrations of H_2 produce less CO and with pure H_2, no CO is generated. Similarly, Fig. 7.6 shows the effect of fuel composition for CH_4/N_2 blends. Higher concentrations of N_2 both reduce the flame temperature and the concentration of carbon available to make CO, which both reduce CO generation.

Creighton [3] has shown the importance of the cooling rate on the level of CO produced from a combustion process. Exhaust gas products cooled rapidly can produce several orders of magnitude CO greater than equilibrium predictions due to rapid quenching. In general, the faster the cooling, the more CO produced. While these results are particularly important for internal combustion engines, they may also have applicability in cases where the exhaust gases are rapidly quenched. This is not typically the case for most industrial combustion systems where CO levels are usually close to equilibrium conditions.

7.2.2 Environmental Concerns

The primary environmental concern for CO is that it is an asphyxiant that can cause death in high enough concentrations. Unfortunately, there are few obvious warning signs such as smell, sound, or sight. Persons begin to become drowsy and sluggish

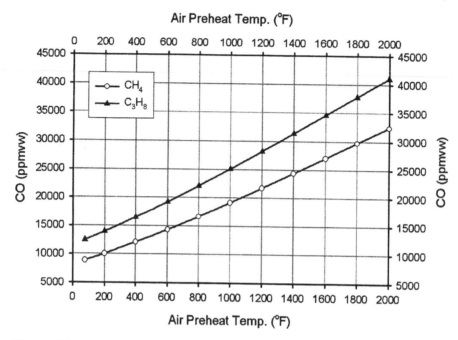

Figure 7.3 Adiabatic equilibrium CO as a function of air preheat temperature for stoichiometric air/fuel flames.

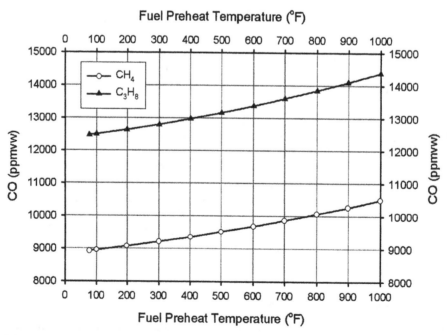

Figure 7.4 Adiabatic equilibrium CO as a function of fuel preheat temperature for a stoichiometric air/CH_4 flame.

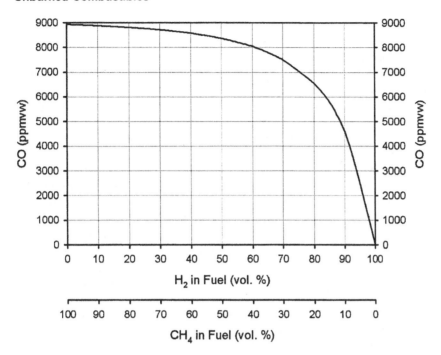

Figure 7.5 Adiabatic equilibrium CO as a function of fuel composition (CH$_4$–H$_2$) for a stoichiometric air/fuel flame. (From Ref. 14. Courtesy of John Zink Co., Tulsa, OK.)

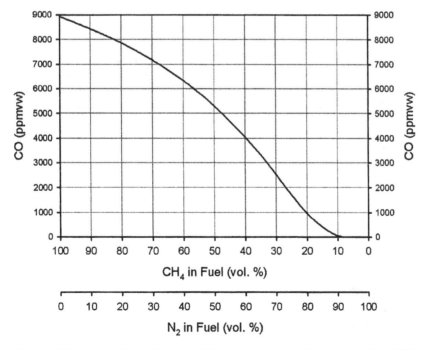

Figure 7.6 Adiabatic equilibrium CO as a function of fuel composition (CH$_4$–N$_2$) for a stoichiometric air/fuel flame. (From Ref. 14. Courtesy of John Zink Co., Tulsa, OK.)

and may die without ever realizing what happened. This is why CO detectors are recommended wherever combustion exhaust gas products may come into contact with personnel because CO is sometimes referred to as the silent killer. Furnaces running under positive pressure could potentially leak combustion gases into the surrounding area where personnel may be present. Even if there is measured excess O_2 in the exhaust stack, that does not guarantee there are no CO emissions. Even further, even if little or no CO is detected in the exhaust products, there could be considerable quantities of CO in the furnace until it is fully combusted. This means that a leak of gases out of the furnace could be potentially dangerous.

Another potential environmental concern for CO emissions is their flammability. However, there would have to be a very high relative concentration for the mixture to be flammable, especially since they would typically be part of exhaust gases produced by a combustion system with insufficient oxygen. These gases could ignite on contact with air if the temperature is high enough and there is proper mixing, but again this is very unlikely. It is usually more feasible for CO-rich gases to ignite in and around a furnace if the gases leak into the surrounding environment, compared to when the gases are emitted into the atmosphere from the exhaust stack.

The presence of significant quantities of CO is also indirectly an environmental concern because it is an indicator of incomplete combustion and, therefore, reduced fuel efficiency. This normally means that more fuel must be combusted than necessary for a given production rate. More fuel consumption translates into more pollution emissions. Therefore, high CO in the exhaust products signals poor combustion efficiency and, therefore, increased emissions.

7.2.3 Regulations

Figure 7.7 shows CO emissions in the United States since 1970 for some of the applications considered in this book [4]. Eisinger et al. [5] note the dramatic reduction in CO ambient concentrations over the past few decades as a result of increased regulations. They believe this trend will continue into the future. Since controlling CO is generally relatively inexpensive, there is not a compelling reason to

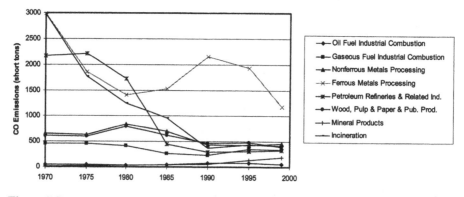

Figure 7.7 CO emissions in the United States since 1970 by application type. (From Ref. 4.)

relax CO emission regulations. The next possible area for regulatory change could be during cold start ups where CO emissions are likely to be higher compared to normal higher temperature operation. This is already happening with automotive exhaust systems where catalysts and other techniques are used to minimize CO emissions during the initial engine warm up.

7.2.4 Measurement Techniques

EPA Method 10 discusses measuring CO emissions from stationary sources from continuous samples extracted from an exhaust stack where the sample is measured with nondispersive infrared (NDIR) analyzer (see Fig. 7.8) [6]. Possible interferences include water, CO_2, and CO. Method 10A tells us how to make certified CO measurements from continuous emission monitoring systems at petroleum refineries. Figure 7.9 shows a typical exhaust gas sampling system including the conditioner and a control unit.

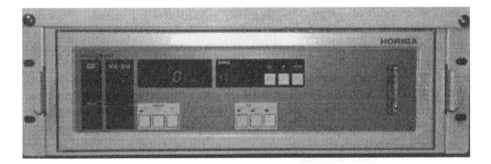

Figure 7.8 Nondispersive IR CO analyzer.

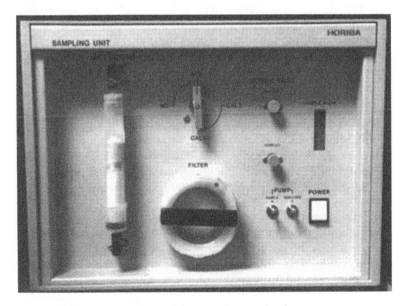

Figure 7.9 Gas sample conditioner and control unit.

7.2.5 Abatement Strategies

de Wit et al. [7] discuss the use of an oxidation catalyst (rhodium) for destroying CO emissions from exhaust gases, particularly in very fuel-lean combustion systems where the gas temperatures are low. The efficiency of the catalytic oxidation increases above 200°C (400°F) and typically operates in flue gas streams at temperatures between 400° and 500°C (750°–930°F). The higher the temperature, the higher the destruction efficiency. If the temperature is high enough, then no catalyst is needed. The catalyst is only needed when it would be more expensive to heat the gases up to a sufficient temperature, for example, using an afterburner. The presence of SO_2 inhibits the destruction efficiency.

7.3 UNBURNED HYDROCARBONS

Unburned hydrocarbons (UHCs), sometimes referred to as products of incomplete combustion (PICs), are not typically present in the exhaust products of most industrial combustion processes because they signal poor operating conditions. Therefore, few emissions monitoring systems include analyzers for measuring unburned hydrocarbons. However, there are some applications where there is a variable amount of hydrocarbons present, usually in the incoming feed materials. For example, scrap metal often contains oil used to facilitate the cutting and drilling of the materials. For those special instances, UHCs can represent a particular emission control challenge because of the transient nature of operating conditions.

7.3.1 Formation Mechanisms

Used beverage cans (UBCs) and other recycled aluminum scrap are typically melted in a rotary furnace. The aluminum processor likes to have an efficient and flexible melter to handle a wide range of scrap materials. An ideal melting process would be able to handle a wide range of scrap and automatically compensate for the contaminants. Scrap aluminum cans commonly have paints and lacquers that must be removed. In some cases, this is done in a pretreatment furnace where the coatings are burned off. Scrap turnings from machining operations typically contain significant quantities of lubrication oils. These oils produce large amounts of smoke during heating and melting in conventional rotary furnaces. This smoke must be removed before the exhaust gases exit the stack.

Stricter regulations are increasing the environmental compliance costs. This includes the costs to clean up contaminated air and water resulting from the melting process, as well as the costs to dispose of any waste products, particularly important for hazardous wastes.

An important environmental factor to be considered is the hydrocarbon emissions. In a conventional rotary furnace, with the burner mounted on the charging door, the burner must be turned off when the door is opened for charging. Fumes are generated by vaporization of any combustibles in the charge, as shown in Fig. 7.10 [8]. Because of the reduced furnace temperature and insufficient free oxygen to burn the newly added combustibles, large quantities of hydrocarbon emissions can be generated. The problem is exacerbated when the door is closed and the burner

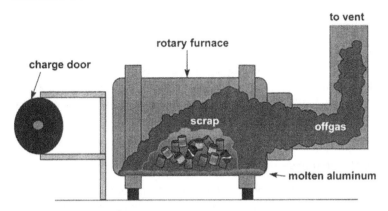

Figure 7.10 Hydrocarbon emissions from a new charge of oily scrap. (From Ref. 8. Courtesy of CRC Press.)

restarted. Even less free oxygen is available when the door is closed, and more combustibles are vaporized by the additional heat from the burner. An afterburner could be used to incinerate these fumes. However, this is costly and does not take advantage of the heating value in the combustibles.

7.3.2 Environmental Concerns

Unburned hydrocarbons represent a fire or explosion hazard due to their flammability. Given proper mixing with an oxidant such as air and an ignition source, these gases could burn in an uncontrolled manner, which is usually the distinction between a fire and controlled combustion in a burner. If a large volume of unburned hydrocarbons suddenly ignites, this could cause an explosion. This is especially dangerous if the unburned hydrocarbons leak out of the combustor in the vicinity of personnel. These gases are often difficult to detect without an analyzer so there may be little or no warning of an impending problem. Any systems with the potential for leaking UHCs in the vicinity of personnel should have detection systems to warn well in advance of reaching a flammable mixture. As previously stated, leaking large quantities of unburned hydrocarbons out of a combustor is not typically an issue for most industrial combustion applications.

Another potential environmental concern with unburned hydrocarbons is smoke formation. This should be readily apparent and easily controlled. This may be done with sufficient oxidant and mixing, or some type of post-treatment system. As with CO, unburned hydrocarbons are usually an indicator of poor combustion operating conditions. This means reduced fuel efficiency and generally higher pollution emissions.

7.3.3 Regulations

For most processes, unburned hydrocarbon emissions are indirectly regulated by limiting CO. CO emissions exceeding regulatory limits would generally be present before any significant quantities of unburned hydrocarbons would start to appear. Therefore, there is usually no need to regulate UHCs because the process would already have exceeded CO limits. Another way that UHCs are indirectly regulated is through smoke or opacity limits. For example, smoke emissions from flares in the petrochemical industry are often regulated (see Chap. 15). Another specific application where unburned hydrocarbons may be an issue is in processing oily aluminum scrap (see Chap. 13).

7.3.4 Measurement Techniques

Figure 7.11 shows an example of a typical unburned hydrocarbon analyzer used in an extractive sampling system. Wilk et al. [9] discuss techniques for measuring polycyclic aromatic hydrocarbons (PAHs) from premixed gas flames [9]. Samples are normally extracted using a probe made of stainless steel or quartz, typically cooled with water to freeze the reactions. The sample may also include solids that may include PAHs adsorbed on their surface. The solids are usually captured by using some type of filter that may be made of quartz or glass fiber. The PAHs must be extracted from the sample to be analyzed. Gas chromatography and mass spectrometry are often used to determine the concentration and composition of PAHs.

7.3.5 Abatement Strategies

A new process was developed to reduce the melting costs and pollution emissions from the conventional rotary melter. It is referred to as the LEAM® process, or *Low Emission Aluminum Melting* process (patent pending) [10]. The LEAM process was first demonstrated in Germany where the environmental regulations are stringent. The first objective of the process was to at least maintain, if not improve, the existing yields in the conventional melting process, which is well known for its high yields. Other objectives included: (1) to lower the emissions, especially hydrocarbons, dioxins, and furans, below the regulated limits, and (2) to reduce the exhaust gas

Figure 7.11 Unburned hydrocarbon analyzer. (Courtesy of California Analytical, Orange, CA.)

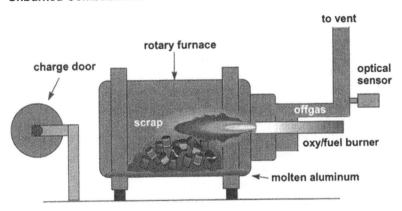

Figure 7.12 LEAM process for reducing hydrocarbon emissions. (From Ref. 8. Courtesy of CRC Press.)

volumes to reduce the cost of the flue-gas cleaning system. This required reductions in the amount of filter dust collected in the cleaning system. All of these objectives had to be achieved at the lowest possible cost.

The conventional rotary melting configuration was shown in Fig. 7.10. As shown in Fig. 7.12, the first major change of the new LEAM process was relocating the burner from the charge door to the exhaust wall. This reversed flame operation offered several benefits. The residence time of the combustion gases in the furnace increased, leading to increased thermal efficiency and higher destruction efficiencies for the volatiles. This configuration allowed the burner to be fired even during charging. Any volatiles coming from the newly charged scrap material had to pass through the flame and were postcombusted within the furnace. The heating value of the volatiles was now used and the volatiles did not have to be removed in the off-gas cleaning system. The overall melting cycle time was reduced because heat was now added during the charging cycle. This new configuration allowed the door to be properly sealed, which better controlled the excess air coming into the furnace.

Another major change was replacing the air/fuel burner in the conventional rotary aluminum melting furnace with an oxy/fuel burner. This provided several advantages. The reduced exhaust gas volume flow made it easier to incinerate the volatiles because they were in higher concentrations. Oxy/fuel burners also have a much wider operating range than air/fuel burners. They can easily operate with several times the stoichiometric amount of O_2 required for complete combustion. This extra O_2 was important for incinerating the volatiles that evolved from the dirty charge materials. Air/fuel burner operation became severely degraded when large quantities of excess air were supplied through the burner.

During furnace charging, the oxy/fuel burner was at low fire, whereas in the conventional process the burner must be off during charging since it is mounted on the door. A major advantage of the LEAM process is that the volatiles emitted during charging of dirty scrap are incinerated by the burner. In the conventional process, all the volatiles emitted during charging are exhausted and must be removed by the exhaust gas-cleaning system.

There are two levels of control in the LEAM process pertaining to course and fine adjustments of the operating conditions. The course adjustments are based on automatic setpoint conditions that are predetermined according to the phase of the melting cycle. The fuel and oxygen flow rates are predefined for low-fire and high-fire conditions.

The fine adjustments are made using a feedback control system. An important element of that system is an optical sensor that is used to control the hydrocarbon emissions. As more volatiles are released into the furnace, the sensor detects the increased emissions and sends a signal to the controller, which increases the oxygen flow to the burner. As the quantity of volatiles being released from the dirty scrap begins to decline, the signal from the sensor automatically reduces the oxygen flow to the burner to the amount required to incinerate the reduced level of volatiles. Only the appropriate flow of oxygen is used. This is important because too much oxygen flow would oxidize aluminum and reduce the yield. Too little oxygen flow would not incinerate all the volatiles, which would increase the load on the off-gas cleaning system and waste the available energy in the volatiles. The optical sensor for the oxygen/fuel ratio correction is both very reliable and inexpensive. The sensor has a fast response time compared to typical gas analyzers which require a sample to be extracted from the system and may have delay times of more than 30 sec.

For the air/fuel system, hydrocarbons were a big problem because they were difficult to control. Using the LEAM process, hydrocarbons were well below the German limit and could even be further reduced by adjusting the sensitivity of the optical sensor. Hydrocarbon emissions dropped by 93% using the LEAM process compared to the conventional air/fuel combustion system it replaced.

de Wit et al. [7] discuss the use of an oxidation catalyst (rhodium) for destroying CH_4 emissions from exhaust gases, particularly in very fuel-lean combustion systems where the gas temperatures are low [7]. The efficiency of the catalytic oxidation increases above 200°C (400°F) typically operates in flue gas streams at temperatures between 400° and 500°C (750°–930°F). The higher the temperature, the higher the destruction efficiency. If the temperature is high enough, then no catalyst is needed. The catalyst is only needed when it would be more expensive to heat the gases up to a sufficient temperature, for example, using an afterburner. The presence of SO_2 inhibits the destruction efficiency.

7.4 VOLATILE ORGANIC COMPOUNDS

Volatile organic compounds (VOCs) are liquids or solids that vaporize easily and contain organic carbon bonded to carbon, hydrogen, nitrogen, or sulfur but do not include carbonate carbon (e.g., $CaCO_3$) nor carbide carbon (e.g., CaC_2, CO, or CO_2). VOCs are generally low molecular weight aliphatic and aromatic hydrocarbons like alcohols, ketones, esters, and aldehydes [11]. The U.S. EPA defines a volatile organic compound as "any organic compound which participates in atmospheric photochemical reactions" [12]. VOCs form a very diverse class of pollutants because of the wide range of chemicals that fall into this category. Typical VOCs include benzene, acetone, acetaldehyde, chloroform, toluene, methanol, and formaldehyde. Shen et al. [13] have written a useful book on many aspects of VOC generation and control. They note that incineration processes are a source of VOC emissions.

Volatility relates to how easily a liquid can vaporize or evaporate. Table 7.1 shows the vapor pressures for some common hydrocarbons [14]. Substances with a normal vapor pressure above atmospheric pressure at room temperature will boil away rapidly unless kept in a closed container. Substances with normal vapor pressures equal to atmospheric pressure at room temperature will evaporate slowly, assuming that the surrounding atmosphere is not saturated. An example is water in an open container, which does not boil at normal room temperature and atmospheric pressure, but it does evaporate slowly into the surrounding air. The evaporation rate increases as the room temperature increases or the humidity decreases. Even substances like mercury that have a very low vapor pressure will still evaporate from an open container but at a slow rate. However, because of the extreme toxicity of mercury it must be kept in a closed container to keep it from entering the ambient environment. Since VOCs readily vaporize and since most industrial combustion processes are at pressures near atmospheric, these chemicals can easily vaporize into the atmosphere if not properly contained and treated. VOCs can be broadly defined as liquid and solid hydrocarbons whose room temperature vapor pressures are greater than about 0.01 psia with atmospheric boiling points up to about 500°F (260°C). Substances with higher boiling points tend to evaporate slowly and are not considered volatile.

Some VOCs (e.g., benzene) are highly toxic and carcinogenic in their own right, and some are only mildly toxic to humans, but most are not toxic. The main problem with most VOCs is not their toxicity but their role in causing photochemical smog and depletion of the ozone layer if they are released into the atmosphere. They can form secondary particles in the atmosphere, which causes smog according to [15]

Hydrocarbons (VOCs) + sulfur oxides + nitrogen oxides → fine particles

$$(7.1)$$

Some VOCs are powerful absorbers of infrared energy and thus contribute to global warming. Therefore, VOCs are regulated pollutants. Table 7.2 shows that the primary sources of most VOC emissions in the United States are solvent usage and motor vehicles, which account for two-thirds of the total. The emissions from the industries considered here include only 4.9% for chemical, metals processing, and petrochemical industries, and 2.5% for other industrial processes. This is primarily because most industrial combustion processes burn clean fuels and are not processing materials containing VOCs. Table 7.3 shows the common types of VOCs emitted from fuel oil and gaseous combustion processes [16].

VOCs are not normally produced in the combustion process, but they may be contained in the material that is being heated, such as in the case of a contaminated hazardous waste in a waste incinerator or in solvents used in cleaning parts. In that case, the objective of the heating process is usually to volatilize the VOCs out of the waste and combust them before they can be emitted to the atmosphere.

There are two strategies for removing VOCs from the off-gases of a combustion process [11]. One is to separate and recover them using techniques like carbon adsorption or condensation. The other method involves oxidizing the VOCs to CO_2 and H_2O. This process includes techniques like thermal oxidation [17], catalytic oxidation, and bio-oxidation. One common way to ensure complete destruction of VOCs in waste incinerators is to add an afterburner or secondary

Table 7.1 Physical Constants for Typical Gaseous Fuel Components

No.	Fuel gas component	Chemical formula	Molecular weight	Boiling point 14.696 psia (°F)	Vapor pressure 100°F (psia)	Specific heat capacity, C_p 60°F & 14.696 psia	Latent heat of vaporization 14.696 psia and boiling point (Btu/lb$_m$)	Specific gravity (Air = 1)	Gas density (lb$_m$/ft²)	Specific volume (ft²/lb$_m$)	LHV (net)	HHV (gross)	LHV (net)	HHV (gross)
								Gas density ideal gas, 14.696 psia, 60°F			Heating value			
Parafin (alkane series C_2H_{2n+2})														
1	Methane	CH_4	16.04	−258.69	—	0.5266	219.22	0.554	0.042	23.651	912	1,013	21,495	23,675
2	Ethane	C_2H_6	30.07	−127.48	—	0.4097	210.41	1.038	0.079	12.618	1,639	1,792	20,416	22,323
3	Propane	C_3H_8	44.10	−43.67	190	0.3881	183.05	1.522	0.116	8.604	2,385	2,592	19,937	21,669
4	n-Butane	C_4H_{10}	58.12	31.10	51.6	0.3867	165.65	2.007	0.153	6.528	3,113	3,373	19,679	21,321
5	Isobutane	C_4H_{10}	58.12	10.90	72.2	0.3872	157.53	2.007	0.153	6.528	3,105	3,365	19,629	21,271
6	n-Pentane	C_5H_{12}	72.15	96.92	15.57	0.3883	153.59	2.491	0.190	5.259	3,714	4,017	19,607	21,095
7	Isopentane	C_5H_{12}	72.15	82.12	20.44	0.3827	147.13	2.491	0.190	6.269	3,706	4,007	19,469	21,047
8	Neopentane	C_5H_{12}	72.15	49.10	36.9	0.3866	136.58	2.491	0.180	6.269	3,692	3,984	19,380	20,978
9	n-Hexane	C_6H_{14}	86.18	166.72	4.966	0.3864	143.96	2.976	0.227	4.403	4,416	4,767	19,415	20,955
Naphthene (cylcoalkene) series (C_nH_{2n})														
10	Cyclopentene	C_5H_{10}	70.13	120.60	9.917	0.2712	167.35	2.420	0.180	5.566	3,512	3,754	19,005	20,358
11	Cyclohexene	C_6H_{12}	84.16	177.40	3.257	0.2901	153.25	2.910	0.220	4.545	4,180	4,482	18,649	20,211
Olefin series (C_nH_{2n})														
12	Ethene (Ethylene)	C_2H_4	28.05	−154.62	—	0.3622	207.57	0.969	0.074	13.525	1,512	1,613	20,275	21,636
13	Propene (propylene)	C_3H_6	42.08	−53.90	226.4	0.3541	188.18	1.453	0.111	9.017	2,185	2,336	19,687	21,048
14	1-Butene (Butylene)	C_4H_8	56.11	20.75	63.05	0.3548	167.94	1.937	0.148	6.782	2,885	3,086	19,493	20,854
15	Isobutene	C_4H_8	66.11	19.58	63.4	0.3701	169.48	1.937	0.148	6.752	2,668	3,069	19,376	20,737
16	1-Pentene	C_5H_{10}	70.13	85.93	19.115	0.3635	154.46	2.421	0.185	5.410	3,565	3,837	19,359	20,720
Aromatic series (C_nH_{2n-6})														
17	Benzene	C_6H_6	78.11	176.17	3.224	0.2429	169.31	2.597	0.206	4.857	3,595	3,745	17,451	16,184
18	Toluene	C_7H_8	92.14	231.13	1.032	0.2698	154.84	3.181	0.243	4.118	4,795	4,497	17,672	18,501
19	o-Xylene	C_8H_{10}	106.17	291.97	0.264	0.2914	148.1	3.665	0.280	3.574	4,970	5,222	17,734	18,533
20	m-Xylene	C_8H_{10}	106.17	282.41	0.326	0.2782	147.2	3.665	0.280	3.574	4,970	5,222	17,734	18,533
21	p-Xylene	C_8H_{10}	106.17	281.05	0.342	0.2769	144.52	3.665	0.280	3.574	4,970	5,222	17,734	18,533
Additional fuel gas components														
22	Acetylene	C_2H_2	26.04	−119	—	0.3966	—	0.899	0.069	14.572	1,448	1,499	20,769	21,502
23	Methyl alcohol	CH_3OH	32.04	148.1	4.63	0.3231	473	1.106	0.084	11.841	757	858	8,066	10,258
24	Ethyl alcohol	C_2H_5OH	46.07	172.92	2.3	0.3323	367	1.590	0.121	8.236	1,449	1,600	11,918	13,161
25	Ammonia	NH_3	17.03	−28.2	212	0.5002	587.2	0.688	0.046	22.279	364	441	7,986	9,667
26	Hydrogen	H_2	2.02	−423.0	—	3.4080	193.8	0.070	0.006	186.217	274.6	326.0	51,525	61,096
27	Oxygen	O_2	32.00	−297.4	—	0.2188	91.6	1.105	0.064	11.868	—	—	—	—
28	Nitrogen	N_2	28.16	−320.4	—	0.2482	87.8	0.972	0.074	13.473	—	—	—	—
29	Carbon monoxide	CO	28.01	−313.6	—	0.2484	92.7	0.967	0.074	13.546	321.9	321.9	4,347	4,347
30	Carbon dioxide	CO_2	44.01	−109.3	—	0.1991	238.2	1.519	0.116	8.621	—	—	—	—
31	Hydrogen sulfide	H_2S	34.09	−76.6	394.04	0.2380	236.6	1.177	0.080	11.133	695	545	5,537	7,097
32	Sulfur dioxide	SO_2	64.06	14.04	88	0.1450	186.7	2.212	0.169	5.923	—	—	—	—
33	Water vapor	H_2O	18.02	212.0	0.9492	0.4446	970.3	0.622	0.047	21.061	—	—	—	—
34	Air	—	28.97	317.6	—	0.7400	92	1.000	0.076	13.099	—	—	—	—

| | | Unit volume per unit volume of combustible | | | | | | | Unit mass per unit mass of combustion | | | | | | | Theoretical air required (lbm/10 000 Atm) | Flammability limits (vol% in air mixture) | |
| | | Required for combustion | | | Flue gas products | | | | Required for combustion | | | Flue gas products | | | | | | |
No.		O_2	N_2	Air	CO_2	H_2O	N_2	SO_2	O_2	N_2	Air	CO_2	H_2O	N_2	SO_2		Lower	Upper
	Paraffin (alkane series) (C_nH_{2n+2})																	
1	Methane	2.0	7.547	9.547	1.0	2.0	7.547	—	3.989	13.246	17.235	2.743	2.246	13.245	—	7.219	5.0	15.0
2	Ethane	3.6	13.205	15.705	2.0	3.0	13.206	—	3.724	12.367	16.092	2.927	1.797	12.357	—	7.209	2.9	13.0
3	Propane	5.0	18.888	23.855	3.0	4.0	18.866	—	3.628	12.047	15.676	2.994	1.634	12.047	—	7.234	2.1	9.5
4	n-Butane	6.5	24.526	31.026	4.0	5.0	24.526	—	3.576	11.882	15.480	3.029	1.550	11.682	—	7.261	1.8	8.4
5	Isobutane	6.5	24.526	31.026	4.0	5.0	24.526	—	3.576	11.882	15.480	3.029	1.550	11.682	—	7.268	1.8	8.4
6	n-Pentane	8.0	30.186	38.186	5.0	6.0	30.185	—	3.548	11.761	15.329	3.090	1.490	11.781	—	7.267	1.4	8.3
7	Isopentane	8.0	30.186	38.186	5.0	6.0	30.185	—	3.548	11.781	15.329	3.090	1.498	11.781	—	7.283	1.4	8.3
8	Neopentane	8.0	30.186	38.186	5.0	6.0	30.185	—	3.548	11.781	15.329	3.090	1.498	11.781	—	7.307	1.4	8.3
9	n-Hexane	9.6	35.846	45.345	8.0	7.0	35.485	—	3.527	11.713	15.240	3.064	1.453	11.713	—	7.269	1.2	7.7
	Naphthene (cycloalkene series) (C_nH_{2n})																	
10	Cyclopentene	7.5	27.939	35.810	5.0	5.0	27.939	—	3.980	11.166	14.793	3.146	1.283	11.155	—	7.262		
11	Cyclohexene	9.0	33.528	42.970	6.0	6.0	33.528	—	4.620	13.386	17.750	3.146	1.283	11.155	—	7.848	1.3	8.4
	Olefin series (C_nH_{2n})																	
12	Ethene (Ethylene)	3.0	11.320	14.320	2.0	2.0	11.320	—	3.422	11.352	14.784	3.138	1.284	11.362	—	6.833	2.7	34.0
13	Propene (propylene)	4.5	16.960	21.480	3.0	3.0	16.980	—	3.422	11.352	14.784	3.138	1.284	11.362	—	7.024	2.0	10.0
14	1-Butene (Butylene)	6.0	22.640	28.640	4.0	4.0	22.640	—	3.422	11.352	14.784	3.138	1.284	11.362	—	7.069	1.6	9.3
15	Isobutene	6.0	22.640	28.640	4.0	4.0	22.640	—	3.422	11.352	14.784	3.138	1.284	11.362	—	7.129	1.6	—
16	1-Pentene	7.5	28.300	36.800	5.0	5.0	28.300	—	3.422	11.352	14.784	3.138	1.284	11.362	—	7.135	1.4	8.7
	Aromatic samples (C_nH_{2n-6})																	
17	Benzene	7.5	28.300	35.800	6.0	3.0	28.300	—	3.072	10.201	13.274	3.380	0.692	10.201	—	7.300	1.38	7.98
18	Toluene	9.0	33.959	42.959	7.0	4.0	33.959	—	3.126	10.378	13.504	3.343	0.782	10.378	—	7.299	1.28	7.18
19	o-Xylene	10.5	39.619	50.119	8.0	5.0	39.619	—	3.164	10.508	13.573	3.316	0.848	10.508	—	7.338	1.18	6.48
20	m-Xylene	10.5	39.619	50.119	8.0	5.0	39.619	—	3.164	10.508	13.573	3.316	0.848	10.508	—	7.338	1.18	6.48
21	p-Xylene	10.5	39.619	50.119	8.0	5.0	39.619	—	3.164	10.508	13.573	3.316	0.848	10.508	—	7.388	1.18	6.48
	Additional fuel gas components																	
22	Acetylene	2.5	9.433	11.933	2.0	1.0	9.433	—	3.072	10.201	13.274	3.380	0.892	10.201	—	7.300	2.5	80
23	Methyl alcohol	1.5	5.880	7.180	1.0	2.0	5.660	—	1.496	4.974	6.472	1.373	1.124	4.974	—	6.309	6.72	36.5
24	Ethyl alcohol	3.0	11.320	14.320	2.0	3.0	11.320	—	2.084	6.919	9.003	1.911	1.173	5.919	—	6.841	3.28	18.95
25	Ammonia	0.75	2.830	3.580	—	1.5	3.330	—	1.409	4.679	6.086	—	1.587	5.502	—	6.298	15.50	27.00
26	Hydrogen	0.5	1.887	2.387	—	1.0	1.887	—	7.936	26.363	34.290	—	0.937	26.353	—	5.613	4.00	74.20
27	Oxygen	—	—	—	—	—	—	—	—	—	—	—	—	—	—	—	—	—
28	Nitrogen	—	—	—	—	—	—	—	—	—	—	—	—	—	—	—	—	—
29	Carbon monoxide	0.5	1.877	2.387	1.0	—	1.887	—	—	1.697	2.468	1.571	—	1.870	—	5.677	12.50	74.20
30	Carbon dioxide	—	—	—	—	—	—	—	—	—	—	—	—	—	—	—	—	—
31	Hydrogen sulfide	1.5	5.650	7.160	—	1.0	5.660	1.0	1.410	4.692	5.093	—	0.529	4.682	1.890	8.585	4.30	45.50
32	Sulfur dioxide	—	—	—	—	—	—	—	—	—	—	—	—	—	—	—	—	—
33	Water vapor	—	—	—	—	—	—	—	—	—	—	—	—	—	—	—	—	—
34	Air	—	—	—	—	—	—	—	—	—	—	—	—	—	—	—	—	—

Table 7.2 Estimated VOC Emissions (short tons) in the United States in 1999 by Source Category

Source category	1999 estimated VOC emissions (short tons)	% of total
Fuel combustion	904	5.0
Industrial processes	7996	44.1
chemical manufacturing	395	2.2
metals processing	77	0.4
petrochemical	424	2.3
other	449	2.5
solvent utilization	4825	26.6
storage and transport	1240	6.8
waste disposal and recycling	586	3.2
Transportation	8529	47.0
Miscellaneous	716	3.9
Total	18,145	100.0

Source: Ref. 4.

combustion chamber, which may or may not have a catalyst, after the main or primary combustion chamber [18].

7.4.1 Formation Mechanisms

Demayo et al. [19] showed experimentally that VOCs can be produced under certain conditions for ultra-lean premixed combustion typically used to reduce NO_x emissions. However, in most cases the amount of VOCs formed was low (< 100 ppbv). This suggests that VOCs may be present in industrial combustion applications, but have never been measured.

In diffusion-type industrial burners, rapid mixing and lean conditions produce both locally lean and rich fuel–air mixture packets that coexist in turbulent eddies. When these packets are too lean or too rich to combust, they can pass through the flame zone without reacting and result in emissions of unburned fuel that can be converted into VOCs in the postcombustion region under certain conditions including rapid quenching.

Demayo and Samuelsen [20] experimentally studied VOC formation in low-NO_x burners. While VOC emissions were generally low for the two burners studied, under certain conditions the emissions were significant. For the generic burner operating under fuel-lean conditions, this occurred at the high-swirl stability limits. For the industrial boiler burner, this occurred near stoichiometric conditions near the lean blowout limit.

7.4.2 Environmental Concerns

There are a number of potential environmental problems with VOCs, which are somewhat dependent on the specific VOC. As described above, the main concern is

Table 7.3 Types of VOCs Emitted from Fuel Oil and Gaseous Fuels

	Fuel oil	Gas
Alkanes		
Methane	10	75
Ethane		12
Propane		4
Butane		4
Alkanes > C_4		
Alkanes, not specified	65	
Alkanes		
Ethene		1.8
Propene		0.6
Butene		0.6
Pentene		
Alkanes > C_3		
Alkanes, not specified	10	
Alkynes		
Ethyne		
Propyne		
Aromatic hydrocarbons		
Benzene		
Toluene		
o-Xylene		
m-Xylene		
p-Xylene		
Ethylbenzene		
Aromatics, not specified	5	
Aldehydes		
Formaldehyde		1.8
Acetadehyde		0.12
Propionaldehyde		0.04
Acrolein		
Aldehydes ~ C_4		0.04
Aldehydes, not specified	10	
Other VOCs		
Ketones, not specified		
VOCs, not specified		

Source: Ref. 16.

the participation of VOCs in photochemical smog formation. Another concern is that some VOCs like benzene are carcinogenic while others may be toxic, although most are not directly harmful to humans by themselves. Most VOCs are highly flammable, which could lead to uncontrolled fires or possible explosions if they are sufficiently mixed with an oxidant like air and come into contact with an ignition source.

7.4.3 Regulations

Figure 7.13 shows VOC emissions in the United States since 1970 for some of the applications considered in this book. As with most other pollution emissions, regulations continue to become more stringent for VOC emissions. There is particular concern with carcinogenic VOCs such as benzene, which are strictly regulated.

7.4.4 Measurement Techniques

EPA Method 25 describes the recommended procedure for determining the total gaseous nonmethane organic (VOCs) emissions from a stack [21]. These measurements are made on a continuous gas sample with a flame ionization detector (FID). Carbon dioxide and water vapor can potentially interfere with the measurements. Specific equipment is required to ensure proper handling and treatment of the sample for valid measurements. Method 25A concerns measuring VOCs with a flame ionization analyzer (FIA) for determining the concentrations of samples containing primarily alkanes, alkenes, and/or arenes. Method 25B concerns the use of a nondispersive infrared (NDIR) analyzer to measure VOCs consisting primarily of alkanes.

Fabrellas et al. [22] note the importance of characterizing VOCs prior to choosing the appropriate control strategy. They recommend the combination of thermal desorption, coupled with chromatographic analysis and a selective ion detector for identifying and quantifying VOCs in waste streams. The U.S. EPA has developed a manual to aid in estimating the cost of air pollution control equipment, including a section on VOCs [23]. The manual includes chapters on adsorbers, refrigerated condensers, and flares.

7.4.5 Abatement Strategies

A variety of strategies are used to control VOC emissions. The actual choice depends on the specific VOCs and on the application. In general, the preferred choice is to

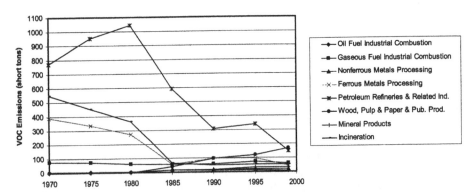

Figure 7.13 VOC emissions in the United States since 1970 by application type. (From Ref. 4.)

combust the VOCs in the combustor so that not only are they destroyed, but their heating value can be used to improve the overall process thermal efficiency, which indirectly reduces other pollution emissions as less auxiliary fuel needs to be burned for a given unit of production.

7.4.5.1 Pretreatment

Minimizing or removing the source of the VOCs is a strategy that seeks to remove the VOCs before they enter the combustion process. This may involve substituting another substance for the VOCs where the VOCs were previously added to the process or removing the VOCs from the process prior to heating if the VOCs do not participate in the production of the end product. For example, water-based paints may be substituted for oil-based paints containing VOCs. Another example is replacing fuels containing VOCs such as gasoline, with fuels that do not contain VOCs such as natural gas. In the previous two examples, the VOCs were active participants in the process. In other cases, the VOCs may be present in the incoming raw materials but may not be required or may even be undesirable in the process. For example, when recycled aluminum parts are fed into an aluminum melter, these parts may be coated with a paint containing VOCs. When these parts are heated, the solid VOCs are vaporized and go into the exhaust gas stream. If they are not incinerated in the combustor, then these VOCs would need to be removed from the exhaust gas stream prior to emission into the atmosphere. An alternative pretreatment strategy might be to sandblast these parts to remove the coatings from the scrap parts prior to putting them into the melting furnace. In some cases it may be possible to separate potential VOC-generating materials and recycle them back into the process. However, the economics need to be evaluated to determine if these are the best courses of action.

7.4.5.2 Process Modification

This treatment strategy involves modifying the process in such a way as to reduce or eliminate VOC emissions. For example, if gasoline is used as a fuel in the process, then making the combustor more efficient indirectly reduces VOC emissions per unit of output. If the car is considered as the combustor, then making cars that get more mileage per gallon of fuel used reduces the amount of pollutants generated per mile driven. Increasing efficiency is always a good strategy, but it may not be difficult or expensive to substantially improve. Another example of a process modification is to use different materials in the process so that VOC usage can be reduced or eliminated. A specific example might be making a part out of aluminum or stainless steel so that it does not need to be painted with a coating containing VOCs. Again, the economics need to be evaluated to determine the proper course of action.

7.4.5.3 Post-Treatment

Figure 7.14 shows some general guidelines for choosing a VOC control system, depending on the gas flow rate and the VOC concentration level [16]. Most of the

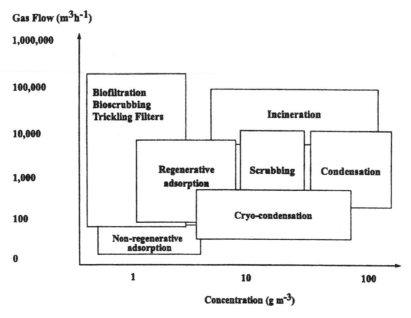

Figure 7.14 Selection guidelines for choosing a VOC control system. (From Ref. 16. Courtesy of Lewis Publishers.)

techniques in the graph are briefly described next. Table 7.4 compares the advantages and disadvantages of the common techniques for controlling VOC emissions [16]. Shen et al. [13] provide a good description of the common control techniques used to minimize VOC emissions.

Thermal Oxidation (Incineration).   The simplest way to destroy VOCs in an industrial combustion process is to burn them in a combustor sometimes referred to as an incinerator, thermal oxidizer, or afterburner (see Fig. 7.15). Figure 7.16 shows a schematic of a commercial thermal oxidizer system, and Fig. 7.17 shows a photograph of a commercial thermal oxidizer. Incineration is generally understood to mean destruction of a waste by combustion. Thermal oxidation is often the only viable method for controlling certain toxic and otherwise harmful pollutants [24]. Incineration is often viewed negatively by the general public and is usually associated with handling solid-waste materials, especially when ash is produced that must be buried in a landfill. However, incineration of gases and vapors normally produces little or no residual solid-waste ash. To avoid the negative public perception associated with solid-waste incineration, vapor incineration is often called thermal oxidation or afterburning.

In a thermal oxidizer (see Fig. 7.18), one or more burners are used in the combustor to ensure that the temperature is above the ignition points of the VOCs to be destroyed. Hannum [25] describes the use of a lean premix burner used in thermal oxidation that also produces ultra-low NO_x emissions of < 10 ppm. Temperatures above about 1800°F are normally sufficient to ensure complete combustion of any VOCs and converting them into primarily CO_2 and H_2O. Not only are most VOCs very flammable, but also the combustor is already at an elevated temperature well

Table 7.4 Comparison of VOC Control Techniques

Control technology	Advantages	Disadvantages
Biofiltration	Low operating and capital costs Effective removal of compounds Low pressure drop No further waste streams produced	Large footprint requirement Medium deterioration will occur Less suitable for high concentrations Moisture and pH difficult to control Particulate matter may clog medium

Biotrickling filters	Medium operating and
	  capital costs
	Effective removal of compounds
	Treats acid-producing
	  contaminants
	Low pressure drop
		Clogging by biomass
		More complex to construct
		  and operate
		Further waste streams produced

Wet scrubbing	Low capital costs
	Effective removal of odors
	No medium disposal required
	Can operate with a
	  moist gas stream
	Can operate with a
	  moist gas stream
	Can handle high flow rates
	Ability to handle variable loads
		High operating costs
		Need for complex chemical
		  feed systems
		Does not remove all VOCs
		Water softening often required
		Nozzle maintenance often required

Carbon adsorption	Short retention time/small unit Effective removal of compounds Suitable for low/moderate loads Consistent, reliable operation	High operating costs Moderate capital costs Carbon life reduced by moist gas stream Creates secondary waste streams

Incineration	System is simple
	Effective removal of compounds
	Suitable for very high loads
	Performance is uniform
	  and reliable
	Small area required
		High operating and
		  capital costs
		High flow/low concentrations
		  not cost effective
		Creates a secondary
		  waste stream
		Scrutinized by public

Source: Ref. 16.

above the typical ignition points of most VOCs. In fact, depending on the process, the combustion of VOCs can actually be an additional source of fuel in the process. For example, scrap metal fed into some type of melting furnace may contain solvents or coatings. These would normally contaminate the final metal product if they were not removed. The heat within the metal causes these solvents and coatings to vaporize and burn, which accomplishes two things. First, it cleans the metal parts of these contaminants, and second, it produces additional heat to aid in the melting process.

The three Ts of incineration are time, temperature, and turbulence. Time refers to adequate residence time in the reactor for the reaction kinetics to go to

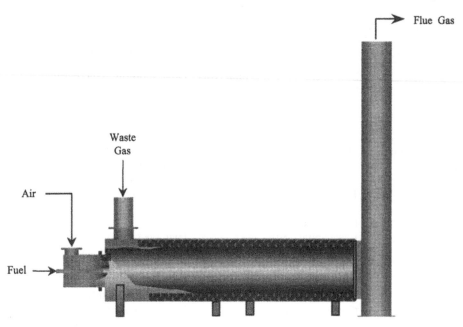

Figure 7.15   Incinerator. (From Ref. 16. Courtesy of Lewis Publishers.)

completion. The residence time in the combustor is determined by the combination
of its length and the gas velocity through the combustor. Since most combustion
reactions occur rapidly, long residence times are not typically required. For example,
fume incinerators usually consist of a refractory-lined chamber sized to produce a
1 to 2 sec retention time to ensure complete destruction [13]. There needs to be
sufficient temperature in the reactor to ensure the pollutant destruction reactions are
initiated and go to completion. While most hydrocarbons can be destroyed at
temperatures of about 1400°F (760°C), chlorinated compounds may require
temperatures up to 2200°F (1200°C) [13]. However, if temperatures are too high,
then other pollutants such as NO_x may increase in the process of reducing VOCs.
Catalytic thermal oxidation systems (see Fig. 7.19) are designed to destroy VOCs at
lower temperatures. For example, the Shell DeVOx process has demonstrated
destruction efficiencies up to 100% at 350°C (662°F) [26]. This reduces the amount
of auxiliary heat that may be required, often eliminates the need for refractory
linings, and can reduce NO_x formation compared to noncatalytic systems.
Turbulence refers to the mixing between the fuel, in this case the VOCs, and the
oxidizer which is usually air. The destruction and removal efficiency or DRE,
requires adequate mixing to ensure the VOCs come into contact with enough
oxidizer to be fully combusted. If insufficient oxygen gets to the VOCs, then other
pollutants like CO may be generated.

Ross and Snape [27] categorize three types of waste streams to be treated in
thermal oxidizers:

1.   Type 1: waste gas stream will sustain combustion on its own without an
     auxiliary fuel but with the addition of sufficient oxidizer (e.g., air); no
     chance for flashback.

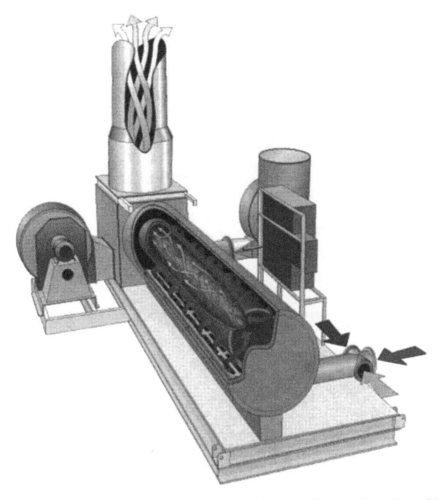

Figure 7.16 Schematic of a thermal oxidizer. (Courtesy of Alzeta, Santa Clara, CA.)

2. Type 2: waste gas stream will sustain combustion on its own without an auxiliary fuel and without additional oxidizer; there is a possibility of flashback so a safety device such as a flame arrestor must be included in the system.
3. Type 3: waste gas stream will not sustain combustion on its own and requires an auxiliary fuel and sufficient oxidizer; no chance of flashback.

As long as sufficient oxygen is available in the combustion chamber, then most VOCs will be easily combusted, producing primarily CO_2 and H_2O, although it is noted that high levels of CO may inhibit destruction efficiencies [13]. For example, the global stoichiometric reaction of benzene and air can be written as

$$C_6H_6 + 7.5(O_2 + 3.76N_2) \rightarrow 6CO_2 + 3H_2O + 28.2N_2 \qquad (7.2)$$

Figure 7.17   Thermal oxidizer. (Courtesy of Alzeta.)

Figure 7.18   Photograph of a thermal oxidizer. (From Ref. 14. Courtesy of John Zink Co.)

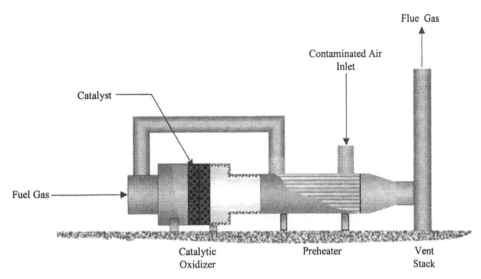

Figure 7.19 Typical horizontal catalytic thermal oxidation system. (From Ref. 14. Courtesy of John Zink Co., Tulsa, OK.)

The potential problem is maintaining the proper amount of O_2 in the combustor when the VOC loading varies. If insufficient O_2 is available, then the fuel and VOCs will not be fully combusted. This leads to CO and unburned hydrocarbons in the exhaust gas stream, which are both pollutants that must be removed prior to emission into the atmosphere, typically with some type of afterburner system to combust fully these flammable gases. If too much O_2 is present in the combustor, the process will be thermally inefficient as the extra O_2 (and N_2 that comes along with the O_2 in air) is a heat load that carries sensible heat out of the combustor. Feedback control systems are available to modulate the combustion system parameters to maintain a given O_2 level in the combustor. In most industrial combustion systems, an O_2 level of approximately 2–3% gives the best balance of high thermal efficiency and high combustion efficiency. Theoretically, no O_2 in the exhaust gas should produce the optimum thermal efficiency assuming perfect mixing between the fuel and the oxidizer. However, in practical combustors perfect mixing is not achievable so some excess O_2 is necessary to ensure all of the fuel and VOCs are fully combusted. Benítez gives a detailed discussion with examples on setting up material and energy balances on thermal incinerators to ensure proper performance [28]. The U.S. EPA has prepared a manual for estimating the costs of air pollution control systems, which includes a chapter on thermal oxidizers [29].

The O_2 control system is more of a problem in batch or discontinuous processes where the amount of hydrocarbons to be combusted can vary significantly. Two examples will illustrate this challenge. If oily scrap metal is loaded into a melting furnace, a significant amount of extra O_2 is needed at the beginning of the melting cycle as the oil vaporizes off the scrap and burns. However, after the oil is removed, much less O_2 is needed. Another example is when there are wide variations in the incoming raw material composition. Waste materials, some containing VOCs are sometimes fed into a waste incinerator in 55-gallon drums. An example might be soil contaminated with liquid hydrocarbons. As the drum heats up inside the incinerator, the volatiles evaporate and are incinerated. This in itself is a batch process where the O_2 requirements to fluctuate. Further complicating the problem is when the volatile concentration in the soil varies widely itself so that each drum is different from the next. This is a demanding problem for the O_2 control system as the O_2 level must be measured and rapidly adjusted as necessary, usually in a nonsteady-state manner. Because the residence time in the combustor is often on the order of seconds, adjustments often need to be made in feed-forward control as there may not be enough time for feedback control. Depending on the nature of the process, the operator may need to enter information (e.g., volatile concentration) into the control system about the incoming raw materials. When there are large variations in the O_2 requirements, these combustion systems are generally operated at higher levels of excess O_2 to ensure complete combustion but at the cost of some thermal efficiency. However, this is an economical tradeoff compared to adding some type of post-treatment system to ensure that no unburned hydrocarbon pollutants are emitted into the atmosphere.

As previously stated, the benefit of combusting the VOCs in the combustor is that additional heat and the associated increase in thermal efficiency can be achieved. Another benefit is that no additional equipment is required to remove the VOCs from the exhaust stream prior to emitting the exhaust gases into the atmosphere. A potential disadvantage is that the combustor may need to be run at a higher excess O_2 level to ensure complete destruction of the VOCs under a wide range of concentrations. This may mean the installation of a fairly sophisticated O_2 control that may be expensive. Another potential disadvantage could be that significant quantities of NO_x could be generated by combusting VOCs containing organically bound nitrogen. Depending on the specific conditions, it may be preferable to remove, rather than destroy, those VOCs as described in some of the techniques below.

There are various configurations that are commonly used in thermal oxidizers (see Fig. 7.20). Figure 7.21 shows a schematic of a thermal oxidation system with heat recovery to improve the thermal efficiency. Figure 7.22 shows a thermal oxidation system with a bypass recuperative system to improve the thermal efficiency.

Flaring. Flares are commonly used as safety device to burn large volumes of waste gases on an infrequent basis, typically during a plant upset or some type of emergency condition. A manual prepared by the U.S. EPA [30] describes flaring as

> "a volatile combustion control process for organic compounds (VOC) in which the VOCs are piped to a remote, usually elevated, location and burned in an open flame in the open air using a specially designed burner tip, auxiliary fuel, and steam or air to promote mixing for nearly complete (>98%) destruction."

One reason to use a flare is when it is not feasible to capture the gases for use in the plant or for resale. Another reason is that the heating value of the gases going to the flare may be too low for them to be usable in most industrial burners without significant design modifications. Flares are often much cheaper than designing special burners and combustors to handle the wide range of conditions that may be encountered including high and low flow rates and gases that may be highly

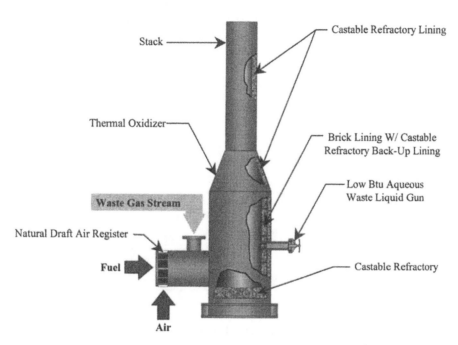

Figure 7.20 Schematic of a thermal oxidizer. (From Ref. 14. Courtesy of John Zink Co., Tulsa, OK.)

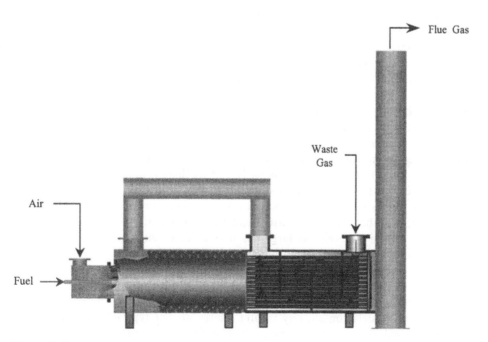

Figure 7.21 Thermal oxidizer system with heat recovery. (From Ref. 14. Courtesy of John Zink Co., Tulsa, OK.)

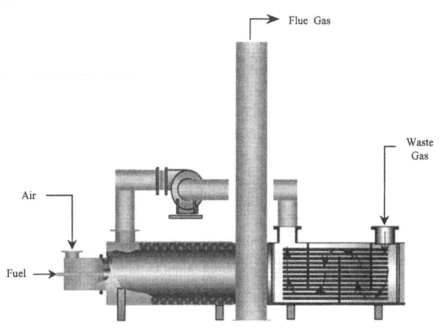

Figure 7.22 Thermal oxidizer system with a bypass recuperative system. (From Ref. 14. Courtesy of John Zink Co., Tulsa, OK.)

flammable or barely flammable. Flares are often very large in size to handle very large flammable gas flows. Flares are often elevated or protected in such a way as to minimize the radiation heat load on surrounding personnel and equipment. The U.S. EPA gives some guidelines for the use of flares as emission control devices [31]. One of the requirements is that flares should have no visible emissions such as smoke.

A flare is simply a very large burner (see Fig. 7.23) or array of burners (see Fig. 7.24) that has a constant ignition source in the event that a large flow of gases containing hydrocarbons is suddenly vented through the flare. Flares have a very wide turndown ratio and rapid turndown response. They are typically used as a safety device to ensure that flammable gases are not released into the atmosphere where they could be reignited elsewhere. However, there is a limit to how dilute the flammable concentration can be to sustain a flame.

There are some potential pollutants that may be generated during the flaring process. These include thermal radiation heat loading of the surrounding area, noise caused by both the large gas flow rates through the flare and by the combustion reactions during flaring, and smoke and odor generated by the incomplete combustion of the flammables in the vent stream. The U.S. EPA has developed emission factors for industrial flares [32]. Besides noise and heat, flares also emit carbon particles (soot), CO and other unburned hydrocarbons, NO_x, and SO_x (if the fuels contain any sulfur). Flares do not typically lend themselves to conventional emission testing because of their configuration and open flames. Therefore, relatively little information is available on pollution emissions from flares.

Figure 7.23   Single flare. (From Ref. 14. Courtesy of John Zink Co., Tulsa, OK.)

Figure 7.24   Multinozzle flare. (From Ref. 14. Courtesy of John Zink Co., Tulsa, OK.)

Some of the advantages of flaring compared to thermal oxidizers include [33,34]:

- Can be an economical way to dispose of sudden large releases of gases.
- Usually do not require auxiliary fuel to support combustion.
- Can be used to control intermittent or fluctuating waste streams.

Some of the problems associated with flare operation include [35]:

- Can produce undesirable noise, smoke, heat radiation, and light.
- Can be a source of SO_x, NO_x, and CO.
- Cannot be used to treat waste streams with halogenated compounds.
- The heat from the combustion process is not profitably used.

Flares are also discussed in Chap. 15 because their primary use has been in petrochemical applications.

Regenerative Thermal Oxidation.   Regenerative thermal oxidation is a popular technique for controlling VOC emissions from industrial processes [36]. Melton and Graham have written a chapter in a recent book on thermal oxidizers used to destroy unwanted hydrocarbon emissions, especially VOCs [37]. Figure 7.25 shows a generic schematic of a regenerative thermal oxidizer (RTO), and Fig. 7.26 shows a schematic of a particular commercial RTO. Figures 7.27 and 7.28 show photographs of RTOs.

The basic principle of RTOs is to have multiple beds containing a packing made of some type of heat-retaining inert material such as a ceramic. The exhaust gas stream enters one or more of the beds containing a heated ceramic where the VOCs burn in this type of preheat chamber. However, the cooler gas stream entering

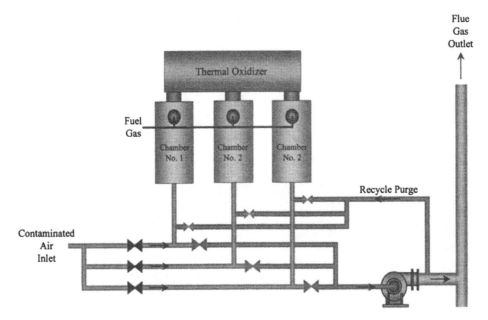

Figure 7.25   Generic regenerative thermal oxidizer system. (From Ref. 14. Courtesy of John Zink Co., Tulsa, OK.)

Figure 7.26 Schematic of a specific regenerative thermal oxidizer. (Courtesy of Alstom, Wellsville, NY.)

Figure 7.27 Regenerative thermal oxidizer. (Courtesy of Alstom.)

the preheat chamber cools down the bed with time. The exhaust gas stream from the preheat chamber then passes through the other set of beds that are being heated by the hot gases. This set of beds is then being thermally regenerated. After the preheat chamber bed has cooled down to a certain level, the incoming gas stream is then shifted to the set of beds that was previously being thermally regenerated and are at a much higher temperature. The cooled bed then becomes the bed to be regenerated.

Figure 7.28   Regenerative thermal oxidizer. (Courtesy of United McGill, Columbus, OH.)

The process continues to regularly switch back and forth between the sets of beds so that one set is being thermally regenerated while the other acts as the preheat chamber for the VOC destruction process. RTOs can be significantly more thermally efficient than nonregenerative thermal oxidizers. Stevanovic and Fassbinder [38] describe a new RTO using pebble-heater technology, which has the potential to be less capital intensive than conventional technology. A pebble heater refers to a regenerator filled with bulk material mostly of a spherical shape that looks like pebbles. This development has the potential to reduce dioxin/furan emissions.

Catalytic and Regenerative Catalytic Oxidation. In a catalytic oxidation system, a catalyst is used to modify the chemical kinetics of the reaction process. A catalyst may be used to combust a flammable at a much lower temperature than without the catalyst, typically at 1000°F (540°C) or less compared to 1800°F (980°C) or more for traditional thermal oxidizers. Figure 7.29 shows that complete destruction of some organic waste gases can be as low as 250°C (480°F) using a platinum catalyst. This would be useful where the flammable is at a lower temperature and it is more economical to combust the material at a lower temperature than to heat the gas stream up to higher temperatures. However, this is rarely a problem in industrial combustion processes where the exhaust gases are typically at temperatures well above the ignition temperature for most hydrocarbons. However, it still may be preferable to use a catalyst to combust flammables at a lower temperature because this may significantly reduce NO_x emissions. For example, if the normal ignition temperature of a flammable is above 1000°F (540°C), but the ignition temperature may be only 500°F (260°C) in the presence of a particular catalyst, then NO_x emissions may be reduced as thermal NO_x production is highly temperature dependent (see Chap. 6). By operating at much lower temperatures, the oxidizer does not need to be refractory lined, which can save on

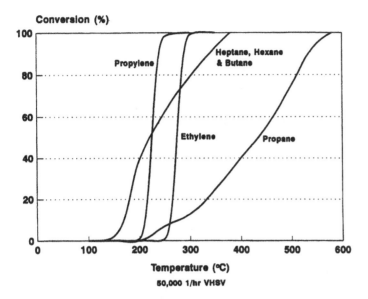

Figure 7.29 Performance of platinum monolithic catalysts for olefins and paraffins. (From Ref. 46. Courtesy of Van Nostrand Reinhold.)

weight; this can be important if the unit needs to be elevated. Another reason to use a catalyst is to speed up the chemical reactions, which can significantly shorten the residence time in the oxidizer and therefore reduce the size and cost of the oxidizer. Catalytic oxidation is often the preferred technology for treating relatively low VOC concentrations (< 10 ppmv) [39].

A regenerative catalytic oxidizer (RCO) typically consists of a fixed bed of catalyst, typically platinum or palladium, between two beds of chemically inert but heat retentive packing [40]. The flow enters through one of the porous inert beds while exhausting through the other. The inlet inert bed is at an elevated temperature and preheats the incoming VOC-laden stream. Oxidation of the VOCs is exothermic and heats up the outlet inert bed. The inlet inert bed cools down with time while the outlet inert bed gains heat over time. After a given period of time, the flow is then reversed and enters through the warmer inert bed that was previously the outlet and exits through the cooler inert bed that was previously the inlet. This transient process is very energy efficient because the liberated heat from VOC oxidation is captured and reused. Figure 7.30 shows catalytic systems without and with heat recovery to further enhance energy efficiency.

There are several advantages that catalytic oxidation systems may have compared to other technologies [41]. These include low operating costs, low capital equipment costs, high VOC destruction efficiency, long-life expectancy due to lower operating temperatures, quick start-up/warm-up times, and minimal maintenance requirements. The equipment is often sold as a prepackaged unit, which minimizes site installation time and costs.

A general problem with catalytic systems is the high initial cost of the catalyst, which is often a precious metal like platinum (e.g., [42]) or palladium (e.g., [43]).

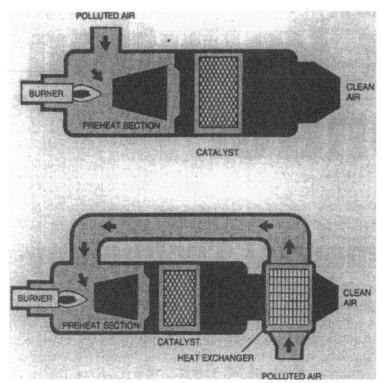

Figure 7.30 Catalytic regenerative systems without and with heat recovery. (From Ref. 46. Courtesy of Van Nostrand Reinhold.)

Other common catalysts include metal oxides like chromia/alumina, cobalt oxide, and copper oxide/manganese oxide. Mazza et al. [44] describe the use of titania loaded with platinum or palladium for VOC oxidation. Metal catalysts are optimized for higher oxidation temperatures than metal oxide catalysts. Figure 7.31 shows an example of a commercial catalyst used in catalytic oxidation systems.

Another general problem with catalytic systems is that the catalyst may need to be regenerated or cleaned on a regular basis depending on the nature of the process. If the catalyst eventually becomes unusable, it may be very expensive to dispose of the spent material. The efficiency of the catalytic process may degrade over time as the catalyst becomes contaminated. Catalytic systems may become quickly contaminated in industrial processes that are particularly dirty or corrosive. The catalyst is often put on some type of substrate such as a honeycomb structure to increase the surface area of contact between the contaminated gas and the catalyst. The structure introduces an additional pressure drop in the system, which may require additional fan power. A RCO system is generally less attractive than an RTO because RCOs usually have higher operating costs, related to the life of the catalyst [36]. Precious metal catalysts can be easily poisoned by contaminants such as lead, zinc, mercury, arsenic, phosphorus, bismuth, antimony, iron oxide, and tin [45]. Halogens, sulfur compounds, and NO_2 are potential chemical inhibitors that can

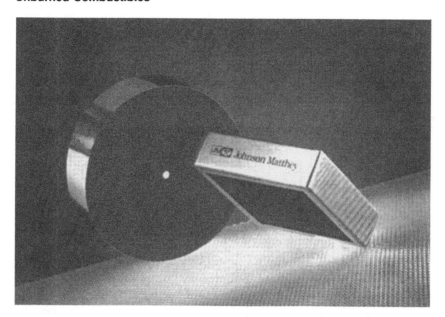

Figure 7.31 Typical catalyst used in catalytic oxidation systems. (Courtesy of Johnson Matthey, Malvern, PA.)

reduce the effectiveness of the catalyst. Relatively low concentrations of sulfur (> 50 ppmv) can degrade the performance of many catalysts [46].

Condensation. This post-treatment strategy attempts to collect VOC vapors so that they can preferably be recycled back into the process and is typically employed where concentrations are greater than 5000 ppmv. The removal efficiency is generally 50–90% so this technique would not be used alone if greater cleaning was required. Then, another VOC removal technique would need to be included further downstream for the final clean-up. The general benefit of this technique is that the VOCs can be reused in the process or sold for use elsewhere, which can be a dramatic savings compared to disposal costs. Gupta and Verma [47] developed a computer model of this technique and noted that relatively little work has been done to understand further the detailed physics of this technology.

The basic principle behind this technique is to cool the gases sufficiently with some type of heat exchanger to condense out the VOCs, which are converted into a liquid that can be simply removed from the gas stream (see Fig. 7.32). Unfortunately, relatively cold temperatures are often required. In many cases, normal mechanical refrigeration may not get the stream sufficiently cold to condense out the VOCs efficiently. In fact, these temperatures may be low enough that normal materials must be replaced with materials capable of going to lower temperatures. For example, carbon steel may need to be replaced with copper or brass, which is often more expensive. Another significant problem in removing VOCs from exhaust gas streams is that the gases often contain large quantities of water that will freeze out well before the VOCs. This causes ice and frost to form on the cooling equipment, which reduces its efficiency and therefore frequent defrosting is often required. Another major obstacle of this technique for cleaning exhaust gas streams is that the

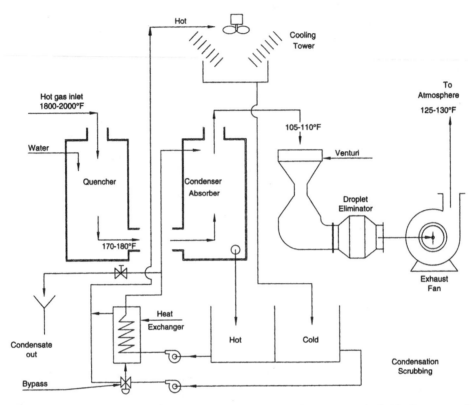

Figure 7.32 Schematic of a condensation scrubbing system. (From Ref. 55. Courtesy of Lewis Publishers.)

gases are often fairly hot to begin with—up to more than 2000°F (1100°C) in some processes. This puts a huge extra load on the condensing system, which must first cool the gases down to ambient conditions and then well below ambient to condense the VOCs. This makes this technique generally impractical for removing VOCs from hot combustion exhaust gases.

There are two predominant types of VOC condensers: surface and contact. In surface condensers, the coolant does not directly contact the gas stream. These are generally some type of heat exchanger, such as shell and tube or plate and frame exchangers, where the gases are cooled on contact with the surface of the exchanger. The coolant is on one side of the exchanger and the VOC-laden gas is on the other. The advantage of this strategy is that the VOCs are not diluted or contaminated by direct contact with the coolant. This technique is commonly used if the condensed VOCs are to be collected and either reused in the process or for use elsewhere including selling to other users.

In a contact condenser, a coolant such as water is sprayed into the gas stream to cool the gas below the condensation temperature of the VOCs. The challenge with this technique is to ensure proper liquid spray distribution so that gases do not pass through the condenser without contacting the coolant.

Adsorption. When a solid medium is surrounded by a gas, the surface of the solid tends to attract molecules of the gas. Depending on the solid and the gas, several layers of gas molecules are tightly held at the surface of the solid. Energy is liberated during the adsorption process, which is normally reversible; this means that energy needs to be supplied to release the gas molecules from the solid surface. Then, adsorption is exothermic and desorption is endothermic. The amount of gas that can be adsorbed at a surface is limited and decreases with temperature.

The solid surface adsorbing the gas molecules is referred to as the *adsorbent* while the gas being adsorbed is referred to as the *adsorbate*. The carrier gas containing the gas to be removed is referred to as the *solvent*. Not all gases adsorb equally. Gases with higher molecular weights and lower boiling temperatures are more readily adsorbed [48]. This is particularly useful for removing VOCs that more readily adsorb on solids than one of the main components in exhaust gas streams—nitrogen. This preferential adsorption is then useful in gas separation processes, including separating air into its major constituents of nitrogen and oxygen.

Not all solids are good adsorbents. Some of the more common adsorbents include charcoal, silica gel, and bauxite. Activated carbon is one of the most popular for air pollution control. For example, Prasertmanukitch et al. [49] discuss the use of activated carbon to eliminate VOC emissions from waste-incineration processes. The term activated refers to increased internal and external surface area produced by special treatment processes. The carbon is activated by heating to a high temperature in the absence of oxygen to drive off the volatiles. The carbon is then activated to increase the surface area by using steam, air, or carbon dioxide at high temperatures. The structure of the adsorbent is also important. Since the adsorption process occurs at the surface of the solid, the adsorbent should have as high a surface area per unit volume as possible to maximize the adsorption sites. Capillary pores, common in porous structures, are desirable as they increase the available adsorption surface area. The adsorption bed can be stationary or moving, but stationary if far more popular [13].

An important characteristic of an adsorbent is its surface polarity, which determines the types of vapors that it would be most efficient in adsorbing [50]. The effective rate of adsorption is also affected by a number of diffusional transport processes [28]:

1. Diffusion in the sorbed state
2. Reaction at the phase boundary
3. Pore diffusion in the fluid phase within the particles
4. Mass transfer from the fluid phase to the external surfaces of the adsorbent particles
5. Mixing throughout the various parts of the contacting equipment

This technique for removing VOCs from gas streams involves sending the dirty gas into a vessel containing a packed bed of an adsorbent designed to remove VOCs. An example of an industrial vapor-recovery system is shown in Fig. 7.33. The VOCs adsorb or attach to the surface of the bed while the rest of the gas passes through the bed. Several beds in series may be required to achieve sufficient removal levels. The problem with this technique is that the beds must be regularly cleaned to maintain an average design removal efficiency. Steam is

Figure 7.33   Industrial vapor-recovery system. (Courtesy of John Zink Co., Tulsa, OK.)

often used to clean the accumulated VOCs from the adsorbent bed. This means that multiple beds are required where some may be in service and some out of service for cleaning at any given time. VOCs build up on the adsorbent until it is saturated. The removal efficiency is highest right after cleaning and lowest right before cleaning. If the bed is not taken out of service for cleaning before it becomes saturated, then VOCs will "break through" the bed without being adsorbed. Activated carbon is a common adsorbent used in these systems. The VOCs are then separated from the cleaning fluid and either recycled back into the process where possible or disposed of properly. The main function of the adsorbent beds is to concentrate the VOCs to make their treatment simpler. This technique is best applied to large gas volumes containing dilute pollution levels [45].

All solids have some adsorptive capability, but only certain materials have sufficient adsorption characteristics to be economically feasible and industrially robust enough for removing pollutants. The shape of the adsorbent material is also important. The packing must not produce too much pressure drop. The material cannot be so fine that it is carried away by the exhaust gases. It must be strong enough so that it is not crushed under the weight of a column of packing. Activated carbon, alumina, silica gel, and molecular sieves are commonly used as adsorbents.

There are two types of adsorption: physical and chemical. Physical adsorption is analogous to condensation and occurs at lower temperatures. It is sometimes referred to as surface condensation. Chemical adsorption, sometimes referred to as chemisorption or surface reaction, occurs at higher temperature and relies on chemical reactions. Therefore, it can only take place when the chemical to be adsorbed can react with the adsorbent surface. In physical adsorption the process can take place between any surface and any gas, needs no energy to start the process, and has a rapid rate of adsorption. In chemical adsorption, there must be a chemical affinity between the adsorbate and adsorbent, and the energy of activation must be supplied before the process can begin. Chemical adsorption liberates more heat than does physical adsorption.

There are a number of factors that affect the retention rates (sometimes referred to as retentivity) of adsorbents. Figure 7.34 shows that the effectiveness of an adsorption system decreases with temperature. This is due to the increase in kinetic energy with temperature, which overcomes the weak Van der Waal forces holding the adsorbate to the adsorbent. Figure 7.35 shows schematically that the adsorption retentivity increases with pressure. Figure 7.36 shows that as the adsorbent internal surface area per unit volume increases, the adsorption effectiveness increases. Figure 7.37 shows that adsorbents become more effective as the molecular weight of the solvent increases.

Hesketh [51] notes the following factors that affect adsorption:

1. Operating conditions: adsorption is improved by low temperature, high pressure, and high adsorbate gas concentration.
2. Gas being adsorbed: molecule size, boiling point, molecular weight, and polarity.
3. Adsorber: surface polarity, pore size, and pore spacing.
4. Adsorber design: adequate surface area and holdup time, good gas distribution, adequate pretreatment to remove adsorbent contaminants, adequate regeneration and cool down, and replacement of unusable adsorbent.

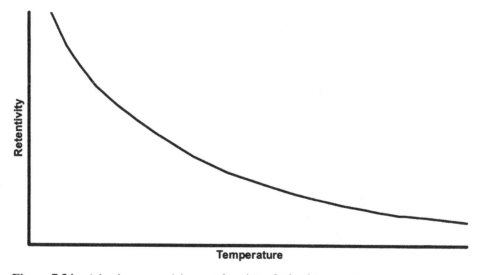

Figure 7.34 Adsorbent retentivity as a function of adsorbent temperature.

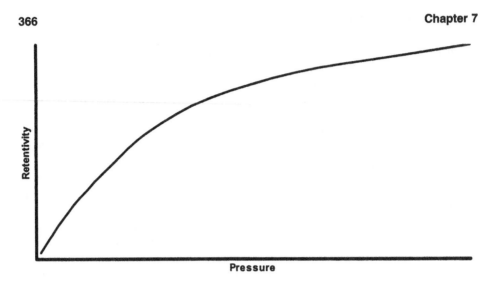

Figure 7.35   Adsorbent retentivity as a function of pressure.

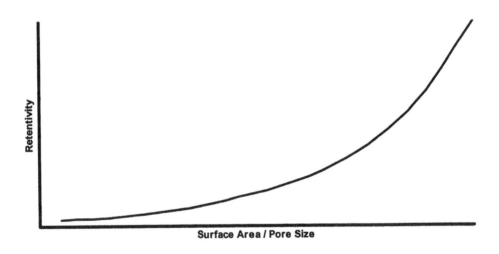

Figure 7.36   Adsorbent retentivity as a function of adsorbent surface area/pore size.

Some of the requirements for optimizing adsorption system design include [52]:

1. Adequate contact time between the gas stream and the adsorbent bed.
2. Sufficient sorption capacity for the given conditions.
3. Minimized pressure drop through the adsorbent bed.
4. Uniform gas flow distribution to utilize fully the adsorbent capacity.
5. Adequate pretreatment of the gas stream to remove nonadsorbable particles.
6. Ability to regenerate the adsorbent on an ongoing basis without significantly degrading its adsorption capability.

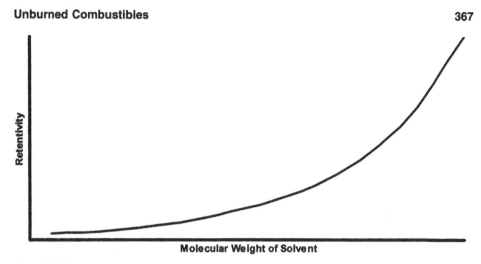

Figure 7.37 Adsorbent retentivity as a function of the molecular weight of the solvent.

Contaminants such as particulates can significantly reduce the effectiveness by clogging up the porosity and increasing the pressure drop through the system. Moisture and carbon dioxide in the gas stream can be undesirably adsorbed in certain types of adsorbents and reduce the effectiveness of the adsorption system.

This technique has several problems related to removing VOCs from high-temperature combustion exhaust gas streams. One problem is that the activated carbon bed is not able to handle some of the high gas temperatures encountered in industrial combustion processes. Since adsorption is exothermic, it may be possible to heat a carbon adsorbent bed above its autoignition temperature. Another problem is that additional fan power would be necessary to draw the exhaust gases through the packed beds. A third problem is the shear size of a system that would be required for typical industrial processes. The cycle for putting the adsorber beds in and out of service can be problematic if the composition or concentration of VOCs varies considerably over time. These systems are optimally designed for a specific set of conditions so that the efficiency can be significantly reduced if the conditions vary too much from the original specifications.

Davis et al. [53] list the following advantages of adsorption systems:

1. Product recovery may be possible.
2. Excellent control and response to process changes.
3. No chemical disposal problem when pollutant is recovered and recycled.
4. Collection system can be automated.
5. Capability to remove pollutants to very low levels.

and the following disadvantages:

1. Product recovery may require an exotic, expensive distillation scheme.
2. Adsorption efficiency deteriorates over time.
3. Adsorption regeneration requires a steam or vacuum source.
4. Relatively high capital cost.
5. Prefiltering may be required to prevent plugging of the adsorbent bed.

6. Needs fairly low gas stream temperatures.
7. High steam requirements for desorbing high-molecular-weight hydrocarbons.

Absorption (Wet Scrubbing).   Absorption scrubbing is sometimes referred to as packed-bed or packed-tower wet scrubbers or as spray-chamber or spray-tower wet scrubbers. This technique captures the VOCs with a liquid solvent. The VOC-laden gas stream flows into the bottom of the scrubber while the solvent flows in from the top of the scrubber. The cleaned gas exits the scrubber from the top while the VOC-laden solvent flows out of the bottom of the scrubber. Therefore, the exhaust gases flow counter to the liquid solvent. The scrubber is designed to optimize the contact between the exhaust gas stream and the liquid solvent through the use of bubble caps, sieve trays, or some type of packing in the scrubber column. The objective of the column packing is to have a high wetted area per unit volume, minimal weight, and sufficient chemical resistance, low liquid holdup, low pressure drop and low cost [54]. Again, the VOCs must be stripped from the solvent in a liquid–liquid separator, similar to the adsorption cleaning process.

Some of the advantages of this technique include low pressure drop, operation in highly corrosive atmospheres with the appropriate materials of construction, relatively high mass-transfer efficiencies, easily expandable mass transfer, relatively low capital cost and space requirements, and the ability to collect particulates as well. An advantage of absorption scrubbing over adsorption beds is that the scrubber does not need to cycle on and off for cleaning as it is designed for continuous operation.

Similar disadvantages are encountered with this technique as with adsorption where the systems may be uneconomically large for the gas streams from industrial combustion processes, high gas temperatures greatly complicate the design of the scrubber, and a suitable solvent must be available for the VOCs to be removed, which may be challenging if a wide range of hydrocarbons are present. Period cleaning of the packing in the column may be necessary, depending on the composition of VOCs being removed. A liquid waste product may need to be treated and particulate matter may plug or clog the mass-transfer equipment. The maintenance costs may be relatively high compared to those of other techniques.

Biofiltration.   This technique is sometimes referred to as biological oxidation. It involves the use of microorganisms that consume pollutants and turn them into benign organic materials. The pollutants are typically water soluble and can be captured by biological organisms. Biofiltration is often referred to as a "green technology" because it is environmentally friendly [55]. Garner [56] refers to biofiltration as a "disruptive technology" that requires only one-fourth to one-tenth the energy of conventional control technologies. A disruptive technology is one that initially underperforms comparable technologies but has some special features that are valued, in this case the environmentally friendly nature of the technology. Removal efficiencies are often up to 90%. Advances in technology have dramatically reduced the size of previous biofiltration systems.

The gas stream containing the VOCs to be treated is passed through the bioreactor containing a bed packed with damp, porous organic particles [57]. A biofilter may be thought of as a packed tower where the biomass support is the packing and the biofilm is the absorbing liquid. Table 7.5 shows that certain classes of VOCs are easier to biodegrade than others [16]. The most important design considerations for biofilters are the media type, moisture control, and air distribution [13].

A schematic of a typical open biofiltration system is shown in Fig. 7.38 [45]. A schematic of an industrial biofiltration system referred to as biovent trickling is shown in Fig. 7.39. A schematic of another industrial biofiltration system with only a single reaction tank is shown in Fig. 7.40. Figure 7.41 shows a photograph of an industrial biofiltration system used to remove methanol and formaldehyde in particleboard manufacturing. The VOCs are sorbed by the biologically active filter bed containing microorganisms that attach to and aerobically degrade the sorbed chemical compounds. The VOCs are considered as food to the microorganisms, which convert the VOCs into oxidation end-products like CO_2, H_2O, and mineral salts. A key aspect of biofiltration is that it is generally considered to be a safe, environmentally friendly air pollution control technology. It may be thought of as an active absorption scrubber where the contaminants (VOCs) are not only captured

Table 7.5 Biodegradability of Individual and Classes of VOCs

Inorganic Rapid	Organic			
	Rapid	Good	Slow	Very slow
Hydrogen sulfide	Alcohols methanol	Esters ethylacetate	Aliphatic hydrocarbons	Many halogenated hydrocarbons
Ammonia	butanol	Ketones	methane	1,1,1-trichloroethane
Sulfur dioxide	Aldehydes formaldehyde acetaldehyde	acetone Phenols benzene	pentane cyclohexane	Polyaromatic hydrocarbons
	Amines	styrene		
	Organic acids butyric acid	Mercaptans methyl mercaptan		

Source: Ref. 16.

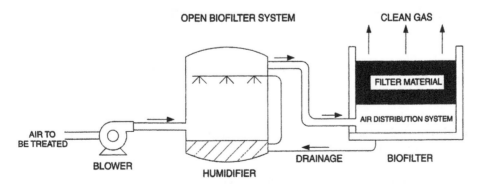

Figure 7.38 Schematic of a typical open biofiltration system. (From Ref. 45. Courtesy of CRC Press.)

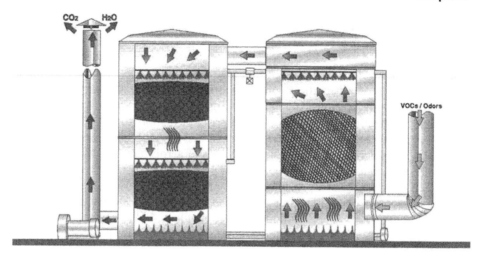

Figure 7.39 Schematic of an industrial biofiltration system with two reactors. (Courtesy of Bioreaction, Tualatin, OR.)

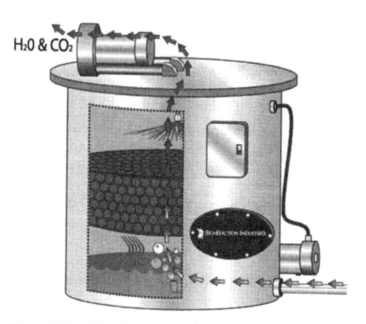

Figure 7.40 Schematic of an industrial biofiltration system with a single reactor. (Courtesy of Bioreaction.)

but also treated in the same device. Devinny et al. [16] have written a comprehensive book covering all aspects of biofiltration including theory, application, design, and operation.

The microorganisms are sometimes divided into two categories: bacteria and fungi. Typical bacteria used in bioreactors include *micrococcus albus*, *proteus vulgarus*, and streptomyces. Typical fungi include penicillium, mucor, and

Figure 7.41 Industrial biofiltration system for removing methanol and formaldehyde in particleboard manufacturing. (Courtesy of Bioreaction.)

stemphilium. Microorganisms are assembled into a type of filter, often in the form of a packed-tower arrangement where the microbial communities grow on the surface of the packing [58]. Some type of liquid layer is used which captures the pollutants for the microbes to consume. While biofiltration is more well known for VOC and odor removal, it can also remove NO, N_2O, and a variety of other pollutants. However, at this time there are more viable methods for removal of most non-VOC pollutants.

A number of factors are important to optimize biofiltration performance. Pretreatment of the gas stream is important to remove particulates that would not be treated but would clog the biofilter. Gas distribution is important for achieving uniform flow. Nonuniform flow may reduce VOC removal efficiency where too much flow is going through a section of the biofilter. Proper gas-distribution systems can solve this problem. Another important factor is humidification where too little moisture can cause dry zones and too much moisture can inhibit gas transport. Insufficient moisture can also limit microbial activity [59]. Proper pH control maximizes microbial activity where conditions at or neutral are generally preferred. Temperature control is important where a narrow temperature range is preferred. Temperatures above or below that range reduce microbial activity. The type of VOC to be treated is an important factor in the performance of a biofilter. Some VOCs are more biodegradable than others as shown in Table 7.6. Residence time is also an important factor as fairly long times may be required to achieve high removal efficiencies.

Another new development is a device called a fungal vapor-phase bioreactor which is designed to remove NO and VOCs from waste gas streams [60]. In the case

Table 7.6 Biodegradability of Volatile Organic Compounds

Contaminant	Biodegradability	Contaminant	Biodegradability
Aliphatic hydrocarbons		*Sulfur-containing carbon[b] compounds*	
Methane	1	Carbon disulfide	2
Propane	?	Dimethyl sulfide	2
Butane	?	Dimethyl disulfide	2
Pentane	1	Methyl mercaptan	1
Isopentane	1	Thiocyanates	1
Hexane	2		
Cyclohexane	1	*Oxygenated carbon*	
Acetylene	1	*compounds*	
		Alcohols	3
Aromatic hydrocarbons		methanol	3
Benzene	2	ethanol	3
Phenol	3	butanol	3
Toluene	3	2-butanol	3
Xylene	2	1-propanol	3
Styrene	2	2-propanol	3
Ethylbenzene	3	Aldehydes	3
		formaldehyde	3
Chlorinated[b] hydrocarbons		acetaldehyde	3
Carbon tetrachloride	1	Carbonic acids (esters)	3
Chloroform	1	butyric acid	3
Dichloromethane	3	vinyl acetate	2
Bromodichloromethane	?	ethyl acetate	3
1,1,1-Trichloroethane	?	butyl acetate	3
1,1-Dichloroethane	?	isobutyl acetate	3
Tetrachloroethene	1[a]	Ethers	1
Trichloroethene	1[a]	diethyl ether	1
1,2-Dichloroethane	?	dioxane	1
1,1-Dichloroethene	?	methyl *tert*-butyl ether	1
Vinyl chloride	1	tetrahydrofuran	3
1,2-Dichlorobenzene	?	Ketones	3
Chlorotoluene	1	acetone	3
		methyl ethyl ketone	3
Nitrogen-containing carbon compounds		methyl isobutyl ketone	3
Amines	3		
aniline	3	*Inorganic[b] compounds*	
Nitriles	1	Ammonia	3
acrylonitrile	?	Hydrogen sulfide	3
pyridine	1	Nitrogen oxide	1

[a]Indicates that cometabolism or anerobic treatment has been identified within a biofilter.
[b]Indicates that a change in filter-bed pH may occur with treatment of these compounds. This change may negatively affect performance.
Note: 1 = some biodegradability; 2 = moderate biodegradability; 3 = good biodegradability; ? = unknown.
Source: Ref. 57.

of VOCs, microorganisms are used to oxidize the hydrocarbons in (or near) ambient conditions through organic processes. The general principle is to provide an environment for the microorganisms that oxidize the VOCs to thrive. The microorganisms act as an organic filter to remove the VOCs from the gas stream through adsorption, absorption, and diffusion processes. This technology shows a resistance to adverse operating conditions and is able to maintain high removal efficiencies over an extended period of time, which are problems with some other post-treatment technologies. Removal efficiencies exceeding 90% for both NO_x and VOCs have been demonstrated on a bench-scale apparatus.

The general advantages of biofiltration include low installation and operating costs, low maintenance, long life for the biofilter, and environmentally friendly operation. Not all VOCs are amenable to this process as some are relatively biologically inactive. However, this process is not likely to be practical for most industrial processes because of the long residence time requirements, high capital costs, high fuel use compared to other technologies, temperature limitations, potentially significant pressure drop through the biofilter, and significant handling constraints.

Michelsen [57] lists the following advantages of biofiltration:

1. Safe, environmentally friendly technology.
2. Generally lower capital and operating costs for treating large waste streams containing low concentrations of biodegradable pollutants.
3. Can achieve high destruction efficiencies in treating certain pollutants.
4. Relatively simple technology that does not require large quantities of chemical additives.

and the following disadvantages:

1. Not as effective for treating waste streams with high pollutant concentrations, pollutants that are difficult to degrade, and for intermittent volume flow rates.
2. Require an acclimation time of the microorganisms to the pollutants before reaching optimum efficiency.
3. Substantial filter material often required to achieve long residence times.
4. Not as generally accepted yet as other types of pollutant control technologies.

Ergas and Kinney [61] list the following advantages of biofiltration:

1. Low operating costs
2. Absence of residuals and by-products
3. Produces less CO_2 and NO_x emissions than does thermal oxidation

and the following disadvantages:

1. High area requirements
2. Moderate to high operating costs
3. Not all gas streams amenable to this technique

They also list a number of factors to be considered concerning biological control systems:

1. Pollutant biodegradability
2. Compound solubility
3. Pollutant loading rate and concentration.
4. Gas temperature (very cold or hot gases are not usually amenable).
5. Incomplete biodegradation.

6. Acidic by-product formation during the biodegradation process.
7. Acclimation time required for the microbial population to acclimate to the pollutant.
8. Transient conditions are encountered when the process is shut down for maintenance or other reasons that require reacclimation of the microbes.
9. Mixtures of compounds usually mean some compounds will be removed before others so that there must be adequate treatment space to remove all desired compounds.
10. Dust and grease can clog biofilters.

Won et al. [62] discuss a unique method for controlling biomass accumulation, which can be a major obstacle for long-term, stable operation of biotrickling filters. Clogging reduces pollutant removal and increases the pressure drop through the filter. The method involves the use of flies and fly larvae literally to consume the accumulated biomass and achieve zero net growth. The flies can be eliminated by short periods of oxygen deficiency in the bioreactor, causing them to be asphyxiated.

Membrane Separation. The basic principle of this technology is the use of a membrane that selectively passes certain chemicals while preventing others from passing through. In the case of VOCs, a polymeric membrane is developed to pass VOCs selectively while remaining impermeable to gases like nitrogen, which is the primary constituent in most exhaust gases. The membranes are best suited for VOC concentrations of 1000 ppmv or higher. This technology has the advantage of collecting the VOCs for reuse in the process or for sale to other industrial users. This is particularly advantageous where the cost of the VOCs is a significant portion of the raw feed materials in the process.

This is a relatively new technology as applied to VOCs and is not generally applicable to most industrial combustion processes. There are a number of potential problems that limit its use at this time. A significant issue is the relatively low-temperature limit of the membranes, which may mean the exhaust gas stream would need to be cooled before reaching the membrane. Depending on the required removal efficiency and the VOC concentration in the gas stream, the pressure drop through the membrane may be significant, which could mean that additional fan power could be required. The membrane structure could be plugged in dirty gas streams that have substantial particle loadings. For very high VOC removal requirements, a secondary technology may need to be used to obtain the higher levels of removal efficiencies. The technology may not be of much interest if there is no demand for the collected VOCs or if the economics of reusing or selling the collected VOCs do not offset the cost of the membrane separation costs.

Other Techniques. A less common technique for destroying VOCs is to expose them to a high-voltage electric field or corona in a device referred to as a corona reactor [63]. Electrons contact the hydrocarbon molecules and break them down into simpler elements and compounds like H_2O, CO_2, and CO. A bed of ferroelectric pellets is charged with an alternating current. Since the pellets are poor electrical conductors, a corona field is generated between the pellets. In general, the higher the voltage field, the more complete the VOC destruction. Another related technique is to use an electric arc between two electrodes.

REFERENCES

1. K Ahlberg (ed.). *AGA Gas Handbook*. Lidingö, Sweden: AGA AB, 1985.
2. K Wark, CF Warner. *Air Pollution: Its Origin and Control*, 2nd edn. New York: Harper & Row, 1981.
3. JR Creighton. Dependence of CO emissions on the rate of product cooling. *Combust. Flame*, Vol. 123, pp. 402–411, 2000.
4. J Elkins, N Frank, J Hemby, D Mintz, J Szykman, A Rush, T Fitz-Simons, T Rao, R Thompson, E Wildermann, G Lear. National Air Quality and Emissions Trends Report, 1999. Washington, DC: U.S. Environmental Protection Agency, Rep. EPA 454/R-01-004, 2001.
5. DS Eisinger, K Dougherty, DPY Chang, T Kear, PF Morgan. A reevaluation of carbon monoxide: past trends, future concentrations, and implications for conformity "hot-spot" policies. *J. Air Waste Mgmt. Assoc.*, Vol. 52, pp. 1012–1025, 2002.
6. U.S. EPA. Method 10: Determination of Carbon Monoxide Emissions from Stationary Sources. Code of Federal Regulations 40, Part 60, Appendix A-4. Washington, DC: U.S. Environmental Protection Agency, 2001.
7. J de Wit, K Johansen, PL Hansen, H Rossen, NB Rasmussen. Catalytic Emission Control with Respect to CH_4 and CO for Highly Efficient Gas Fueled Decentralised [sic] Heat and Power Production. *Proceedings of 5th European Conference on Industrial Furnaces and Boilers*, Lisbon, Portugal, Vol. II, pp. 587–595, 2000.
8. CE Baukal (ed.). *Oxygen-Enhanced Combustion*. Boca Raton, FL: CRC Press, 1998.
9. R Wilk, A Ksiadz, A Szlek. Measurement Methods and Preliminary Experimental Investigations of Polycyclic Aromatic Hydrocarbons (PAHs) Formation in Premixed Flames of Gaseous Fuels. *Fifth International Conference on Technologies and Combustion for a Clean Environment*, Lisbon, Portugal, Vol. 2, The Combustion Institute—Portuguese Section, pp. 855–858, 1999.
10. CE Baukal, R Scharf, PB Eleazer. Low Emission Aluminum Melting Process. Presented at the 1996 Spring AFRC Meeting, Orlando, FL, May 6–7, 1996.
11. S Setia, VOC emissions—hazards and techniques for their control. *Chem. Eng. World*, Vol. XXXI, No.9, pp. 43–47, 1996.
12. Office of the Federal Register. 60.2 Definitions. U.S. Code of Federal Regulations Title 40, Part 60. Washington, DC: U.S. Government Printing Office, 2001.
13. TT Shen, CE Schmidt, TR Card. *Assessment and Control of VOC Emissions from Waste Treatment and Disposal Facilities*. New York: Van Nostrand Reinhold, 1993.
14. CE Baukal (ed.). *The John Zink Combustion Handbook*. Boca Raton, FL: CRC Press, 2001.
15. N de Nevers. *Air Pollution Control Engineering*. New York: McGraw-Hill, 2000.
16. JS Devinny, MA Deshusses, TS Webster. *Biofiltration for Air Pollution Control*. Boca Raton, FL: Lewis Publishers, 1999.
17. P Melton, K Graham. Thermal Oxidizers. In *The John Zink Combustion Handbook*, Boca Raton, FL: CRC Press, 2001.
18. WR Niessen. *Combustion and Incineration Processes*, 2nd edn. New York: Marcel Dekker, 1995.
19. TN Demayo, MM Miyasato, GS Samuelsen. Hazardous Air Pollutant and Ozone Precursor Emissions from a Low-NO_x Natural Gas-Fired Industrial Burner. *Twenty-Seventh Symposium (International) on Combustion*. Pittsburgh, PA: The Combustion Institute, pp. 1283–1291, 1998.
20. TN Demayo, GS Samuelsen. Hazardous Air Pollutant and Ozone Precursor Emissions from "Low-NO_x" Natural Gas-Fired Burners: a Parametric Study. *Fifth International Conference on Technologies and Combustion for a Clean Environment*, Lisbon, Portugal, Vol. 2, The Combustion Institute—Portuguese Section, pp. 663–671, 1999.

21. U.S. EPA. Method 25: Determination of Total Gaseous Nonmethane Organic Emissions as Carbon. Code of Federal Regulations 40, Part 60, Appendix A-7. Washington, DC: U.S. Environmental Protection Agency, 2001.

22. B Fabrellas, Y Benito, P Sanz. Characterization of VOC's [sic] Emissions from Wastes Advanced Oxidation and Combustion Processes by TD/GC/MS System. In N Piccinini, R Delorenzo (eds.). *Chemical Industry and Environment II*, Vol 1. Turin, Italy: Politecnico Di Torino, 1996.

23. U.S. EPA. Section 3: VOC Controls. In DC Mussatti (ed.). *Air Pollution Control Cost Manual*, 6th edn. Rep. EPA.452/B-02-001. Washington, DC: U.S. Environmental Protection Agency, Jan. 2002.

24. CD Cooper. Vapor Incineration. Air Pollution Control Engineering. In WT Davis (ed.). *Air Pollution Engineering Manual*. New York: John Wiley, 2000.

25. MC Hannum. Benefits of Lean Premix Combustion Technology to Thermal Oxidation. Paper 45470. *Proceedings of the Air & Waste Management Association's 95th Annual Conference & Exhibition*, Baltimore, MD, June 23–27, 2002.

26. M Crocker, RJM Groenen, BM Veldmeijer. The Shell DeVOx Process for VOC Abatement. In *Controlling Industrial Emissions—Practical Experience*, pp. 169–177. *IChemE Symposium Series 143*. Warwickshire, UK: Institution of Chemical Engineers, 1997.

27. RD Ross, TH Snape. Thermal Incineration. In PN Cheremisinoff (ed.). *Air Pollution Control and Design for Industry*. New York: Marcel Dekker, 1993.

28. J Benítez. *Process Engineering and Design for Air Pollution Control*. Englewood Cliffs, NJ: Prentice Hall, 1993.

29. WM Vatavuk, DR van der Vaart, JJ Spivey. Section 3.2: VOC Destruction Controls, Chap. 2: Incinerators. In DC Mussatti (ed.). *Air Pollution Control Cost Manual*, 6th edn. Rep. EPA/452/B-02-001. Washington, DC: U.S. Environmental Protection Agency, Jan. 2002.

30. LB Evans, WM Vatavuk, DK Stone, SK Lynch, RF Pandullo, W Koucky. Section 3.2: VOC Destruction Controls, Chap. 1: Flares. In DC Mussatti (ed.). *Air Pollution Control Cost Manual*, 6th edn. Rep. EPA/452/B-02-001. Washington, DC: U.S. Environmental Protection Agency, Jan. 2002.

31. Office of the Federal Register. 60.18 General Control Device Requirements. U.S. Code of Federal Regulations Title 40, Part 60. Washington, DC: U.S. Government Printing Office, 2001.

32. U.S. EPA. Compilation of Air Pollutant Emission Factors, Vol. I: Stationary Point and Area Sources, Section 13.5: Industrial Flares. 5th edn. Washington, DC: U.S. Environmental Protection Agency Rep. AP-42, 1995.

33. U.S. EPA. Control Technologies for Hazardous Air Pollutants. Rep. EPA/625/6-91/014. Washington, DC: U.S. Environmental Protection Agency, 1991.

34. U.S. EPA. Control Techniques for Volatile Organic Emissions from Stationary Sources. Rep. EPA-453/R-92-018. Washington, DC: U.S. Environmental Protection Agency, 1992.

35. JM Tyler. Combustion Systems. In JC Mycock, JD McKenna, L Theodore (eds.). *Handbook of Air Pollution Control Engineering and Technology*, Boca Raton, FL: Lewis Publishers, 1995.

36. K Tierney. RTOs: A new look at an established technology. *Pollut. Eng.* Vol. 32 No.13, pp. 28–31, 2000.

37. P Melton, K Graham. Thermal Oxidizers. In C Baukal (ed). *The John Zink Combustion Handbook*. Boca Raton, FL: CRC Press, pp. 638–689, 2001.

38. D Stevanovic, H-G Fassbinder. Regenerative Thermal Oxidizers Based on the Pebble-Heater Technology. *Proceedings of 5th European Conference on Industrial Furnaces and Boilers*, Lisbon, Portugal, Vol. II, pp. 607–616, 2000.

39. R Cooley. Burning questions: catalytic oxidation Q and A. *Environ. Protect.*, Vol. 13, No. 2, pp. 12–17, 2002.

40. VO Strots, GA Bunimovich, CR Roach, YS Matros. Regenerative Catalytic Oxidizer Technology for VOC Control. In MA Abraham, RP Hesketh (eds.). *Reaction Engineering for Pollution Prevention.* Amsterdam: Elsevier, 2000.

41. CM Martinson. A breath of fresh air for your oxidizer system. *Process Heat.*, Vol. 9, No. 8, pp. 43–46, 2002.

42. Y Wang, H Shaw, RJ Farrauto. Catalytic Oxidation of Trace Concentrations of Trichlorethylene over 1.5% Platinum on γ-Alumina. In RG Silver, JE Sawyer, JC Summers. (eds.). *Catalytic Control of Air Pollution.* Washington, DC: American Chemical Society, 1992.

43. T-C Yu, H. Shaw, RJ Farrauto. Catalytic Oxidation of Trichlorethylene over PdO Catalyst on γ-Al_2O_3. In RG Silver, JE Sawyer, JC Summers (eds.). *Catalytic Control of Air Pollution*, Washington, DC: American Chemical Society, 1992.

44. D Mazza, I Mazzarino, S Bodoardo, S Ronchetti, M Lucco-Borlera. Evaluation of Pt, Pd Loaded Titania as Catalyst for VOC Oxidation. In N Piccinini, R Delorenzo (eds.). *Chemical Industry and Environment II*, Vol 1. Turin, Italy: Politecnico Di Torino, 1996.

45. KB Schnelle, CA Brown. *Air Pollution Control Technology Handbook.* Boca Raton, FL: CRC Press, 2002.

46. RM Heck, RJ Farrauto. *Catalytic Air Pollution Control: Commercial Technology.* New York: Van Nostrand Reinhold, 1995.

47. VK Gupta, N Verma. Removal of volatile organic compounds by cryogenic condensation followed by adsorption. *Chem. Eng. Sci.*, Vol. 57, pp. 2679–2696, 2002.

48. M Crawford. *Air Pollution Control Theory.* New York: McGraw-Hill, 1976.

49. S Prasertmanukitch, MJ Tierney, I Richeh. The Elimination of VOCs and Other Toxic Gaseous Emissions from Waste Incinerator Plant. *Proceedings of 5th European Conference on Industrial Furnaces and Boilers*, Lisbon, Portugal, Vol. II, pp. 641–650, 2000.

50. KE Noll. Adsorption. In WT Davis (ed.). *Air Pollution Engineering Manual*, New York: John Wiley, 2000.

51. H Hesketh. *Air Pollution Control: Traditional and Hazardous Pollutants.* Lancaster, PA: Technomic, 1991.

52. PN Cheremisinoff. Gas Phase Adsorption for Air Pollution Control. In PN Cheremisinoff (ed.). *Air Pollution Control and Design for Industry.* New York: Marcel Dekker, 1993.

53. WT Davis, AJ Buonicore, L Theodore. Air Pollution Control Engineering. In WT Davis (ed.). *Air Pollution Engineering Manual.* New York: John Wiley, 2000.

54. CD Cooper, FC Alley. *Air Pollution Control: A Design Approach.* Prospect Heights, IL: Waveland Press, 1994.

55. KC Schifftner. *Air Pollution Control Equipment Selection Guide.* Boca Raton, FL: Lewis Publishers, 2002.

56. LG Garner. Biofiltration—A Disruptive Technology for Sustainable Air Pollution Control, Paper 42633. *Proceedings of the Air & Waste Management Association's 95th Annual Conference & Exhibition*, Baltimore, MD, June 23–27, 2002.

57. RF Michelsen. Biofiltration. In JC Mycock, JD McKenna, L Theodore (eds.). *Handbook of Air Pollution Control and Technology.* Boca Raton, FL: Lewis Publishers, 1995.

58. JB Eweis, SJ Ergas, DPY Chang. ED Schroeder. *Bioremediation Principles.* New York: McGraw-Hill, 1998.

59. SJ Hwang, S-J Wu, C-M Lee. Water transformation in the media of biofilters controlled by *Rhodococcus fascians* in treating an ethyl acetate-contaminated airstream. *J. Air Waste Mgmt. Assoc.*, Vol. 52, pp. 511–520, 2002.

60. JR Woertz, KA Kinney, PJ Szaniszlo. A fungal vapor-phase bioreactor for the removal of nitric oxide from waste gas streams. *J. Air Waste Mgmt. Assoc.*, Vol. 51, pp. 895–902, 2001.

61. SJ Ergas, KA Kinney. Biological Control Systems. Air Pollution Control Engineering. In WT Davis (ed.). *Air Pollution Engineering Manual*. New York: John Wiley, 2000.
62. Y-S Won, JHJ Cox, WE Walton, MA Deschusses. An Environmentally Friendly Method for Controlling Biomass in Biotrickling Filters for Air Pollution Control, Paper 42633. *Proceedings of the Air & Waste Management Association's 95th Annual Conference & Exhibition*, Baltimore, MD, June 23–27, 2002.
63. WL Heumann. Miscellaneous Technologies. Chap. 15 in *Industrial Air Pollution Control Systems*, ed. WL Heumann. New York: McGraw-Hill, 1997.

8

Sulfur Oxides (SO$_x$)

8.1 INTRODUCTION

Elemental sulfur is relatively inert and harmless to human beings and in fact is needed in some quantity for life. It occurs naturally in the environment, mostly in the form of sulfates like CaSO$_4$. However, sulfur oxides are recognized pollutants that are harmful to the environment.

Sulfur oxides, usually referred to as SO$_x$, include SO, S$_2$O, S$_n$O, SO$_2$, SO$_3$, and SO$_4$ of which SO$_2$ and SO$_3$ are of particular importance in combustion processes [1]. Sulfur dioxide (SO$_2$) tends to be preferred at higher temperatures while sulfur trioxide (SO$_3$) is more preferred at lower temperatures [2]. Since most combustion processes are at high temperatures, SO$_2$ is the more predominant form of SO$_x$ emitted from systems containing sulfur. Interestingly, SO$_x$ reactions may inhibit fuel NO$_x$ formation [3].

Sulfur dioxide is a colorless nonflammable acidic gas with a pungent odor that is detectable by the human nose and is used in a variety of chemical processes. SO$_2$ can be very corrosive in the presence of water and is highly soluble in water. While SO$_2$ is produced naturally from, for example, volcanic eruptions, the largest source is from industrial combustion processes. SO$_2$ can injure plants under certain conditions although this is not generally of significant concern compared to its contribution to acid rain [4]. Dullien [5] believes SO$_2$ is the most important gaseous pollutant.

It is often assumed that any sulfur in a combustor will be converted into SO$_2$ that will then be carried out with the exhaust gases [6]. The sulfur may come from the fuel or from the raw materials used in the production process. Fuels like heavy oil and coal generally contain significant amounts of sulfur while gaseous fuels like natural gas tend to contain little or no sulfur. The two strategies for minimizing or eliminating SO$_x$ are: (1) removing the sulfur from the incoming fuel or raw materials, and (2) removing the SO$_x$ from the exhaust stream using a variety of dry and wet scrubbing techniques [7]. One dry scrubbing technique is limestone injection. After use, the combined limestone and sulfur can be used in gypsum board. New membrane separation technologies is another reduction technique being developed.

Figure 8.1 shows that more than two-thirds of SO$_x$ emissions come from coal combustion, mostly for power generation. Less than 10% of SO$_x$ emissions come from industrial processes like chemicals, metals, and petroleum production.

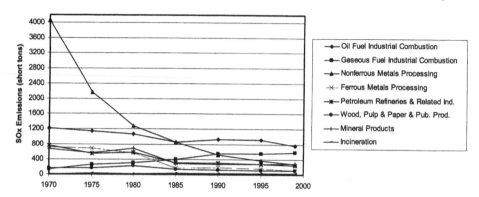

Figure 8.1 Sources of SO$_x$ emissions in the United States (From Ref. 11.)

Interestingly, SO$_2$ emissions were among the first identified as so-called "long-range transboundary air pollutants" that migrate from a source in one geographic location to other surrounding locations over long distances [8]. This phenomenon has necessitated cooperation between the governments of adjoining areas where pollutants from one impact another.

8.2 FORMATION MECHANISMS

One of the common sources of sulfur in industrial combustion processes is in the fuel. Solid fuels like coal and coke may contain 1% or more sulfur by weight. Liquid fuels like gasoline and heavy fuel oil may contain 2% or more sulfur by weight. Gaseous fuels like natural gas and propane typically contain little or no sulfur. In a high-temperature combustion process, the sulfur is oxidized by oxygen from the combustion air to form SO$_x$. The first oxidation step is

$$S + O_2 \rightarrow SO_2 \tag{8.1}$$

This step takes place inside the high-temperature combustor. The second oxidation step is

$$SO_2 + 0.5O_2 \rightarrow SO_3 \tag{8.2}$$

This step can occur in the combustion system but the ratio of SO$_2$/SO$_3$ is typically on the order of 40–80:1 [4]. This step may also occur in the atmosphere after leaving the combustor if the conditions are right. At ambient temperatures, equilibrium strongly favors SO$_3$ rather than SO$_2$ formation. At elevated temperatures, SO$_2$ formation dominates and at temperatures below about 700°F, the reaction rate for SO$_3$ formation declines dramatically so that most of the SO$_x$ in flue gases is in the form of SO$_2$. Further oxidation steps are possible but the most common SO$_x$ pollutants are SO$_2$ and SO$_3$.

The problem in the atmosphere is that SO$_3$ reacts with H$_2$O (e.g., rain) to form sulfuric acid:

$$SO_3 + H_2O \rightarrow H_2SO_4 \tag{8.3}$$

Sulfuric acid by itself is very corrosive and damaging to the environment, including plants and animals. It is also damaging to buildings and other structures like bridges and railroads. This can cause a serious safety problem if structures are not periodically checked for integrity. Another problem with the sulfuric acid is that it may further react with some other compounds in the atmosphere to form particles that contribute to smog.

Another source of SO_x is from the processing of sulfur-bearing ores commonly used in obtaining elemental copper. This process can be described by the overall global reaction:

$$CuFeS_2 + 2.5O_2 \rightarrow Cu + FeO + 2SO_2 \tag{8.4}$$

One of the treatment strategies involves reducing sulfur with hydrogen to form H_2S which can be relatively easily scrubbed out of an exhaust gas stream:

$$S + H_2 \rightarrow H_2S \tag{8.5}$$

Another common treatment strategy is to react SO_x with limestone ($CaCO_3$) to form $CaSO_4$ which is known as anhydrite (or with H_2O is known as gypsum):

$$CaCO_3 + SO_2 + 0.5O_2 \rightarrow CaSO_4 + CO_2 \tag{8.6}$$

Example 8.1

Given: A No. 6 fuel oil containing 2% sulfur by weight is used in an industrial burner at the rate of 10 gpm.

Find: Mass flow rate of SO_x produced.

Solution: It is normally assumed that all sulfur in a high temperature combustion reaction will be converted to SO_2. First, find the mass flow rate of sulfur produced. From Appendix F-3 for No. 6 oil, the typical density is 8.08 lb/gal. The mass flow rate of sulfur is then:

$$\dot{m}_S = (10\,\text{gpm/oil})\left(8.08\,\frac{\text{lb oil}}{\text{gal oil}}\right)\left(0.02\,\frac{\text{lb s}}{\text{lb oil}}\right) = 1.6\,\frac{\text{lb s}}{\text{min}}$$

Then assume there is enough available oxygen to convert all of the S to SO_2:

$$S + O_2 \rightarrow SO_2$$
$$\dot{m}_{SO_2} = \dot{m}_S\left(\frac{64\,\text{lb}\,SO_2}{32\,\text{lb}\,S}\right) = 1.6\,\frac{\text{lb S}}{\text{min}}\left(2\,\frac{\text{lb}\,SO_2}{\text{lb S}}\right) = 3.2\,\frac{\text{lb}\,SO_2}{\text{min}}$$

8.3 SOURCES

The primary source of sulfur needed to produce SO_x comes from the fuel. This is particularly a problem with heavy-oil firing and with coal combustion. Another source of sulfur for producing SO_x is from the incoming raw materials. This is usually limited to some specific applications. Both sources are briefly discussed next.

8.3.1 Sulfur in the Fuel

Oil, coal, and coke are examples of fuels that naturally contain sulfur. In furnace heating operations in the steel industry such as soaking pits and reheat furnaces, the main source of sulfur is from the fuel, which may be oil or coke oven gas [9]. In petroleum refineries, SO_2 emissions from refinery combustion processes can result from sulfur contained in the refinery gas fuel, which is a by-product from the refining process. The oils being processed usually contain some level of sulfur. Sometimes these oils are fired directly while in other cases the resulting refinery gases are used as fuel for the process heaters. In lime and cement manufacturing, all of the fuels commonly used, except natural gas, contain some sulfur.

8.3.2 Sulfur in the Incoming Raw Materials

Since essentially any sulfur going into an industrial combustion system will be converted into SO_x, then the amount of sulfur contained in the incoming raw materials is an important factor in the amount of SO_x produced. Some examples will be given here for illustration purposes. In copper smelting, some of the incoming raw materials may include sulfur such as chalcocite (Cu_2S), bornite (Cu_5FeS_4), tetrahedrite ($Cu_5Sb_2S_7$), or chalcopyrite ($CuFeS_2$). In lead smelting, some ores may contain high concentrations of lead sulfide. Raw materials used in zinc smelting may have as much as one-third of the weight as sulfur. These are in contrast to aluminum smelting where little if any sulfur is contained in the incoming raw feed materials.

In pulp mills, there is a significant amount of sulfur in the kraft in the form of sulfate. Various reactions in the mill during the process of the chemicals generate a variety of sulfur-containing gases including methyl mercaptan (CH_3SH), hydrogen sulfide (H_2S), dimethyl sulfide (CH_3SCH_3), dimethyl disulfide (CH_3SSCH_3), and SO_2 and SO_3 in small amounts.

In the glass manufacturing process, SO_x emissions may result from sodium sulfate (salt cake) used to condition the glass in the manufacture of soda-lime glass and wool fiberglass. Sulfur is present in limestone used in the production of lime and cement. In the production of ethylene from a liquid feed, the incoming raw materials may consist of gas oils, and heavier fractions often contain sulfur that can be removed by hydrodesulfurization. Some of the sulfur for the SO_x emissions from waste-incineration processes comes from the incoming waste materials usually in the form of sulfates or sulfides from, for example, paper, food waste, garden waste, and rubber. Sewer sludge incinerated in waste combustors also usually contains some sulfur.

8.4 ENVIRONMENTAL AND HEALTH CONCERNS

There are two primary adverse affects of SO_x, which are discussed next. These include its adverse respiratory affects on humans and its contribution to acid rain, which damages the environment.

8.4.1 Health Effects

Sulfur dioxide is considered to be a criteria pollutant because of the choking effect it can cause on the human respiratory system at high enough concentrations. It is referred to as a pulmonary irritant and is especially troublesome for children and the elderly. It is particularly dangerous for those who have asthma and other respiratory diseases. SO_2 can react with moisture in the respiratory system to form sulfuric acid. At lower concentrations, it can cause a taste sensation. These affects have been particularly acute at certain times in history and in certain places. Great Britain had a large problem with this during the initial stages of the industrial revolution. The former Soviet-bloc countries struggled with this issue just within the last decade or two because of the lack of sufficient post-treatment systems to remove SO_x from combustion exhaust gases.

8.4.2 Acid Rain

Emissions of SO_x are also damaging to green plants, which are more sensitive than people and animals to SO_2. When SO_2 is released into the atmosphere, it can produce acid rain by combining with water to produce sulfuric acid (H_2SO_4) as shown in the diagram in Fig. 8.2. Sulfur oxides are precursors to acid rain, which is corrosive to the environment, particularly plant and aquatic life. SO_2 may damage leaves by causing bleached spots, bleached areas between veins, chlorosis, and insect injury [10]. Sulfuric acid is very corrosive and can cause considerable damage to man-made structures such as buildings and bridges. In fact, sulfuric acid is deliberately produced in some chemical plants for sale in cleaning and etching applications.

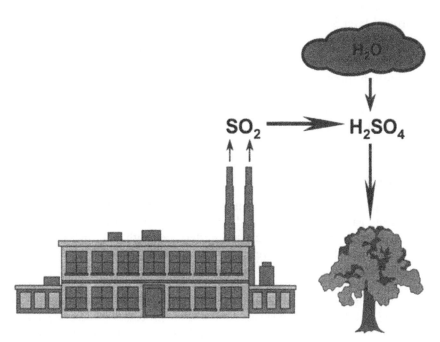

Figure 8.2 Sulfuric acid rain schematic.

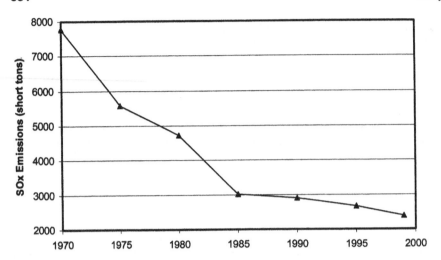

Figure 8.3 SO_x emissions in the United States since 1970 for certain industrial combustion sources (From Ref. 11.).

8.5 REGULATIONS

Regulations for SO_x are commonly written in one of five different forms:

1. Mass of SO_x emitted per hour (e.g., lb SO_x/hr)
2. Mass of sulfur feed (e.g., lb S/hr)
3. Mass of SO_x per unit of product (e.g., lb SO_x/ton of cement)
4. Concentration of SO_x in the flue gas (e.g., ppmvd SO_x in the exhaust gases)
5. Ground-level SO_x concentration (e.g., ppm)

Since essentially any sulfur coming into the system will be converted into SO_x, limiting the mass of incoming sulfur directly limits the amount of SO_x produced in the combustion process. The regulations vary by application, geographic location, fuel type, individual source size, and overall plant size, among other things. Figure 8.3 shows total SO_x emissions in the United States since 1970 for the industrial combustion sources shown in Fig. 8.1 [11].

8.6 MEASUREMENT TECHNIQUES

EPA Method 6 provides procedures for measuring SO_2 emissions from stationary sources where the gas sample is extracted from the exhaust stack [12]. Ammonia, water-soluble cations, and fluorides cause interferences with SO_x measurements. Method 6A concerns SO_2, moisture, and CO_2 measurements from fossil-fuel combustion sources by chemically separating the SO_2 and CO_2 components where different reagent chemicals are used. Method 6C discusses the use of instrument analyzers to measure SO_2 emissions from stationary sources. This is the most commonly used method in industrial combustion processes. These analyzers typically use ultraviolet (UV), nondispersive infrared (NDIR), or fluorescence techniques. A schematic of an accepted sampling system is shown in Fig. 5.21.

8.7 ABATEMENT STRATEGIES

The four basic strategies of pretreatment, process modification, combustion modification, and post-treatment apply to SO_x emissions and are briefly considered next. The specific choice of a technique depends primarily on the application, the potential SO_x concentration in the exhaust gas, and the economics of each technique.

8.7.1 Pretreatment

Since essentially any sulfur present in high-temperature combustion reactions forms SO_x under normal conditions, an effective strategy is to remove as much of the incoming sulfur as possible since SO_x formation is directly proportional to the sulfur concentration present in the process. This strategy for reducing or eliminating SO_x emissions then involves treating the incoming fuel, oxidizer, or feed materials in some way. This is usually referred to as fuel desulfurization. Removing sulfur from coal is commonly done to reduce SO_x emissions in power generation [13]. An alternative is to replace a sulfur-containing fuel with one that has less or no sulfur. The problem with this approach is that it may be uneconomical to treat or replace the fuel as high-sulfur fuels are often much less expensive than low- or no-sulfur fuels. The economics of using a more expensive fuel need to be compared to those of removing SO_x from the exhaust gas stream.

Since sulfur in the fuel is a common source of SO_x from industrial combustion applications, then removing some or all of the sulfur in the fuel prior to combustion would reduce SO_x emissions. One example would be removing H_2S that is sometimes present in natural gas. This can be done in a wet scrubber where the H_2S is absorbed by the solvent (usually water plus a weak alkali such as an ethanolamine) while the rest of the gas passes through the scrubber [14]. Once the H_2S has been separated from the rest of the fuel, it can be further oxidized to form elemental sulfur that can be sold for the production of sulfuric acid or other chemical products:

$$H_2S + 0.5O_2 \rightarrow S + H_2O \tag{8.7}$$

which is referred to as the Claus process. Sulfuric acid is an inexpensive industrial acid that is used in many processes. It is also used in the production of phosphate fertilizer.

There are a wide range of technologies for reducing the sulfur content in gasoline products including catalytic distillation hydrotreating, hydrofining with octane recovery and selective catalytic naphtha hydrofining, dual-catalyst reactors, low-pressure fixed-bed hydroprocessing, olefinic alkylation, sorbents, and extractive mass transfer [15]. Improving hydrotreater performance [16] and gasoline fractionation [17] have been shown to be effective in reducing the sulfur content in gasoline products. The technology choice depends on the initial capital cost, ongoing operating costs, and removal efficiency. Catalysts are commonly used to reduce the sulfur content in petroleum products to meet increasingly more stringent regulatory requirements. Considerable research is being devoted to developing more advanced catalysts to improve the removal and energy efficiency in minimizing the sulfur content of fuels, including increasingly more complex sulfur compounds [18].

In some cases, removing sulfur from the incoming feed streams may be the only economic option. For example, while it may be economical for a refinery to have post-treatment equipment for removing SO_x from exhaust gases, it is clearly not economical to have such treatment equipment on every automobile. In that case, it makes more sense to remove the sulfur from the gasoline used in the cars to avoid making SO_x in the first place. However, the regulations governing allowable sulfur content in automotive fuels continue to get tougher and it is becoming more complex and expensive to attain these very low sulfur contents [19]. This will require refineries to be operationally excellent, including increased training of plant personnel.

8.7.2 Process Modification

There are several strategies related to process modification to reduce or eliminate SO_x emissions. One strategy is to use an alternative method of generating the energy needed to process the materials. A common alternative is to use electrical energy instead of burning a fossil fuel. An example of a common industrial heating process used to make steel from scrap metal is known as an electric arc furnace or EAF (see Chap. 13). In an EAF, three graphite electrodes arranged in a triangle are inserted into a vessel containing a load of scrap metal. A very high voltage is applied which arcs between the electrodes through the scrap metal causing it to melt. The appropriate chemicals are added to the molten bath to produce the desired grade of steel. While electricity is often much more expensive than fossil fuels, the heating efficiency of the EAF is high, which makes it economical. However, this particular alternative is only a possibility for materials that are electrically conductive, which is not the case for most industrial combustion processes. A related alternative is to convert electrical energy into thermal energy in radiant heaters. The two challenges of this approach are the very high radiant power output density required in most industrial material processes and the overall system economics where electrical power is often several times more expensive than fossil fuels for the same heat input.

Another strategy is to modify the materials process by replacing the incoming sulfur-bearing feed materials with other materials that have less or no sulfur. This approach only applies to those processes where the incoming feed materials contain sulfur, which are fairly limited. It is also only viable if there is a suitable substitute for the materials containing the sulfur.

8.7.3 Combustion Modification

One alternative to burning high-sulfur coal in a traditional combustor is to use a fluidized bed combustor where the bed contains limestone particles. The limestone reacts with the SO_x formed to form $CaSO_4$, which is then scrubbed out in the exhaust gas treatment system. However, a fluidized-bed combustor is not generally a viable alternative for most industrial combustion processes. It is an alternative for power generation to replace traditional coal-fired boilers. This technology is still in the development phase and has not yet proven to be better than conventional boilers.

Another combustion modification technique under development is to convert sulfur-containing coal into a synthetic fuel through coal gasification where the sulfur is converted into H_2S, which is relatively easy to remove. The synthetic gaseous fuel produced by burning the coal in a gasifier passes through an H_2S removal process where the H_2S is converted into elemental sulfur in a Claus process. The cleaned gas then goes to a gas turbine to generate power and then to a steam boiler to generate steam. While the overall system is relatively complicated, the sulfur removal is simplified and the thermal efficiency is high compared to other conventional systems.

Most of the combustion modification alternatives for handling SO_x emissions are not viable alternatives for industrial combustion processes and are designed for power generation where the large scale makes them economically competitive. These systems are too complex at this time to be viable for the scale common to the industries considered here. This makes SO_x emissions fairly unusual among the air pollutants from industrial combustion processes because they cannot be practically reduced through combustion modification techniques.

8.7.4 Post-Treatment

Removing SO_x from exhaust gases is often referred to as flue-gas desulfurization or FGD. There are four major categories of SO_x post-treatment techniques: nonregenerable, regenerable, wet, and dry [9]. Nonregenerable processes produce either sludge or some type of waste liquor that must be disposed of properly. Regenerable processes incorporate a mechanism to regenerate the absorbent and produce elemental sulfur or a sulfur compound that can be sold. Wet processes may require stack gas reheating to ensure adequate buoyancy and to avoid corrosion problems. Gas reheating is not required in dry removal processes. Wet sodium-based nonregenerable FGD is the most widely used in larger industrial combustion applications. Space constraints and other site-specific factors impact the choice of an appropriate FGD process in retrofit situations.

If the combustion gases are too hot prior to reaching the FGD system, they need to be cooled down to acceptable temperatures with some type of quench system. Some types of FGD systems incorporate scrubbers that not only remove SO_x, but also particulates. Materials of construction may also be important if the SO_x reacts with water in the FGD system to form sulfuric acid, which is very corrosive. This must be considered in water-spray absorption systems.

One type of FGD is the lime process which is a wet, nonregenerable SO_2 absorption process. An alkaline slurry is recirculated through a scrubber/absorber tower. Calcium sulfite and sulfate are formed by the reaction, which are then separated in settlers or clarifiers and filters. A sludge is produced that can be inerted and then landfilled. The systems generally have four major operations: scrubbing or absorption to capture the gaseous SO_2, flue-gas handling including ductwork and fans, lime handling and preparation, and sludge processing. The scrubber/absorber may be a tray absorber, a packed scrubber, a mobile bed scrubber, a venturi scrubber, or a spray tower. The limestone process is very similar to the lime process except that the feed preparation equipment is different. Limestone is less reactive than lime so some of the process parameters are different. The so-called double alkali process is basically an indirect lime/limestone process where some of the plugging and scaling associated with lime/limestone processes can be avoided. Many of these systems are large, capital intensive, have significant operating costs, and produce a sludge that needs to be disposed of properly. They may only be economically feasible for larger SO_x emission sources.

Another type of post-treatment FGD system is an ammonia-based wet absorption process. The flue gas is pretreated to remove particulates using an electrostatic precipitator, fabric filtration, or some other technique (see Chap. 9). The flue gas is then water-quenched to its adiabatic saturation temperature. The conditioned and humid gas is then brought into contact with an aqueous solution containing ammonia, which rapidly absorbs the SO_2. The by-products are saleable and contain either ammonium sulfate or elemental sulfur/sulfuric acid. This technique has commonly been used to remove SO_2 from pulp mills.

The Wellman–Lord process uses sodium bisulfite to absorb SO_2 from the exhaust gas stream. The SO_2 is then concentrated in a stripping step while the absorbent is regenerated and returned to the absorber. The concentrated SO_2 stream with water vapor enters a condenser where most of the water is removed. If required, the SO_2 can be further dried in a drying tower. Sulfur in the form of liquid SO_2, liquid SO_3, sulfuric acid, or elemental sulfur may be recovered and potentially sold or used from the resulting stream. A similar process involves the use of magnesium oxide (MgO) slurry scrubbing.

Common dry removal processes include spray drying, dry injection, and combustion of fuel/limestone mixtures. An important advantage of dry techniques is that there is no sludge to be disposed of after the SO_x removal. Another advantage is that special construction materials are often not required compared to wet techniques because there is usually not enough water present to form sulfuric acid that could corrode the system. Dry systems may be regenerable or nonregenerable. Sodium carbonate and lime slurries are common sorbents.

Benítez [20] notes that one of the unique challenges of flue-gas desulfurization is the combination of very large gas flow rates and very low SO_x pollutant concentrations. de Nevers [21] notes that unlike some other pollutants such as particulates, which can be removed by physical means (e.g., scrubbing), SO_x removal from an exhaust gas stream is a more chemically oriented process. Johnsson and Kiil [22] note that FGD is likely to be the most important technology to reduce SO_x emissions well into the future. Dullien [5] lists numerous flue-gas desulfurization processes under development: aqueous carbonate, catalytic oxidation, copper oxide absorption, dual alkali, magnesium oxide, sodium carbonate, dry adsorption, and a variety of dry collection techniques. Schnelle and Brown [23] present a summary in Table 8.1 of some common processes for acid gas control.

One method of classifying FGD techniques is based on whether the resulting end product must be disposed of or can be reused (regenerated). A process is considered regenerative if the captured sulfur can be recovered in a usable form, but does not necessarily have to be reused in the process at the given plant. Regenerative techniques are generally more expensive, but this depends on many factors that should be analyzed prior to reaching any final conclusions. The FGD economics are greatly impacted by whether the recovered products can be sold or if there is a cost to dispose of them. Because the cost of the saleable products (e.g., gypsum) are relatively low per unit of weight, they cannot usually be transported very far before the process becomes uneconomical due to the cost of transportation. Therefore, the ability to sell the product is usually limited to a certain radius around the plant.

Table 8.1 Summary of Common Processes for Acid Gas Control

Process	Efficiency	Capital Cost	Reagent Cost	Complexity	Comments
Wet limestone	High	High	Low	High	Low reagent cost offsets high cost of operation and maintenance in very large systems
Wet soda ash/caustic	High	Moderate	High	Moderate	No slurry or solids handling. Very effective for smaller systems
Lime spray dryer	Moderate	Moderate	Moderate	Moderate	Mature technology widely used for industrial applications. Subject to deposits accumulation during upset conditions
Circulating lime reactor	Moderate to high	Moderate	Moderate	Moderate	High-solids circulation rate prevents deposits accumulation, allowing slightly higher reactivity or lower reagent cost than that of lime spray dryer
Sodium bicarbonate/trona injection	Moderate	Low	High	Low	Lower cost of trona an advantage in HCl applications, but mostly offset by lower reactivity in SO_2 applications

Source: Ref. 23.

Other FGD processes in the vicinity of this plant also attempting to sell products will greatly impact the overall economics as well. Srivastava and Jozewicz have reviewed state-of-the-art flue gas desulfurization techniques for coal-fired boilers [37].

8.7.4.1 Scrubbers

One common technique for removing SO_x from exhaust gas streams is to use some type of scrubbing system. These are typically either dry scrubbers or wet scrubbers. In either case, the objective is to impinge on the unwanted pollutants, separate them from the main gaseous combustion products like CO_2, H_2O, and N_2, and then collect those pollutants for disposal or reuse. There is sometimes a tendency to assume that dry scrubbers are used for dry pollutants and wet scrubbers for wet pollutants, but wet scrubbers are often more cost effective on dry pollutants depending on the application [24]. Table 8.2 shows a comparison of SO_x scrubbing techniques.

Wet Scrubbers. Wet scrubbers (see Fig. 8.4) utilize the process of gas absorption into a liquid for separation and purification of gas streams, as product recovery devices, or as pollution control devices. The suitability of this technique depends on the availability of a suitable solvent to absorb the desired gas, the required removal efficiency, and the concentration of the gas to be removed. The physical absorption process depends on the properties of both the gas to be removed and the liquid stream (e.g., diffusivity, equilibrium solubility), which are often highly temperature dependent. The ideal solvent would have a high solubility for the desired gas to be removed, low vapor pressure, low viscosity, high availability, and low cost.

Wet scrubbers are used as pollution control devices to remove SO_2 emissions from exhaust gas streams. The SO_2 in the exhaust gas stream is absorbed into the liquid in the scrubber where it can be either separated from the liquid, which can then be reused or where the liquid containing the absorbed SO_2 can be properly disposed. Wet scrubbers are typically relatively simple to operate and are often similar to other processes within production plants such as wastewater treatment pumps and process towers. This aids operators in becoming familiar with how to use wet scrubbers. In addition to removing SO_2, wet scrubbers are also effective for removing particulates if they happen to be present in the exhaust stream. The U.S. EPA has written a useful manual for estimating the cost of various air pollution control techniques, which includes a chapter on wet scrubbers for acid gas control [25].

These scrubbers are generally of two types: nonregenerative and regenerative. These refer to whether or not the reagent used in the scrubbing process is regenerated. Commonly used reagents are typically caustic, magnesium oxide, or soda ash. In a nonregenerative system, the reagent cannot be reused and is spent after its initial use. In a regenerative system, the reagent is used over and over again by being regenerated after each use. Figure 8.5 shows a schematic of a nonregenerative wet scrubbing system [26], and. Fig. 8.6 shows a schematic of a regenerative wet scrubbing system. One example of a regenerative scrubbing system is a process known as LABSORB™ developed by Belco Technologies (Parsippany, NJ).

Nonregenerable processes produce some type of liquid or solid waste that must be disposed of or sold, where the absorbent is not reused. Savu et al. [27] describe the use of limestone scrubbers in industrial combustion applications in Romania that demonstrated SO_2 removal efficiencies up to 98%. Bravo et al. [28] have studied the absorption of SO_2 from mixtures of SO_2 and N_2 into limestone slurries to determine

Table 8.2 Comparison of SO$_x$ Scrubbing Techniques

Technology		Capital cost to install	Efficient (reagent utilization)	Neutralizing reagent cost	Energy cost	Process/ maintenance complexity	Ease of retrofit	Waste water	Visible plume
Wet scrubbing	Calcium based	High	Medium	Low	High	High	Low	Yes	Yes
	Sodium based	High	High	High	High	High	Low	Yes	Yes
Semi-dry scrubbing	Calcium based	High	Medium	Medium	High	High	Low	No	No
Dry scrubbing	Calcium based	Low–medium	Low–medium	Medium	Low	Medium	Medium	No	No
	Sodium based	Low	Low–medium	Medium–high	Low	Low	High	No	No

Source: Ref. 33.

Figure 8.4   Example of a wet scrubber for removing SO_x. (Courtesy of Belco Technologies, Parsippany, NJ.)

appropriate mass-transfer coefficients. Finely ground limestone or $CaCO_3$ is mixed with water for use as the scrubbing agent. The overall stoichiometry can be written as:

$$CaCO_3(s) + H_2O + 2SO_2 \rightarrow Ca^{2+} + 2HSO_3^- + CO_2(g) \qquad (8.8)$$

$$CaCO_3(s) + 2HSO_3^- + Ca^{2+} \rightarrow 2CaSO_3 + CO_2 + H_2O \qquad (8.9)$$

This is often done in some type of packed-tower arrangement that has trays or sieves to improve the mass transfer between the exhaust gas containing the SO_x and the liquid scrubbing agent. A specific process for this technique is sometimes referred to as forced-oxidation limestone wet scrubbing. The water/limestone slurry is sprayed from the top while the exhaust gas is fed from the bottom in a counter-flow geometry. The SO_x that is captured in the slurry is then pumped to a liquid/solid separator like a hydroclone. The resulting semidry gypsum is then often sold as a raw material for making plasterboard for building construction. Because there is so much gypsum produced from a large power plant, there must be a fairly large demand for the gypsum or else the economics of the process can be very unfavorable. The gypsum must be landfilled if no commercial use can be found for the product. The water resulting from the hydroclone separation must be treated in a wastewater plant to remove any remaining impurities. This process is currently only economical for very large sources of SO_x, typically from power generating plants.

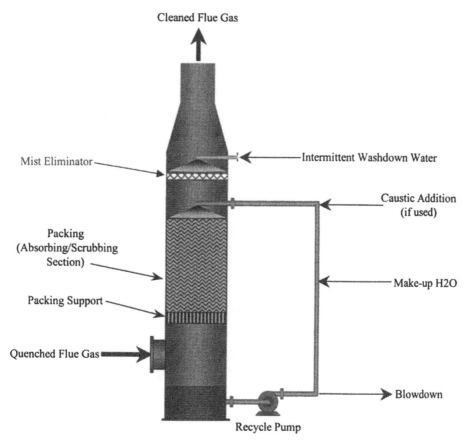

Cleaned Flue Gas

Mist Eliminator

Intermittent Washdown Water

Caustic Addition
(if used)

Packing
(Absorbing/Scrubbing
Section)

Packing Support

Make-up H2O

Quenched Flue Gas

Blowdown

Recycle Pump

Figure 8.5 Simple packed-column wet scrubber. (From Ref. 26. Courtesy of John Zink Co., Tulsa, OK.)

Many of the initial problems with limestone scrubbers have been greatly reduced or eliminated due to advances in technology after years of development [21]. These problems included corrosion, solids deposition, scaling, plugging, poor reagent utilization, and poor solid–liquid separation.

Another variation of the limestone scrubber is to use hydrated lime or $Ca(OH)_2$, sometimes called quicklime, which is more reactive than limestone but requires additional processing. The overall stoichiometry can be written as:

$$CaO + H_2O \rightarrow Ca(OH)_2 \tag{8.10}$$

$$SO_2 + H_2O \longleftrightarrow H_2SO_3 \tag{8.11}$$

$$H_2SO_3 + Ca(OH)_2 \rightarrow CaSO_3 + 2H_2O \tag{8.12}$$

$$CaSO_3 + 2H_2O + 0.5O_2 \rightarrow CaSO_4 + 2H_2O \tag{8.13}$$

This is a throwaway system because the reagent is only used once and then thrown away. The additional reactivity of the lime was an initial benefit because of the early

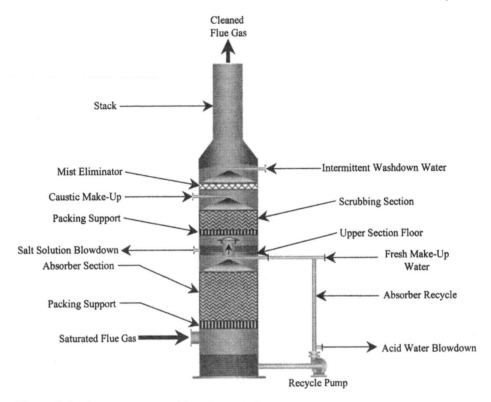

Figure 8.6   Two-stage wet scrubber. (From Ref. 26. Courtesy of John Zink Co., Tulsa, OK.)

problems with limestone scrubbers, but as most of these have been solved, the added cost of the lime and system complexity are no longer economically justified in most cases.

Sodium-based materials are also used in wet scrubbers to remove SO_x because sodium salts are highly soluble in water. The overall reaction of sodium hydroxide with SO_2 is

$$2NaOH + SO_2 \rightarrow \cdot Na_2SO_3 + H_2O \tag{8.14}$$

The overall reaction of sodium carbonate with SO_2 is

$$Na_2CO_3 + SO_2 \rightarrow \cdot Na_2SO_3 + CO_2 \tag{8.15}$$

An example of a regenerative wet scrubbing system is known as the Wellman–Lord process, which consists of the following five subprocesses:

1. Flue-gas pretreatment to remove particulates and to cool and humidify the flue gas.
2. SO_2 and SO_3 absorption by a sodium sulfite solution:

$$Na_2SO_3 + SO_2 + H_2O \rightarrow 2NaHSO_3 \tag{8.16}$$

$$2Na_2SO_3 + SO_3 + H_2O \rightarrow Na_2SO_4 + 2NaHSO_3 \tag{8.17}$$

3. Purge treatment.
4. Sodium sulfite regeneration where concentrated SO_2 is produced:

$$2NaHSO_3 \rightarrow 2Na_2SO_3 + SO_2 + H_2O \tag{8.18}$$

5. Sulfur recovery.

Vanderschuren et al. [29] describe a process for removing SO_x from exhaust streams by absorption using magnesium hydroxide [$Mg(OH)_2$] slurries in a three-stage scrubber. The basic process can be described by [30]

$$SO_2 + MgCO_3 \rightarrow MgSO_3 + CO_2 \tag{8.19}$$

$$SO_2 + Mg(OH)_2 \rightarrow MgSO_3 + H_2O \tag{8.20}$$

Removal efficiencies of up to 96% were demonstrated. A primary advantage of this process compared to other wet processes is that the resulting effluent can be used in other commercial applications compared to the effluents from traditional wet processes that create disposal issues.

Dry Scrubbers. One of the main problems with wet scrubbing systems is that the combination of water and SO_x forms sulfuric acid, which is highly corrosive. It would be very advantageous if a dry scrubbing system could be used to avoid the problem of corrosion. Such systems have been developed where an electrostatic precipitator (ESP) is used to separate the SO_x out of the exhaust stream instead of a solid–liquid separator in the case of a wet scrubbing system. Figure 8.7 shows some schematics of common dry flue gas desulfurization processes [31].

The principle of most dry scrubbers is to inject dry alkaline particles into the flue gas stream to react with the SO_x. Depending on the process, the particles may be injected into the combustor and/or the exiting exhaust gas stream. Figure 8.8 shows an example of a limestone flue-gas desulfurization process. These particles are then removed by the ESP where the collected dry solids can be disposed of along with other waste solids such as fly ash, usually in a landfill. This is often simpler and less expensive than dealing with the sludge produced by wet scrubbers. One reagent that is used is CaO:

$$CaO + SO_2 \rightarrow CaSO_3 \tag{8.21}$$

where the $CaSO_3$ is oxidized to $CaSO_4$ prior to reaching the ESP where it is removed. However, the efficiency of the process is such that considerable excess CaO may need to be injected, which increases the reagent costs. An alternative is to use more efficient reagents like $NaHCO_3$ or Na_2CO_3, which are also much more expensive.

Garea et al. [32] have shown that the concentration of CO_2 in flue gases at low and medium temperatures can affect the desulfurization process in dry scrubbing. No effect was observed on the desulfurization process by the CO_2 concentration at lower temperatures (<100°C or 212°F). However, CO_2 concentration was

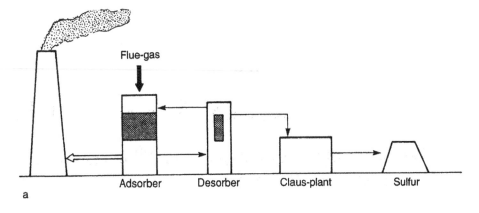

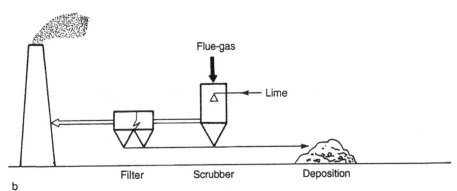

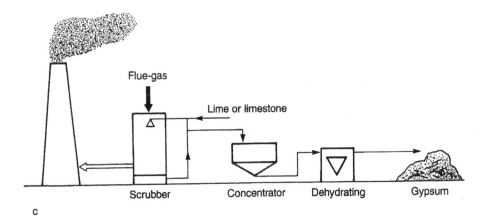

Figure 8.7   Flue-gas desulfurization processes. (From Ref. 31. Courtesy of Springer.)

found to have a significant impact on desulfurization at medium gas temperatures (350°–450°C or 660°–840°F) where $CaCO_3$ is formed and reacts differently with SO_x compared to the reaction at lower temperatures. It is preferred to minimize $CaCO_3$ in order to increase desulfurization.

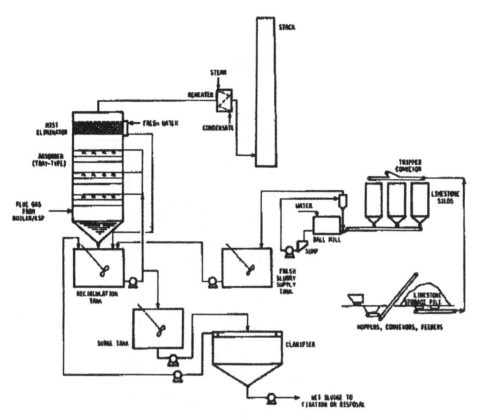

Figure 8.8   Limestone flue-gas desulfurization process. (From Ref. 9.)

Maziuk and Kumm [33] note that, while dry hydrated lime is often the reagent of choice in dry scrubbers, sodium-based sorbents such as sodium bicarbonate (NaHCO3), nacholite (which is naturally occurring sodium bicarbonate), and trona ($Na_2CO_3 \cdot NaHCO_3 \cdot 2H_2O$) can be preferred in certain applications. In certain cases, sodium-based sorbents can be less costly to implement, operate, and maintain than calcium-based sorbents, especially in areas where a source of natural sodium-based materials is close by. Calcium hydroxide is considered a hazardous material whereas sodium bicarbonate and trona are only considered to be nuisance dusts from the occupational exposure standpoint.

Wet–Dry Scrubbers.   Wet–dry scrubbing systems, also referred to as semidry scrubbing, use features from both wet and dry scrubbers. The most common wet–dry scrubber is a spray dryer where a slurry consisting of reagent dissolved in water is injected into the flue-gas stream through some type of atomizing nozzle to generate a fine mist (see Fig. 8.9). Another version of this type of scrubber is where a hydrated lime slurry is used as the absorbing medium and is formed by "slaking" lime:

$$CaO + H_2O \rightarrow Ca(OH)_2 \tag{8.22}$$

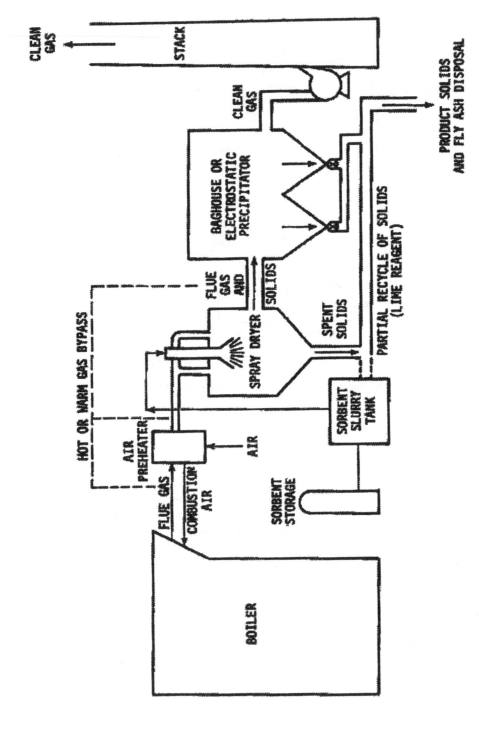

Figure 8.9 Spray dryer flue-gas desulfurization process diagram. (From Ref. 9.)

The absorbed SO_x is then neutralized in the suspended alkaline material according to the overall reaction:

$$Ca(OH)_2 + SO_2 \rightarrow CaSO_3 + H_2O \qquad (8.23)$$

The heat from the hot exhaust products vaporizes the water in the reagent slurry leaving the dry reagent particles, which can then react with the SO_x. In this type of scrubbing system, a wet slurry is used to add the reagent to the flue-gas stream but dry solids are collected for disposal; hence, its name as a wet–dry scrubber. Any unreacted dry reagent particles fall to the bottom of the spray-dryer chamber where they are collected and then recycled for reuse. In this case, using excess reagent is less of a problem because whatever does not react with SO_x is reused.

Zheng et al. [34] studied the conversion of a spray-dry absorption product (SDAP) into gypsum in a wet flue-gas desulfurization process. They showed it was possible to reuse the SDAP and that residual products generated in the process could be used to reduce limestone consumption and increase SO_2 removal rates.

One of the limitations of the wet–dry scrubber is how much water can be sprayed into the spray dryer. This depends on how much energy is available in the flue-gas exhaust stream being treated to vaporize the water added in the reagent slurry. If too much water is injected then not all will be vaporized, which will tend to clog the collection system at the bottom of the dryer and cause corrosion in the scrubber. Another issue in wet–dry scrubbers is that the reaction conditions are multiphase involving liquids and solids in a slurry and, therefore, not as ideal as in a wet scrubber.

Gas Treatment and Sulfur Recovery. In certain applications it may be preferable and more economic to recover SO_x emissions where they can either be reused in the given process, sold as a chemical, or more easily disposed of in the captured form.

Claus Process. Petrochemical plants produce off-gas streams containing high concentrations of sulfur that must be treated prior to emission into the atmosphere. The sulfur-containing gases are commonly referred to as "sour gas" and are typically by-products from high-sulfur content crude oils often referred to as "sour crude." The sulfur-containing gases are typically in the form of hydrogen sulfide (H_2S). In the Claus process, the H_2S-rich gas stream is reacted with one-third of the stoichiometric quantity of air, which produces SO_2 with some of the H_2S remaining. These are then reacted in the presence of a catalyst (e.g., bauxite) to produce elemental sulfur. This process can be significantly enhanced by using pure O_2 instead of air since the high concentration of N_2 in air dilutes the process and makes separation more expensive. However, there is a cost for using pure O_2 so this must be weighed against the process improvements. This technique can have up to 97% sulfur recovery.

Desulfurization. Martin et al. [35] describe a process called ELSA to be used in conjunction with Claus units. SO_2 removal efficiencies of up to 90% have been demonstrated on a laboratory and pilot scale with the ELSA process, which features a sorbent (magnesium sulfate) that can be regenerated. Longer residence times significantly increase removal efficiencies. The sorbent showed better activity levels

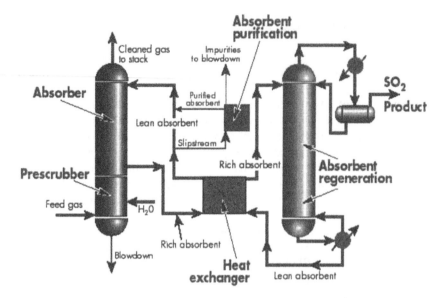

Figure 8.10 SO$_2$-absorption system schematic. (Courtesy of Cansolv, Montreal, Canada.)

than those of conventional sorbents after many cycles of use. Figure 8.10 shows an example of a commercial absorption system for SO$_x$ removal [36].

REFERENCES

1. ED Weil. Sulfur Compounds. *Kirk–Othmer Encyclopedia of Chemical Technology*, 3rd edn. Vol. 22. New York: John Wiley, 1983.
2. CT Bowman. Chemistry of Gaseous Pollutant Formation and Destruction. *Fossil Fuel Combustion*. In W Bartok, AF Sarofim (eds.). New York: John Wiley, 1991.
3. JE Johnsson, P Glarborg. Sulphur Chemistry in Combustion I: Sulphur in Fuels and Combustion Chemistry. In C Vovelle (ed.). *Pollutants from Combustion: Formation and Impact on Atmospheric Chemistry*. Dordrecht, The Netherlands: Kluwer, 2000.
4. K Wark, CF Warner. *Air Pollution: Its Origin and Control*, 2nd edn. New York: Harper & Row, 1981.
5. FAL Dullien. *Introduction to Industrial Gas Cleaning*. San Diego, CA: Academic Press, 1989.
6. CR Bruner. *Handbook of Incineration Systems*. New York: McGraw-Hill, 1991.
7. SR Turns. *An Introduction to Combustion*. New York: McGraw-Hill, 1996.
8. G Sundqvist, M Letell, R Lidskog. Science and policy in air pollution abatement strategies. *Environ. Sci. Policy* Vol. 5, pp. 147–156, 2002.
9. U.S. EPA. Control Techniques for Sulfur Oxides Emissions from Stationary Sources, 2nd edn. Rep. EPA-450/3-81-004. Washington, DC: Environmental Protection Agency, 1981.
10. AC Stern, RW Boubel, DB Turner, DL Fox. *Fundamentals of Air Pollution*, 2nd edn. Orlando, FL: Academic Press, 1984.
11. J Elkins, N Frank, J Hemby, D Mintz, J Szykman, A Rush, T Fitz-Simons, T Rao, R Thompson, E Wildermann, G Lear. National Air Quality and Emissions Trends Report, 1999. Washington, DC: U.S. Environmental Protection Agency, Rep. EPA 454/R-01-004, 2001.

12. U.S. EPA. Method 6: Determination of Sulfur Dioxide Emissions from Stationary Sources. Code of Federal Regulations 40, Part 60, Appendix A-4. Washington, DC: U.S. Environmental Protection Agency, 2001.

13. RC Eliot (ed.). *Coal Desulfurization Prior to Combustion*. Park Ridge, NJ: Noyes Data Corp., 1978.

14. JC Mycock, JD McKenna, L Theodore. *Handbook of Air Pollution Control Engineering and Technology*. Boca Raton, FL: Lewis Publishers, 1995.

15. C Fredrick. Sulfur reduction: What are the options? *Hydrocarbon Process.*, Vol. 81, No. 2, pp. 45–50, 2002.

16. RR Bharvani, RS Henderson. Revamp your hydrotreater for deep desulfurization. *Hydrocarbon Process.*, Vol. 81, No. 2, pp. 61–64, 2002.

17. SW Golden, DW Hanson, SA Fulton. Use better fractionation to manage gasoline sulfur concentration. *Hydrocarbon Process.*, Vol. 81, No. 2, pp. 67–72, 2002.

18. WK Shiflett, LD Krenzke. Consider improved catalyst technologies to remove sulfur. *Hydrocarbon Process.*, Vol. 81, No. 2, pp. 41–43, 2002.

19. NL Gilsdorf, TS Ratajczak, SE Zillman, WH Keesom. What does it take to meet the clean fuels challenge. *Hydrocarbon Process.*, Vol. 81, No. 2, pp. 37–39, 2002.

20. J Benítez. *Process Engineering and Design for Air Pollution Control*. Englewood Cliffs, NJ: Prentice Hall, 1993.

21. N de Nevers. *Air Pollution Control Engineering*. New York: McGraw-Hill, 2000.

22. JE Johnsson, S Kiil. Sulphur Chemistry II: Flue Gas Desulphurization. In C Vovelle (ed.). *Pollutants from Combustion: Formation and Impact on Atmospheric Chemistry*. Dordrecht, The Netherlands: Kluwer, 2000.

23. KB Schnelle, CA Brown. *Air Pollution Control Technology Handbook*. Boca Raton, FL: CRC Press, 2002.

24. RJ Chironna. Dry/wet scrubbers for clean air compliance. *Pollut. Eng.*, Vol. 24, No. 19, pp. 56–58, 1992.

25. W Barbour, R Oommen, GS Shareef, WM Vatavuk. Section 5.2: Post-Combustion Controls, Chap. 1: Wet Scrubbers for Acid Gas Control. In DC Mussatti (ed.). *Air Pollution Control Cost Manual*, 6th edn. Rep. EPA/452/B-02-001. Washington, DC: U.S. Environmental Protection Agency, Jan. 2002.

26. P Melton, K Graham. Thermal Oxidizers. In CE Baukal (ed.). *The John Zink Combustion Handbook*. Boca Raton, FL: CRC Press, 2001.

27. A Savu, L Dragos, N Scarlat, M Girjoaba. Researches for SO_2 Pollutant Emission Reducing. *Proceedings of 5th European Conference on Industrial Furnaces and Boilers*, Portugal, Vol. II, pp. 633–640, 2000.

28. RV Bravo, RF Camacho, VM Moya, LAI Garcia. Desulphurization of SO_2–N_2 mixtures by limestone slurries. *Chem. Eng. Sci.*, Vol. 57, pp. 2047–2058, 2002.

29. J Vanderschuren, A Laudet, X Pettiau. Reduction of Acid Pollutants by Means of Magnesium Hydroxide Slurries in a Three-Stage Scrubber. In N Piccinini, R Delorenzo (eds.). *Chemical Industry and Environment II*, Vol 1. Turin, Italy: Politecnico Di Torino, 1996.

30. A Ersoy-Meriçboyu, N Karatepe, U Beker, S Küçükbayrak. Removal of Sulphur Oxides Resulting from Coal Combustion by Different Methods. In N Piccinini, R Delorenzo (eds.). *Chemical Industry and Environment II*, Vol. 2. Turin, Italy: Politecnico di Torino, 1996.

31. G Baumbach. *Air Quality Control*. Berlin: Springer, 1996.

32. A Garea, I Fernández, JR Viguri, JA Irabien. Influence of CO_2 in Flue Gas Desulfurization at Low and Medium Temperatures. In N Piccinini, R Delorenzo (eds.). *Chemical Industry and Environment II*, Vol. 2. Turin, Italy: Politecnico di Torino, 1996.

33. J Maziuk, JH Kumm. Comparison of Dry Injection Acid–Gas Control Technologies. Paper 43151. *Proceedings of the Air & Waste Management Association's 95th Annual Conference & Exhibition.* Baltimore, MD, June 23–27, 2002.

34. Y Zheng, S Kiil, JE Johnsson, Q Zhong. Use of spray dry absorption product in wet flue gas desulphurisation [sic] plants: pilot-scale experiments. *Fuel*, Vol. 81, pp. 1899–1905, 2002.

35. G Martin, E Lebas, L Nougier. Flue Gas Desulfurization by Regenerative Sorbents. *Fifth International Conference on Technologies and Combustion for a Clean Environment.* Lisbon, Portugal. Vol. 1. The Combustion Institute — Portuguese Section, pp. 97–101, 1999.

36. L Hakka, J Sarlis. Scrubbing sulfur dioxide. *Hydrocarbon Eng.*, Vol. 7, No. 12, pp. 53–54, 2002.

37. RK Srivastava, W. Jozewicz. Flue Gas Desulfurization: The state of the art. *J. Air & Waste Manage. Assoc.*, Vol. 51, pp. 1676–1688, 2001.

9

Particulates

9.1 INTRODUCTION

Particulate matter (PM) is a broad classification of material consisting of either tiny solid particles or fine liquid droplets found in the atmosphere. The U.S. EPA defines particulates as "any finely divided solid or liquid material, other than uncombined water" [1]. The particulates of interest here are the fine solid particles that may be emitted from industrial combustion processes. The chemical composition of these particulates can vary widely, depending on their source and they can further react in the atmosphere to form other compounds. Many forms of PM exist naturally and include spores, dust, volcanic ash, and combustion products from wildfires. By their very nature, particulates are heterogeneous because of their varying size, shape, and chemical composition. Lee et al. [2] have experimentally investigated particulates from coal and No. 4 fuel oil combustion to study the composition and size distribution in the light of increasingly more stringent emission limits. Violi et al. [3] developed a new kinetic model for the fundamental formation of carbon from rich premixed ethylene flames.

Particulates in the atmosphere typically range in size from 0.1 to 50 μm, where larger sizes typically settle out of the air. Figure 9.1 shows some common terms and sizes used to categorize particulates [4]. PM less than 10 μm and smaller is called PM_{10} and is considered inhalable. These present a health risk because they can enter the respiratory system and become lodged in the tiny sacs of the lungs. In addition, the particulates may be toxic, depending on their chemical composition, or they may act as a carrier to bring toxic materials into the body that have attached to the surface of the particulates. Particulates can be particularly troublesome for those with respiratory problems or susceptibility such as the very young and the very old. The majority of natural PM is larger than 1 μm while most combustion-generated PM is in the size range 0.1–10 μm.

There are many forces that may act on particles including inertial (impingement, scrubbing, scavenging), electrostatic, magnetic, phoretic or indirect forces (diffusiophoresis, thermophoresis, photophoresis), diffusion, acoustic, adsorption, and other forces such as centrifugal and combinations of the previously mentioned forces [5]. Some of these forces are used in the various techniques discussed later for capturing particulates. There are also a number of forces for the interaction between particles including adhesion, coalescence, evaporation and condensation, and

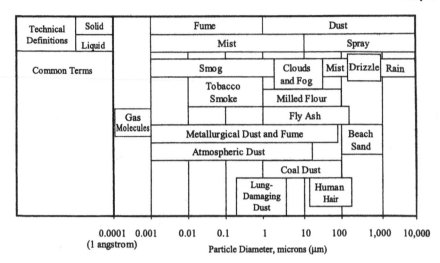

Figure 9.1 Common particulate terms and size ranges. (From Ref. 4. Courtesy of CRC Press.)

adsorption. These are also important in many of the collection techniques. Friedlander [6] has written an extensive textbook on all aspects of particle dynamics including particle transport properties, light scattering, experimental methods, coalescence, and agglomeration, among others.

A number of factors are often important in determining the best way to control particulate emissions. These include the loading density, size distribution, composition, shape, density, stickiness, corrosivity, reactivity, and toxicity. Therefore, it is very important to sample the particulates in a given exhaust gas stream to determine these properties and to determine if emissions are within regulatory limits. The properties of the gas carrying the particles are also often important. These properties include the temperature, pressure, composition, viscosity, velocity, humidity, and flammability.

Visibility problems in the atmospheres around cities are generally caused by particulates, which produce a haze that obscures the clarity, color, texture, and form of what we see. The particles cause light to be scattered, absorbed, and reflected, which distorts and reduces visibility. Smaller size particles can be particularly troublesome, not only because they are more easily inhaled, but because they can also remain airborne for much longer than larger particles. The fluid dynamics of particulates in the atmosphere are complicated by many factors including the climatic conditions like wind speed and direction and by the size, shape, and composition of the particles.

There are three primary sources of particulates that may be carried out of an industrial combustion process with the exhaust gases. One is entrainment and carryover of incoming raw materials. A second is particulates generated during the combustion process. The third is fuel carryover where some portion of a solid fuel, for example, coal, is not fully combusted before leaving the exhaust stack. A particular health concern regarding particulate emissions is the hazardous materials that can condense on the particle surfaces and be carried into the atmosphere [7]. For example, heavy metals vaporized during high-temperature combustion processes can condense on solid particles and be carried out with the exhaust products.

9.2 SOURCES

The three principle sources of particulates in most industrial combustion applications are: (1) dry fine particles being carried out of the process from the raw materials being processes, (2) particles generated in the combustion process, and (3) fuel carryover where some of a solid fuel passes through the combustor essentially unreacted. The first mechanism is not usually a problem in most hydrocarbon and petrochemical applications.

9.2.1 Particle Entrainment/Carryover

The gas flow through the combustor may entrain particles from the raw materials used in the process (Fig. 9.7). This is often referred to as carryover. An example of this would be in the glass-making process where fine dust materials like sand are used to make the glass and can be carried out of the glass furnace if the gas velocity in the combustion space is high enough. This type of particulate is expensive because not only must the particles be captured by some type of flue-gas scrubbing equipment, but some of the raw materials needed for the process are also lost.

A variation of particle entrainment is where incoming feed materials fed through the burner go through the system unreacted. For example, if a sludge is cofired in an incinerator, the inorganic solids remaining after the removal of the moisture from the sludge pass through the combustion system unreacted. These particulates must be captured prior to emission into the atmosphere.

9.2.2 Combustion-Generated Particles

The second method by which particles may be emitted from the combustion system is through the production of particles in the combustion process. For example, in the combustion of solid fuels, like coal for example, ash is normally produced. The airborne portion of the ash, usually referred to as fly ash, may be carried out of the combustor by the exhaust gases. Heavy-oil flames also tend to generate particulates due to the high carbon contents and greater difficulty in fully oxidizing those particles prior to exiting the exhaust stack.

Wark and Warner [8] note that four types of particles are formed during combustion by the following mechanisms:

1. Heat from the combustion process vaporizes some of the materials being processed in the combustor. These vaporized materials then condense in the much cooler atmosphere to produce particles in the size range 0.1–1 µm.
2. Relatively short-lived and unstable molecular clusters may be formed during the combustion reactions that produce particles smaller than 0.1 µm.
3. Mechanical processes such as coal grinding may release large fuel particles into the atmosphere greater than 1 µm in size.

4. Liquid fuel sprays may produce very fine ash particles that escape into the exhaust gas stream.
5. Soot may be produced by the incomplete combustion of fossil fuels. This is particularly more likely for some solid and liquid fuels, but may also take place for very fuel-rich combustion of gaseous hydrocarbons.

Another source of combustion-generated particles is soot that may be produced in a flame. Under certain conditions, even gaseous fuels may produce soot. To a certain extent, soot is desirable because it generally enhances the radiant heat transfer between the flame and the load. Fuels that have a higher carbon to hydrogen mass ratio tend to produce more soot than fuels with a lower ratio. For example, propane (C_3H_8), which has a C:H mass ratio of about 4.5, is more likely to produce soot than methane (CH_4), which has a C:H mass ratio of about 3.0. For clean burning fuels like natural gas, it is much more difficult to produce sooty flames compared to other fuels, like oil and coal, which have little or no hydrogen and a high concentration of carbon. Flames containing more soot are more luminous and tend to release their heat more efficiently than flames containing less soot, which tend to be nonluminous. Soot particles generally consist of high molecular weight polycyclic hydrocarbons and are sometimes referred to as "char."

Ideally, soot would be generated at the beginning of the flame so that it could radiate heat to the load and then it would be destroyed before exiting the flame so that no particles would be emitted. Soot can be produced by operating a combustion system in a very fuel-rich mode or by incomplete combustion of the fuel due to poor mixing. If the soot particles are quenched or "frozen," they are more difficult to incinerate and more likely to be emitted with the exhaust products. The quenching could be caused by contact with much colder gases or possibly by impingement on a cool surface, such as a boiler tube. Soot particles tend to be sticky and can cling to the exhaust ductwork, clogging the ductwork and other pollution treatment equipment in the system. If the soot is emitted into the atmosphere, it can contribute to smog in addition to being dirty. The emitted soot particles become a pollutant because they produce a smoky exhaust that has a high opacity. Most industrial heating processes have a regulated limit for opacity. Ammouri et al. [9] discuss the effects of soot on heat transfer and NO_x emissions in industrial-scale experiments using propane and natural gas combusted with high-purity oxygen.

Linak et al. [10] experimentally and numerically studied fine particulate emissions from residual fuel oil combustion [9]. Tests were conducted in a small-scale laboratory combustor and a small commercial fire-tube boiler. The ultrafine particles from both combustors were consistent with the nucleation of vapor-phase metal sulfates accumulating into an evolving aerosol. The large particles were consistent with large porous carbonaceous particles produced by incomplete burnout. The fine particles are becoming increasingly important as particulate emissions standards regulate smaller and smaller particle sizes.

9.2.3 Solid Fuel Carryover

This mechanism for generating particulates involves some of the solid fuel passing through the combustor without being completely reacted. This is particularly a concern in applications where coal or coke is used as fuel. These typically produce some type of residual ash, even after being completely combusted. The ash residue must be collected and properly disposed of or treated. In some cases, the ash may even contain hazardous heavy metals, which makes the treatment and disposal even more critical and expensive. There are also some applications where a solid may be cofired with an auxiliary fuel, particularly in waste-incineration applications. There may be inorganic components in the waste fuel that are carried through the combustion system in the form of particulates. Fortunately, solid-fuel carryover is not usually a concern in most industrial combustion applications, but it does need to be considered in certain applications.

9.3 ENVIRONMENTAL CONCERNS

9.3.1 Atmospheric Degradation

A primary environmental concern for particulate emissions is reduced visibility, which is a particular problem in more urban areas. The degree of impact on visibility depends on a number of factors including loading density, particle size, weather conditions (temperature, wind speed and direction, humidity, angle of the sun, cloud conditions, etc.), source location, and time of day (e.g., reduced visibility is more of a concern during daylight hours when sunlight is blocked). Related to the reduced visibility is the reduction in solar radiation caused by absorption of the sun's rays by the particulates. This reduces the temperature in areas where the absorption is high, which can have significant environmental effects. It can also reduce plant growth rates and cause things like depression in certain individuals due to the reduction in sunlight.

9.3.2 Environmental Damage

Another environmental concern of particulates is deposition. This can reduce the aesthetic appeal of structures and monuments. This can lead to frequent cleaning or painting, depending on how severe the deposition. If the particles are corrosive, or become corrosive on contact with water (e.g., rain) then further damage can result, including reducing the structural integrity of infrastructure like roads and bridges.

Particles can damage vegetation and waterways if they are in high concentrations, are toxic, or are large enough in size. Particulates can reduce crop production by blocking the natural processes of plants. If the particles are alkaline or acidic they can alter the pH of waterways. If they are toxic, then the plant and animals in contact with the particles can become toxic and adversely impact the food chain for humans.

Particulates can enhance chemical reactions in the atmosphere by acting as a catalyst, especially if the particulates contain some catalytic metals. Particulates have a high surface area per unit volume, which is needed for enhancing catalytic reactions such as the conversion of VOC emissions to photochemical smog. In most cases, the enhanced chemical reactivity promoted by catalytic activity on particulates is undesirable.

Table 9.1 Air Quality Guide for Particulate Matter

Air quality index	Air quality	Fine particles (PM$_{2.5}$)	Coarse particles (PM$_{10}$)
0–50	Good	None	None
51–100	Moderate	None	None
101–150	Unhealthy to sensitive groups	People with respiratory or heart disease, the elderly, and children should limit prolonged exertion	People with respiratory disease, such as asthma, should limit outdoor exertion
151–200	Unhealthy	People with respiratory or heart disease, the elderly, and children should avoid prolonged exertion; everyone else should limit prolonged exertion	People with respiratory disease, such as asthma, should avoid outdoor exertion; everyone else, especially the elderly and children, should limit prolonged outdoor exertion
201–300	Very unhealthy (alert)	People with respiratory or heart disease, the elderly, and children should avoid any outdoor activity; everyone else should avoid prolonged exertion	People with respiratory disease, such as asthma, should avoid any outdoor exertion; everyone else, especially the elderly and children should limit outdoor exertion

Source: Ref. 11.

9.3.3 Health Hazards

Another environmental concern related to particulates is the health hazard they present to the human respiratory system. Small particles can be captured on the mucous membranes in the nasal passages and in the lungs. This can cause premature death in the elderly and the ill. Particles can cause inflammation, cellular damage, increased permeability of the lungs, and increased heart rate and blood pressure. Long-term exposure can cause a decrease in lung function and an increased risk of respiratory illness. The severity of the health hazards associated with particulates depend on many factors including loading density, particle surface area, particle acidity, particle surface chemistry, electrical charge, composition, exposure duration, temperature, humidity, activity level, age, and general physical condition. Table 9.1 shows guidelines established by the U.S. EPA for the health hazards associated with different particulate loadings [11].

9.4 REGULATIONS

In 1997, the U.S. EPA issued a revision to the National Ambient Air Quality Standard for Particulate Matter, reducing the particulate size of interest from 10 to 2.5 µm because of recent studies indicating health concerns for the smaller particles. While that standard refers to ambient air quality, it indirectly affects the permissible

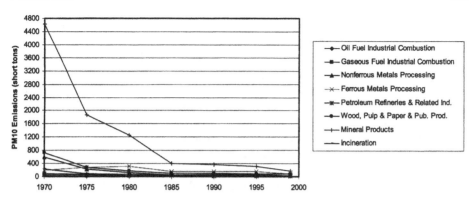

Figure 9.2 PM$_{10}$ emissions in the United States since 1970 by source type. (From Ref. 12.)

particulate emissions from an industrial combustion process. If a geographic area has a high concentration of particulates in the atmosphere and problems with visibility, it would be expected that regulated particulate emissions limits would be more stringent compared to an area with a low concentration of particulates and not a significant problem with visibility. Figure 9.2 shows particulate (PM$_{10}$) emissions in the United States since 1970 for some of the industrial combustion applications considered in this book [12].

The emission of toxic particulates is of particular concern. This usually occurs when the material being treated in the combustor contains toxic materials. For example, lead particles are added to the raw batch feed materials to make leaded glass. When some of these lead particles are carried out of the combustor in the exhaust products, they must be scrubbed out to prevent lead from being emitted into the atmosphere. This is a problem not only because of the toxicity of lead, but also because the incoming lead in the batch is an expensive raw material. Lead particles escaping out of the glass-melting process are in effect paid for twice: once as incoming raw materials and again to be removed from the exhaust products.

Other toxic particulates include heavy metals that may be present in certain fuels like coal or heavy fuel oil, or that may be present in the raw materials such as in contaminated soil being processed in a waste incinerator.

9.5 MEASUREMENT TECHNIQUES

EPA Method 5 is used to determine particulate emissions from stationary sources [13]. Method 5B refers to the determination of nonsulfuric acid particulate matter emissions from stationary sources. Method 5D concerns determining particulate matter emissions from positive pressure fabric filters, also known as baghouses, where this method has particular emphasis on the sample location. Method 5E describes the procedure for determining particulate emissions from fiberglass wool manufacturing. Method 5F concerns nonsulfate particulate emissions from stationary sources, which includes a modified method for separating the sample from the filter. Method 5I concerns the determination of low concentrations of particulate emissions from stationary sources. This method differs from Method 5 with the following modifications: improved sample-handling procedures, use of a lightweight sample filter assembly, and the use of low-residue grade acetone to recover the particulate matter collected on the filter.

For particulate measurements, the gas sample must be withdrawn isokinetically (see Chap. 5) to ensure that a representative sample of particles are withdrawn from the gas stream. If the extraction rate is too high, then the measured particulate concentration will be too high. If the extraction rate is too low, then the measured particulate concentration will be too low. A specially designed probe is used to sample gases from particle-laden flows. The construction of the sampling system is also important to ensure that water and particulates do not come out of the extracted sample until it reaches the measurement location in the system. For example, the probe should be lined with borosilicate or quartz glass tubing and heated to prevent condensation and to prevent the particulates from coating the walls.

The extracted sample containing particulates passes through a glass fiber filter maintained at a temperature of $120° \pm 14°C$ ($248 \pm 25°F$). The sample is extracted for a known amount of time, long enough to collect a sample of sufficient mass. Moisture is removed from the particulate sample captured on the filter. The dried sample and filter are then weighed. This is compared to the weight of the clean filter prior to sampling. The difference in the weights then gives the weight of particulates captured. Using the measured sample extraction time, sample flow rate, and total exhaust gas flow rate, the particulate matter emission rate can then be calculated. Figure 9.3 shows an example of an in situ particulate sampling system, and Fig. 9.4 shows an example of a Method 5 extractive sampling system.

A number of state-of-the-art commercially available technologies are available for monitoring particulate matter including opacity monitors, light-scattering technologies (see Fig. 9.5), beta gauges, acoustic-energy monitoring, tapered-element oscillating microbalance (TEOM), and triboelectric technologies [14]. The problem with most of these techniques is that they do not directly measure the mass of particulates generated. Table 9.2 shows a comparison of some of the techniques used to measure particulate emissions [15].

Figure 9.3 Particulate analyzer mounted at the sample location. (Courtesy of Environmental Systems, Knoxville, TN.)

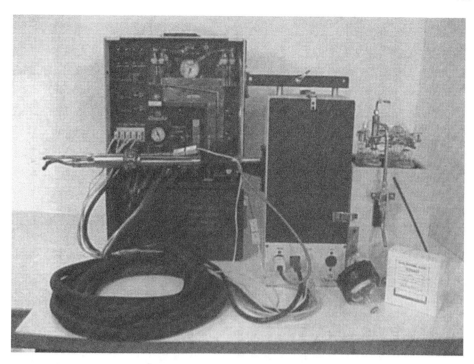

Figure 9.4 Extractive Method 5 particulate sample system. (Courtesy of Clean Air Engineering, Palatine, IL.)

Figure 9.5 Backscatter in-situ dust monitor. (Courtesy of Monitor Labs, Denver, CO.)

Table 9.2 Overview of Particulate Measurement Techniques

Type	Sensitivity	Influence	Problems/advantages
Transmissivity (opacity)	• Range limited to approximately 100 mg/m³ • Output is related to concentration not mass emission	• Particle size • Particle shape • Particle color • Ambient light	• Dirt on lens windows • Misalignment of the transmitter and receiver reflector units owing to thermal expansion of duct
Scattering	• Output is related to concentration not mass emission • Not suitable for high-opacity applications such as smoke	• Particle size • Particle shape • Particle color • Ambient light	• Dirt on lens windows although less sensitive to this than is the transmissivity type
Optical scintillation	• AC measurement • Higher sensitivity than transmissivity type • Response is related to mass emission i.e., velocity-dependent	• Particle size • Particle shape • Particle emission reflection effects • Dynamic flow conditions	• Dirt on lens windows although less sensitive than the transmissivity type • Suitable for small—and large—diameter ducts
Triboelectric	• DC or AC measurement • High sensitivity • DC type: response is related to mass emission (velocity dependent)	• Particle size • Particle shape • Particle electrostatic characteristics • Duct electrostatic characteristics • Dynamic flow conditions	• Not suitable for damp or wet gas • Influenced by ionization and humidity variations • Will tolerate superficial contaminations (AC type: high tolerance) • Applicable to small-diameter ducts down to 100 mm
Acoustic	• Moderate sensitivity • Response is related to mass emission • Alarm duty only	• Particle mechanical characteristics • Dust geometry • Dynamic flow conditions • Ambient noise	• Complex and application dependent

Source: Ref. 15.

EPA Method 9 concerns the visual determination of opacity from emissions from stationary sources [16]. The opacity may be caused by particulates and/or by steam. This technique requires a certified observer trained in this measurement technique. An alternative technique known as LIDAR [17] based on light backscattering can be used to determine opacity remotely as described in Alternate Method 1. Method 22 discusses the visual determination of smoke emissions from flares. This method does not require a certified observer as in Method 9 as the opacity is not quantified. It is merely a visual determination to see if smoke is present from flaring.

One technique that is sometimes used to quantify smoke based on visual observations was developed by the U.S. Bureau of Mines and is referred to as the Ringelmann Smoke Chart [18]. A series of cards are visually compared against the smoke from a stack or flare to determine the approximate smoke density. The cards (see Fig. 9.6) were designed by Professor Ringelmann as follows:

Card 0: all white
Card 1: black lines 1 mm thick, 10 mm apart, leaving white spaces 9 mm square
Card 2: lines 2.3 mm thick, spaces 7.7 mm square
Card 3: lines 3.7 mm thick, spaces 6.3 mm square
Card 4: lines 5.5 mm thick, spaces 4.5 mm square
Card 5: all black

It is noted that this is an empirical technique, although many regulatory bodies have adapted its use. There are many factors that can affect the apparent smoke density including the particulate concentration and size distribution in the exhaust gases, the depth of the smoke plume, the natural lighting conditions (especially the position of the sun relative to the plume and the viewer), and the color of the particles. The chart is designed to be supported at eye level at a distance from the observer where the lines on the chart merge into shades of gray, as closely in line with the stack as possible. The observer compares the smoke from the stack to the Ringelmann cards to find the one that most closely resembles the shading of the smoke. Clear gases would be recorded as No. 0, while pure black smoke would be recorded as No. 5. These observations can be averaged over some time period if desired.

9.6 TREATMENT TECHNIQUES

There are a variety of techniques used to control particulate emissions from combustion processes. The U.S. EPA has prepared a document [19] that considers particulate control techniques for approximately 80 industrial stationary sources including most of those considered in this book. The specific method chosen will depend on many factors including economics, particle size distribution and composition, volumetric flow rate, exhaust stream temperature, and particle moisture content. In general, it is often more cost effective to minimize the formation of particulates in the first place rather than remove them at the back end of the process before being emitted into the atmosphere.

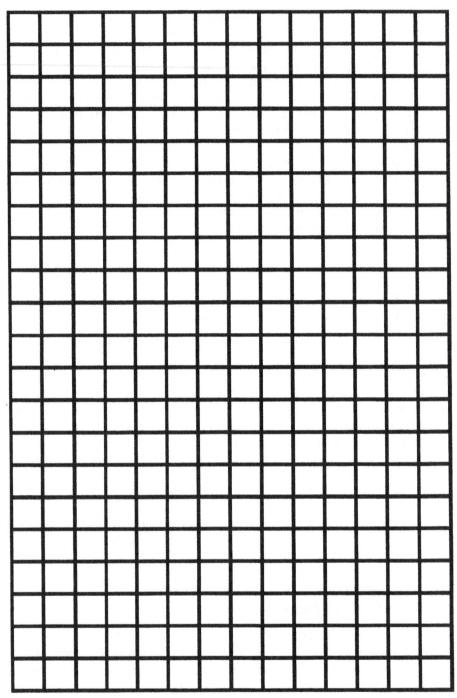

1. EQUIVALENT TO 20 PERCENT BLACK

Figure 9.6 Ringelmann smoke cards 1–4 which are equivalent to smoke densities of 20, 40, 60, and 80%, respectively. (From Ref. 18.)

2. EQUIVALENT TO 40 PERCENT BLACK

Figure 9.6 Continued.

3. EQUIVALENT TO 60 PERCENT BLACK

Figure 9.6 Continued.

4. EQUIVALENT TO 80 PERCENT BLACK

Figure 9.6 Continued

Wark and Warner [8] note the following strategies for handling particulate emissions:

1.   Gas cleaning
2.   Source relocation
3.   Fuel substitution
4.   Process changes
5.   Good operating process
6.   Source shutdown
7.   Dispersion

Some of these are considered next, but others like source relocation, source shutdown, and dispersion are either not viable options or do not solve the problem but merely redistribute the particulates so that the concentration is reduced.

9.6.1 Pretreatment

Pretreatment involves modifying the fuel, oxidizer, or raw materials in the process in some way as to reduce particulate emissions. One example would be either to grind a solid fuel like coal into a finer size or to filter out the larger fuel particles in order to increase the probability of complete combustion as the finer the solid fuel size, the easier it is to combust fully.

Another example is removing the finer particles in the incoming batch materials that are more likely to become airborne inside the combustor. In the glass industry, sand is one of the primary batch materials in the process (see Chap. 14). Fine grains of sand may be entrained into the exhaust products and carried out of the stack. The problem with these sand particles is that they will not combust in the exhaust stream and unless they are captured before entering the atmosphere, they will enter the environment essentially unchanged. While sand is inert and not specifically hazardous to the environment, it increases the opacity of the exhaust stream and can be a real nuisance to residents living in the vicinity of the plant who could get a fine coating of sand dust on their property.

Air is the most common oxidizer used in industrial combustion processes. It has been shown that replacing some or all of the combustion air with high-purity oxygen can significantly reduce particulate emissions under certain circumstances. The global stoichiometric reaction of air with methane can be written as

$$CH_4 + 2(O_2 + 3.76N_2) = CO_2 + 2H_2O + 7.52N_2 \tag{9.1}$$

where air is approximately 21% O_2 and 79% N_2 by volume (which is represented by $O_2 + 3.76N_2$). This equation shows that for every one volume of methane, 10.52 volumes of exhaust products are produced, most of which is in the form of nitrogen. This process can be enhanced by removing some or all of the nitrogen from the air to produce a higher concentration of O_2 in the oxidizer. This process is usually referred to as *oxygen-enhanced combustion* or OEC [20]. In the extreme case of using pure O_2, the global stoichiometric reaction of methane can be written as

$$CH_4 + 2O_2 = CO_2 + 2H_2O \tag{9.2}$$

where it can be seen now that only three volumes of exhaust products are produced for each volume of methane. The gas velocity inside the combustor is directly related to the gas flow rate. By reducing the exhaust products from 10.52 to 3 volumes (a reduction of almost two-thirds), the resulting gas velocity will be decreased proportionally (assuming that the pressure and temperature of the gases are the same and the combustor geometry remains the same). This dramatic reduction in velocity causes fewer particles in the raw batch materials to be entrained into the flue gases (see Fig. 9.7). In a mobile incinerator (see Chap. 16) soil cleanup project, OEC was used in part to reduce particulate emissions dramatically [21]. In a transportable incinerator (see Chap. 16) used to clean contaminated soil, replacing air/fuel combustion with oxy/fuel combustion reduced particulate emissions by more than 50% while simultaneously increasing soil processing rates by 22% [22]. However, separating the nitrogen from air comes at a cost. Each project should be evaluated on its own merits to determine if the added cost of the higher purity oxidizer offset the costs of a post-treatment system. It is unlikely that using OEC strictly for reducing particulate emissions will be cost effective. However, there are often numerous other benefits of using OEC, that would need to be evaluated [23].

Another example of pretreatment is to remove possible particulate sources from incoming raw materials. An example is cleaning of scrap metal prior to its introduction into a melting furnace. There are several ways that may be done. One way is some type of spray washing system where the scrap metal is cleaned. The problem with this technique is that if the parts are not dried prior to putting them into the furnace, then there are some safety concerns where rapid vaporization of water on the scrap could produce a vapor explosion under the molten metal. Another potential problem is that the water in the cleaning system presents a significant added heat load that reduces the overall thermal efficiency of the process. A specific challenge of scrap metal is paint and other coatings on the parts that can become a source of particulates in the combustor. Spraying a cryogenic like liquid nitrogen on to the scrap parts causes the outer surface to freeze preferentially. The coating contracts and cracks off the surface of the metal. The parts are then cleaned

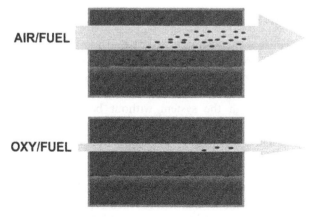

Figure 9.7 Comparison of particle entrainment in a furnace using air/fuel and oxy/fuel combustion. (From Ref. 20. Courtesy of CRC Press.)

to remove the coating flakes before the scrap is put into the melter. This can be an expensive process and is limited by how well the cryogenic spray can get into the interstices of the scrap parts. The parts can be mechanically cleaned with wire wheels, brushes, or similar devices but these techniques are limited as to their removal efficiencies. Complete pretreatment systems are available for cleaning scrap materials.

Ferrara [24] describes the use of a novel chemical additive to improve combustion efficiency and reduce particulates during the firing of a heavy fuel oil. In some cases, NO_x also decreased slightly with the use of the proprietary additive. Filho et al. [25] experimentally demonstrated similar results using proprietary chemical additives that reduced particulate emissions from heavy oil firing.

9.6.2 Process Modification

The preferred method in most cases for reducing particulate emissions is to minimize particulate formation in the first place by modifying the process. For example, substituting a gaseous fuel or lighter fuel oil for a heavy fuel oil can significantly reduce particulate emissions resulting from the fuel. Another example would be using raw materials that are less dusty and, therefore, have less tendency to be entrained into the flue gas products. The dust produced from existing raw materials may be reduced by increasing the moisture content although this also increases the heat load on the combustion system. If there are any undesired components from an incoming waste fuel that cause particulate emissions, these could be separated out of the waste fuel prior to firing if this is economically feasible.

Another process modification strategy for reducing particulate emissions is to capture the particles inside the combustion system for recycling back into the process. One example is a fluidized-bed reactor where the majority of the particles are deliberately recirculated back into the process. This not only reduces the load on any downstream post-treatment equipment for removing particulates, but it also reduces raw materials costs and enhances process economics. The specifics of the application will dictate the viability of this option.

9.6.3 Combustion Modification

Modifying the combustion system is sometimes the easiest and most economical solution for certain applications where, for example, dust from incoming raw materials is not an issue. The basic premise is to optimize the conditions for combustion to minimize the tendency for either forming particulates, or allowing combustible particulates to pass through the system without being completely burned. One of the more important conditions includes sufficient oxygen for combustion. While sufficient oxygen is not generally a problem in most applications, there are some where the incoming hydrocarbon composition and quantity may vary with time (e.g., waste fuel firing in incineration). A control system must be in place to adjust the available oxygen to meet any changing demands because particulates are much more likely to form during combustion when there is insufficient oxygen.

Another set of conditions important for minimizing particulate emissions due to combustion are the so-called three Ts of time, temperature, and turbulence. This means that there must be adequate residence time in the combustor for the combustion reactions to go to completion. This is particularly important in the combustion of liquid and solid fuels that require more steps, such as vaporization and devolatilization, compared to gaseous fuels. The combustion chamber must be hot enough for heat to transfer to the incoming fuel for proper combustion. There must be sufficient mixing (turbulence) to ensure that the fuel is adequately blended with the oxidizer for proper combustion. As discussed in Chap. 1 and elsewhere [20], increasing the oxygen content in the oxidizer can significantly enhance the combustion process and reduce the tendency for particulate formation due to the fuel.

9.6.4 Post-Treatment

Three basic collection mechanisms used to remove particulates from exhaust gas streams are impaction, interception, and diffusion (Fig. 9.8) [4]. In impaction, the particles are sufficiently large that they travel in a straight line toward some type of collection target without being carried around the target with the rest of the gases. Interception is where smaller particles are carried around the target with the main gas flow, but the particles travel close enough to the target to be intercepted at the surface. In diffusion, very small particles travel to the target surface by Brownian motion. Other mechanisms include electrostatic diffusion, gravity, centrifugal force, thermophoresis, and diffusiophoresis. Table 9.3 shows how the collection efficiency of various devices depends on the particle diameter [26].

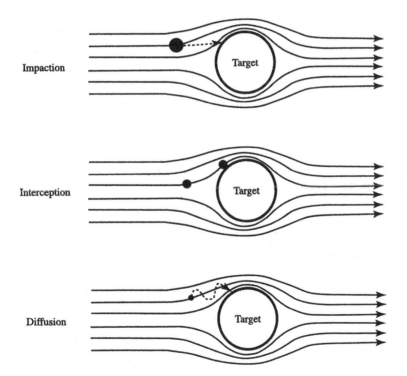

Figure 9.8 Basic particle collection mechanisms. (From Ref. 4. Courtesy of CRC Press.)

Table 9.3 Particulate Capture Dependence on the Particle Diameter d_p

Control device	Principal particle capture mechanism	Particle size dependence[a]
Settling chamber	Gravity settling	d_p^2
Momentum separator	Gravity settling	d_p^2
	Inertial separation	d_p^2
Large-diameter single cyclone	Inertial separation	d_p^2
Small-diameter multiple cyclones	Inertial separation	d_p^2
Fabric filters	Impaction on dry surfaces	d_p^2
	Interception	d_p
	Diffusion to dry surfaces	$1/d_p$
Electrostatic precipitator	Electrostatic attraction	d_p^2 and $1/d_p$
	Gravity settling	d_p^2
Wet scrubber	Impaction on surfaces	d_p^2
	Impaction on liquid droplets	d_p^2
	Diffusion to wetted surfaces	$1/d_p$
	Diffusion to liquid droplets	$1/d_p$
Incinerator	Particle oxidation	$1/d_p$

[a]Based on particle capture mechanism.
Source: Ref. 26.

Because of typically higher costs, the last choice is usually to remove the particles from the exhaust stream before they are emitted into the atmosphere. This may be done with electrostatic precipitators (wet or dry), filters, venturi scrubbers, or cyclones [6,27]. de Nevers [28] discusses the mathematics behind most of these particulate removal systems. Wark and Warner [8] note the basic classes of particulate collection equipment:

1. Gravity settling chambers
2. Cyclone (centrifugal) separators
3. Wet collectors
4. Fabric filters
5. Electrostatic precipitators

These are discussed below. The various techniques have advantages and disadvantages and generally rely on different technologies depending on the particle size. Figure 9.9 shows how particulate removal efficiency depends on both the collection technique and on the particle size, where settling chambers are the least efficient and bag filters are the most efficient [29]. However, settling chambers are also typically the least expensive so both capital and operating costs need to be considered as well.

9.6.4.1 Gravity Settler

Gravity settlers, sometimes referred to as elutriators, are fairly crude and unsophisticated but effective in removing larger particulates if properly designed.

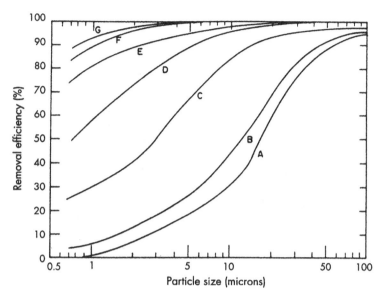

Figure 9.9 Removal efficiencies for common particulate collection techniques as a function of particle size. (From Ref. 29. Courtesy of Butterworth-Heinemann.)

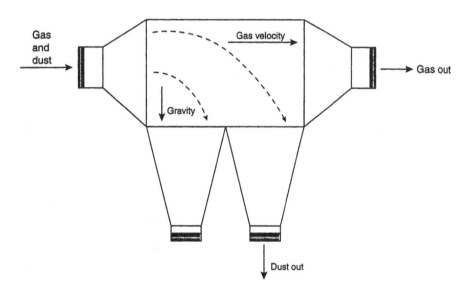

Figure 9.10 Schematic of a gravity settling device. (From Ref. 50. Courtesy of Technomic Publishing.)

These are sometimes referred to as precleaners, which reduce the particulate loading going into downstream particulate removal devices like fabric filters. The settling chamber is basically a long chamber with a large cross-sectional area and baffles that allow the flue gases to slow down enough for the particulates to fall to the bottom simply by gravity (see Fig. 9.10). The baffles are designed to distribute the flow evenly and prevent the bulk of the gases from going through the center of the device. Even flow distribution improves the particulate removal efficiency of the settling device. The chamber is then cleaned at regular intervals to remove the particles that have accumulated. The removal efficiency of a gravity settling chamber increases with (1) particle size and/or density, (2) a decrease in flow velocity, and (3) the number of vertical tubes or towers.

Some of the benefits of this device include: it is very simple to build, it usually requires very low maintenance, has no moving parts, low energy requirements, low pressure drop, excellent reliability, resistant to abrasive particles, temperature and pressure limitations only dependent on materials of construction, and the collection is dry, which makes disposal easier compared to wet methods. A potential problem is the large amount of space it requires, which is often difficult to find in existing plants that are being retrofitted with pollution control equipment. Another problem is the efficiency of these devices is reduced as the size of the particulates decreases because of their reduced weight, which makes it easier for them to remain airborne. This method is not suitable for sticky or tacky materials.

9.6.4.2 Cyclone

Another type of particulate separating device is known as a cyclone or centrifugal separator (see Fig. 9.11). These are low-cost and low-maintenance and are widely used to remove medium-sized and coarse particles. Particle-laden gas streams enter a conically shaped chamber tangentially to the walls and close to the top of the vessel. This causes the gases to swirl downward toward the bottom of the vessel. The higher specific gravity particles are thrown against the wall and slide down the wall by gravity into a collection hopper. The heavier the particle, the easier to remove from the gas stream. The collected particles can then be disposed of or recycled back into the process, depending on the application. On reaching the bottom of the cone, the gases then turn and travel back up inside the downward-swirling outer vortex. The cleaned gases then exit the top of the cyclone.

As noted above, the problem with settling chambers is their relative ineffectiveness for capturing smaller size particles. The cyclone solves that problem by using a stronger separating force than gravity—a centrifugal force that throws the particles against the walls of the cyclone. This force can be calculated from

$$F_{\text{centrifugal}} = \frac{mV_c^2}{r} \tag{9.3}$$

where m is the mass of the particle, V_c is the velocity of the particle in a circular path, and r is the radius of the circular path.

To show why cyclones are more effective than gravity settlers in collecting small particles, the ratio of the centrifugal force to the gravity force can be calculated from

$$\frac{F_{\text{centrifugal}}}{F_{\text{gravity}}} = \frac{mV_c^2/r}{mg} = \frac{V_c^2}{rg} \tag{9.4}$$

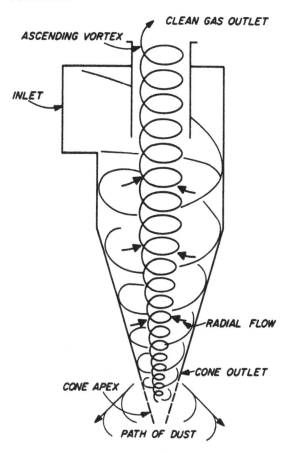

ASCENDING VORTEX

CLEAN GAS OUTLET

INLET

RADIAL FLOW

CONE OUTLET

CONE APEX

PATH OF DUST

Figure 9.11 Basic cyclone steps. (From Ref. 30. Courtesy of Marcel Dekker.)

For a particle traveling at 50 ft/sec in a cyclone 3 ft in diameter, the ratio of these two forces is

$$\frac{F_{\text{centrifugal}}}{F_{\text{gravity}}} = \frac{V_c^2}{rg} = \frac{(50\,\text{ft/s})^2}{(3\,\text{ft})(32.2\,\text{ft/s}^2)} = 25.9 \tag{9.5}$$

which shows how much more powerful the centrifugal force can be under the proper conditions of velocity and radius. A schematic of a cyclone separator is shown in Fig. 9.12 [30]. The separation efficiency is related to the centrifugal force in the cyclone. There are two ways this force can be increased: increase the velocity or reduce the radius. The problem with reducing the radius is that a large volume of gas often needs to be treated and the pressure drop through a small cyclone would be unacceptably high. The pressure drop is related to the square of the velocity and the velocity is related to the cross-sectional area of the cyclone, which is proportional to the square of the radius of the cyclone. Therefore, a small reduction in radius can cause a dramatic increase in pressure drop. However, multiple smaller cyclones can be joined in parallel to be able to treat large volumes of gas without an undue penalty

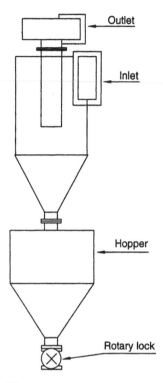

Figure 9.12   Schematic of a cyclone. (From Ref. 30. Courtesy of Marcel Dekker.)

in pressure losses. Figure 9.13 shows a schematic of a device, called a multicyclone, consisting of many cyclones.

Some of the factors that influence the efficiency of cyclones include secondary effects (e.g., when particles are entrained into the inner upward-swirling vortex, the physical dimensions of the cyclone, the properties of both the carrier gas and of the particles, and changes in process variables, (e.g., dust loading). Some specific geometric dimensions produce higher efficiency collection as shown in Fig. 9.14.

There are generally two main classifications for cyclones: (1) high-efficiency cyclones and (2) high-throughput cyclones [30]. High-efficiency cyclones have higher inlet velocities to increase the centrifugal force, but at the cost of throughput. High-throughput cyclones are generally larger in diameter to handle high gas flow rates with less pressure drop, but at the cost of collection efficiency. Figure 9.14 shows the recommended dimensional proportions for a high-efficiency cyclone. The four general types of cyclones include tangential inlet, axial discharge; axial inlet, axial discharge; tangential inlet, peripheral discharge; and axial inlet, peripheral discharge. The first two types are the most common.

Some advantages of cyclones include [31]:

1.   Inexpensive to purchase
2.   No moving parts
3.   Can usually withstand harsh operating conditions
4.   Nearly maintenance free

Figure 9.13 Schematic of a multiclone or multiple cyclones. (Courtesy of Research-Cottrell, Somerville, NJ.)

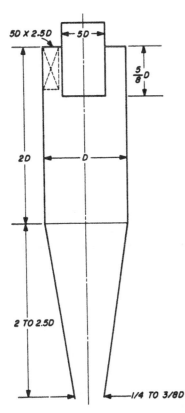

Figure 9.14 Dimensional proportions for a high-efficiency cyclone. (From Ref. 30. Courtesy of Marcel Dekker.)

Other advantages are that higher gas temperatures can be treated compared to other devices, such as fabric filters, and relatively little space is required. Heumann considers cyclones the most efficient and economical separator for industrial control systems [32]. The collection efficiency normally varies considerably with the particle size. Most cyclones are very efficient at collecting larger particles, but may be very inefficient at collecting the smaller sizes. Some factors that affect collection efficiency include the gas flow rate and inlet velocity, the cyclone diameter and geometry, the gas viscosity, the difference in density between the gas and the particles, and the particle-loading concentration and size distribution. A cyclone is sometimes used as a precleaner for other particulate removal systems because cyclones are inexpensive and can remove the larger particles easily and economically compared to other techniques like electrostatic precipitation which is discussed next.

There are certain advantages and disadvantages of wet removal techniques in a cyclone [33]. One advantage is that liquid droplets can prevent particles from being re-entrained into the exhaust gas flow. The liquid can often be more easily drained, collected, and transported compared to dry dust. Air entrainment from the outside can be prevented with a simple liquid seal. Wet cyclones often have higher removal efficiencies and less erosion and plugging problems compared to dry cyclones. The main disadvantage of wet cyclones is corrosion if the dust particles are corrosive where the presence of water exacerbates the problem. Another disadvantage is the added cost of the water and the necessary water treatment equipment needed to clean the water prior to disposal or to recycle the water back to the cyclone in a continuous closed loop.

9.6.4.3 Electrostatic Precipitator

When neither a gravity separator nor a cyclone is effective, an electrostatic precipitator (ESP) may be used. This relies on electrostatic forces to drive the particles out of the stream and on to a collector. While ESPs also have difficulty collecting smaller particles, they are more effective than gravity settlers and cyclones because the electrostatic force is less dependent on the diameter of the particle than those techniques. Some of the advantages of ESPs include large gas volume capacities, high collection efficiencies (up to 90% or more) for a wide range of particle sizes, low energy consumption and draft loss, and the ability to work with relatively high-temperature gases. A unique feature of ESPs is that the energy input works only on the particles being collected and not on the entire gas stream. The U.S. EPA has written a manual to assist in estimating the cost of various air pollution control techniques, which includes a chapter on electrostatic precipitators [34]. There are a variety of ESP designs including plate-wire, flat plate, tubular, wet, and two-stage precipitators.

The basic principle is to charge the particles electrostatically and then put them in an electrostatic field to drive the charged particles toward a collector. This two-step process of first charging the particles and then driving them towards the collector may be carried out either in the same part of an ESP on in consecutive parts of an ESP, depending on the design. The latter design is more common in industrial applications. In many cases, particulates have some electrical charge prior to reaching the ESP, but the charge is not sufficient to drive the particles toward the oppositely charged plate for removal. The charge must be sufficient for the particle to migrate in a direction perpendicular to the main gas flow, over to the collection plate.

An ESP consists of a series of high-voltage electrodes (often wires) and grounded collector plates (see Fig. 9.15) [35]. The high-voltage electrodes (on the order of −40 kV) are located in the flow stream and create a corona discharge that releases electrons into the gas stream flowing past the electrodes. Typical electrodes consist of small diameter wires, coiled wires, flat rectangular shapes, flat rectangular shapes with spikes, barbed wires, square twisted rods, star wires, or rigid tubes with spikes [36]. Electrons striking particles in the gas stream cause the particles to become negatively charged. The charged particles are then electrically driven toward the grounded plates (0 V). For this process to work efficiently, the particle must be sufficiently charged, the spacing between the collector plates must not be too large, and there must be enough time for the particles to migrate through the gas to reach the collector plates, which is a function of both travel distance and gas velocity.

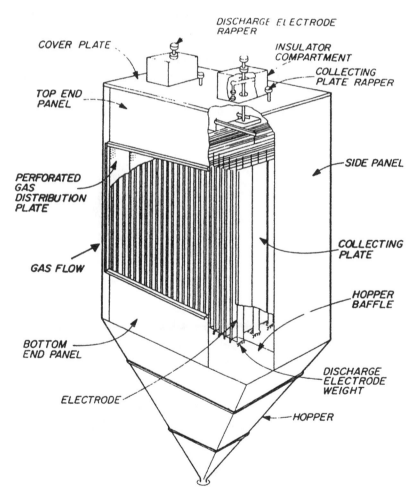

Figure 9.15 Schematic of an electrostatic precipitator. (From Ref. 35. Courtesy of Marcel Dekker.)

Particle buildup on the collector electrodes can reduce collection efficiency. The migration velocity is proportional to the square of the electrical field strength, directly proportional to the particle diameter, and inversely proportional to the gas viscosity. The particles lose their electrical charge on striking the collection plate and tend to agglomerate into a cake.

There are two types of collection mechanisms, depending on the design of the ESP. In dry ESPs (see Fig. 9.16), the particles are removed from the plate by mechanically vibrating it at regular intervals, for example, with some type of rapper, causing the particles to fall to the bottom of the ESP where they are then removed. The problem is that some of these particles are re-entrained into the exhaust stream, which reduces the overall removal efficiency of the ESP; this is sometimes referred to as rapper re-entrainment. In a wet ESP, the particles are captured in a thin film of water that flows down the collection electrode. An advantage of the wet ESP is that particles are not re-entrained into the exhaust gas stream, which can be a problem in dry ESPs. Altman et al. [37] notes that compared to wet ESPs, dry ESPs cannot effectively handle moist or sticky particulates that would cling to the collector plates, that dry ESPs require more real estate for multiple fields due to particle re-entrainment problems, and that the mechanical methods used for cleaning of the collector plates require maintenance and periodic shutdowns. Wet ESPs are particularly effective when the gas stream has a high moisture content, the particles are sticky, some of the particles are submicrometer in size, the gas stream has acid droplets or mist-SO_3, or the temperature of the gas stream approaches the gas dew point.

Figure 9.16 Industrial dry electrostatic precipitator. (Courtesy of Wheelabrator, Pittsburgh, PA.)

Dunkle [36] lists the following process parameters that affect the design of an ESP:

1. Particulate loading
2. Gas temperature, composition, and dew point
3. Electrical resistivity of the particulates
4. Particle size distribution
5. Shape of the particles
6. Cohesion and adhesion properties of the particles
7. Electric field intensity
8. General cleanliness of the discharge and collecting electrodes

An ESP is a type of gravity separator where the force driving the particles to the collector is electrostatic instead of gravitational. As with other particulate removal devices previously discussed, the electrode and plate devices can be constructed in parallel to increase the amount of exhaust gas that can be treated. A typical dry industrial ESP is shown in Fig. 9.17 where internal baffles are used to direct the flow to the various electrode-collector sections and to achieve uniform flow distribution [38]. A schematic of a typical wet ESP, often referred to as a WESP, is shown in Fig. 9.18. Figure 9.19 shows a schematic of an actual industrial wet electrostatic precipitator. Figure 9.20 shows a detail of that wet ESP, and Figs. 9.21 and 9.22 show photograph of some industrial wet ESPs.

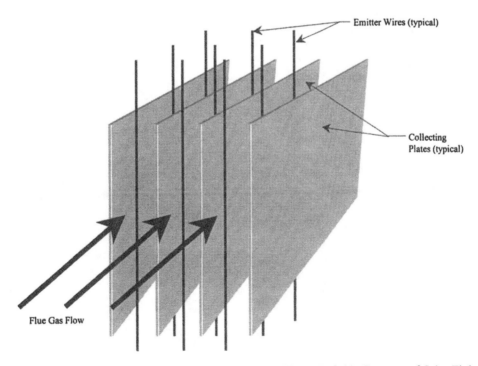

Figure 9.17 Typical dry electrostatic precipitator. (From Ref. 30. Courtesy of John Zink Co., Tulsa, OK.)

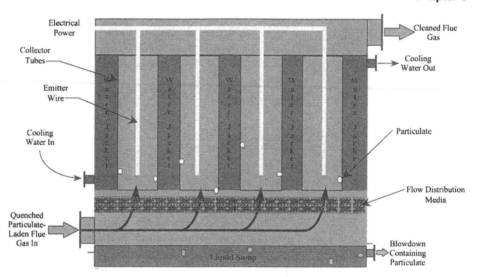

Figure 9.18 Typical wet electrostatic precipitator. (From Ref. 30. Courtesy of John Zink Co., Tulsa, OK.)

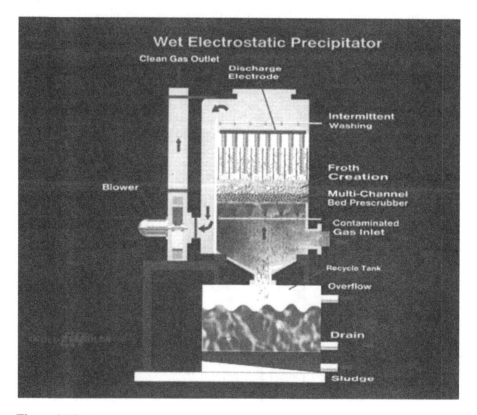

Figure 9.19 Schematic of an industrial wet electrostatic precipitator. (Courtesy of Croll, Westfield, NJ.)

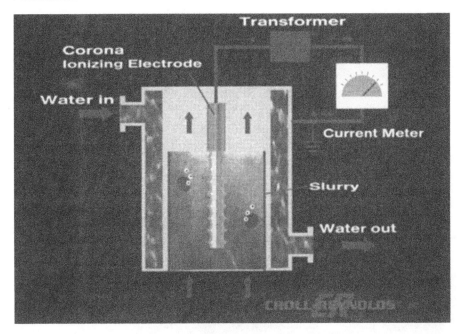

Figure 9.20 Detail of industrial wet electrostatic precipitator. (Courtesy of Croll.)

Stern et al. [39] list the following advantages of ESPs:

1. Can handle high gas temperatures, which makes them attractive in many industrial combustion processes.
2. They have extremely small pressure drops, which minimizes fan costs.
3. If properly operated, they can have very high collection efficiencies.
4. They can handle a wide range of particulate sizes and dust concentrations.
5. They can be more efficient to operate and maintain compared to other particulate collection systems if properly designed.

and the following disadvantages:

1. They often have the highest initial capital cost.
2. They may require a large amount of space.
3. They are not suitable for combustible particles.

Cooper and Alley [31] list the following advantages of ESPs:

1. Very high removal efficiencies, even for very small particles.
2. Can handle very large gas volumes with low pressure drop.
3. Can collect wet or dry materials.
4. Can operate over a wide range of gas temperatures.
5. Low operating costs, except at very high removal efficiencies.

and the following disadvantages:

1. High capital costs.
2. Cannot control gaseous emissions (e.g., NO_x).

Figure 9.21   Industrial wet electrostatic precipitator. (Courtesy of Croll.)

3. Somewhat inflexible to changes in operating conditions once installed.
4. Requires a significant amount of space.
5. May not work on particles with very high electrical resistivities.

Davis et al. [40] list the following advantages of ESPs:

1. Extremely high collection efficiencies (both coarse and fine particulates).
2. Dry collection and disposal.
3. Low pressure drop.
4. Designed for continuous operation with minimum maintenance.
5. Relatively low operating costs.
6. Can operate at high pressures (up to 150 psi) or in a vacuum.
7. Can operate at high temperatures (up to 1300°F).
8. Can handle relatively large flow rates.

Figure 9.22 Industrial wet electrostatic precipitator. (Courtesy of United McGill.)

and the following disadvantages:

1. High capital cost.
2. Very sensitive to fluctuations in gas stream conditions (e.g., flow rates, temperatures, particulate loading).
3. High or low resistivity particulates difficult to collect.
4. Relatively large space requirements.
5. Explosion hazards when treating combustible gases or particulates.
6. High voltage needs to be safeguarded for personnel protection.
7. Ozone is produced during gas ionization.
8. Sophisticated maintenance personnel required.

ESPs are effective for collecting particles of medium electrical resistivity, but are less effective or ineffective on particles that have higher or lower resistivities. The particles remain at the plate because of the continuous flow of electrons from the electrode to the collector plate. However, particles of lower resistivity have very little voltage gradient across the collected particles, which produces very little electrostatic force to keep the particles on the plate. These lower resistivity particles then easily fall off the plate and are re-entrained into the gas stream. High-resistivity particles have the opposite problem where most of the voltage drop is across the particles at the plate, thus reducing the strength of the corona discharge at the electrode, which is a function of voltage gradient. High-resistivity cakes formed on the collector can actually produce a back corona discharge because of the high-voltage gradient across the cake, which can cause minor gas explosions that drive the cake off the collector. Less than optimal electrical resistivities can be a problem for some coal and fly ashes.

Increasing the temperature of the particles can sometimes lower the electrical resistivity of particles that have higher resistivities at normal temperatures. Unfortunately, no such simple solution exists for particles of low electrical resistivity.

Another potential problem is called sneakage where some of the particle-laden gas gets through the ESP without being treated. It may have gotten past the collector plates in the gap between the plates and the walls of the ESP, it may have gone into the collection hopper, or it may have gotten around the collector some other way. Whatever the cause, this problem can be minimized or eliminated with proper baffling to ensure that all gases to be treated flow through the collection system. ESPs are now being designed using computational fluid dynamics (see Chap. 4) to optimize the flow characteristics of the system to increase the collection and operating efficiency. Physical modeling, usually at a reduced scale, is another tool used to improve the design of ESPs.

Flue-gas conditioning is sometimes used to improve the effectiveness of ESPs by [4]:

- Increasing the potential gradient between the collectors.
- Reducing particle re-entrainment by increasing the cohesive properties of the particles.
- Improving the agglomeration properties of the particles to coalesce small particles into larger ones that are easier to collect.
- Making the potential gradient more uniform so that higher voltages can be used, which increases collection efficiency.

The conditioning is done in a variety of ways. Cooling the flue gas with a water spray increases the effective gas density, decreases the volumetric flow rate, increases the water content of the gas, and reduces particle resistivity. This cooling increases collection efficiency and increases the potential field strength limit. It is important the water evaporates in the gas stream, otherwise water impacting the collector plates and ESP housing could cause corrosion or particle agglomeration in the wrong places. Adding SO_3 or sulfuric acid can lower particle resistivity; however, if done improperly can create other pollutant problems or corrosion. Ammonia and ammonium sulfate are other additives that can improve performance or create other problems if improperly applied.

Ray [41] believes that advanced electrostatic precipitators may be the control technology of choice for capturing fine particles in the light of the $PM_{2.5}$ limits proposed by the U.S. EPA. While fabric filters have very high removal efficiencies for smaller particles sizes, they are somewhat limited by the chemical and physical nature of the particles. Venturi scrubbers are more efficient for removing larger particles. Both dry and wet electrostatic precipitators have been used most often for many types of small particle removal applications.

9.6.4.4 Filters

Filters are a mechanical separation device used to remove particles from gas streams. The principle behind filters is simple—the holes in the filter are large enough for gas to pass through but too small for particles to go through (although as discussed below sometimes even the particles could pass through the holes); hence, the particles get trapped on the surface of the filter. The filter is composed of any suitable porous material including granular or fibrous materials. The filter material must be compatible with the gas and particles. Common filter fabrics include wool, cotton, nylon, glass fiber, and polyester. The filters may be arranged in deep beds, mats, or fabric, all having large void spaces. Donovan [42] has written an entire book on using filters to remove pollutants from combustion sources. He notes the difference between fabric and textile filters, including structural and material composition differences. The U.S. EPA has prepared a manual for use in estimating the cost of various air pollution control techniques, which includes a chapter on baghouses and filters for particulate emission control [43].

Fabric filters use a variety of mechanisms to collect particulates including sieving, interception, impingement, diffusion, gravitational settling, and electrostatic attraction [44]. Sieving is where large particles will not fit through small holes and are therefore trapped by the fabric. Interception is where particles flowing around cloth fibers are attracted to the fibers by van der Waals forces. Impingement or inertial impact occurs when particles directly impinge on a cloth fiber. Diffusion occurs when particles impact the filter by random Brownian motion. Electrostatic attraction occurs when charged particles are attracted to the cloth, which is oppositely charged.

Dullien classified filters into three categories: (1) woven or felted fabrics in the form of bags, envelopes, or sleeves, (2) fibrous filters fabricated in the form of mats or screens composed of fibers, filaments, or ribbons, and (3) bulk granular material in a packed column [45]. Jorgenson offered the following general categories of filter media: needle-punched felt, woven media, filter paper, plastic media, ceramic filter candles, and sintered metal [46]. A variety of surface treatments are used including singeing, glazing, sprayed-on coatings, membranes, multilayer construction, and abraded surfaces. There are a number of options for high-temperature filters including metallic fibers (up to 1000°F), graphite fibers (up to 600°F), quartz filters, and ceramic filters (up to 1600°F). Some of the important factors to consider in fabric selection include [26]:

- Dust penetration
- Continuous and maximum operating temperatures
- Chemical degradation
- Abrasion resistance
- Cake release
- Pressure drop
- Cost
- Cleaning method
- Fabric construction (woven or felted)

Three of the primary performance criteria for fabric filters include pressure loss, collection efficiency, and lifetime [47]. These are all related to technical feasibility and economics. For example, the higher the pressure loss the higher the operating costs and the larger the fan needed. The collection efficiency is important to ensure that particulate emissions are below the allowable limits. The efficiency is dependent on the particle size distribution and on the particulate loading. If higher efficiencies are required then more filters may be required, which also means more fan horsepower and higher capital and operating costs.

In industrial-scale dust collection, the filters are typically arranged and mounted in a housing referred to as a baghouse (see Fig. 9.23). Some of the basic design parameters to consider for a fabric filter baghouse include [4]:

- Cleaning mechanism (shake/deflate, reverse air, pulse jet)
- Size (air-to-cloth ratio, velocity)
- Pressure drop (fan power, vacuum/pressure rating)
- Fabric (material, weave)
- Bag life (cleaning frequency, gas composition, inlet design)

In some cases, while the holes in the filter may initially be larger than the particles, the particles get stuck in the sides of the holes forming bridges that effectively make the holes smaller, thus trapping more particles. In surface filters, the trapped particles form a cake on the entrance side of the filter, while in depth filters the particles are trapped inside the filter material. Gas is still permitted to flow through the cake although the pressure drop increases with the thickness of the cake. In surface filters, the cake is removed by taking the filter out of service (removing the flow of dirty gas or even putting a slight negative pressure on the filter) and then

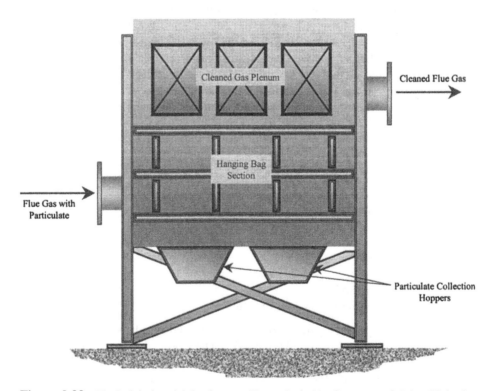

Figure 9.23 Typical industrial baghouse. (From Ref. 38. Courtesy of John Zink Co., Tulsa, OK.)

mechanically shaking the filter to dislodge the particles that fall down into a hopper where they can be removed and either disposed of or recycled back into the process if they are reusable. If the supply of exhaust gases is continuous, then multiple filters are needed so that at least one is always available while others may be getting cleaned. The removal efficiency of the filter varies continuously as the particle cake builds up and then is cleaned off, with the highest efficiency right after each cleaning. However, the filter system is designed so that the average removal efficiency is suitable for the application.

There are several types of filters that are used to remove particulates from exhaust streams. One type is a fabric filter, which typically has a high removal efficiency. Fabric filters are assembled into units referred to as baghouses (see Fig. 9.24) because the filters are often formed in the shape of a bag for easier mounting. An additional fan is usually required to pull the exhaust gases through the baghouse because of the high pressure drop caused by the filters, especially when they contain particulate matter. The filter bags are cleaned through mechanical-shaking, reverse-flow, or pulse-jet technologies. The pulse-jet type can be cleaned while in service while mechanical-shaking and reverse-flow types must be taken out of service to be cleaned.

A benefit of surface filter systems is that at low gas velocities, they are effective at capturing both large and small particles. The filter system effectiveness is not dependent on particle resistivity as in the case of ESPs. A problem with filter systems is that they can sometimes require a significant amount of maintenance, replacing clogged or damaged filter bags. Another major limitation of baghouses is the

Figure 9.24 Typical industrial baghouse. (Courtesy of Wheelabrator.)

temperature limit those of the fabric. These units can be fairly large and expensive. Ceramic filters, which have much higher temperature limits than those of fabric filters, are also used in some cases. Low-density ceramic filters, can have pressure drops comparable to those of fabric filters.

Depth filters, sometimes referred to as high-efficiency particle-arresting (HEPA) filters, trap particulates inside a porous matrix where they impinge on to and adhere to the material. These are most commonly used to trap sulfuric acid mist, which is not commonly encountered in typical industrial combustion applications. There are several problems with this type of filter in industrial service including the difficulty in cleaning and reusing.

There are several common techniques used to clean fabric filters. One is simply manual cleaning where the filter is removed from service and cleaned by hand. Another technique is referred to as shake and deflate where the structure holding the structures is mechanically shaken so that the particles fall to the bottom of the baghouse where they can be collected in one or more hoppers. This may require stopping the gas flow or even deflating the bags, which both increase the effectiveness of the shaking action to clean the filters. Another technique is to reverse the air flow so that the gas flows backward through the filters, which knocks the accumulated particles off into collection hoppers at the bottom of the baghouse. Again, some or all of the baghouse must be temporarily taken out of service during the cleaning. A third common technique is to use high-pressure short-duration pulse jets of compressed air that deforms the bags and dislodges the dust cake that has accumulated on the filters. This technique can be done without taking the filters out of service, which is a significant advantage compared to the first two techniques. However, the high-pressure jets can significantly shorten the life of the fabric filters. An alternative to the high-pressure pulse jets is to use low-pressure jets that distend the filters enough to cause the dust cake to fragment for removal. Another technique for cleaning the fabric filters is to use a sonic or acoustic horn that creates a low-frequency vibration that loosens the filter cake and promotes cleaning. Another cleaning alternative is where the fabric is mounted on a cartridge or piston that periodically moves up and down to shake off the accumulated particles.

The basic problem with fabric filters is that they must be periodically cleaned, which sometimes means being taken out of service. In some cases, the cleaning cycle may be only a few seconds, while in the worst case, the filter must be completely replaced. Between cleanings, there are two competing effects that impact the efficiency of the filtration system. The removal efficiency actually increases because the filter cake catches more particles as it builds up. On the other hand, particle re-entrainment may increase with the thickness of the filter cake if the particles do not agglomerate sufficiently and fall off the filter to be carried through. Another problem is that the system is inherently in an unsteady state. As the filters get dirty, more energy is needed to pass a certain flow rate of gas through them. Right after cleaning, the least amount of energy in the cycle is required. Another problem is the possibility of a dust explosion if any sparks occur in the vicinity of the dust-laden filters.

Some advantages of fabric filters include [31]:

1. Very high collection efficiencies even for very small particles.
2. Operable on a wide variety of particle types.
3. Modular design minimizes costs.
4. Operable over an extremely wide range of volumetric flow rates.
5. Relatively low pressure drops.

and some disadvantages include:

1. Large floor area requirement.
2. Temperature and corrosion limitations.
3. Can become clogged in moist environments.
4. Potential for fire or explosion depending on particle composition.

Davis et al. [40] list the following advantages of fabric filters:

1. Extremely high collection efficiency on both coarse and fine particulates.
2. Relatively insensitive to gas stream fluctuations.
3. Filter outlet air can be recirculated within the plant for energy conservation.
4. Collected material is dry for easy subsequent handling and treatment.
5. No wastewater produced that needs to be disposed of or that could freeze.
6. Corrosion not an issue.
7. No high-voltage hazards.
8. Selected fibrous or granular filter aids permits high-efficiency collection of submicrometer smokes and gaseous contaminants.
9. Many filter configurations available to fit wide range of specifications.
10. Simple operation.

and the following disadvantages:

1. Lower temperature limits for standard fabrics.
2. Certain dusts may require special fabric treatments.
3. Concentrations of some dusts can present a fire or explosion hazard.
4. Relatively high maintenance requirements.
5. High gas temperatures with acid or alkaline gases or particulates can shorten filter life.
6. Moisture or tar in the collected particles can require special cleaning procedures.
7. Replacing fabrics usually requires respiratory protection.
8. Medium-pressure drop requirements.

9.6.4.5 Absorption Scrubbers

Absorption scrubbers are sometimes referred to as packed-bed or packed-tower wet scrubbers. Absorption is a mass transfer process where a gas or solid phase is transferred into a liquid phase. This is shown schematically in Fig. 9.25 [48]. The absorbent is the liquid phase, usually water, designed to absorb the contaminant gas or solid phase. The absorbate or solute is the contaminant being absorbed.

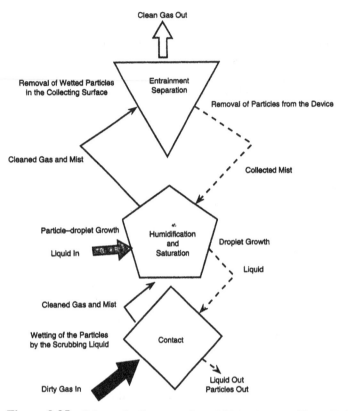

Figure 9.25 Schematic diagram of scrubbing process. (From Ref. 48. Courtesy of CRC Press.)

The carrier gas is the gas stream carrying the contaminants. Note that, while absorption scrubbers are capable of collecting both solid and gaseous contaminants, a single absorber design is not efficient at collecting both types. A key in designing absorption systems is to optimize the mass transfer by maximizing the contact area between the absorbent and the absorbate, providing good mixing, allowing sufficient residence time, and ensuring that the absorbate is highly soluble in the absorbent. Table 9.4 shows some typical applications for wet scrubbers [49]. Schifftner and Hesketh [50] have written an entire book on wet scrubbers that includes theory, applications to a wide range of problems, maintenance and control, and testing.

Scrubbers are often defined as devices used to collect particles by contacting the exhaust gas stream to be cleaned with liquid drops, usually water. The principal factor determining scrubber performance is the solubility of the pollutants in the absorbing liquid [51]. These drops strike the particles and carry them out of the gas stream. For example, a cyclone may not be effective for capturing smaller particles but if those particles are contacted by larger droplets, thus effectively making the particles larger, then the cyclone can be used to collect these larger particles that were previously difficult to collect because of their small size. The higher the velocity of the gas relative to the liquid, the more

Table 9.4 Typical Applications for Wet Scrubbers

Source of gas	Contaminants
Iron and steel industry	
Gray iron cupola	Iron, coke, silica dust
Oxygen steel converter	Iron oxide
Steel open-hearth furnace (scrap)	Iron and zinc oxide
Steel open-hearth furnace (oxygen lanced)	Iron oxide
Blast furnace (iron)	Iron ore and coke dust
Electric furnace	Ferromanganese fume
Electric furnace	Ferrosilicon dust
Rotary kiln—iron reduction	Iron, carbon
Crushing and screening	Taconite iron ore dust
Scarfer	Iron oxide fume
Chemical industry	
Acid—humidified SO_2, scrub with water scrub with 40% acid	H_2SO_4 mist
Acid concentrator	H_2SO_4 mist
Copper roasting kiln	H_2SO_4 mist
Chlorosulfonic acid plant	H_2SO_4 mist
Dry ice plant	Amine fog
Wood distillation plant	Tar and acetic acid
$TiCl_4$ plant, TiO_2 dryer	TiO_2–HCl fumes
Spray dryers	Detergents, fume, and odor
Flash dryer	Furfural dust
Phosphoric acid plant	H_3PO_4 mist
Phthalic anhydride	Benzoic, maleic, phthalic anhydride
Nonferrous metals industry	
Blast furnace (secondary: lead)	Lead compounds
Reverberatory lead furnace	Lead and tin compounds
Ajax furnace—aluminum alloy	Aluminum chloride
Zinc sintering	Zinc and lead oxide dusts
Reverberatory brass furnace	Zinc oxide fume
Aluminum pot lines	Fluorides
Briquetting processes	Nonferrous dusts
Mineral products industry	
Lime kiln	Lime dust, soda fume
Asphalt stone dryer	Limestone and rock dust
Cement kiln	Cement dust
Coal dryers	Coal dust
Petroleum industry	
Catalytic reformer	Catalyst dust
Acid concentrator	H_2SO_4 mist
FCC catalyst regenerator	Oil fumes

(*continued*)

Table 9.4 Continued

Source of gas	Contaminants
Fertilizer industry	
Fertilizer dryer	Ammonium chloride fumes, dust
Superphosphate den and mixer	Fluorine compounds
Phosphate rock dryer	Phosphate rock dust
Curing building ventilation	SiF_4
Pulp and paper industry	
Lime kiln	Lime dust, soda fume
Black liquor recovery boiler	Salt cake
Dissolving stack	Smelt and green liquor
Power/bark boiler	Fly ash, SO_2

Source: Ref. 49.

effective the scrubber at collecting smaller particles. The liquid droplets increase the size of the particles, which makes it easier to remove them. The particulates attach themselves to larger liquid droplets so that they can be collected and removed.

There are actually several sections in a typical scrubber. The first is sometimes referred to as the gas–liquid contactor where the dirty gas flows through a continuous stream of liquid droplets, usually water. The mixture of gas, clean droplets, and droplets encapsulating particles to be removed is then taken through a gas–liquid separator, often a cyclone, where the clean gas stream goes out of the top while the dirty liquid is collected at the bottom due to gravity. If possible, the dirty liquid can be used somewhere else in the plant. If not, the dirty liquid then flows through a liquid–solid separator of some type so that the solid can be disposed of or recycled if possible, and the clean liquid is recycled back into the scrubber to the gas–liquid contactor. The liquid atomization system including the spray nozzles is a critical element in the overall design of a scrubber. A schematic of a typical scrubbing system is shown in Fig. 9.25.

There are several variations for the flow geometry of the dirty gas stream relative to the liquid droplets in the gas–liquid contactor. In a crossflow scrubber, the gases flow perpendicular to the liquid droplets where the gas typically flows horizontally, while the droplets flow vertically downward due to gravity. The problem with crossflow scrubbers is that the contact time with the gas is often small and too many particles can get through without being contacted by a liquid droplet.

In a counterflow scrubber, the liquid droplets again flow vertically downward but the dirty gas flows vertically upward. The optimum collection efficiency for counterflow scrubbers occurs when the downward liquid velocity is equal to the upward dirty-gas velocity. This maximizes the contact time as the particle appears to be hovering in the scrubber. Unfortunately, equal velocities cause the scrubber to become flooded with liquid unless the gas velocity is higher than the liquid velocity.

For that reason, counterflow scrubbers are not commonly used in industrial applications.

A third type of geometry is referred to as a coflow scrubber where both the liquid droplets and the dirty gas stream enter at the same location and exit at the same location. This configuration allows much higher gas velocities and low liquid droplet velocities. A specific design for a coflow scrubber is a venturi scrubber where the gas enters a venturi where the liquid is injected into the gas stream. This design uses less fan power and thus uses less energy to operate. In a cyclone scrubber, the dirty gas is swirled around a cyclone while liquid is sprayed toward the walls of the cyclone to knock the particles out of the gas stream.

There are also several types of scrubbers, depending on the design. These include spray chamber scrubbers, cyclonic scrubbers sometimes referred to as wet cyclones, and venturi scrubbers (see Fig. 9.26). In the spray scrubber, water or some other liquid is atomized and sprayed into the exhaust gas stream being treated. The fine mist impinges on the particles to be removed, causing them to fall to the bottom of the scrubber for collection. There must be sufficient spray coverage and residence time to ensure that no particles go through the scrubber without being impinged on by the liquid spray. Sometimes these spray dryers have baffles to divert the gas towards the sprays and increase the residence time in the chamber to increase the removal efficiency.

Cyclonic scrubbers use a variation of the spray scrubbers where the spray nozzles are arranged in a ring and spray toward the cyclone wall. The exhaust gas stream is swirled in the cyclone, causing the particles to go toward the wall. These cyclones are as effective on small particles because their mass is not sufficient to cause them to fall to the bottom of the cyclone. By adding the liquid spray, the liquid impinges on the dust particles, which makes them heavy enough to fall to the bottom of the cyclone for collection.

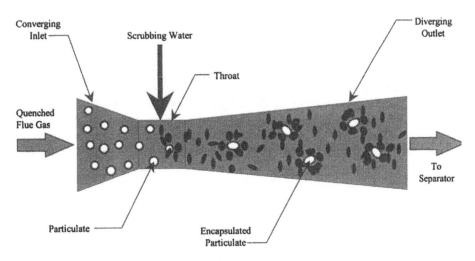

Figure 9.26 Venturi scrubber. (From Ref. 38. Courtesy of John Zink Co., Tulsa, OK.)

Table 9.5   Major Types of Wet Scrubbers

Category	Particle capture mechanism	Liquid collection mechanism	Types of scrubbers
Preformed spray	Inertial impaction	Droplets	Spray towers Cyclonic spray towers Vane-type cyclonic towers Multiple-tube cyclones
Packed-bed scrubbers	Inertial impaction	Sheets, droplets (moving bed scrubbers)	Standard packed-bed scrubbers Fiber-bed scrubbers Moving-bed scrubbers Cross-flow scrubbers Grid-packed scrubbers
Tray-type scrubbers	Inertial impaction Diffusional impaction	Droplets, jets, and sheets	Perforated-plate Impingement-plate scrubbers Horizontal impingement-plate (baffle) scrubbers
Mechanically aided scrubbers	Inertial interception	Droplets and sheets	Wet fans Disintegrator scrubbers
Venturi and orifice scrubbers (gas atomized scrubbers)	Inertial impaction Diffusional impaction	Droplets	Standard venturi scrubbers Variable throat venturi scrubbers; flooded disk, plumb bob, movable blade, radial flow, variable rod Orifice scrubbers

Source: Ref. 26.

In the venturi scrubber, the exhaust gas stream containing the particles is accelerated through a converging–diverging section. Liquid is injected either in the throat or just before the throat entrance. The high–velocity gas flow through the throat helps to atomize the liquid to yield better droplet distribution. The droplets impact the particles, which are then collected at the bottom of the scrubber. Table 9.5 shows a comparison of the major types of wet scrubbers. Figure 9.27 shows a typical industrial scrubbing system.

One of the problems with wet scrubbers is how to dispose of the wet sludge produced by the impingement of the liquid on the solid particles that are then collected. While in some cases the sludge may be preferred to fine dry dust particles, in other cases the water may create other problems such as freezing, where ambient temperatures are below freezing, and corrosion of the metal components of the scrubber that may need to be made out of a noncorrosive material such as stainless steel. Another problem is that higher power may be required due to the need to spray water and then collect the heavier water-laden particles.

Cooper and Alley [31] list the following advantages of scrubbers:

1. Low risk while handling flammable or explosive dusts
2. Gas absorption and dust collection in one unit

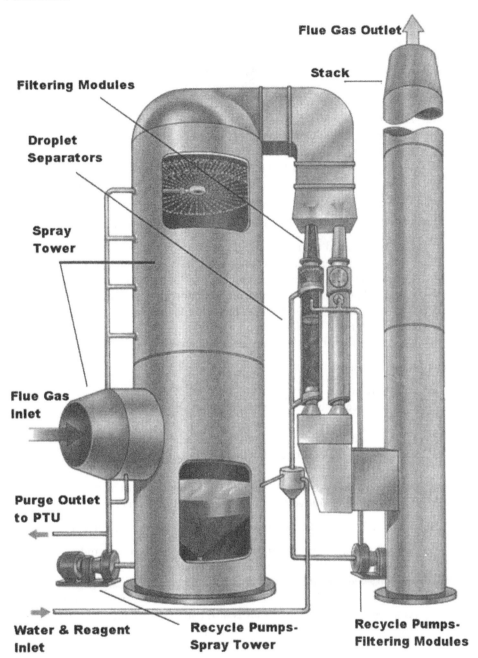

Figure 9.27   Typical industrial scrubber. (Courtesy of Belco, Parsippany, NJ.)

3. Can handle mists
4. Provides hot gas cooling
5. Variable collection efficiency
6. Can neutralize corrosive gases and dusts

and the following disadvantages:

1. High potential for corrosion problems
2. Effluent can be a water-treatment problem
3. Potential freezing problems
4. Cleaned gas may need to be reheated to avoid a visible plume
5. Collected particles may be contaminated and therefore not recyclable
6. Waste sludge disposal can be expensive.

Mycock et al. [49] list the following advantages of wet scrubbers:

1. Can be designed to fit into small spaces.
2. No re-entrainment problems once particles have been collected in hopper.
3. Can collect both particulates and gaseous pollutants.
4. Handles high-temperature and high-humidity gas streams.
5. Can humidify a gas stream, which can make downstream gas moving equipment more efficient.
6. Minimizes fire and explosion hazards associated with dry dusts.

and the following disadvantages:

1. Can have corrosion problems.
2. Can produce an undesirable visible plume by humidifying the exhaust gas stream.
3. It is more difficult to recover reusable materials that have been captured by the scrubber and formed into a sludge containing the dust particles.
4. High collection efficiencies require high power requirements.
5. The by-product sludge can be a water-treatment issue.

Heumann and Subramania [52] list the following advantages of wet scrubbers:

1. High-temperature gases can be simultaneously quenched and scrubbed.
2. The capital cost of scrubbers is often less than that of comparable baghouses and ESPs.
3. Both gaseous and solid (particulates) pollutants can be simultaneously removed in a single device that is also capable of reducing odors.
4. Combustible and explosive materials are often more safely removed with scrubbers compared to filters and ESPs.
5. Sticky and/or hygroscopic materials can be handled.
6. Scrubbers are often more compact compared to dry systems like baghouses.

and the following disadvantages:

1. Treating the liquid and solid wastes produced in the scrubber may be expensive.

2. Energy consumption may be higher than that of other comparable methods.
3. Detailed specifications, including aerodynamic particle size distribution, is necessary for optimum system design and performance.
4. Corrosion, which is not a problem for dry collection techniques, can be an issue under certain circumstances.
5. The liquid spray systems are subject to freezing in cold temperatures.
6. The exhaust gases can be saturated with water, which may produce a steam plume and/or condensation in the exhaust stack; these are both usually undesirable.

Davis et al. [40] note the following advantages of wet scrubbers:

1. No secondary dust sources
2. Relatively small space requirements
3. Ability to collect both gases and particulate pollutants
4. Can handle high-temperature and high-humidity gases
5. Low capital cost (if wastewater treatment system not required)
6. For high-pressure gas streams, pressure drop is not significant
7. Can have high collection efficiencies on fine particulates

and the following disadvantages:

1. May create water disposal problems
2. Product is collected wet
3. Corrosion may be a problem
4. Steam plume opacity may be an issue
5. Pressure drop and horsepower requirements may be high
6. Solids buildup at the wet–dry interface may be a problem
7. Relatively high maintenance costs

9.6.4.6 Collection Recommendations

Wark and Warner [8] suggest the following guidelines for choosing the appropriate particulate collection system:

1. Use cyclones when the dust is coarse, the concentrations are fairly high, classification of the particles is required, and/or very high efficiency is not required.
2. Use wet scrubbers when fine particles need to be removed efficiently, cooling is desirable and moisture is not objectionable, gases are combustible, and/or gaseous pollutants also need to be removed.
3. Use fabric filters when very high removal efficiencies are required, valuable material needs to be collected dry, the gas is always above its dew point, the gas volumes are relatively low, and/or the gas temperature is relatively low.
4. Use electrostatic precipitators when very high efficiencies are required for removing fine particles, very large gas volumes are being treated, and/or valuable particulate materials need to be recovered.

de Nevers [28] gives the following guidelines for selecting the appropriate type of collection device for removing particles from gas streams:

1. If the gas stream to be cleaned is relatively small or only occasional, a throwaway filter may be the best option.
2. If the particles are sticky and easily agglomerate as in the case of tars, a throwaway device or a device using a liquid, such as in a scrubber, may be the best alternative to keep the sticky particles from becoming a maintenance headache.
3. Particles that adhere well to each other but not to surfaces are generally easy to collect. However, particles that do not adhere well to each other but do adhere well to typical surfaces may require the surfaces to be treated with some type of nonstick material such as Teflon to make particle removal easier.
4. The electrical properties of the particles are particularly important in devices like ESPs that work best on particles having intermediate electrical resistivities.
5. Cyclone separators are the best option for larger particles (over about 5 μm) that are nonsticky.
6. For smaller particles (under about 5 μm), ESPs, filters, and scrubbers are usually the preferred separating devices.
7. Scrubbers can be very expensive to operate for large gas flows due to the high pumping costs.
8. Corrosion resistance and acid dew point must be considered in collecting certain types of particles.

Stern et al. [39] compared a variety of particulate removal systems, as shown in Table 9.6. Different techniques are preferred for different particle sizes, removal efficiencies, and exhaust gas temperatures. Table 9.7 shows a particulate removal selection guide prepared by the U.K. Environment Agency.

Table 9.6 Comparison of Particulate Removal Systems

Collector type	Particle size range (μm)			Space requirement	Max. temp. (°C)
	0.1–1.0	1.0–10.0	10.0–50.0		
Baghouse (cotton bags)	Fair	Good	Excellent	Large	80
Baghouse (Dacron, nylon, orlon)	Fair	Good	Excellent	Large	120
Baghouse (glass fiber)	Fair	Good	Good	Large	290
Baghouse (Teflon)	Fair	Good	Excellent	Large	260
Electrostatic precipitator	Excellent	Excellent	Good	Large	400
Standard cyclone	Poor	Poor	Good	Large	400
High-efficiency cyclone	Poor	Fair	Good	Moderate	400
Spray tower	Fair	Good	Good	Large	540
Impingement scrubber	Fair	Good	Good	Moderate	540
Venturi scrubber	Good	Excellent	Excellent	Small	540
Dry scrubber	Fair	Good	Good	Large	500

Adapted from Ref. 39.

Table 9.7 Particulate Removal Selection Guide

Particle Size range (µm)	Type of collector	Pressure drop (mbar)	Collection efficiency (%)				Maximum temperature (°C)	Condition of effluent	Dew-point sensitivity	Effect of particle density
			at 5 µm	at 2 µm	at 1 µm					
> 150	Settling chamber	< 2.5		150 µm–50% > 300 µm–95%			500	Dry	Fairly sensitive	Efficiency increases with density
> 10	Cyclones	2.5–10		5 µm–40% 20 µm–80% > 50 µm–97%			500	Dry	Sensitive to plugging and corrosion	Efficiency increases with density
> 5	Irrigated cyclone	5–20		5 µm–50% 20 µm–95% > 50 µm–95%			350	Wet	Not sensitive	Efficiency increases with density
> 5	Spray tower	5–17.5	94	87	55		350	Wet	Not sensitive	Little effect
> 3	Impingement scrubber	15–30	97	95	80		350	Wet	Not sensitive	Little effect
> 0.1	Ceramic filter	15–20	99.9	99.5	99		900	Dry	Sensitive	Little effect
> 0.5	High pressure venturi scrubber	35–90	99.8	99	97		350	Wet	Not sensitive	Little effect
> 1	Medium pressure drop wet scrubber	30–100	97	95	80		350	Wet	Not sensitive	Little effect
> 0.1	Fabric bag filter	8–18	99.8	99.5	99		220 (with special material)	Dry	Very critical	Little effect
> 0.5	Wet disintegrator	10–20	98	95	91		350	Wet	Not sensitive	Little effect
> 0.01	Dry electrostatic precipitator	0.6–2	99	95	86		375	Dry	Critical	Little effect
> 0.01	Wet electrostatic precipitator	0.6–4	98	97	92		375	Wet	Not sensitive	Little effect

Source: Ref. 15.

REFERENCES

1. Office of the Federal Register. 60.2 Definitions. U.S. Code of Federal Regulations. Title 40, Part 60. Washington, DC: U.S. Government Printing Office, 2001.

2. SW Lee, B Kan, R Pomalis. Characterization of Fine Particulates Formation in Oil and Coal Combustion. *Fifth International Conference on Technologies and Combustion for a Clean Environment*, Lisbon, Portugal. Vol. 2, The Combustion Institute—Portuguese Section, pp. 1167–1174, 1999.

3. A Violi, A D'Anna, A D'Alessio. A Kinetic Model of Particulate Carbon Formation in Rich Premixed Flames of Ethylene. *Fifth International Conference on Technologies and Combustion for a Clean Environment*, Lisbon, Portugal. Vol. 2, The Combustion Institute—Portuguese Section, pp. 1185–1191, 1999.

4. KB Schnelle, CA Brown. *Air Pollution Control Technology Handbook*. Boca Raton, FL: CRC Press, 2002.

5. HE Hesketh. *Fine Particles in Gaseous Media*, 2nd edn. Chelsea, MI: Lewis Publishers, 1986.

6. SK Friedlander. *Smoke, Dust, and Haze*, 2nd edn. New York: Oxford University Press, 2000.

7. I Ray. Particulate emissions: evaluating removal methods. *Chem. Eng.*, Vol. 104, (No. 6), pp.135–141, 1997.

8. K Wark, CF Warner. *Air Pollution: Its Origin and Control*, 2nd edn. New York: Harper & Row, 1981.

9. F Ammouri, B Labegorre, JM Samaniego, A Coppalle, M Talbaut-Haudiquert. Soot Effects on Heat Transfer and NO_x Formation in Industrial Scale Furnace—Experimental Study on Turbulent Diffusion Oxygen Gas Flames. *Fifth International Conference on Technologies and Combustion for a Clean Environment*, Lisbon, Portugal, Vol. 2. The Combustion Institute—Portuguese Section, pp. 1097–1100, 1999.

10. WP Linak, CA Miller, JOL Wendt. Fine Particulate Emissions from Residual Fuel Oil Combustion: Characterization and Mechanisms of Formation. *Proceedings of the Combustion Institute*, Pittsburgh, PA, Vol. 28, pp. 2651–2658, 2000.

11. U.S. EPA. Air Quality Guide for Particulate Matter. EPA draft report. Washington, DC, 2001.

12. J Elkins, N Frank, J Hemby, D Mintz, J Szykman, A Rush, T Fitz-Simons, T Rao, R Thompson, E Wildermann, G Lear. National Air Quality and Emissions Trends Report, 1999. Washington, DC: U.S. Environmental Protection Agency, Rep. EPA 454/R-01-004, 2001.

13. U.S. EPA, Method 5: Determination of Particulate Matter Emissions from Stationary Sources. Code of Federal Regulations 40, Part 60, Appendix A-4. Washington, DC: U.S. Environmental Protection Agency, 2001.

14. E Eiseman. *Monitoring for Fine Particulate Matter*. Washington, DC: Rand, 1998.

15. U.K. Environment Agency. Iron and Steel Making Processes. IP Guidance Note S2 2.01. London, 1999.

16. U.S. EPA. Method 9: Visual Determination of the Opacity of Emissions from Stationary Sources. Code of Federal Regulations 40, Part 60, Appendix A-4. Washington, DC: U.S. Environmental Protection Agency, 2001.

17. J Grabowski. The Evaluation of a Backscattering Lidar for Measurements of Air Pollution Concentration Profiles and Particulate Emissions from Single Stacks—Computer Simulations. In N Piccinini, R Delorenzo (eds.). *Chemical Industry and Environment II*, Vol 1. Turin, Italy: Politecnico Di Torino, 1996.

18. U.S. Bureau of Mines. Ringelmann Smoke Chart. Information Circular 8333. Washington, DC: U.S. Department of the Interior, 1967.

19. U.S. EPA. Control Techniques for Particulate Emissions from Stationary Sources—Vol. 2. Rep. EPA-450/3-81-005b. Washington, DC: U.S. Environmental Protection Agency, Sept. 1982.

20. CE Baukal. *Oxygen-Enhanced Combustion*. Boca Raton, FL: CRC Press, 1998.

21. CR Griffith. PCB and PCP Destruction Using Oxygen in Mobile Incinerators. *Proceedings of 1990 Incineration Conference*, San Diego, CA, May 14–18, 1990.

22. FJ Romano, BM McLeod. The Use of Oxygen to Reduce Particulate Emissions Without Reducing Throughput. *Proceedings of 1995 International Incineration Conference*, p. 637. Bellevue, WA, May 1995.

23. CE Baukal. Basic Principles. In CE Baukal (ed.). *Oxygen Enhanced Combustion*, Boca Raton, FL: CRC Press, 1998.

24. M Ferrara. Fuel Oil Additives Improve Combustion and Reduce Emissions. *Fifth International Conference on Technologies and Combustion for a Clean Environment*, Lisbon, Portugal, Vol. 1. The Combustion Institute—Portuguese Section, pp. 425–433, 1999.

25. RV Filho, FDA de Sousa, LS Messias, MM dos Santos. Experimental Investigation of Efficacy of Fuel Oil Additives on the Emission of Atmospheric Pollutants from Burning Ultra-Viscous Oils. *Fifth International Conference on Technologies and Combustion for a Clean Environment*, Lisbon, Portugal, Vol. 1. The Combustion Institute—Portuguese Section, pp. 443–452, 1999.

26. U.S. EPA. Control Techniques for Particulate Emissions from Stationary Sources—Vol. 1. Rep. EPA-450/3-81-005a. Washington, DC: Environmental Protection Agency, 1982.

27. G Elliot, A Startin. Controlling particulate emissions: reducing pollution using ceramic filters. *Chem. Process.*, Vol. 64, No. 9: 69–70, 2001.

28. N de Nevers. Control of Primary Particulates. Chap. 9 in *Air Pollution Control Engineering*. New York: McGraw-Hill, 2000.

29. JJ Peirce, RF Weiner, PA Vesilind. *Environmental Pollution and Control*, 4th edn. Boston, MA: Butterworth-Heinemann, 1998.

30. MV Bhatia, PN Cheremisinoff. Cyclones. In PN Cheremisinoff (ed.). *Air Pollution Control and Design for Industry*. New York: Marcel Dekker, 1993.

31. CD Cooper, FC Alley. *Air Pollution Control: A Design Approach*, 2nd edn. Prospect Heights, IL: Waveland Press, 1994.

32. WL Heumann. Cyclones, Chap. 8 in *Industrial Air Control Systems*, ed. WL Heumann. New York: McGraw-Hill, 1997.

33. KJ Caplan. 3 Source Control by Centrifugal Force and Gravity. In AC Stern (ed.). *Air Pollution*, 3rd edn. Vol. 4. New York: Academic Press, 1977.

34. JH Turner, PA Lawless, T Yamamoto, DW Coy, JD McKenna, GP Greiner, JC Mycock, WM Vatavuk. Sec. 6: Particulate Matter Controls. Chap. 3: Electrostatic Precipitators. In DC Mussatti (ed.). *EPA Air Pollution Control Cost Manual*, 6th edn. Rep. EPA/452/B-02-001. Washington, DC: U.S. Environmental Protection Agency, Jan. 2002.

35. MJ Freeman, PN Cheremisinoff, RW Ziminski. Electrostatics and Electrostatic Precipitation. In PN Cheremisinoff (ed.). *Air Pollution Control and Design for Industry*. New York: Marcel Dekker, 1993.

36. SG Dunkle. Electrostatic Precipitators. Chap. 11 in *Industrial Air Pollution Control Systems*, ed. HL Heumann. New York: McGraw-Hill, 1997.

37. R Altman, W Buckley, I Ray. Wet electrostatic precipitation demonstrating promise for fine particulate control—Part 1. *Power Eng.*, Vol. 105, No. 1, pp. 37–39, 2001.

38. P Melton, K Graham. Thermal Oxidizers. In CE Baukal (ed.). *The John Zink Combustion Handbook*, Boca Raton, FL: CRC Press, 2001.

39. AC Stern, RW Boubel, DB Turner, DL Fox. *Fundamentals of Air Pollution*, 2nd edn. Orlando, FL: Academic Press, 1984.

40. WT Davis, AJ Buonicore, L Theodore. Air Pollution Control Engineering. In WT Davis (ed.). *Air Pollution Engineering Manual*, New York: John Wiley, 2000.
41. I Ray. The quest for a better submicron particle trap. *Environ. Technol.*, Vol. 7, No. 3, pp. 22–26, 1997.
42. RP Donovan. *Fabric Filtration for Combustion Sources*. New York: Marcel Dekker, 1985.
43. JH Turner, JD McKenna, JC Mycock, AB Nunn, WM Vatavuk. Sec. 6: Particulate Matter Controls. Chap. 1: Baghouses and Filters. In DC Mussatti (ed.). *EPA Air Pollution Control Cost Manual*, 6th edn. Rep. EPA/452/B-02-001. Washington, DC: U.S. Environmental Protection Agency, Jan. 2002.
44. MG Kennedy, PN Cheremisinoff, L Bergmann. Fabric Filters. In PN Cheremisinoff (ed.). *Air Pollution Control and Design for Industry*, New York: Marcel Dekker, 1993.
45. FAL Dullien. *Introduction to Industrial Gas Cleaning*. San Diego, CA: Academic Press, 1989.
46. RL Jorgenson. Media Filtration, Chap. 9 in *Industrial Air Pollution Control Systems*, ed. HL Heumann. New York: McGraw-Hill, 1997.
47. K Iinoya, C Orr. 4 Filtration. In AC Stern (ed.) *Air Pollution*, 3rd edn. Vol. 4. New York: Academic Press, 1977.
48. PC Reist. Particulate Controls: Wet Collectors. In DHF Liu, BG Liptak (eds.). *Environmental Engineers' Handbook*, 2nd edn. Boca Raton, FL: Lewis Publishers, 1997.
49. JC Mycock, JD McKenna, L Theodore. *Handbook of Air Pollution Control Engineering and Technology*, Boca Raton, FL: Lewis Publishers, 1995.
50. KC Schifftner, HE Hesketh. *Wet Scrubbers*, 2nd edn. Lancaster, PA: Technomic Publishing, 1996.
51. RJ Heinsohn. Absorption. In WT Davis (ed.). *Air Pollution Engineering Manual*, New York: John Wiley, 2000.
52. WL Heumann, V Subramania. Chap. 10: Particle Scrubbing, In HL Heumann (ed.). *Industrial Air Pollution Control Systems*, New York: McGraw-Hill, 1997.

10

Noise and Vibration

10.1 INTRODUCTION

Combustion systems have the potential to be very noisy. The noise may come from a variety of sources. The turbulent combustion process itself may generate a loud roaring sound. The high-velocity jets of fuel and oxidizer may cause jet noise. The fluid flow through the piping system usually makes some noise. Most combustion systems have fans and blowers that also generate noise. The combustion system may also cause vibrations. While in many cases, these vibrations may be very small and nearly undetectable without the aid of sophisticated sensors, there are some occasions where the vibration may be significant. This vibration may be caused by flame instability or by some acoustic resonance between the combustion process and the combustion chamber.

Noise and vibration are considered to be "pollutants" in the broad sense of the word because they disturb the environment, particularly for the workers in the vicinity. Vibration may also damage equipment in the surrounding area and has the potential for catastrophic damage if the vibration persists over a long enough time without being mitigated or eliminated. Both noise and vibration from combustion processes have received relatively little attention compared to other pollutants like NO_x, SO_x, and particulates. For both noise and vibration in industrial combustion systems, further research is recommended to understand what causes them and how to control them.

There are a number of good general references on the subjects of noise and vibration. Barber [1] has edited the sixth edition of a comprehensive handbook on noise and vibration control. While it has sections on engine and gas turbine noise, there are no sections specifically dedicated to industrial combustion noise and vibration. However, it has much useful information for the interested reader. Liu and Roberts [2] have written a useful chapter on noise pollution in an environmental engineering handbook. While the material does not specifically address combustion-generated noise, it does cover a wide range of issues including characterizing noise sources, reactions to noise, acoustic trauma, psychological effects of noise, noise assessment, and noise control among other things. Harris [3] has edited a very large handbook on acoustical measurements and noise control, which includes the subject of vibration. A number of books are available on the general subject of noise [4–13] and the subjects of noise and vibration [14,15].

10.2 NOISE

Sound is a physical disturbance, measured in a frequency unit, known as hertz (Hz), that can be detected by the human ear, which is normally capable of hearing from approximately 20 Hz to 20 kHz. The human ear is most sensitive to sound between 2 and 5 kHz and is less sensitive at higher and lower frequencies. Frequencies that are too low to be heard by humans are referred to as infrasound while those too high to be heard are referred to as ultrasound. Table 10.1 lists the 10 octave bands that cover the human range of hearing and the center frequencies that represent each octave band. Each octave band extends over seven fundamental musical notes. Acoustics is the science of sound, including its production, transmission, and effects [16].

Sound is to be distinguished from noise, which is often thought of as disagreeable or unwanted sound [17]. Figure 10.1 shows a comparison of the thresholds

Table 10.1 Ten Octave Bands for Human Hearing

Octave band (Hz)	"Center" frequency (Hz)
22–44	31.5
44–88	63
88–177	125
177–355	250
355–710	500
710–1420	1000
1420–2840	2000
2840–5680	4000
5680–11,360	8000
11,360–22,720	16,000

Source: Ref. 42.

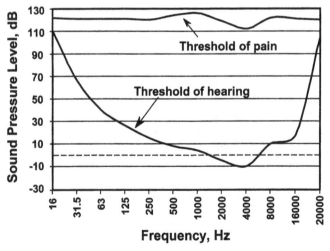

Figure 10.1 Threshold of hearing compared to the threshold of pain in humans. (From Ref. 42. Courtesy of John Zink Co., Tulsa, OK.)

of hearing and pain for humans as a function of frequency and sound pressure level. These two curves nearly intersect at the lowest and highest frequencies. The threshold of pain is relatively flat at about 120 dB. Noise is a common agent in most industrial facilities that can often be mitigated if addressed in the design phase [18]. It is a very complex phenomenon that is not fully understood, but has the potential to cause permanent hearing loss to personnel working around noisy equipment if adequate hearing protection is not used. Reed [19] has written a book on furnace operations in the petrochemical industry, which starts out with a chapter on noise. Interestingly, there is very little discussion of NO_x in that book as there was much less emphasis at that time on that pollutant. Harris [20] defines noise control, separate and distinct from noise reduction, as

> "the technology of obtaining an acceptable noise environment, consistent with economic and operational considerations."

The main source of noise from combustion processes is typically combustion roar, which is considered in this section. Other sources, also considered briefly here, may play important roles, depending on the specific installation.

10.2.1 Basics

Noise is the transfer of energy without transfer of mass [21].

The speed of sound in a gas is given by:

$$c = \left(\frac{\gamma P}{\rho}\right)^{0.5} = \left(\frac{\gamma RT}{M}\right)^{0.5} \tag{10.1}$$

where γ is the ratio of specific heats, P is the ambient absolute pressure, ρ is the gas density, R is the universal gas constant, T is the absolute gas temperature, and M is the gas molecular weight.

Example 10.1

Given: The temperature of air is 70°F.
Find: The speed of sound in air at that temperature.
Solution: For air at ambient conditions at sea level, $\gamma = 1.4$, $P = 14.7\,\text{lb/in}^2$, and $\rho = 0.074\,\text{lb/ft}^3$; substituting into the equation above:

$$c = \left(\frac{\gamma P}{\rho}\right)^{0.5} = \left(\frac{(1.4)(14.7\,\text{lb/in.}^2)}{0.074(\text{lb/ft}^3)}\frac{144\,\text{in.}^2}{\text{ft}^2}32.2\frac{\text{lb m} - \text{ft}}{\text{lb ft} - \text{sec}^2}\right)^{0.5}$$

$$= \left(1.290 \times 10^6 \frac{\text{ft}^2}{\text{sec}^2}\right)^{0.5} = 1136\frac{\text{ft}}{\text{sec}}$$

where the appropriate conversion factors have been used. Therefore, mach one for air under the above conditions is 1136 ft/sec.

Sound is dependent on the wavelength of the waves, which is measured in meters. Wavelength is related to frequency as follows:

$$\lambda = \frac{c}{\nu} \tag{10.2}$$

where λ is the wavelength (m or ft), c is the velocity of sound in the given medium (m/sec or ft/sec), and v is the frequency (cycles/sec or hertz).

Example 10.2

Given: The primary sound frequency is 10 kHz.
Find: The wavelength in ft for the conditions from the previous example.
Solution: From the previous example, $c = 1136$ ft/sec; substitute into equation:

$$\lambda = \frac{c}{v} = \frac{1136\,\text{ft/sec}}{10,000\,\text{cycles/sec}} = 0.1136\,\text{ft}$$

Sound intensity level (L_I) is measured in decibels (dB) as

$$IL_b = \log_{10}\left(\frac{I}{I_0}\right) \tag{10.3}$$

where I is the sound intensity ($10^{-12}\,\text{W/m}^2$), and I_0 is the intensity of least audible sound, usually given as $10^{-12}\,\text{W/m}^2$.

Sound pressure level (L_p) is measured in dB as

$$L_p = 20\log_{10}\left(\frac{P}{P_0}\right) \tag{10.4}$$

where P is the pressure of the sound wave (N/m^2), and P_0 is the reference pressure, usually chosen as the threshold of hearing, $2 \times 10^{-5}\,\text{N/m}^2$.

Table 10.2 shows a comparison of sound pressure levels for a variety of sources [1]. Figure 10.2 shows the relationship between decibels and watts, and Figure 10.3 shows the effect of distance r from the source on sound pressure level.

Table 10.2   Comparison of Sound Pressure Levels for Different Sources

Sound pressure level (dB)	Condition	Sound pressure (N/m^2)
120	Threshold of plain	2 000 000
110	Thunder, artillery	
100	Steel riveter at 4.5 m	200 000
90	Noisy factory	
80	Tube train (open window)	20 000
70	Average factory	
60	Loud conversation	2000
50	Average office	200
40	Average living room	200
30	Private office	
20	Whisper	20
10	Soundproof room	
0	Threshold of audibility	2

Adapted from Ref. 1.

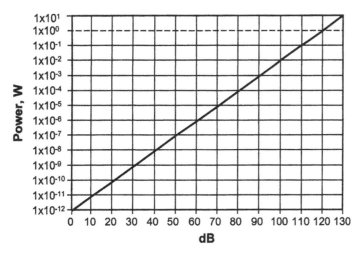

Figure 10.2 Relationship between decibels and watts. (From Ref. 42. Courtesy of John Zink Co., Tulsa, OK).

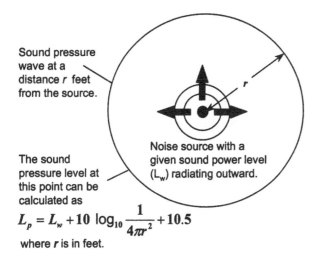

Sound pressure wave at a distance *r* feet from the source.

Noise source with a given sound power level (L$_w$) radiating outward.

The sound pressure level at this point can be calculated as

$$L_p = L_w + 10 \ \log_{10} \frac{1}{4\pi r^2} + 10.5$$

where *r* is in feet.

Figure 10.3 Sound pressure level at a distance *r*. (From Ref. 42. Courtesy of John Zink Co., Tulsa, OK.).

Sound power level (L_w) is measured in dB as

$$L_w = 10 \log_{10}\left(\frac{W}{W_0}\right) \tag{10.5}$$

where W is the sound power of the source (10^{-12} W), and W_0 is the reference sound power, usually chosen as 10^{-12} W. Figure 10.4 shows a comparison of sound power levels for various noise sources [1].

Since humans do not hear equally well at all frequencies, correction factors are normally used to modify noise measurements to quantify sounds that really affect

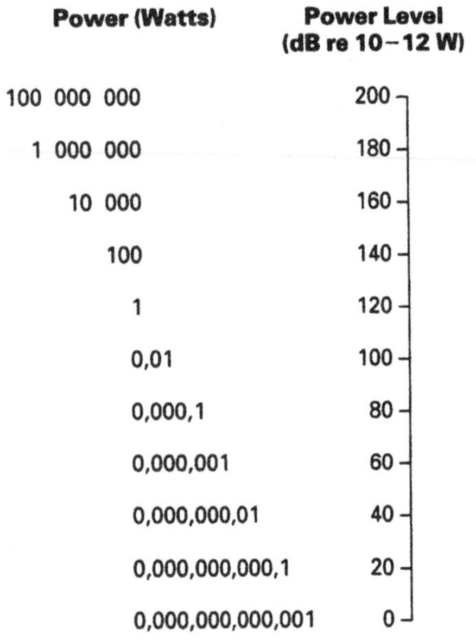

Power (Watts)	Power Level (dB re 10−12 W)
100 000 000	200
1 000 000	180
10 000	160
100	140
1	120
0,01	100
0,000,1	80
0,000,001	60
0,000,000,01	40
0,000,000,000,1	20
0,000,000,000,001	0

Figure 10.4 Comparison of different sound power levels for various sources. (Adapted from Ref. 1.)

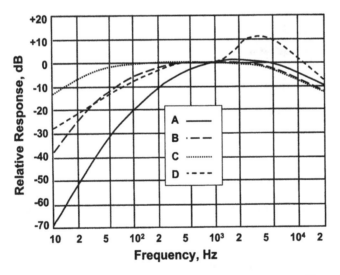

Figure 10.5 Correction scales for noise measurements. (From Ref. 42. Courtesy of John Zink Co., Tulsa, OK.)

humans. The four correction scales commonly used are shown in Figure 10.5. The most common correction is the A scale which represents an idealized inverse, except for level, of the threshold of the hearing curve (see Figure 10.6). The A-weighted sound level correlates reasonably well with hearing damage risk. Figure 10.7 shows a comparison of the raw and corrected noise data from a burner.

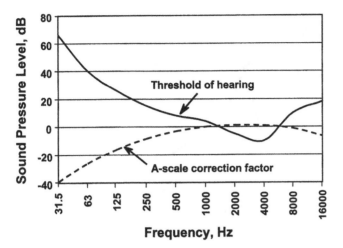

Figure 10.6 Comparison of human threshold of hearing with A-weighted correction scale. (Adapted from Ref. 42.)

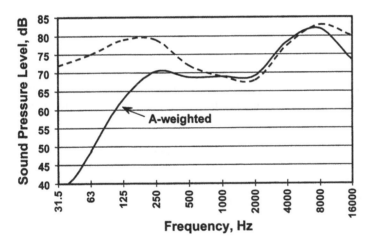

Figure 10.7 Raw noise data for a burner compared to the A-weighted noise curve. (From Ref. 42. Courtesy of John Zink Co., Tulsa, OK.)

The thermoacoustic efficiency of a process is a measure of the fraction of the thermal input energy that is converted into sound power. For turbulent flames, a general rule of thumb is that the thermoacoustic efficiency is approximately 1×10^{-7} times the energy input.

10.2.2 Sources

Noise from industrial combustion processes may come from a variety of sources that are briefly discussed below. As an example, Figure 10.8 shows a noise spectrum produced by shock waves and by mixing noise. There are three pronounced peaks in

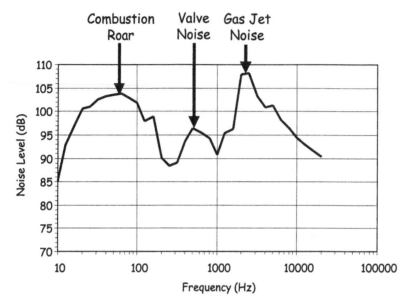

Figure 10.8 Noise spectrum from a high pressure flare. (Courtesy of John Zink Co., Tulsa, OK.)

the noise signature, caused by different sources: combustion roar, valve noise, and gas jet noise. The presence and relative magnitude of each type of noise varies by the type of combustion process.

Lee et al. [22] discuss the formation of pulsations through the use of externally applied tones that were used to study their effects on flames. While these tones were artificially generated and applied, they demonstrate what can happen in a combustion system that has some type of harmonic resonance. The study showed that flames can be made unstable with the addition of an excited tone.

10.2.2.1 Blowers and Fans

Blowers are commonly used to supply the combustion air required by the burners in the heating system. Fans are commonly used to move exhaust gas streams through the exhaust duct system. In many cases, the pressure drop through pollution control equipment requires additional fan power. Both blowers and fans generate noise primarily from the high-speed rotation of the blades and impellors [23]. There are a number of factors that impact the noise produced by fans and blowers. Some of these include how the equipment is mounted and to what, the type of enclosure (e.g., room size and insulation material) where the equipment is located, the gas flow rate and pressure, and the downstream ducting system [24]. The noise emitted by industrial fans typically consists of two noise components: broadband and discrete tones. The broadband noise comes from the interaction of turbulence with the solid construction parts of the fan. The discrete tones are created by the periodic interactions of the rotating blades and both upstream and downstream surfaces.

Fan noise can emanate from both the intake and exhaust. The noise can also radiate downstream through the ductwork.

One major source of noise in a conventional air/fuel combustion system is the air-handling system [25]. The blower, which moves the air, is typically noisy and may need to be located outside the building or acoustically insulated in order to meet noise regulations. Alternatively, workers may need to wear hearing protection when in the vicinity of the blowers. This source of noise can be completely eliminated if the air is replaced by pure oxygen since a blower is not typically required for oxygen. If a low-pressure oxygen source like a vacuum swing adsorption (VSA) unit [26] is used, an oxygen blower, used to boost the supply pressure, may be a source of noise.

10.2.2.2 Gas Flow and Jet Noise

There are two common types of noise caused by gas flow. The first is from the high volumetric flow of gas through the piping system. The actual flow of gas through a pipe is commonly given as "acfh" or actual cubic feet of gas per hour. This is the flow rate of gas at the pressure and temperature in the pipe. This actual flow is usually corrected to a standard temperature and pressure (STP) level and reported as "scfh" or standard cubic feet of gas per hour. Although there are various definitions for STP, they are usually at or about 70°F (21°C) and atmospheric pressure (14.7 psia or 760 mmHg). Since air is typically supplied at a low pressure, the actual flow of gas through the piping is high, which may produce a significant amount of noise. Lower gas velocities generate less noise.

The second type of noise caused by gas flow is from gas-jet mixing from high-speed gases exiting a nozzle, sometimes referred to as jet noise. The noise source begins at the nozzle exit and extends several diameters downstream. Near the nozzle exit, the scale of the turbulent eddies is small and is responsible for producing the higher frequency component of the noise. The lower frequency component of the noise is produced predominantly further downstream from the nozzle exit by the larger scale eddies. While this noise source is often easily reduced by reducing the jet exit velocity, that solution may radically alter the performance of the combustion system especially in terms of other pollutant emissions such as NO_x. High-velocity jets are often used in burner designs to entrain furnace gases and control mixing to reduce NO_x emissions. Reducing the gas velocities may reduce the noise but simultaneously increase NO_x.

There are several factors that affect the noise produced by gas-jet mixing. These include the diameter and smoothness of the nozzle exit, the gas exit velocity, the angle of the observer's position relative to the exit plane of the nozzle, and the temperature ratio of the fully expanded jet to the ambient conditions. Figure 10.9 shows a comparison of noise produced by shock waves and by mixing noise.

10.2.2.3 Combustion Roar

Another source of noise in a combustion system comes from the burner and is sometimes referred to as "combustion roar" [27]. This noise is a combination of the gas flow through the burner nozzles and also from the combustion process itself. It is broadband noise (noise distributed over a wide range of frequencies with no pronounced peaks) generated by the combustion processes in the flame and can be a

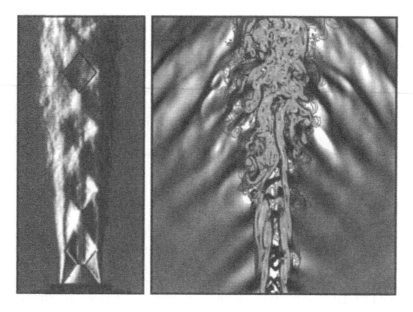

Figure 10.9 Shock vs. mixing noise. (Courtesy of John Zink Co., Tulsa, OK.)

dominant noise source for high-velocity burners. It is present in essentially all turbulent combustion processes, which are the predominant type in industrial combustion systems. There are many factors that affect the noise level produced by the combustion system. These include the firing rate, oxidizer-to-fuel ratio, turbulence intensity of the gas flows, combustion or mixing intensity, amount of swirl, preheat of the oxidizer or fuel, type of fuel and oxidizer, number of burners, geometry of the combustion chamber, insulation used in the combustor, and even the dampening effects of the material being heated. High-hydrogen fuels, for example, can increase noise levels due to the increased intensity of the combustion process [28].

It is difficult to generalize a comparison of the noise between air/fuel and oxy/fuel burners, because of the wide variety of designs that are available. The older style oxy/fuel burners that were used in the steel industry in the 1970s and 1980s typically had flames that were of very high intensity and high momentum, which made them very noisy. One study reported noise levels up to 125 dB for one such high-intensity oxy/fuel burner [29]. The same study also reported noise levels as low as 85 dB for a lower intensity oxy/fuel burner at the lower end of the firing rate range for the burner. Many of the burners commonly used today for OEC are lower momentum and lower intensity and therefore significantly quieter than the older style high-momentum burners. In many cases, new design oxy/fuel burners can be significantly quieter than most air/fuel burners.

10.2.2.4 Combustion Instability

Combustion instability is characterized by a high-amplitude, low-frequency noise resembling the puffing sound of a steam locomotive, which is sometimes referred to

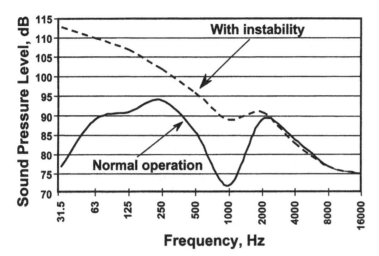

Figure 10.10 Sound pressure level for a burner operated normally and with instability. (From Ref. 42. Courtesy of John Zink Co., Tulsa, OK.)

as *huffing*. This type of noise can create significant pressure fluctuations in a furnace, which can cause equipment damage. Figure 10.10 shows how the noise levels for the same burner can increase significantly when it is operated unstably.

Combustion-driven oscillations occur when there is a match between a characteristic frequency of the flame and of the combustion chamber. This can produce very high noise levels, combustion instability, and vibration. Higher frequencies tend to produce undesirable noise while lower frequencies can cause equipment damage.

10.2.3 Environmental Concerns

There are two groups that are affected by noise: those working in and around the equipment and those outside the so-called fenceline of the plant often referred to as the local community. The workers may routinely wear protective equipment such as earplugs and earphones while working around noisy equipment. However, the local community will not normally have hearing protection and, therefore, the noise levels at the fenceline must be below levels that would cause injury or disturbance. It is assumed here that noise levels heard by the surrounding community are below harmful levels so that the primary focus is on the workers in the plant. Two of the primary concerns are hearing loss and performance reduction, which are both discussed next. There are other concerns including sleep disturbance and possible mental health problems that are not discussed here. There is also another important concern, which is communication impairment that can have serious safety consequences. If workers have difficulty communicating with each other due to high noise levels, then operation mistakes could be made that may lead to dangerous situations.

10.2.3.1 Hearing Loss

The principle environmental concern related to high noise levels is possible hearing loss up to and including permanent deafness [30]. Hearing loss can be divided into three categories: acoustic trauma, noise-induced temporary threshold shift, and noise-induced permanent threshold shift. Acoustic trauma usually results from a single exposure that causes permanent hearing damage, as in the case of an explosion, which causes parts of the ear, for example, the eardrum, to break down. Noise-induced temporary threshold shift is a temporary loss of hearing sensitivity following noise exposure where the loss is reversible. This might occur at a rock concert, for example, where the audience has difficulty hearing lower tones immediately after leaving the concert but regains all hearing by the next day. Noise-induced permanent threshold shift is a permanent and nonreversible hearing loss that could result from a single traumatic noise event or from repeated exposure to excessive noise conditions. The latter is particularly dangerous as the loss is very gradual and often undetectable until it is too late. Older factory workers often now have permanent hearing loss due to repeated exposure to high-noise conditions prior to the days of the U.S. Occupational Safety and Health Administration (OSHA) noise limits for workers.

In the case of a single traumatic noise exposure, the degree of hearing loss is determined more by the noise level. In the case of repeated noise exposure, the degree of hearing loss is determined by a combination of noise level and exposure time. In this case, hearing damage is cumulative and may occur over a period of many years. This points to the need to use appropriate noise control measures even if noise levels do not "seem to be that bad," especially for those used to attending raucous sporting events or rock concerts. In certain types of jobs, for example, flagmen for jet aircraft, regular hearing checks may be required to check for threshold shifts and hearing loss. These checks are done by qualified personnel using specific equipment and under guidelines and standards developed for this purpose [31].

10.2.3.2 Reduced Performance

While some types of sound, for example, relaxing music, can have a positive effect on humans, noise can create some other problems, especially related to worker performance, which can be significantly reduced by either excessive noise levels or by erratic changes in noise level [32]. Reduced performance can range from a slight temporary decline to a substantial and longer-lasting reduction, depending on the noise level and how it changes over time. Under certain conditions, it is even possible to enhance worker performance temporarily through the appropriate use of sound. For example, so-called white background noise has been shown to slightly increase heart rate and blood pressure, which is often only temporary [33]. This may occur when the sounds serve to stimulate workers to be more attentive, which often improves their efficiency and productivity, although the effects are usually relatively short-lived. However, noise generally increases worker stress and therefore is usually undesirable.

Noise can interfere with communications between workers where understanding speech is impaired. It can cause stress among workers, lower morale, reduce efficiency and productivity, and cause fatigue among other things.

10.2.4 Regulations

The American National Standards Institute (ANSI) has some standards regarding noise:

S1.4   Specification for sound level meters.
S1.13   Methods for the measurement of sound pressure levels.
S1.23   Method for the designation of sound power emitted by machinery and equipment.
S3.19   Methods for the measurement of real-ear protection of hearing protectors.
S12.6   Methods for the measurement of real-ear attenuation of hearing protectors.

The American Society of Testing & Materials (ASTM) also has some standards regarding noise:

C384   Test method for impedance and absorption of acoustical materials by the impedance tube method.
E413   Classification for rating sound insulation.
E1124   Standard test method for free-field measurement of sound power level by the two-surface method.

In most industrialized nations, noise is a "pollutant" that is regulated in the work environment. The U.S. OSHA has set maximum permissible industrial noise levels, as shown in Table 10.3 [34]. The table shows that as the length of exposure increases, the permitted sound level decreases. Noise is commonly measured with a sound-level meter, which is an instrument designed to respond to sound in approximately the same way as the human ear. The meter is designed to give objective, repeatable measurements compared to the human ear. Human hearing varies from person to person and even varies for the same person as they age. Different weighting systems are used over the frequency spectrum to closely match the frequency response of a typical human ear. The A-weighting network is the most

Table 10.3 U.S. Occupational Safety and Health Administration Maximum Permissible Industrial Noise Levels

Duration per day (h)	Sound level (dBA), slow response
8	90
6	92
4	95
3	97
2	100
1 ½	102
1	105
½	110
¼ or less	115

Source: Ref. 34.

widely used in noise work so that sound measured with the A-weighting is given as dB(A) or sometimes simply as dBA. Zegel [35] has written a helpful overview of noise standards.

10.2.5 Measurement Techniques

Figure 10.11 shows a commonly used device for measuring industrial noise. A noise measurement system consists primarily of some type of microphone to receive the sound, some type of transducer to convert the sound waves into a usable signal, typically digital, and a meter to convert the signals into the proper weighting system and to display and record the signals. Figure 10.12 shows an engineer measuring sound from a flare.

American Petroleum Institute Recommended Practice 531M gives information on measuring noise from process heaters [36]. Recommendations are given for

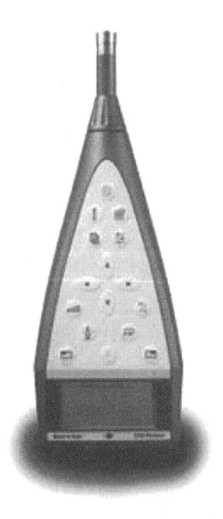

Figure 10.11   Typical noise meter. (Courtesy of Bruel & Kjaer, Naerum, Denmark.)

Figure 10.12 Measuring sound from a flare. (From Ref. 42. Courtesy of John Zink Co., Tulsa, OK.)

making field measurements, and procedures for sound-level measurement including correcting for background noise. Measurements are suggested to be made at 1 m (3.3 ft) from the outside heater walls at various locations, depending on the configuration.

Hassall [37] lists the following procedural steps when making noise measurements:

1. Determine what quantities are to be measured.
2. Select the instruments including the type of microphone to be used.
3. Determine the minimum number of microphone positions and their locations including a diagram of the arrangement (multiple microphones are necessary to measure the sound field).
4. Check the system sensitivity.

5. Measure the acoustical and electrical noise level.
6. Measure the source sound levels.
7. Correct the measurements as appropriate (e.g., subtracting background noise).
8. Record the data.

Lang [38] discusses how to make sound power level (typically used as a descriptor for stationary equipment) measurements. These measurements can be used to calculate the sound pressure level at a given distance from equipment, compare the noise radiated by equipment of the same type and size, compare the noise radiated by equipment of different types and sizes, determine whether equipment complies with a specified upper sound emission limit, determine the amount of noise reduction required for a particular circumstance, and develop quieter equipment. This type of measurement requires a three-dimensional array of microphones and is, therefore, not commonly used when measuring combustion noise.

Measuring sound intensity can be challenging as multiple types of probes are required to measure both the sound pressure and the particle velocity [39]. This is often done using two microphones and performing appropriate calculations to determine the particle velocity. This type of measurement is also not commonly done for industrial combustion systems.

There are a number of factors that can affect noise measurements and instrument performance including temperature, humidity, atmospheric pressure, wind or dust, and even magnetic fields [40]. Instruments should be calibrated according to manufacturer's instructions before and after each day of use and whenever temperature or relative humidity changes significantly. Zahringer et al. [41] experimentally demonstrated the effectiveness of using microphones combined with an intensified camera system to measure acoustic oscillations in a boiler.

10.2.6 Abatement Strategies

There are several strategies that can be used to reduce combustion noise. One strategy is either to move the source of the undesirable sound away from the people or to move the people away from the sound. However, this may not be practical for many industrial applications. Another strategy is to put some type of sound barrier between the noise and the people. The barrier can be either reflective or absorptive to minimize the noise. The noise source might be surrounded by an enclosure or the operators may be located inside a sound-proofed enclosure. In some cases, it may be possible to use a silencer, which would act as a barrier, to reduce the noise. For example, the exhaust from a car is reduced by the muffler, which acts like a silencer. The barrier could also be in the form of earplugs, ear phones, or some other sound-reducing safety device worn by people in the vicinity of the noise. Another technique is to reduce the exposure time to the noise since noise has a cumulative effect on human hearing. In some cases, it may be possible to replace noisy equipment with new equipment that has been specifically designed to produce less noise, or to retrofit existing equipment to produce less noise. For example, old combustion air blowers and fans could be replaced by new, quieter blowers and fans. Another way to reduce noise is to increase the pipe size and reduce the number of bends in the pipe to reduce the jet noise of the fluids flowing through the pipe. Resonance and instabilities can usually be designed out of a system if they are a problem. Noisy burners can be replaced by quieter burners. The burner noise is a function of the burner design, fuel, firing rate, stoichiometry, combustion intensity, and aerodynamics of the combustion chamber. These strategies are discussed in some detail next.

10.2.6.1 Source Reduction

The primary source of noise in most industrial combustion systems is from the burner(s). The design of the burner nozzle and the burner tile or quarl are important factors in noise generation in combustion processes. One method to reduce noise from a burner is to use larger exit ports that produce lower gas velocities. However, there are limits to how low gas velocities can be, depending on the fuels used and the burner type. For example, the exit gas velocities in a premix burner must be greater than the flame speed of the fuel or else flashback will result. Fuels containing high concentrations of hydrogen will necessarily require higher exit velocities because of the high flame speed of hydrogen.

Another method for reducing the sound from a combustion process is to add a pipe or tube, often referred to as a quarter-wave tube, to the resonance system to cancel out the harmonics causing the noise. In this technique, a specially designed tube is attached to the chamber where sound of a predominant frequency is causing resonance. The quarter-wave tube is designed to cancel out this resonance and therefore reduce the noise levels. Figure 10.13 shows a quarter-wave tube installed on the side of an exhaust stack for a gas-fired propylene vaporizer that previously exhibited a low-frequency rumble during operation, producing excessive noise levels. The quarter-wave tube was built with some adjustability to determine the best geometry to maximize noise reduction. Figure 10.14 shows the noise octave bands for the vaporizer with and without the quarter-wave tube. The total noise level for the vaporizer without the tube was 95.3 dBA. The total noise level with the tube dropped to 83.8 dBA for a noise reduction of more than 10 dBA. This brought the noise levels within acceptable limits. Note that adding a quarter-wave tube is not always a practical option because, depending on the combustion chamber geometry and the frequency of the harmonic, the tube diameter and length may be excessive.

Other techniques to mitigate noise caused by combustion instability include modifying the [42]:

- Furnace stack height
- Internal volume of the furnace
- Acoustical properties of the furnace lining
- Pressure drop through the burner by varying the damper positions
- Location of the pilot
- Flame-stabilization techniques

Figure 10.15 shows some of the common silencers used to reduce the source of noise from industrial processes [43]. Figure 10.16 shows a natural draft burner without an air inlet muffler. Figure 10.17 shows a common type of muffler used on a natural draft burner used in the floor of refinery heaters where the air inlet has a baffle lined with sound-deadening insulation. Figure 10.18 shows a comparable standard muffler for a radiant wall burner used in the side of ethylene cracking furnaces. Figure 10.19 shows a nonstandard muffler used on a floor-fired natural draft burner where the

Figure 10.13 Quarter-wave tube installed on the exhaust stack of a propylene vaporizer: (top) front view; (bottom) back view. (Courtesy of John Zink Co., Tulsa, OK.)

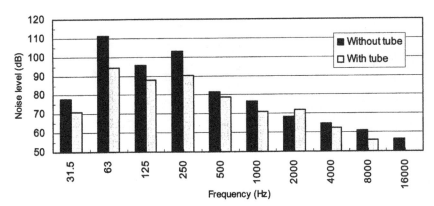

Figure 10.14 Noise octave bands for the propylene vaporizer with and without the quarter-wave tube. (Courtesy of John Zink Co., Tulsa, OK.)

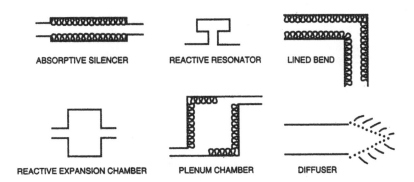

Figure 10.15 Silencers. (From Ref. 43. Courtesy of Lewis Publishers.)

muffler is larger than the burner because of the very low noise requirements for the particular application. Figure 10.20 shows a nonstandard muffler for low noise requirements on wall-fired burners.

10.2.6.2 Sound Transmission Mitigation

This technique involves mitigating the transmission of the sound from the source to the receiver. This can be done in a variety of ways. One rather simple but not always practical method is simply to increase the distance between the source and the receiver, which reduces the sound levels at the receiver. Another strategy is to put some type of barrier between the source and the receiver. For example, a concrete wall could be built around the source. People commonly plant trees and shrubs on their property to mitigate some of the sound from their neighbors and from road traffic noise. Figure 10.21 shows enclosed flares with walls around the enclosures to help mitigate the sound produced by the flares. Some type of sound-absorptive material could be placed between the source and the receiver. Figure 10.22 shows an

Figure 10.16   Burner with no muffler. (Courtesy of John Zink Co., Tulsa, OK.)

Figure 10.17   Typical muffler for a floor-fired natural draft burner. (Courtesy of John Zink Co., Tulsa, OK.)

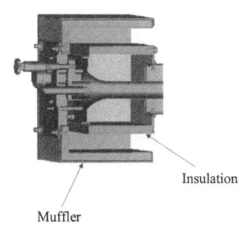

Insulation

Muffler

Figure 10.18 Typical muffler for a radiant wall-fired natural draft burner. (Courtesy of John Zink Co., Tulsa, OK.)

Figure 10.19 Nonstandard mufflers for low-noise requirements for a floor-fired natural draft burner. (Courtesy of John Zink Co., Tulsa, OK.)

example of soundproof curtains used to shield workers from noisy equipment and machinery. Different materials have different sound-absorbing characteristics, depending on both the composition and configuration [44]. Another way could be to use a medium that is less transmissive for sound. For example, water is less transmissive than air.

Noise generated by the burners in a combustion system may be greatly mitigated by the combustion chamber, which is usually a furnace of some type. The refractory linings in most furnaces generally significantly reduce any noise emitted from the burners. Noise is not commonly considered in many industrial

Figure 10.20 Nonstandard mufflers for low-noise requirements for a radiant wall-fired natural draft burner. (Courtesy of John Zink Co., Tulsa, OK.).

heating applications for a variety of reasons. This is evidenced by the general lack of information available on the subject. It is difficult to predict the noise levels before installing the equipment due to the wide variety of factors that influence noise. Often, there are many other pieces of machinery that are much noisier than the combustion system so that the workers are already required to wear hearing protection. In the future, noise reduction may become more important and OEC may be one way to minimize the noise produced by the combustion system.

Mufflers are commonly used on burners to reduce noise levels (see Figure 10.23). Figure 10.24 shows how effective a muffler is at mitigating the noise produced by a burner. These mufflers typically go on the combustion air inlet to the burner and usually have some type of noise-reducing insulation on one or more sides. Reed gives an example of a natural draft burner producing 107 dB of noise before any mitigation [19]. The addition of a primary muffler reduced the noise to 93 dB and the addition of a secondary muffler further reduced the noise to 84 dB.

Figure 10.25 shows a custom-made cylindrical muffler on the air outlet of a velocity thermocouple, often called a suction pyrometer, used to measure higher temperatures. These type of temperature-measuring devices rely on high air flow rates across a venturi to induce furnace gases to flow through the pyrometer and across a shielded thermocouple. This minimizes the effects of thermal radiation on the thermocouple and produces a more accurate measure of the true temperature.

Figure 10.21 Two enclosed flares with a wall around the bottom helps to reduce noise (Courtesy of John Zink Co. LLC) [42].

Figure 10.22 Soundproof curtains. (Courtesy of Kinetics Noise Control, Dublin, OH.)

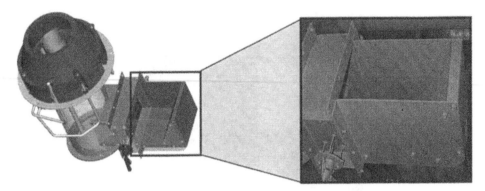

Figure 10.23 Typical muffler used on a natural draft burner. (Courtesy of John Zink Co., Tulsa, OK.)

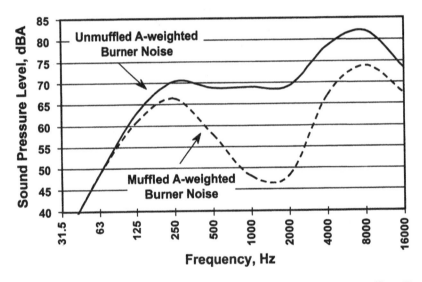

Figure 10.24 Sound pressure level for a burner with and without a muffler. (From Ref. 42. Courtesy of John Zink Co., Tulsa, OK.)

Measurements using bare-wire thermocouples can be as much as 200°F or more too low. Compressed air is often used as the motive gas to induce furnace gas inspiration. The high exit gas flow rates into the atmosphere can produce relatively high noise levels that are a concern for workers in the vicinity. The cylindrical muffler shown in Figure 10.25 reduced noise levels by more than 10 dBA down to an acceptable level.

10.2.6.3 Personnel Protection

There are two levels of protection commonly used by industrial workers to reduce noise levels [45]: plugs (see Figure 10.26) and muffs (see Figure 10.27). One or both types may be used, depending on the noise levels. A third type of protection device is

Figure 10.25 Muffler on a velocity thermocouple used to measure furnace temperatures (Courtesy of John Zink Co., Tulsa, OK.).

Figure 10.26 Typical ear plugs.

a helmet, used commonly by motorcycle drivers, which provides relatively little hearing protection and is rarely used in industry for hearing protection. Therefore, this is not considered further here.

Plugs can reduce noise levels by 5–45 dB, depending on the plug type and sound frequency. They come in a variety of forms including disposable and reusable. Disposable plugs are typically made of some type of moldable material

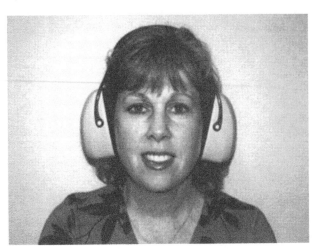

Figure 10.27 Typical ear muffs.

(e.g., foam) that can be inserted into a variety of ear sizes and shapes. These are very inexpensive and are typically bought in large quantities. They are especially convenient for visitors who do not have their own ear plugs. Reusable ear plugs can be cleaned and used multiple times. Custom-molded ear plugs are available that are made to fit exactly in a specific person's ears and are designed to be reusable.

Ear muffs are designed to cover the entire ear and typically reduce levels by 5–50 dB depending on the muff type and sound frequency. These can be used in lieu of or in addition to ear plugs. When both plugs and muffs are worn, noise protection is greater than either individually but is not additive. They may be more comfortable to some compared to ear plugs. However, they may also interfere with other personal protective equipment such as hard hats. Special muffs are made to attach to certain types of hard hats where the muffs can be folded down or up as needed (see Figure 10.28).

Convenience and comfort are important factors when choosing appropriate noise protection for a given environment. If too much effort is required to use or maintain the hearing-protection devices, or if they are uncomfortable, then they are less likely to be used. Proper training and education is essential to maximize the effectiveness of any hearing-conservation program [46].

While not possible in many cases, a simple way to protect workers is either to increase their distance from the sound source or to put them in a sound-proofed enclosure such as a control room or building. However, it is almost impossible to keep all workers away from high-noise sources all the time so hearing protection will probably be necessary.

10.3 VIBRATION

Harris [17] defines vibration as "an oscillation wherein the quantity is a parameter that defines the motion of a mechanical system." Vibrations caused by combustion

Figure 10.28   Ear muffs designed to be used with hard hats.

processes have been given many names including: combustion-driven oscillations, combustion resonance, pulsations, and combustion hum to name a few. These combustion-induced vibrations can be even more difficult to analyze compared to noise sources because the vibrations result from a coupling between the flame and the combustor. Therefore, both must be considered when reducing these vibrations which are typically low frequency.

To show how specialized combustion-induced vibration is, a recent three-volume encyclopedia of vibration contains nothing on this topic [47]. Shabana [48] has written a general text on vibration that does not mention combustion-induced vibration. Fuller et al. [49] have written a book on the active control of vibration. Although it does not consider combustion-induced vibration, it gives a good introduction to control techniques such as feedforward control, distributed transducers for active control, and active isolation of vibrations. de Silva [50] has written an extensive handbook on vibration that is primarily focused on controlling vibration rather than on eliminating vibration sources. It does not include any discussions on combustion-induced vibration.

The American National Standards Institute (ANSI) has some standards regarding vibration:

S2.8   Guide for describing the characteristics of resilient mountings.
S2.61   Guide to the mechanical mounting of accelerometers.
S3.40   Guide for the measurement and evaluation of gloves which are used to reduce exposure to vibration transmitted to the hand.

10.3.1   Sources

The primary sources of vibration related to industrial combustion systems are due to instability and thermoacoustic coupling of the combustion processes with the furnace. Here, it is assumed that vibration caused by, for example, combustion air blowers can be easily mitigated with proper vibration-dampening mounts and is, therefore, not considered here. Blevins [51] has written a book on flow-induced vibrations that could be present in the auxiliary equipment in a combustion system. Some of the topics discussed include vortex-induced vibration, galloping and flutter, vibrations caused by oscillating flow, and vibration induced by turbulence and sound. As an example of techniques used to mitigate a specific type of flow-induced vibration, vortex-suppression devices may be used to minimize or eliminate vibration caused by vortices. These devices include helical strakes, perforated shrouds, axial slats, streamlined fairings, splitters, ribbon or hair cables, guiding vanes, spoiler plates, and stepped cylinders.

Combustion-induced vibration is very undesirable as it poses safety and equipment issues. Furnace vibrations caused by combustion instability could lead to temporary flame extinguishment and then reignition, possibly causing an explosion. This should not be a concern on a properly designed system unless the equipment is operated outside the normal design regime. For example, if certain burners are turned down below their lowest design firing rate, they could become unstable. Another possibility is that a change in fuel composition could cause flame instability. Therefore, it is important to check with the burner equipment supplier when operating burners outside their design operating conditions.

Thermoacoustic coupling occurs when the acoustics of the flame couple with the geometry and characteristics of the combustion chamber to cause resonance. This occurs when a fundamental frequency of the combustion system matches a fundamental frequency of the combustion chamber, causing a harmonic resonance. This is a difficult and challenging problem to model and determine a priori because of the complications of modeling transient-combustion physics and the geometry of the combustion chamber. This resonance may only occur under certain conditions, for example, at a particular firing rate. This resonance can also be violent enough to cause equipment damage.

Buoyant jet diffusion flames in the open air are known to produce a constant low-frequency (10–20 Hz) oscillation under certain conditions [52]. While these flames are not common in industrial combustion applications where the flames are usually highly turbulent, buoyant diffusion flames are often studied to gain further insight into combustion instability. The coherent flow structure produced by these buoyant flames is attributed to a modified Kelvin–Helmholtz instability.

Froud et al. [53] studied Helmholtz resonance in swirl burners. Experimental results in pilot-scale tests showed that the air/fuel ratio, the mode of fuel entry, and the length of the exhaust pipe attached to the exit of the furnace were all important variables. The experiments showed that the resonance can become severe enough virtually to extinguish the flame from the burner and cause highly complex transient flows in the furnace.

Excessive heat loss from a flame and/or a large Lewis number (ratio of the thermal diffusivity to the mass diffusivity) can induce spontaneous oscillations and instability [54]. For premixed flames, pulsations occur for Lewis numbers sufficiently greater than one. For diffusion flames, instability may occur for large Lewis numbers, and also for Lewis numbers near unity with large heat losses. The resulting oscillations may lead to flame extinction.

Füri et al. [55] experimentally studied the regular axisymmetric flame pulsations at the anchoring base of diluted propane and methane jet diffusion flames operating near the extinction limit. They found that Lewis numbers and mixture strengths are both important factors in predicting flame pulsations.

10.3.2 Environmental Concerns

While there are safety and equipment damage concerns with vibration caused by the combustion system, the primary environmental concern is for the effects on humans around the equipment. These effects can range from mild discomfort to motion sickness. Prolonged vibration can produce fatigue, and reduced efficiency and perception.

10.3.2.1 Thermoacoustic Instability

Cammarata et al. [56] experimentally studied combustion instabilities in a lean premixed combustion system. These systems have the promise of low NO_x emissions, but are more susceptible to instabilities because of constraints like preventing flashback and autoignition. While other studies have examined this phenomenon intrusively, the goal of their study was to study it nonintrusively with external microphones. The measurement technique may be applicable for controlling instabilities in industrial combustors.

10.3.2.2 Combustion-Induced Vibration

Combustion-induced vibration occurs when the combustion system causes the combustor, supporting structure, and nearby equipment to vibrate, typically at a low frequency. Depending on the frequency and amplitude of the vibration, this is usually a problem that is cumulative over time. The longer the vibration, the more likely equipment damage will occur. Most industrial combustion systems are not specifically designed for high-amplitude vibration. Besides the obvious movement of the equipment, one early sign of damage is often the hard ceramic refractory inside the combustor, which begins to fracture and fall to the floor. This can result in overheating in those spots where there is insufficient insulation due to damaged refractory. Combustion-induced vibration can be an insidious problem in that there is often no immediate damage. Equipment problems may not show up for some time, which may cause personnel to ignore the vibration. However, catastrophic damage could occur after continued equipment vibration. This could result in fires, explosion, furnace collapse, or other significant damage.

Besides the damage to equipment, there is likely to be some harmful effects on personnel working in the vicinity of the vibrating equipment for long periods. This is somewhat akin to the "jackhammer" affect. One may not realize how violent a jackhammer is until after using it for a while and then stopping. It feels like the body is still vibrating. Continuous excessive vibration usually reduces worker performance and increases fatigue.

10.3.3 Measurement Systems

A typical vibration-measurement system consists of an accelerometer, which is a transducer (see Figure 10.29) that converts the mechanical motion to an electrical signal, a preamplifier to increase the signal level, a signal conditioner, a detector to perform appropriate calculations (e.g., RMS), and a meter for displaying and recording results [57]. The transducer can be affected by temperature, humidity and acoustic noise [58]. Figure 10.30 shows an example of a vibration data-acquisition system.

Figure 10.29   Vibration transducer on a furnace. (Courtesy of John Zink Co., Tulsa, OK.)

Figure 10.30   Vibration data-acquisition system. (Courtesy of John Zink Co., Tulsa, OK.)

American Petroleum Institute Recommended Practice 531 M gives information on measuring vibration from process heaters [36]. Harris [59] recommends the following procedure for making vibration measurements:

- Select the appropriate transducer (e.g., acceleration, velocity, or displacement)
- Select the proper mounting arrangement
- Determine field calibrations
- Select appropriate wire and cabling
- Choose appropriate noise suppression techniques

10.3.4 Abatement Strategies

There are three common strategies for mitigating vibration: source reduction, transmission path modification, and receiver protection [60]. However, in reality, for most industrial combustion installations it is not practical to modify the transmission path. Therefore, only source reduction and receiver protection are considered here.

10.3.4.1 Source Reduction

There are two strategies for reducing the source of vibration. One is to modify the equipment so that there is little or no vibration. The other is to control the vibrating equipment using some type of isolator, for example, a special vibration isolation mount under a blower, to prevent it from affecting other equipment or personnel in the vicinity. The latter is commonly used, for example, to isolate rotating equipment. However, this is not usually the type of vibration problem caused by industrial combustion systems, which are typically more difficult to identify and then control. Vibration isolators are, therefore, not considered further here and are discussed in detail elsewhere for the interested reader [61,62].

The first source reduction strategy involves modifying the equipment to minimize vibration generation. This is often the most desirable vibration control strategy as substantial combustion-induced vibration is often an indicator of other problems, such as combustion instability, that need to be addressed anyway. Some of the techniques to reduce vibration generation may be relatively simple such as changing the overall firing rate or adjusting the firing rates of individual burners. Some strategies are more complicated such as modifying the combustor geometry, modifying the burner design especially the mixing, or using some advanced control technique discussed below. The nature of this problem usually means that one solution does not fit all problems, and sometimes multiple solutions may need to be applied.

10.3.4.2 Receiver Protection

This refers to protecting equipment and personnel subject to vibration. It does not seek to reduce the source of the vibration, but rather its effects. This is not generally preferred except if it is only a temporary solution until a more permanent solution can be found. It may also be used when the vibration is either relatively mild (which is somewhat subjective) or only occurs under limited circumstances that only happen very infrequently. One example might be separating the vibrating equipment from other equipment and personnel. If the furnace is shaking, then elevated walkways and platforms can be built in such a way that they do not attach to the vibrating furnace. Flexible hoses can be used close to the furnace so that downstream piping and flow control equipment is not significantly impacted by the vibrations. There is not really any personal protective equipment that can be worn to mitigate substantial equipment vibration.

10.3.4.3   Advanced Controls

One of the means of controlling vibration caused by combustion instability is by the use of advanced control techniques. The primary purpose is to eliminate the combustion instability, which then in turn eliminates the combustion-induced vibration. Advanced control techniques have received more attention in recent years. The combustion system is continuously monitored and adjusted according to feedback from sensors such as microphones or vibration transducers. Kemal and Bowman [63] describe a real-time adaptive feedback control system for controlling combustion instability. The primary elements of the system include sensors, actuators, and a control strategy using a least-mean-square algorithm. The damping of the pressure oscillations is acoustically driven. The system was shown experimentally to reduce the normalized oscillations by 15%. Neumeier and Zinn [64] experimentally and theoretically demonstrated control of combustion instabilities using an active control technique and secondary fuel injection. The system consists of an electronic "observer" to determine the characteristics of the instabilities, an injector to control the flow of secondary fuel into the combustor, and a control strategy. The observer determines the frequency, amplitude, and phase of the oscillations while the fuel injector compensates for the oscillations over a 0–1000 Hz frequency range. Hantschk et al. [65] describe an active control system for controlling combustion instabilities in liquid-fuel fired systems using direct-drive servo motors. The servo motors drive valves that control the fuel-injection flow rate to dampen oscillations.

Hathout et al. [66] developed a low-order heat-release model that accounts for the impact of flame surface area and equivalence rate oscillations in conjunction with system acoustics to design control strategies to mitigate combustion instabilities. The model results compared well with those of practical and experimental results of other researchers. Blonbou et al. [67] discuss the use of neural networks in a closed-loop system to control combustion instabilities in Rijke-tube combustion. The authors note the importance of control systems as the a priori prediction of combustion instability is currently not possible. Johnson et al. [68] describe a technique to estimate experimentally the stability margin in an actual combustor based on the response to artificially imposed oscillations and to fuel-injector actuation.

Hong et al. [69] describe an advanced robust system for controlling combustion instabilities. The system has a slow-time supervisory controller and a fast-time flame controller. The slow-time controller is used to optimize overall combustor performance including minimizing pollutant emissions. The fast-time controller is used to suppress combustion instabilities. Kulsheimer and Buchner [70] describe the development of predictive methods for studying combustion instabilities in turbulent swirl flames. These instabilities may result from advanced burner and combustion designs attempting to minimize pollution emissions. The model can be used to develop suppression techniques.

REFERENCES

1. A Barber (ed.). *Handbook of Noise and Vibration Control*, 6th edn. Oxford, UK: Elsevier Science, 1992.
2. DHF Liu, HC Roberts. 6 Noise Pollution. In DHF Liu, BG Lipták (eds.). *Environmental Engineers' Handbook*, 2nd edn. Boca Raton, FL: Lewis Publishers, 1997.
3. CM Harris (ed.). *Handbook of Acoustical Measurements and Noise Control*, 3rd edn. Woodbury, NY: Acoustical Society of America, 1998.
4. JS Anderson. *Noise: Its Measurement, Analysis, Rating, and Control*. Aldershot, Hants, UK: Avebury Technical, 1993.
5. NS Kamboj. *Control of Noise Pollution*. New Delhi, India: Deep & Deep Publications, 1993.
6. HK Pelton. *Noise Control Management*. New York: Van Nostrand Reinhold, 1993.
7. LH Bell, DH Bell. *Industrial Noise Control: Fundamentals and Applications*. New York: Marcel Dekker, 1994.
8. DA Bies. *Engineering Noise Control: Theory and Practice*. New York: E&FN Spon, 1996.
9. NP Cheremisinoff (ed.). *Noise Control in Industry*. Westwood, NJ: Noyes Publications, 1996.
10. R Smith. *Noise*. Erin, Ontario: Porcupines' Quill, 1998.
11. CA Davis. *Noise Abatement*. London: Bloodlines, 2000.
12. SD Snyder. *Active Noise Control Primer*. New York: AIP Press, 2000.
13. G Rosenhouse. *Active Noise Control*. Boston, MA: Wit Press, 2001.
14. LL Beranek, IL Ver. *Noise and Vibration Control Engineering: Principles and Applications*. New York: John Wiley, 1992.
15. CE Wilson. *Noise Control: Measurement, Analysis, and Control of Sound and Vibration*. Malabar, FL: Krieger, 1994.
16. AD Pierce. *The Wave Theory of Sound*. asa.aip.org/pierc.html, 2002.
17. CM Harris. Definitions, Abbreviations, and Symbols. In CM Harris (ed.). *Handbook of Acoustical Measurements and Noise Control*, 3rd edn. Woodbury, NY: Acoustical Society of America, 1998.
18. PD Owens. Health Hazards Associated with Pollution Control and Waste Minimization. In DL Wise, DJ Trantolo (eds). *Process Engineering for Pollution Control and Waste Minimization*. New York: Marcel Dekker, 1994.
19. RD Reed. *Furnace Operations*, 3rd edn. Houston, TX: Gulf Publishing, 1981.
20. CM Harris. Introduction. In CM Harris (ed.). *Handbook of Acoustical Measurements and Noise Control*, 3rd edn. Woodbury, NY: Acoustical Society of America, 1998.
21. JJ Peirce, RF Weiner, PA Vesilind. *Environmental Pollution and Control*. Boston, MA: Butterworth-Heinemann, 1998.
22. KM Lee, TK Kim, WJ Kim, SG Kim, J Park, SI Keel. A visual study on flame behavior in tone-excited non-premixed jet flames. *Fuel*, Vol. 81, pp. 2249–2255, 2002.
23. JB Graham, RM Hoover. Fan Noise. In CM Harris (ed.). *Handbook of Acoustical Measurements and Noise Control*, 3rd edn. Woodbury, NY: Acoustical Society of America, 1998.
24. PN Cheremisinoff. Fans and Blowers. In PN Cheremisinoff (ed.). *Air Pollution Control and Design for Industry*. New York: Marcel Dekker, 1993.

25. R.D. Bruce. Noise Pollution. *Kirk–Othmer Encyclopedia of Chemical Technology*, 3rd edn. Vol. 16, New York: John Wiley, 1983.

26. RM McGuinness, WT Kleinberg. Oxygen Production. In CE Baukal (ed.). *Oxygen-Enhanced Combustion*. Boca Raton, FL: CRC Press, 1998.

27. A Putnam, L Faulkner. An overview of combustion noise. *J Energy*, 458–469, Nov.–Dec., 1983.

28. API. Publication 535: Burners for Fired Heaters in General Refinery Services. Washington, DC: American Publication Institute, 1995.

29. X-RC Song. Experimental Study of Combustion Noise Generated by Oxygen–Fuel Burners. *Proceedings of the 1993 National Conference on Noise Control Engineering: Noise Control in Aeroacoustics*, Williamsburg, VA, pp. 97–102, 1993.

30. W Melnick. Hearing Loss from Noise Exposure. In CM Harris (ed.). *Handbook of Acoustical Measurements and Noise Control*, 3rd edn. Woodbury, NY: Acoustical Society of America, 1998.

31. MH Miller, LA Wilber. Hearing Evaluation. In CM Harris (ed.). *Handbook of Acoustical Measurements and Noise Control*, 3rd edn. Woodbury, NY: Acoustical Society of America, 1998.

32. DM Jones, DE Broadbent. Human Performance and Noise. In CM Harris (ed.). *Handbook of Acoustical Measurements and Noise Control*, 3rd edn. Woodbury, NY: Acoustical Society of America, 1998.

33. G Jansen. Physiological Effects of Noise. In CM Harris (ed.). *Handbook of Acoustical Measurements and Noise Control*, 3rd edn. Woodbury, NY: Acoustical Society of America, 1998.

34. U.S. Department of Labor, Occupational Safety and Health Administration. "Occupation. Noise Exposure," OSHA Standard 1910.95, 1995.

35. WC Zegel. Standards. In DHF Liu, BG Lipták (eds.). *Environmental Engineers' Handbook*, 2nd edn. Boca Raton, FL: Lewis Publishers, 1997.

36. API. Measurement of Noise from Fired Process Heaters. Recommended Practice 531 M. Washington, DC: American Petroleum Institute, reaffirmed Aug. 1995.

37. JR Hassall. Noise Measurement Techniques. In CM Harris (ed.). *Handbook of Acoustical Measurements and Noise Control*, 3rd edn. Woodbury, NY: Acoustical Society of America, 1998.

38. WW Lang. Measurement of Sound Power. In CM Harris (ed.). *Handbook of Acoustical Measurements and Noise Control*, 3rd edn. Woodbury, NY: Acoustical Society of America, 1998.

39. MJ Crocker. Measurement of Sound Intensity. In CM Harris (ed.). *Handbook of Acoustical Measurements and Noise Control*, 3rd edn. Woodbury, NY: Acoustical Society of America, 1998.

40. U.S. Occupational Safety and Health Administration. OSHA Technical Manual — Sec. III: Chap. 5 Noise Measurement. www.osha.gov/dts/osta/otm/otm_iii/otm_iii_5.html, 2002.

41. K Zahringer, J-C Rolon, J-P Martin, S Candel, O Gicquel, S Arefi. Optical Diagnostics for Analysis of Acoustic Coupling in Domestic Gas Boilers. *Proceedings of 5th European Conference on Industrial Furnaces and Boilers*, Portugal, Vol. 1, pp. 679–688, 2000.

42. W Bussman, JD Jaykaran. Noise. In CE Baukal(ed.). *The John Zink Combustion Handbook*. Boca Raton, FL: CRC Press, 2001.

43. DHF Liu. 6.7 Noise Control in the Transmission Path. In DHF Liu, BG Lipták (eds.). *Environmental Engineers' Handbook*, 2nd edn. Boca Raton, FL: Lewis Publishers, 1997.

44. R Moulder. Sound-Absorptive Materials. In CM Harris (ed.). *Handbook of Acoustical Measurements and Noise Control*, 3rd edn. Woodbury, NY: Acoustical Society of America, 1998.

45. CW Nixon, EH Berger. Hearing Protection Devices. In CM Harris (ed.). *Handbook of Acoustical Measurements and Noise Control*, 3rd edn. Woodbury, NY: Acoustical Society of America, 1998.

46. LH Royster, JD Royster. Hearing Conservation Programs. In CM Harris (ed.). *Handbook of Acoustical Measurements and Noise Control*, 3rd edn. Woodbury, NY: Acoustical Society of America, 1998.

47. SG Braun, DJ Ewins, SS Rao. *Encyclopedia of Vibration* (3 vols.). New York: Academic Press, 2002.

48. AA Shabana. *Theory of Vibration: An Introduction*, 2nd edn. New York: Springer-Verlag, 1996.

49. CR Fuller, SJ Elliott, PA Nelson. *Active Control of Vibration*. London: Academic Press, 1996.

50. CW de Silva. *Vibration: Fundamentals and Practice*. Boca Raton, FL: CRC Press, 2000.

51. RD Blevins. *Flow-Induced Vibration*, 2nd edn. New York: Van Nostrand Reinhold, 1990.

52. A Lingens, K Neemann, J Meyer, M Schreiber. Instability of Diffusion Flames. *Twenty-Sixth Symposium (Intenational) on Combustion*. Pittsburgh, PA: The Combustion Institute, pp. 1053–1061, 1996.

53. D Froud, A Beale, T O'Doherty, N. Syred. Studies of Helmholtz Resonance in a Swirl Burner/Furnace System. *Twenty-Sixth Symposium (International) on Combustion*. Pittsburgh, PA: The Combustion Institute, pp. 3355–3362, 1996.

54. S Cheatham, M Matalon. Heat Loss and Lewis Number Effects on the Onset of Oscillations in Diffusion Flames. *Twenty-Sixth Symposium (International) on Combustion*. Pittsburgh, PA: The Combustion Institute, pp. 1063–1070, 1996.

55. M Füri, P Papas, PA Monkewitz. Non-Premixed Jet Flame Pulsations Near Extinction. *Proceedings of the Combustion Institute*, Pittsburgh, PA, Vol. 28, pp. 831–838, 2000.

56. G Cammarata, A Fichera, C Losenno, F Romanello. An Experimental Study of Combustion Instabilities in Premixed Combustor. *Fifth International Conference on Technologies and Combustion for a Clean Environment*. Lisbon, Portugal. Vol. 1, The Combustion Institute—Portuguese Section, pp. 501–505, 1999.

57. RB Randall. Vibration Measuring Instruments. In CM Harris (ed.). *Handbook of Acoustical Measurements and Noise Control*, 3rd edn. Woodbury, NY: Acoustical Society of America, 1998.

58. EE Eller, RM Whittier. Vibration Transducers. In CM Harris (ed.). *Handbook of Acoustical Measurements and Noise Control*, 3rd edn. Woodbury, NY: Acoustical Society of America, 1998.

59. CM Harris. Vibration Measurement Techniques. In CM Harris (ed.). *Handbook of Acoustical Measurements and Noise Control*, 3rd edn. Woodbury, NY: Acoustical Society of America, 1998.

60. EE Ungar, DH Sturz. Vibration Control Techniques. In CM Harris (ed.). *Handbook of Acoustical Measurements and Noise Control*, 3rd edn. Woodbury, NY: Acoustical Society of America, 1998.

61. RH Racca. Part 1: Types and Characteristics of Vibration Isolators. In CM Harris (ed.). *Handbook of Acoustical Measurements and Noise Control*, 3rd edn. Woodbury, NY: Acoustical Society of America, 1998.

62. HL Hain, JJ Heintzel, CJ Leingang. Part 2: Selection and Applications of Vibration Isolators. In CM Harris (ed.). *Handbook of Acoustical Measurements and Noise Control*, 3rd edn. Woodbury, NY: Acoustical Society of America, 1998.

63. A Kemal, CT Bowman. Real-Time Adaptive Feedback Control of Combustion Instability. *Twenty-Sixth Symposium (International) on Combustion*. Pittsburgh, PA: The Combustion Institute, pp. 2803–2809, 1996.

64. Y Neumeier, BT Zinn. Experimental Demonstration of Active Control of Combustion Instabilities Using Real-Time Modes of Observation and Secondary Fuel Injection.

Twenty-Sixth Symposium (*International*) *on Combustion*. Pittsburgh, PA: The Combustion Institute, pp. 2811–2818, 1996.

65. C Hantschk, J Hermann, D Vortmeyer. Active Instability Control With Direct-Drive Servo Valves in Liquid-Fueled Combustion Systems. *Twenty-Sixth Symposium* (*International*) *on Combustion*. Pittsburgh, PA: The Combustion Institute, pp. 2835–2841, 1996.

66. JP Hathout, M Fleifil, AM Annaswamy, AF Ghoniem. Heat-Release Actuation for Control of Mixture-Inhomogeneity-Driven Combustion Instability. *Proceedings of the Combustion Institute*, Pittsburgh, PA, Vol. 28, pp. 721–730, 2000.

67. R Blonbou, A Laverdant, S Zaleski, P Kuentzmann. Active Control of Combustion Instabilities on a Rijke Tube Using Neural Networks. *Proceedings of the Combustion Institute*, Pittsburgh, PA, Vol. 28, pp. 747–755, 2000.

68. CE Johnson, Y Neumeier, TC Lieuwen, BT Zinn. Experimental Determination of the Stability Margin of a Combustor Using Exhaust Flow and Fuel Injection Rate Modulations. *Proceedings of the Combustion Institute*, Pittsburgh, PA, Vol. 28, pp. 757–763, 2000.

69. B-S Hong, A Ray, V Yang. Wide-range robust control of combustion instability. *Combust Flame*, Vol. 128, pp. 242–258, 2002.

70. C Kulsheimer, H Buchner. Combustion dynamics of turbulent swirling flames. *Combust. Flame*, Vol. 131, pp. 70–84, 2002.

11

Other Pollutants

11.1 INTRODUCTION

This chapter contains discussions of pollutants that generally receive less attention than NO_x, unburned combustibles, SO_x, particulates, and noise. Specifically, the chapter discusses carbon dioxide, thermal radiation, and dioxins and furans. There are a number of other pollutants that are of interest in waste incineration, but are rarely encountered in other types of industrial combustion processes. Therefore, discussion of those pollutants is given in the chapter on waste incineration (Chap. 16). In a limited number of industrial combustion processes, odor may also be considered a pollutant, such as when H_2S is present. However, odor pollution is not considered here as it is not generally due to the industrial combustion process, but rather due to chemicals in the incoming feed materials.

11.2 CARBON DIOXIDE

Carbon dioxide (CO_2) has recently been gaining prominence as an important factor in global warming. At this time, its affect on the environment is somewhat disputed, but most agree the rapid rise in the ambient levels of CO_2 in the atmosphere has been a leading cause in the rise in atmospheric temperatures. The rise in CO_2 levels has been caused by both a rise in the burning of hydrocarbon fuels and the cutting down of vast forests of trees that naturally remove CO_2 from the atmosphere via photosynthesis. Because of the prominence of hydrocarbon fuels in nearly all industrial processes, either directly or indirectly, CO_2 emissions may need to be considered as another pollutant to be regulated. Paul and Pradier [1] have edited a book dedicated to the environmental issues associated with CO_2.

Carbon dioxide is somewhat usual compared to other pollutants for a variety of reasons. First, it is inert and not directly harmful to humans and in fact is required by plants. Second, there is disagreement as to whether it is actually a pollutant. Third, it is the natural by-product of hydrocarbon combustion processes and therefore is very difficult to reduce without dramatically affecting throughput rates. Fourth, the primary treatment after CO_2 is produced is sequestration where the pollutant is captured and then stored elsewhere. It is likely that this "pollutant" will continue to gain prominence in the future due to global warming.

11.2.1 Introduction

Carbon dioxide is a colorless, odorless, inert gas that does not support life since it can displace oxygen and act as an asphyxiant. CO_2 is found naturally in the atmosphere at concentrations averaging 0.03% or 300 ppmv. Concentrations of 3–6% can cause headaches, dyspnea, and perspiration. Concentrations of 6–10% can cause headaches, tremors, visual disturbance, and unconsciousness. Concentrations above 10% can cause unconsciousness eventually leading to death.

Carbon dioxide is a naturally occurring element that is part of the natural cycle in the atmosphere. CO_2 enters the global atmosphere from vegetative decay and methane oxidation. The CO_2 is removed from the atmosphere by photosynthesis and solution in bodies of water. This balance of sources and sinks has been thrown out of balance by the increase in CO_2 generation from fossil-fuel combustion.

Carbon dioxide emissions are produced when a fuel containing carbon is combusted near or above stoichiometric conditions. Some studies indicate that CO_2 is a greenhouse gas that may contribute to global warming. Some believe that CO_2 is the main culprit causing global warming although that is disputed by some scientists who argue other gases such as chlorofluorocarbons, methane (CH_4), and nitrous oxide (N_2O) are the key culprits [2]. The petrochemical industry is particularly interested in this debate because of the implications for its primary products, which contain hydrocarbons that produce CO_2 when burned [3]. The veracity of the historical global CO_2 and temperature data is questionable at best. There does not appear to be a satisfactory explanation for the cooling trend observed during the early 1900s as no significant pollution control methods were introduced, nor were any large-scale climatic changes identified.

Many schemes have been suggested for "disposing" of CO_2, including injection deep into the ocean [4–6] or deep-well injection for oil recovery. In Europe, CO_2 emissions are considered a pollutant and as such are regulated. Any technique that improves the overall thermal efficiency of a process can significantly reduce CO_2 emissions because less fuel needs to be burned for a given unit of available heat output. Some predict that reductions in CO_2 emissions will become increasingly important for the petrochemical industry [7].

11.2.2 Formation Mechanisms

This section discusses some of the important variables that impact CO_2 formation including fuel composition and temperature, oxidizer composition and temperature, equivalence ratio, and exhaust gas temperature.

11.2.2.1 Fuel Composition and Temperature

There are many factors that affect the amount of CO_2 produced during the combustion of fossil fuels in industrial processes. The global simplified reaction of a hydrocarbon with air can be written as follows:

$$C_xH_y + a(O_2 + 3.76N_2) \rightarrow xCO_2 + 0.5yH_2O + bO_2 + 3.76aN_2 + \text{minor species}$$

$$(11.1)$$

where x and y are determined by the fuel, and the minimum a for complete combustion can be calculated from an O atom balance where $b = 0$:

$$2a = 2x + 0.5y$$
$$a = x + y/4 \tag{11.2}$$

In reality, there will be some excess O_2 ($b > 0$) to ensure complete combustion. For simplicity, b and the minor species will be assumed to be zero here. Then, the global reaction can be rewritten as

$$C_xH_y + (x + y/4)(O_2 + 3.76N_2) \rightarrow xCO_2 + 0.5yH_2O + 3.76aN_2 \tag{11.3}$$

so there will be x scf of CO_2 produced per scf of C_xH_y. Table 11.1 shows the calculated amount of CO_2 produced per 10^6 Btu for some common gaseous fuels. There are two extremes for hydrogen and carbon monoxide where either no CO_2 is produced or 3226 scf $CO_2/10^6$ Btu, respectively. In between these two fuels, the amount of CO_2 produced is approximately in the range of 1200 scf $CO_2/10^6$ Btu for the fuels considered. This is relatively independent of the hydrogen:carbon ratio (x/y) in the fuel.

Figure 11.1 shows the predicted volume per cent of CO_2 in the exhaust gas from the adiabatic equilibrium combustion of CH_4/H_2 and C_3H_8/H_2 fuel blends with ambient air. As the hydrogen content in the blend increases, the fraction of CO_2 in the exhaust gas decreases. The graph also shows that more CO_2 is produced for propane compared to methane as discussed above. Figure 11.2 shows that the predicted fraction of CO_2 in the exhaust stream decreases slightly with fuel preheat temperature. This is primarily due to dissociation.

11.2.2.2 Oxidizer Composition and Temperature

The oxidizer composition affects the percentage of CO_2 in the exhaust stream. This is important when considering ways to capture the CO_2 for other uses or for storage

Table 11.1 Flow Rate of CO_2 Produced per 10^6 Btu

Fuel	Formula (C_xH_y)	x	y	$\frac{x}{y}$	Heat of combustion (Btu/scf)	scf Fuel/10^6 Btu	scf CO_2/scf fuel	scf $CO_2/10^6$ Btu
Hydrogen	H_2	0	2	0.000	320	3125	0.0	0
Methane	CH_4	1	4	0.250	985	1015	1.0	1015
Ethane	C_2H_6	2	6	0.333	1712	584	2.0	1168
Propane	C_3H_8	3	8	0.375	2450	408	3.0	1224
Butane	C_4H_{10}	4	10	0.400	3300	303	4.0	1212
Ethylene	C_2H_4	2	4	0.500	1480	676	2.0	1351
Propylene	C_3H_6	3	6	0.500	2310	433	3.0	1299
Butene	C_4H_8	4	8	0.500	3150	318	4.0	1270
Acetylene	C_2H_2	2	2	1.000	1450	690	2.0	1379
Carbon monoxide	CO	1	0	∞	310	3226	1.0	3226

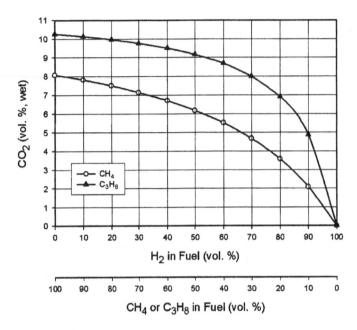

Figure 11.1 Adiabatic equilibrium CO_2 (wet basis) as a function of the hydrogen content in CH_4/H_2 or C_3H_8/H_2 blends combusted with ambient air at an equivalence ratio of one.

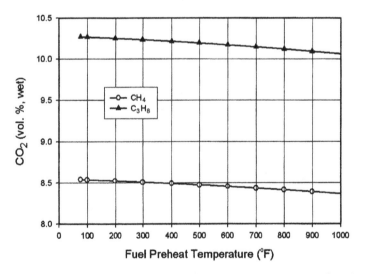

Figure 11.2 Adiabatic equilibrium CO_2 (wet basis) as a function of the fuel preheat temperature for the combustion of CH_4 or C_3H_8 with ambient air at an equivalence ratio of one.

(sequestration). The higher the concentration of CO_2, the easier it is to separate the CO_2 from the other components in the exhaust stream. Figure 11.3 shows the CO_2 concentration (wet basis) in the exhaust gases for methane and propane combusted adiabatically with oxidizers consisting of oxygen and nitrogen, where the oxygen concentration varies from 21% (air) to 100% (pure oxygen). The CO_2 concentration

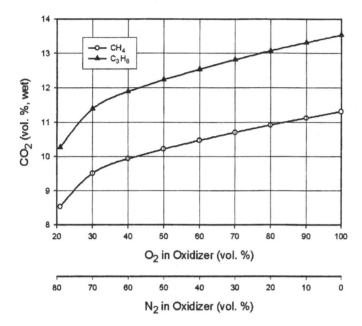

Figure 11.3 Adiabatic equilibrium CO_2 concentration (wet basis) as a function of the oxygen content in the oxidizer for the combustion of CH_4 or C_3H_8 with the oxidizer in ambient conditions at an equivalence ratio of one.

increases with the oxygen concentration in the oxidizer. Figure 11.4 shows a similar effect for the CO_2 concentration on a dry basis (water removed from the exhaust gases).

Since real combustion processes are not adiabatic, the exhaust gases will be at a much lower temperature than the predicted adiabatic flame temperature. For illustration purposes, the exhaust gas temperature will be assumed to be 2400°F (1300°C). The actual exhaust gas temperature will be a function of many factors including the burner design, the operating conditions, the flame heat-transfer characteristics, the combustor design, the geometrical configuration between the flame and the load, the load heat-transfer characteristics, and the exhaust gas ducting configuration and properties among other things. Figure 11.5 shows there is a significant increase in the wet-basis concentration of CO_2 as a function of the oxidizer composition when the flue gas temperature is assumed to be 2400°F (1300°C). The peak CO_2 concentration on a wet basis in Fig. 11.5 is over 40% compared to less than 14% in Fig. 11.4. Even more striking is that the dry-basis concentration in Fig. 11.5 is approximately 100% when the oxidizer is pure O_2. This is important because the exhaust gas stream can be used in applications requiring high-purity CO_2. Figure 11.6 shows that there is little difference in the dry-basis concentration of CO_2 in the exhaust stream produced by combusting methane or propane with varying oxidizer compositions. Since removing water from an exhaust gas stream is straightforward and relatively inexpensive, using oxidizers with higher O_2 concentrations makes it easier to capture the CO_2 for other uses or for storage because it is more concentrated.

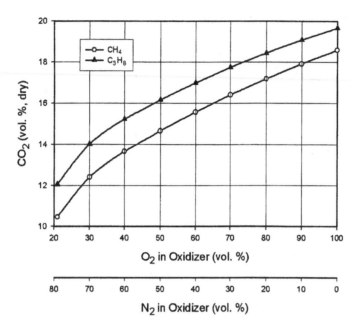

Figure 11.4 Adiabatic equilibrium CO_2 concentration (dry basis) as a function of the oxygen content in the oxidizer for the combustion of CH_4 or C_3H_8 with the oxidizer in ambient conditions at an equivalence ratio of one.

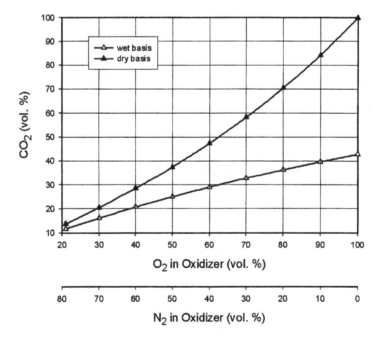

Figure 11.5 Equilibrium CO_2 concentration (wet and dry basis) as a function of the oxygen content in the oxidizer for the combustion of C_3H_8 with the oxidizer in ambient conditions for a gas temperature of 2400°F (1300°C) at an equivalence ratio of one.

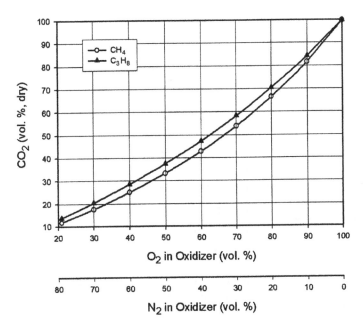

Figure 11.6 Equilibrium CO_2 concentration (dry basis) as a function of the oxygen content in the oxidizer for the combustion of CH_4 or C_3H_8 with the oxidizer in ambient conditions for a gas temperature of 2400°F (1300°C) at an equivalence ratio of one.

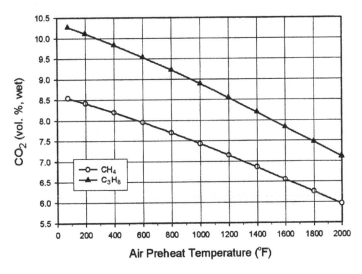

Figure 11.7 Adiabatic equilibrium CO_2 concentration (wet basis) as a function of air preheat temperature for the combustion of CH_4 or C_3H_8 at an equivalence ratio of one.

Figure 11.7 shows that the predicted concentration of CO_2 on a wet basis declines with air preheat temperature. This is due primarily to dissociation where more CO and less CO_2 are produced at the higher adiabatic equilibrium temperatures caused by air preheating. Figure 11.8 shows that the predicted CO_2

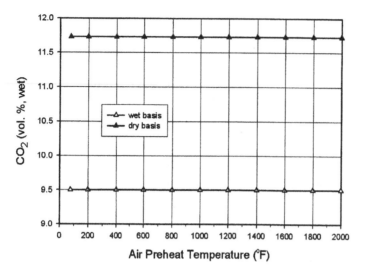

Figure 11.8   Equilibrium CO_2 concentration (wet and dry basis) as a function of the air preheat temperature for the combustion of CH_4 with air in ambient conditions for a gas temperature of 2000°F (1100°C) at an equivalence ratio of one.

concentration on either a wet or dry basis is independent of the air preheat temperature for an assumed gas temperature of 2000°F (1100°C) where the effects of dissociation are negligible.

11.2.2.3   Mixture Ratio

Figure 11.9 shows the effect of the equivalence ratio on the predicted CO_2 concentration in the exhaust stream resulting from the combustion of methane or propane with ambient air. As the air/fuel mixture becomes more fuel rich (equivalence ratio increasing above 1.0), the CO_2 concentration decreases. There, more CO is formed due to the lack of sufficient oxygen to combust fully the carbon in the fuel to CO_2. As the air/fuel mixture becomes more fuel lean (equivalence ratio decreasing from 1.0), the CO_2 concentration again decreases. In that case, excess oxygen and more diluent nitrogen are present in the exhaust gases, which dilute the concentration of CO_2. In either the fuel rich or lean case, the lower concentration of CO_2 makes it more expensive to separate the CO_2 for other uses or for storage.

11.2.2.4   Exhaust Gas Temperature

Figure 11.10 shows how the gas temperature affects the predicted CO_2 concentration in the exhaust gas stream for the combustion of methane or propane with ambient air at stoichiometric conditions (equivalence ratio equals one). As previously shown above, the adiabatic equilibrium predictions compared to those at an assumed gas temperature of 2400°F (1300°C) were significantly different where all other factors were the same. This is graphically shown in Fig. 11.10 where the predicted CO_2 concentration is essentially constant for gas temperatures up to about 2400°F

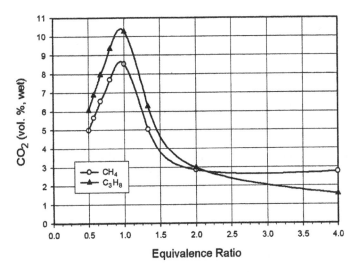

Figure 11.9 Adiabatic equilibrium CO_2 concentration (wet basis) as a function of equivalence ratio for the combustion of CH_4 or C_3H_8 with ambient air.

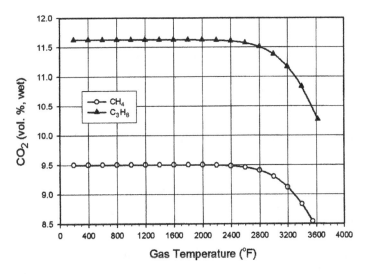

Figure 11.10 Equilibrium CO_2 concentration (wet basis) as a function of the exhaust gas temperature for the combustion of CH_4 or C_3H_8 with ambient air at an equivalence ratio of one.

(1300°C), but then significantly decline for higher temperatures due to dissociation. Since the thermal efficiency declines as the exhaust gas temperature increases, most combustion systems operate at exhaust gas temperatures below 2400°F (1300°C) for the combustion of fuels like methane or propane with ambient air in or near stoichiometric conditions. Therefore, at typical exhaust gas temperatures below 2400°F (1300°C), the CO_2 concentration is essentially constant as a function of gas temperature.

11.2.3 Environmental Concerns

The primary environmental concern for CO_2 emissions is their impact on global warming as CO_2 is a greenhouse gas [8]. However, as stated in Chap. 1, this is a controversial topic that does not have universal acceptance in the scientific community because of problems both with the data and their interpretation [9]. In fact, some scientists argue that increased levels of CO_2 are actually beneficial, particularly to plant life [10].

One of the effects of global warming would be a potential yield loss in agricultural production [11]. While yield production would initially improve due to the higher temperatures caused by global warming, yield loss would eventually result from the more rapid depletion of nutrients in the soil, causing lower plant fertility [12]. Related to this would be the possible use of fragile and marginal land to make up for the crop yield loss. This could mean the transformation of forests and wetlands into cropland. There could also be a significant increase in the insect population, due to the warmer temperatures, that would be a problem for both plants and animals. Soil erosion could be accelerated by the increased rainfall associated with warmer temperatures.

11.2.4 Regulations

One strategy to reduce CO_2 emissions is through government legislation. One method is through a carbon tax so that any usage of fuels that produce CO_2 (e.g., natural gas) would be taxed and therefore cost more. This would encourage fuel conservation to reduce fuel usage and therefore CO_2 emissions. It would also encourage investigating alternative fuels and energy sources. It would indirectly make alternative energy sources like solar and wind power more cost competitive as these would not be taxed. In fact, these alternative energy sources often receive tax credits to encourage their use.

Another strategy for using legislation to control CO_2 emissions is through emissions trading [13]. Willis et al. [14] discuss preparation for an internal emissions trading within a cement company if and when carbon emissions trading would become a reality. This encourages free-market forces to develop new technologies that can be employed to reduce emissions. However, these new technologies may only be economically feasible in certain applications. For example, maybe they are only economic on a large scale, which limits their use to fewer end-users. Those large end-users can improve the economics of the new process by reducing their own emissions and selling the credits to other, smaller users who cannot justify employing some of the newer reduction technologies themselves. Since regulations normally get stricter over time, the available emission credits will decline and become more valuable. This in turn gives economic incentive to continue developing new technologies. Therefore, this free-market system can lead to reduced emissions based on a combination of stricter regulations and new technology development. Flavin and Lenssen [15] note that CO_2 and global warming are particularly challenging legislative problems that need to include policies governing energy efficiency and continued development of renewable energy resources.

11.2.5 Measurement Techniques

EPA Method 3 [102] is used to determine the dry molecular weight of an exhaust gas stream. This is needed, for example, to calculate the exhaust gas flow rate used in calculating the pollutant emission rates. The method specifically refers to determining the CO_2 concentrations in a gas stream. The technique uses single-point grab sampling (batch extractive sample), single-point integrated sampling, or multipoint integrated sampling. The sample is typically analyzed using an Orsat analyzer. This method is not commonly used in industrial combustion applications where instrument analyzers are typically used.

Method 3A concerns the determination CO_2 using instrument analyzers that are commonly used in industrial combustion systems. This method discusses test and calibration procedures. Method 3B is specific to gas streams that contain significant quantities of gases other than O_2, CO_2, CO, and nitrogen (excluding water which can be easily removed from the stream), which could affect the gas analysis results. For example, significant quantities of SO_2 or HCl can interfere with CO_2 measurements using an Orsat analysis.

11.2.6 Control Strategies

Most of the abatement and control strategies for CO_2 involve either improving the thermal efficiency of the combustion system so that less fuel needs to be consumed for a given level of production, or to sequester the resulting CO_2 in some form so that it does not contribute to the growing increase in CO_2 concentration in the atmosphere. A key factor when improving thermal efficiency is to ensure that there is an overall reduction in fossil-fuel energy consumption. For example, if some fossil fuel is replaced by electrical energy, there may not be an overall fuel savings as fuel must be burned at the power plant to generate the electricity. The methods to control CO_2 from combustion fall into the following broad categories [16]:

1. Increased system efficiency
2. Revolutionary new combustion processes
3. Sequestration

11.2.6.1 Pretreatment

One of the most effective pretreatment strategies for controlling CO_2 emissions is to remove as much of the diluent from the system as is feasible. The primary diluent in most industrial combustion processes is the nitrogen in the combustion air. Removing some or all of this N_2 can dramatically increase the thermal efficiency of the system [17]. As shown in Chap. 6, using a high-purity O_2 oxidant can have the added benefit of reducing NO_x emissions as well.

Diluent in the combustion process absorbs heat, which is then carried out with the exhaust gas stream. Nitrogen in the combustion air puts a very large heat load on the system. Depending on the exhaust gas temperature, half or more of the energy released during combustion may be carried out of the exhaust stack. Available heat is defined as the gross heating value of the fuel minus the energy carried out of the system by the exhaust gases. Not all of the available heat normally goes into the heat load as some is usually lost by heat conduction through the walls and by radiation and

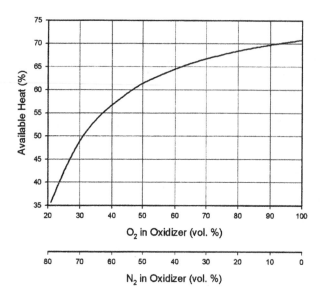

Figure 11.11 Available heat as a function of the oxygen content in the oxidizer where exhaust gases are at 2400°F (1600 K).

convection through openings in the combustor. Ideally, these heat losses are minimized to maximize the thermal efficiency of the system. The available heat is a convenient way of showing trends related to thermal efficiency. Figure 11.11 shows how removing nitrogen from the combustion air can significantly increase the available heat in a combustion system. As the oxygen content in the oxidant increases, the available heat also increases because less energy is carried away by the diluent nitrogen.

Reducing the diluent then reduces the sensible heat load on the combustion process. One problem with nitrogen is that it does not radiate its heat back into the combustor [18]. It does transfer heat by convection but this is often not as efficient or desirable in many types of industrial heating processes, especially those that are not designed with integral heat recovery systems.

Removing some or all of the nitrogen from the oxidant comes at some energy cost, normally in the form of electricity used in the separation process. An overall energy analysis must be done to show that less fossil fuel is used both in the process and at the power plant to generate the higher O_2 oxidant.

Another strategy for reducing CO_2 emissions is to use fuels that produce less CO_2 per unit of heat output. For example, replacing a fossil fuel with hydrogen can completely eliminate CO_2 as the products of combustion from a hydrogen fuel are H_2O and N_2 and some O_2. Liakos et al. [19] showed a reduction in CO_2 emissions when methane was used as a fuel instead of propane.

11.2.6.2 Process Modification

The primary process modification strategy to minimize CO_2 emissions is reduced combustion of fossil fuels. Reduced fuel burning is advocated for many reasons, not just CO_2 emission reductions. There are other pollutants like CO and NO_x that are also reduced if less fossil fuel is consumed. Reserves of important fuels like natural gas and oil continue to dwindle, so reducing fuel consumption will help conserve these limited resources. The problem is that demand for energy continues to rise around the world. Therefore, fuel conservation would make other energy sources, particularly renewable sources like solar, wind, and hydroelectric energy, more attractive and cost competitive as usage increases. More research is likely to be funded for these alternative sources if fossil fuel use is restricted.

In some countries, particularly in Europe, CO_2 emissions are being reduced by taxing fuels containing carbon. This is commonly referred to as a carbon tax. Another indirect method of reducing CO_2 emissions is by imposing minimum efficiency regulations on equipment that directly or indirectly uses the energy produced by burning fossil fuels. This includes making electrical devices more efficient as most electricity in most countries is produced by burning fossil fuels. The more efficient the electrical devices, the less electricity is needed, which means less fuel is consumed at the power plant. Improving efficiency reduces fuel consumption, which then reduces CO_2 emissions. This is always a preferred method to reduce emissions as it simultaneously conserves precious resources. Improving the thermal efficiency of an industrial combustion process reduces the amount of fossil fuel burned for a given production rate and therefore reduces CO_2 emissions indirectly. Brune et al. [20] note that using heat recuperation is a method to increase system efficiency and reduce CO_2 generation. A complication arises when dealing with existing low-efficiency equipment and how to impose and enforce standards in that case. Some recommend a combination of mandatory CO_2 emissions reductions combined with a market-based emission trading system, where the emissions would be controlled in part based on permitting fuel sources [8].

Another suggested combustion modification strategy is to use fuels with less carbon than other fuels so less CO_2 is generated per unit of energy produced. One recommendation is to replace some or all of the coal used in large power plants for producing electricity with natural gas [21]. The simplified global reaction for the combustion of methane, which is the primary component in most natural gases, can be written as

$$CH_4 + 2O_2 \rightarrow CO_2 + 2H_2O \tag{11.4}$$

where the diluent nitrogen in the combustion air has been ignored as it carries through on both sides of the equation. A variant of this technique is to use pure O_2 as the oxidizer instead of air, which makes it easier and simpler to recycle, capture, and/or separate CO_2 from the exhaust gas stream [22]. Notice that the combustion products (excluding nitrogen) consist of one-third CO_2 and two-thirds water on a volume basis. Compare this to the simplified global combustion reaction of coal:

$$C + O_2 \rightarrow CO_2 \tag{11.5}$$

where again the diluent nitrogen has been ignored. Notice that, excluding the nitrogen, all of the combustion products are CO_2. An even better fuel for reducing CO_2 emissions is hydrogen:

$$2H_2 + O_2 \rightarrow 2H_2O \tag{11.6}$$

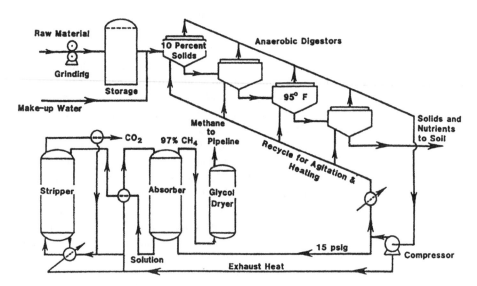

Figure 11.12 Process for production of CH_4 from biomass. (From Ref. 23. Courtesy of CRC Press.)

where no CO_2 is produced. However, it must be realized that some type of hydrocarbon fuel is normally burned to produce the hydrogen in the first place so a system analysis is required to determine how much CO_2 is generated overall.

A variation of fuel switching is to use a biomass fuel, such as corn husks, or to use fuels derived from plants, such as methanol or methane (see Fig. 11.12) [23]. The idea here is that using biofuels or fuels derived from biomass balances CO_2 generation with photosynthetic consumption so there is no net gain in CO_2 production. At this time, the technology has not progressed sufficiently, nor are the economics favorable yet, to make this a viable option at this time for industrial combustion processes. However, this is likely to change in the future.

A more dramatic process modification strategy is to use completely different sources of energy such as nuclear, wind, or solar, which do not generate any CO_2 emissions. However, in the case of nuclear power, there are other environmental issues such as the disposal of the used fuel rods that must be considered. At this time, "green" sources of energy like wind and solar have not been developed to the point to be economic or practical for the large quantities of energy that would be needed for large industrial users. Continued advances in technology should make these options more viable in the future.

11.2.6.3 Combustion Modification

A major objective in combustion modification is to increase the overall thermal efficiency so that less fuel will need to be burned for a given level of production. Reducing fuel consumption simultaneously reduces CO_2 emissions. The U.S. Department of Energy has funded major research projects designed to mitigate

CO_2 generation by combustion modification processes [24]. Most of the research is targeted at power generation, which produce the largest portion of CO_2 emissions.

One common method for increasing thermal efficiency is flue-gas recirculation. Weber et al. [25] experimentally demonstrated a furnace design methodology for using air preheating by external flue-gas recirculation to increase thermal efficiency and substantially decrease CO_2 emissions.

11.2.6.4 Post-Treatment

Pradier [26] believes that catalysis is the only viable option for moderating the accumulation of CO_2 generated by fossil-fuel combustion. Seo et al. [27] describe the absorption of CO_2 into aqueous mixtures of 2-amino-2-methyl-1-propanol with monoethanolamine.

ThermoEnergy with the support of the Department of Energy and the EPA has developed a patented process called TIPS (ThermoEnergy Integrated Power System), which is designed to capture CO_2 emissions from large power plants [28]. The process involves separating the carbon from the fuel and the nitrogen from the oxidizing air and combusting the combination in such a way as to produce high-purity CO_2 that can be more easily captured and used in a variety of applications such as enhanced oil recovery. The power needed for the various separation and scrubbing operations is self-generated in the process. However, this method will only be economical for the foreseeable future in very large processes such as in power plants.

11.2.6.5 Alternative Uses

The basic problem with most post-treatment processes designed to capture CO_2 is what to do with the collected gas. Most strategies involve some type of sequestration [29] where the CO_2 is permanently stored, rather than used. Some examples include deep-well injection and deep-water storage in the ocean [30]. The world's first commercial-scale storage of CO_2 is 800 m under the Norwegian North Sea in a sandstone aquifer referred to as the Sleipner West field run by Statoil [31]. Another variation is referred to as mineral sequestration or mineral carbonation where CO_2 is reacted with silicates of alkaline earth metals to form thermodynamically stable carbonates that are environmentally benign [32]. White et al. (2003) have written an extensive review on the capture and sequestration of CO_2 in geological formations, particularly unmineable coalbeds and deep saline aquifers [32a]. There are many potential capture techniques, including physical adsorption, cryogenic separation, membrane separation, and electrochemical pumps. Much work remains before large-scale CO_2 capture and sequestration is economically and environmentally feasible.

One unique strategy for handling CO_2 emissions compared to other pollutants is to recover the CO_2 and use it in other processes or convert it into another chemical. In order to do that, it is usually necessary to separate the CO_2 out of the exhaust gas stream, which typically contains large quantities of nitrogen and water vapor. The water is very easy to remove and requires very little energy, but separating the CO_2 from the nitrogen is not as simple and requires a significant amount of energy. Exxon has developed a process called FLEXSORB® that can separate CO_2 using sterically hindered amines [33]. Monoethanolamine has been successfully demonstrated to separate CO_2 from combustion flue gases [34]. While membranes have been demonstrated for CO_2 separation, commercial membrane systems are not yet available on a large scale to be viable for industrial combustion processes [35]. Zeolite adsorbents are being developed for separating CO_2 [36]. A hollow-fiber gas–liquid contactor (microporous membrane) is being developed for CO_2 removal from flue gas streams [37]. Regenerable polymeric sorbents have been demonstrated for separating CO_2 from gas streams [38].

One example of an alternative use of CO_2 is in preparing new chemicals such as the hydrogenation of CO_2 to make methanol [39]:

$$CO_2 + 3H_2 \rightarrow CH_3OH + H_2O \tag{11.7}$$

Palladium-promoted Cu/ZnO catalysts promote the reactions [40]. Again, it must be remembered that there is an energy cost to make the hydrogen for this process. CO_2 can also be reacted with hydrogen to make formic acid in the presence of a transition metal catalyst [41].

$$CO_2 + H_2 \rightarrow HCO_2H \tag{11.8}$$

CO_2 can be reacted with methane to form a synthesis gas [42]:

$$CO_2 + CH_4 \rightarrow 2CO + 2H_2 \tag{11.9}$$

Rhodium [43] and nickel-based [44] catalysts promote this reaction. Schmidt [45] describes a process for converting CO_2 into methane in an electrochemical cell. CO_2 can also be converted into methane in a catalytic reaction with hydrogen [46]:

$$CO_2 + 4H_2 \rightarrow CH_4 + 2H_2O \tag{11.10}$$

using a rare earth alloy catalyst. CO_2 can also be used in deep-well injection to help in the recovery of natural gas and oil.

Another example of using CO_2 in the production of other chemicals is for the production of hydrogen peroxide (H_2O_2) [47]. This process requires the use of liquid or supercritical CO_2, which is used as a solvent in the traditional anthraquinone/anthrahydroquinone technique. Use of CO_2 eliminates the organic solvent normally used, produces less waste due to the minimization of side reactions, eliminates the gas phase in the hydrogenation reactor along with the associated safety hazards, and significantly reduces pollutant emissions. There are also some significant economic benefits of this new process.

There are a number of emerging techniques for converting CO_2 into other forms. Some of these are biological. CO_2 is naturally converted by plants into carbohydrates through the process of photosynthesis where the energy is provided by the sun. There are a variety of other biological processes where CO_2 can be converted into organic matter [48–51]. There are also some emerging techniques for converting CO_2 including electrochemical [52], photon-assisted reduction [53], photoelectrochemical [54], and using CO_2 as both a solvent and a monomer in copolymerizations [55].

11.3 THERMAL RADIATION

Thermal radiation is a necessary and important phenomenon in nearly all industrial combustion processes [56]. It is typically the predominant form of heat transfer from the flame to the combustor walls and heat load. Most combustion engineers design systems to optimize the radiation heat transfer to get the proper temperature distribution in the flame. Burners are deliberately designed to maximize the radiation from the flame by creating luminous flames [57].

11.3.1 Formation Mechanisms

While thermal radiation inside the combustor is normally highly desirable, thermal radiation outside of the combustor is usually undesirable and considered a pollutant. The three primary sources of this radiation are flames occurring outside the combustor, energy radiating through openings in the furnace, and from the outside combustor walls being too hot. The first source is usually a symptom of a significant problem that needs to be corrected. The second source may be from radiation coming through viewports (see Fig. 11.13). The last source is usually due to improper insulation, a defect in the combustor design, or some type of failure in the combustor. In some cases, the nature of the high-temperature processes necessarily lead to high external wall temperatures. For example, speaking from personal experience, two large glass furnaces next to each other inside a building (not air-conditioned) located in a plant in Texas in August produce high ambient temperatures in the vicinity of the furnaces. Workers doing maintenance on top of these high-temperature glass furnaces have to wear special insulated clothing to

Figure 11.13 Radiation from a viewport.

protect them from excessive radiation. In general, this source of radiation is limited to specific applications or circumstances. It is usually either easily fixed or personnel can be properly protected from it.

The source of external thermal radiation, caused by flames outside the combustor, is also generally rare except in the case of flares that are designed to burn large quantities of fuel in the open atmosphere [58]. Thermal radiation from flares can be a significant environmental problem and is discussed further in Chap. 15. Exterior flames normally indicate a major problem and usually necessitate immediate action.

11.3.2 Environmental Concerns

Excessive external thermal radiation can cause equipment damage and injure personnel. Electronic equipment in particular is susceptible to damage from high heat loadings. External radiation can melt plastic parts such as valve seals, dry out lubricated parts such as motors, and make equipment operation more difficult where, for example, operators may need to wear gloves to open or close valves. An even more serious concern is premature ignition of fuels prior to reaching the burners, where the fuel and oxidant are premixed upstream of the burners.

Personnel injury from excessive thermal radiation should be considered in most industrial combustion applications. Heat stress can cause illnesses ranging from behavioral disorders to heat stroke and even death [59]. Heat stress is the mildest form and is temporary in nature. It could include heat rashes sometimes referred to as "prickly heat" and to dehydration. High-humidity ambient conditions can exacerbate the problem for personnel in the vicinity of high thermal radiation conditions. Skin can be damaged from excessive heat (see Chap. 15). Even lower thermal radiation loadings can cause worker fatigue and reduce performance. Older, overweight, or out-of-shape workers are particularly at risk from high thermal radiation loads.

11.3.3 Regulations

Thermal radiation is not generally directly regulated, except in the case of radiation from flares, which may have a specific permitted limit. However, general worker safety laws require that personnel be adequately protected from excessive heat. In some cases, this is somewhat subjective as workers respond differently to thermal radiation loads. Therefore, allowable radiation limits may vary according to the specific needs of the operating personnel. While not strictly regulatory, equipment manufacturers often have specifications for maximum allowable temperatures, which are indirectly determined by thermal radiation loading.

11.3.4 Measurement Techniques

Thermal radiation is typically measured with some type of radiometer [56]. Several different radiometers may be used. One is referred to as an ellipsoidal radiometer which has a hollow, gold-plated cavity in the shape of an ellipse [60]. Radiation enters through a small opening located at one of the foci of the ellipsoidal cavity. The radiation is reflected inside the ellipsoid on to a thermopile located at the other focal

Figure 11.14 Radiometer.

point. The thermopile converts the radiant flux into an electrical signal. Proper calibration of the radiometer equates the signal to a thermal flux. Another technique for measuring radiation is with heat-flux gauges. One common transducer used to measure thermal radiation is referred to as a Gardon gauge [61]. These are sometimes positioned in water-cooled housings with specific view angles and a glass window to prevent convection heating of the probe. They are calibrated in a similar way to ellipsoidal radiometers. Figure 11.14 shows a radiometer with a hemispherical plastic dome over it to prevent convective heating of the probe in order to capture only thermal radiation. Highly sophisticated measurements can be made with spectral radiometers; however, these are usually not required for external radiation measurements.

Thermal radiation can also be calculated based on surface temperatures and geometric view angles. The surface temperature of the radiating surface, for example, an external furnace wall, can be measured. The radiation from that surface can then be calculated using the surface temperature, surface emissivity, and configuration view factor [62].

11.3.5 Abatement Strategies

There are two general abatement strategies used to mitigate and control external thermal radiation loading: reducing the source of the radiation or shielding personnel and equipment from the source.

11.3.5.1 Source Reduction

This technique involves reducing the external radiation source. One way is to insulate the source, for example, a furnace wall, to reduce the external surface temperature and hence the radiation. Another way is to reduce the energy source causing the radiating surface to be heated. The easiest way to do that is to reduce the firing rate; however, this will also reduce the production rate which is normally not desirable. Another way to reduce the radiation source is to cool it, for example, by water cooling the external furnace walls. Any unplanned external flames should be immediately eliminated because of the safety risks. Viewports should have some type of shutter to minimize the heat escaping through them when they are not in use (see Fig. 11.15). Furnace leaks should be repaired, not only to reduce external thermal radiation, but also to improve the process efficiency.

11.3.5.2 Shielding and Cooling

This technique involves shielding personnel and equipment from the external radiation source. This can be done with some type of physical barrier such as a wall. It can be done with screens around the radiation source to prevent personnel from getting too close. Individual pieces of equipment can be shielded with insulation, a reflective surface, or some type of solid material. Personnel can be shielded by

Figure 11.15 Viewport with shutter.

Figure 11.16   Protective clothing for high-heat environments.

wearing appropriate protective clothing designed for high heat environments (see Fig. 11.16).

11.4 DIOXINS AND FURANS

This class of pollutants includes the carbon–hydrogen–oxygen–halogen compounds, chemically classified as halogenated aromatic hydrocarbons, and has received considerable attention from both the general public and from regulatory agencies because of the potential health hazards associated with them. Dioxins generally refer to compounds that are structurally and chemically related to chlorinated congeners (members of the same chemical family):

- Polychlorinated dibenzo-*para*-dioxins (PCDDs)
- Polychlorinated dibenzofurans (PCDFs)
- Coplanar polychlorinated biphenyls (co-PCBs)

There are over 400 types of dioxin compounds that have been identified so far, with about 30 having significant toxicity, with 2,3,7,8-TCDD (tetrachlorodibenzo-*para*-dioxin) being the most toxic. Many regulations are written in terms of 2,3,7,8-TCDD equivalent toxicity, commonly referred to as TEQ for toxic equivalent. Table 11.2 shows toxic equivalency factors developed by the World Health Organization for

Table 11.2   Toxic Equivalency Factors (TEFs) Established by the World Health
Organization for Various Dioxins and Furans

Congener	TEF value	Congener	TEF value
Dibenzo-*p*-dioxins		Non-*ortho* PCBs	
2,3,7,8-TCDD	1	PCB 77	0.0001
1,2,3,7,8-PnCDD	1	PCB 81	0.0001
1,2,3,4,7,8-HxCDD	0.1	PCB 126	0.1
1,2,3,6,7,8-HxCDD	0.1	PCB 169	0.01
1,2,3,7,8,9-HxCDD	0.1		
1,2,3,4,6,7,8-HpCDD	0.01		
OCDD	0.0001		
Dibenzofurans		Mono-*ortho* PCBs	
2,3,7,8-TCDF	0.1	PCB 105	0.0001
1,2,3,7,8-PnCDF	0.05	PCB 114	0.0005
2,3,4,7,8-PnCDF	0.5	PCB 118	0.0001
1,2,3,4,7,8-HxCDF	0.1	PCB 123	0.0001
1,2,3,6,7,8-HxCDF	0.1	PCB 156	0.0005
1,2,3,7,8,9-HxCDF	0.1	PCB 157	0.0005
2,3,4,6,7,8-HxCDF	0.1	PCB 167	0.00001
1,2,3,4,6,7,8-HpCDF	0.01	PCB 189	0.0001
1,2,3,4,7,8,9-HpCDF	0.01		
OCDF	0.0001		

Source: Ref. 63.

some of the more common dioxins and furans [63]. Compared to 2,3,7,8-TCDD,
most of the other dioxins and furans are orders of magnitude less toxic.

Furans generally refer to polychlorinated dibenzofuran (PCDF) compounds.
Some of the potential health risks include toxicity because of the poisoning effect on
cell tissues, carcinogenicity because cancerous growth may be stimulated, mutageni-
city because of possible mutations in cell structure or function, and teratogenicity
because of the potential changes to fetal tissue [64]. Over 200 dioxin/furan
compounds are regulated in certain industries, particularly in waste incineration,
and also in certain geographical locations for a wide range of applications, especially
in Europe.

Polychlorinated biphenyls (PCBs) are a special class of dioxins that have
received considerable attention recently. PCBs were first manufactured in 1929 and
were particularly popular because of their nonflammability, chemical stability, and
electrical insulating characteristics [65]. They were used in numerous industrial
applications requiring fluids with these properties and were even used as an additive
to paints, plastics, and rubber products. The problem arose when laboratory animal
studies seem to suggest that PCBs may be carcinogenic although there are not
enough data yet to be irrefutable.

In the vast majority of cases, dioxin/furan emissions result from some
contaminant in the load materials being heated in the combustor. A quick scan of
most of the textbooks on combustion shows that these emissions are essentially
ignored because they are not generally produced in the flame, except in certain
limited cases. This is primarily because there are not usually any halogens in either

the fuel or the oxidizer to produce dioxins or furans. An exception is the case when waste materials are burned as a fuel by direct injection into a flame. One example is the destruction of waste solvents that may be injected into an incinerator through the burner.

Koshland [66] has written a comprehensive review article on air toxics produced by combustion reactions including dioxins and furans. Much more research is needed to better understand dioxin/furan formation, artificial sources, long-term effects on humans and the environment, what parts of the population are at highest risk, and the ultimate fate in order to determine appropriate regulations and control techniques.

11.4.1 Formation Mechanisms

Dioxins and furans are not typically found at the end of combustion processes in waste incineration, but may be formed downstream in the exhaust gases at temperatures between 250° and 350°C [67]. They are formed from precursor compounds (predioxins) such as chlorinated benzenes, phenols, biphenyls, or chlorinated biphenyl ether. Lenoir et al. [68] experimentally showed that hexachlorobenzene is a likely precursor to chlorinated phenols and dioxins/furans. A generally accepted mechanism for dioxin/furan formation is by heterogeneous reactions in the cooler regions of a combustor involving gas-phase organic precursors, a chlorine donor (e.g., HCl), and a flyash-bound metallic catalyst such as copper chloride [69]. Dioxins/furans may be formed from nonchlorinated organic substances plus chlorine, which is referred to as de novo synthesis [70,71]. Altwicker [72] discussed the formation of dioxins and furans both from precursors and from de novo reactions.

Dioxins/furans may also be formed from the incomplete combustion of substances already containing dioxins. Dioxins and furans break down at higher temperatures. Lighty and Veranth [73] argue that there is little or no correlation between dioxin formation and chlorine in the feed or between dioxin formation and furnace conditions. They showed the three major pathways for the formation of dioxin-like compounds in Fig. 11.17. Instead, they argue that dioxin formation occurs from heterogeneous synthesis reactions on fly ash surfaces downstream of the flame in a low-temperature window around 300°C (600°F). Precursor synthesis is theorized to occur when carbon from unburned fuel combines with chlorine in the gas phase to form precursors and then the precursors condense on particle surfaces and undergo catalytic recombination reactions to form aromatics. Another formation mechanism, referred to as de novo synthesis, is theorized to occur when a carbon source is bound to particulate carbon and chlorination occurs on the particulate surface. Ring structures are liberated from the unburned carbon matrix and may undergo recombination reactions to form larger ring structures. Further research was recommended to understand dioxin formation better. Everaert and Baeyens [74] showed experimentally that de novo synthesis is the predominant formation mechanism for dioxins in large-scale thermal processes where conditions that favor precursor formation are not present.

A review of dioxin formation and minimization in municipal solid-waste incinerators discusses three primary formation theories [75]. The first is from dioxins

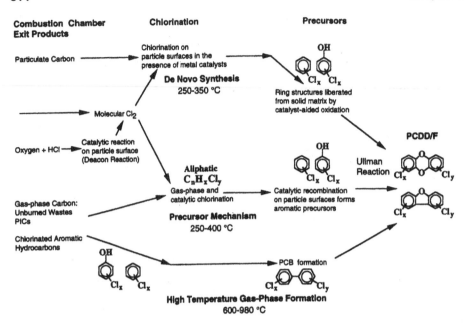

Figure 11.17 Three major pathways of dioxin-like compounds based on recent literature. (From Ref. 73. Courtesy of the Combustion Institute, Pittsburgh, PA.)

and furans already present in the feedstock. The second is from precursor compounds in the feed that can lead to dioxin and furan formation. The third is by de novo synthesis, from smaller, relatively innocuous, chemical molecules combining together, of dioxins and furans. The precursors may already exist and be carried through the combustor adsorbed or absorbed on materials such as soot or dust. The precursors can also be formed during the cooling process in conjunction with heterogeneous catalytic reactions. Maximum formation rates for dioxins occur in the temperature range 300°–400°C (600°–800°F). While most dioxins are formed from combustion processes, they have also been shown to form naturally in soils and sediments through biological processes. Tan and coworker [76,77] have assembled three thermodynamic databases for use in simulating PCDD/F generation in industrial combustion processes.

Blumenstock et al. [78] showed that the conditions in the postcombustion chamber (650°–900°C or 1200°–1650°F) of a pilot-scale waste incinerator strongly affected the formation of chlorinated and nonchlorinated aromatics including dioxins and furans. Nonoptimal combustion conditions increased dioxins, furans, and polycyclic aromatic hydrocarbons, while they had little effect on chlorinated benzenes and PCBs, and higher chlorinated dioxins/furans were only weakly affected or even decreased. It has been found that dioxin/furan formation is inhibited by the presence of sulfur so that the higher the sulfur-to-chlorine ratio, the lower the generation of dioxins/furans. The key parameter in dioxin/furan formation is residence time in the postcombustion zone [79].

Hatanaka et al. [80] experimentally studied the effect of gas temperature on dioxin/furan formation in a laboratory-scale fluidized-bed incinerator. The results showed that dioxin/furan formation increased with temperature in the primary combustion zone and decreased with temperature in the secondary combustion zone. They used electrical heaters to control the temperatures accurately and used polyvinyl chloride or sodium chloride as the chlorine sources and copper chloride as a catalyst. The increase in formation with temperature in the primary zone was explained by the devolatilization of the solid surrogate waste. Ohta et al. [81] experimentally showed that gas temperatures above about 800°C (1500°F) produced far less dioxins/furans than those produced at temperatures below about 750°C (1400°F) .

Chang and Chen [82] developed a numerical tool based on genetic programming and neural network modeling to predict the potential emissions of dioxins/furans from municipal waste incinerators. The tool is especially suited to this problem, which is highly nonlinear. Stanmore [83] also developed a model to simulate dioxin/furan formation in solid-waste incinerators. Predicted results compared favorably with experimental data.

11.4.2 Sources

In the vast majority of industrial combustion applications, there is no source to produce dioxins and furans. The combustion products contain nitrogen, water vapor, carbon dioxide, and trace pollutants like NO_x. The main source of dioxins and furans from industrial combustion applications is when hazardous wastes are burned in some type of combustor. This is due to the incoming feed materials, which may contain chlorine, especially from plastics. Katami et al. [84] showed a clear correlation between the chloride content of incoming feed materials and the levels of dioxins generated in the exhaust gases. On the other hand, an exhaustive report done for the American Society of Mechanical Engineers, using data from numerous actual commercial incinerators of all types, showed no clear and convincing relationship between dioxin/furan formation and the amount of chlorine in the waste stream [85]. The waste materials may be either solid or liquid. Preto et al. [86] discuss dioxin and furan formation from solid-waste incineration from straw or wood contaminated with salt water. They also showed a direct correlation between the amount of dioxins/furans formed and both the chlorine content in the waste feed and the amount of sulfur in the system. The higher the chlorine content the more dioxins/furans and the lower the chlorine/sulfur ratio the lower the dioxins/furans.

The most common potential source of dioxins and furans from an industrial combustion application is from waste incineration. This is discussed in more detail in Chap. 16. Other potential industrial combustion sources of dioxins are from cement kilns, light aggregate kilns, sewage sludge treatment, ferrous and nonferrous metal smelting operations, scrap metal recovery furnaces, high-temperature steel production, kraft black liquor boilers, and boilers burning hazardous wastes. For example, Buekens et al. [87] showed that dioxins/furans may be present in iron and steel manufacturing, copper smelters, and aluminum plants. Anderson and Fisher [88] showed that iron and steel production in the U.K. is a source of dioxins. Birat et al. [89] showed that dioxins may be found in the exhaust gases from electric arc furnaces. Zheng et al. [90] showed that dioxins/furans may be present in pulp mills. All of these processes are typically highly regulated, continuously monitored, and carefully designed to minimize the emissions of these carcinogens. Interestingly, Baker and Hites [91] argue that combustion is not the major source of dioxins/furans in the environment. They believe that photochemical synthesis of a specific dioxin congener (octachlorinated dibenzo-*p*-dioxin or OCDD) from pentacholorophenol in atmospheric condensed water is the most significant source of OCDD in the environment.

McKay [75] lists the following factors that affect dioxin emissions from combustion sources:

1. PCDD in the feed (e.g., wood preservatives or pesticides).
2. Precursors in the feed (e.g., chlorinated phenols, chlorinated benzenes, and PCBs that may be present in wood preservatives, herbicides, solvents, dyes, and dielectric fluids formerly used in electrical transformers).
3. Chlorine in the feed.
4. Combustion temperature (from 500°–800°C or 900°–1500°F promotes PCDD formation, while temperatures greater than 900°C or 1700°F destroy PCDDs).
5. Residence time (the higher the gas temperature, the less residence time needed to destroy PCDDs and vice versa).
6. Oxygen availability (insufficient oxygen for complete combustion of the hydrocarbons or poor air/fuel mixing will promote poor combustion conditions and PCDD formation).
7. Feed processing (liquid and gaseous fuels tend to be easier to mix with air and combust so solid fuels may need to be dried, shredded, separated, or treated in some other way to promote complete combustion to minimize PCDD formation).
8. Supplemental fuel (when the heating value of the waste material being burned is not sufficient and supplemental fuel is required, the combustion process tends to be more efficient, and the temperature, is higher which is less favorable for PCDD formation).

11.4.3 Environmental Concerns

A report issued by the U.S. EPA notes that dioxins are a serious public health threat and that there do not appear to be any "safe" levels of exposure [92]. Dioxins are known carcinogens that can cause severe reproductive and developmental problems. Nursing infants, those eating fish as part of their basic subsistence, and those eating their own livestock for subsistence appear to be at greater risk of having higher levels of dioxins in their bodies than the average person and therefore at greater risk from developing cancer.

Dioxins bind to sediment and organic material in the environment and tend to be absorbed rapidly in animal and human fatty tissue. The most common method for dioxins/furans to enter humans is through the food supply. A problem with dioxins/furans is that they are very resistant to chemical and biological transformation processes (in or near ambient conditions) and therefore they persist in the environment and accumulate in the food chain. As they are toxic in sufficient quantity, they present a health risk for both humans and animals. The most noted health effect in people besides cancer is that large amounts of dioxin/furan exposure can cause chloracne, which is a severe skin disease with acne-like lesions occurring mainly on the face and upper body. Other effects include skin rashes, skin discoloration, excessive body hair, and possibly mild liver damage.

The eventual fate of dioxins that get into the environment is not well understood. They exhibit little potential for significant leaching or volatilization once they have been adsorbed on to particulate matter. The only environmentally significant transformation process for them is believed to be photodegradation of nonsorbed species in the gaseous phase, at the soil–air interface or water–air interface, or in association with organic cosolvents. Dioxins and furans entering the atmosphere are either removed by photodegradation or by deposition. Burial in-place, resuspension back into the air, or erosion of soil to water bodies appears to be their predominant fate. The ultimate environmental sink for them appears to be aquatic sediments. Since there is no current evidence that they eventually degrade into benign species, but instead appear to accumulate, it is important to minimize their generation because of their toxic characteristics.

11.4.4 Regulations

Dioxin emissions are highly regulated because of their harmful effects on humans. Since there are relatively few applications where dioxins are an issue, these are closely monitored. The main application where there may be dioxin emissions is waste incineration (see Chap. 16), which tends to be highly regulated by nature anyway. Exhaust gas emissions must be regularly, if not continually, monitored to ensure permit limits are not exceeded. For this particular pollutant, European regulations appear to be more well developed compared to U.S. regulations. This may have to do, in part, to higher processing of contaminated materials that may contain dioxins and furans.

11.4.5 Measurement Techniques

EPA Method 23 discusses how to measure the emissions of dioxins and furans from stationary sources [93]. A sample is withdrawn isokinetically and collected on a glass fiber filter. The sample is then separated using high-resolution gas chromatography and measured by high-resolution mass spectrometry. A schematic of an approved gas-sampling train is shown in Fig. 5.42. The condenser and adsorbent trap must be specially designed for this application. Specific filters, reagents, adsorbents, and sample recovery are required. The sample analysis is also extensive. These types of measurements must be done by qualified professionals to ensure high accuracy, especially because of the toxic nature of the constituents and the strict regulatory requirements.

Everaert and Baeyens [94] note that it is not currently possible to measure dioxins/furans in real time. They collected literature data from large-scale municipal solid-waste incinerators to assess statistically possible correlations with dioxin/furan emissions for use in controlling the emissions. They found that the temperature in the electrostatic precipitator can be an important indicator of dioxin/furan levels.

11.4.6 Abatement Strategies

The most obvious way to prevent the formation of dioxins and furans is to avoid the conditions that cause them to form. The most important would be the exclusion of chlorine from the combustion system if this is possible. Other ways include minimizing particulate formation where dioxins are believed to form, having high combustion temperatures and long residence times to destroy any PCDDs that may be present, and provide adequate oxygen and mixing to ensure destruction of any PCDDs. Hunsinger et al. [95] recommend efficient flue-gas burnout and effective combustor cleaning to minimize dioxin/furan generation.

Milosavljevic and Pullumbi [96] have shown that the use of oxygen-enriched combustion air in a waste incinerator can reduce dioxin and furan emissions. Modeling and experiments showed this was due to an improvement in the combustion process, leading to higher destruction efficiencies of the dioxins and furans. The overall global oxygen level due to enrichment was 23.8%, compared to the typical 20.9% found in normal air.

Samaras et al. [97] experimentally demonstrated dioxin/furan prevention through the use of inorganic sulfur and nitrogen solid compounds in the fuel (refuse-derived fuel in this case). Reduction efficiencies of up to 98% were measured.

Liljelind et al. [98] experimentally showed in a laboratory-scale incinerator that a titanium/vanadium catalyst can effectively destroy ($>99.9\%$ destruction efficiency) dioxins/furans and also polycyclic aromatic hydrocarbons at temperatures at or above 150°C (300°F) and gas hour space velocities of 8000 ft^3/hr-ft^3. This low-temperature effectiveness is important because of the possibility of forming dioxins/furans at low temperatures in electrostatic precipitators (ESPs). This catalyst could be located after the ESP to clean up any dioxins/furans that may have formed.

Ruokojarvi et al. [99] demonstrated the use of urea as a dioxin/furan inhibitor in a pilot-scale waste incinerator. An aqueous solution of urea was injected into the flue gas, which was at a temperature of about 730°C (1350°F). The results showed that proper urea concentrations and injection points can inhibit dioxin/furan formation. Licht [100] describes the use of ammonia injection to control dioxin and HCl emissions from secondary aluminum processes that use chlorine as a fluxing agent and aluminum scrap that may contain chlorine compounds. A secondary benefit is reduction of NO$_x$ emissions.

McKay [75] recommends high burnout of particulates, with low residual carbon and low residence time for particulates in the low-temperature zone of 300°–400°C (572–752°F) to minimize PCDD formation. This assumes that the appropriate chemical species containing chlorine are present in the system so that PCDDs could be formed if the conditions were right. Everaert et al. [101] discuss the use of activated carbon to adsorb dioxins and furans, and then to capture the carbon by fabric filtration. The carbon may be injected as particulates into the flue gas stream.

REFERENCES

1. J Paul, C-M Pradier (eds.). *Carbon Dioxide Chemistry: Environmental Issues*. Cambridge, UK: The Royal Society of Chemistry, 1994.
2. K Tierney. Carbon Dioxide—the only cause of global warming? Pollut. Eng., Vol. 32, No. 11, pp. 12–13, 2000.

3. TH Standing. Climate change projections hinge on global CO_2, temperature data. *Oil & Gas J*, Vol. 99, No. 46, pp. 20–26, 2001.

4. D Dyrssen, DR Turner. Uptake of Carbon Dioxide by the Oceans. In J Paul, C-M Pradier (eds.). *Carbon Dioxide Chemistry: Environmental Issues*. Cambridge, UK: The Royal Society of Chemistry, 1994.

5. HJ Herzog, J Edmond. Disposing of CO_2 in the Ocean. In J Paul, C-M Pradier. (eds.). *Carbon Dioxide Chemistry: Environmental Issues*. Cambridge, UK: The Royal Society of Chemistry, 1994.

6. Y Shindo, Y Fujioka, K Takeuchi, T Hakuta, H Komiyama. Deep Sea CO_2 Sequestration. In J Paul, C-M Pradier (eds.). *Carbon Dioxide Chemistry: Environmental Issues*. Cambridge, UK: The Royal Society of Chemistry, 1994.

7. M Thorning. How climate change policy could shrink the federal budget surplus and stifle US economic growth. *Oil & Gas J.*, Vol. 97, No. 50, pp. 22–26, 1999.

8. P Barnes. The right way to cut carbon emissions. Environ. Protect. Vol. 12, No. 6, pp. 22–26, 2001.

9. SF Singer, R Revelle, C Starr. What to do About Greenhouse Warming: Look Before You Leap. In RA Geyer (ed.). *A Global Warming Forum: Scientific, Economic, and Legal Overview*. Boca Raton, FL: CRC Press, 1992.

10. SB Idso. Carbon Dioxide Can Revitalize the Planet. In MA Kraljic (ed.). *The Greenhouse Effect*. New York: H.W. Wilson Co., 1992.

11. H Önal. Global Warming and Agriculture: A Survey of Economic Analyses and Preliminary Assessments. In RA Geyer (ed.). *A Global Warming Forum: Scientific, Economic, and Legal Overview*. Boca Raton, FL: CRC Press, 1992.

12. NC Bhattacharya. Prospects of Agriculture in a Carbon Dioxide-Enriched Environment. In RA Geyer (ed.). *A Global Warming Forum: Scientific, Economic, and Legal Overview*. Boca Raton, FL: CRC Press, 1993.

13. DA Wirth, DA Lashof. Beyond Vienna and Montreal: A Global Framework Convention on Greenhouse Gases. In RA Geyer (ed.). *A Global Warming Forum: Scientific, Economic, and Legal Overview*. Boca Raton, FL: CRC Press, 1993.

14. D Willis, E Willis, M Harris, D Lieberman. Internal Trading of Greenhouse Gas Emissions. *Proceedings of the Air & Waste Management Association's 95th Conference & Exhibition*, Paper 43073. Baltimore, MD, June 23–27, 2002.

15. C Flavin, N Lenssen. Global Warming: The Energy Policy Challenge. In RA Geyer (ed.). *A Global Warming Forum: Scientific, Economic, and Legal Overview*. Boca Raton, FL: CRC Press, 1993.

16. U.S. Department of Energy (DOE). *Industrial Combustion Vision: A Vision by and for the Industrial Combustion Community*. Washington, DC: U.S. DOE, 1998.

17. CE Baukal (ed.). *Oxygen-Enhanced Combustion*. Boca Raton, FL: CRC Press, 1998.

18. B Ji, CE Baukal. Spectral Radiation Properties of Oxygen-Enhanced/Natural Gas Flames. In D Dolenc (ed.). *Proceedings of 1998 International Gas Research Conference*, San Diego, CA. Vol. 5, pp. 422–433, 8–11 Nov 1998.

19. HH Liakos, MA Founti, NC Markatos. Energy Savings and Environmental Impacts from Fuel Substitution in Premixed Flame Processes. *Fifth International Conference on Technologies and Combustion for a Clean Environment*, Lisbon, Portugal, Vol. 1. The Combustion Institute—Portuguese Section, pp. 459–466, 1999.

20. M Brune, M Bob, M Flamme, J Heym, A Lynen, JA Wunning, JG Wunning, HJ Dittman. Improvement of Efficiency of Industrial Furnaces with New Ceramic Self-Recuperative Burners. *Fifth International Conference on Technologies and Combustion for a Clean Environment*, Lisbon, Portugal, Vol. 1. The Combustion Institute—Portuguese Section, pp. 513–523, 1999.

21. PR Dey, EE Berkau, KB Schnelle. Environmental and Economic Benefits of Natural Gas Use for Pollution Control. In RA Geyer (ed.). *A Global Warming*

Forum: Scientific, Economic, and Legal Overview. Boca Raton, FL: CRC Press, 1993.

22. RW Bilger. The Future for Energy from Combustion of Fossil Fuels. *Fifth International Conference on Technologies and Combustion for a Clean Environment*, Lisbon, Portugal, Vol. 1. The Combustion Institute—Portuguese Section, pp. 617–623, 1999.

23. MD Ackerson, EC Clausen, JL Gaddy. The Use of Biofuels to Mitigate Global Warming. In RA Geyer (ed.). *A Global Warming Forum: Scientific, Economic, and Legal Overview*. Boca Raton, FL: CRC Press, 1993.

24. AC Bose, SI Plasynski, HM Ness, DJ Seery. Carbon Dioxide Mitigation via Combustion Modification: An Overview of U.S. Department of Energy's Power Systems Technology R&D Program. In MA Abraham, RP Hesketh (eds.). *Reaction Engineering for Pollution Prevention*. Amsterdam: Elsevier, 2000.

25. R Weber, AL Verlaan, S Orsino, N Lallemant. On Emerging Furnace Design Methodology that Provides Substantial Energy Savings and Drastic Reductions in CO_2, CO and NO_x Emissions. *Proceedings of 5th European Conference on Industrial Furnaces and Boilers*, Portugal, Vol. 1, pp. 43–57, 2000.

26. C-M Pradier. Introduction. In J Paul, C-M Pradier (eds.). *Carbon Dioxide Chemistry: Environmental Issues*. Cambridge, UK: The Royal Society of Chemistry, 1994.

27. DJ Seo, WH Hong, S-M Yang, H Lee. Absorption of Carbon Dioxide into Aqueous Mixtures of 2-Amino-2-Methyl-1-Propanol with Monoethanolamine. *Fifth International Conference on Technologies and Combustion for a Clean Environment*, Lisbon, Portugal, Vol. 1. The Combustion Institute—Portuguese Section, pp. 405–407, 1999.

28. N Moretti. TIPS for Clean Air. Pollut. Eng., Vol. 33, No.8, pp. 10–14, 2001.

29. P Mathieu, R Dubuisson, S Houyou, R Nihart. Combination of Quasi-Zero Emission Power Cycles and CO_2 Sequestration. *Fifth International Conference on Technologies and Combustion for a Clean Environment*, Lisbon, Portugal, Vol. 1. The Combustion Institute — Portuguese Section, pp. 391–398, 1999.

30. A Yamasaki, H Teng, M Wakatsuki, Y Yanagisawa, K Yamada. An Analysis of the Disposal of Anthropogenic CO_2 in the Ocean Via a Submerged Hydrate Crystallizer. In MA Abraham, RP Hesketh (eds.). *Reaction Engineering for Pollution Prevention*. Amsterdam: Elsevier, 2000.

31. www.statoil.com

32. H-J Ziock, DP Butt, KS Lackner, CH Wendt. The Need and Options Available for Permanent CO_2 Disposal. In MA Abraham, RP Hesketh (eds.). *Reaction Engineering for Pollution Prevention*. Amsterdam: Elsevier, 2000.

32a. CM White, BR Strazisar, EJ Granite, JS Hoffman, HW Pennline. Separation and Capture of CO_2 from Large Stationary Sources and Sequestration in Geological Formations—Coalbeds and Deep Saline Aquifers. *J. Air & Waste Manage. Assoc.*, Vol. 53, pp. 645–715, 2003.

33. G Sartori, WS Ho, WA Thaler, GR Chludzinski, JC Wilbur. Sterically-hindered Amines for Acid Gas Absorption. In J Paul, C-M Pradier (eds.). *Carbon Dioxide Chemistry: Environmental Issues*. Cambridge, UK: The Royal Society of Chemistry, 1994.

34. T Suda, M Fujii, T Miura, S. Shimojo, M. Iijima, S Mitsuoka. Development of Flue Gas Carbon Dioxide Recovery Technology. In J Paul, C-M Pradier (eds.). *Carbon Dioxide Chemistry: Environmental Issues*. Cambridge, UK: The Royal Society of Chemistry, 1994.

35. PHM Feron. Membranes for Carbon Dioxide Recovery from Power Plants. In J Paul, C-M Pradier (eds.). *Carbon Dioxide Chemistry: Environmental Issues*. Cambridge, UK: The Royal Society of Chemistry, 1994.

36. LVC Rees, JA Hampson. CO_2–Zeolite Reactions for Gas Separations. In J Paul, C-M Pradier (eds.). *Carbon Dioxide Chemistry: Environmental Issues*. Cambridge, UK: The Royal Society of Chemistry, 1994.

37. H Matsumoto, T Kamata, H Kitamura, M Ishibashi, H Ohta, N Nishikawa. Fundamental Study on CO_2 Removal from the Flue Gas of Thermal Power Plant by Hollow-fiber Gas–Liquid Contactor. In J Paul, C-M Pradier (eds.). *Carbon Dioxide Chemistry: Environmental Issues*. Cambridge, UK: The Royal Society of Chemistry, 1994.

38. A Diaf, J Garcia, EJ Beckman. Thermally-reversible Polymeric Sorbents for Acid Gases. In J Paul, C-M Pradier (eds.). *Carbon Dioxide Chemistry: Environmental Issues*. Cambridge, UK: The Royal Society of Chemistry, 1994.

39. B Eliasson. CO_2 Chemistry: An Option for CO_2 Emission Control? In J Paul, C-M Pradier (eds.). *Carbon Dioxide Chemistry: Environmental Issues*. Cambridge, UK: The Royal Society of Chemistry, 1994.

40. Y Kanai, T Watanabe, M Saito. Catalytic Conversion of Carbon Dioxide to Methanol over Palladium-promoted Cu/ZnO Catalysts. In J Paul, C-M Pradier (eds). *Carbon Dioxide Chemistry: Environmental Issues*. Cambridge, UK: The Royal Society of Chemistry, 1994.

41. E Dinjus, W Leitner. Transition Metal Catalysed Activation of Carbon Dioxide. In J Paul, C-M Pradier (eds). *Carbon Dioxide Chemistry: Environmental Issues*. Cambridge, UK: The Royal Society of Chemistry, 1994.

42. K Seshan, JA Lercher. Challenges in $CH_4 + CO_2$ Reforming. In J Paul, C-M Pradier (eds.). *Carbon Dioxide Chemistry: Environmental Issues*. Cambridge, UK: The Royal Society of Chemistry, 1994.

43. F Solymosi. Activation of and Reactions of CO_2 on Rh Catalysts. In J Paul, C-M Pradier (eds.). *Carbon Dioxide Chemistry: Environmental Issues*. Cambridge, UK: The Royal Society of Chemistry, 1994.

44. T Inui. Selective Synthesis of Gaseous and Liquid Fuels from CO_2. In J Paul, C-M Pradier (eds.). *Carbon Dioxide Chemistry: Environmental Issues*. Cambridge, UK: The Royal Society of Chemistry, 1994.

45. M Schmidt. The Thermodynamics of CO_2 Conversion. In J Paul, C-M Pradier (eds.). *Carbon Dioxide Chemistry: Environmental Issues*. Cambridge, UK: The Royal Society of Chemistry, 1994.

46. Y Souma, H Ando, M Fujiwara. Hydrogenation of Carbon Dioxide to Hydrocarbons. In J Paul, C-M Pradier (eds.). *Carbon Dioxide Chemistry: Environmental Issues*. Cambridge, UK: The Royal Society of Chemistry, 1994.

47. D Hâncu, EJ Beckman. Production of Hydrogen Peroxide in CO_2. In MA Abraham, RP Hesketh (eds.). *Reaction Engineering for Pollution Prevention*. Amsterdam: Elsevier, 2000.

48. GH Lorimer. Interactions of Rubisco, Nature's Most Abundant Enzyme, with CO_2. In J Paul, C-M Pradier (eds.). *Carbon Dioxide Chemistry: Environmental Issues*. Cambridge, UK: The Royal Society of Chemistry, 1994.

49. K Palmqvist. Inorganic Carbon Fluxes in Lichens and their Photosynthesizing Partners. In J Paul, C-M Pradier (eds.). *Carbon Dioxide Chemistry: Environmental Issues*. Cambridge, UK: The Royal Society of Chemistry, 1994.

50. RH Brown. Biological Uptake of CO_2. In J Paul, C-M Pradier (eds.). *Carbon Dioxide Chemistry: Environmental Issues*. Cambridge, UK: The Royal Society of Chemistry, 1994.

51. DN Silverman. The Hydration of CO_2 Catalysed by Carbonic Anhydrase. In J Paul, C-M Pradier (eds.). *Carbon Dioxide Chemistry: Environmental Issues*, Cambridge, UK: The Royal Society of Chemistry, 1994.

52. WM Ayers. An Overview of Electrochemical Carbon Dioxide Reduction. In J Paul, C-M Pradier (eds.). *Carbon Dioxide Chemistry: Environmental Issues*, Cambridge, UK: The Royal Society of Chemistry, 1994.

53. K Ravindranathan Thampi, AJ McEvoy, M Grätzel. Photon Assisted Reduction of CO_2. In J Paul, C-M Pradier (eds.). *Carbon Dioxide Chemistry: Environmental Issues.* Cambridge, UK: The Royal Society of Chemistry, 1994.

54. K Hashimoto, A Fujishima. Electrochemical and Photoelectrochemical Carbon Dioxide Reduction. In J Paul, C-M Pradier (eds.). *Carbon Dioxide Chemistry: Environmental Issues.* Cambridge, UK: The Royal Society of Chemistry, 1994.

55. MS Super, KL Parks, EJ Beckman. Carbon Dioxide as both Solvent and Monomer in Copolymerizations. In J Paul, C-M Pradier (eds.). *Carbon Dioxide Chemistry: Environmental Issues.* Cambridge, UK: The Royal Society of Chemistry, 1994.

56. CE Baukal. *Industrial Combustion Heat Transfer.* Boca Raton, FL: CRC Press, 2000.

57. AG Slavejkov, CE Baukal, ML Joshi, JK Nabors, Advanced Oxygen/Natural Gas Burner for Glass Melting. *Proceedings of 1992 International Gas Research Conference,* Orlando, FL, Nov. 16–19, 1992, published by Government Institutes, Inc., Rockville, MD, 1993.

58. R Schwartz, J White, W Bussman. Flares. In CE Baukal (ed.). *The John Zink Combustion Handbook.* Boca Raton, FL: CRC Press, 2001.

59. PD Owens. Health Hazards Associated with Pollution Control and Waste Minimization. In DL Wise, DJ Trantolo (eds.). *Process Engineering for Pollution Control and Waste Minimization.* New York: Marcel Dekker, 1994.

60. J Chedaille, Y Braud. *Industrial Flames.* Vol. 1: *Measurements in Flames.* New York: Crane, Russak & Co., 1972.

61. R Gardon. A transducer for the measurement of heat flow rate. *J. Heat Transfer,* Vol. 82, pp. 396–398, 1960.

62. R Siegel, JR Howell. *Thermal Radiation Heat Transfer,* 2nd. edn. Washington, DC: Hemisphere, 1981.

63. www.who.int/pcs/docs/dioxin-exec-sum/exe-sum-final.doc

64. WR Niessen. *Combustion and Incineration Processes,* 2nd edn. New York: Marcel Dekker, 1995.

65. T Himes. Polychlorinated biphenyls (PCBs): industrial miracle to environmental concern. *Indust. Heat.,* Vol. LXVIII, No. 6, p.16, 2001.

66. CP Koshland. Impacts and Control of Air Toxics from Combustion. *Twenty-Sixth Symposium (Int'l) on Combustion.* Pittsburgh, PA: The Combustion Institute, pp. 2049–2065, 1996.

67. G Baumbach. *Air Quality Control.* Berlin: Springer, 1996.

68. D Lenoir, A Wehrmeier, SS Sidhu, PH Taylor. Formation and inhibition of chloroaromatic micropollutants formed in incineration processes. *Chemosphere,* Vol. 43, No. 1, pp. 107–114, 2001.

69. PM Lemieux, CW Lee, JV Ryan, CC Lutes. Bench-scale studies on the simultaneous formation of PCBs and PCDD/Fs from combustion systems. *Waste Mgmt.,* Vol. 21, No. 5, pp. 419–425, 2001.

70. H Huang, A Buekens. De novo synthesis of polychlorinated dibenzo-*p*-dioxins and dibenzofurans—Proposal of a mechanistic scheme. *Sci. Total Environ.,* Vol. 193, pp. 121–141, 1996.

71. H Matzing. A simple kinetic model of PCDD/F formation by de novo synthesis. *Chemosphere,* Vol. 44, pp. 1497–1503, 2001.

72. ER Altwicker. Relative rates of formation of polychlorinated dioxins and furans from precursor and de novo reactions. *Chemosphere,* Vol. 33, No. 10, pp. 1897–1904, 1996.

73. JS Lighty, JM Veranth. The Role of Research in Practical Incineration Systems—A Look at the Past and the Future. *Twenty-Seventh Symposium (International) on Combustion.* Pittsburgh, PA: The Combustion Institute, pp. 1255–1273, 1998.

74. K Everaert, J Baeyens. The formation and emissions of dioxins in large scale thermal processes. *Chemosphere,* Vol. 46, No. 3, pp. 439–448, 2002.

75. G McKay. Dioxin characterisation [sic], formation and minimisation [sic] during municipal solid waste (MSW) incineration: review. *Chem. Eng. J.*, Vol. 86, pp. 343–368, 2002.

76. P Tan, I Hurtado, D Neuschutz, G Eriksson. Thermodynamic modeling of PCDD/Fs formation in thermal processes. *Environ. Sci. Technol.*, Vol. 35, No. 9, pp. 1867–1874, 2001.

77. PF Tan, I Hurtado, D Neuschutz. Predictions for isomer distributions of toxic dioxins and furans in selected industrial combustion processes. *Chemosphere*, Vol. 46, Nos. 9–10, pp. 1287–1292, 2002.

78. M Blumenstock, R Zimmerman, KW Schramm, A Kettrup. Influence of combustion conditions on the PCDD/F-, PCB-, PCBz- and PAH-concentrations in the post-combustion chamber of a waste incineration pilot plant. *Chemosphere*, Vol. 40, Nos. 9–11, pp. 987–993, 2000.

79. I Fangmark, B van Bavel, S Marklund. Influence of combustion parameters on the formation of polychlorinated dibenzo-*p*-dioxins, benzene, and biphenyls and polyaromatic hydrocarbons in a pilot incinerator. *Environ. Sci. and Technol.*, Vol. 27, pp. 1602–1610, 1993.

80. T Hatanaka, T Imagawa, A Kitajima, M Takeuchi. Effects of combustion temperature on PCDD/Fs formation in laboratory-scale fluidized-bed incineration. *Environ. Sci Technol.*, Vol. 35, No. 24, pp. 4936–4940, 2001.

81. M Ohta, S Oshima, T Iwasa, N Osawa, K Kumatoriya, A Yamazaki, T Takasuga, M Matsushita, N Umedzu. Formation of PCDDs and PCDFs during the combustion of polyvinylidene chloride. *Chemosphere*, Vol. 44, No. 6, pp. 1389–1394, 2001.

82. NB Chang, WC Chen. Prediction of PCDDs/PCDFs emissions from municipal incinerators by genetic programming and neural network modeling. *Waste Mgmt. & Res.*, Vol. 18, No. 4, pp. 341–351, 2000.

83. BR Stanmore. Modeling the formation of PCDD/F in solid waste incinerators. *Chemosphere*, Vol. 47, No. 6, pp. 565–573, 2002.

84. T Katami, A Yasuhara, T Shibamoto. Formation of PCDDs, PCDFs, and coplanar PCBs from polyvinyl chloride during combustion in an incinerator. *Environ. Sci. Technol.*, Vol. 36, No. 6, pp. 1320–1324, 2002.

85. HG Rigo, AJ Chandler, WS Lanier. The Relationship Between Chlorine in Waste Streams and Dioxin Emissions from Waste Combustor Stacks. American Society of Mechanical Engineers' Report CRTD-Vol. 36, New York, 1995.

86. F Preto, R McCleave, P Gogolek, D McLaughlin. Factors Affecting Formation of Dioxins and Furans from Combustion of High Chloride Waste Wood. *Proceedings of 5th European Conference on Industrial Furnaces and Boilers*, Portugal, Vol. II, pp. 565–578, 2000.

87. A Buekens, E Cornelis, H Huang, T Dewettinck. Fingerprints of dioxin from thermal industrial processes. *Chemosphere*, Vol. 40, Nos. 9–11, pp. 1021–1024, 2000.

88. DR Anderson, R Fisher. Sources of dioxins in the United Kingdom: the steel industry and other sources. *Chemosphere*, Vol. 46, No. 3, pp. 371–381, 2002.

89. JP Birat, A Arion, M Faral, F Baronnet, PM Marquaire, R Rambaud. Abatement of organic emissions in EAF exhaust flue gas. *Rev. Metall. Cahiers Inform. Tech.*, Vol. 98, No. 10, pp. 839–854, 2001.

90. MH Zheng, ZC Bao, XB Xu. Polychlorinated dibenzo-*p*-dioxins and dibenzofurans in paper making from a pulp mill in China. *Chemosphere*, Vol. 44, No. 6, pp. 1335–1337, 2001.

91. JI Baker, RA Hites. Is combustion the major source of polychlorinated dibenzo-*p*-dioxins and dibenzofurans to the environment? A mass balance investigation. *Environ. Sci. Technol.*, Vol. 34, No. 14, pp. 2879–2886, 2000.

92. U.S. EPA. Estimating Exposure to Dioxin-Like Compounds. Rep. EPA/600/6-88/ 005Cb. Washington, DC: Office of Research and Development, 1994.

93. U.S. EPA. Method 23: Determination of Polychlorinated Dibenzo-*p*-Dioxins and Polychlorinated Dibenzofurans from Stationary Sources. Code of Federal Regulations 40, Part 60, Appendix A-7. Washington, DC: U.S. Environmental Protection Agency, 2001.

94. K Everaert, J Baeyens. Correlation of PCDD/F emissions with operating parameters of municipal solid waste incinerators. *J. Air Waste Mgmt. Assoc.* Vol. 51, No. 5, pp. 718–724, 2001.

95. H Hunsinger, K Jay, J Vehlow. Formation and destruction of PCDD/F inside a grate furnace. *Chemosphere*, Vol. 46, Nos. 9–10, pp. 1263–1272, 2002.

96. I Milosavljevic, P Pullumbi. Reduction of Dioxins and Furans in Incineration. In MA Abraham, RP Hesketh (eds.). *Reaction Engineering for Pollution Prevention*. Amsterdam: Elsevier, 2000.

97. P Samaras, M Blumenstock, D Lenoir, KW Schramm, A Kettrup. PCDD/F prevention by novel inhibitors: Addition of inorganic S- and N-compounds in the fuel before combustion. *Environ. Sci. Technol.*, Vol. 34, No. 24, pp. 5092–5096, 2000.

98. P Liljelind, J Unsworth, O Maaskant, S Marklund. Removal of dioxins and related aromatic hydrocarbons from flue gas streams by adsorption and catalytic destruction. *Chemosphere*, Vol. 42, Nos. 5–7, pp. 615–623, 2001.

99. P Ruokojarvi, A Asikainen, J Ruuskanen, K Tuppurainen, C Mueller, P Kilpinen, N Yli-Keturi. Urea as a PCDD/F inhibitor in municipal waste incineration. *J. Air Waste Mgmt. Assoc.*, Vol. 51, No. 3, pp. 422–431, 2001.

100. CA Licht. Use of Ammonia Gas to Control Hydrogen Chloride Gas Emissions and Limit the Formation of Dioxins in Secondary Aluminum Melting Operations. *Proceedings of the Air & Waste Management Association's 95th Conference & Exhibition*, Paper 41775. Baltimore, MD, June 23–27, 2002.

101. K Everaert, J Baeyens, J Degreve. Removal of PCDD/F from incinerator flue gases by entrained-phase adsorption. *J. Air Waste Mgmt. Assoc.* Vol. 52, pp. 1378–1388, 2002.

102. EPA 40CFR 60—Standards of performance for New Stationary Sources, Washington, D.C.: U.S. Environmental Protection Agency, 2001.

12

Pollution from Burners

12.1 INTRODUCTION

The U.S. Department of Energy has identified a number of goals for superior industrial burner performance [1]. Those related to environmental quality include reducing emissions of criteria pollutants by 90% and reducing CO_2 emissions to levels agreed upon by the international community. Other nonenvironmental goals that impact burner design include: ability to use multiple fuels, faster development of advanced designs, reduced specific fuel consumption, reduced product loss, safe and reliable operation, and reduced cost. Some of the identified barriers to achieving burner performance goals include:

- Burner stability limits for low NO_x burners.
- Poor understanding of mixing momentum characteristics of fuels and oxidants.
- Poor understanding of noise as a pollutant.
- Limited ability to predict property changes.
- Poor understanding of liquid-fuel vaporization.
- Poor understanding of multiburner interactions.
- No techniques for converting fuel nitrogen into N_2.
- Poor understanding of particulate physical processes.
- Difficulties in scaling.
- Inability to calculate theoretical limits (e.g., of NO_x).
- Poor understanding of radiation in furnaces.

12.2 BURNER TYPE

While there are many different types of burners and many ways of classifying burners, two broad categories are considered in this section: open flame and radiant. This section is not intended to be comprehensive, but merely representative of some of the general parameters that influence emissions from these two types of burners.

12.2.1 Open-Flame Burners

Figure 12.1 shows an example of multiple open-flame hearth burners in an ethylene cracking furnace. Figure 12.2 shows an example of a single open-flame burner used in reformer furnaces [2]. There are many factors that affect the emission performance of these burners, which may also be referred to as hearth burners, freestanding burners, or simply burners. This section will briefly consider the effects of some key burner design factors on emission performance. However, even for a given design, there are numerous other factors that significantly influence emissions. The most obvious is the overall system geometry. Figure 12.3 shows two vertical cylindrical furnaces with the same total firing rate. One has four burners firing at a higher rate while the other has six smaller burners firing at a lower rate. The momentum and flame shapes are different for the two burner sizes. This usually changes the emissions, depending on the burner design. Figure 12.4 shows two vertical cylindrical furnaces with four burners each where the burners are spaced differently. In one furnace the burners are farther apart from each other and closer to the wall. In the other furnace, the burners are farther from the wall, but closer to each other. Again, emissions are often affected by the spacing between burners and from the furnace walls. The next subsections discuss some specific burner design factors on emissions.

Figure 12.1 Hearth burners in an ethylene cracking furnace. (Courtesy of John Zink Co. LLC, Tulsa, OK.)

Figure 12.2 Example of a downfired, forced-draft, preheated air burner used in reformers. (From Ref. 2. Courtesy of John Zink Co. LLC, Tulsa, OK.)

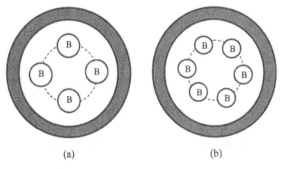

Figure 12.3 Cross-sectional view of a vertical cylindrical furnace with (a) four larger burners and (b) six smaller burners.

12.2.1.1 Momentum Effects

Momentum is the product of mass times velocity. Then, for a given mass flow rate, momentum is directly proportional to velocity. Figure 12.5 shows a diagram of two cabin heaters of different geometries. The cabin heater on the left is tall and narrow while the one on the right is short and wide. A long narrow flame is preferred on the left heater while a short wide flame is preferred on the right. For a given firing rate, the momentum from the left burner would generally be higher than for the burner on

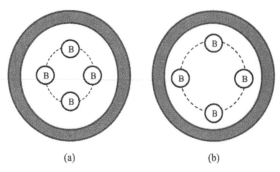

Figure 12.4 Cross-sectional view of a vertical cylindrical furnace with four burners: (a) closer to the wall; (b) farther from the wall.

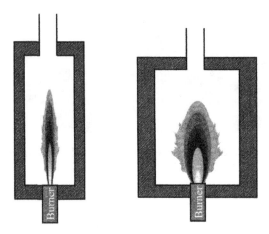

Figure 12.5 Different flame momentums in cabin heaters.

the right, although this does not necessarily have to be the case, depending on the burner design. The flame momentum is an important factor in determining furnace gas recirculation. The higher the gas-jet momentum from the burner, the more gases tend to be entrained into the flame. This flue-gas entrainment cools the flame and tends to reduce NO_x emissions (see Chap. 6). On the other hand, high-momentum gas jets can increase turbulence and mixing in the flame that may increase NO_x emissions. Therefore, the specific effect on NO_x is dependent on how the burner is designed and especially on the gas-jet velocities (momentum).

12.2.1.2 Flame Luminosity

Figure 12.6 shows a diagram of soot formation in flames where the soot forms on the interface between the fuel and the oxidizer. Soot formation is needed to make a flame luminous, which is desirable to increase flame radiation that can increase the overall system thermal efficiency. Excessive soot formation in the flame could produce smoke, unburned hydrocarbon (UHC) emissions, and high CO. Smoke is particularly an issue in flares (see Chap. 15). However, excessive soot is normally

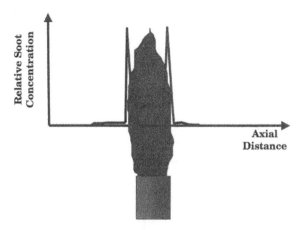

Figure 12.6 Relative soot concentration in flames.

easily controlled in burners so that smoke, UHCs, and CO are not generally problems. While soot may be formed in the initial part of the flame to enhance radiation heat transfer, the burner can be designed to burn out the soot before the exhaust gases exit the combustor. Besides increasing flame radiation, the increased flame luminosity can help decrease NO_x emissions by reducing the temperatures in the flame because of the increased heat transfer from the flame. Therefore, increasing flame luminosity may indirectly reduce NO_x emissions. Liquid and solid fuels tend to produce luminous flames, but gaseous flames generally produce more nonluminous flames (see Chap. 3). Carefully controlled mixing and burner design can enhance flame luminosity for NO_x reduction.

12.2.1.3 Firing Rate Effects

To assess the validity of the theoretical NO_x reduction using oxygen enrichment as previously discussed, a comprehensive R&D program was conducted, with partial funding from the Gas Research Institute [3]. The two regimes of low- and high-level oxygen enrichment were studied. Low-level enrichment typically involves adding pure oxygen to air to increase the total O_2 concentration from 21% to as high as 35%. In high-level enrichment, air is replaced with oxygen of varying purity, depending on the oxygen production method. These two regimes are important because they encompass most industrial applications.

These two regimes were studied by conducting an extensive set of experiments in a pilot-scale furnace (see Fig. 12.7). Several parameters were varied to study their effects on NO_x. The furnace pressure was positive to prevent air infiltration. This was done for two reasons. Many industrial furnaces are run under positive pressure to prevent infiltrating air from affecting the process. In addition, excluding air-infiltration effects isolates the effect of the burner on NO_x production. The natural gas, used for the tests, consisted of 96.6% CH_4, 0.4% N_2, plus higher hydrocarbons. Low-level enrichment was studied using a standard North American Mfg. Co. (Cleveland, OH) model 4425 air/fuel burner, as shown in Fig. 12.8 [4]. High-level

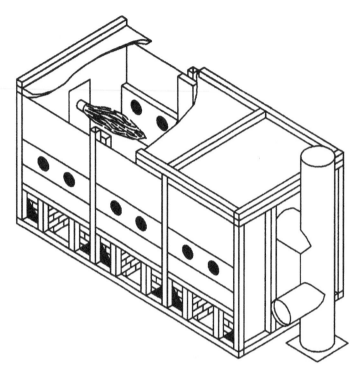

Figure 12.7 Pilot-scale industrial test furnace.

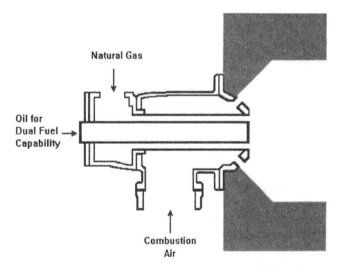

Figure 12.8 North American (Cleveland, OH) 4425-8A air/fuel burner. (From Ref. 4. Courtesy of CRC Press.)

enrichment was studied using a standard Air Products and Chemicals, Inc. (Allentown, PA) model KT-3 oxy/fuel burner, as shown in Fig. 12.9. This burner consisted of three concentric tubes with the fuel going through the inner and outer passages; O_2 flowed through the middle passage.

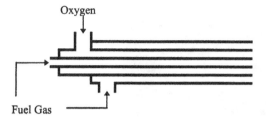

Figure 12.9 Schematic of Air Products (Allentown, PA) KT-3 oxy/fuel burner. (From Ref. 4. Courtesy of CRC Press.)

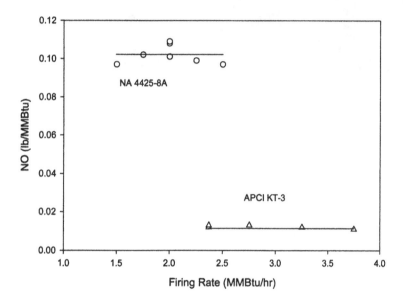

Figure 12.10 Measured NO as a function of firing rate for air/fuel and oxy/fuel burners. (From Ref. 4. Courtesy of CRC Press.)

Figure 12.10 shows a comparison of the normalized flue NO as a function of the burner firing rate. For both burners, the normalized NO_x emissions were not dependent on the firing rate for the firing rate ranges that were tested. Note that with pure oxygen and no air infiltration there was still some flue NO for the oxy/fuel burner. This was caused by the small amount of N_2 in the natural gas.

The above effect does not hold for all burner designs and all firing rates within the given operating range of a burner. Because the flame momentum is directly affected by the firing rate, the mixing and flue gas entrainment are also affected, which can cause the pollution emissions to change as a result. Very little has been published on the effects on pollution emissions as a function of the firing rate for industrial burners.

12.2.1.4 Flame Shape Effects

Figure 12.11 shows identical heaters with the same number of burners, but with different burner shapes: round flame and flat flame. For the sake of comparison,

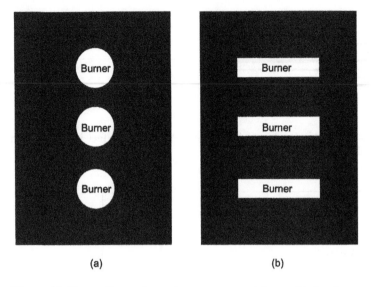

(a)   (b)

Figure 12.11   Different flame shapes: (a) round flame; (b) flat flame.

assume that the firing rate and burner cross-sectional areas are the same for both:

$$A_r = \pi r^2 = lw = A_f \tag{12.1}$$

where r is the radius of the round-flame burner, l is the length of the flat-flame burner, and w is the width of the flat-flame burner. Define the aspect ratio of the flat shape as the length divided by the width:

$$\alpha = l/w \tag{12.2}$$

Then, solve the area equation in terms of r and α:

$$w = \sqrt{\frac{\pi}{\alpha}}r \tag{12.3}$$

Calculate the perimeters of each flame in terms of r:

$$P_r = 2\pi r \tag{12.4}$$

$$P_f = 2l + 2w = 2\alpha w + 2w = 2w(\alpha + 1) = 2\sqrt{\frac{\pi}{\alpha}}r(\alpha + 1) \tag{12.5}$$

Then, divide the perimeter of the flat-flame burner by the perimeter of the round-flame burner:

$$\frac{P_f}{P_r} = \frac{2\sqrt{(\pi/\alpha)}r(\alpha + 1)}{2\pi r} = \sqrt{\frac{\pi}{\alpha}}\frac{\alpha + 1}{\pi} \tag{12.6}$$

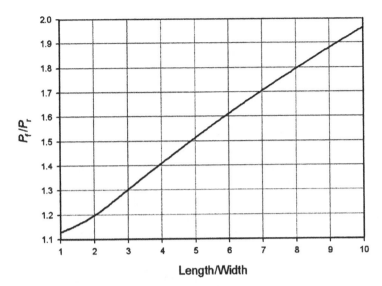

Figure 12.12 Ratio of perimeter areas for flat (rectangular) and round flame shapes as a function of the aspect ratio (length/width) of the flat flame shape.

Figure 12.12 shows a plot of this relationship. The graph shows that the ratio of perimeter lengths increases with the aspect ratio. If, for example, the length is five times the width ($\alpha = 5$), then the ratio of perimeter lengths is 1.51. This would mean that a rectangularly shaped burner would have 51% more perimeter than a round-shaped burner with the same area. Even a square burner ($\alpha = 1$) has a longer perimeter than a round burner of the same cross-sectional area. This is important in flue gas entrainment because there is more surface area for the flue gases to be entrained. This increased surface area typically enhances flue gas recirculation, which typically reduces NO_x emissions. The flat-flame shape may also enhance the radiation from the flame, which that can indirectly reduce NO_x emissions as discussed above. Therefore, in certain cases flat-shaped burners may produce emissions lower than those from comparable round burners.

12.2.1.5 Swirl Effects

Swirl means that some angular motion is imparted to the flame. This is typically achieved by angling some or all of either the combustion air or the fuel or both with respect to the main axis or normal of the burner. Swirl is often used to intensify mixing, which usually shortens the flame and increases the flame temperatures near the burner. This enhances fuel burnout and reduces or eliminates CO emissions. Swirl is often used in liquid and solid fuel combustion because these fuels are more difficult to mix with the combustion air and burn compared to gaseous fuels. Therefore, swirl can be used to reduce certain emission such as UHCs and particulates.

However, swirl may have a detrimental effect on NO_x emissions because of the increased combustion intensity and flame temperatures. Terasaki and Hayashi [5] discussed the effect of air–fuel mixing on NO_x for nonpremixed swirl burners with air preheat. Conventional swirl and double-swirl burners were compared. For the burners tested, the double-swirl burner achieved the lowest NO_x. This was primarily due to better control of the mixing and equivalence ratio profile in the upstream flame region compared to conventional swirl burners.

In some cases, swirl may actually reduce NO_x under certain conditions. Cheng et al. [6] experimentally studied the effects of fuel–air mixing on NO_x for two types of swirling burners referred to as fuel-jet dominated and strongly recirculating. The swirl number, fuel-air momentum flux ratio, and fuel injection were varied. The data showed that the strongly recirculating swirl burner produced lower NO_x. This was attributed to the strong and rapid mixing, which increases the mixture homogeneity and shortens the characteristic time for NO_x formation, resulting in lower NO_x.

Bizzo et al. [7] studied the effect of swirl on pollutant emissions in an experimental waste incinerator. They found that the amount of swirl had relatively little impact on the temperature inside the nearly adiabatic combustor, but did influence the emissions of CO and UHCs which tended to reduce with swirl due to the enhanced mixing.

Demayo and Samuelsen [8] experimentally studied VOC formation in low-NO_x generic research burners incorporating swirl. While VOC emissions were generally low under most operating conditions, the emissions were significant when the burner operated under fuel-lean conditions near the high-swirl stability limits. As this is a condition that can be controlled, VOC emissions can be minimized by operating outside of the high-swirl stability limits.

Therefore, the effect of burner swirl on pollutant emissions is dependent on the burner design, the operating conditions, and the application. Swirl usually reduces hydrocarbon and CO emission and may or may not increase NO_x emissions.

12.2.2 Radiant Burners

There are many types of radiant burners including perforated ceramic, wire mesh, flame impingement, and porous radiant to name a few (see Fig. 3.29). Figure 12.13 shows an example of radiant wall burners used in an ethylene cracking furnace. These burners are used in a wide range of applications ranging from lower temperature applications such as paper drying (see Chap. 17), medium temperature applications such as ethylene cracking furnaces (see Chap. 15), and higher temperature minerals processing (see Chap. 14). The emissions from these burners are widely variable, but in general the only pollutant of significant concern is NO_x which typically increases as the application temperature increases. In some low-temperature drying applications, NO_x emissions may be extremely low [9].

One type of porous radiant burner is made from a ceramic foam. Lammers et al. [10] have shown numerically and experimentally that this type of burner can reduce NO_x emissions by producing lower flame temperatures due to the high radiation of heat from the gases. Bouma et al. [11] numerically and experimentally studied similar burners and their effect on CO emissions. The model results for CO were slightly different than the measurements, which was attributed to the intrusive nature of the probe.

Bingue et al. [12] experimentally studied NO_x and CO emissions from methane/air flames in inert porous media. NO_x and CO emissions were found to be very low

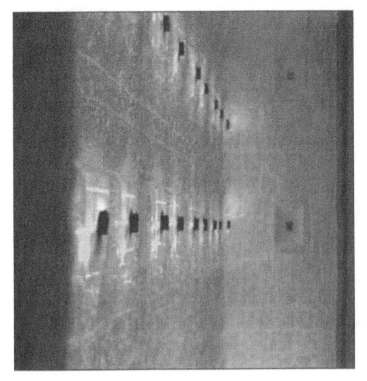

Figure 12.13 Radiant wall burners in an ethylene cracking furnace. (Courtesy of John Zink Co. LLC, Tulsa, OK.)

in ultra fuel-lean regions where NO_x levels were less than 4 ppm and CO levels were below 10 ppm. NO_x increased with equivalence ratio. Lim and Chung [13] studied liquid-fuel combustion in a porous ceramic burner. This type of burner produces clean burning and high combustion efficiency. Both CO and NO_x emissions were lower than those from a comparable burner without the porous ceramic.

12.3 EFFECTS ON POLLUTION

There are a variety of factors that affect the emissions produced by burners. Some of the more important parameters are considered next including the fuel type, mixing effects, oxidizer, and heat recuperation.

12.3.1 Fuel Type

Figure 12.14 shows an example of a liquid fuel-fired flame. Figure 12.15 shows emissions from industrial combustion processes (not including power generation) firing fuel oil [14]. Figure 12.16 shows the effect of the nitrogen content in oil fuels on NO_x emissions [15]. As can be seen, NO_x emissions increase with nitrogen content in the fuel.

Figure 12.14   Downfired burner with liquid naphtha fuel. (Courtesy of John Zink Co. LLC, Tulsa, OK.)

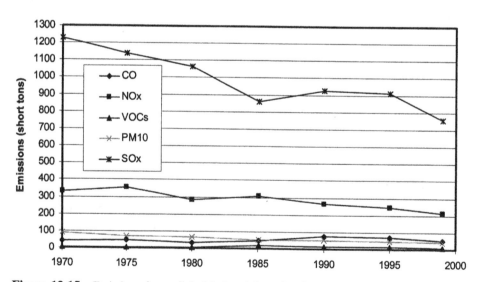

Figure 12.15   Emissions from oil-fuel industrial combustion processes (From Ref. 14.)

Allouis et al. [16] discuss soot and polycyclic aromatic hydrocarbons (PAHs) formation from rapeseed fuel oil made from vegetable oils. While this fuel is not typically used in industrial combustion processes, the effects on pollution are similar to those of other liquid fuels. The rapeseed was fired in a vertical

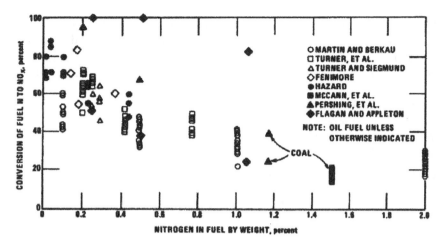

Figure 12.16 Fuel nitrogen effects on NO$_x$ from oil and coal flames. (From Ref. 15.)

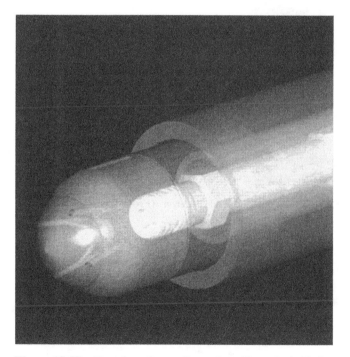

Figure 12.17 Drawing of new oil atomizer. (From Ref. 17. Courtesy of John Zink Co. LLC, Tulsa, OK.)

cylindrical furnace using low NO$_x$ burners. Fewer particulates and PAHs were formed compared to those formed by heavier fuel oils as the rapeseed was found to be cleaner burning.

Chung et al. [17] discuss the development of a new oil atomizer (see Fig. 12.17) for firing heavy residual fuel oils. The objective of the development was to maximize combustion efficiency and minimize pollutant emissions compared to those from conventional oil atomizers. The new atomizer (patent pending) is referred to as the HERO (high-efficiency residual oil) gun. Figure 12.18 shows this atomizer operating in a DEEPstar™ burner designed for firing both gas and oil. Figure 12.19 shows

Figure 12.18 Burner firing heavy oil with new oil atomizer. (From Ref. 17. Courtesy of John Zink Co. LLC, Tulsa, OK.)

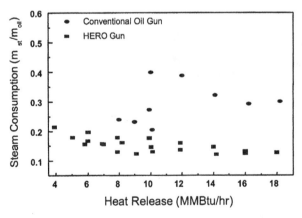

Figure 12.19 Steam consumption comparison for conventional and new oil atomizers. (From Ref. 17. Courtesy of John Zink Co. LLC, Tulsa, OK.)

that the steam consumption of the HERO gun is significantly lower than that from a conventional oil gun. This represents significant cost savings to the end-user. Figure 12.20 shows that NO_x emissions are dramatically less for the HERO gun compared to a conventional oil atomizer where both were tested in the same burner under identical conditions. Table 12.1 shows the measured particulate emissions for the HERO gun during pilot-scale testing. Table 12.2 shows a comparison of the performance of five different types of oil gun where the HERO had the lowest NO_x emissions.

Figure 12.21 shows NO_x emissions from industrial combustion processes (excluding power generation) firing on gaseous fuels [14]. This plot shows that there can be a wide range of emissions, depending on the burner design and excess O_2 in the exhaust gases. As discussed in Chap. 6, air and fuel staging are proven techniques for reducing NO_x. Internal flue-gas recirculation (FGR) is another proven technique for reducing NO_x that has the potential for very low emissions. Some of these are considered further below. The latest burner designs for low NO_x emissions often combine multiple techniques. There are typically more options available for reducing NOx with gaseous fuels compared to liquid and solid fuels where getting the fuel to burn efficiently is usually the primary concern. This is not usually a concern for gaseous fuels, which typically burn readily.

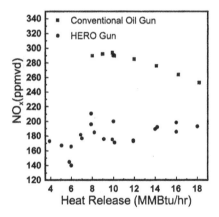

Figure 12.20 NO_x comparison for conventional and new oil atomizers. (From Ref. 17. Courtesy of John Zink Co. LLC, Tulsa, OK.)

Table 12.1 Particulate Emissions from the HERO Gun Measured in the Exhaust from a Pilot-Scale Test Furnace

Oil flow rate (lb/min)	Combustion air temperature (°F)	Furnace temperature (°F)	Stack O_2 (vol%)	Particulate emissions (mg/Nm³)
14	450	1592	3.9	47.5
14	457	1555	4	31.0

Source: Ref. 17. (Courtesy of John Zink Co. LLC.)

Table 12.2 Performance Comparison of Five Different Oil Guns

	Oil gun				
	Type A	Type B	Type C	Type D	HERO
Oil pressure (psig)	80	60	89	78	80
Steam pressure above oil pressure (psig)	20	30	19	20	20
Furnace temperature (°F)	1427	1360	1400	1379	1459
NO_x at 3% O_2 (ppmvd)	269	238	287	252	200
NO_x at 3% O_2 and 1400°F furnace temperature (ppmvd)	262	248	287	257	179
lb Steam/lb fuel	0.263	0.216	0.271	0.218	0.145
Flame length (ft)	14.5	16	14	17.5	10

All data based on 0.4 wt% fuel-bound nitrogen.
Source: Ref. 17. Courtesy of John Zink Co. LLC.

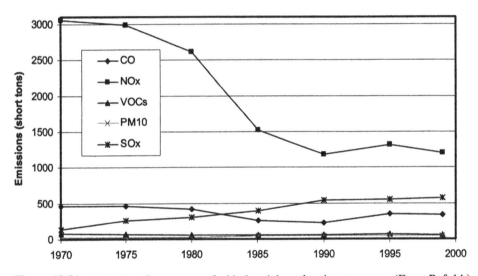

Figure 12.21 Emissions from gaseous fuel industrial combustion processes. (From Ref. 14.)

Figure 12.22 Primary and secondary flame zones.

12.3.2 Mixing Effects

Figure 12.22 shows a schematic of multiple flame zones that result from staging either the fuel or the oxidizer. Table 12.3 shows a summary of some of the common combustion modification techniques used to reduce NO_x emissions from stationary sources [15]. These are discussed in some detail in Chap. 6. Some of these are discussed here as examples of how pollution emissions are impacted by the burner design. Many of these directly or indirectly involve controlling the mixing between the fuel and the oxidant.

12.3.2.1 Fuel Staging

A schematic of fuel staging is shown in Fig. 12.23 [18] and is discussed in Chap. 2. Andrews et al. [19] experimentally and numerically showed that very low NO_x emissions (as low as 15 ppmv at 0% O_2) can be achieved with fuel staging. NO_x increased with equivalence ratio, firing rate, and with radial distance from the centerline. NO_x decreased when using reburn and with using two stages instead of one. High thermal efficiencies, low CO and low UHCs were achievable under conditions of low NO_x emissions. Lee et al. [20] studied the effect of central fuel injection with partial premixing on NO_x and CO emissions. Experimental results showed that both NO_x and CO increased with the equivalence ratio; CO was reduced by staging the fuel in the center. However, NO_x was not significantly reduced for the conditions tested by separately injecting the fuel into the center of the flame, except at lower equivalence ratios.

A specialized type of fuel staging is to pulse the incoming fuel to create alternating zones of fuel-rich and fuel-lean combustion, both of which generally produce less NO_x compared to near-stoichiometric combustion. The overall time-averaged oxidizer/fuel mixture is stoichiometric. Joshi et al. [21] describe a pulse combustion process involving oxy/fuel burners in an industrial furnace as shown in Fig. 12.24. This may include some or all of the burners depending on a variety of factors. Table 12.4 shows a comparison of nonoscillating and oscillating combustion applied to a glass-melting furnace. The oscillating case produced 45% less NO_x than the nonoscillating case with no change in the operating parameters of the furnace.

12.3.2.2 Oxidizer Staging

A schematic of air staging is shown in Fig. 12.25 and is discussed in some detail in Chap. 2. Figure 12.26 shows NO_x emissions from various burner types including conventional (no staging), fuel staging, air staging, and internal furnace gas recirculation. This technique works best for forced draft burners as natural draft burners do not usually have sufficient motive force to be effective. The technique also works for higher purity oxidizers such as pure O_2, which are often available at fairly high pressures. The primary objective of this technique is to reduce the peak flame temperatures in the flame, which reduces NO_x emissions. If improperly done, oxidizer staging can increase other emissions such as CO and particulates. Oxidizer staging may be combined with fuel staging for even deeper NO_x reductions.

12.3.2.3 Fuel/Oxidant Mixture Ratio

Figure 12.27 shows how the fuel-to-oxidant mixture ratio affects NO_x emissions [4]. The stoichiometric ratio (see Chap. 2), defined here as the ratio of the volumetric fuel flow rate to the volumetric flow rate of oxygen in the oxidizer, for the fuel in the

Table 12.3 Summary of Combustion Modification Process for Reducing NO_x Emissions

Combustion conditions	Control concept	Applicable equipment	Effect on thermal NO_x	Effect on fuel NO_x	Primary applicable controls		
					Operational adjustments	Hardware modification	Major redesign
Decrease primary flame zone O_2 level	Decrease overall O_2 level	Boilers, furnaces	Reduces O_2-rich, high NO_x pockets in the flame	Reduces exposure of fuel nitrogen intermediaries to O_2	Low excess air firing	Flue gas recirculation (FGR)	
	Delayed mixing of fuel and air	Boiler, furnaces	Flame cooling and dilution during delayed mix reduces peak temp.	Volatile fuel N reduces to N_2 to the absence of oxygen	Burner adjustments	Low NO_x burners	Optimum burner/firebox design
	Increased fuel/air mixing	Gas turbines	Reduces local hot stoichiometric regions in over all fuel lean combustion	Increases			New can design; premix, prevap.
	Primary fuel-rich flame zone	Boilers, furnaces, IC	Flame cooling in low-O_2, low-temp. primary zone reduces peak temp.	Volatile fuel N reduces to N_2 in the absence of oxygen	Burners out of service; biased burner firing	Overfire air ports, stratified charge	Burner/firebox design for two-stage combustion

Objective	Method	Application	Mechanism		Reduced air preheat	Water injection, FGR	
Decrease peak flame temperature	Decreases adiabatic flame temperature	Boilers, furnaces, IC, gas turbines	Direct suppression of thermal NO_x mechanism	Ineffective			Enlarged firebox, increased burner spacing
	Decrease combustion intensity	Boilers, furnaces	Increased flame zone cooling yields lower peak temp.	Minor direct effect; indirect effect on mixing	Load reduction		
	Increased flame zone cooling/reduce residence time	Boilers, furnaces	Increased flame zone cooling yields lower peak temp.	Ineffective			Redesign heat transfer surfaces, firebox aerodynamics
Chemically reduce NO_x in post-flame region	Inject reducing agent	Boilers, furnaces				Ammonia injection possible on some units	Redesign convective section for NH_3 injection

Source: Ref. 15.

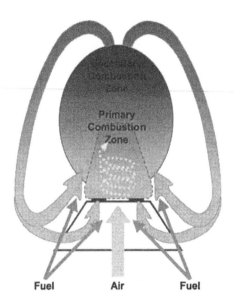

Figure 12.23 Schematic of fuel staging. (From Ref. 18. Courtesy of John Zink Co. LLC, Tulsa, OK.)

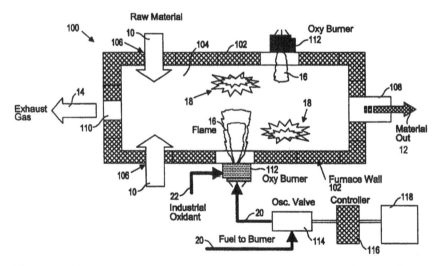

Figure 12.24 Oxy/fuel pulse combustion applied to an industrial furnace. (From Ref. 21.)

figure is approximately two when the natural gas consisted almost entirely of methane. For oxygen:fuel ratios less than two, the mixture is fuel rich and for ratios greater than two, the mixture is fuel lean. As discussed in Chap. 6, fuel-rich combustion tends to minimize NO_x emissions while fuel-lean combustion tends to increase NO_x (except in the case of ultra-lean mixtures). The data in Fig. 12.27 show this same phenomenon.

The fuel-to-oxidant ratio is very important in determining NO_x emissions. Figure 12.28 shows typical NO_x curves for premix and diffusion flames as a function

Table 12.4 Comparison of Nonoscillating and Oscillating Combustion Applied to a Glass Furnace

Parametric tests	Nonoscillating operation	Oscillating combustion
Pull rate (tons/day)	(nominal)	(nominal)
Cullet (%)	7	7
Elec. boost (kW·hr)	600	600
Fuel use (scfh NG)	20,000	19,000
Oxygen-to-fuel ratio	2.10	2.00
Oxygen use (scfh)	42,000	38,000
Furnace pressure (in. W.C.)	0.03	0.03
Air infiltration (scfh)	11,830	12,300
Oscillating parameters		
Oscillating frequency (f) (Hz)	—	0.8
Oscillating amplitude (A) (%)	—	50
Oscillating duty cycle (D) (%)	—	50
Avg. emissions (dry)		
O_2 (%)	5.0	3.6
CO_2 (%)	70.5	70.1
CO (ppmv in flue)	< 100	< 100
CO (ppmv, in stack)	< 5	< 5
NO_x (ppmv)	1,225	717
NO_x (lbs/ton of glass)	1.0	0.55
NO_x reduction (%)	—	45

Source: Ref. 21.

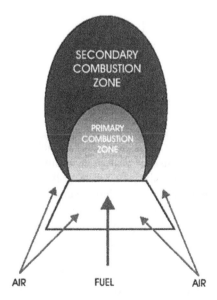

Figure 12.25 Schematic of air staging. (From Ref. 18. Courtesy of John Zink Co. LLC, Tulsa, OK.)

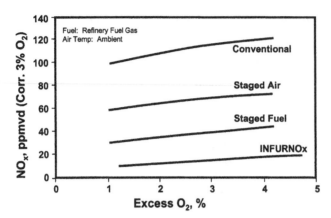

Figure 12.26 NO$_x$ emissions for various burner types. (From Ref. 18. Courtesy of John Zink Co. LLC, Tulsa, OK.)

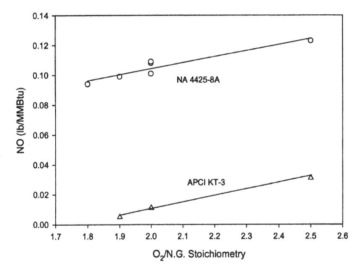

Figure 12.27 Measured NO as a function of O$_2$/natural gas stoichiometry for air/fuel and oxy/fuel burners. (From Ref. 4. Courtesy of CRC Press.)

of fuel-rich and fuel-lean mixtures. The curves are very similar on the fuel-rich side but very different on the fuel-lean side. Fuel-lean, premix flames produce much less NO$_x$ compared to fuel-rich, diffusion flames. The premix NO$_x$ curve follows the general shape of the adiabatic equilibrium predictions of NO$_x$ that assume perfect mixing (see Fig. 6.28). Peak NO$_x$ occurs near stoichiometric conditions. For diffusion flames, NO$_x$ peaks at higher excess air levels due to imperfect mixing. Figure 12.29 shows the operating regime for so-called lean premix combustion, which can produce very low NO$_x$ levels.

As previously discussed in Chap. 2, gas recirculation reduces the flame temperature and therefore NO$_x$ emissions. Figure 12.30 shows that flue/furnace gas recirculation can significantly reduce NO$_x$ in diffusion flames. Figure 12.31 shows a drawing of a diffusion flame burner, incorporating furnace gas recirculation, that

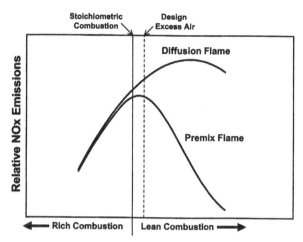

Figure 12.28 NO$_x$ vs. stoichiometry for diffusion and premix flames.

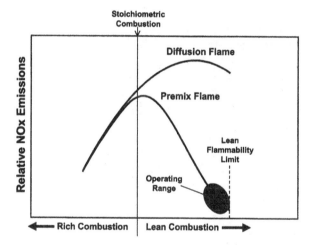

Figure 12.29 Operating regime for lean premix flames.

produces low NO$_x$ emissions. Figure 12.32 shows that burner operating in a hot furnace. Figure 12.33 shows that flue/furnace gas recirculation reduces NO$_x$ in premix flames. Figure 12.34 shows NO$_x$ emissions for diffusion and premix flames with and without flue/furnace gas recirculation. It also shows one operating scheme where the primary combustion zone is fuel-lean premixed, the secondary zone is fuel-rich premixed, and the overall operation is at the design excess air level. Note that the overall NO$_x$ emissions for this scheme are much lower than if all of the fuel and air were premixed in a single zone at the design excess O$_2$.

Figure 3.16 shows a drawing of a burner incorporating lean premixing and furnace gas recirculation. Figure 12.35 shows a picture of these burners installed on the outside of a process heater. Figure 12.36 shows an example of an ultra-lean premix burner firing in a furnace that produced very low NO$_x$ emissions.

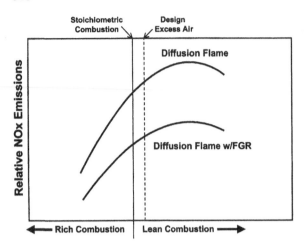

Figure 12.30 NO$_x$ vs. stoichiometry for diffusion flames with and without flue gas recirculation.

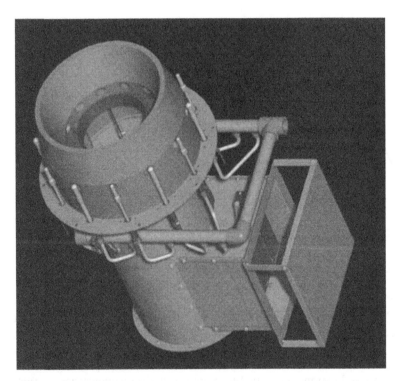

Figure 12.31 Drawing of a John Zink LM-300 burner incorporating furnace gas recirculation. (Courtesy of John Zink Co. LLC, Tulsa, OK.)

Figure 12.37 shows a crude vacuum furnace equipped with standard nonlean premix burners firing across the floor [22]. The NO$_x$ emissions were 180 ppm. Figure 12.38 shows the same crude furnace equipped with lean premix burners. The burners were installed while the furnace was in operation. NO$_x$ emissions after the conversion

Figure 12.32 John Zink LM-300 burner incorporating furnace gas recirculation operating in a furnace. (Courtesy of John Zink Co. LLC, Tulsa, OK.)

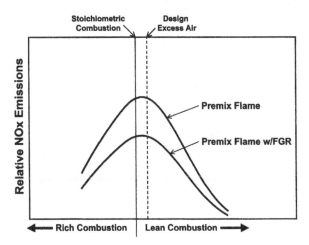

Figure 12.33 NO_x vs. stoichiometry for premix flames with and without flue gas recirculation.

to ultra-low NO_x burners were reduced to 20 ppm. Figure 12.39 shows a drawing of a radiant wall burner incorporating lean premix and furnace gas recirculation. Figure 12.40 shows a photograph of these burners operating in a process heater.

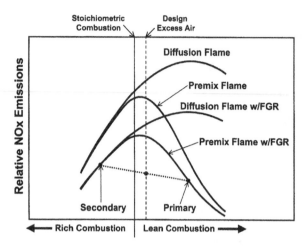

Figure 12.34 NO$_x$ vs. stoichiometry for diffusion and premix flames with and without flue gas recirculation.

Figure 12.35 Ultra-lean premix burners installed in a process heater. (Courtesy of John Zink Co. LLC, Tulsa, OK.)

12.3.2.4 Burner Design

Some specific examples of the effects of burner design on NO$_x$ emissions will be used to illustrate the importance of the burner itself, in addition to its operating conditions. A series of burner designs known as EZ-Fire™ were developed by

Figure 12.36 Flat-shaped flame burner incorporating lean premixing and furnace gas recirculation. (Courtesy of John Zink Co. LLC, Tulsa, OK.)

Figure 12.37 Process heater with conventional burners before installation of lean premix burners. (Courtesy of John Zink Co. LLC, Tulsa, OK.)

Figure 12.38 Process heater after installation of lean premix burners. (Courtesy of John Zink Co. LLC, Tulsa, OK.)

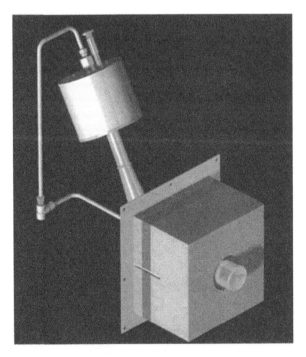

Figure 12.39 Drawing of a radiant burner incorporating lean premixing and furnace gas recirculation. (Courtesy of John Zink Co. LLC, Tulsa, OK.)

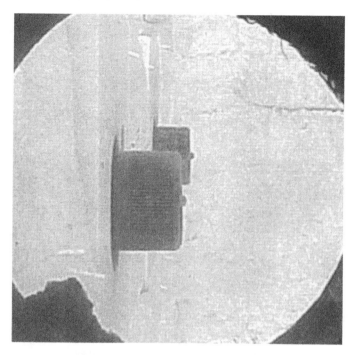

Figure 12.40 Radiant burner incorporating lean premixing and furnace gas recirculation. (Courtesy of John Zink Co. LLC, Tulsa, OK.)

Air Products and Chemicals, Inc. (Allentown, PA). Figure 12.41 shows a drawing of a first-generation model (referred to here as EZ-0) designed to use a gaseous fuel like natural gas and a combined oxidant of air and pure oxygen [23]. This design was developed for easy retrofitting of an existing air/fuel burner to increase the throughput of industrial combustion processes, particularly aluminum production. It was not designed for low NO_x emissions and in fact increased NO_x compared to air/fuel-only combustion. The increased emissions were unacceptable, which led to the development of a second-generation model (referred to here as EZ-1) shown in Fig. 12.42 [24]. Two variants of this design are shown in Figs. 12.43 and 12.44, referred to here as EZ-1A and EZ-1B, respectively.

During the development of the EZ-1, another series of burner designs (referred to here as EZ-3) was developed and tested. As will be shown, these did not reduce NO_x emissions as much as did the EZ-1 configuration. Figure 12.45 is a drawing of the EZ-3A configuration consisting of an annular design with a full flare between the middle oxygen passage and the outer air passage. Figure 12.46 shows the effect of the length of the burner block on NO_x emissions. For this design, the shorter the block the lower the NO_x. Figure 12.47 shows a similar burner design with only a half-flare between the oxygen and air passages. Figure 12.48 generally shows that the larger internal burner block diameter and the shortest internal burner block length produced lower NO_x emissions. Figure 12.49 is a drawing of the same burner shown in Fig. 12.47, except with (17) 0.5 in. holes in the half-flare. Figure 12.50 again shows that the burner with no burner tile produced less NO_x than the 7 in. long tile. In Fig. 12.51, the flare was completely removed. Fig. 12.52 again shows that for a

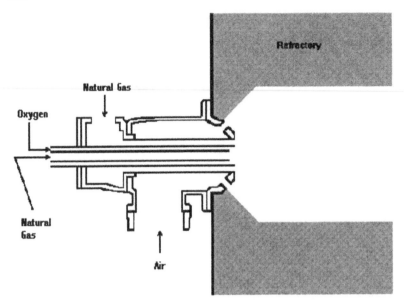

Figure 12.41   EZ-0 burner drawing. (From Ref. 23.)

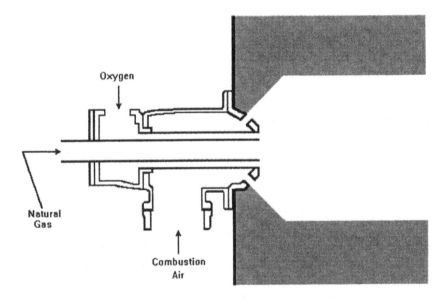

Figure 12.42   EZ-1 burner drawing. (From Ref. 24.)

fixed internal burner block diameter of 8 in., the shorter the burner block the lower the NO_x emissions. Figure 12.53 shows that the EZ-1 design produced less NO_x than did any of the EZ-3 designs. This figure also shows that the addition of the flare increased NO_x emissions.

A third-generation EZ-Fire design (referred to here as EZ-4) was developed and tested [25]. A drawing of this model is shown in Fig. 12.54. As can be seen, the half-moon-shaped air passages are significantly different from the previous designs.

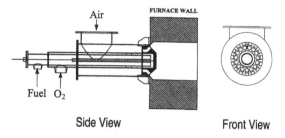

Figure 12.43 EZ-1A burner drawing. (From Ref. 24.)

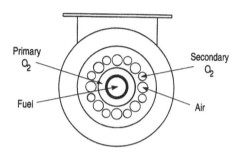

Figure 12.44 EZ-1B burner drawing. (From Ref. 24.)

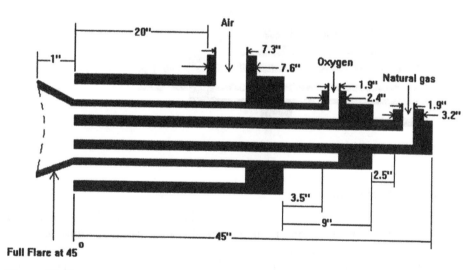

Figure 12.45 EZ-3A burner drawing. (From Ref. 24.)

In addition, oxygen lances were added in the middle of the air passages. This design was tested to vary the amount of oxygen injected through these lances compared to the amount through the central oxygen annulus. Figure 12.55 shows field measurements of the NO_x from the EZ-1A (Fig. 12.43), EZ-1B (Fig. 12.44), and EZ-4 (Fig. 12.54) designs. The EZ-4 produced considerably less NO_x compared to the EZ-1 designs. Figure 12.56 shows that lancing some of the oxygen through the air passage can significantly reduce NO_x compared to no O_2 lancing.

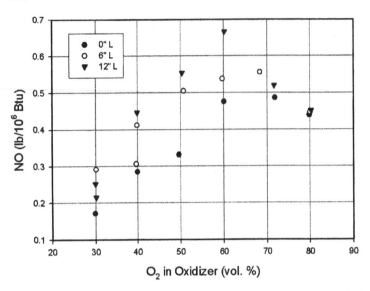

Figure 12.46   Measured NO vs. oxidant composition and burner block length for an EZ-3A with an 8 in. ID burner block. (From Ref. 24.)

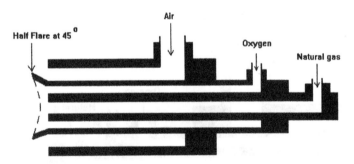

Figure 12.47   EZ-3B burner drawing. (From Ref. 24.)

12.3.3   Oxidizer Effects

The composition and temperature of the oxidizer are significant factors in determining pollutant emissions from burners. In some cases, multiple oxidizers may be used in a single burner (see Sec. 12.3.2.4 for a discussion of air–oxy/fuel burners that are used in aluminum melting). This section briefly discusses how the O_2 content in the oxidizer and the temperature of the oxidizer impact pollution emissions.

12.3.3.1   Oxidizer Composition

Figure 12.57 shows a diagram of air/fuel and oxy/fuel burners. Figure 12.58 shows how the oxygen content significantly impacts the NO_x emissions for the air/fuel (see Fig. 12.8) and oxy/fuel (see Fig. 12.9) burners previously discussed [3]. At relatively low levels of O_2 enrichment, the NO_x emissions increase rapidly (see Fig. 6.4).

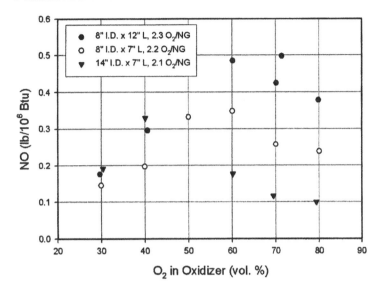

Figure 12.48 Measured NO vs. oxidant composition and burner block shape for an EZ-3B with no holes in the half-flare. (From Ref. 24.)

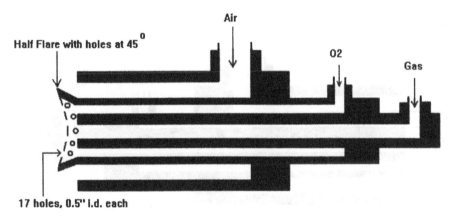

Figure 12.49 EZ-3B burner with holes in the flare. (From Ref. 24.)

Similarly, as only a small amount of nitrogen is added to pure oxygen as the oxidant, the NO_x increases rapidly. These are both important because low-level oxygen enrichment is commonly used in industrial combustion applications to improve incrementally the throughput and because oxygen purities less than 100% are often used in combustion applications as they are often much less expensive and perform nearly the same as pure oxygen. In both cases, NO_x emissions are likely to increase and must be taken into consideration.

12.3.3.2 Oxidizer Temperature

As discussed in Chap. 2, the oxidizer is typically preheated for two primary reasons. The first is to increase the overall system efficiency by using some type of heat

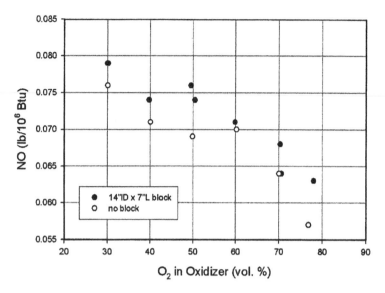

Figure 12.50 Measured NO vs. oxidant composition and burner block shape for an EZ-3B with holes in the half flare. (From Ref. 24.)

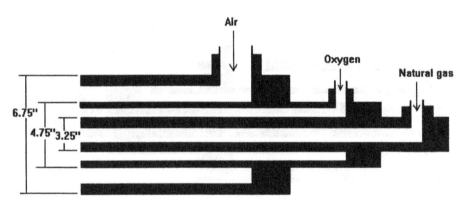

Figure 12.51 EZ-3C burner drawing. (From Ref. 24.)

exchanger to capture some of the heat in the exhaust gases to preheat the incoming oxidizer. This is generally preferable to preheating the incoming fuel because the fuel flow rate is much lower than the combustion air flow rate and because preheating the fuel could cause the hydrocarbons to deposit soot on the inside of the burner, which hinders its performance. The other reason for preheating the oxidizer is to increase the flame temperatures for high-temperature heating and melting applications such as glass production. The flame temperatures from nonpreheated air combustion are generally insufficient to melt, for example, ferrous metals and glass batch materials. However, as was shown in Chap. 6, preheating the oxidizer, particularly when it is air, can significantly increase NO_x emissions. Preheating the incoming oxidizer may also complicate the burner design because of the much higher gas temperatures inside the burner.

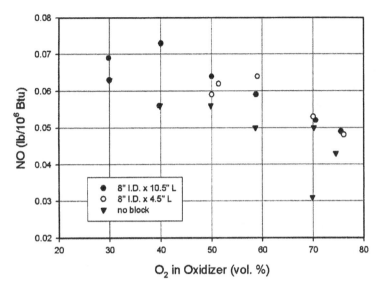

Figure 12.52 Measured NO vs. oxidant composition and burner block length for an EZ-3C with an 8 in. ID burner block. (From Ref. 24.)

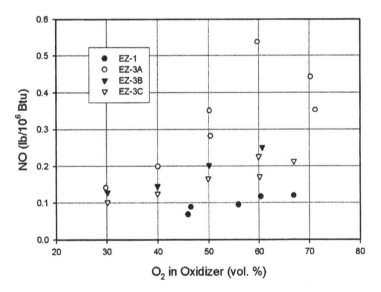

Figure 12.53 Measured NO vs. oxidant composition and burner design where the burners had no burner block, the O_2–natural gas (NG) stoichiometry was approximately 2.3, and the firing rate was 3×10^6 Btu/hr (From Ref. 24.)

Lim et al. [26] experimentally and numerically showed that NO_x emissions increase significantly with air preheat temperature. This was shown theoretically in Figs. 6.33 and 34. CO emissions increased slightly with preheat temperature. Combustion air temperatures from 300 to 560 K (81°–550°F) were tested. The experimental data and numerical predictions were in good agreement, especially for NO_x.

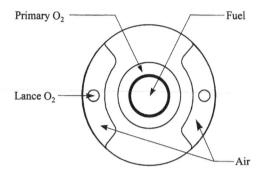

Figure 12.54   EZ-4 burner drawing. (From Ref. 25.)

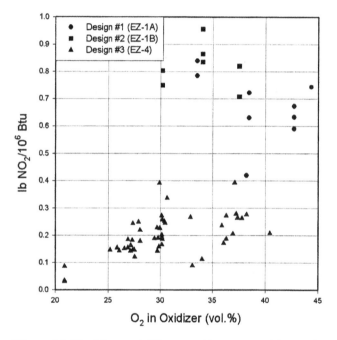

Figure 12.55   Measured NO$_2$ vs. oxidant composition and burner design (EZ-1A, EZ-1B, EZ-4) where the burners had no burner block and the O$_2$:NG stoichiometry was approximately 2.0; the firing rate and the O$_2$ flow through the lances of EZ-4 varied. (From Ref. 25.)

12.3.4   Heat Recuperation

As discussed above, heat recuperation is used both to increase the overall system thermal efficiency and sometimes to increase the flame temperatures for high-temperature applications. The most common way of recuperating heat is with the use of external heat exchangers (see Figs.1.12 and 3.1). Large regenerative heat exchangers are commonly used in the glass industry (see Figs. 14.11 and 14.12). However, there are also some burners designed with built-in heat exchangers. While fewer and larger external heat exchangers may be cheaper per unit of heat recovery

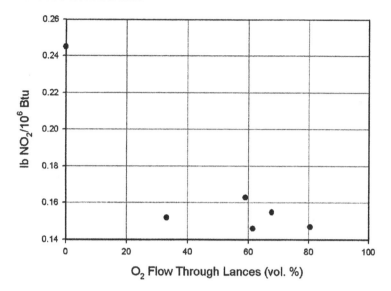

Figure 12.56 Measured NO_2 vs. the fraction of oxygen through the lances in the EZ-4 burner for an overall O_2:NG stoichiometry of approximately 2.0 and approximately 27% O_2 in the combined oxidizer. (From Ref. 25.)

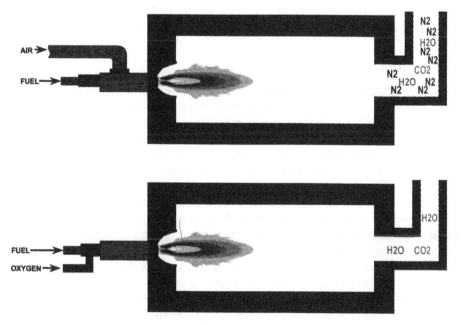

Figure 12.57 Schematic of air/fuel and oxy/fuel flames.

compared to many smaller internal heat exchangers built into the burners, the large external heat exchangers require a significant amount of large insulated ductwork. Therefore, it is sometimes more economical to have the heat exchangers built into the burners, although this complicates the burner design. The two common types of

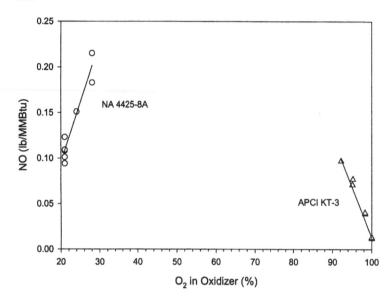

Figure 12.58 Measured NO as a function of O_2 in the oxidant for air/fuel and oxy/fuel burners (From Ref. 4. Courtesy of CRC Press.)

heat recuperation in burners are discussed briefly next. Exhaust gas recirculation through the burner is also discussed as it impacts the design and pollutant emissions.

12.3.4.1 Recuperative Burners

Recuperative burners have an internal heat exchanger of some type where the exhaust gases from the combustion process are pulled through the burner to heat typically the incoming combustion air. An example of a recuperative burner is shown in Fig. 3.35. Unlike regenerative burners, recuperative burners are in continuous operation rather than cycling between heating and cooling. While the burner design is complicated by both the internal heat exchanger and by the temperature differences of the ambient fuel gases and preheated combustion air, there is no external ductwork with the associated costs and heat losses.

However, increased emissions with preheating are an important issue. With an external heat exchanger, it is somewhat easier to reduce the preheat temperature without too many major modifications. It is not always as simple to modify a recuperative burner to reduce the preheat temperature without designing a new burner. Advances are being made in recuperative burner design to maintain the benefit of increased thermal efficiency, but without the penalty of excessive emissions. One example is by including a catalyst in the burner's recuperator to reduce emissions. Changing the aerodynamics and mixing produced by the burner is also being used to reduce emissions.

Scherello et al. [27] experimentally studied NO_x emissions from high-temperature systems incorporating recuperation to preheat the incoming combustion air to the burner. In this case the recuperator was external to the burner, but the preheated air flowed through the burner designed for high-temperature glass-melting furnaces. While the original NO_x emissions from the burner were very high, it was shown that they could be reduced substantially by changing the injection angle, changing the momentum of the combustion air, and by reducing the air preheat temperature.

12.3.4.2 Regenerative Burners

Regenerative burners incorporate some type of thermal storage medium inside the burner to increase the system fuel efficiency and in some cases the flame temperature for higher temperature applications. These burners differ from recuperative burners because they cycle between heating and cooling. For the first part of the cycle, half of the burners are firing while the other half of the burners are off. The exhaust gases from the first set of burners flow through the second set of burners. The thermal storage media is heating up in the second set of burners and cooling down in the first set of burners. After a given cycle time, typically on the order of 10–20 min, the process reverses and the first set of burners turn off while the second set begin firing. The process continues the reversal after each cycle is completed. An example of a regenerative burner is shown in Fig. 3.37.

Reducing emissions from these burners is complicated by both the cycling and by the transient temperature fluctuations during each cycle. It is often less practical to use a catalyst in the burner because of the rapidly changing temperatures. Usually the best approach is to design the burner to control the mixing carefully through the use of, for example, oxidizer and fuel staging, to minimize emissions. In some cases, it may be more practical and economic to use some type of post-treatment system such as selective catalytic reduction (see Chap. 6) to reduce emissions because the higher thermal efficiencies and flame temperatures produced by regenerative burners offset the costs of the post-treatment equipment where sufficient emission reductions cannot be easily achieved in modifying the burner design.

12.3.4.3 Exhaust Gas Recirculation

Recirculating combustion exhaust gases through a flame is a proven technique for reducing emissions such as NO_x (see Figs. 6.36–38). The exhaust gas recirculation is commonly accomplished in one of the two ways. The first, referred to here as external flue-gas recirculation (FlGR), involves taking the exhaust gases from the stack and recirculating them back to the burners with external fans and ductwork (see Figs. 1.12 and 3.1). The second technique, referred to here as internal furnace gas recirculation (FuGR), uses the burner motive force to cause the exhaust gases inside the combustor to recirculate back through the flame. Figure 12.59 shows a schematic of furnace gas recirculation. Figure 12.31 showed an example of a burner designed to recirculate furnace gases into the flame to moderate the temperature and reduce NO_x.

Figure 12.60 shows experimental data on NO_x reduction as a result of flue-gas recirculation [15]. The more exhaust gases recirculated through the flame increase, the lower the NO_x, although at a diminishing reduction rate. Figure 12.26 showed that internal FuGR was very effective at reducing NO_x compared to air and fuel staging. Rogaume et al. [28] showed that recirculating the products of combustion from a pilot-scale incinerator burning polyurethane back to the burner and mixing with the incoming ambient combustion air can reduce NO_x emissions by up to 45%

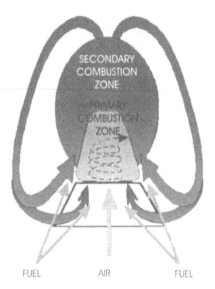

Figure 12.59 Schematic of furnace gas recirculation. (From Ref. 18. Courtesy of John Zink Co. LLC, Tulsa, OK.)

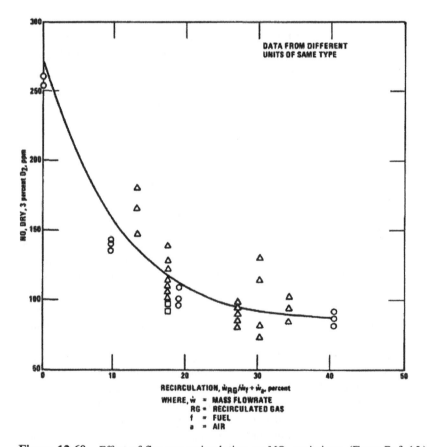

Figure 12.60 Effect of flue gas recirculation on NO_x emissions. (From Ref. 15.)

and reduce CO emissions by up to 15%. Many new low NO_x burner designs incorporate some type of FuGR because of the significant emissions reduction without the cost of external fans and ductwork to recirculate the gases.

REFERENCES

1. U.S. Department of Energy (DOE). Industrial Combustion Roadmap: A Roadmap by and for the Industrial Combustion Community. Washington, DC: U.S. DOE, 1999.
2. CE Baukal. *Heat Transfer in Industrial Combustion*. Boca Raton, FL: CRC Press, 2000.
3. AI Dalton and DW Tyndall. Oxygen Enriched Air/Natural Gas Burner System Development. Final Report, July 1984–Sept. 1989, Gas Research Institute Rep. GRI-90/ 0140, Chicago, IL, 1989.
4. CE Baukal (ed.). *Oxygen-Enhanced Combustion*. Boca Raton, FL: CRC Press, 1998.
5. T Terasaki, S Hayashi. The Effects of Fuel-Air Mixing on NO_x Formation in Non-Premixed Swirl Burners. *Twenty-Sixth Symposium (International) on Combustion*. Pittsburgh, PA: The Combustion Institute, pp. 2733–2739, 1996.
6. TS Cheng, Y-C Chao, D-C Wu, T Yuan, C-C Lu, C-K Cheng, J-M Chang. Effects of Fuel–Air Mixing on Flame Structures and NO_x Emissions in Swirling Methane Jet Flames. *Twenty-Seventh Symposium (International) on Combustion*. Pittsburgh, PA: The Combustion Institute, pp. 1229–1237, 1998.
7. WA Bizzo, L Goldstein, PR Tardin. Temperature and Gas Distributions in an Experimental Incinerator Combustion Chamber. *Fifth International Conference on Technologies and Combustion for a Clean Environment*. Lisbon, Portugal, Vol. 1. The Combustion Institute—Portuguese Section, pp. 329–335, 1999.
8. TN Demayo, GS Samuelsen. Hazardous Air Pollutant and Ozone Precursor Emissions from "Low-NO_x" Natural Gas-Fired Burners: a Parametric Study. *Fifth International Conference on Technologies and Combustion for a Clean Environment*. Lisbon, Portugal, Vol. 2. The Combustion Institute—Portuguese Section, pp. 663–671, 1999.
9. S Longacre. Using infrared to dry paper and its coatings. *Process Heat.*, Vol. 4, No. 2, pp. 45–49, 1997.
10. FA Lammers, PH Bouma, TAM Althuizen, LPH de Goey. Stability and Performance of Ceramic Foam Surface Burners in a Hot Environment. *Fifth International Conference on Technologies and Combustion for a Clean Environment*. Lisbon, Portugal, Vol. 1. The Combustion Institute—Portuguese Section, pp. 533–539, 1999.
11. PH Bouma, FA Lammers, LPH de Goey. Influence of Gas Radiation on the Combustion Behaviour (sic) of Radiant Surface Burners. *Fifth International Conference on Technologies and Combustion for a Clean Environment*. Lisbon, Portugal, Vol. 1. The Combustion Institute—Portuguese Section, pp. 541–546, 1999.
12. JP Bingue, AV Saveliev, AA Fridman, LA Kennedy. NO_x and CO Emissions of Lean and Ultra-Lean Filtration Combustion of Methane/Air Mixture in an Inert Porous Media. *Fifth International Conference on Technologies and Combustion for a Clean Environment*. Lisbon, Portugal, Vol. 1. The Combustion Institute—Portuguese Section, pp. 551–554, 1999.
13. IG Lim, KH Chung. Active Control of Combustion Process in a Liquid Fuel Burner Using Porous Ceramic Materials. *Fifth International Conference on Technologies and Combustion for a Clean Environment*. Lisbon, Portugal, Vol. 1. The Combustion Institute—Portuguese Section, pp. 563–565, 1999.
14. J Elkins, N Frank, J Hemby, D Mintz, J Szykman, A Rush, T Fitz-Simons, T Rao, R Thompson, E Wildermann, G Lear. National Air Quality and Emissions Trends Rep. 1999. Washington, DC: U.S. Environmental Protection Agency, Report EPA 454/ R-01-004, 2001.

15. U.S. EPA. Control Techniques for Nitrogen Oxides Emissions from Stationary Sources. EPA Rep. 450/1-78-001. Washington, DC: U.S. Environmental Protection Agency, 1978.

16. C Allouis, B Apicella, R Barbella, F Beretta, A Treegossi, A Ciajolo. Soot and PAH Formation in the Rapeseed Oil Spray Combustion. *Fifth International Conference on Technologies and Combustion for a Clean Environment*. Lisbon, Portugal, Vol. 1. The Combustion Institute—Portuguese Section, pp. 225–231, 1999.

17. IP Chung, A Patel, P Singh, C Strupp, CE Baukal. Innovations in atomisation [sic]. *Hydrocarbon Process.*, Vol. 7, No. 3, pp. 103–106, 2002.

18. CE Baukal (ed.). *The John Zink Combustion Handbook*. Boca Raton, FL: CRC Press, 2001.

19. GE Andrews, MC Mkpadi, JMN Mohd, M Pourkashanian, Y Yang. Lean/Lean Staged Low NO_x Combustion for Near Stoichiometric Burners. *Fifth International Conference on Technologies and Combustion for a Clean Environment*. Lisbon, Portugal, Vol. 2. The Combustion Institute—Portuguese Section, pp. 769–777, 1999.

20. T-W Lee, A Mitrovic, T Wang. Temperature, velocity, and NO_x/CO emission measurements in turbulent flames: effects of partial premixing with central fuel injection. Combust. Flame, Vol. 121, pp. 378–385, 2000.

21. ML Joshi, H Borders, O Charon. Oxy–Fuel Combustion Firing Configurations and Methods. U.S. Patent 6 398 547 B1, issued 4 June 2002.

22. JG Seebold, RG Miller, GW Spesert, DE Beckley, DJ Coutu, RT Waibel, D Venizelos, RR Hayes, W Bussman. Developing and Retrofitting Ultra Low NO_x Burners in a Refinery Furnace. *Proceedings of Joint International Combustion Symposium*, AFRC/JFRC/IEA 2001, Sec. 2C, Paper 2, Kauai, Hawaii, Sep. 2001.

23. ER Bazarian, JF Heffron, CE Baukal. Method for Reducing NO_x Production During Air–Fuel Combustion Processes. U.S. Patent 5 308 239 issued 3 May 1994.

24. CE Baukal, AG Slavejkov, LW Monroig. Method and Apparatus for Reducing NO_x Production During Air–Oxygen–Fuel Combustion. U.S. Patent 5 611 683, issued 18 Mar. 1997.

25. CE Baukal, VY Gershtein, JF Heffron, RC Best, PB Eleazer. Method and Apparatus for Reducing NO_x Production During Air–Oxygen/Fuel Combustion. U.S. Patent 5 871 343, issued 16 Feb. 1999.

26. J Lim, J Gore, R Viskanta. A study of the effects of air preheat on the structure of methane/air counterflow diffusion flames. *Combust. Flame*, Vol. 121, pp. 262–274, 2000.

27. A Scherello, M Flamme, H Kremer. New Approaches to NO_x Control for Glass Melting Furnaces resulting from advanced Burner and Process Designs. *Fifth International Conference on Technologies and Combustion for a Clean Environment*. Lisbon, Portugal, Vol. 2. The Combustion Institute—Portuguese Section, pp. 705–708, 1999.

28. Y Rogaume, F Jabouille, M Auzanneau, J-C Goudeau. Thermal Degradation and Incineration of Polyurethane: Parameters to Reduce NO_x Emissions. *Fifth International Conference on Technologies and Combustion for a Clean Environment*. Lisbon, Portugal, Vol. 1. The Combustion Institute—Portuguese Section, pp. 345–351, 1999.

13

Metal Industries

13.1 INTRODUCTION

In the metals industries, there are generally three types of processes: metal production, parts production from that metal, and heat treatment of the parts. Final finished parts are not usually made directly in a single process starting with raw materials. In the case of metals production, many of these processes are batch in nature. Solid raw ore materials or scrap metals are heated and melted to the liquid state. The molten metal is then blended with the appropriate ingredients to obtain desired grade of metal. Chemical reactions may occur in the molten metal. In the second type of high-temperature metal heating process, a specific grade of metal is melted and then formed into a specific shape or part. For example, a given grade of steel may be cast into sheet, plate, I-beams, bar stock, or other shapes. Another example would be taking a specific grade of aluminum and casting it to make a part like an engine component. The batch nature of many high-temperature metals processes complicates the heat-transfer analysis because of the transient nature and the high-temperature gradients and nonuniformity caused by the discontinuous processing. A variety of melting furnaces are used in the metals industry, as shown in Fig. 13.1 [1].

The metallurgical industry offers some of the more challenging air pollution control problems because the gas volumes and temperatures are very large and often contain particulates [2]. In addition, heavy metals and other highly toxic pollutants may be present in the gas streams. The U.S. EPA ranks metallic mineral processing as number 14 on its prioritized list of 59 major source categories where the lower the number the higher the priority [3]. Steel and gray iron foundries rank number 17 and secondary aluminum production ranks number 30.

Many metallurgical processes use some form of combustion, but not all of those are considered here. For example, a blast furnace is used in iron making. However, traditional burners are not used in this process where coke, iron ore, limestone, and other feed materials are heated in a very large vessel where they react to form iron. The coke is the fuel, but it is not fired through burners—it is actually in intimate contact with the iron ore in the vessel. This type of process is not specifically considered here.

The heating and melting of metal has been around for centuries. Many of the techniques still being used today were developed years ago and are often little

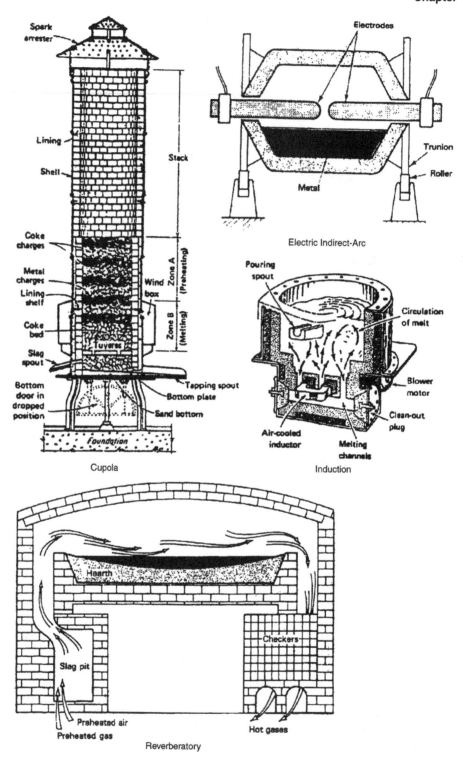

Figure 13.1   Sectional views of metal melting furnaces. (From Ref. 1.)

changed. As is customary, the metals industry has been broken into two categories: ferrous and nonferrous. The ferrous industry includes iron and steel making. The nonferrous industry includes essentially all other metals, but the more common ones include aluminum, brass, copper, and lead.

13.2 FERROUS METAL PRODUCTION

Iron is one of the more abundant and widely distributed elements in the earth's crust. It is the fourth most abundant after aluminum, silicon, and oxygen. Because raw iron readily combines with other elements, it is rarely found by itself in nature. Iron oxides are the most common naturally occurring form of iron. Steel is a generic name for a group of ferrous metals composed primarily of iron, some carbon, and other elements. Steel is used in a very wide range of applications because of its low cost, high strength, durability, formability, versatility, and flexibility. It is contained in a wide range of products such as cars, bridges, buildings, household appliances, and piping to name a few. Nearly all of the furnaces and heaters discussed in this book are constructed from some type of steel.

Figure 13.2 shows a conceptual process flow diagram for the key steps in steelmaking [4], and Fig. 13.3 shows an overview of the iron and steel manufacturing

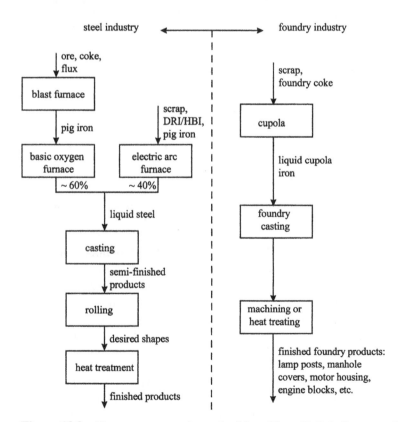

Figure 13.2 Key process steps in steelmaking. (From Ref. 4. Courtesy of CRC Press.)

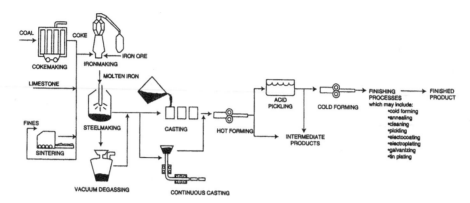

Figure 13.3 Iron and steel manufacturing process flow diagram. (From Ref. 5.)

process [5]. There are many processes in the iron and steel industries that use large quantities of fossil-fuel energy. Examples include the blast furnace, the open-hearth furnace, and the basic oxygen furnace (BOF). However, these processes do not use burners. The combustion occurs inside the vessels in conjunction with processing the materials inside where the fuel may consist of coal or coke mixed with the raw materials. Therefore, these processes are not considered here as they are not industrial combustion processes in the sense of the definition used in this book. Although a furnace like the electric arc furnace, which is discussed next, uses electricity as the primary energy source, it also often uses supplemental burners and is, therefore, considered here. Lehrman et al. [6] list the following applications in steel mills that use gaseous fuels: coke-oven heating, blast-furnace stoves, gas turbines for power generation, boilers, soaking pits, reheating furnaces, forge and blacksmith furnaces, normalizing and annealing furnaces, controlled-cooling pits, foundry core ovens, blast furnace and steel ladle drying, drying of blast-furnace runners, hot-top drying, ladle preheating, and oxy/fuel burners. Several of those that are "conventional" combustion applications as defined here are considered next. A relatively recent batch steelmaking process, known as the energy optimizing furnace (EOF), uses oxy/fuel burners to assist in melting a blend of approximately 50% scrap and 50% hot metal [7]. However, since the EOF is primarily used only in Brazil, it has not been included here.

Steel is made through a variety of processes. The preferred method in the past to produce steel was in very large vessels capable of making many tons of finished product in a so-called integrated steel mill that took raw iron ore, refined it, and then made finished products such as I-beams and piping. These large mills were located close to large users such as automotive producers. They supplied many other industries over a wide geographical area. Some of the large vessels used to make steel are open-hearth furnaces and BOFs. Figure 13.4 shows some of the products made from continuous casters.

While there are certain economies of scale associated with an integrated mill, many other factors have changed the industry to more smaller and less integrated mills, sometimes called mini-mills. These mini-mills usually produce a narrower range of products and supply a smaller geographical area. However, this specialization

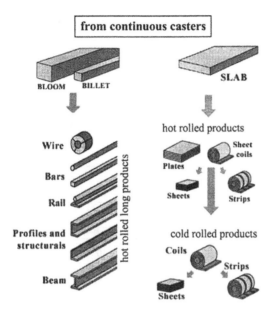

Figure 13.4 Major semifinished and finished steel products. (From Ref. 4. Courtesy of CRC Press.)

compared to large integrated mills has reduced production costs due to the focus on fewer products and generally reduced transportation costs. In general, much more scrap metal is used in mini-mills compared to integrated mills. Only a fraction of the energy required to produce steel from iron ore is needed to process scrap steel into a usable end product. While the per unit energy costs in a mini-mill may be considerably higher than those in an integrated mill, far less energy is needed per unit of steel produced, resulting in an overall cost reduction. These mills are also much more flexible in terms of what alloys can be made and in what quantities as it is not economic to make small runs of anything in the very large vessels used in integrated mills. Therefore, different production techniques are used in mini-mills to achieve this flexibility. One of these is the electric arc furnace, which is discussed in Sec. 13.2.1.

Figure 13.5 shows pollution emissions from ferrous metal production processes in the United States since 1970 [8]. All have dropped significantly since that time. CO emissions went up from 1980 to 1990 before dropping again since then. Particulate emissions increased from 1970 to 1980 before declining since then. Note that not all of these are from combustion. Many are from the high-temperature production processes such as BOFs and argon oxygen decarburizers (AODs), which are not industrial combustion applications as defined here.

13.2.1 Electric Arc Furnace

Even before the arrival of the BOF, a major alternative to conventional ore-based steelmaking had begun to gain in acceptance [9]. The electric arc furnace or EAF was conceived as a unit to melt scrap steel and recycle the iron units back to useful

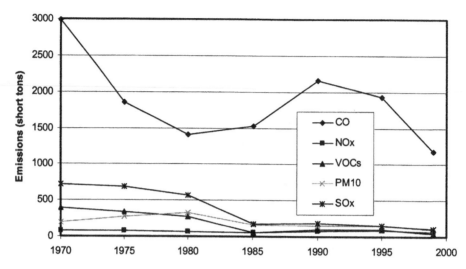

Figure 13.5   Pollution emissions from ferrous metal production processes. (From Ref. 8.)

service. A schematic of an EAF is shown in Fig. 13.6, and a photograph of an EAF is shown in Fig. 13.7. It did not rely on molten iron from a blast furnace, but rather solid scrap was melted by energy input from electrical arcs passed between three carbon electrodes. Originally this furnace was conceived as an all-electric melter. Freedom from the blast furnace and comparatively cheap electric power fueled the growth of the EAF, particularly in what came to be known as mini-mills. They offer several important advantages including low construction cost, flexibility in the use of raw materials, ability to produce a wide range of grades of steel, improved process control, and the ability to run at reduced capacity. Their popularity is increasing in the steel industry, at the expense of open-hearth furnace production, because of these attributes [10].

Electric arc furnaces are used to produce various grades of steel either from molten metal produced in an open-hearth furnace or more commonly from scrap steel with little or no iron oxide (see Fig. 13.8). This includes various grades of carbon steel, stainless steel, and other steel alloys. EAFs are cylindrically shaped, refractory-lined, vertical vessels with a sealed bottom and open top that has a lid which can be swung open or closed. There is normally a tap hole on the side of the vessel near the bottom for pouring out the refined molten metal. There is also a door on the side for removing slag, checking bath composition, adding chemicals, and blowing gases into the molten metal for further processing such as oxygen for decarburization. The vessel is mounted in a trunion device so that it can be tilted for tapping. Three graphite electrodes are inserted through the roof, in a triangular arrangement. Current is passed from one electrode to another by arcing through the metal, which may be solid, liquid, or a combination of the two, depending on the phase of the refining process. These electrodes are raised and lowered as the height of the scrap changes after a new charge is loaded down to when the scrap has been completely melted to a bath of molten metal. Most EAFs are powered by alternating current (AC furnaces) but there are some direct current (DC furnaces) EAFs in

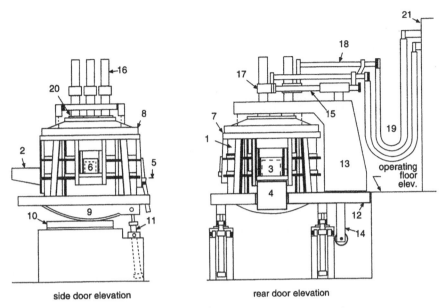

side door elevation   rear door elevation

Fig. 1.5 Schematic of a typical AC electric arc furnace. Elements are identified as follows:

1. shell	8. roof ring	15. electrode mast arm
2. pouring spout	9. rocker	16. electrode
3. rear door	10. rocker rail	17. electrode holder
4. slag apron	11. tilt cylinder	18. bus tube
5. sill line	12. main (tilting) platform	19. secondary power cables
6. side door	13. roof removal jib structure	20. electrode gland
7. bezel ring	14. electrode mast stem	21. electrical equipment vault

Figure 13.6   Schematic of an electric arc furnace. (From Ref. 15. Courtesy of AISI.)

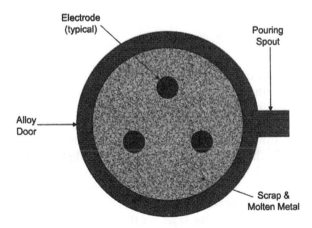

Figure 13.7   Plan view of an electric arc furnace.

certain geographical locations where electrical supplies are poor. DC furnaces only have two electrodes. A relatively newer advancement is the use of water-cooled panels for the walls to reduce refractory costs and to make the surrounding environment cooler.

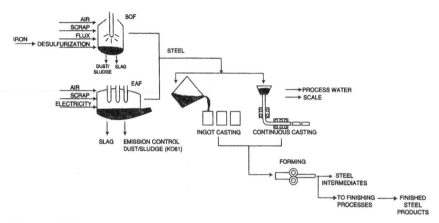

Figure 13.8 Electric arc furnace process schematic. (From Ref. 5.)

One of the advantages of the EAF is that low-grade scrap metal can be refined into a variety of steel alloys. The EAF process is very transient as multiple charges of molten metal or scrap are loaded into the vessel at various times during the production cycle. EAFs have rapid and accurate heat control compared to the large vessels used in integrated mills. While it is preferred to keep EAFs running round-the-clock, it is economically feasible to run them intermittently. This is not economically feasible with large vessels like BOFs that need to run continuously as too much energy would be wasted heating them up and cooling them down.

The Environmental Protection Agency (EPA) has established emission guidelines for electric arc furnaces [11,12] and for EAFs producing ferroalloys such as ferrosilicon [13]. A schematic of an EAF and its associated equipment is shown in Fig. 13.9 [14].

13.2.1.1 Melting Cycle

The EAF is a batch process where heats or cycles may take anywhere from 1 to 5 h, depending on the size and quality of the charge materials, the power input to the furnace including supplemental fossil-fuel firing, and the quality of the steel being produced. An initial charge of scrap metal is placed in the furnace. Often a pool of molten metal, referred to as a heal, from the previous charge is left behind. One of the strengths of the EAF process is the wide range of scrap metal that can be processed. The scrap may be pretreated by removing contaminants such as paints and varnishes, preheating the scrap with a heat recuperation system, and/or shredding the scrap to increase the loading density. Fluxes such as lime and dolomite are added to the scrap to provide a protective slag layer on the top of the molten metal, which also pulls contaminants out of the metal. For example, lime is used as a flux to reduce the sulfur and phosphorus content of the molten metal.

After a charge has been added to the furnace, the roof is swung back over the furnace after the charge and the electrodes are lowered until they contact the scrap metal. The arcing then begins and causes the scrap metal to melt. Once the electrodes are shielded by surrounding scrap, the power is increased to complete melting. Radiation from the high-intensity arcs is the primary mode of heating and melting

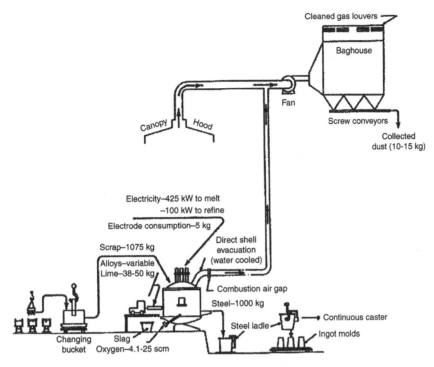

Figure 13.9 Electric arc furnace system diagram. (From Ref. 14.)

the charge. Radiation from the hot furnace walls also contributes to the heating and melting. When most or all of the scrap is molten, the roof is opened and the next charge of scrap is added to the vessel and the meltdown cycle is repeated. There may be up to five scrap charges for a complete heat.

Once all of the scrap has been melted, it is sampled to determine its composition. The metal is then refined by adding alloying agents in the proper proportions to achieve the desired composition. If the carbon levels are too high, then high-purity oxygen may be injected into the melt to oxidize the carbon to CO and CO_2, depending on the conditions. This process is referred to as decarburization. Any CO in the off-gases is usually combusted inside the gas space of the furnace both to prevent this pollutant from exiting the furnace and to gain additional heating efficiency from the heating value of the CO. This is referred to as postcombustion. Argon or other inert gases may be injected into the molten bath to promote stirring and temperature homogeneity. The slag layer is skimmed off the top of the molten metal prior to tapping. Once the desired chemistry has been achieved, the molten metal is tapped out of the furnace, either into a transfer ladle to be taken to another location or directly into molds.

13.2.1.2 Supplemental Burners

EAFs would not be considered here because the primary fuel source is electricity. However, many EAFs use supplemental fossil-fuel fired burners mounted either on the side or the roof of the vessel to assist in heating and melting the scrap metal.

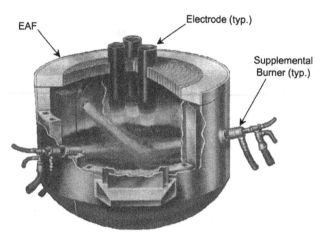

Figure 13.10 Electric arc furnace with supplemental burners.

The most common fuel in the United States for these supplemental burners is natural gas. These burners also reduce the melting cycle time and can reduce the overall production costs as natural gas is often much less expensive than electricity.

Due to the very high temperature requirements in most ferrous metal processes, oxy/fuel burners have become standard equipment on most EAFs [15]. Initially, firing through the slag door, oxy/fuel burners mounted on a boom or a carriage were introduced to increase scrap melting. Later, to target the cold spots efficiently in the EAF, the burner location was moved to either the roof or sidewalls, to aim at the cold spots associated with the spaces between the electrodes, with one to four burners in a furnace, providing supplemental energy. Figure 13.10 shows a sketch of a three-burner, wall-fired supplemental system, firing between the electrodes. In addition to productivity improvements of 5 to 20%, the burners provide economical energy for melting scrap at a lower cost.

Burners must have the flexibility of operating in either the fuel-rich or fuel-lean mode. They can be fired fuel-rich with a bushy, luminous flame early in the heat to preheat scrap without excessive bridging or welding of the scrap. They are then switched to an oxygen-rich mode to cut away scrap and assist with uniform melt-in. Operating in an oxygen-rich mode is beneficial for slag door installations to assist with cutting away the scrap to allow earlier use of an oxygen/carbon lance manipulator to produce a foamy slag.

13.2.1.3 Pollutants

The air emissions from an EAF depend on a variety of factors including the furnace size, the type and composition of the scrap, the quality (e.g., cleanliness) of the scrap, the melt rate, the number of backcharges, the refining procedure, and tapping temperature and duration [16]. These emissions are in addition to those created by supplemental fossil-fuel firing. One of the major air pollution problems associated with EAFs is the large volume of particulates generated during the melting of the scrap, which may be contaminated with oils, paints, and other coatings.

The particulates may contain oxides or iron, manganese, aluminum, calcium, magnesium, or silicon [17]. Some of these are incinerated prior to leaving the EAF, but much of these must be removed from the exhaust gas stream. Draft hoods are normally used to control the particulate emissions from the EAF process. Fabric filters (baghouses) are commonly used to capture these emissions. NO_x emissions from supplemental burners can be significant because of the high gas temperatures in an EAF. Another pollutant generated by EAFs that is somewhat unique is a massive amount of vibration and noise in the vicinity of the vessel, depending on its size. The arcing between the electrodes causes the vibrations and noise, which can be cumulatively disturbing to the operators and personnel in the vicinity.

The common air pollutants from EAFs include NO_x, CO, ozone, and metal dusts [18]. EAF dust is considered a hazardous waste, but is not considered further here as it is a by-product of the manufacturing process rather than from the combustion process. Birat et al. [19] showed that dioxins may be present in the exhaust gases from EAFs. Dioxins, furans, SO_x, and noise may also be emissions from EAFs [20]. Overcamp et al. (2003) discuss the pollutants from a DC electric melter used to treat soils contaminated with metal compounds [20a]. Typical gaseous emissions included N_2, CO_2, CO, O_2, CH_4, and particulates.

13.2.2 Smelting

Smelting is a bulk production process where solid metal ore mixtures are heated, which causes a chemical change that results in liquid metal. This process is done in a furnace often referred to as a smelter. The liquid phase may be either a liquid metal sulfide (matte) common in the production of copper and nickel, or a metallic phase common for the production of iron, steel, lead, zinc, and aluminum. The two most common smelting processes are flash smelting and bath smelting. Depending on the specific production method, large quantities of heat are normally required to melt the metal ore, which is where heat transfer from a combustion process is important.

Davila et al. [21] described an oxy/fuel rotary furnace process for smelting white cast-iron and low alloy cast carbon steels. The main parameters studied were the furnace lining, raw material composition, load protection (slag), flame types and temperature, and furnace rotation speed. The main measurements included the chemical analysis of each heat, analysis of the gas content, heat performance, metal yield, and slag analysis. The flames were only slightly oxidizing to minimize metal oxidation that would reduce yields. This process reduced energy consumption by more than 35% compared to electric induction melting.

The high temperatures necessary for smelting cause trace elements like arsenic, lead, and cadmium that may be present in the metal to volatilize. If they are not captured prior to exiting the exhaust stack then they will enter the atmosphere and deposit in the environment where they can enter plants, animals, and aquatic life and then ultimately end up in humans who eat the contaminated food sources. Smelters are equipped with fume collection and control equipment to minimize emission of heavy metals and other harmful chemicals into the atmosphere. Another potential product from smelting is SO_x, which is easily captured with the proper post-treatment equipment (see Chap. 8).

13.2.3 Ladle Preheating

Another major use for burners in the steel industry's melt shops is for ladle preheating, as shown schematically in Fig. 13.11 [22]. Figure 13.12 [23] shows a ladle that is used to transport molten metal in steel mills from the melting furnace to the casting station. Figure 13.13 shows a photograph of a ladle preheater in use. Ladles are refractory-lined cylindrical vessels closed at one end. The molten metal is poured into the top and then the ladle is usually transported via an overhead crane for pouring. There are four primary reasons for preheating ladles, prior to pouring molten metal into them. The first is to minimize the cooling effect on the molten metal. A cold ladle could cool the liquid metal enough to cause the metal to solidify inside the ladle. This reduces the metal yield and increases maintenance costs to clean out the frozen metal. The relatively high thermal conductivity of the refractories used in ladles means that excessive heat losses would result if the ladles were not preheated [24]. The second reason for preheating a ladle is to minimize the thermal shock on the refractory, which could go from ambient conditions to molten-metal temperatures (> 2000°F or 1100°C) in a matter of seconds. This rapid temperature rise often damages the hard ceramic refractory used to line ladles. A third reason to

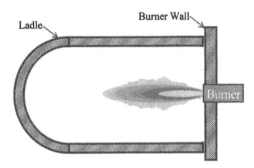

Figure 13.11 Schematic of a ladle preheating system. (From Ref. 22. Courtesy of CRC Press.)

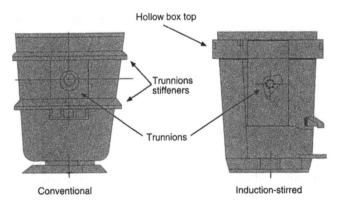

Figure 13.12 Conventional ladle used in steel melting processes. (From Ref. 23. Courtesy of AISE Steel Foundation, Pittsburgh, PA.)

Figure 13.13 Ladle preheater.

preheat ladles is to remove any moisture that may have accumulated in the vessel. As any metal producer knows, water is anathema to molten metal. Any liquid water in the ladle could quickly turn to steam with the addition of molten metal. The rapid expansion from liquid water to steam often produces violent explosions that can damage equipment and threaten the safety of personnel. The fourth reason to preheat ladles is to make the process more consistent. Without preheating, the temperature of a given ladle would vary significantly from one heat to the next, depending on how long between uses, the ambient conditions in the plant, how long the last heat was held in the vessel, how much metal was poured into the vessel, etc. The benefits of ladle preheating include lower tap temperatures, increased ladle lining life, lower refractory maintenance costs, and increased productivity and quality in casting due to more consistent steel temperatures [23].

Today's refractories and steelmaking processes often require ladle preheat temperatures above 2000°F (1100°C), which can be difficult to obtain with the relatively low temperatures generated by conventional air/fuel flames. Because of their higher flame temperature, oxy/fuel burners provide the energy required for ladle preheating to reach the desired temperature in a shorter time with minimum flue losses and maximum fuel efficiency. Even with high-temperature flames, Tomazin et al. [25] showed that, although the inside ladle refractory temperature (hot face) of an initially cold dry ladle heats up to 2000°F (1100°C) in about 8 hr, the outside ladle refractory temperature next to the steel shell (cold face) only heats up to about 300°F (150°C). Steady state may not be reached in the ladle refractory until 18 hr after the start of the heating cycle. Heating the refractory too quickly should be avoided because of the nonequilibrium temperature profile in the refractory, excessive stresses on the shell, and excessive shock on the refractory, which may cause damage [24].

Oxy/fuel ladle preheating installations provide the following benefits: faster heating times, hotter ladle bottoms, fuel savings of over 70%, decreased off-gas of 90% compared to air/fuel combustion, reduced maintenance, and no water-cooling requirements. Although fuel savings are a major economic benefit, other advantages associated with the oxy/fuel ladle preheating system are just as important. For example, ladle bottom preheat improves. The firewall life also improves dramatically. In one reported example, the firewall in which an oxy/fuel system lasted more than 70% longer than a typical air–fuel firewall in this particular shop. This result is due to the lower amount of off-gas volume flowing across the firewall.

13.2.4 Casting

The term "metalcasting" applies to the process of melting metal and pouring it into a mold to achieve a desired shape. Virtually any metal that can be melted can be cast. A casting is a metal part formed by pouring molten metal into a sand mold or metal die [26]. Sand molds must be remolded for each casting while metal dies can be used over and over again. In the case of die casting, the molten metal is typically injected at high pressure into the die, compared to mold castings where the metal normally enters the mold by gravity. The mold or die is usually made of two parts that when mated together form a cavity into which the molten metal is poured. One part is fixed and the other part is movable. The mated parts are clamped together while the molten metal cools. Internal cavities are made by putting a core inside the mold cavity. After the molten metal solidifies, the mold or die is broken down, the cores are removed, and the part is ready to be finished.

Ferrous metal castings make up about 85% of all metal castings in the United States, which is the world's leading casting producer with 21% of the world casting market. The most common cast ferrous metals include gray iron, ductile iron, malleable iron, compacted graphite iron, carbon and low-alloy steel, and corrosion-resistant steel. The metalcasting industry is highly diverse and segmented. It includes sand casting, investment casting, lost foam casting, permanent mold casting, centrifugal casting, and diecasting among others. The term "foundry" is applied to all of these except diecasting. Metal casting is an environmentally friendly process from the perspective of the high percentage of used metal parts that can be recycled and reused in new castings.

Cast metal products are found in many industries including transportation, aerospace, defense, energy exploration and conversion, mining, construction, maritime, fluid power, instrumentation, and computers among others. Examples of cast metal parts include valves, pumps, faucets, pipes, fittings, surgical equipment, and components for many of the household devices we use every day. Over 90% of manufactured goods contain at least one metal cast part. Die castings have greater utility and are used in more applications than metal parts produced by almost any other metal-forming process [27]. Die casting is used to produce accurately dimensioned, sharply defined parts with either a smooth or textured surface.

The U.S. Department of Energy has developed a vision [28,29] and roadmap [30] for the future of the metalcasting industry. The metalcasting industry consumes over 300 trillion Btu/year. Melting accounts for 55% of the energy consumption and heat treatment accounts for another 6%. Both of these are of interest here. The vision includes producing castings in an efficient and environmentally friendly manner. The roadmap identified that pollutants such as particulates and off-gases (e.g., CO) present major environmental problems for metalcasters. A high-priority research need identified in the roadmap was to develop an emissions database for foundries to use in educating regulators. One of the major technology barriers related to emissions that was identified was the inability to control deleterious elements (e.g., lead, phosphorus, and sulfur) found in recycled materials. These elements lead to pollutant generation. Another goal is to reduce energy consumption significantly, which also reduces pollutant emissions when less fuel is combusted.

Schifftner and Hesketh [31] describe the use of wet scrubbers to control particulate emissions from foundry cupolas. They list some of the key requirements for the wet air pollution control system as:

(1) Provide for the safe ignition of the CO to CO_2 and adequate residence time for the combustion of the hydrocarbons to CO_2 and H_2O.
(2) Quench the gas stream quickly to permit particle growth among the small particles present.
(3) Avoid bends or other changes in direction that may cause abrasive wear.
(4) Avoid or accumulate the effects of the use of fluxes for slag formation that could produce corrosive conditions in the scrubber.
(5) Provide adjustable pressure drop control to compensate for volume changes during the melt cycle.
(6) Provide adequate draft at the cupola charge door for worker safety if the cupola is hand charged or if workers are in the area.
(7) Remove the particulate to opacity levels of less than Ringleman 1 (20% opacity—see Figure 9.6a) or an outlet loading of 0.1 grains per dry standard cubic foot (gr/dscf).
(8) Neutralize the normally acidic scrubbing liquid stream.
(9) Provide reasonable horsepower requirements in performing these tasks.

13.2.5 Reheating

Soaking pits and reheat furnaces are used to raise the temperature of steel during processing so that the products can be hot worked or shaped. They are designed to heat the products uniformly and hold them at the desired temperature until they are ready to be further processed. The main objective of a reheating furnace is to deliver properly reheated slabs, blooms, or tubes to a rolling mill [32]. The desired characteristics of a reheating furnace are:

- Correct stock discharge temperature
- Proper temperature distribution along the length of the stock
- Low temperature difference between the surface and the core of the stock
- Thermally efficient
- Minimum scale formation on the stock surface
- Low maintenance and high availability
- Minimized operating and capital costs

There are a wide range of reheating furnaces, but they are generally classified as either batch or continuous. The latter type is usually further subdivided as to the mechanism used to move the parts including: roller, pusher, walking beam, and rotary hearth.

A standard steel reheat furnace consists of the following components: heating chamber or furnace, hearth or support for carrying the charge, controls to maintain a specified temperature, distribution system for heating and waste gas removal, and a materials' handling system for moving the charge into and out of the furnace [33]. The required material outlet temperature depends on its composition and geometry and can be as high as 2350°F (1290°C).

Steiner [34] notes that SO_x is one of the potential significant pollutant emissions from reheat furnaces that burn a fuel containing sulfur such as coke oven gas that is produced in the steel mill from the coke ovens [34]. The obvious way to minimize or eliminate the SO_x emissions is to switch to a fuel that contains little or no sulfur, such as most natural gases.

Handa and Tomita [35] discussed the development of a highly luminous natural gas-fired burner for use in large reheat furnaces. One application for the technology is replacing oil with natural gas as the fuel for reheat furnaces. A major problem with switching these fuels is the large disparity that often exists in the flame luminosity since oil flames are normally very luminous and natural gas flames are normally fairly nonluminous. Handa and Tomita showed that while the thermal efficiency increases with flame emissivity, there is very little increase in efficiency for emissivities above about 0.5. Therefore, the goal of their technology development was to make a natural gas flame with an emissivity of at least 0.5. The resulting burner design had staged air combustion with the natural gas injected through the center of the burner. The high-luminosity natural gas burner had an emissivity higher than that of both an oil and a conventional lower luminosity natural gas burner. The high luminosity natural gas burner had a flame shape about 10% larger than the oil flame. The calculated heat flux to 140 mm (5.5 in.) wide × 140 mm (5.5 in.) high × 900 mm (35 in.) long steel billets was 76.4 kW/m² (Btu/ft²-hr). The thermal efficiency increased by 8% by converting a furnace from oil firing to the high-luminosity natural gas burner technology.

Walsh et al. [36] discuss the use of oxygen/fuel burners in a continuous steel reheat furnace. The furnace, which was about 100 ft (33 m) long and 34 ft (11 m) wide, had four top-fired zones and one bottom-fired zone. The design production rate for 8 in. × 6 in. × 30 ft (3.1 cm × 2.4 cm × 9.8 m) blooms was 200 ton/hr (180 mton/hr). An overall fuel savings of 28% was demonstrated compared to the base case using ambient air for combustion.

Blanco and Sala [37] used computer modeling to optimize the performance of steel reheat furnaces used for heating round billets for making chains. One of the problems they were trying to solve was excessive energy consumption. It was estimated that the furnace efficiency was reduced by as much as 10% by air infiltration alone. As a result of the modeling, the cross-section of the furnace was narrowed and the burners above the load were tilted down 20° from the horizontal toward the load. A heat-recovery system was added to preheat the combustion air to 400°C (750°F). These and other modifications resulted in improving the overall thermal efficiency from 40 to 64%. Lisin and Marino [38] have also developed a simplified model to predict the heat-transfer performance of direct-fired, batch, steel reheat furnaces.

Biro and Sandor [39] discuss modernizing pusher-type reheat furnaces to meet environmental regulations. Techniques like reducing excess air, flameless combustion, and reburn can be applied to these furnaces to reduce NO_x. Li [4] discusses the use of oxygen-enhanced combustion to increase throughput by 20% and reduce NO_x emissions by 50% in a reheat furnace (see Fig. 13.14). Baukal [40] shows flame impingement heating in a reheat furnace (see Fig. 13.15).

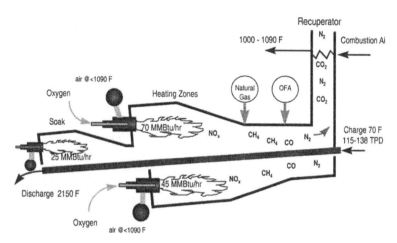

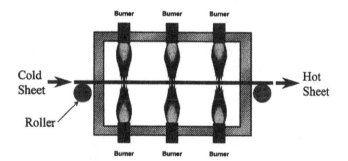

Figure 13.14 Use of oxygen and gas reburn in a reheat furnace to increase throughput and reduce NO$_x$; OFA = over-fire air for reburn. (From Ref. 4. Courtesy of CRC Press.)

Figure 13.15 Elevation view of flame impingement inside a continuous reheat furnace. (From Ref. 40. Courtesy of CRC Press.)

13.2.6 Forging

Forging is a metal-working process used to change the shape of a metal part by using compressive forces, often with the use of a die to obtain the desired shape. This can be done in either a batch or continuous process, depending on the original and final part shapes, the material temperature, and the material composition. Forging may be done when the part is hot or cold. Hot parts may require energy to heat them up if they started cold, but require lower compressive forces to do the shaping. Cold forging does not require any energy to heat the part, but does require higher compressive forces for shaping. Hot forging is of interest here because it usually requires fossil-fuel combustion to do the heating, since electric induction heating is often expensive.

Ward et al. [41] showed how the thermal efficiency and productivity of a forge furnace can be increased by using a high-temperature metallic recuperator for heat recuperation. Combustion air preheat temperatures up to 700°F (370°C) were achieved. The fuel consumption was reduced by 28% compared to the base case without heat recuperation. The heating time was reduced by 8% and scale formation on the outside of the billets was reduced by one-third.

Salama and Desai [42] studied electric induction heating and gas metal heating in forging applications. The following market factors were considered: workpiece size (temperature uniformity of large workpieces and the need to reheat partially forged products from large workpieces), product run (frequency of changes in shape and size of the workpieces for short-run products), quality tolerance (surface defects), and energy price. The first two factors were considered to be the most important. The two extremes of gas furnace sophistication were chosen to compare against electric induction: the slot furnace with low efficiency (4–25%) and low capital cost, and the rotary hearth furnace with high energy efficiency (20–45%), but high capital cost (almost five times higher than that of the slot furnace). Electric induction furnaces have high efficiency (45–55%), but even higher capital costs (nearly 6.5 times the cost of a slot furnace). The gas-fired rotary hearth furnace is comparable to the electric induction furnace. A great advantage of the electric induction furnace is better product quality because of reduced surface defects. Therefore, the most important recommended area for research is to reduce scale formation in gas-fired systems. One method would be to use indirect heating with radiant tube burners.

Takamichi [43] described the development of a high-performance forging furnace utilizing twin regenerative burners. Energy savings of 40% were reported for heating tips of round steel in a forge furnace and 25% for heating aluminum in a roller hearth furnace. The combustion air was preheated to 1000°C (1800°F), compared to 150°–350°C (300°–630°F) for conventional preheat systems on forge furnaces. Better temperature uniformity was also measured in the furnace using the regenerative burners.

13.2.7 Emissions

Cupolas, reverb furnaces, and electric arc furnaces used in the production of iron and steel products may emit particulate matter, carbon monoxide, unburned hydrocarbons (e.g., from oily scrap), sulfur oxides, nitrogen oxides, small quantities of chloride and fluoride compounds, and metallic fumes [44–46]. Particulate emissions will vary according to the furnace type, fuel, metal type, melting temperature, and operating practices. Table 13.1 shows the uncontrolled NO_x emissions from typical

Table 13.1 Uncontrolled NO_x Emissions from Iron and Steel Heating Processes

Process facility	Firing type	Uncontrolled emissions (ppm at 3% O_2)	Uncontrolled emissions (MMBtu)
Soaking pits	Preheat	—	0.14
Reheat furnace	Regenerative	650	0.79
	Recuperative	220	0.20
	Cold-air	120	0.14
Annealing furnace	Regenerative	780	1.15
	Recuperative	330	0.40
	Cold-air	120	0.15
Galvanizing furnace	Regenerative	9240	1.2
	Recuperative	330	0.40
	Cold-air	120	0.14

Source: Ref. 47.

heating and melting processes in iron and steel mills [47]. The exhaust gases from some metallurgical production processes may contain hazardous air pollutants (HAPs), such as VOCs, heavy metals, chlorides, and fluorides that may be present in the incoming raw materials, from the fluxes added to control the slags used to minimize metal oxidation, or from the finishing and cleaning operations. These are not considered further here as they are not a result of the combustion process.

13.2.8 Regulations

The EPA has set minimum standards for lowest achievable emission rates (LAER) and best available control technologies (BACT) for iron and steel mills in its new source performance standards 40 CFR 60 [11,12]. Subpart AA regulates opacity and particulate matter emissions for electric arc furnaces.

13.2.9 Emission Control Technologies

One of the best ways to reduce pollution emissions from metallurgical processes is to improve furnace efficiencies. The less fuel that is combusted per unit of metal production, the less pollution that will be generated. Increased thermal efficiencies also reduce unit operating costs as less fuel is consumed.

Wet scrubbers are commonly used to control emissions from iron and steel processes. However, wastewater emissions result, which must be dealt with accordingly. Catalytic afterburners are used to reduce VOC emissions. Scrubbers, fabric filters, electrostatic precipitators, cyclones, and settling chambers are used to control particulate emissions. Table 13.2 shows the types of NO_x control techniques that have been applied to various heating and melting processes in iron and steel mills including how effective the technique is at reducing NO_x.

Table 13.2 NO_x Control Techniques Applied to Iron and Steel Heating and Melting Processes

Furnace type	Control	Emissions (lb/(MMBtu) regenerative	Emissions (lb/MMBtu) recuperative	Emissions (lb/MMBtu) cold-air	Percentage reduction
Reheat	LEA	0.69	0.17	0.12	13
	LNB	0.27	0.068	0.046	66
	LNB+FGR	0.18	0.046	0.031	77
Annealing	LNB	0.48	0.20	0.07	50
	LNB+FGR	0.38	0.16	0.07	60
	SNCR	0.38	0.16	0.07	60
	SCR	0.14	0.06	0.02	85
	LNB+SNCR	0.19	0.08	0.03	80
	LNB+BCR	0.095	0.04	0.015	90
Galvanizing	LNB	0.57	0.20	0.07	50
	LNB+FGR	0.46	0.16	0.06	60

Source: Ref. 47.

13.3 NONFERROUS METAL PRODUCTION

Nonferrous metal production includes everything but iron and steel. Common nonferrous processes incorporating industrial combustion equipment include aluminum, copper, brass, and lead production. Only some typical nonferrous metals are considered here as they are representative of the other processes. Some of this section has been adapted from Saha and Baukal [48]. The nonferrous industry is very broad and diverse. It includes aluminum, copper and copper alloys, beryllium, lead, mercury, zinc, and other lesser-used metals such as arsenic, cadmium, zirconium, and titanium.

Figure 13.16 shows emissions from nonferrous metal production processes in the United States since 1970 [8].

13.3.1 Aluminum Production

Aluminum is one of the most essential materials for U.S. transportation, packaging, and construction industries [49]. Aluminum has some unique properties, which make it particularly valuable including its high strength-to-weight ratio, low maintenance, and ease of recycling. It is used in the transportation industry (automobiles, trucks, rail cars, ships, and commercial aircraft), the packaging industry (aseptic food containers and beverage cans), the construction industry (windows, doors, and siding), and for infrastructure (bridges, airports, highways, and wiring). Aluminum recycling dramatically reduces pollutant emissions as only 5% of the energy required to produce aluminum from ore is required to produce aluminum from recycled aluminum parts.

Primary production of aluminum involves the refining of alumina from bauxite ore followed by electrolysis or smelting of the alumina. This is typically done by the Hall–Heroult process, which is very energy intensive. This process is not considered here as it does not involve a traditional combustion system with fossil-fuel-fired burners. Secondary production of aluminum involves recovering aluminum from recycled aluminum products and scrap aluminum. Secondary production is far less

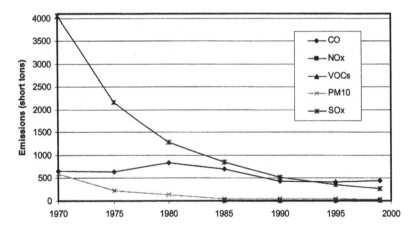

Figure 13.16   Pollution emissions from nonferrous metal production processes. (From Ref. 8.)

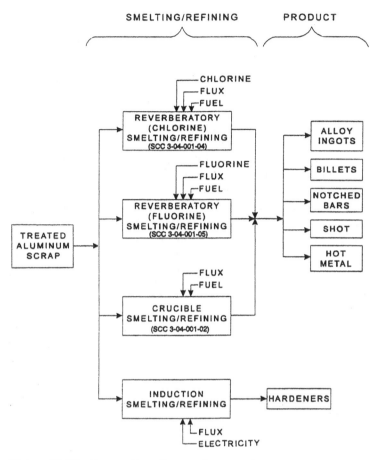

Figure 13.17   Process flow diagram for secondary aluminum production. (From Ref. 50.)

energy intensive than primary production. In the United States in 1998, 33% of the total supply of aluminum was from secondary production. Figure 13.17 shows the process flow for secondary aluminum production [50].

The aluminum industry is generally divided into three segments: raw materials (e.g., alumina, primary molten metal, secondary molten metal, and ingot), semifabricated products (sheet, plate, foil, forgings, castings, wire, rod, bar, extrusions, powders, and aluminum-based chemical products), and finished products (aircraft, automobiles, windows, doors). It is predicted that future demand for aluminum will continue to be strong. The U.S. aluminum industry is the largest in the world.

The U.S. Department of Energy (DOE), in conjunction with the Aluminum Association, has identified the following environmental targets for the aluminum industry [51]:

- Recycle and treat all types of aluminum wastes.
- Increase recyclability of aluminum scrap which often contains considerable contamination.

- Achieve 80% wrought recycling of autos by 2004.
- Reduce NO_x emissions on existing furnaces while simultaneously increasing capacity.

All of these would reduce pollution emissions as less fuel would be needed for a given amount of aluminum production.

Semifabricated products are made from aluminum raw materials. Molten primary or secondary aluminum is typically cast into ingots which are used as the raw material feed for subsequent remelting or mechanical operations. It is also possible to cast continuously molten primary or secondary aluminum to be fed into a rolling mill. Hot or cold rolling is used to roll aluminum into sheet, plate, or foil. Tube, wire, and similar products are made from extrusion processes. The DOE recommends developing an optimum furnace design for the future for the casting of semifabricated products where the furnace would be environmentally benign among other things [51].

Improving the energy efficiency in aluminium-melting operations reduces fuel consumption and simultaneously pollutant emissions [52]. A primary pollutant from secondary aluminum production is CO_2. Other common air pollutants from the combustion process include particulates, unburned combustibles including carbon monoxide, SO_x, and NO_x. Chloride and fluoride emissions may result from the fluxing and demagging operations.

13.3.1.1 Furnaces Used in Aluminum Melting

The general method for secondary aluminum production includes the following processes: charging, melting, fluxing, demagging, degassing, alloying, skimming, and pouring. Charging involves adding pretreated scrap (to remove impurities, shredding, etc.) and flux to a small pool (referred to as a heel) of molten aluminum left behind from the previous heat. The flux materials combine with contaminants and cause them to float so that they can be easily skimmed off and removed. The materials floating on the top of the aluminum also provide a protective coating to minimize oxidation of the molten aluminum. The layer is called slag. Demagging reduces the magnesium content from approximately 0.5 to 0.1%. This is usually done with liquid chlorine injected into the melt at high pressure to combine with the magnesium to form magnesium chloride. The demagging process also helps remove any entrained gases in the melt as the injected liquid forms bubbles that rise to the top of the melt causing stirring and entrainment. The gases are then released when they reach the melt surface. Alloying involves adding the appropriate agents (e.g., zinc, copper, manganese, or silicon) to achieve the desired grade of aluminum. Skimming involves removing the slag layer, which contains the contaminants from the top of the melt just prior to pouring. The molten aluminum is then poured out of the furnace to make the desired products (e.g., ingots, billets, or bars).

Rotary Furnaces. Rotary furnaces have traditionally been used for melting dross. The furnace sizes range from 8000 to 30,000 lb (3600–14,000 kg), with fuel consumption of 1800–2400 Btu/lb (0.77–1.0 J/kg). A typical combustion system for rotary furnaces can be a single or double pass. The single-pass system has a burner at one end of the furnace and a flue at the other. A typical single-pass rotary aluminum melting furnace is shown in Fig. 13.18. The combustion gases flow straight through

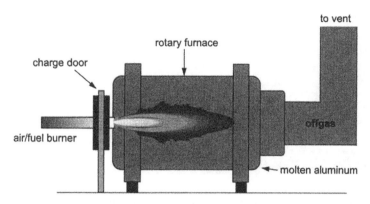

Figure 13.18 Conventional single-pass, rotary aluminum melting furnace. (From Ref. 48. Courtesy of CRC Press.)

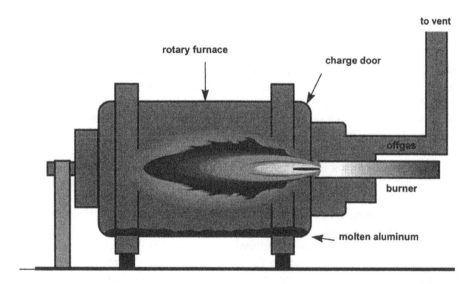

Figure 13.19 Conventional double-pass rotary aluminum-melting furnace. (From Ref. 53. Courtesy of CRC Press.)

the combustion space exiting through the flue. With the double-pass system the flue is located on the same side of the furnace as the burner. A typical double-pass rotary aluminum melting furnace is shown in Fig. 13.19 [53]. The combustion gases make a "U" turn inside the furnace, which increases the gases residence time and improves furnace efficiency. This furnace type is often referred to as a concrete-mixer type furnace. Radiation and convection from the hot combustion products are transferred to both the furnace walls and the charge materials. The heat in the furnace wall is then transferred to the charge material as the furnace rotates on its horizontal axis. On the burner side is the charge door for adding the feed materials to the furnace. The exhaust is located at the opposite end of the furnace. A typical charge contains mixed scrap such as aluminum foils, castings, turnings, and dross.

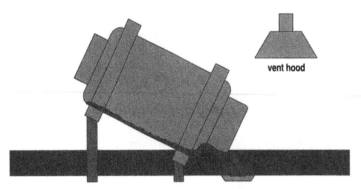

Figure 13.20   Typical tilting rotary aluminum-melting furnace. (From Ref. 53. Courtesy of CRC Press.)

The aluminum producer would like to recover as much of that aluminum as possible to maximize the yield.

Rotary-furnaces can also be divided into fixed axis or tilters. A bottom taphole is required for the fixed-axis furnaces to pour the molten aluminum. A delivery trough is needed for molten metal transport to a crucible, to the sow molds, or to a casting furnace. The fixed-axis furnace also requires a procedure for saltcake removal. A towmotor with a rake or boom assembly mounted on the front is typically used to extract the saltcake. Such a procedure is time consuming and also can damage the furnace refractory because of the raking process. The alternative is to remove the saltcake through the taphole when the "wet process" is used.

In the tilting furnaces, the aluminum is poured from the furnace front, once the melting process is completed. A typical tilting rotary aluminum melting furnace is shown in Fig. 13.20. The back of the furnace is tilted up gradually until all the metal is poured from the furnace lip. When the metal is removed completely, the furnace is rotated again and tilted more to remove the saltcake. The benefits of the tilting furnaces are their ability to remove contaminants at any given time, to control the chemistry of the melt, and to remove the metal effectively. The melt and saltcake removal processes reduce the poring time, increase furnace production, and reduce maintenance cost.

Another significant distinction of the rotary furnaces is the melting process selection. The melting process can be "wet" or "dry." The wet process uses twice as much salt compared to the dry one. It also requires more energy since the flux has to be molten throughout the process. When the dry process is used, a saltcake is formed above the molten metal bath. This saltcake has to be removed from the furnace at the end of the melting cycle by tilting the furnace or raking it out. The removal procedure depends on the type of furnace. More dross can be processed per cycle in the rotary furnaces using the dry process because less flux is used.

Flux is used for aluminum melting in both the wet and dry processes. It serves several purposes. First, the flux melts at a temperature lower than that of the aluminum. Therefore, it is used as an oxidation protective blanket over the metal surface. Second, the salt flux acts as a separation agent between the molten metal and the oxides. A typical salt flux consists of sodium chloride (NaCl) and potassium chloride (KCl). Sometimes Cryolite is added to the flux to reduce its melting temperature.

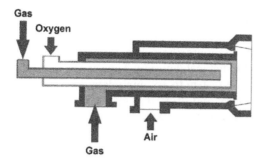

Figure 13.21 Combination air/fuel–oxy/fuel burner. (From Ref. 48. Courtesy of CRC Press.)

The rotary furnace lends itself very well to the use of oxygen-assisted combustion. The reduced flue gas volume also provides for a longer residence time of the hot combustion gases in the furnace, thus allowing more heat to be transferred to the charge material. Also, because of the reduced flue gas velocity, less fines are carried into the dust-collection system.

The concept of using an oxygen/fuel burner within an air/fuel burner provides the aluminum melter with the flexibility of using either burner, or a combination of both burners (see Fig. 13.21). This innovative approach, developed by Air Products under the trade name of EZ-Fire™, has proved to be very successful in the aluminum industry [54]. This differs considerably from both conventional oxygen-enriched techniques and the oxy/fuel burners typically used for melting in the metals industries. This technology is retrofittable on existing air/fuel combustion systems. An oxy/fuel burner is installed in the center of the conventional air/fuel burner. This approach marries the existing air/fuel burner with a retrofitted nonwater-cooled oxy/fuel burner, providing the more efficient heat transfer and available heat benefits of oxy/fuel, while maintaining the external flame characteristics of air/fuel. The result is a low-capital cost, air–oxygen/fuel combustion system that improves the furnace efficiency and lowers production costs.

During meltdown, the oxygen/fuel burner is used along with the air/fuel burner to provide maximum melt rates and minimum flue gas volumes, while during holding and casting, only the air/fuel burner is used. For the melting of aluminum, a combination of air–oxy/fuel combustion provides the best results. The outer air/fuel flame envelope provides a shielding effect for the hot oxygen/fuel inner flame. By firing the oxygen/fuel inner burner on a fuel-rich mixture, flame luminosity is increased as combustion radicals and gas particles are heated to incandescence by the hot oxy/fuel flame. This in turn improves the radiative heat transfer of the flame to the charge material and furnace wall while preventing excess oxygen from contacting the load and causing metal oxidation.

This air/oxy/fuel burner concept was first tried in 1989 at a Canadian aluminum dross reclamation plant where drosses and skimmings are smelted [55]. The dross, with 10% salt, is charged through the burner end of a 25,000 lb (11,000 kg) rotary furnace, fired by a dual-fuel burner of 20×10^6 Btu/hr (6 MW) capacity for melting. The rotary furnace flue exhaust as well as dross crushers and cooling tables are all connected to the same baghouse. When the dross reclamation unit was used at the same time the rotary furnace was firing, fume collection efficiency was reduced and fugitive emissions became a problem.

Because of the draft limitations on the baghouse, the melt furnace firing rate had to be maintained at a reduced rate, which limited metal output. Time constraints and economics precluded upgrading the baghouse system to accommodate the combined dust and fume loads from the melting furnace and dross reclamation systems. The most logical approach was to address the melt furnace, which generated the largest amount of hot gases and fines processed by the baghouse. The use of oxygen-assisted combustion to alleviate the baghouse problems while simultaneously increasing furnace productivity became the preferred choice. By substituting air with pure oxygen, flue gas output from the furnace would be reduced and combustion efficiency would increase. The oil tube/atomizer section was removed from the dual fuel burner and an oxy/fuel burner was inserted in its place. The burner retrofit package including the natural gas and oxygen flow control trains were prefabricated to prevent furnace downtime. The oxygen–fuel flow controls and safety switches were interlocked to the conventional air/fuel controls, thus creating a hybrid air–oxy/fuel combustion system.

Used beverage cans (UBCs) and other recycled aluminum scrap are typically melted in a rotary furnace. A new process has been developed that reduces the melting costs and pollution emissions from the conventional rotary melter. It is referred to as the LEAM® process, or *Low Emission Aluminum Melting* process (patent pending) [56].

The aluminum processor likes to have an efficient and flexible melter to handle a wide range of scrap materials. An ideal melting process would be able to handle a wide range of scrap and automatically compensate for the contaminants. Scrap aluminum cans commonly have paints and lacquers that must be removed. In some cases, this is done in a pretreatment furnace where the coatings are burned off. Scrap turnings from machining operations typically contain significant quantities of lubrication oils. These oils produce large amounts of smoke during heating and melting in conventional rotary furnaces. This smoke must be removed before the exhaust gases exit the stack.

Stricter regulations are increasing the environmental compliance costs. This includes the costs to clean up contaminated air and water resulting from the melting process, as well as the costs to dispose of any waste products, particularly important for hazardous wastes.

A slag layer covers the top of the molten aluminum in the melter to minimize aluminum oxidation. The slag is produced by adding salt that contains 10–20% potassium chloride and 80–90% sodium chloride. The slag is skimmed off before tapping the furnace. This slag, in certain geographic regions, is classified as a hazardous waste. It is expensive either to reprocess the slag to recover further more aluminum or to dispose of the slag, usually in a landfill. The exhaust gases from the rotary furnace are usually ducted to a baghouse that removes any particulate matter. In some cases, this filter dust may also be classified as hazardous waste, which again increases the disposal costs.

Another important factor is the hydrocarbon emissions. In a conventional rotary furnace, with the burner mounted on the charging door, the burner must be turned off when the door is opened for charging. Fumes are generated by vaporization of any combustibles in the charge, as shown in Fig. 7.10. Because of the reduced furnace temperature and insufficient free oxygen to burn the newly added combustibles, large quantities of hydrocarbon emissions can be generated. The problem is exacerbated when the door is closed and the burner restarted. Even less free oxygen is available when the door is closed, and more combustibles are vaporized by the additional heat from the burner. An afterburner could be used to incinerate these fumes. However, this is costly and does not take advantage of the heating value in the combustibles.

Another problem related to the hydrocarbon fumes is the formation of dioxins and furans. The hydrocarbons react with the halogens, typically contained in the salts that are used to produce the slag covering on the molten bath, to form dioxins and furans (see Chap. 11). These are carcinogens that are currently regulated in several countries. Methods to reduce dioxin and furan emissions include: (1) to postcombust the hydrocarbons, (2) to quench the off gases quickly within a certain temperature window, and (3) to use an activated carbon to adsorb the dioxins and furans. However, any of these methods would be at some additional cost to the aluminum producer.

The LEAM process, developed to minimize the pollutant emissions from rotary aluminum melting furnaces, was first demonstrated in Germany where the environmental regulations are stringent. The first objective of the process was to at least maintain, if not improve, the existing yields in the conventional melting process, which is well known for its high yields. Other objectives included: (1) to lower the emissions, especially hydrocarbons, dioxins and furans, below the regulated limits, and (2) to reduce the exhaust gas volumes to reduce the cost of the flue gas cleaning system. This required reductions in the amount of filter dust collected in the cleaning system. All of these objectives had to be achieved at the lowest possible cost.

As shown in Fig. 7.12, the first major change was relocating the burner from the charge door to the exhaust wall. This reversed flame operation offered several benefits. The residence time of the combustion gases in the furnace increased, leading to increased thermal efficiency and higher destruction efficiencies for the volatiles. This configuration allowed the burner to be fired even during charging. Any volatiles coming from the newly charged scrap material had to pass through the flame and were postcombusted within the furnace. The heating value of the volatiles was now used and the volatiles did not have to be removed in the off-gas cleaning system. The overall melting cycle time was reduced because heat was now added during the charging cycle. This new configuration allowed the door to be properly sealed, which better controlled the excess air coming into the furnace.

Another major change was replacing the air/fuel burner in the conventional rotary aluminum melting furnace with an oxy/fuel burner. The reduced exhaust gas volume flow made it easier to incinerate the volatiles because they were in higher concentrations. Oxy/fuel burners also have a much wider operating range than air/fuel burners. They can easily operate with several times the stoichiometric amount of oxygen required for complete combustion. This extra O_2 was important for incinerating the volatiles that evolved from the dirty charge materials. Air/fuel burner operation became severely degraded when large quantities of excess air were supplied through the burner.

During furnace charging, the oxy/fuel burner was at low fire, whereas in the conventional process the burner must be off during charging since it is mounted on the door. A major advantage of the LEAM process is that the volatiles emitted during charging of dirty scrap are incinerated by the burner. In the conventional process, all the volatiles emitted during charging are exhausted and must be removed by the exhaust gas cleaning system.

Many process improvements were achieved by converting to the LEAM process, as shown in Table 13.3. Three options were tested and compared: (1) conventional air/fuel burner in the charge door, (2) oxy/fuel burner in the charge door, and (3) the LEAM process. Using oxy/fuel (options 2 and 3), the tap-to-tap time went from 10 to 7 hr. This was primarily a consequence of the higher flame temperatures with oxy/fuel burners. The energy consumption using the LEAM process, was less than half that of the conventional process, which was again a consequence of using the more efficient oxy/fuel burner.

Aluminum yields increased an average of 1% due to better control of the furnace atmosphere. Slag formation went down by 20%. There was a dramatic reduction in the filter dust generated during the melting process, which is especially significant where the dust is classified as a hazardous waste. In Germany in 1996, the disposal costs were about \$0.50/lb (\$0.23/kg) so this represents a large cost savings for the aluminum producer.

There were also significant reductions in the pollutant emissions. The regulated and measured values are shown in Table 13.4. For the air/fuel system, hydrocarbons were a great problem because they were difficult to control. NO_x was not a problem because of the low melting temperature of the aluminum and the low flame temperatures. Dust emissions were effectively controlled by the baghouse. However, dioxins and furans were a serious problem. Using oxy/fuel (option 2), the hydrocarbon, dioxin, and furan emissions were similar to the base case air/fuel system. However, NO_x was higher because of high levels of air infiltration due to the poor door seal in option 2. The infiltrated air passed through the very hot oxy/fuel flame, which generated high NO_x levels.

Table 13.3 Process Improvements using the LEAM Process

	Air/fuel	Oxy/fuel	LEAM	Improvement (%)
Tap-to-tap time (hr)	10	7	7	30
Consumption (kWh/ton)	850	550	420	52
Thermal efficiency (%)	35	55	71	100
Relative metal yield (%)	0	0–0.5	0.5–2.0	
Slag (lb/ton)	770	770	620	20
Filter dust (lb/ton)	55	33	13	76

Source: Ref. 48. Courtesy of CRC Press.

Table 13.4 Pollutant Emissions on a Volumetric Basis

	Air/fuel	Oxy/fuel	LEAM	German limits
Hydrocarbons (mg/m^3)	0–1000	0–1000	0–30	50
NO_x (mg/m^3)	50–200	200–3000	100–500	500
Dust (mg/m^3)	2–5	2–5	2–5	10
Dioxin/furan (ng-TE/m^3)	0.2–10	0.2–10	0.0–0.4	1

Source: Ref. 48. Courtesy of CRC Press.

Using the LEAM process, hydrocarbons were well below the German limit and could even be further reduced by adjusting the sensitivity of the optical sensor. The NO_x levels were reduced below the regulated limit by sealing the charge door and by putting a pressure control system on to the furnace to minimize air infiltration. Dust levels were again low. There was a significant reduction in the dioxin and furan emissions, which are typically reported on a toxic-equivalent (TE) basis. There is currently some discussion in Germany about reducing the existing regulation for dioxins and furans down to $0.1 \, ng \, TE/m^3$. Some type of post-treatment system would have to be added to the existing air/fuel system to meet that limit. The LEAM process may be able to meet that limit without any additional equipment.

The pollutant emissions in Table 13.4 are given on a volumetric basis. This does not reflect the dramatic reduction in the flue gas volume that occurs when replacing an air/fuel system with oxy/fuel. The pollutant emissions have been normalized in Table 13.5. This shows the dramatic reductions in hydrocarbons, dust, dioxins and furans, and CO_2. CO_2 is considered to be a greenhouse gas that contributes to global warming and as such is regulated in many locations (see Chap. 11). There was also a slight reduction in NO_x compared to the base case.

Figure 13.22 shows that the LEAM process can reduce the costs for the aluminum producer. In Germany, the costs to dispose of the filter dust and to reclaim the aluminum in the slag are high. By reducing those emissions, the cost of the oxygen is more than offset by the LEAM process improvements. Not included in the figure are the costs that could be incurred if a post-treatment system were

Table 13.5 Pollutant Emissions per Ton of Scrap Input

	Air/fuel	Oxy/fuel	LEAM	Improvement (%)
Hydrocarbons (g)	87	60	6	93
NO_x (g)	222	1360	214	4
Dust (g)	24	17	3	88
Dioxins and furans (mg)	36	25	0.2	99
CO_2 (kg)	238	154	114	52

Source: Ref. 48. Courtesy of CRC Press.

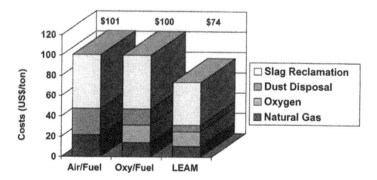

Figure 13.22 Production costs including emissions disposal. (From Ref. 48. Courtesy of CRC Press.)

required to remove dioxins and furans. The actual savings for any given aluminum producer will depend on a variety of factors and are somewhat site specific. Other important factors to consider are the increase in metal yield and the ability to use cheaper grades of scrap.

Side-Well Reverberatory Furnaces. Reverberatory furnaces, often referred to as reverbs, have a much wider use in the aluminum industry than rotary furnaces. These furnaces are typically used for primary or secondary aluminum melting. The operation procedures for both rotary and reverb furnaces is similar. Side-well reverberatory furnaces are discussed next.

A typical side-well reverberatory furnace is shown schematically in Fig. 13.23. The furnace consists of a number of burners typically firing parallel to the hearth, against the furnace hot wall or against the furnace door (see Fig. 13.24). Figure 13.25 shows a picture of the inside of a hot reverb furnace with the burners turned off. A charging well and a pump well containing a molten-metal pump (see Fig. 13.26) are attached to the furnace hot wall on the outside of the furnace. Both wells are connected to each other and with the furnace hearth by the arches, which are used for aluminum circulation in the furnace. Two types of arches are used between the wells and the hearth. The first type is the tall arch. These are built to stretch from

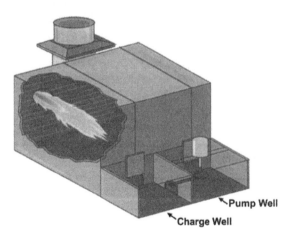

Figure 13.23 Side-well-fed reverberatory furnace. (From Ref. 22. Courtesy of CRC Press.)

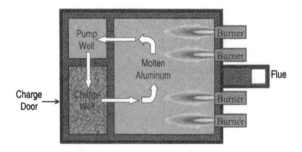

Figure 13.24 Plan view of side-well-fed reverberatory furnace.

Figure 13.25 View inside a reverb furnace with the burners off.

Figure 13.26 Molten-metal pump.

the furnace bottom to well above the metal line. These arches are typically used in side-well furnaces with no metal pump [57] and, therefore, no pump well. The second type is the submerged arch. These arches are built from the furnace floor to a level below the metal line. A typical height of this type of arch varies from 10 to 15 in. (25–38 cm), depending on the furnace design. Reverb furnaces with submerged arches typically use a metallic pump.

A material is charged into the furnace charging (side) well, which already has a certain level of molten aluminum in it. The material melts into the existing molten metal in the charge well. The molten aluminum circulates from the hearth into the charging well and back into the hearth for reheating. As the metal level reaches its maximum, the furnace is tapped into a casting line, crucible, ladle, or a holding furnace. Often, the same melting furnace is used for degassing and alloying. In that case, the total furnace cycle consists of two subcycles: the melting cycle and the holding cycle. A typical holding cycle time varies from 25 to 50% of the total furnace cycle time.

The efficiency of the side-well furnaces is determined by the ability of the molten metal to receive the energy from the combustion chamber. There are several important factors influencing heat transfer inside the aluminum melter. First, the layer of dross at the molten bath surface inside the furnace hearth can be viewed as a thermal insulation, restricting heat transfer between the combustion space and the melt. Figure 13.27 shows a photograph of slag being removed from the surface of the molten aluminum being produced in a reverb furnace. This process is referred to as "slagging off." At the same time, a thin layer of dross is essential to prevent excessive aluminum oxidation at the gas–melt interface. Second, the furnaces with a metallic pump are more efficient than the ones without the pump. The melt circulation helps to enhance heat transfer to the melt compared to the process with no melt circulation. Convective heat transfer is added to conductive heat transfer in the case with a nonstagnant melt (the furnace with the pump). Third, flame characteristics have a great influence on the heat transfer to the melt. For example, high-momentum air/fuel burners with relatively transparent (nonluminous) flames deliver less energy to the melt compared to highly luminous, high-temperature oxygen-enriched flames. Figure 13.28 shows a picture of molten aluminum being

Figure 13.27   Slagging off a reverb furnace.

Figure 13.28 Pouring molten metal into a ladle.

Figure 13.29 Pouring molten metal into casting machine.

poured into a ladle for transfer to another location in the plant for further processing. Figure 13.29 shows molten aluminum being poured into a casting machine to produce ingots.

Furnace design and operating practices are of great importance for the aluminum makers. An optimized ratio of the furnace length to its width, molten bath height, combustion space height, shape and cross-sectional area of the arches,

Table 13.6 Oxygen Enrichment of an Aluminum Melting Reverberatory Furnace

	Steady-state melt rate (tons/hr)	Corrected specific consumption (Btu/lb)	Efficiency during melt
Without recuperator	9.30 (base case)	1974 (base case)	25.1 (base case)
3% Enrichment (Δ%)	10.99 (+18.2%)	1628 (−17.5%)	30.6 (+21.9%)
5% Enrichment (Δ%)	11.38 (+22.4%)	1569 (−20.5%)	31.7 (+26.3%)
7% Enrichment (Δ%)	11.76 (+26.5%)	1507 (−23.7%)	33.0 (+31.5%)

(Δ%) = Percentage change from base case of no recuperator and no enrichment.
Source: Ref. 48. Courtesy of CRC Press.

burner firing rates, and burner position and orientation usually determine the furnace performance. Until recently, furnace designs and operating practices were based on the experience and limited field data available from aluminum makers. Computational fluid dynamic (CFD) modeling (see Chapter 4) is an excellent tool that offers a significant help to the furnace designers and operators [58–61].

In the mid-1980s, oxygen-enrichment tests were conducted in a nonrecuperative reverberatory aluminum remelt furnace. Pure oxygen was introduced into the combustion air stream. Oxygen analyzers were used to assure the proper enrichment level was maintained and the oxygen level in the flue was kept to approximately 2%. The results obtained are shown in Table 13.6 [62]. During the test, no significant change in melt loss was found from the use of oxygen. However, productivity improvements of up to 27% were obtained with 7% oxygen enrichment. Although emissions were not measured, assuming that the pollutant generation per unit of fuel was the same, the overall emissions would be reduced significantly due to the increased efficiency.

A technique of retrofitting conventional air/fuel burners with oxygen capability allows operators of reverberatory melting furnaces to obtain easily the advantage of oxygen-assisted melting. In most cases, the air inlet of the conventional burner is retained and a new internal assembly including the oxygen connection is fabricated for installation on the hot furnace. The existing air/fuel controls are retained and the oxygen controls are interconnected. The ignition and flame supervision components are also retained although sometimes they must be relocated on the burner body. The resulting burner retains the capabilities of the air/fuel burner and can be operated as the original air/fuel burner or as an oxygen-assisted air/fuel burner.

The inclusion of some air in the combustion process helps to moderate the flame temperature to eliminate the possibility of localized overheating of the refractory or metal bath. Air surrounding the oxy/fuel flame cools the burner tile, prolonging the life of the tile. More importantly, the overall cost of operation is reduced compared with a 100% oxy/fuel burner for a desired production increase. Since the burner can also operate as an air/fuel burner without oxygen, the operating cost can be further reduced by switching to air/fuel when holding, alloying, or because of production delays. In many cases the conventional burner has an oil tube and atomizing air section in the center of the burner. This can be removed and replaced with an independent oxy/fuel burner. In other cases, the gas tube is replaced with a combined oxygen–gas nozzle.

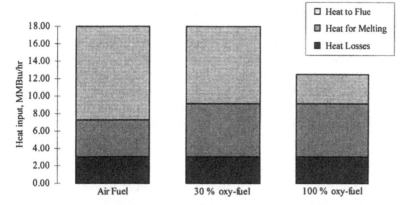

Figure 13.30 Heat-balance comparison for aluminum reverberatory furnace. (From Ref. 48. Courtesy of CRC Press.)

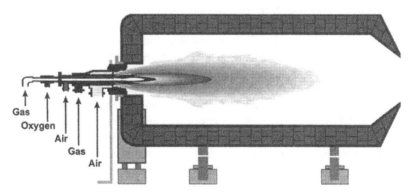

Figure 13.31 Combination air/fuel–oxy/fuel burner in a rotary furnace. (From Ref. 48. Courtesy of CRC Press.)

Using the available heat calculated for the flue temperature and the heat contained in the metal at the pouring temperature (475 Btu/lb or 1100 kJ/kg) the heat lost to the environment from the furnace is calculated. Assuming that the heat loss is constant, the amount of fuel required at various oxygen participation levels is calculated to give the desired 30% increase in production rate. The resulting heat balances are shown in Fig. 13.30.

An aluminum smelter, with three open-well reverberatory furnaces melting scrap, who needed increased production, implemented oxygen-assisted combustion for increased furnace productivity, reduced cost per pound of material melted, and improved metal yield [63]. Two furnaces of 150,000 lb (68,000 kg) capacity each and one 70,000 lb (32,000 kg) capacity were side-well-fed reverberatories. Four dual-fuel burners fired along one furnace wall in the larger furnaces, while the smaller furnace was equipped with two burners. For the first furnace, the EZ-Fire burners (see Fig. 13.31) were installed within the existing air/fuel burners. Once the oxy/fuel burners were operating, the major change in furnace practice involved increasing the charge rate to keep up with the higher melt rate. By monitoring melt temperature, the furnace operators found that the normal charge rate could be increased by 25%.

During charging and melting, the EZ-Fire burners were fired 25% from the air/fuel burners and 75% from the oxy/fuel burners. During extended waiting periods, the oxy/fuel burners were shut off, but otherwise the burners were kept at high fire throughout the charging/melting cycle and reduced to low fire during tapping. As a result of the faster melting rate the total cycle time was reduced. This smelter, to better accommodate work shifts and casting line utilization, adopted new pouring schedules. The measured daily output of ingot and sow was, therefore, increased by 15 to 20 % for each of the three furnaces. Natural gas consumption was reduced from $2.8 \, ft^3$ $(0.08 \, m^3)$ to an average $1.8 \, ft^3$ $(0.05 \, m^3)$ per pound of aluminum melted. The oxygen requirement was $1.4 \, ft^3$ $(0.04 \, m^3)$ per pound of aluminum melted. Overall, the average cost of processing a pound of aluminum was reduced about 0.75 cent when labor and overhead costs were taken into account.

Another aluminum producer of secondary ingots, billets and slabs was interested in evaluating the EZ-Fire technology for both production enhancements and emissions reductions [64]. For the initial installation, a direct-charge reverberatory furnace was retrofitted. The furnace was charged with billet ends and primary and secondary sows. The furnace combustion system contained two air/fuel type burners firing at $14 \times 10^6 \, Btu/hr$ (4.1 MW), with air preheated to $500°F$ ($260°C$). The existing control system was designed to regulate firing rates to control the roof and bath temperature at the desired setpoint. This system could meter the rate down below $5 \times 10^6 \, Btu/hr$ (1.5 MW) when the roof or bath temperature was attained. The EZ-Fire controls were incorporated into the existing furnace controls enabling the systems to work concurrently. Three oxy/fuel firing rates corresponding with three setpoint temperatures were utilized while the air/fuel firing rate was fixed to $5 \times 10^6 \, Btu/hr$ (1.5 MW). When the maximum setpoint temperature was reached, the oxygen system would turn off and the air/fuel system would take over.

As a result of the increased efficiency of the EZ-Fire system the melt cycle was reduced from over 8 to 5.5 hr for the EZ-Fire system (a 30% melt rate increase). The reduction in melt time enabled the customer to optimize the work load for the entire billet casting operation. The optimized EZ-Fire system has been able to demonstrate emissions levels below some of the most stringent state and local emissions factors in the United States. The system has demonstrated emissions levels under $0.3 \, lb \, NO_2/ton$ ($0.15 \, g \, NO_2/kg$) aluminum melted. Multiple emissions tests have seen levels consistently below $0.19 \, lb \, NO_2/ton$ ($0.085 \, g \, NO_2/kg$) aluminum melted.

Direct-Charge Furnaces.   Another type of a reverberatory furnace is a direct-charge furnace [65,66]. A photograph of a typical direct-charge furnace is shown in Fig. 13.32. Figure 13.33 shows an elevation view of a schematic of a direct-charge melting furnace, and Fig. 13.34 shows a plan view of the same furnace. This furnace type differs from the side-well furnace by the absence of a side-well. The metal is charged directly into the furnace hearth where it is exposed to the open flames. Therefore, three phases are present in this melting process. Combustion gases, molten aluminum, and the solid metal are all together inside the furnace. The metal recovery of such a system is usually lower compared to that of the

Figure 13.32 Photograph of a typical direct-charge furnace. (From Ref. 53. Courtesy of CRC Press.)

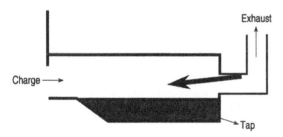

Figure 13.33 Elevation view of a direct-charge melting furnace.

Figure 13.34 Plan view of a direct-charge melting furnace.

side-well operation, but nevertheless, these furnaces are widely used in the aluminum industry.

Figure 13.35 shows an example of a round-top furnace used for melting large charges of aluminum. The roof is swung over to add each charge. Multiple burners are fired toward one side of the furnace. The exhaust gases then turn back around

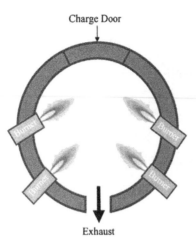

Figure 13.35   Cross-section of a round-top aluminum-melting furnace.

and go to the other end, exhausting through an opening on the other end. The burner operating conditions are often varied during each phase of the melting cycle. The burners are at low fire during the charging phase. After the roof is closed, the burners are on high fire while melting the scrap charge. After the full load is melted, it is checked for chemical composition, and alloys are added or subtracted as required. The burners are at a lower fire to maintain the temperature without causing excessive oxidation and metal loss. Pollution emissions vary during the production cycle. Scrap contaminants such as oils and greases burn off during the initial phases. The scrap may also contain plastics and other waste materials that could produce pollutants such as dioxins and furans. NO_x emissions typically increase with the furnace temperature.

13.3.1.2   Oxygen-Enhanced Aluminum Production

Conventional practice is to use air/fuel burners in these rotary furnaces. The concept of using an oxygen/fuel burner within an air/fuel burner provides the aluminum melter with the flexibility of using either burner, or a combination of both burners (see Fig. 13.31). This innovative approach has proved to be very successful in the aluminum industry [67]. This differs considerably from both conventional oxygen-enriched techniques and the oxy/fuel burners typically used for melting in the metals industries. During meltdown, the oxygen/fuel burner is used along with the air/fuel burner to provide maximum melt rates and minimum flue gas volumes, while during holding and casting, only the air/fuel burner is used. For the melting of aluminum, a combination of air–oxy/fuel combustion provides the best results. The outer air/fuel flame envelope provides a shielding effect for the hot oxygen/fuel inner flame. By firing the oxygen/fuel inner burner on a fuel-rich mixture, flame luminosity is increased as combustion radicals and gas particles are heated to incandescence by the hot oxy/fuel flame. This in turn improves the radiative heat transfer of the flame to the charge material and furnace wall while preventing excess oxygen from contacting the load and causing metal oxidation.

This air–oxy/fuel burner concept was first tried in 1989 at a Canadian aluminum dross reclamation plant where drosses and skimmings are smelted [55]. The dross, with 10% salt, was charged through the burner end of a 25,000 lb (11,000 kg) rotary furnace, fired by a dual-fuel burner of 20×10^6 Btu/hr (6 MW) capacity for melting. By firing the oxy/fuel burner at 9×10^6 Btu/hr (2.6 MW) and the air/fuel burner at 4×10^6 Btu/hr (1.2 MW), the combined air–oxygen/fuel input of 13×10^6 Btu/hr (3.8 MW) provided a meltdown reduction of 50–65% from 3 to 4 hr per heat to 1 to 2 hr per heat, depending on the charge material. Phillips et al. [69] also reported the benefits of using an air–oxy/fuel burner for aluminum dross processing. Compared to conventional air/fuel operation, the melting time was reduced by 50%, productivity was increased by 38%, and natural gas consumption was reduced by 45%.

An aluminum smelter, with three open-well reverberatory furnaces melting scrap, who needed increased production, implemented oxygen-assisted combustion for increased furnace productivity, reduced cost per pound of material melted, and improved metal yield [64]. Two furnaces of 150,000 lb (68,000 kg) capacity each and one 70,000 lb (32,000 kg) capacity were side-well-fed reverberatories (see Fig. 13.23). Four dual-fuel burners fired along one furnace wall in the larger furnaces, while the smaller furnace was equipped with two burners. For the first furnace, the EZ-Fire burners (see Fig. 13.31) were installed within the existing air/fuel burners without interrupting production. Adding molten-metal recirculating pumps can significantly improve melt rates by up to 15% and reduce fuel consumption by more than 20%, with melt yields up to 96% [70]. This is due to the increased heat-transfer rates by mixing the metal, which has a more uniform temperature instead of much higher surface temperatures as is often the case in naturally circulating systems.

Stala and Hindman [71] described a gas-fired immersion tube for melting aluminum. One of the problems with the conventional aluminium-melting processes is that they rely primarily on radiation heat transfer. This mode of heat transfer is often not optimal for a material like aluminum, which can be highly reflective when it is polished or when it is molten, both of which may exist at certain times in the aluminium melting cycle. Another possible problem with radiant heating is overheating of the surface, which can produce dross that lowers yields. A heating method that utilizes the high thermal conductivity of aluminum could increase the thermal efficiency of the aluminium-melting process. An immersion-type burner not only heats primarily by thermal conduction, but it also separates the combustion products from the gases evolving from the aluminium-melting process. This makes heat recuperation easier, especially in melting processes where chlorine is injected into the aluminum for contaminant removal because the excess chlorine gases coming out of the molten aluminum are highly corrosive and can damage recuperative burners. The key to the development of an immersion burner is to find the proper ceramic material that has high thermal conductivity, good thermal shock resistance, good corrosion resistance, and is easy to fabricate. Stala and Hindman found that α silicon carbide had the best characteristics during testing in an aluminium-holding furnace, which showed the feasibility of the technology despite some of the problem encountered during the testing.

Kimura and Taniwaki [71a] described a similar immersion-type burner for melting smaller quantities of aluminum used in some segments of the aluminium-casting industry. The burner tubes only survived up to 6 months in the aggressive environment of aluminium-melting processes. Both super-alumina and hi-alumina materials showed similar performance characteristics in the single-ended radiant tube burners. Melting efficiencies up to 60% were measured. Other benefits included: low metal loss, more uniform aluminum temperature, higher product quality, reduced exhaust volume, and smaller equipment size.

Ward et al. [72] described the development of a new, modular, ceramic recuperator for use in an aluminium-remelt furnace. With thermal efficiencies ranging from 10 to 30% for nonrecuperated furnaces, this development was designed to increase efficiencies in reverberatory furnaces by 40% with heat recuperation. The difficulty with conventional metal recuperators is that the atmosphere in aluminum furnaces is moderately corrosive so the life of the metal recuperators is shortened.

13.3.1.3 Pollution Control

A variety of techniques are used to control particulate emissions from secondary aluminum processes [17]. Wet scrubbers are used not only to capture particulates but also to absorb any chlorine gases that may have gotten into the exhaust gas stream from the chlorination demagging process when chlorine is injected into the molten-aluminum bath. The use of a caustic solution in the wet scrubber has been found to be effective for maximizing the collection efficiency. Fabric filters or baghouses are also commonly used to remove particulates in this application. However, these do not have the added benefit of capturing other pollutant fumes like chlorine gases, unless the filters are coated with an absorbent material that neutralizes the acid gases while simultaneously removing particulates. Wet scrubbers are sometimes used in combination with baghouses where the primary function of the scrubber is to remove the fumes and the primary function of the baghouse is to capture the particulates.

Afterburners are sometimes used in this application to combust fully any hydrocarbon emissions that may still be in the exhaust gases. This can happen, for example, if a large quantity of especially oily aluminum borings or turnings are charged into the furnace beyond the normal capability of the furnace system to combust the oil fully. This could be due to a lack of oxygen, poor mixing, or insufficient residence time. The afterburner then finishes the combustion process that was started but not completed in the furnace. These are designed to provide adequate oxygen for combustion, turbulence for mixing, and residence time to complete combustion to achieve very high destruction efficiencies. Afterburners are not usually required if the incoming scrap is sufficiently cleaned to remove excess oils or if the amount of oily scrap is properly controlled to ensure that the furnace system is capable of handling the added hydrocarbon load.

13.3.2 Copper, Brass, and Bronze Production

Copper is made through both primary and secondary processes. Primary processes take raw ore and convert it into copper while secondary processes take recycled metal containing copper and reprocesses it to make copper. Since primary processes do not involve traditional industrial combustion processes, these are not considered here. Figure 13.36 shows a typical flow diagram for the production of copper from both primary and secondary processes [73]. Figure 13.37 shows a process flow diagram for the production of high-grade brass and bronze [74].

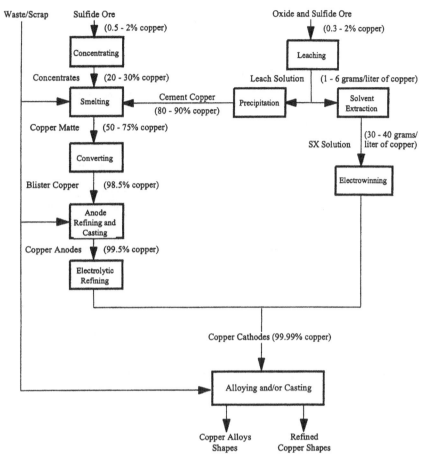

Source: *Office of Technology Assessment.*

Figure 13.36 Copper-production process flow diagram. (From Ref. 73.)

In secondary copper processing, the scrap is often pretreated to remove the large contaminants and to make the scrap easier to feed into the furnace where, for example, the scrap may be shredded or made into briquettes. Smelting of low-grade copper is done first in a blast or rotary furnace. This is followed by melting in a fired furnace, such as a reverb, to increase further the purity. Fluxes are added to remove impurities, which are then removed as slag. The furnace may then be run in a reducing mode to convert oxidized copper (CuO) into pure copper. Alloys are added to make the appropriate products including brass and bronze. The alloying agents may include tin, zinc, silver, lead, aluminum, or nickel. The final product is then poured into the desired shape such as shot, wirebar, anodes, cathodes, ingots, or other cast shapes. Copper alloyed with zinc is usually termed brass, and when alloyed with tin is termed bronze. Many other copper alloys are possible including, for example, aluminum bronze and silicon bronze. In most cases the combustion process is very similar for all of these.

ENTERING THE SYSTEM   LEAVING THE SYSTEM

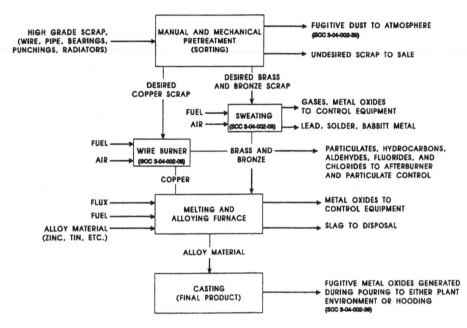

Figure 13.37 Process flow diagram for the production of brass and bronze. (From Ref. 74.)

The Environmental Protection Agency (EPA) has established emission guidelines for secondary brass and bronze production plants [75]. The major sources of emissions from secondary copper processes are from the furnaces [76]. Common pollutants from secondary copper processes include particulates, NO_x, VOCs, CO, unburned hydrocarbons, and heavy metals. Cupola furnaces are usually the largest sources of emissions and particulates are usually the pollutant of most concern. The particulates may contain zinc, lead, tin, copper, chlorine, or sulfur [17]. Baghouses (see Chap. 9) are commonly used to control particulate emissions. SO_x emissions are not typically a problem because the fuels and scrap used generally contain little sulfur. However, some of the incoming raw materials, such as chalcocite (Cu_2S), bornite (Cu_5FeS_4), tetrahedrite ($Cu_5Sb_2S_7$), or chalcopyrite ($CuFeS_2$), may include sulfur.

A variety of equipment is commonly used to control pollutant emissions in copper smelting [77]. For particulates, cyclones, venturi scrubbers, electrostatic precipitators, mist precipitators, spray chambers, baghouses, settling chambers, and scrubbing towers are used to control particulate emissions. Both single- and double-contact sulfuric acid plants are used to control SO_x emissions. Afterburners may be used to control CO and unburned hydrocarbon emissions.

A rotary-type copper-alloy melting furnace, rated for 1.2 tons per hour, was used to experiment with oxy/fuel combustion where productivity improvements and energy savings were realized [78]. The results of the test are presented in Table 13.7. Reducing fuel consumption directly reduces pollution emissions.

Table 13.7. Performance of a Rotary Copper Melting Furnace

	Air/fuel	Oxy/fuel	Difference (%)
Specific fuel consumption (Nm^3/t)	102.1	40.2	−60.7
Specific oxygen consumption (Nm^3/t)	—	72.9	—
Total energy consumption (kg/kg)	3633	2224	−38.8
Productivity (kg/hr)	1175	2595	+121

Source: Ref. 48. Courtesy of CRC Press.

In a 20 metric ton copper rotary furnace, where the dual-fuel burner was retrofitted with a nonwater-cooled oxy/fuel burner that provided the flexibility of firing air/fuel or oxy/fuel or any combination in between, furnace productivity improvements and fuel savings were realized. The oil leg of the dual-fuel burner was replaced with an oxy/natural gas burner. The firing rate was reduced from an equivalent of $600\,m^3/hr$ ($21,000\,ft^3/hr$) natural gas firing (1:1 gas:oil) to $400\,m^3/hr$ ($14,000\,ft^3/hr$) natural gas firing. The heat time was reduced from 7.5 to 6.5 hr, with a fuel savings of 33%. In addition, fugitive emissions from the furnace during the early part of the cycle were significantly reduced and refractory damage was controlled.

Several major copper smelters use oxygen in their reverberatory furnaces for increasing matte production in order to utilize excess downstream converter capacity. Alternatively, producing the same amount of matte from fewer reverberatory furnaces or fewer furnace hours can also be cost effective.

Conventional pyrometallurgical smelting in reverberatory furnaces for recovery of copper from its sulfide concentrates uses large quantities of hydrocarbon fuels because the process makes little use of the energy available from oxidation of the sulfide charge. The use of tonnage oxygen in this process is a first step in retrofitting the furnaces to reduce overall energy costs and increase production [79]. In some cases, it may be desirable to enrich selectively some of the end-wall burners to higher levels than others, in order to concentrate the heat in colder or slower moving portions of the melt. This selectivity is easily accomplished with either premixing oxygen and combustion air or undershooting oxygen enrichment. Both techniques are relatively simple to install. A production increase of 20–30% may be attained on a routine basis using O_2 enrichment. Generally, 9% enrichment or 30% total O_2 in the combustion air is considered to be a practical maximum. In addition to a production increase and fuel savings, O_2 enrichment can provide additional cost savings in the gas cleaning and sulfur fixation systems.

After successful experience with oxygen enrichment of wall-mounted conventional air/fuel burners, several smelters have successfully tried roof-mounted oxy–fuel burners, firing directly on to the charge. Substantial improvements in efficiency can be achieved with these burners because the heat is transferred to the unmelted charge by both radiation and convection. These burners can be used to supplement oxygen-enriched conventional burners to increase production further.

When using oxy/fuel burners substantially greater benefits can be achieved. Data from Inco's Copper Cliff Smelter indicates a 45% increase in smelting rate with simultaneous reduction in fuel consumption of 55% [80]. The Noranda Horne Smelter reports a 60% increase in smelting rates and an 18% reduction in fuel consumption. The fuel rates are gradually being reduced further. Codelco's Caletones Smelter reports a smelting rate improvement over 100% and a 55% reduction in fuel. Their flue gas contains SO_2 concentrations of 6–8%.

In oxygen-sprinkle smelting, a reverberatory furnace is retrofitted with roof-mounted burners that combust dry feed and oxygen. This concept, developed by Queneau and Schuhmann [81], transforms an existing reverberatory furnace into an oxygen flash smelting unit with minimal capital investment. It is possible to establish autogenous smelting conditions at a certain oxygen-to-concentrate ratio. Based on heat-balance calculations, the autogenous level can be attained when smelting to a matte grade of 64% Cu with 42% SO_2 in the off-gas. In order to reach autogenous conditions [81], the feed rate needs to be almost doubled compared with conventional reverberatory smelting (matte grade 35% Cu). The oxygen-to-concentrate mass ratio should be 0.22:1. Small-scale oxygen-sprinkle smelting test conducted at Phelps Dodge Corporation at its Morenci smelter [82] indicated that furnace throughput increases of 100% were possible with two-thirds reduction in total smelting energy requirements per unit of charge (including electricity and energy for oxygen production). SO_2 concentrations of 20–30% were expected.

The copper smelting process consists of two major operations: melting the charge materials and oxidation of the molten bath. Melting requires a supply of external heat energy, such as the burning of hydrocarbon fuels. For many years, melting has been done by supplying heat via combustion of fossil fuels. The oxidation of a molten bath is a chemical phenomenon where an oxidant is required to remove sulfur from concentrates. Although oxidation is exothermic, external fuel is also required here unless the sulfur level is high enough to make the process autogenous.

In flash smelting, the concentrate is dispersed in an air or oxygen stream, and smelting and converting occur while the particles are in suspension. The major reasons that prompted the use of oxygen in flash smelting include: increased matte production from an existing smelter, use of a more efficient and lower cost SO_2 recovery system due to increased SO_2 content of the furnace off-gas, and an autogenous process resulting in energy savings [83].

The INCO flash smelting process uses pure oxygen instead of air, and is an autogenous operation. Concentrates are injected through two burners in each end-wall of the furnace and combusted in a horizontal stream of oxygen. The oxygen flash smelting is very flexible and can treat feeds of various compositions to produce a wide range of matte grades. For producing higher grades, the oxygen-to-concentrate ratio is increased and a coolant is added, whereas for lower grades the ratio is decreased with addition of a supplemental external fuel burner. Metallurgical advantages from the use of tonnage oxygen include the absence of magnetite buildup as a result of a very low oxygen potential in the furnace, and the rapid establishment of slag–matte equilibrium [84]. Slag can be discarded directly as it contains about 0.8% Cu at a matte grade of 55% Cu. Other benefits related to oxygen use in the autogenous process are low off-gas volume (20% of other processes), low dust carryover due to low gas velocity in the furnace, high SO_2 content (80%) in the waste gas offering increased flexibility in choosing SO_2 recovery system, and relatively small size and cost for gas cleaning and treatment facilities.

The Mitsubishi continuous smelting process is a multistep process where dried, fluxed concentrates and 9–14% oxygen-enriched air enter the smelting furnace through nonsubmerged vertical lances. Fuel is required in the smelting furnace to maintain the heat balance. In the converting step, 5–7% oxygen-enriched air is blown through vertical lances above the bath to oxidize continuously the matte to blister copper. Fuel is also required in the converting furnace to compensate for the heat deficit [85]. The smelting unit can be made autogenous with 65% O_2 in air [86]. Trials at Naoshima [87] have demonstrated that throughput can be increased by 50% when the oxygen enrichment level is increased from 7 to 18%. A concurrent fuel reduction of 70% of the normal requirement has also been achieved. The SO_2 content of off-gas was increased from 14 to 24% with a volume reduction of 88%.

Oxygen enrichment of Pierce–Smith copper converters has been practiced on a production scale since the late 1950s. Most major copper smelters currently use oxygen enrichment in their side-blown Pierce–Smith converters. Usually, oxygen is added to the tuyere (nozzle for injecting gases into molten metal) air to a maximum enrichment limit of 9%. Above this level, refractory wear at the tuyere line becomes excessive unless sufficient cold dope is added to control bath temperature. In the side-blown Hoboken converter, oxygen is top blown through a lance directly into the copper bath to prevent refractory wear at the tuyere area. The incentives for oxygen enrichment of the converter blowing air or direct oxygen injection into the converter bath are [88]:

(1) To conserve heat in the charge to melt additional cold reverts, precipitates, or concentrates (during slag blow) and copper scrap (during copper blow).
(2) To increase converter capacity by reducing the blowing time or increasing charge size.
(3) To increase the SO_2 content in the off-gas, which can result in more efficient and lower cost SO_2 recovery systems. Higher SO_2 concentrations will not only allow smaller pollution control equipment with associated lower investment, but also reduce operating costs for the SO_2 recovery system.

13.3.3 Lead Production

Lead is made through primary and secondary processes. Since traditional industrial combustion processes are not used in primary processes, only secondary smelting is considered here where recycled metal is used in a smelting furnace. There are three major operations in secondary lead smelting: scrap pretreatment, smelting, and refining [89]. Scrap pretreatment includes removal of large contaminants, shredding large pieces, draining fluids, and possibly drying wet materials. Smelting is used to separate the lead from metallic and nonmetallic contaminants and reduce lead oxides to elemental lead. This is usually done in a blast, reverberatory, or rotary furnace. The pretreated scrap, slag to be reprocessed, scrap iron, coke, recycled dross, flue dust, and limestone are charged into the furnace. As the charge melts, the limestone and iron float to the top of the molten bath and form a flux that retards oxidation of the elemental lead. The molten lead is then tapped into a holding pot before being cast into large ingots called pigs or sows. A typical process flow schematic for the refining process is shown in Fig. 13.38. If required, further refining is done in a refining furnace to remove either copper and antimony for soft lead production or to remove arsenic, copper, and nickel for hard lead production.

SMELTING

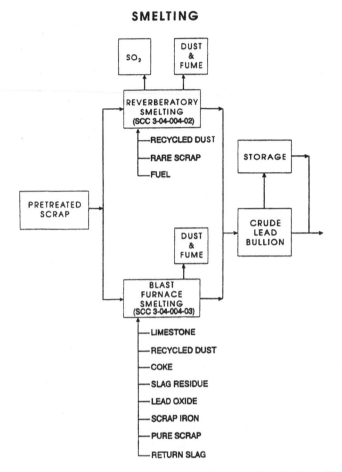

Figure 13.38 Process flow diagram for secondary lead smelting. (From Ref. 89.)

Sulfur may be added to remove copper while aluminum chloride may be added to remove copper, antimony, and nickel. Air can also be bubbled through the molten metal to remove antimony.

The Environmental Protection Agency (EPA) has established emission guidelines for secondary lead smelters [90]. Typical emissions from secondary lead processing include particulates, NO_x, SO_x, CO, and heavy metals. Fabric filters (baghouses), dry cyclones, wet cyclones, settling chambers, venturi scrubbers, and combinations of these are used to control particulate emissions that may contain PbO, SnO, ZnO, tar, flyash, coke dust, sulfates, and sulfides [17].

Lead rotary furnaces are used either for melting battery scrap or smelting concentrates. Oxy/fuel technology was introduced about 20 years ago in lead scrap melting furnaces. Most of the lead rotary furnaces have been converted to oxy/fuel in Europe. The benefits of oxy/fuel burners in secondary plants include: up to 100% production increases, energy savings of up to 60%, lower CO, CO_2 and NO_x, 85% lower combustion products, 50% furnace off-gas, and up to 60% lower particulate emissions.

When an air/oil burner of a 15-ton concentrate smelting rotary furnace is replaced with an oxy/oil burner, the following benefits have been realized: 27% reduction in smelting time, 30% reduction in tap-to-tap time, 59% oil savings per batch, 34% reductions in flue dust, higher tapping temperature, better working environment, less smoke exiting the furnace, lower noise, and overall furnace productivity improvements of 42%.

No major changes, other than installing the oxy/oil burner (which was much smaller in size) and an oxy–oil flow train, were needed to convert the furnace from air/oil to an oxy/oil system. This furnace was charged with oxide concentrates with coke and other fluxes. The increased furnace productivity allowed the plant to operate fewer rotaries and still exceeded the plant production obtained with the old air/fuel system.

In a lead reverberatory furnace, when 2% oxygen enrichment was practiced, the production increased by 6.8%. When the polypropylene-rich scrap content of the charge was increased from 30 to 60%, the burner firing rate was reduced from 18 to 14.4×10^6 Btu/hr (5.3–4.2 MW), a 20% reduction in fuel. The increased use of polypropylene-rich scrap is an important benefit of the oxygen-enrichment practice. Although high temperatures were generated by oxidation of the plastic-rich scrap, the increased antimony content of the lead produced indicated the presence of reducing conditions in the furnace.

In lead smelting, the oxygen-enriched blast has been used in lead cupola and slag fuming furnaces. Oxygen enrichment of the copper blast furnace is also common. The cupola charge mix consists of reverberatory furnace slag, battery plates, coke, iron, and limestone. Injection of 2.5% pure oxygen into the cupola wind results in a 20% increase in lead production. Reductions in coke consumption with enrichment are normally in the range 4–7% on a dry weight basis. At another location, 3% enrichment increased production 30% and reduced coke consumption by 33%. Good slag fluidity, and a higher metal temperature resulting from oxygen enrichment, facilitated tapping operations. Enrichment also allowed the tuyeres to remain cleaner. Care should be taken to maintain a CO-to-CO_2 ratio of 1:4 at the exit gas analysis to ensure smooth operation of a lead blast cupola. The results of the use of oxygen enrichment of the lead blast furnaces are shown in Table 13.8 [91].

One of the primary concerns in lead production is the potential for lead emissions, particularly during smelting operations where high temperatures cause the solid lead to volatilize. There is also the potential for dust, SO_2, antimony, cadmium,

Table 13.8 Effect of Oxygen Enrichment of a Lead Blast Furnace

	Base	3% Enrichment	5% Enrichment
Production increase (%)	Base	25.5	52.9
Coke savings (%)	Base	9.9	16
Dust production, reduction (%)	Base	36.4	45.5
Pb in slag (%)	1.54	1.40	
	0.96		1.3

Source: Ref. 48. Courtesy of CRC Press.

zinc, and other volatile metals. Baghouses or electrostatic precipitators are commonly used in most smelting processes.

13.4 HEAT TREATING

Heat treating involves the thermal treatment of metal to produce some type of enhanced performance characteristic. Typical material improvements include surface hardening, strengthening the part, relieving stresses, and improving ductility. Typical thermal treatments include annealing, brazing, carburizing, normalizing, sintering, and tempering.

Annealing is a heat treatment used to remove internal stresses and to make a material less brittle by heating a part and then cooling it. This may be done under a protective atmosphere, commonly nitrogen, to prevent surface reactions if the application is particularly sensitive or critical. Sometimes a reducing agent like hydrogen may be added to the protective atmosphere to produce clean, bright surfaces.

Brazing is a method of joining metals together using a nonferrous filler material such as brass or a brazing alloy. The melting point of the braze material is lower than that of the metals being joined, which dictates the furnace temperature. Again a protective atmosphere of nitrogen, possibly with small amounts of hydrogen or a hydrocarbon, may be used to produce high-quality parts. Dewpoint control in the atmosphere is often important in certain applications to minimize surface reactions with water at higher temperatures that would reduce the surface quality.

Carburizing and *neutral hardening* refer to surface hardening of steels by adding carbon just to the surface of a part at a temperature high enough for the carbon to convert the surface from a lower carbon steel to a higher carbon steel. This must be done under an atmosphere, sometimes referred to as an endothermic gas, that contains carbon. The endothermic gas may have a composition of approximately 20% CO, 40% H_2, and 40% N_2.

Normalizing involves taking a ferrous alloy above its upper transformation temperature and cooling it well below that temperature. This refines the grain structure of the metal and is usually done after a part has been hot worked to soften the material.

Sintering is used to produce solid metal parts from metal powders by heating them to a high enough temperature, often under a protective atmosphere, depending on the metals being used, that the powder fuses together without melting.

Tempering is the heat treatment of ferrous alloys by heating them below the lower transformation temperature to obtain certain mechanical properties. Although it is similar to annealing and stress relieving, the resulting material properties are distinct in all three processes.

One method that has been used to classify heat-treating furnaces is based on the heat-transfer medium: gaseous (inert, oxidizing, or reducing), vacuum (no medium), liquid (molten metal or molten salt), and solid (fluidized bed) [92]. This method focuses on the heating process rather than on the metal-treatment process.

Heat-treating furnaces may be either continuous or batch [93]. The major types of continuous furnaces include rotary retort, rotary hearth, pusher, roller hearth, conveyor, and continuous strand. A rotary retort furnace has a gas-tight cylindrical

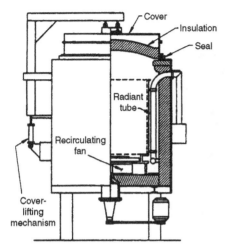

Figure 13.39 Pit furnace. (From Ref. 92. Courtesy of American Society of Metals, Metals Park, OH.)

retort that revolves inside a firebrick and steel-enclosed shell. The rotary hearth furnace has a cylindrical gas-tight shell that rotates. In a pusher furnace the load is usually loaded on trays and pushed into the furnace. The major types of batch furnaces include box, rotary retort, rotary hearth, car bottom, bell, pit, liquid bath (molten lead and salt), fluidized bed, and vacuum. A pit furnace is located below the floor and extends up to or slightly beyond the floor. An example of a pit furnace is shown in Fig. 13.39. The challenge in either batch or continuous processes requiring a specific atmosphere composition is to maintain that atmosphere during the heat-treating cycle. For example, in vacuum furnaces (batch operation), fairly elaborate door seals are needed to maintain high levels of vacuum inside the furnace. In belt furnaces, gas curtains are usually employed at both ends of the furnace to prevent ambient air infiltration into the process that would reduce product quality. Heat-treat furnaces are either direct or indirect fired. If a special atmosphere is required, the furnace is indirect fired. If no special atmosphere is required, the furnace is often direct fired.

13.4.1 Standard Atmosphere

There are some heat-treating processes that do not require a special atmosphere, where the gas around the parts being treated is air and the products of combustion. One example is a car bottom furnace that is often used to relieve stresses in large welded parts. Furnaces used for annealing also do not usually require any specific gas composition around the parts being treated. Another example is a rotary-hearth continuous heat-treating furnace, shown in Fig. 13.40, which operates like a revolving door with parts entering on one part of the circle and exiting at another part after rotating some angular distance. Rotary-hearth furnaces are often used for heating larger workpieces.

Ferguson (1979) compared the energy requirements for indirect-fired (radiant tube) and direct-fired (with recuperation) furnaces for batch annealing furnaces. The

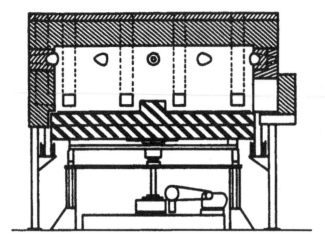

Figure 13.40 Rotary hearth furnace. (From Ref. 92. Courtesy of American Society of Metals, Metals Park, OH.)

direct-fired furnace used about one-third less energy than the indirect-fired furnace. In the radiant tube furnace, over half of the energy is lost in the flue gases, compared to only 30% for the direct-fired recuperative furnace. This shows the importance of the heating method in energy consumption. The analysis assumed that no special atmosphere was required for the heating application.

13.4.2 Special Atmosphere

In this type of process, the combustion products cannot come into contact with the metal load as they would reduce the product quality. There are two basic types of special atmospheres that are commonly used, depending on the type of heat-treating process. In the first type of special atmosphere, the objective is to prevent certain chemicals from reaching the part that could contaminate the surface. One such type is sometimes referred to as a neutral atmosphere, where, for example, an inert gas like helium is used to protect metal products from oxidation, which causes scale formation on the metal surface that reduces the metal yield and product quality. Another type of protective atmosphere is the absence of any gases, created by a vacuum, contacting the part surface. Some specific contaminants that can detrimentally affect a part's surface quality include oxygen (air), water, carbon-containing materials, hydrogen, and nitrogen. Oxygen and water can cause surface oxidation, carbon-containing materials can cause surface hardening, hydrogen can cause embrittlement, and nitrogen can cause nitriding. The last two are sometimes desirable, but they can also be detrimental, depending on the desired part characteristics.

The second type of special atmosphere is where a specific reaction is desired in which the atmosphere contacts the part surface at an elevated temperature. One example is a carburizing atmosphere that contains a specific amount of gases containing carbon, which diffuses into the metal surface for hardening. This is used, for example, in tool steels, which require hard outer surfaces. An atmosphere containing nitrogen can be used to nitride a metal. Table 13.9 shows common special furnace atmosphere compositions and their applications [95]. Many of these

Table 13.9 Special Atmosphere Compositions and Applications

Class Description	Nominal composition (% by volume)					Application
	N_2	CO	CO_2	H_2	CH_4	
101 Lean exothermic	86.8	1.5	10.5	1.2	0	Oxide coating of steel
102 Rich exothermic	71.5	10.5	5.0	12.5	0.5	Bright annealing; copper brazing; sintering
201 Lean prepared nitrogen	97.1	1.7	0	1.2	0	Neutral heating
202 Rich prepared nitrogen	75.3	11.0	0	13.2	0.5	Annealing and brazing of stainless steels
301 Lean endothermic	45.1	19.6	0.4	34.6	0.3	Clean hardening
302 Rich endothermic	39.8	20.7	0	38.7	0.8	Carburizing (enriched 301)
402 Charcoal	64.1	34.7	0	1.2	0	Carburizing
501 Lean exothermic–endothermic	63.0	17.0	0	20.0	0	Clean hardening
502 Rich exothermic–endothermic	60.0	19.0	0	21.0	0	Carburizing
601 Dissociated ammonia	25.0	0	0	75.0	0	Brazing; sintering
621 Lean combusted ammonia	99.0	0	0	1.0	0	Neutral heating
622 Rich combusted ammonia	80.0	0	0	20.0	0	Sintering of stainless powders

Adapted from Ref. 95.

atmospheres are directly or indirectly produced by combusting hydrocarbon fuels. One example of a furnace that has a protective atmosphere is a bell furnace. The inner part of the furnace has a heated muffle that contains a protective atmosphere, often hydrogen or hydrogen blends. These furnaces are commonly used for annealing steel coils. Another example of a furnace using a protective atmosphere is a roller hearth as shown in Fig. 13.41. The picture shows an entrance vestibule that helps contain the protective atmosphere. Furnace pressure control is especially important in furnaces containing special atmospheres not only to ensure that ambient air does not leak into the furnace, but also that not much special atmosphere leaks out of the furnace, which would be wasted cost [96].

Thekdi et al. [97] described an indirect, natural gas-fired heating system used where controlled atmospheres are needed for heat treating. The heat source was a high-intensity radiant tube burner capable of temperatures greater than 2000°F (1100°C). Heat-flux improvements of up to three times that of conventional heating systems were measured.

Liang et al. [98] discussed the application of ceramic tube burners in indirect-fired heat-treat furnaces. They described developments of advanced materials to increase the maximum operating temperatures of metallic radiant tubes that were limited to 1800°F (1000°C). Furnaces operating at higher temperatures typically used electric heating, which is often more expensive than using natural gas.

Hemsath [99] discussed a new gas-fired heating system for indirect heating under a protective or reactive atmosphere, or under a vacuum. Important advantages

Figure 13.41   Photograph of a roller hearth furnace. (Courtesy of Electric Furnace, Salem, OH.)

of the system include lower operating costs compared to electrically heated indirect furnaces and uniform heating rates. Conventional gas-fired indirect furnaces use radiant tubes. The system described by Hemsath used forced convection to heat the vacuum vessel. The impinging jets consist of the exhaust products from a single burner. The maximum reported temperature for a pilot-scale furnace was 1250°F (677°C), which was expected to be raised to 1750°F (954°C) by subsequent development. The application of this novel furnace design was discussed for batch coil annealing and as a thermal cleaner for aluminum borings and scrap.

Copes and Brooks [100] tested a ceramic composite material called Siconex™ for use in a controlled-atmosphere furnace. The material, which is a fiber-reinforced composite, was tested in endothermic, exothermic, and nitrogen-based atmospheres. Tubular shapes were tested with and without coatings (23 different coatings were tested) to determine their properties under the different atmospheres. The performance tests for the coatings included burst strength, permeability, adhesion, stability, cracking, and maintenance of emissivity. Only two of the coatings performed well over a wide range of conditions. The inert atmosphere proved to be the most detrimental to the strength of the ceramic samples.

Schultz et al. [101] studied heat transfer in a gas-fired, high-temperature, "soft" vacuum (1–100 torr) furnace with a design operating temperature of 2350°F (1290°C). The furnace was designed for hardening tool steels, annealing stainless steel, and for sintering and brazing applications. The furnace was heated with single-ended radiant tube burners. The object of the study was to model the heat transfer in the furnace with various silicon carbide radiant tube sizes ranging from 3.25 to 6 in. (8.26–15 mm) o.d., quantities, and locations. The following parameters were used in the evaluation: empty furnace temperature uniformity, surface temperature uniformity during heating, process or cycle time, and temperature uniformity within the load. A combination of modeling and experiments showed that for the given furnace configuration, four 6 in. (15 mm) o.d. single-ended radiant tubes provided the most optimized results.

Erinov et al. [102] described the development of a low-inertia, high-efficiency heat-treating furnace using indirect heating with contained, flat, gas-fired, infrared burners. The claimed benefits included fast heat-up and cool-down, better heating uniformity, higher thermal efficiency, increased productivity, and low NO_x emissions. The burner design incorporated flue gas heat recuperation and had a high surface temperature uniformity.

Kurek et al. [103] presented the development of high-temperature uniformity, natural gas fired, flat radiant panels for use in indirect-fired heat-treating furnaces. These panels were designed as a potential replacement for radiant tube burners. They had increased radiating surface area, which means that they could be operated at lower temperatures to achieve the same heat-flux output. The lower radiating temperature and high surface area improved the temperature uniformity to the load, which improved product quality. Because of the higher surface area, less refractory was required in the furnace, which meant faster heat-up and cool-down. Self heat recuperation increased the thermal efficiency to over 70%.

REFERENCES

1. U.S. EPA. EPA Office of Compliance Sector Notebook Project: Profile of the Metal Casting Industry. Washington, DC: U.S. Environmental Protection Agency, Oct. 1998.
2. AC Stern, RW Boubel, DB Turner, DL Fox. *Fundamentals of Air Pollution*, 2nd edn., Orlando, FL: Academic Press, 1984.
3. Office of the Federal Register. 60.16 Priority List. U.S. Code of Federal Regulations Title 40 Part 60. Washington, DC: U.S. Government Printing Office, 2001.
4. XJ Li. CFD Modeling for the Steel Industry. In CE Baukal, VY Gershtein, X Li. (eds.). *Computational Fluid Dynamics in Industrial Combustion*. Boca Raton, FL: CRC Press, 2001.
5. U.S. EPA. EPA Office of Compliance Sector Notebook Project: Profile of the Iron and Steel Industry. Washington, DC: U.S. Environmental Protection Agency, Sept. 1995.
6. A Lehrman, CD Blumenschein, DJ Doran, SE Stewart. Steel Plant Fuels and Water Requirements, In RJ Fruehan, (ed.). *The Making, Shaping and Treating of Steel*, 11th edn., *Steelmaking and Refining Volume*, Pittsburgh, PA: AISE Steel Foundation, pp. 311–412, 1998.
7. RJ Fruehan, CL Nassaralla. Alternative Oxygen Steelmaking Processes. In RJ Fruehan (ed.). *The Making, Shaping and Treating of Steel*, 11th edn., *Steelmaking and Refining Volume*. PA: AISE Steel Foundation, Pittsburgh, pp. 743–759, 1998.
8. J Elkins, N Frank, J Hemby, D Mintz, J Szykman, A Rush, T Fitz-Simons, T Rao, R Thompson, E Wildermann, G Lear. National Air Quality and Emissions Trends Report, 1999. Washington, DC: U.S. Environmental Protection Agency, Rep. EPA 454/R-01-004, 2001.
9. MD Kistler, JS Becker. Ferrous Metals. CE In Baukal (ed.). *Oxygen-Enhanced Combustion*. Boca Raton, FL: CRC Press, pp. 165–180, 1998.
10. U.S. EPA. Electric Arc Furnaces and Argon-Oxygen Decarburization Vessels in Steel Industry—Background Information for Proposed Revisions to Standards. Rep. EPA-450/3-82-020a. Washington, DC: Environmental Protection Agency, July 1983.

11. Office of the Federal Register. Subpart AA: Standards of Performance for Steel Plants: Electric Arc Furnaces Constructed After October 21, 1974, and On or Before August 17, 1983. U.S. Code of Federal Regulations Title 40, Part 60. Washington, DC: U.S. Government Printing Office, 2001.

12. Office of the Federal Register. Subpart AAa: Standards of Performance for Steel Plants: Electric Arc Furnaces and Argon–Oxygen Decarburization Vessels Constructed After August 17, 1983. U.S. Code of Federal Regulations Title 40, Part 60. Washington, DC: U.S. Government Printing Office, 2001.

13. Office of the Federal Register. Subpart Z: Standards of Performance for Ferroalloy Production Facilities. U.S. Code of Federal Regulations Title 40, Part 60. Washington, DC: U.S. Government Printing Office, 2001.

14. U.S. EPA. Electric Arc Furnaces and Argon–Oxygen Decarburization Vessels in Steel Industry—Background Information for Proposed Revisions to Standards. Rep. EPA-450/3-82-020a. Washington, DC: U.S. Environmental Protection Agency, July 1983.

15. JAT Jones, B Bowman, PA Lefrank. Electric Furnace Steelmaking. In RJ Fruehan (ed.). *The Making, Shaping and Treating of Steel*, 11th edn. *Steelmaking and Refining Volume*. Pittsburgh, PA: AISE Steel Foundation, pp. 525–660, 1998.

16. PF Fennelly, PD Spawn. Air Pollutant Control Technologies for Electric Arc Furnaces in the Iron and Steel Industry. Rep. EPA 450/2-78-024. Washington, DC: U.S. Environmental Protection Agency, 1978.

17. U.S. EPA. Control Techniques for Particulate Emissions from Stationary Sources— Volume 2. Rep. EPA-450/3-81-005b. Washington, DC: U.S. Environmental Protection Agency, Sept. 1982.

18. U.S. EPA. Profile of the Iron and Steel Industry. Washington, DC: U.S. Environmental Protection Agency Rep. EPA/310-R-95-005, 1995.

19. JP Birat, A Arion, M Faral, F Baronnet, PM Marquaire, R Rambaud. Abatement of organic emissions in EAF exhaust flue gas. Rev. Metall.-Cahiers Inform. Tech. Vol. 98, No. 10, pp. 839–854, 2001.

20. U.K. Environment Agency. Iron and Steel Making Processes. IP Guidance Note S2 2.01. London, 1999.

20a. TJ Overcamp, MP Speer, SJ Griher, DM Cash. Gaseous and particulate emissions from a DC arc melter. *J. Air & Waste Manage. Assoc.*, Vol. 53, pp. 13–20, 2003.

21. JR Davila, A Garrote, AM Gutierrez, P Carnicer, L Cobos, I Erauskin. New Applications of the Oxygas Rotary Furnace as a Means of Production in Melting. *Proceedings of 1995 International Gas Research Conference*, ed. DA Dolenc. Govt. Institutes, Rockville, MD, pp. 2764–2773, 1996.

22. CE Baukal. *Heat Transfer in Industrial Combustion*. Boca Raton: CRC Press, 2000.

23. DH Hubble, RO Russell, HL Vernon, RJ Marr. *Steelmaking Refractories*. In RJ Fruehan (ed.). *The Making, Shaping and Treating of Steel*, 11th edn., *Steelmaking and Refining Volume*. Pittsburgh, PA: AISE Steel Foundation, pp. 227–290, 1998.

24. GJW Kor, PC Glaws. Ladle Refining and Vacuum Degassing. In RJ Fruehan (ed.). *The Making, Shaping and Treating of Steel*, 11th edn. *Steelmaking and Refining Volume*. Pittsburgh, PA: AISE Steel Foundation, pp. 661–713, 1998.

25. CE Tomazin, EA Upton, RA Willis. Effect of ladle refractories and practices on steel temperature control. *Iron & Steelmaker*, Vol. 13, No. 6, pp. 28–34, 1986.

26. www.afsinc.org/Trends/FactsandFigures.htm

27. www.diecasting.org/faq/introduction/background.htm

28. U.S. Department of Energy. Beyond 2000: A Vision of the American Metalcasting Industry. Washington, DC: U.S. Department of Energy, 1995.

29. U.S. Department of Energy. A Vision of the U.S. Metal Casting Industry: 2002 and Beyond. Washington, DC: U.S. Department of Energy, 2002.

30. U.S. Department of Energy. Metalcasting Industry Technology Roadmap. Washington, DC: U.S. Department of Energy, 1998.

31. KC Schifftner, HE Hesketh. *Wet Scrubbers*, 2nd edn. Lancaster, PA: Technomic Publishing, 1996.

32. R Klima. Improved knowledge of gas flow and heat transfer in reheating furnaces. *Scand. J. Metall. (Suppl.)*, Vol. 26, pp. 25–32, 1997.

33. GS Koch and JH Williams, Characterization of Natural Gas-Fired and Alternative Steel Reheat Technologies. In MA Lukasiewicz (ed.). *Industrial Combustion Technologies*. Warren, PA: American Society of Metals, pp. 295–301, 1986.

34. BA Steiner. Ferrous Metallurgical Operations. In AC Stern (ed.). *Air Pollution*, 3rd edn., Vol. IV. New York: Academic Press, 1977.

35. K Handa and Y Tomita. Development of Highly Luminous Gas Burner and Its Application to Save Energy in the Steel Industry. *Proceedings of 1984 International Gas Research Conference*, Govt. Institutes, Rockville, MD, pp. 729–739, 1985.

36. LT Walsh, M Ho, MG Ding. Demonstrated Fuel Savings and Uniform Heating with 100% Oxygen Burners in a Continuous Steel Reheat Furnace. In MA Lukasiewicz (ed.). *Industrial Combustion Technologies*, Warren, PA: American Society of Metals, pp. 259–265, 1986.

37. JM Blanco, JM Sala. Improvement of the efficiency and working conditions for reheating furnaces through computational fluid dynamics. *Indus. Heat.*, Vol. LXVI, No. 5, pp. 63–67, 1999.

38. F Lisin, JA Marino. Heat transfer analysis in a batch steel reheat furnace. *Indus. Heat.*, Vol. LXVI, No. 10, pp. 99–102, 1999.

39. A Biro, P Sandor. Modernization of Pusher-Type Furnaces in Order to Meet the Requirements of Environmental Protection. *Proceedings of 5th European Conference on Industrial Furnaces and Boilers*, Portugal, Vol. 1, pp. 587–600, 2000.

40. CE Baukal. Modeling Impinging Flame Jets. In CE Baukal, VY Gershtein, X Li (eds.). *Computational Fluid Dynamics in Industrial Combustion*. Boca Raton, FL: CRC Press, 2001.

41. ME Ward, D Knowles, SR Davis, J Bohn. Effect of Combustion Air Preheat on a Forge Furnace Productivity. *Proceedings of 1984 International Gas Research Conference*, Govt. Institutes, Rockville, MD, pp. 763–776, 1985.

42. SY Salama, TM Desai. Electric Induction versus Gas Metal Heating: A Case Study. *Proceedings of 1984 International Gas Research Conference*, Govt. Institutes, Rockville, MD, pp. 777–786, 1985.

43. S Takamichi. Development of High Performance Forging Furnaces. *Proceedings of 1998 International Gas Research Conference*, In D Dolenc (ed.). Vol. V: *Industrial Utilization and Power Generation*. Rockville, MD: Govt. Institutes, pp. 100–112, 1998.

44. U.S. EPA. Compilation of Air Pollutant Emission Factors. Vol. I: Stationary Point and Area Sources, Sect. 12.10: Gray Iron Foundries, 5th edn. Washington, DC: U.S. Environmental Protection Agency Rep. AP-42, 1995.

45. U.S. EPA. Compilation of Air Pollutant Emission Factors. Vol. I: Stationary Point and Area Sources, Sect. 12.13: Steel Foundries, 5th edn. Washington, DC: U.S. Environmental Protection Agency Rep. AP-42, 1995.

46. U.S. EPA. Profile of the Metal Casting Industry. Washington, DC: U.S. Environmental Protection Agency Rep. EPA/310-R-97-004, 1998.

47. B Neufer. Alternative Control Techniques Document—NO_x Emissions from Iron and Steel Mills. Rep. EPA-453/R-94-065. Washington, DC: Environmental Protection Agency, Sept. 1994.

48. D Saha, CE Baukal. Nonferrous Metals. In CE Baukal (ed.). *Oxygen-Enhanced Combustion*, Boca Raton, FL: CRC Press, pp. 181–214, 1998.

49. The Aluminum Association. *Partnerships for the Future*. Washington, DC: The Aluminum Association, 1996.

50. U.S. EPA. Compilation of Air Pollutant Emission Factors. Vol. I: Stationary Point and Area Sources, Sect. 12.8: Secondary Aluminum Operations, 5th edn. Washington, DC: U.S. Environmental Protection Agency Rep. AP-42, 1995.

51. The Aluminum Association. *Aluminum Industry Technology Roadmap*. Washington, DC: The Aluminum Association, 1997.

52. HW Edwards, DC Paulus. Reducing Remelting in Manufacturing Extruded Aluminum Products. *Proceedings of the Air & Waste Management Association's 95th Conference & Exhibition*, Paper 42919. Baltimore, MD, June 23–27, 2002.

53. VY Gershtein, CE Baukal. Aluminum Industry. In CE Baukal (ed.). *Heat Transfer in Industrial Combustion*, Boca Raton, FL: CRC Press, 2000.

54. ER Bazarian, JF Heffron, CE Baukal. Method for Reducing NO_x Production During Air–Fuel Combustion Processes. U.S. Patent 5,308,239, issued 3 May 1994.

55. MW Paget, JF Heffron, M Lefebre, C. Bazinet. A Novel Burner Retrofit Used to Increase Productivity in an Aluminum Rotary Furnace and Reduce Baghouse Loading. *TMS Conference*, San Francisco, CA, 1994.

56. CE Baukal, R Scharf, PB Eleazer. Low Emission Aluminum Melting Process. Presented at the 1996 Spring AFRC Meeting, Orlando, FL, May 6–7, 1996.

57. DV Neff. The Use of Gas Injection Pumps in Secondary Aluminum Metal Refining. *Proceedings of AIME The Metallurgical Society*, Fort Lauderdale, FL, Dec. 1–4, 1985.

58. VY Gershtein, CE Baukal, RJ Hewertson. Oxygen-enrichment of side well aluminum furnaces: Part I. *Indus. Heat.*, Vol. LXVII, No. 5, pp. 41–44, 2000.

59. VY Gershtein, CE Baukal, RJ Hewertson. Oxygen-enrichment of side well aluminum furnaces: Part II. *Indus. Heat.*, Vol. LXVII, No. 9, pp. 124–126, 2000.

60. BS Brewster, BW Webb, MQ McQuay, M D'Agostini, CE Baukal. Combustion measurements and modeling in an oxygen-enriched aluminum recycling furnace. *J. Inst. Energy*, Vol. 74, pp. 11–17, 2001.

61. VY Gershtein, CE Baukal. CFD in Burner Development. In CE Baukal, VY Gershtein, X Li (eds.). *Computational Fluid Dynamics in Industrial Combustion*. Boca Raton, FL: CRC Press, 2001.

62. JH Boneberg. Oxygen Enrichment Test on a Non-Recuperative Reverberatory Remelt Furnace. *Aluminum Industry Energy Conservation Workshop IX*, Washington DC: Aluminum Association, 1986.

63. DJ Krichten, WJ Baxter, CE Baukal. Oxygen Enhancement of Burners for Improved Productivity, *EPD Congress 1997, Proceedings of the 1997 TMS Annual Meeting*. ed. B Mishra. Orlando, FL, pp. 665–672, February 9–13, 1997.

64. DJ Krichten, WJ Baxter, CE Baukal. Oxygen Enhancement of Burners for Improved Productivity. *EPD Congress 1997, Proceedings of the 1997 TMS Annual Meeting*, ed. B Mishra. Orlando, FL, pp. 665–672, Feb. 9–13, 1997.

65. JS West. US Experience with High Efficiency Aluminum Melting Techniques. *IHEA-AGA Energy Seminar*, Oct. 4, 1983.

66. JS Becker, JF Heffron. The changing role of oxygen-based combustion in aluminum melting. *Light Metal Age*, June 1994.

67. ER Bazarian, JF Heffron, CE Baukal. Method for Reducing NO_x Production During Air-Fuel Combustion Processes. U.S. Patent 5 308 239, issued 3 May 1994.

69. B Phillips, D Pakulski, M Mazzei, E Lepoutre, H Spoon. Use of air/oxy/fuel burners for aluminum dross processing. *Indus. Heat.*, Vol. LX, No. 3, pp. 65–69, 1993.

70. LD Areaux, P Corio. Aluminum Scrap Melting System Provides High Metal Quality and Melt Yield. *Indus. Heat.*, Vol. LXVI, No. 8, pp. 39–41, 1999.

71. C Stala, DL Hindman. Gas-Fired Immersed Ceramic Tube Aluminum Melter. *Proceedings of 1989 International Gas Research Conference*, ed. TL Cramer. Govt. Institutes, Rockville, MD, pp. 1453–1461, 1990.

71a. Y Kimura, K Taniwaki. Development of Immersion Type Aluminum Melting Furnace. *Proceedings of 1989 International Gas Research Conf.*, Vol 3: *Industrial Utilization*, ed. TL Cramer. Govt. Institutes, Rockville, MD, pp. 202–214, 1990.

72. ME Ward, RE Gildersleeve, SJ Sikirica, WW Liang. Application of a New Ceramic Recuperator Technology for Use on an Aluminum Remelt Furnace. *Proceedings of 1986 International Gas Research Conference*, ed. TL Cramer. Govt. Institutes, Rockville, MD, pp. 864–874, 1987.

73. U.S. EPA. Profile of the Nonferrous Metals Industry. Washington, DC: U.S. Environmental Protection Agency. Rep. EPA/310-R-95-010, 1995.

74. U.S. EPA. Compilation of Air Pollutant Emission Factors, Vol. I: Stationary Point and Area Sources, Section 12.9: Secondary Copper Smelting, 5th edn. Washington, DC: U.S. Environmental Protection Agency Rep. AP-42, 1995.

75. Office of the Federal Register. Subpart M Standards of Performance for Secondary Brass and Bronze Production Plants. U.S. Code of Federal Regulations Title 40, Part 60. Washington, DC: U.S. Government Printing Office, 2001.

76. MK Snyder, FD Shobe. Source Category Survey: Secondary Copper Smelting and Refining Industry. EPA Rep. EPA-450/3-80-011. Washington, DC: U.S. Environmental Protection Agency, 1980.

77. U.S. EPA. Control Techniques for Sulfur Oxides Emissions from Stationary Sources, 2nd edn. Rep. EPA-450/3-81-004. Washington, DC: Environmental Protection Agency, 1981.

78. M De Lucia. Oxygen enrichment in combustion processes: comparative experimental results from several application fields, *Trans. ASME*, Vol. 113, pp. 122, June 1991.

79. D Saha, JF Heffron, KJ Murphy, JS Becker. The Application of Tonnage Oxygen in Copper Smelting. *Fall Meeting of the Arizona Section of AIME*, Dec. 7, 1987.

80. TN Antonioni, JA Blanco, GJ Dayliw, CA Landolt. Oxy/fuel Smelting in Reverberatory Furnace at Inco's Copper Cliff Smelter. Paper presented at the *50th Congress of the Chilean Institute of Mining and Metallurgical Engineers*, Santiago, Chile, Nov. 1980.

81. PE Queneau, R Schuhmann. Metamorphosis of the copper reverberatory furnace: oxygen sprinkle smelting. *J. Metals*, pp. 12–15, Dec. 1979.

82. Anon. Successful test encourage helps Dodge Corporation to modify its copper smelter. *Chem. Eng.*, Vol. 89, No. 3, pp. 18–19, 1982.

83. EJ Harbison, JA Davidson. Oxygen in Copper Smelting, Paper presented by Air Products & Chemicals, Inc. at the *Fall Meeting of the Arizona Section of the American Institute of Mining, Metallurgical and Petroleum Engineers*, Dec. 1972.

84. TN Antonioni, et al. Control of the INCO Oxygen Flash Smelting Process, Copper Smelting—An Update. *Conference Proceedings, TMS-AIME Meeting*, Dallas, TX, Feb. 1982.

85. JG Eacott. The Role of Oxygen Potential and Use of Tonnage Oxygen in Copper Smelting. *Conference Proceedings of 1983 Fall Extractive and Process Metallurgy Meeting of the Metallurgical Society of AIME*, Livermore, CA, pp. 583–634, 1983.

86. T Suzuki et al. Recent Operation of Mitsubishi Continuous Copper Smelting and Converting Process at Naoshima, *CIM Bulletin*, Sept., pp 51–76, 1982.

87. T Suzuki et al. Test Operation for Smelting more Tonnages of Copper Concentrates at the Mitsubishi Continuous Copper Smelting and Converting Process. Paper presented at the *112th AIME Annual Meeting*, Atlanta, GA, Mar. 1983.

88. EJ Harbison, JA Davidson. Oxygen in Copper Smelting, Paper presented by Air Products & Chemicals, Inc. at the *Fall Meeting of the Arizona Section of the American Institute of Mining, Metallurgical and Petroleum Engineers*, Dec. 1972.

89.   U.S. EPA. Compilation of Air Pollutant Emission Factors, Vol. I: Stationary Point and Area Sources, Sec. 12.11: Secondary Lead Processing. 5th edn. Washington, DC: U.S. Environmental Protection Agency Rep. AP-42, 1995.

90.   Office of the Federal Register. Subpart L: Standards of Performance for Secondary Lead Smelters. U.S. Code of Federal Regulations Title 40, Part 60. Washington, DC: U.S. Government Printing Office, 2001.

91.   EA Hase. Experiments with Oxygen-Enriched Blast at the ASARCO East Helena Lead Smelter. *First Operating Metallurgy Conference and Symposium*, Pittsburgh, PA.

92.   HE Boyer, *Practical Heat Treating*, American Society for Metals, Metals Park, OH, 1984.

93.   ASM Committee on Gas Carburizing, Case Hardening of Steel. *Metals Handbook*, 8th edn. Vol. 2: *Heat Treating, Cleaning and Finishing*, Metals Park, Ohio: American Society for Metals, pp. 93–114, 1964.

94.   NT Ferguson. Energy and furnace design for batch annealing and heat treating furnaces. *Iron & Steel Engineer*, Vol. 56, No. 6, pp. 31–33, 1979.

95.   ASM Committee on Furnace Atmospheres, Furnace Atmospheres and Carbon Control. *Metals Handbook*, 8th edn., Vol. 2: *Heat Treating, Cleaning and Finishing*. Metals Park, OH: American Society for Metals, pp. 67–84, 1964.

96.   W Trinks, MH Mawhinney. *Industrial Combustion*, Vol. II, 4th edn. New York: John Wiley, 1967.

97.   AC Thekdi, SR Huebner, MA Lukasiewicz. Development of an Indirect Gas-Fired High Temperature Heating System. *Proceedings of 1984 International Gas Research Conference*, Govt. Institutes, Rockville, MD, pp. 709–718, 1985.

98.   WW Liang, ME Schreiner, SJ Sikirica, ES Tabb. Application of Ceramic Tubes in High Temperature Furnaces. *Proceedings of 1986 International Gas Research Conference*, ed. TL Cramer. Govt. Institutes, Rockville, MD, pp. 875–888, 1987.

99.   KH Hemsath. A Novel Gas Fired Heating System for Indirect Heating. *Fossil Fuel Combustion Symposium 1990*, ed. SN Singh. ASME PD-Vol. 30, New York, pp. 155–159, 1990.

100. JS Copes, DL Brooks. Siconex^{TM} Coupon Testing in Controlled-Atmosphere Furnaces. *Fossil Fuel Combustion—1991*, ed. R Ruiz. ASME PD-Vol. 33, New York, pp. 155–160, 1991.

101. TJ Schultz, RA Schmall, I Chan. Selection of a Heating System for a High Temperature Gas Fired Soft Vaccum Furnace. *Fossil Fuels Combustion—1992*, ed. R Ruiz. ASME PD-Vol. 39, New York, pp. 23–30, 1992.

102. AE Erinov, AM Semernin, VA Povarenkov, MJ Khinkis, HA Abbasi. Development of a Gas-Fired, Low-Inertia, High-Efficiency Heat Treating Furnace. *Proceedings of 1995 International Gas Research Conference*, ed. DA Dolenc. Govt. Institutes, Rockville, MD, pp. 2774–2782, 1996.

103. HS Kurek, M Khinkis, W Kunc, A Touzet, A de La Faire, T Landais, A Yerinov, O Semernin. Flat Radiant Panels for Improving Temperature Uniformity and Product Quality in Indirect-Fired Furnaces. *Proceedings of 1998 International Gas Research Conference*, Vol. V: *Industrial Utilization and Power Generation*, ed. D Dolenc. Govt. Institutes, Rockville, MD, pp. 100–112, 1998.

14

Minerals Industries

14.1 INTRODUCTION

The U.S. EPA has defined a mineral processing plant as any facility that processes or produces any of the following minerals or their concentrates: alumina, ball clay, bentonite, diatomite, feldspar, fire clay, fuller's earth, gypsum, industrial sand, kaolin, lightweight aggregate, magnesium compounds, perlite, roofing granules, talc, titanium dioxide, and vermiculate [1]. There are a wide range of mineral products including cement, limestone, asphalt, glass, bricks, and refractory to name some of the more common ones. Figure 14.1 shows the emissions from minerals production processes in the United States since 1970 [2]. The U.S. EPA ranks nonmetallic mineral processing as number 13 on its prioritized list of 59 major source categories where the lower the number the higher the priority [3]. Glass ranks number 33, fiberglass ranks number 42, and brick and related clay products ranks number 46.

Calciners and dryers are commonly used in minerals processing. Typical calciners include rotary flash, kettle calciners, and multiple hearth furnaces. Table 14.1 shows what types of calciners are used for a given type of mineral. Typical dryers include rotary (direct or indirect), fluid bed, vibrating-grate, flash, and spray dryers. Figure 14.2 shows an example of two direct-fired rotary dryers used in minerals processing. Table 14.2 shows what types of dryers are used for each type of mineral. High-temperature furnaces are used to make glass, including regenerative and recuperative melters.

This chapter has basically been divided into three types of mineral processing. Section 14.2 discusses the production of glass, which is done in very high-temperature furnaces that are usually rectangular in shape with no moving parts. Section 14.3 considers common minerals (lime and cement) produced in high-temperature rotary furnaces, which are rotating refractory-lined cylinders. Section 14.4 considers mineral products produced in high-temperature furnaces that are long and rectangular, where the material is conveyed through the furnace with some type of moving mechanism.

14.2 GLASS

The glass industry consists primarily of four segments: container, flat, pressed and blown, and fiberglass. Glass is formed from raw materials that are fed into a

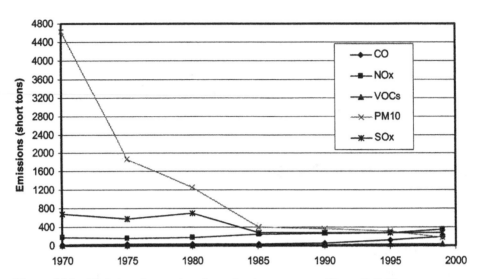

Figure 14.1 Emissions from minerals production processes. (From Ref. 2.)

Table 14.1 Types of Calciners Used in Minerals Processing

Industry[a]	Rotary	Flash	Multiple hearth furnace	Kettle	Expansion furnace
Alumina	x	x			
Diatomite	x				
Fire clay	x				
Fuller's earth	x				
Gypsum		x		x	
Kaolin	x	x	x		
Lightweight aggregate	x				
Magnesium compounds	x		x		
Perlite					x
Talc	x				
Titanium dioxide	x				
Vermiculite					x

[a]Calciners are not used in the ball clay, bentonite, feldspar, industrial sand, and roofing granules industries.
Source: Ref. 1.

high-temperature melting furnace to produce glass, which may be clear or colored. A schematic of the basic glass manufacturing process is shown in Fig. 14.3 [4]. The specific furnace and heat-transfer issues vary slightly by segment. There are four common furnace designs used to make glass: direct-fired furnaces, regenerative side port, regenerative end port, and all-electric furnaces [5].

The major raw materials used in all types of glass manufacturing include sand, soda ash, limestone, and cullet, which is recycled crushed glass. A variety of additives may also be included in the incoming raw materials. Fluxes are added to lower the

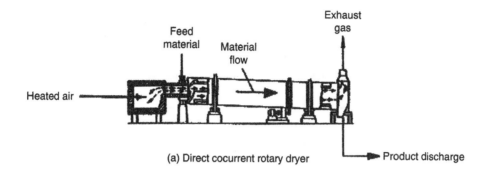

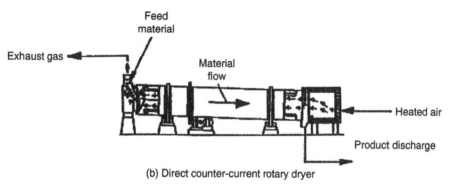

Figure 14.2 Direct-fired rotary kilns: (a) co-current and (b) counter-current. (From Ref. 1.)

Table 14.2 Types of Dryers Used in Minerals Processing

Industry[a]	Rotary (direct)	Rotary (indirect)	Fluid bed	Vibrating grate	Flash	Spray
Ball clay		x		x[b]		
Bentonite	x		x			
Diatomite	x				x	
Feldspar	x		x			
Fire clay	x			x		
Fuller's earth	x		x			
Gypsum	x					
Industrial sand	x		x			
Kaolin	x					x
Perlite	x					
Roofing granules	x		x			
Talc	x				x	
Titanium dioxide	x	x	x		x	x
Vermiculite	x		x			

[a]Dryers are not used in the alumina, lightweight aggregate, and magnesium compounds industries.
[b]Indirect.
Source: Ref. 1.

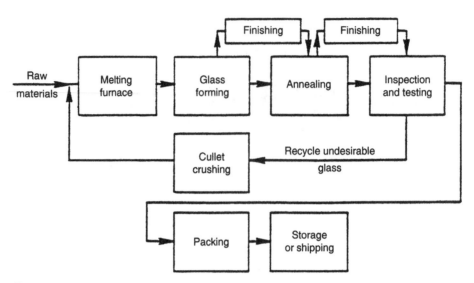

Figure 14.3 Flow diagram of a typical glass manufacturing process. (From Ref. 4.)

melting points and the working temperatures of the raw material mixture. Stabilizers are added to improve the chemical durability of the glass. Borates are added to increase thermal durability. Lead increases the refractive index and density. Aluminum increases the glass strength. Sodium accelerates the melting process. Trace amounts of various metal oxides may be added to change the color of the glass. The raw materials are mixed to produce a homogeneous mixture that is processed at high temperatures in some type of melting furnace.

One of the unique aspects of glass melting is that the load not only absorbs radiation, but it also transmits radiation. Gardon [6] refers to such materials as diathermanous and argues that the absorption and emission in glass can be treated as bulk phenomena. For glass melting, the heat input should be prompt, intense, and uniformly distributed, which favors luminous flames [7].

It has been shown by Russian scientists that the heat transfer in a glass-melting furnace with firing ports can be improved by injecting the fuel under the port instead of on the side of the port as is the common practice [8–10]. The specific efficiency depends on the design of the fuel nozzle, such as the angle of attack. Another Russian paper determined the optimum flame length for both luminous and nonluminous flames to be one-quarter of the length of an end-fired recuperative glass-melting furnace [11]. Computer modeling results in another paper by Russian scientists suggested an optimum flame length of between one-half and three-quarters of the furnace width on side-port regenerative glass melters [12].

Because of the high temperatures required to melt glass, two common approaches have been used in designing the combustion equipment. The conventional approach has been to use high air preheat temperatures to increase significantly the adiabatic flame temperature (see Fig. 2.30). The second and more recent approach has been to use oxygen-enhanced combustion. Industrial oxygen has been used to enhance combustion in the glass industry for several decades [13]. Most of these installations utilized supplemental oxy/fuel burners, premixed oxygen enrichment of the combustion air, or undershot lancing of oxygen to the port or burner. Supplemental oxy/fuel is the practice of installing one or more oxy/fuel burners into an air/fuel furnace. Premixed oxygen enrichment is the practice of introducing oxygen into the combustion air to a level of up to 27% total contained oxygen (i.e., 6% oxygen enrichment). The amount of oxygen enrichment is limited by materials compatibility issues in highly oxidizing environments. Undershot lancing is the practice of strategically injecting oxygen through a lance into the combustion region.

Pincus [14] has compiled a series of articles that have appeared in the magazine *The Glass Industry*, which are related to combustion in the glass industry [14]. It contains major sections on fuels, combustion, heat balances, and emissions control. It is noted that glass melting requires the heat input to a batch composed of raw materials plus cullet to bring about the desired chemical and physical changes to yield a chemically homogenous liquid that can be formed into the desired shape.

Glass is an environmentally friendly and important material because it is 100% recyclable. However, the production of glass generates pollutant emissions such as NO_x, SO_x, and particulates. Glass chemistry is beyond the scope of this book. The interested reader is referred to other books on that subject [15,16].

14.2.1 Furnace Types

The type of furnace for melting glass typically depends on the type and quantity of glass being produced, and the local fuel and utility costs. Figure 14.4 shows a schematic of a typical glass tank [17]. The main parts of the glass tank are the melter section, throat, refiner, and forehearth. The throat connects the melter section to the refiner. The forehearth transports the glass to the forming machines. For some types of glasses (e.g., fiberglass), a refiner is not used and the glass goes directly from the melter to the forehearth. Figure 14.5 shows a schematic of a furnace used for flat-glass production that is slightly different from the furnace shown in Fig. 14.4.

While there are exceptions, the following discussion describes the primary furnace types and the glass segments that most commonly use each style. Scholes [18] has written one of the classic books on the manufacturing of glass that includes an interesting historical discussion of the development of the furnaces used to make glass.

14.2.1.1 Unit Melter

Figure 14.6 shows a schematic of a typical unit melter. The term "unit melter" is generally given to any fuel-fired glass-melting furnace that has no heat recovery device. Typically, the air/fuel-fired unit melters are relatively small in size and are fired with two to 16 burners. Furnaces range in production from as large as 40 tons (36 metric tons) of glass per day to as small as 500 lb (230 kg) of glass per day. Larger air/fuel unit melters are found in areas where fuel is extremely cheap. Frit, tableware,

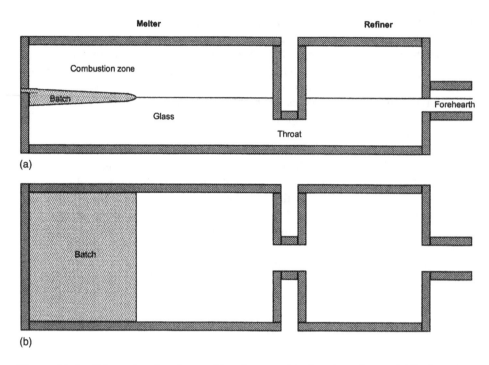

Figure 14.4 Schematic of a glass-melting furnace: (a) elevation view and (b) plan view. (From Ref. 17. Courtesy of CRC Press.)

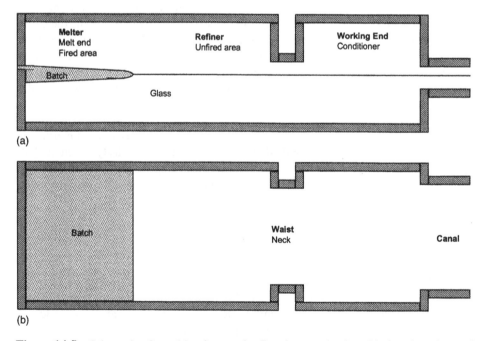

Figure 14.5 Schematic of a melting furnace for flat-glass production: (a) elevation view and (b) plan view. (From Ref. 17. Courtesy of CRC Press.)

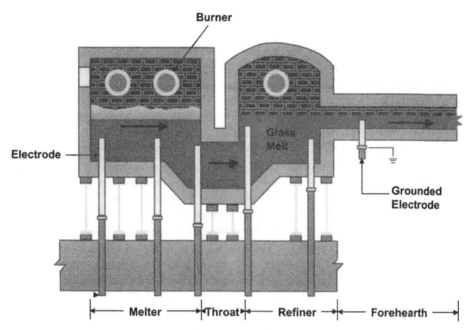

Figure 14.6 Unit melter.

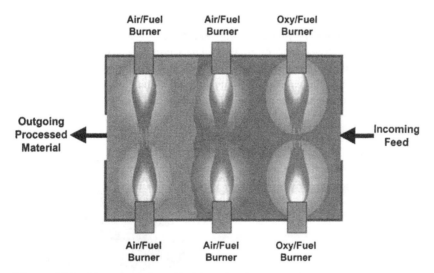

Figure 14.7 Plan view of oxy/fuel boosting in a glass furnace. (From Ref. 19. Courtesy of CRC Press.)

ophthalmic glass, fiberglass, and specialty glasses with highly volatile and corrosive components are produced in unit melters.

Due to the very low energy efficiency and the use of individual burners, the air/fuel unit melters are very amenable to oxygen-enhanced combustion techniques including supplemental oxy/fuel boosting (see Fig. 14.7 [19]), premix oxygen enrichment, and full oxy/fuel combustion. Oxy/fuel unit melters have been built as large as 350 tons (320 metric tons) per day of glass to as small as 500 pounds (230 kg) of glass per day.

The Gas Research Institute (Chicago, IL) funded Avco Research Lab [20] and Vortec Corp. (Collegeville, PA) [21] worked on an advanced glass melter (AGM) design that promised to reduce pollution emissions, increase the thermal efficiency, reduce capital costs, and improve operational flexibility compared to conventional furnace designs. The major steps include: (1) rapid suspension of the glass-forming ingredients (batch) in a high-intensity combustor, (2) acceleration of the gas–solids suspension in a converging nozzle that which directs the gas–solids flow at a vertically oriented center (bluff) body, (3) impact and separation of the batch on the center body, which serves as a site for the glass-forming reactions, and (4) homogenization and fining of the thin glass layer that moves down the center body. Both radiation and convection heat the glass-forming ingredients as they are in intimate proximity of the exhaust products generated by the combustion of the fuel (natural gas) and the preheated air. The AGM process takes on the order of minutes, compared to conventional glass-melting processes, which take on the order of hours up to days. The AGM claims 20–25% higher thermal efficiencies than those of conventional melting processes due to the improved heat transfer.

Abbasi et al. [22] described a new process that has been developed known as oxygen-enriched air staging (OEAS). A schematic of the process for an end-port furnace is shown in Fig. 14.8. A schematic for application of the process on a side-port furnace is shown in Fig. 14.9. While the primary objective of the technology is to reduce NO_x emissions (50–70% reduction was measured), there are also some effects on heat transfer. Since the preheated air flames are now initially fuel rich, they are more luminous and therefore more radiative. The secondary oxygen added to the exhaust products completes the combustion before the gases exit the furnace. This secondary combustion zone helps to spread out the heat transfer over a wider region in the furnace, which helps to make the heat flux more uniform.

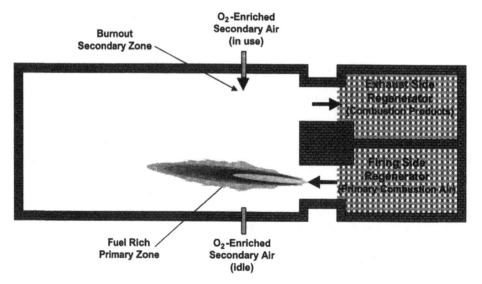

Figure 14.8 Oxygen-enriched air staging in an end-port regenerative furnace.

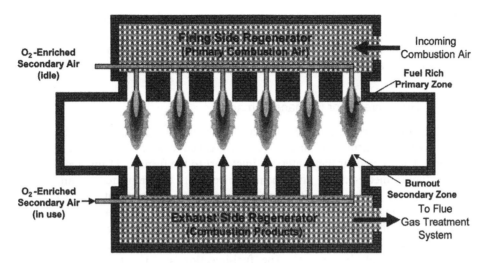

Figure 14.9 Oxygen-enriched air staging in an side-port regenerative furnace.

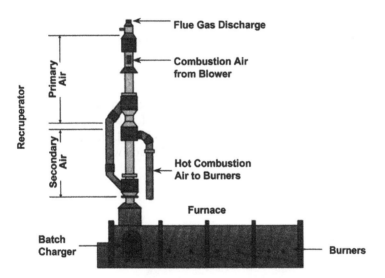

Figure 14.10 Recuperative melter. (From Ref. 13. Courtesy of CRC Press.)

14.2.1.2 Recuperative Melter

Figure 14.10 shows a schematic of a typical recuperative melter. A recuperative melter is a unit melter equipped with a recuperator. Typically, the recuperator is a metallic shell and tube style heat exchanger that preheats the combustion air to 1000°–1400°F (540°–760°C). The furnace is fired with four to 20 individual burners. These furnaces range in size from as large as 280 tons (250 metric tons) per day of glass to as small as 20 tons (18 metric tons) per day of glass. These furnaces are common in fiberglass production, but can also be used to produce frit. Some recuperative furnaces are used in the container industry, though this is not common.

Furnace life is a function of glass type being produced. For example, a 6-year furnace life is typical for wool fiberglass. Booker [23] noted that production rates were increased by 25% by using oxy/fuel instead of air/fuel combustion in an end-fired recuperative furnace used to make mineral fiber.

14.2.1.3 Regenerative or Siemans Furnace

In a regenerative or Siemans furnace, air for combustion is preheated by being passed over hot regenerator bricks, typically called checkers (sometimes spelled chequers). This heated air then enters an inlet port to the furnace. Using one or more burners, fuel is injected at the port opening, mixes with the preheated air, and burns over the surface of the glass. Products of combustion exhaust out of the furnace through nonfiring ports and pass through a second set of checkers (regenerator), thereby heating them [24]. After a period of 15–30 min, a reversing valve changes the flow and the combustion air is passed over the hot checkers that were previously on the exhaust side of the process. The fuel-injection system also reverses. After reversing, the exhaust gases pass through and heat the checkers that had previously heated the combustion air.

The Siemans furnace is the workhorse of the glass industry. Most flat glass and container glass are produced in this furnace type. Regenerative furnaces are also used in the production of TV products, tableware, lighting products, and sodium silicates. There are two common variants of the Siemans furnace: the side-port regenerative melter and the end-port regenerative melter.

End-Port Regenerative Furnace. End-port regenerative furnaces are typically used for producing less than 250 tons (230 metric tons) of glass per day. In an end-port furnace, the ports are located on the furnace back wall. Batch is charged into the furnace near the back wall on one or both of the side walls. Figure 14.11

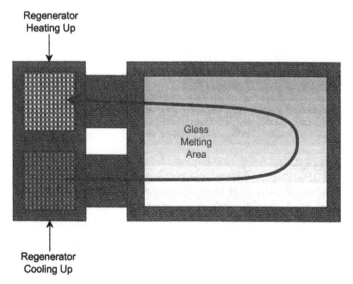

Figure 14.11 Schematic of an end-port regenerative furnace. (From Ref. 24. Courtesy of CRC Press.)

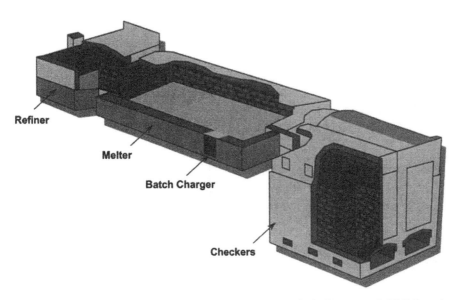

Refiner

Melter

Batch Charger

Checkers

Figure 14.12 End-port regenerative melter. (From Ref.13. Courtesy of CRC Press.)

shows a schematic of the operation of a typical end-port regenerative furnace, and Fig. 14.12 shows the layout of a typical end-port furnace. These furnaces are commonly used for producing container glass, but are also used for producing tableware and sodium silicates. For container production, a furnace campaign typically lasts 8 years. Undershot of oxygen through lances and supplemental oxy/fuel have been used successfully on this type of furnace [25].

Side-Port Regenerative Furnace. Side-port regenerative furnaces have ports located on the furnace side walls. Batch is charged into the furnace from the back wall. Figure 14.13 shows a schematic of a side-port regenerative furnace [26], and Fig. 14.14 shows the layout of a typical side-port furnace. Side-port regenerative furnaces are typically used for producing greater than 250 tons (230 metric tons) of glass per day. A side-port furnace for float glass commonly produces 500–700 tons (460–630 metric tons) of glass per day. For container glass, side-port furnaces ordinarily produce between 250 and 350 tons (230–320 metric tons) of glass per day. These furnaces are commonly used in container and float glass production, but are also used for the production of tableware and sodium silicates. Undershot and supplemental oxy/fuel oxygen enrichment have been successfully used on this type of furnace [27]. These furnaces have also been converted to full oxy/fuel.

14.2.1.4 Forehearth Furnace

Multiple forehearth furnaces (see Fig. 14.15) are located after the melting furnaces to prepare the glass for further processing. Many small burners line both sides of the forehearth to homogenize the temperature of quantities of glass much smaller than those of the upstream melting furnace. At the exit of the forehearth, the glass is ready to be formed into the desired shapes. Temperature uniformity is important to making good quality glass products. Nonuniformities can cause defects in the glass, which

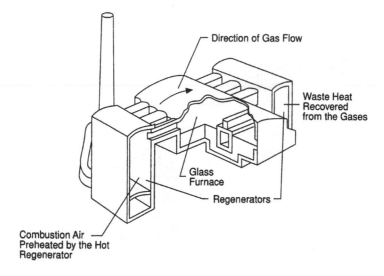

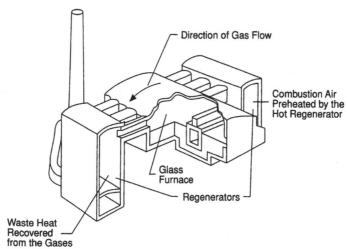

Figure 14.13   Schematic of a side-port regenerative furnace. (From Ref. 26. Courtesy of CRC Press.)

reduces the overall efficiency of the process. This indirectly increases the pollution emissions because fuel is being combusted to produce unusable product.

14.2.2   Industry Segments

Nearly all glass produced commercially is one of five basic types: soda–lime, lead, fused silica, borosilicate, and 96% silica [28]. Silica is used in most commercial glasses, but its high melting point (3133°F or 1723°C) and high viscosity make it difficult to melt and work. Soda in the form of sodium carbonate or nitrate is typically added to the silica to reduce its melting temperature. Lime is also added to increase the chemical durability and reduce the solubility of the glass to make it

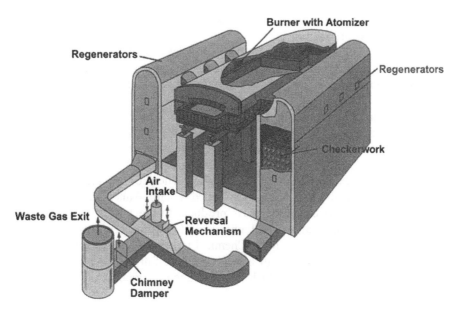

Figure 14.14 Side-port regenerative melter. (From Ref. 13. Courtesy of CRC Press.)

Figure 14.15 Forehearth furnace.

impermeable to liquids like water. This forms the basic soda–lime–silica mixture used in common glass production.

There are five main steps in the manufacture of glass: mixing, melting, forming, annealing, and finishing. Mixing is either wet or batch agglomeration, depending on the type. Both minimize dusting and the resulting particulate emissions, ensure homogeneity, and increase melting efficiency and glass quality. Melting of small quantities is done in pot furnaces or crucibles. Larger quantities of glass are melted in refractory-lined furnaces that usually incorporate some type of heat recuperation. Forming varies by the type of product. Glass bottles and jars are typically formed in automatic machines. Pressing is used to form flat items such as lenses and plates. Drawing and casting are used to mold molten glass. The products are then slowly cooled or annealed, usually in a long oven called a lehr. Annealing reduces the internal stresses that could cause the glass products to crack. The two types of finishing processes are mechanical and chemical. Mechanical processes include cutting, drilling, grinding, and polishing. Chemical treatments are used to alter the strength, appearance, and durability of a product. Once finished, the products are cleaned with some type of solvent, which may be aqueous, organic, or halocarbon.

14.2.2.1 Container Glass

This segment includes glass used to make bottles, jars, and other similar containers used in packaging. Figure 14.16 shows a photograph of glass bottle making. Bottle

Figure 14.16 Glass bottle making.

glass plants are sometimes located directly on the site of the beverage (e.g., soda, wine, or beer) production. This segment has received a great deal of pressure from the plastics and aluminum industries. Increased emphasis on reducing pollution emissions has put on further pressure. The main quality concern in bottle production is seeds, which can cause the bottle to explode under certain conditions. The containers may also hold materials under some pressure, for example, carbonated beverages. Glass homogeneity is very important for achieving good quality. Container glass also comes in a variety of colors, depending on the beverage manufacturer's requirements.

14.2.2.2 Flat Glass

This segment includes flat glass used in windows, car windshields, instrumentation windows, mirrors, and tabletops. The four main products are tempered glass, laminated glass, glass mirrors, and insulating units. Tempered glass is a type of safety glass produced by a thermal process where the glass is made stronger by heating and then subsequent rapid cooling, which produces surface and interior stresses. Laminated glass consists of two or more layers of glass separated by and bonded to thin layers of plastic that prevent the glass from shattering when broken. These are commonly used in automobile windshields and windows. Glass mirrors are produced by putting a reflective coating on one side. Insulating units are produced by putting two or more parallel panes of glass together separated by a gap where the unit is completely sealed to prevent air from leaking through the unit. Low thermal conductivity gases like xenon may also be used to fill the space between the glass panes to reduce the heat losses through the window. Insulating units reduce surface condensation, sound transmission, and heat loss compared to single pane windows.

Flat glass is made by the float process. The raw materials used in this process include silica sand, soda ash, limestone, dolomite, cullet (scrap glass), and small amounts of other materials. These components are mixed together in the proper proportions and melted together in a furnace at temperatures of approximately 1600°C (2900°F). Coloring agents may be added to produce different glass tints. The molten glass is fed as a continuous ribbon into a bath of molten tin where the glass floats on top of the tin because it is lighter. The glass is fire polished before leaving the float glass tank where it then enters an annealing lehr where it is gradually cooled to prevent stresses that could cause flaws. The glass is then cut. It may then be cooled, packaged, and shipped to the customer, further processed, or sent to storage for future processing, which might include adding coatings to absorb or reflect certain wavelengths of light.

14.2.2.3 Fiberglass

This refers to glass fibers used to making insulation for buildings and piping, textile fabrics, and as reinforcements for plastics and other materials. This segment is sometimes broken into two subsegments: glass wool and textile fibers. Glass wool is used to make insulation both to minimize heat losses in buildings and sound transmission through walls. Textile fiberglass is used in the production of fireproof cloth.

Glass-fiber manufacturing involves the high-temperature processing of raw materials into a homogenous melt followed by fabrication of the melt into glass

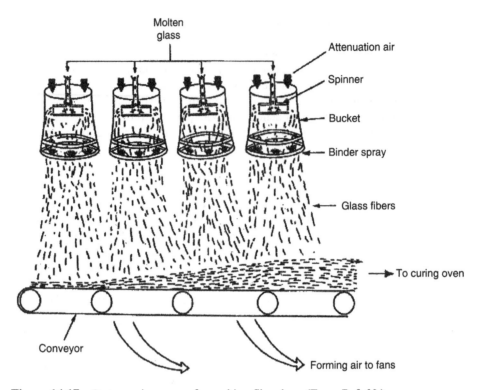

Figure 14.17   Rotary spin process for making fiberglass. (From Ref. 30.)

fibers. There are three phases of the production process: raw materials' handling, glass melting and refining, and glass-fiber forming and finishing. The raw materials include sand, feldspar, sodium sulfate, boric acid, and other components. These are fed into the melting furnace using belts, screws, or bucket elevators. The materials are melted and chemically react in the furnace to temperatures ranging from 2700 to 3100°F (1500°–1700°C) [29]. The glass is melted in one of four types of melting furnaces: recuperative, regenerative, unit, and electric (discussed later). The molten glass is then made into fibers through either a rotary spin process (see Fig. 14.17) or through a flame attenuation process (see Fig. 14.18). Figure 14.19 shows a process flow diagram of the wool fiberglass manufacturing process including typical emissions from each step [30].

14.2.2.4   Frit

Frit is used to put glazes and enamels on materials for both decorative and protective purposes. Frit furnaces generally produce less than 30 tons per day per furnace and are not continuous, with frequent startups and shutdowns of the furnace. Therefore, heat recovery/air preheat is uncommon. Without heat recovery, fuel savings alone have justified the use of full oxy/fuel. In parallel with the conversions in frit, Corning converted a large number of their smaller specialty glass furnaces, mainly ones that had no heat recovery, to 100% oxy/fuel firing [31]. In general, emissions from batch

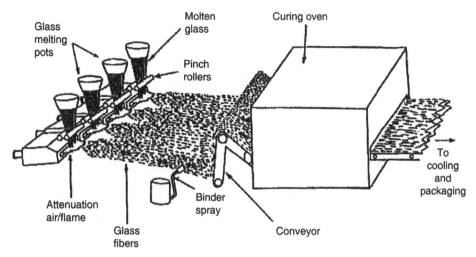

Figure 14.18 Flame attenuation process for making fiberglass. (From Ref. 30.)

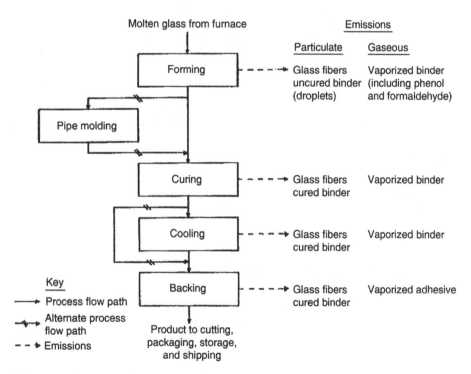

Figure 14.19 Fiberglass manufacturing process. (From Ref. 30.)

processes vary compared to continuous processes. CO emissions are often higher and NO_x emissions lower at the beginning of the heating cycle when the furnace and gas temperatures are lower. As these temperatures increase, the CO emissions usually decrease while the NO_x emissions increase.

14.2.2.5 Specialty Glass

This segment includes a wide range of glass products including table- and oven-ware, flat panel displays, lighting, TV tubes, fiber optics, scientific and medical equipment, and hand-blown glass products. Because of the specialized nature of these products, they are typically produced in smaller furnaces. Quality is usually especially important in this segment.

14.2.3 Emissions

Glass production is a significant source of emissions because of a variety of factors. The EPA estimates that approximately 99% of the total emissions from a glass plant come from the melting furnaces [4]. The materials used to produce glass are a major source of particulate emissions because of the fine nature of the sand and other batch materials. The raw materials can also be a source of lead emissions, for example, in the production of leaded glass. The incoming batch often contains niter, which is a source of NO_x emissions. Glass production is a high-temperature process that requires large quantities of fossil fuels, which both often lead to high NO_x emissions. Other common emissions include CO, SO_x, VOCs, and fluorides.

The U.S. Department of Energy Office of Industrial Technologies sponsored a consortium of glass industry experts to develop a vision for the year 2020 [32]. One of the goals in the vision related to environmental concerns was to reduce air and water emissions by 20% through sound environmental practices. To achieve this goal, research and development of the entire melting process was recommended to improve emissions control. It was noted that glass is one of the most environmentally benign materials available and is completely recyclable, does not release leachates, and does not decompose. Environmental costs are particularly important in the flat-glass segment because of the significant costs to meet emissions regulations. Environmental technology challenges related to air pollution include:

- Development of combustion technologies that reduce air emissions
- Expanded use of 100% oxygen combustion
- Development of alternate materials or furnace designs that reduce particulate emissions

A U.S. Department of Energy [33] roadmap workshop identified the following environmental research needs:

- Develop predictive emissions modeling tools
- Develop integrated control systems to link production with emissions
- Develop new heating mechanisms to melt glass economically without noxious emissions
- Develop improved oxy–fuel firing to lower particulate and gaseous emissions

As shown in Table 14.3, many technology barriers were identified to achieving superior environmental performance. The report recommends achieving environmental goals through research and development in emissions characterization and control, alternative materials, and oxy–fuel firing. An important element is the development of accurate models for quantitative prediction and control. A better understanding of the general melting process and fundamental mechanisms in

Table 14.3 Technology Barriers to Achieving Superior Environmental Performance

Process	Fundamental understanding of emissions	Regulation	Recovery and reuse
Lack of process control and instrumentation	Fuels used (energy)—link to process	Government regulations act to inhibit some technologies	Lack of tracking technology for recovery
Theoretical limits of process efficiency	Lack of understanding of process mechanisms influencing particulates	—minimum emission standards	Composition dependence of glass
Lack of selective process sensors	Lack of predictive emissions modeling tools	—state vs. federal (non-uniform)	Separation and sorting of post consumer waste
Fuel efficiency and adaptability	Modeling inadequacies for combustion	Lack of central industry group to interact with government	—implications for other industries under the Industries of the Future initiative
Inefficient heat recovery	Composition—dependence of glass	Moving regulatory drivers	Economics of recycle
Modeling inadequacies for combustion	Nature of hazardous materials	Uneven enforcement of regulations	Reverse distribution system—collection
High cost of oxygen production for oxy–fuel firing	—carcinogenic	Misguided regulations	Technology for characterizing chemical and physical properties of glass
Production of by products and contamination	—fibrous		Cost effectiveness of beneficiation
Broader compositional range	—air		Industry and consumer attitudes to recycle vs. reuse.
Lack of *cost-effective* water treatment technology	—etc.		
Technologies to control CO_2 emissions	Chemistry of treating materials (lack of understanding)		
Lack of measurement technology for emissions (low cost)	Lack of understanding of mold release chemistry/emissions		
Current melting and refining technology produces emissions	Measurement technology for emissions		
	Understanding of melting and refining technology		

Source: Ref. 33.

Table 14.4 Proposed Emission Limits for the Soda–Lime–Silica
Glass Industry in Italy

Concentration in mg/Nm^3 (at 8%O_2 dry, in dry waste gases)	Decree limits	Typical emissions
Particulates	100	150
SO_x	1800	600
NO_x	3500	2500
Chloride	30	60
Fluoride	5	6

Source: Ref. 34. Courtesy of Institution of Chemical Engineers.

emission formation are needed. One specific technology recommended for further investigation is the laser destruction of NO_x in glass-melting furnaces.

In recent years in Europe, there has been more of an emphasis on controlling particulate emissions from glass production processes, while allowing relatively generous limits for NO_x and SO_x as shown in Table 14.4 [34]. However, it can be shown that NO_x from glass-making processes is a much higher percentage of total NO_x than the contribution from the glass industry to particulate emissions. Therefore, greater emphasis on reducing NO_x emissions is recommended.

14.2.4 Regulations

The EPA has established emission guidelines for glass manufacturing plants [35] and for fiberglass wool insulation manufacturing plants [36].

14.2.5 Pollution Control Strategies

A variety of techniques are used to control emissions from glass processes, depending on the specific application and pollutant.

14.2.5.1 Pretreatment

The primary pretreatment technique used in glass production is to modify the oxidizer. Virtually all segments of the glass industry have implemented oxygen-enhanced combustion (OEC) [37] technology to reduce a variety of pollutants [13]. OEC is used to reduce particulate emissions in the fiberglass, container glass, and specialty glass industries [38]. It is used to reduce NO_x in the container glass industry. Table 14.5 shows how converting a glass-melting furnace from air/fuel to oxy/fuel can dramatically reduce NO_x emissions. This is not surprising as most glass-melting furnaces incorporate heat recuperation to preheat the incoming combustion air to high levels. This significantly increases the flame temperature, which correspondingly increases thermal NO_x generation.

OEC has been used in a variety of ways in the glass industry. Figure 14.20 shows the use of OEC in a glass furnace by lancing oxygen into the flames of the burners. Figure 14.21 shows OEC in a glass furnace. Figure 14.22 shows the effect on the economics of using OEC to control NO_x and particulate emissions [13].

Table 14.5 No$_x$ Emission Reductions from Furnace Converted from Air/Fuel to Oxy/Fuel Combustion

Furnace	Air/fuel (lb NO/ton)	Oxy/fuel (lb NO/ton)	Percentage change	Excess oxygen in the stack (%)
A				
B	4.5	1.5	−67	1.5
C	16.3	8.8	−46	12.0
D	18.0	0.5	−97	0.5
E				
F	13.1	0.7	−95	1.0

Source: Ref. 13. Courtesy of CRC Press.

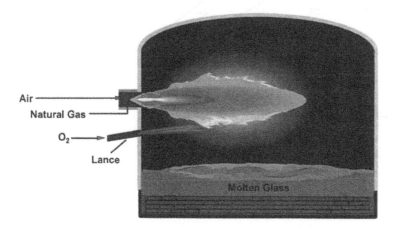

Figure 14.20 Oxygen lancing into the flames in a glass furnace.

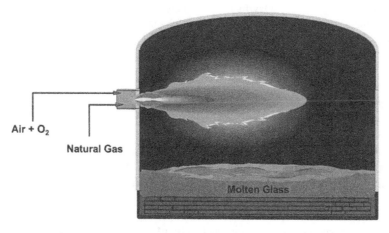

Figure 14.21 Oxygen-enhanced combustion in a glass furnace.

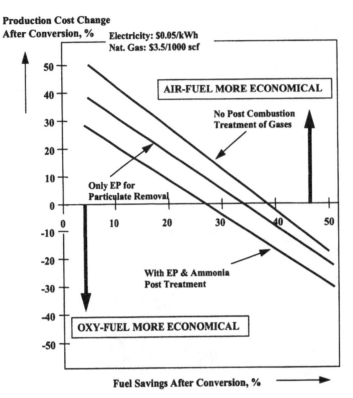

Figure 14.22 Effect of NO$_x$ and particulate controls on process economics for air/fuel vs. oxy/fuel combustion. (From Ref. 13. Courtesy of CRC Press.)

Table 14.6 Particulate Emission Reductions from Four Furnaces Converted from Air/Fuel to Oxy/Fuel Combustion

Furnace	Air/fuel (lb particulates/ton)	Oxy/fuel (lb particulates/ton)	Percentage change
B	3.3	1.3	−60
C	5.2	2.6	−50
D	3.7	0.8	−78
F	1.0	0.3	−70

Source: Ref. 13. Courtesy of CRC Press.

There are two primary sources for particulate emissions from glass furnaces: (1) physical entrainment of the incoming raw materials into the exhaust gas products, and (2) volatilization of glass and batch components that recombine and condense on cooling. As previously shown in Figure 2.41, using OEC dramatically reduces the exhaust gas velocity in a furnace originally designed for air/fuel combustion which in turn reduces particle entrainment. Volatilization is reduced per unit production rate because of the significant reduction in fuel consumption using OEC. Table 14.6 shows how converting a glass-melting furnace from air/fuel to oxy/fuel can dramatically reduce NO$_x$ emissions. An added benefit of

Table 14.7 Summary of Oxy/Fuel Conversions of Glass Manufacturing Processes in North American up to Approximately 1997 on Furnaces Larger than 30 tons/day

Segment	Number of conversions	Most common driving force	Second most common driving force	Factors against
Fiberglass	21	Particulate	Capital reduction	Oxygen cost
Container	24	NO_x	Particulate	Oxygen cost
Lighting/table ware/TV	25	Particulate	Capital reduction	Oxygen cost
Float	3	Production flexibility	Capital reduction	Risk, oxygen cost, glass quality
Frit	21	Fuel savings	Production increase	Oxygen cost
Sodium silicate	1			Need for large production swings
Specialty	15	Fuel savings	Production increase	Oxygen cost

Source: Ref. 13. Courtesy of CRC Press.

OEC is that the pollutants in the exhaust gas stream are much more concentrated due to the reduction of the diluent nitrogen present in air/fuel combustion systems. This concentration makes post-treatment equipment more efficient. Jankes et al. [39] showed as much as a 90% reduction in NO_x emissions per ton of glass by using oxygen instead of air for combustion compared to conventional preheated air combustion. Particulates were reduced by 80% and specific energy consumption was reduced by 50%, which is directly proportional to CO_2 emissions. However, oxygen enrichment of the combustion air in a glass furnace increased NO_x emissions.

Table 14.7 shows a summary of the glass manufacturing processes, by market segment, that had been retrofitted with OEC up to approximately 1997 in North America on furnaces larger than 30 tons/day [13]. The table shows that reducing particulate or NO_x emissions was the most common driving force or the second most common driving force in many cases. This was balanced against the added cost of using oxygen compared to air for combustion.

14.2.5.2 Combustion Modification

As with virtually all industrial combustion applications, the burner design is a critical element in determining pollution emissions. Caldeira-Pires et al. [40] note the importance of the burner design as the key factor in pollution emission generation in glass melting. They experimentally studied a small (50 kW) air/methane burner with swirl to improve the understanding of how the burner design affects emissions. The tile and burner geometries, flow distribution, and gas velocities were varied. NO_x increased with fuel gas velocity. The burner and tile geometries were important factors in NO_x production.

14.2.5.3 Post-Treatment

A variety of post-treatment techniques are used to control various pollutant emissions from glass production processes. Table 14.8 shows the common post-treatment

Table 14.8 Emission Control Methods for Fiberglass
Manufacturing Processes

Emission sources	Control technology
Forming	Wet electrostatic precipitator
	Wet scrubber
	Process modification—
	forming air recirculation
Curing	High-velocity air filter
	Incinerator
	Process modification—
	special binder formulations
Cooling	High-velocity air filter
Forming + curing	Wet electrostatic precipitator
Curing + cooling	High-velocity air filter
	Wet scrubber
Forming + curing + cooling	Wet electrostatic precipitator

Source: Ref. 30.

emission control techniques for fiberglass manufacturing processes. Wet electrostatic precipitators are used in a wide variety of glass applications because of the significance of particulate emissions (see Chap. 9). Wet scrubbers, cyclones, and fabric filters are also used. Selective noncatalytic reduction is often used to control NO_x emissions from glass furnaces [28]. Thermal incinerators and afterburners are used to destroy volatile organic emissions (see Chap. 7).

14.3 CEMENT AND LIME

Cement is used for a wide range of building and construction purposes and has been used for thousands of years starting at least with the ancient civilizations in Egypt to make the great pyramids. The prosperity of a nation is often determined in part by its ability to use cement for infrastructure (roads, bridges, dams, etc.) and for building. It is the key component in mortar, which consists of cement and sand. It is the key component in concrete, which consists of cement, sand, gravel, and stone. Portland cement is made by the calcination of an alumina-containing limestone. A number of references are available for more information on cement production and uses [41–45]. Limestone is the largest component in cement manufacturing so both are considered together here as they are made by very similar processes.

Cement has some unique properties that have made it so popular throughout history. It can be transported dry and then mixed with water at the use point. In wet form, it is flowable but also rigid enough to maintain many shapes. It remains workable for enough time to form it into the desired shape, but it dries quickly enough to be usable in a relatively short time. Its strength and hardness increase with age. It can withstand the weather literally for centuries as evidenced by the works of the Greeks and Romans that have survived to the present day. While it originally contained water, cured cement is very durable in wet applications including use under water such as supports for bridges. It is relatively inexpensive and is usually easy to repair when damaged. Its adhesive ability makes it stick both to other materials and to itself, especially when joining to previously made cement products. It can be installed even under fairly cold weather conditions as long as the water in the slurry can be kept from freezing.

The components of cement products are often categorized into two groups: active and inactive. The active components are the cement and water. The inactive components are the sand, gravel, stone, and other aggregate materials. During the hardening process, the water and the cement react while the inactive components remain unchanged. The cement, which may only constitute 10% by volume of a concrete mix, binds together the inactive components to form a strong matrix. Lime is one of the products used in cement. It is typically produced by burning limestone in a rotary kiln.

14.3.1 Introduction

A schematic of the cement manufacturing process is shown in Fig. 14.23 [46]. Cement is made by burning together, in a specific proportion, a mixture of naturally occurring argillaceous (containing alumina) and calcareous (containing calcium carbonate or lime) materials to a partial fusion at high temperature (about 2640°F or 1450°C). The resulting product is called clinker, which is cooled and then ground into a fine material known as cement. A relatively small amount of gypsum or plaster of paris is sometime added to the clinker to adjust the setting time of the mixture. Common calcareous materials include limestone, chalk, oyster shells, and marl. Limestone is the predominant material and consists of lime (CaO), silica (SiO_2), alumina (Al_2O_3), and iron oxide (Fe_2O_3). Common argillaceous materials include clay, shale, slate, and slag from blast furnaces used in steel making. Five different types of cement are commonly produced: regular, moderate heat of hardening and sulfate resisting, high early strength, low heat, and sulfate resisting.

The pyroprocessing system transforms the raw mix into clinkers, which are gray, glass-hard, spherically shaped nodules up to 2 in. (5.1 cm) in diameter. The chemical and physical processes that occur are very complicated but generally include the following sequence [46]:

1. Evaporation of free water.
2. Evolution of combined water in the argillaceous components.
3. Calcination of the calcium carbonate ($CaCO_3$) to calcium oxide (CaO).
4. Reaction of CaO with silica to form dicalcium silicate.
5. Reaction of CaO with the aluminum and iron-bearing constituents to form the liquid phase.
6. Formation of the clinker nodules.
7. Evaporation of volatile constituents (e.g., sodium, potassium, chlorides, and sulfates).
8. Reaction of excess CaO with dicalcium silicate to form tricalcium silicate.

This sequence can be divided into 4 stages:

1. Evaporation of uncombined water from raw materials as the material temperature increases to 212°F (100°C).
2. Dehydration as the material temperature increases to approximately 800°F (430°C) to form oxides of silicon, aluminum, and iron.

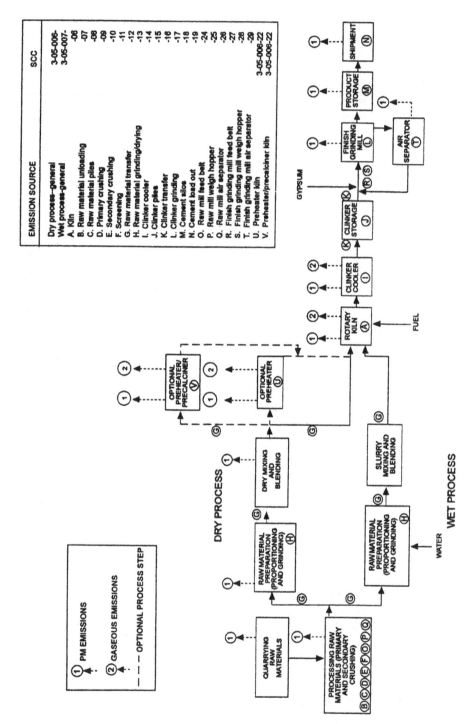

Figure 14.23 Process flow diagram for Portland cement manufacturing. (From Ref. 46.)

3. Calcination, where CO_2 is evolved between 900 and 982°C (1650 and 1800°F) to form CaO.

4. Reaction of the oxides in the burning zome of the rotary kiln to form cement clinker at about 1510°C (2750°F).

Cement manufacturing processes are normally classified as wet, dry, or semi-dry. The same raw materials are used in each type except that water is added in the wet process to minimize dust emissions. Dust emissions from a dry process are often more than double those from a wet process. If the incoming feed materials already contain a high moisture content then wet processes are often employed. In either case, the raw materials are fed into a large refractory-lined kiln (see Fig. 14.24) fired typically with either coal or a gaseous fuel like natural gas. These kilns range in diameter from 10 to 25 ft (3–8 m) and in length from 150 to 750 ft (50–230 m). They typically rotate at 1 to 3 rpm. The longitudinal axis of the kiln is at a slight angle so that the incoming raw materials fed in at the higher end flow naturally toward the lower exit end. Excess water in the mixture is driven off at about 970°F (520°C). Further along the kiln at about 1600°F (870°F) the limestone breaks down into calcium oxide and carbon dioxide. Towards the end of the kiln at about 2600°F (1430°C), the material begins initial melting and sintering takes place to form the clinker.

These large rotary kilns may be either co-current (see Fig. 14.25) or counter-current (see Fig. 14.26), depending on how the exhaust gases flow compared to the direction of the material flow. A single very large burner is located at one end of the kiln. The flame lengths are typically very long, sometimes beyond the midpoint of

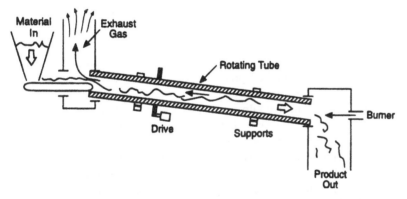

Figure 14.24   Rotary cement kiln. (From Ref. 19. Courtesy of CRC Press.)

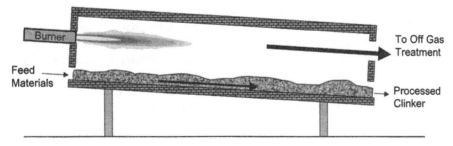

Figure 14.25   Co-current rotary kiln.

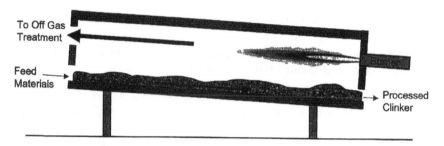

Figure 14.26 Counter-current rotary kiln.

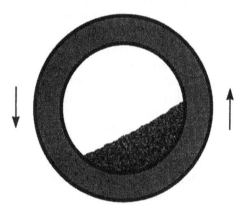

Figure 14.27 Cross-sectional view of a granular load in a rotary kiln. (From Ref. 19. Courtesy of CRC Press.)

very long kilns. The heat distribution needs to be uniform enough that the raw materials are not overheated or melted prematurely. If the flux is too high in one location then the refractory may be damaged.

Cement is manufactured in five kiln types: wet process, dry process, preheater, precalciner, and semidry process. The U.S. EPA defines a calciner as "equipment used to remove combined (chemically bound) water and/or gases from mineral material through direct or indirect heating" [47]. While the materials are the same in each case, the moisture content and processing techniques differ. Wet kilns are longer to dry the higher moisture content mixture. The kiln is an inclined refractory-lined cylinder that slowly rotates to help convey the material from the inlet to the outlet (see Fig. 14.27). A very large burner is located at the exit end of the kiln, which makes the process counter-current where the raw materials flow in one direction and the combustion products flow in the opposite direction. The fuel may be coal, oil, or natural gas, depending on the availability and costs of each. The primary combustion air going through the burner is supplied by a blower while the remaining air comes from the clinker cooler at the exit end of the kiln.

Lime is produced by the high-temperature calcination of limestone, which contains at least 50% calcium carbonate. Two common lime manufacturing processes include [48]:

$$CaCO_3 + heat \rightarrow CO_2 + CaO \tag{14.1}$$

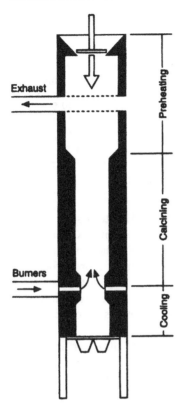

Figure 14.28   Elevation view of a vertical shaft kiln. (From Ref. 49. Courtesy of CRC Press.)

for high calcium lime and

$$CaCO_3 + MgCO_3 + heat \rightarrow 2CO_2 + CaO \cdot MgO \qquad (14.2)$$

for dolomitic lime. A vertical shaft kiln (see Fig. 14.28 [49]) is often used to calcine limestone used in the production of cement. This kiln acts like a counterflow heat exchanger where hot exhaust products from burners at the bottom flow upward while the incoming cold raw materials flow downward.

14.3.2   Emissions

The potential emissions from cement production include: dust emissions, NO_x, SO_x, CO, total hydrocarbons, dioxins and furans, heavy metals (e.g., mercury, lead, and zinc), CO_2, and water-related pollutants such as suspended solids [50]. One of the principle pollutants from cement and lime production is the vast amount of dust that is generated during various phases of the processes. Dust emissions come principally from the kiln/raw mill exhaust, the cooler exhaust, and the cement mill exhaust. The raw materials, such as rock and limestone, are ground down to the appropriate size, which generates dust. Some of the raw materials, like sand, are already fine enough to produce dust emissions. The dust emissions from the incoming raw materials

Table 14.9 Uncontrolled NO$_x$ Emissions from Cement Kilns

Cement kiln type	Heat input requirement (MMBtu/ton of clinker)	NO$_x$ emission rate (lb/ton of clinker)	Range of NO$_x$ emissions (lb/ton of clinker)
Long wet kiln	6.0	9.7	3.6–19.5
Long dry kiln	4.5	8.6	6.1–10.5
Preheater kiln	3.8	5.9	2.5–11.7
Precalciner kiln	3.3	3.4	0.9–7.0

Source: Ref. 51.

cause a variety of health and environmental problems. Health issues include respiratory illnesses and aggravation of existing conditions like asthma. Cement dust is particularly dangerous because when it comes into contact with the moisture in the human respiratory system it can harden and then becomes much more difficult to remove by coughing. Fine cement dust particles can travel long distances in the atmosphere, coating plants, trees, waterways, and man-made structures. However, the pollution generated during the preparation of the raw materials is not of specific concern here as they are not part of the combustion process.

NO$_x$ emissions are typically high compared to those of other industrial combustion processes because of the high temperatures in the cement kiln. Table 14.9 shows typical uncontrolled NO$_x$ emissions from various types of cement kilns [51]. Precalciners tend to have the lowest emissions while long wet kilns tend to have the highest emissions, although there is a fairly large range for each type. There are several possible sources for NO$_x$ in cement and lime production. The primary source is from the high-temperature combustion of hydrocarbon fuels in the presence of large quantities of nitrogen in the combustion air, commonly referred to as thermal NO$_x$. Air preheating further increases emissions. Figure 14.29 shows the effect of the cement kiln temperature on NO$_x$ emissions [52]. Fuel composition, excess O$_2$ in the kiln, and burner design also significantly impact NO$_x$ emissions. Both the fuel (e.g., coal) and the raw feed materials may contain significant quantities of nitrogen that can lead to NO$_x$ emissions, although these are normally much lower quantities compared to those of thermal NO$_x$. The overall thermal efficiency of the process directly impacts all emissions. The higher the thermal efficiency, the less fuel is combusted and, therefore, less emissions are generated.

14.3.3 Regulations

The U.S. EPA has developed emission and operating guidelines for Portland cement plants [53], lime-manufacturing plants [54], and calciners and dryers in mineral industries [55]. Table 14.10 shows emission factors developed by the U.S. EPA for the cement manufacturing process. These factors vary, depending on the pollutant and on the production type. These factors are often used as guidelines in the permitting process to determine the approximate emissions from a new process that has not yet been built or for existing processes that do not have emissions monitoring equipment.

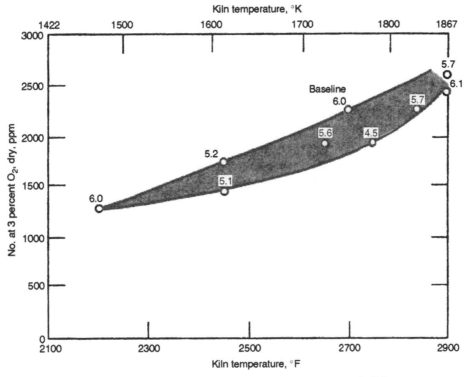

Figure 14.29 Effect of cement kiln temperature on NO_x. (From Ref. 52.)

14.3.4 Pollution Control Strategies

14.3.4.1 Dust Control

All cement and lime kilns are equipped with some type of particulate collection system to remove cement kiln dust. Baghouses are commonly used to control dust emissions in glass manufacturing processes [28]. Bandhu and Garg [56] have written a short book on pollution emissions and control in the cement industry. Much is devoted to the discussion of dust, which is one of the largest air pollution problems in the cement industry. Most of the dust, is generated prior to the actual calcination process that required combustion. Some of these upstream processes include: ball mills, blending silos, discharge of hoppers and feeder tanks, and the transfer of materials around the plant. Dry dust and other particulates are typically controlled by baghouses, cyclones, and electrostatic precipitators (see Chap. 9). Gravity settling chambers, dry cyclones, and electrostatic precipitators are commonly used in cement and lime kilns [57]. Wet dust is a little more difficult because of the possibility of the wet cement hardening, but is generally handled with equipment like packed column scrubbers, vortex scrubbers, and venturi scrubbers.

Dust is also produced from the cement kiln where high volumes of combustion exhaust products flow countercurrently to the incoming feed materials that are being dried prior to being converted into clinker. The high-temperature exhaust gases entrain some of the fine particles and carry them out of the kiln where they must be captured prior to emission into the atmosphere. These particles are dry and can be

Table 14.10 Cement Manufacturing Emission Factors[a]

Process	SO_2[b]	Emission factor rating	NO_x	Emission factor rating	CO	Emission factor rating	CO_2[c]	Emission factor rating	TOC	Emission factor rating
Wet process kiln	8.2	C	7.4	D	0.12	D	2100	D	0.028	D
Long dry process kiln	10	D	6.0	D	0.21	E	1800	D	0.028	E
Preheater process kiln	0.55	D	4.8	D	0.98	D	1800	C	0.18	D
Preheater/precalciner kiln	1.1	D	4.2	D	3.7	D	1800	E	0.12	D
Preheater/precalciner kiln with spray tower	1.0	E	ND		ND		ND		ND	

[a]Factors represent uncontrolled emissions unless otherwise noted. Factors are in lb/ton of clinker produced, unless noted. TOC = total organic compounds; ND = no data.
[b]Mass balance on sulfur may yield a more representative emission factor for a specific facility than the SO_2 emission factors presented in this table.
[c]Mass balance on carbon may yield a more representative emission factor for a specific facility than the CO_2 emission factors presented in this table.
Source: Ref. 4.

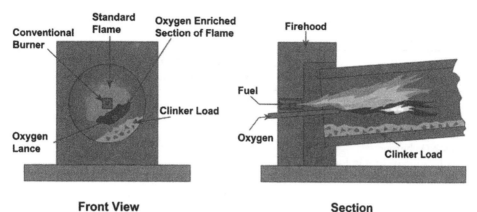

Figure 14.30   Oxygen-enhanced combustion in a rotary kiln.

captured by the standard particulate collection methods. To some extent, the particulate emissions can be controlled by adjusting the operating conditions and the kiln design. Another source of particulate emissions related to the combustion process is from the grinding and pulverization of coal when it is used as the fuel. Fabric filters and electrostatic precipitators are commonly used to control particulate emissions. Coolers are required to reduce the high gas temperatures prior to entering the dust removal equipment. The fabric filters may be made of a high-temperature textile such as Nomex® made by Dupont. This increases the filter cost, but the life is increased as well. Electrostatic precipitators, venturi scrubbers, and cyclones are also used to control particulate emissions [58].

Another technique for controlling dust emissions is through the use of oxygen-enhanced combustion. This is done by replacing a substantial amount of the air used for combustion with high-purity oxygen. Figure 14.30 shows oxygen-enhanced combustion applied to a rotary kiln. Essentially this amounts to removing large quantities of inert nitrogen from the system. This has two positive effects on dust emissions. The first is that the overall system thermal efficiency is improved so that less fuel needs to be fired for a given production rate. Less fuel means less flue gas and, therefore, less particle entrainment. The second positive effect is also related to particle entrainment. By removing large quantities of nitrogen, the flue gas volume decreases, which reduces particle entrainment. Low levels of oxygen enrichment, however, can increase NO_x emissions so there may be a tradeoff of one emission for another. Often, oxygen-enhanced combustion is used in this application to increase throughput for a given size kiln and auxiliary equipment. This means that the firing rate may not be decreased, but the material processing rate is increased. The overall dust emissions may actually increase although they may decrease on a unit production basis (e.g., tons of clinker).

14.3.4.2 SO_x Control

Lime has been a proven absorbent for reducing SO_x emissions (see Chap. 8). Fortunately, lime is produced during cement production and absorbs the vast majority of SO_x formed. Controlling the O_2 levels in the kilns between about 1.5 and 3.5% can minimize SO_x generation [150]. Clemente and Sillero [59] have shown that

Table 14.11 NO$_x$ Reduction with Various Control Technologies in Cement Manufacturing

NO$_x$ control technology	Achievable NO$_x$ emissions reduction (%)
Process modifications	25
Staged combustion in precalciner	30–45
Conversion to indirect firing with a low NO$_x$ burner	20–30
Midkiln firing of tires in long kilns	20–40
SNCR	30–70
SCR	80–90

Source: Ref. 51.

substituting oil coke for coal in a cement kiln can both reduce pollutant emissions, particularly SO$_x$.

14.3.4.3 NO$_x$ Control

The three predominant strategies used to minimize NO$_x$ emissions in these types of processes are process modifications, combustion modifications, and post-treatment. The *first two* minimize NO$_x$ formation while the *third* removes the NO$_x$ that has been formed. Depending on how much reduction is required, these strategies may all be used. Table 14.11 shows a comparison of typical NO$_x$ reductions for various control techniques used in cement manufacturing processes.

Process modifications can reduce NO$_x$ emissions by either increasing the overall energy efficiency or by reducing the nitrogen content in either the incoming raw materials or in the fuel. For example, the lower the required alkali content in the finished cement, the more energy is required, which indirectly increases all emissions. Using raw materials with lower alkali contents or using bypass flue gases for preheating the higher alkali raw materials can significantly reduce NO$_x$ emissions [51]. These are not always possible or economic, but in limited cases they may be viable options. Other ways to improve the overall cement production operation including reducing pollutant emissions are better training of the kiln operators, homogenizing the incoming raw materials, and improving the cooler's operation. A relatively recent process modification is called CemStar, which involves the addition of steel slag to the kiln feed to reduce the amount of limestone calcination per unit of product and, therefore, reduce the energy requirements, which directly reduces emissions [60]. Reported NO$_x$ reductions per ton of product have been as high as 55%.

Because cement kilns are high-temperature processes, thermal NO$_x$ (see Chap. 6) is the predominant mechanism for NO$_x$ generation. Because more fuel is needed in wet processes used to make cement, more NO$_x$ is also generated in those processes, often double those of dry processes. Therefore, there is an incentive both to reduce fuel consumption and NO$_x$ generation by using dry processes instead of wet processes. It had been common practice to run cement kilns at higher than needed temperatures to ensure that no underburned material was produced. However, advanced control equipment has replaced manual kiln operation and has allowed kilns to be run at lower temperatures. Again, this reduces both fuel consumption and pollution generation, in addition to extending the life of the kiln. Switching to a higher volatility coal often produces lower NO_x emissions.

An effective combustion modification technique for minimizing NO_x emissions is to minimize excess air in the kiln. This also increases energy efficiency, which also indirectly reduces all emissions. Control of the gas temperatures in the kiln can also reduce NO_x by minimizing hot spots. By minimizing peak gas temperatures, thermal NO_x emissions are reduced.

Iliuta and coworkers [61,62] describe an in-line low NO_x calciner for use in making cement and related products. The process involves optimizing the primary firing in the rotary kiln and of the secondary firing in the calciner. Fuel is fed into a reducing zone and preheated combustion air is fed into a downstream oxidizing zone. A computer model showed that the most important hydrodynamic parameter is the mixing rate of preheated combustion air into the main flow. NO decreased with increase in the primary CO/CO_2 product ratio. NO increased/decreased with an increase/decrease in the char/limestone particle diameter. Standard combustion modification techniques such as staging, flue gas recirculation, and low NO_x burners may also be used to reduce the amount of NO_x produced by the combustion process. NO_x reductions of up to 47% have been reported using low NO_x burners [60].

A relatively recent development is a low NO_x precalciner where the fuel is burned substoichiometrically to create a reducing atmosphere that inhibits NO_x formation. Another relatively recent development is referred to as mid-kiln firing where waste fuels are added near the middle of the kiln. Fuel is periodically injected into the kiln through pivoting doors. This fuel staging reduces NO_x emissions by up to 59%.

The most common postcombustion control techniques used to reduce NO_x emissions in industrial combustion processes are selective catalytic reduction (SCR) and selective noncatalytic reduction (SNCR). Both of these involve the injection of ammonia, urea, or similar chemicals into the exhaust gas stream to be treated, within a certain temperature range, to convert the NO_x into nitrogen and water (see Chap. 6). SNCRs are most commonly used in cement and lime production processes. The use of SCRs in cement and lime production is still in the commercialization phase, but has been successfully demonstrated in some pilot applications. A recent postcombustion SNCR NO_x control technology is biosolids injection (BSI) where the naturally occurring ammonia content in dewatered biosolids is utilized instead of injecting ammonia or urea as in the conventional technique. Table 14.12 shows the applicability by kiln type for the more common NO_x control technologies in the cement industry.

14.3.4.4 Other Pollutants

While it is possible to have CO and unburned hydrocarbon emissions from cement kilns, these are very easy pollutants to control (see Chap. 7) by proper control of the combustion system. Dioxins and furans have been measured from some cement processes [50]. Keeping the exhaust gas temperature below about 250°C (480°F) and minimizing the residence times in the electrostatic precipitators or baghouses seems to mitigate these pollutants. Heavy metals are often produced in cement processes and tend to end up in the clinker, rather than exiting the exhaust stack. Only thallium, chromium, and mercury are volatile enough that they can exit with the flue gases. Proper selection and treatment of the incoming raw materials is one of the best ways to reduce heavy metals emissions.

Table 14.12   Applicability of NO_x Control Techniques of Different Types of Cement Kilns

NO_x control technique	Applicable kiln type			
	Wet	Long dry	Preheater	Precalciner
Process control systems	Yes	Yes	Yes	Yes
CemStar	Yes	Yes	Yes	Yes
Low-NO_x burner[a]	Yes	Yes	Yes	Yes
Midkiln firing	Yes	Yes	No	No
Tire-derived fuel[b]	Yes	Yes	Yes	Yes
SNCR	No	No	Yes	Yes

[a]Low- NO_x burners can only be used on kilns that have indirect firing.
[b]Tire-derived fuel can be introduced midkiln in a wet or long dry kiln, or at the feed end of a preheater or precalciner kiln.
Source: Ref. 60.

14.4   BRICKS, REFRACTORIES, AND CERAMICS

Bricks and related clay products such as pipe and pottery involve the mining, grinding, screening, and blending of the raw materials, and then forming, cutting, drying or curing, and firing of the final product [63]. Refractories are made from either clays or nonclays. The manufacturing process includes four steps: raw material processing, forming, firing, and final processing. Figure 14.31 shows a process flow diagram for the brick manufacturing process [58]. Firing involves heating the refractory material to high temperatures.

Bricks, refractories, and ceramics are often heated in a tunnel kiln (see Fig. 14.32). The materials to be heated, sometimes referred to as ware, are loaded on to cars and pushed through the kiln at regular intervals. The ware passes through three zones: preheating, firing, and cooling. In the preheating zone, the ware is preheated by the exhaust products from the firing zone. The primary function of the preheating zone is to remove moisture and burn out any organic material that might be present. In the firing zone, roof-mounted burners are fired in the free space around the ware. The function of this zone is to promote high-temperature chemical transformations, such as the decomposition of carbonates, and to heat the ware to the sintering temperature. In the third zone, the cooling zone, the ware is cooled down by the incoming combustion air, which is simultaneously preheated by the hot ware to improve the thermal efficiency of the system. Some of the preheated air is removed from this zone and used in other parts of the plant.

An important aspect of the process is how the ware is stacked on the cars traveling through the tunnel kiln. The ware is arranged to maximize the convective heat and mass transfer of the combustion gases flowing around it. The ware must be arranged so that all pieces receive adequate convection and are properly treated. This precludes dense packing, which would not give adequate exposure to the innermost pieces of ware. Carvalho and Nogueira [64] have modeled a tunnel kiln and shown how the stacking arrangement of the ware affects the load temperature.

The primary air pollutants from the kiln combustion process are CO, NO_x, particulates, and carbon dioxide. SO_x, VOCs, and fluorides may also be emitted if high enough temperatures are reached. Traditional air pollution control equipment

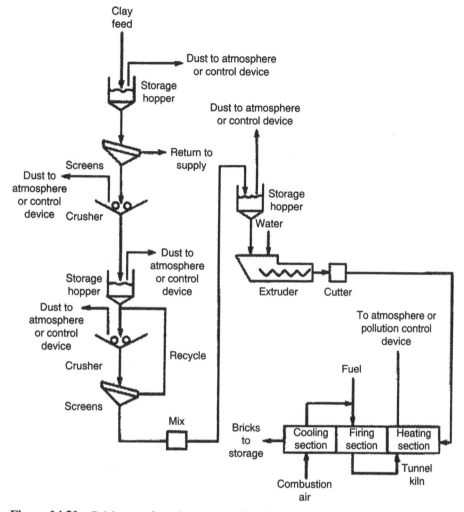

Figure 14.31 Brick manufacturing process flow diagram. (From Ref. 58.)

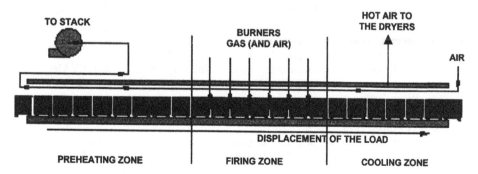

Figure 14.32 Tunnel kiln. (From Ref. 19. Courtesy of CRC Press.)

such as electrostatic precipitators, baghouses, and scrubbers are used to control these pollutants.

REFERENCES

1. U.S. EPA. Calciners and Dryers in Mineral Industries—Background Information for Proposed Standards. Rep. EPA-450/3-85-025a. Washington, DC: Environmental Protection Agency, Oct. 1985.

2. J Elkins, N Frank, J Hemby, D Mintz, J Szykman, A Rush, T Fitz-Simons, T Rao, R Thompson, E Wildermann, G Lear. National Air Quality and Emissions Trends Report, 1999. Washington, DC: U.S. Environmental Protection Agency, Rep. EPA 454/R-01-004, 2001.

3. Office of the Federal Register. 60.16 Priority List. U.S. Code of Federal Regulations Title 40, Part 60. Washington, DC: U.S. Government Printing Office, 2001.

4. U.S. EPA. Compilation of Air Pollutant Emission Factors, Vol. I: Stationary Point and Area Sources, Sect. 11.15: Glass Manufacturing, 5th edn. Washington, DC: U.S. Environmental Protection Agency Rep. AP-42, 1995.

5. RA Drake. Combustion Progress, Problems, Needs in the Glass Industry. In MA Lukasiewicz (ed.). *Industrial Combustion Technologies*. Warren, PA: American Society of Metals, pp. 23–25, 1986.

6. R Gardon. A review of radiant heat transfer in glass. *J. Am. Ceram. Soc.*, Vol. 44, No. 7, pp. 305–312, 1961.

7. RJ Reid. What you should know about combustion. *Glass Indust.*, Vol. 70, No. 7, pp. 24–35, 1989.

8. VK Gegelashvili, VV Zhukovskii, GA Svidzinskii. Separate admission of gas and air to a glass furnace. *Steklo Keram.*, No. 11, pp. 38–39, 1968.

9. AI Kukarkin, VE Dunduchenko, VS Marinskii et al. A rational way to burn fuel in glass tank furnaces. *Steklo Keram.*, No. 6, pp. 4–5, 1972.

10. VB Kut'in, SN Gushchin, VG Lisienko. Heat transfer in the cross-fired glass furnace. *Glass & Ceram.*, Vol. 54, Nos. 5–6, pp. 135–138, 1997.

11. VG Lisienko, VB Kut'in, VYa Dzyuzer. Influence of flame length and luminosity on heat transfer in glass tank furnaces. *Steklo Keram.*, No. 3, pp. 6–8, 1981.

12. VB Kut'in, SN Gushchin, and VG Lisienko. Heat exchange in the cross-fired glass furnace. *Glass & Ceram.*, Vol. 54, Nos. 5–6, pp. 172–174, 1997.

13. PB Eleazer, BC Hoke. Glass. In CE Baukal (ed.). *Oxygen-Enhanced Combustion*, Boca Raton, FL: CRC Press, pp. 215–236, 1998.

14. AG Pincus (ed.). *Combustion Melting in the Glass Industry*. New York: Magazines for Industry, 1980.

15. RH Doremus. *Glass Science*. New York: John Wiley, 1973.

16. A Paul. *Chemistry of Glasses*. London: Chapman and Hall, 1982.

17. BC Hoke, P Schill. CFD Modeling for the Glass Industry. In CE Baukal, VY Gershtein, X Li (eds.). *Computational Fluid Dynamics in Industrial Combustion*. Boca Raton, FL: CRC Press, 2001.

18. SR Scholes, CH Greene (ed.). *Modern Glass Practice*. 7th revised edn. Boston, MA: Cahners Books, 1975.

19. CE Baukal. *Heat Transfer in Industrial Combustion*. Boca Raton, FL: CRC Press, 2000.

20. LF Westra, LW Donaldson. Development of an Advanced Glass Melter System. *Proceedings of 1986 International Gas Research Conference*, ed. TL Cramer. Govt. Institutes, Rockville, MD, pp. 889–897, 1987.

21. J Hnat, D Bender, L Donaldson, A Bendre, D Tessari, J Sacks, A Litka. Development of an Advanced Gas-Fired Glass Melting Furnace. *Proceedings of 1989 International*

Gas Research Conference, ed. TL Cramer. Govt. Institutes, Rockville, MD, pp. 1381–1390, 1990.

22. HA Abbasi, RE Grosman, LW Donaldson, CF Youssef, ML Joshi, SR Hope. A Low-NO_x Retrofit Technology for Regenerative Glass Melters. *Proceedings of 1995 International Gas Research Conference*, Vol. II ed. by DA Dolenc, Govt. Institutes, Rockville, MD, pp. 2322–2331, 1996.

23. PI Booker. Developments in the use of oxygen in glass furnace combustion systems. *Glass*, Vol. 59, No. 5, pp. 172–178, 1982.

24. SC Hill, AM Eaton, LD Smoot. PCGC-3. In CE Baukal, VY Gershtein, X Li (eds.). *Computational Fluid Dynamics in Industrial Combustion*. Boca Raton, FL: CRC Press, 2001.

25. D Ertl, A McMahon. Conversion of a Fiberglass Furnace from 100% Electric to Oxy/fuel Combustion. *Proceedings from 54th Conference on Glass Problems*, pp. 186–190, Champaign, IL, Oct. 1993.

26. AJ Willmot. Regenerative Heat Exchangers. In GF Hewitt, GL Shires, YV Polezhaev. (eds.). *International Encyclopedia of Heat & Mass Transfer*. Boca Raton, FL: CRC Press, 1997.

27. S Hope, S Schemberg. Oxygen-fuel boosting on float furnaces. *Int. Glass Rev.*, pp. 63–66, spring/summer, 1997.

28. U.S. EPA. Profile of the Stone, Clay, Glass, and Concrete Products Industry. Washington, DC: U.S. Environmental Protection Agency. Rep. EPA/310-R-95-017, 1995.

29. U.S. EPA. Compilation of Air Pollutant Emission Factors, Vol. I: Stationary Point and Area Sources, Sect. 11.13: Glass Fiber Manufacturing, 5th edn. Washington, DC: U.S. Environmental Protection Agency Rep. AP-42, 1995.

30. U.S. EPA. Wool Fiberglass Insulation Manufacturing Industry—Background Information for Proposed Standards. Rep. EPA-450/3-82-022a. Washington, DC: U.S. Environmental Protection Agency, Dec. 1983.

31. JT Brown. 100% Oxygen–Fuel Combustion for Glass Furnaces. *Collected Papers from the 51st Conference on Glass Problems*, pp. 202–217, 1990.

32. U.S. Department of Energy. Glass: A Clear Vision for a Bright Future. Washington, DC: U.S. Dept. of Energy, 1996.

33. Energetics, Inc. *Glass Technology Roadmap Workshop*. Columbia, MD, 1997.

34. RM McIntosh. Controlling Emissions in the Glass Industry—A Developing Approach. Controlling Industrial Emissions—Practical Experience. *IChemE Symposium Series 143*. Warwickshire, UK: Institution of Chemical Engineers, pp. 203–209, 1997.

35. Office of the Federal Register. Subpart CC: Standards of Performance for Glass Manufacturing Plants. U.S. Code of Federal Regulations Title 40, Part 60. Washington, DC: U.S. Government Printing Office, 2001.

36. Office of the Federal Register. Subpart PPP: Standards of Performance for Wool Fiberglass Insulation Manufacturing Plants. U.S. Code of Federal Regulations Title 40, Part 60. Washington, DC: U.S. Government Printing Office, 2001.

37. CE Baukal. *Oxygen-Enhanced Combustion*. Boca Raton, FL: CRC Press, 1998.

38. AG Slavejkov, CE Baukal, ML Joshi, JK Nabors. Oxy–Fuel glass melting with a high performance burner. *Ceram. Bull.*, Vol. 71, No. 3, pp. 340–343, 1992.

39. G Jankes, M Stanjevic, M Karan, M Kuburovic, M Adzic. The Use of Technical Oxygen for Combustion Processes in Industrial Furnaces. *Proceedings of 5th European Conference on Industrial Furnaces and Boilers*, Portugal, Vol. 1, pp. 649–658, 2000.

40. A Caldeira-Pires, DP Correia, P Maia, P Lacava, MV Heitor. Influence of burner-port geometry in hydrocarbon oxidation and NO_x formation mechanisms in methane/air flames. *Fuel*, Vol. 81, pp. 771–783, 2002.

41. JP Skalny. *Cement Production and Use*. Engineering Foundation, 1980.

42. B Kohlhaas, ORFZ Labahn (eds.). *Cement Engineers Handbook*, 4th edn. International Philadelphia, PA: Heyden & Sons, Inc., 1982.

43. KE Peray. *Rotary Cement Kiln*, 2nd edn. Chemical Publishing Co., 1986.

44. DK Mittal. *Cement Industry*. New Delhi, India: Anmol Publications, 1994.

45. EM Gartner, H Uchikawa. *Cement Technology*. American Ceramic Society, 1995.

46. U.S. EPA. Compilation of Air Pollutant Emission Factors, Vol. I: Stationary Point and Area Sources, Sect. 11.6: Portland Cement Manufacturing, 5th edn. Washington, DC: U.S. Environmental Protection Agency Rep. AP-42, 1995.

47. Office of the Federal Register. 60.731 Definitions. U.S. Code of Federal Regulations Title 40, Part 60. Washington, DC: U.S. Government Printing Office, 2001.

48. U.S. EPA. Compilation of Air Pollutant Emission Factors, Vol. I: Stationary Point and Area Sources, Sect. 11.17: Lime Manufacturing, 5th edn. Washington, DC: U.S. Environmental Protection Agency Rep. AP-42, 1995.

49. GL Shires. Kilns. GF Hewitt, GL Shires, YV Polezhaev (eds). *International Encyclopedia of Heat & Mass Transfer*. Boca Raton, FL: CRC Press, pp. 651–653, 1997.

50. LP Evans. Cement—Present and Future Environmental Challenges. Controlling Industrial Emissions—Practical Experience. *IChemE Symposium Series 143*. Warwickshire, UK: Institution of Chemical Engineers, pp. 51–66, 1997.

51. U.S. EPA. Alternative Control Techniques Document—NO_x Emissions from Cement Manufacturing. Rep. EPA-453/R-94-004. Washington, DC: Environmental Protection Agency, 1994.

52. U.S. EPA. Control Techniques for Nitrogen Oxides Emissions from Stationary Sources. EPA Rep. 450/1-78-001. Washington, DC: U.S. Environmental Protection Agency, 1978.

53. Office of the Federal Register. Subpart F: Standards of Performance for Portland Cement Plants. U.S. Code of Federal Regulations Title 40, Part 60. Washington, DC: U.S. Government Printing Office, 2001.

54. Office of the Federal Register. Subpart HH: Standards of Performance for Lime Manufacturing Plants. U.S. Code of Federal Regulations Title 40, Part 60. Washington, DC: U.S. Government Printing Office, 2001.

55. Office of the Federal Register. Subpart UUU: Standards of Performance for Calciners and Dryers in Mineral Industries. U.S. Code of Federal Regulations Title 40, Part 60. Washington, DC: U.S. Government Printing Office, 2001.

56. D Bandhu, RK Garg. *Cement Industries and Environmental Management*. New Dehli: Friedrich Ebert Stiftung, 1992.

57. G Ramachandran. 5.16 Particulate Controls: Dry Collectors; 5.17 Particulate Controls: Electrostatic Precipitators. In DHF Liu, BG Lipták (eds.). *Environmental Engineers' Handbook*, 2nd edn. Boca Raton, FL: Lewis Publishers, 1997.

58. U.S. EPA. Control Techniques for Particulate Emissions from Stationary Sources—Vol. 2. Rep. EPA-450/3-81-005B. Washington, DC: U.S. Environmental Protection Agency, Sep. 1982.

59. C Clemente, E Sillero. Optimization of the Use of Fuel in the Cement Industry. In N Piccinini, R Delorenzo (eds.). *Chemical Industry and Environment II*, Vol. 2. Turin, Italy: Politecnico di Torino, 1996.

60. R Battye, S Walsh, J Lee-Greco. NO_x Control Technologies for the Cement Industry. EPA Report, EPA Contract No. 68-D98-026, Washington, DC: Environmental Protection Agency, Sep. 2000.

61. I Iliuta, K Dam-Johansen, A Jensen, LS Jensen. Modeling of in-line low-NO_x calciners—a parametric study. *Chem. Eng. Sci.*, Vol. 57, pp. 789–803, 2002.

62. I Iliuta, K Dam-Johansen, LS Jensen. Mathematic modeling of in-line low-NO_x calciner. *Chem. Eng. Sci.*, Vol. 57, pp. 805–820, 2002.

63. EPA. Compilation of Air Pollutant Emission Factors, Vol. I: Stationary Point and Area Sources, Sect. 11.3: Bricks and Related Clay Products, 5th edn. Washington, DC: U.S. Environmental Protection Agency Rep. AP-42, 1995.

64. M Carvalho, M Nogueira. Improvement of energy efficiency in glass-melting furnaces, cement kilns and baking ovens, *Appl. Therm. Eng.*, Vol. 17, Nos. 8–10, pp. 921–933, 1997.

15

Chemicals Industries

15.1 INTRODUCTION

The chemicals industry encompasses a wide range of end products. It is sometimes divided into two general categories referred to as inorganic and organic chemicals. Inorganic chemicals include things like acids, bases, fertilizers, chlorines, bromines, and phosphates. While mineral products like cement are sometimes included under inorganic chemicals, here they are considered as a separate class of products (see Chap. 14). Organic chemicals include petrochemical and other hydrocarbon products such as carbon black, explosives, paints, and varnishes. Hydrocarbon-based products are the primary focus of this chapter. The petrochemicals industry general refers to chemicals derived from petroleum or natural gas and includes a variety of compounds from acetylene to vinyl chloride. Raw materials include crude petroleum, natural gas, refinery gas, natural gas condensate, light tops or naphtha, and heavy fractions such as fuel oil [1]. The primary products of the petroleum industry fall into three categories [2]:

1. Fuels such as motor gasoline, diesel and distillate fuel oil, liquefied petroleum gas, jet fuel, residual fuel oil, kerosene, and coke.
2. Finished nonfuel products such as solvents, lubricating oils, greases, petroleum wax, petroleum jelly, asphalt, and coke.
3. Chemical industry feedstocks such as naphtha, ethane, propane, butane, ethylene, propylene, butylenes, butadiene, benzene, toluene, and xylene.

By virtue of the chemical production process, the chemicals industry often has many other air pollutants generated during the manufacturing process. For example, in the production of hydrochloric acid, this compound is often a pollutant in the exhaust gas stream, which must be treated prior to emission into the atmosphere. This type of pollutant is not the primary concern of this book as it is not caused by the combustion process, but rather by the manufacturing process.

The purpose of this chapter is to alert the interested reader about the potential effects on pollutant emissions from the combustion processes in the chemicals (sometimes referred to as the CPI or chemical processing industry), petrochemical, and hydrocarbon (sometimes referred to as the HPI or hydrocarbon processing industry) industries. Petroleum refining is usually categorized as separating crude oil into various usable petroleum products such as fuels (e.g., gasoline, jet fuel, and

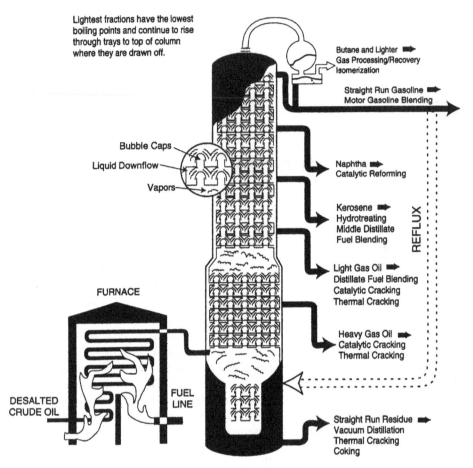

Lightest fractions have the lowest
boiling points and continue to rise
through trays to top of column
where they are drawn off.

Butane and Lighter ➡
Gas Processing/Recovery
Isomerization

Straight Run Gasoline ➡
Motor Gasoline Blending

Bubble Caps

Liquid Downflow

Vapors

Naphtha ➡
Catalytic Reforming

Kerosene ➡
Hydrotreating
Middle Distillate
Fuel Blending

REFLUX

Light Gas Oil ➡
Distillate Fuel Blending
Catalytic Cracking
Thermal Cracking

FURNACE

Heavy Gas Oil ➡
Catalytic Cracking
Thermal Cracking

DESALTED
CRUDE OIL

FUEL
LINE

Straight Run Residue ➡
Vacuum Distillation
Thermal Cracking
Coking

Figure 15.1   Crude-oil distillation schematic. (From Ref. 3.)

residual fuel oil), nonfuel products (e.g., solvents, lubricating oils, asphalt, and coke), and chemical industry feedstocks (e.g., propane, butane, ethylene, and propylene). Figure 15.1 shows a schematic of crude oil distillation in the petrochemical industry [3]. Figure 15.2 shows the products and product yields from typical U.S. refineries. Figure 15.3 shows a process flow diagram for the production of organic chemicals [4]. Liu [5] gives a good general discussion of reducing pollution emissions from chemical manufacturing. He notes the important distinction between source reduction and discharge reduction. The EPA has tended toward the former while industry has tended toward the latter, which can be achieved through source reduction, but also through recycling, treatment, and disposal or some combination of these. Detailed analysis, including life cycle costs, of each process is recommended to ensure pollution emissions are minimized and controlled.

The Responsible Care Code initiated in 1988 by the Chemical Manufacturers' Association requires its members to commit to proper management of chemicals, the highest standards of safety and health, and superior environmental performance. The U.K. Environment Agency has developed a Guidance Note for the organic

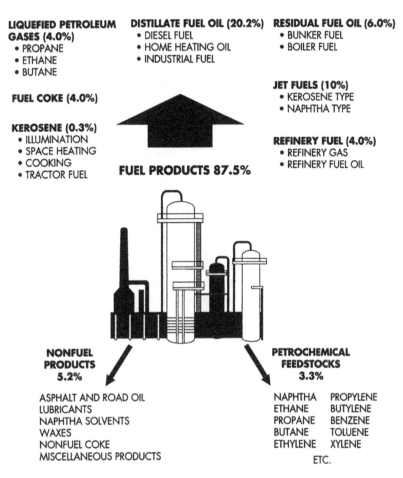

MOTOR GASOLINE (43%)

LIQUEFIED PETROLEUM GASES (4.0%)
• PROPANE
• ETHANE
• BUTANE

DISTILLATE FUEL OIL (20.2%)
• DIESEL FUEL
• HOME HEATING OIL
• INDUSTRIAL FUEL

RESIDUAL FUEL OIL (6.0%)
• BUNKER FUEL
• BOILER FUEL

FUEL COKE (4.0%)

JET FUELS (10%)
• KEROSENE TYPE
• NAPHTHA TYPE

KEROSENE (0.3%)
• ILLUMINATION
• SPACE HEATING
• COOKING
• TRACTOR FUEL

REFINERY FUEL (4.0%)
• REFINERY GAS
• REFINERY FUEL OIL

FUEL PRODUCTS 87.5%

NONFUEL PRODUCTS 5.2%

ASPHALT AND ROAD OIL
LUBRICANTS
NAPHTHA SOLVENTS
WAXES
NONFUEL COKE
MISCELLANEOUS PRODUCTS

PETROCHEMICAL FEEDSTOCKS 3.3%

NAPHTHA   PROPYLENE
ETHANE   BUTYLENE
PROPANE   BENZENE
BUTANE   TOLUENE
ETHYLENE   XYLENE

ETC.

Figure 15.2 U.S. refinery products and yields. (From Ref. 3.)

chemicals industry [6]. This Note includes process descriptions for the production of: petrochemicals like butadiene, ethylene, propylene, and isopropyl alcohol; nitrogen compounds like aniline, methylamines, and nitriles; acids and aldehydes like acetic acid, acetylene, and formaldehyde; sulfur compounds like thiols and thiophene; and monomers and polymers like styrene, acrylates, and vinyl chloride. Scott and Mohan [7] note the importance of developing a plan for the event of an equipment malfunction, particularly pollution control equipment, that could send a plant out of compliance. Singh and Seto [8] discuss how to monitor the performance of air pollution control equipment used in the chemical process industries to ensure continued compliance and minimize operational problems throuh maintenance and troubleshooting [8].

The U.S. Department of Energy sponsored the development of a vision for the chemicals industry for the year 2020 consisting of leading industry experts and led by

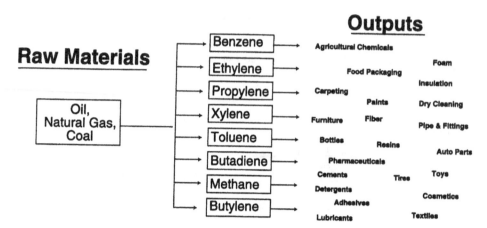

Figure 15.3 Organic chemicals flow diagram. (From Ref. 4.)

the American Chemical Society with the help of several other societies [9]. One of the challenges identified was societal demands for increasing environmental performance. An environmental goal of the vision was to continue playing a leadership role in balancing environmental and economic objectives. One of the steps identified in achieving environmental objectives was to improve the legislative and regulatory climate by the reform of programs to emphasize performance rather than specific methods.

Anderson [10] has written a general introductory book on the petroleum industry, tracing its development from the beginning up to some projections for the future of oil. There is no specific discussion of combustion in petroleum refining. Leffler [11] has written an introductory book on the major processes in petroleum refining, including catalytic cracking, hydrocracking, and ethylene production among many others. The book is written from an overall process perspective and has no discussion of the heaters in a plant. Gary and Handwerk [12] have written a good overview of petroleum refining. The book discusses many of the processes involved in petroleum refining operations, including coking, catalytic cracking, and catalytic reforming among others. However, it does not specifically discuss the combustion processes involved in heating the refinery fluids. Meyers [13] has edited a recently updated handbook on petroleum refining processes. The book is divided into 14 parts, each on a different type of overall process, including catalytic cracking and reforming, gasification and hydrogen production, hydrocracking, and visbreaking and coking, among others. Each part is further divided into the individual subtypes and variations of the given overall process. Companies such as Exxon, Dow–Kellogg, UOP, Stone and Webster, and Foster–Wheeler have written about the processes they developed, which they license to other companies. Many aspects of the processes are discussed including flow diagrams, chemistry, thermodynamics, economics, and environmental considerations, but very little discussion of the combustion systems.

The standard book on the subject of combustion in the hydrocarbon and petrochemical industries that had been used for decades was *Furnace Operations* by Robert Reed [14]. This has been replaced by the *John Zink Combustion*

Figure 15.4   Typical refinery.

Handbook [15], which is a comprehensive reference for the various aspects of combustion related to these industries. The American Petroleum Institute has developed some guidelines for burners and heaters used in refineries [16,17]. The U.K. Environmental Agency has also prepared a useful guide for petroleum processes that includes process descriptions, typical emissions, and pollution control information [18].

The hydrocarbon and petrochemical industries present unique challenges to the combustion engineer, compared to other industrial combustion processes. One of the more important challenges in those industries is the wide variety of fuels, which are usually off-gases from the petroleum refining processes that are used in a typical plant (see Fig. 15.4). Table 15.1 shows how the composition of refinery gas varies, depending on the process in the refinery [19]. This differs significantly from most other industrial combustion systems that normally fire a single purchased fuel such as natural gas or fuel oil. Another important challenge is that many of the burners commonly used in the hydrocarbon and petrochemical industries are natural draft, where the buoyant combustion exhaust products create a draft that induces the combustion air to enter the burners. This is different from nearly all other industrial combustion processes, which utilize a combustion air blower to supply the air used for combustion in the burner. Natural draft burners are not as easy to control as forced-draft burners, and are subject to things like the wind, which can disturb the conditions in a process heater.

Abilov et al. [20] note that furnaces are "the basic and most important industrial units of petroleum refineries and petrochemical process." Not only do they provide the energy needed to process the crude oil, but they also generate most of the pollutant emissions from the plant so they are also important from that perspective.

According to the U.S. Department of Energy, petroleum refining is the most energy-intensive manufacturing industry in the U.S., accounting for about 7% of total U.S. energy consumption in 1994 [21]. Table 15.2 lists the major processes in petroleum refining, most of which require combustion in one form or another. Figure 15.5 shows the process flow through a typical refinery.

Table 15.1 Composition of Typical Refinery Gases

Fuel gas component	Refinery fuel gas source (dry gas)					
	Cracked gas (%)	Coking gas (%)	Reforming gas (%)	FCC gas (%)	Combined refinery gas–sample 1(%)	Combining refinery gas–sample 2(%)
CH_4	65	40	28	32	36	53
C_2H_4	3	3	7	7	5	2
C_2H_6	16	21	28	9	18	19
C_3H_6	2	1	3	15	8	6
C_3H_8	7	24	22	25	20	14
C_4H_8	1	—	—	—	—	—
C_4H_{10}	3	7	7	0	2	1
C_5 and higher	1	—	—	—	—	—
H_2	3	4	5	6	3	3
CO	—	—	—	—	—	—
CO_2	—	—	—	—	—	—
N_2	—	—	—	7	8	3
H_2O	—	—	—	—	—	—
O_2	—	—	—	—	—	—
H_2S	—	—	—	—	—	—
Total	100	100	100	100	100	100

Source: Ref. 19. (Courtesy of John Zink CO. LLC.)

Table 15.2 Major Petroleum Refining Process

Category	Major process
Topping (seperation of crude oil)	Atmospheric distillation
	Vacuum distillation
	Solvent deasphalting
Thermal and catalytic cracking	Delayed coking
	Fluid coking/flexicoking
	Visbreaking
	Catalytic cracking
	Catalytic hydrocracking
Combination/rearrangement of hydrocarbons	Alkylation
	Catalytic reforming
	Polymerization
	Isomerization
	Ethers manufacture
Treating	Catalytic hydrotreating/hydroprocessing
	Sweetening /sulfur removal
	Gas treatment
Speciality product manufacture	Lube oil
	Grease
	Asphalt

Source: Ref. 21.

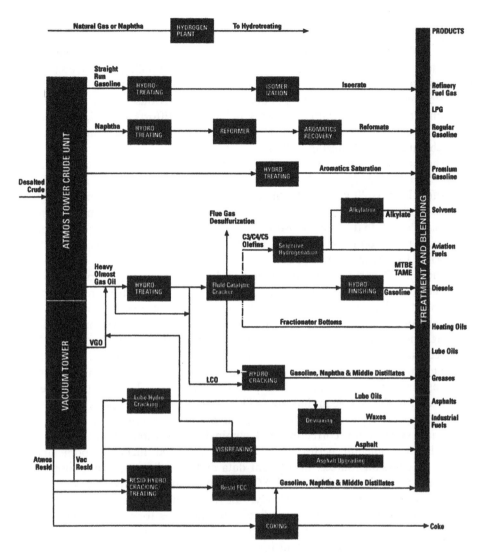

Figure 15.5 Typical refinery flow diagram. (From Ref. 21.)

15.1.1 Environmental Concerns

The U.S. Department of Energy Office of Industrial Technologies has prepared a Technology Roadmap for industrial combustion [22]. For process heating systems, some key performance targets for the year 2020 have been identified for burners and for the overall system. For the burners, the targets include: reducing criteria pollutant emissions by 90%, reducing CO_2 emissions to levels agreed on by the international community, reducing specific fuel consumption by 20–50%, and maximizing the ability to use multiple fuels. For the heating system, the targets include: reducing the total cost of combustion in manufacturing, enhancing system integration, reducing product loss rate by 50%, maximizing system robustness, and zero accidents.

The following were identified as top priority R&D needs in process heating: burner capable of adjusting operating parameters in real time, advanced burner stabilization methods, robust design tools, and economical methods to premix fuel and air.

There continues to be increasing interest in reducing pollutant emissions of all types from all combustion processes. One prognosticator predicts this will continue well into the future [23]. These pollutants have deleterious effects on both the environment and on the health of humans and animals. Efforts are underway from a broad cross-section of organizations to improve existing techniques and to develop new techniques for minimizing pollution.

The American Petroleum Institute has developed a technology vision for the U.S. petroleum industry that identified environmental concerns as one of the five key drivers for the next 20 years [24]. One of the performance targets identified in that report is to demonstrate continuous improvement in environmental impact on air emissions. Part of that target is also to develop appropriate environmental metrics and then to report on those every 5 years. It is recommended that novel methods be investigated for reducing greenhouse gases.

The U.S. Department of Energy prepared a technology roadmap as a follow-up document to the vision [25]. Some of the key industry drivers related to the environment include environmental regulations, increasing pressure to reduce CO_2 emissions, and proactively dealing with public scrutiny, the environment, global warming, and other related issues. It is believed that refiners will continue looking for process improvements in response to growing environmental costs. Some of the environmental performance targets included attaining a leadership position in emissions standards and reducing air pollution emissions. Future environmental performance in the petroleum industry will look to integrate emissions control with production, work toward zero emissions, and look for improved sensing and monitoring systems for correcting and eliminating emissions. Some of the key technical environmental performance barriers identified included: poor understanding of emission sources, insufficient data and modeling for ozone formation, inadequate methods for NO_x and SO_x removal, and an inability to control combustion emissions cost effectively. Environmentally related R&D needs identified included pursuing bio-remediation technologies for removing pollutants, improving ozone modeling, and achieving complete understanding and modeling of combustion chemistry and pollutant formation.

While there are other pollutants potentially produced in the hydrocarbon and petrochemical industries, this chapter is only concerned with the air pollutants resulting from the combustion processes. There is also concern about the pollution generated by the use of the products (e.g., automotive gasoline) produced in petroleum processes; however, this is outside the scope of this work.

The U.S. Environmental Protection Agency has written a booklet entitled "Profile of the Petroleum Refining Industry," which provides some useful information on pollution from this industry [26]. While this booklet is specific to the petrochemical industry, most of the information is relevant to the broader chemical industry.

There are numerous factors that affect the pollutant emissions generated from the combustion of fuels. The U.S. Department of Energy has classified emission factors by fuel type for petroleum refining, as shown in Table 15.3 [27]. An EPA report identified the following heater design parameters that affect NO_x emissions

Table 15.3 Combustion Emission Factors (lb/10^6 Btu) by Fuel Type

Fuel type	SO$_x$	NO$_x$	CO	Particulates	VOCs
Distillate fuel	0.160	0.140	0.0361	0.010	0.002
Residual fuel	1.700	0.370	0.0334	0.080	0.009
Other oils	1.700	0.370	0.0334	0.080	0.009
Natural gas	0.000	0.140	0.0351	0.003	0.006
Refinery gas	0.000	0.140	0.0340	0.003	0.006
LPG	0.000	0.208	0.0351	0.007	0.006
Propane	0.000	0.208	0.0351	0.003	0.006
Steam coal	2.500	0.950	0.3044	0.720	0.005
Petroleum coke	2.500	0.950	0.3044	0.720	0.005
Electricity	1.450	0.550	0.1760	0.400	0.004

Table 1–11 on p. 16 of U.S. Dept. of Energy, Energy & Environmental Profile of the U.S. Petroleum Refining Industry, 1998.
Source: Ref. 27.

from process heaters: fuel type, burner type, combustion air preheat, firebox temperature, and draft type [28]. The important factors that influence pollution are considered here.

15.2 FIRED EQUIPMENT IN THE CHEMICALS PROCESSING INDUSTRIES

Table 15.4 lists the major applications for fired heaters in the chemical industry. These can be broadly classified into two categories: (1) low- and medium-firebox temperature applications such as feed preheaters, reboilers, and steam superheaters, and (2) high-firebox temperature applications such as olefins, pyrolysis furnaces and steam–hydrocarbon reformers. The low- and medium-firebox temperature heaters represent about 20% of the chemical industry requirements and are similar to those in the petroleum refining industry [28]. The high-firebox temperature heaters represent the remaining 80% of the chemical industry heater requirements and are unique to the chemical industry.

15.2.1 Process Heaters

The U.S. EPA defines a process heater as
"a device that transfers heat liberated by burning fuel to fluids contained in tubular coils, including all fluids except water that is heated to produce steam" [29].

Process heaters are designed to heat petroleum products, chemicals, and other liquids and gases. Typical petroleum fluids include gasoline, naphtha, kerosene, distillate oil, lube oil, gas oil, and light ends [30]. Kern noted that refinery heaters may carry liquids at temperatures as high as 1500°F (1100 K) and pressures up to 1600 psig (110 barg). The primary modes of heat transfer in process heaters are radiation and convection. Figure 15.14 shows one common firing configuration where burners fire against a wall (sometimes referred to as a center or target wall) in

Table 15.4 Major Fired Heater Applications in the Chemical Industry

Chemical	Process	Heater type	Firebox temperature (°F)	1985 Fired heater energy requirement, (10^{12} Btu/yr)	Percentage of known chemical industry heater requirements
Low- and medium-temperature applications					
Benzene	Reformate extraction	Reboiler	700	64.8	9.9
Styrene	Ethylbenzene dehydrogenation	Steam superheater	1500–1600	32.1	4.9
Vinyl chloride monomer	Ethylene dichloride cracking	Cracking furnace	N/A	12.6	1.9
p-Xylene	Xylene isomerization	Reactor fired preheater	N/A	13.0	2.0
Dimethyl terephthalate	Reaction of p-xylene and methanol	Preheater, hot oil furnace	480–540	11.1	1.7
Butadiene	Butylene dehydrogenation	Preheater, reboiler	1100	2.6	0.4
Ethanol (synthetic)	Ethylene hydration	Preheater	750	1.3	0.2
Acetone	Various	Hot oil furnace	N/A	0.8	0.1
High-temperature applications					
Ethylene/propylene	Thermal cracking	Pyrolysis furnace	1900–2300	337.9	51.8
Ammonia	Natural gas reforming	Steam hydrocarbon reformer	1500–1600	150.5	23.1
Methanol	Hydrocarbon reforming	Steam hydrocarbon	1000–2000	25.7	4.0
Total known fired heater energy requirement				652.4	100.0

N/A = not applicable.
Source: Ref. 42.

the middle of the heater and the wall reradiates to the process tubes. Figure 15.18 shows another configuration where a horizontally mounted burner fires up along a wall, which then radiates to the process tubes. Other configurations, previously discussed, are shown in Figs. 15.11 and 15.13.

Fired or tubestill heaters are used in the petrochemical and hydrocarbon industries to heat fluids in tubes for further processing. A fired heater consists of three major components: the heating coil, the furnace enclosure, and the combustion equipment. The objective is to transfer heat to the fluids in the heating coil. The heat is produced by the combustion equipment (burners), and is transferred directly to the tubes and also to the furnace enclosure, which in turn also radiates heat to the tubes. The design of all three components is optimized for efficiently and uniformly transferring heat to the fluids in the tubes.

Process heaters are sometimes referred to as process furnaces or direct-fired heaters. They are heat-transfer units designed to heat petroleum products, chemicals, and other liquids and gases flowing through tubes. Typical petroleum fluids include gasoline, naphtha, kerosene, distillate oil, lube oil, gas oil, and light ends [30]. The heating is done to raise the temperature of the fluid for further processing downstream or to promote chemical reactions in the tubes, often in the presence of a catalyst. The initial part of the fluid heating is done in the convection section of the furnace, while the latter heating is done in the radiant section (see Fig. 15.6 [31]). Figure 15.7 shows a diagram of a typical fired heater [32]. Each section has a bank of tubes in it where the fluids flow through, as shown in Fig. 15.8 [33]. Evans [34] has written some general information on the design principles used in fired heaters for the interested reader.

In this type of process, fluids flow through an array of tubes located inside a furnace or heater. The tubes are heated by direct-fired burners that often use fuels that are by-products from processes in the plant and that vary widely in

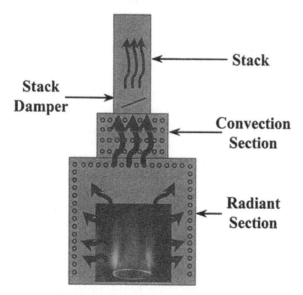

Figure 15.6 Schematic of a process heater. (From Ref. 31. Courtesy of John Zink Co. LLC.)

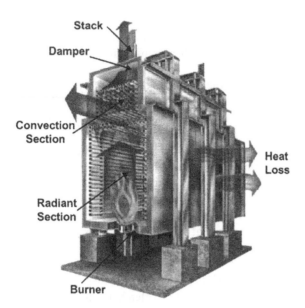

Figure 15.7 Typical fired heater. (From Ref. 32. Courtesy of John Zink Co. LLC.)

composition. Process heaters are used for a variety of processes in refineries. They are used in distillation and fractionation for separating hydrocarbons by boiling point. They are used in:

- Thermal cracking processes like visbreaking and coking.
- Catalytic cracking for producing lighter products than those produced through thermal processes.
- Hydrotreating, which is a catalytic process to remove sulfur components.
- Hydrocracking, which is a high temperature and pressure process in a hydrogen atmosphere for reducing coking.
- Catalytic reforming, which is a high temperature and pressure process to improve the octane number in naphtha for blending in gasoline production.

The initial part of the fluid heating is done in the convection section of the furnace, while the latter heating is done in the radiant section. Each section has a bank of tubes in it where the fluids flow through, as shown in Fig. 15.6 [33]. Early heater designs had only a single bank of tubes that failed prematurely because designers did not understand the importance of radiation in the process [37]. The tubes closest to the burners failed due to overheating that caused the hydrocarbons to coke inside the tube, which further aggravated the problem by reducing the cooling ability of the fluids due to the reduction in thermal conductivity through the coke layer inside the tube. One of the key challenges for the heater designer is to get even heat distribution inside the combustor to prevent coking inside the tubes. Bell and Lowy [35] estimated that approximately 70% of the energy is transferred to the fluids in the radiant section of a typical heater and the balance in the convection section. The tubes in the convection section often have fins to improve convective heat-transfer efficiency. These fins are designed to withstand temperatures up to about 1200°F (650°C).

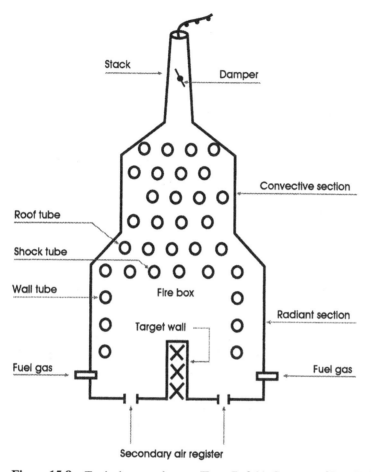

Figure 15.8 Typical process heater. (From Ref. 31. Courtesy of PennWell Books, Tulsa, OK.)

If delayed combustion occurs in the convection section, the fins can be expose to temperatures up to 2000°F (1100°C), which can damage the fins [33].

Garg [36] gives some typical energy consumptions for fired heaters in several common processes: 0.32×10^6 Btu/bbl (94 kW/bbl) of crude oil in the refining industry, 22×10^6 Btu/ton (7.1 MW/m-ton) of ethylene, and 28.5×10^6 Btu/ton (9.21 MW/m-ton) of ammonia. In these types of processes, fluids flow through an array of tubes located inside a furnace or heater. The tubes are heated by direct-fired burners that often use fuels that are by-products from processes in the plant and that vary widely in composition. Using tubes to contain the load is somewhat unusual compared to the other types of industrial combustion applications considered in this book. It was found that heating the fluids in tubes has many advantages over heating them in the shell of a furnace [37]. These include better suitability for continuous operation, better controllability, higher heating rates, more flexibility, less chance of fire, and more compact equipment.

Vertical heaters can be broadly categorized into two types: cylindrical and box heaters [38]. Both have tubes along the walls in the radiant section. In cylindrical heaters, the tubes are installed vertically. In box heaters, the tubes are generally installed horizontally. In both designs the burners are usually installed in the floor and fire vertically upwards. Both designs normally have a convection section near or in the entrance of the stack. A rule-of-thumb principal for the heat distribution is that greater than 60–70% of the total heat duty should be in the radiant section of the heater [39]. The calculated maximum radiant flux ranges from 1.70 to 1.85 times the average radiant flux. A tall, narrow heater is more economical to build, but the radiant flux is often poorly distributed with much higher fluxes in the lower part of the heater. The recommended height-to-diameter ratio for a cylindrical heater is between 1.5 and 2.75 to achieve more uniform heat-flux loading. Garg (1989) notes that fired heater performance can be enhanced by installing a convection section to an all-radiant heater, enlarging the heat transfer area of an existing convection section, converting a natural-draft heater to forced-draft, and adding air preheating or steam-generation equipment [40].

Garg [41] notes a number of factors that need to be considered when specifying burners for fired heaters. Choosing the right burners can increase the heater capacity by 5–10% and increase the efficiency by 2–3%. These factors include burner type, heat release and turndown, air supply, excess air, fuel specifications, firing position, flame dimensions, ignition mode, atomization media for liquid fuel burners, noise, NO_x emission, and waste gas firing. Most of those factors directly influence the heat transfer in the heater.

Kern noted that process heaters are typically designed around the burners [30]. There may be anywhere from one to over 100 burners in a typical process heater, depending on the design and process requirements. In the refinery industry, the average number of burners in a heater varies by the heater type, as shown in Table 15.5 [42,43]. On average, mechanical draft burners have firing rates higher than those of natural draft. For forced-draft systems, burners with air preheat typically have higher heat releases than burners without air preheat. According to one survey, 89.6% of the burners in oil refineries are natural draft, 8.0% are forced draft with no air preheat, and 2.4% are forced draft with air preheat [44]. The mean size of all process heaters is 72×10^6 Btu/hr (21 MW), which are mostly natural draft. The mean size of forced draft heaters is 110×10^6 Btu/hr (32 MW). Figure 15.9 shows the distribution for the overall firing rate for fired heaters. Table 15.6 shows the variety of processes in a refinery that uses fired heaters.

Berman [45] discussed the different burner designs used in fired heaters. Burners may be located in the floor firing vertically upward. In vertical cylindrical (VC) furnaces, those burners are located in a circle in the floor of the furnace. Figure 15.10 shows two burner configurations in VC furnaces [46]. For even larger

Table 15.5 Average Burner Configuration by Heater Type

Heater type	Average number of burners	Average design total heat release (10^6 Btu/hr)	Average firing rate per burner (10^6 Btu/hr)
Natural draft	24	69.4	2.89
Mechanical draft, no air preheat	20	103.6	5.18
Mechanical draft, with air preheat	14	135.4	9.67

Source: Ref. 42.

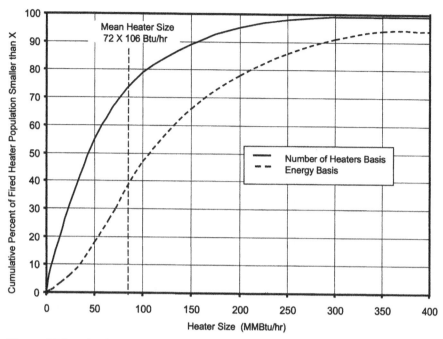

Figure 15.9 Fixed heater size distribution. (From Ref. 42.)

diameter VCs, there may be one burner in the center surrounded by a set of burners at a given radius from the centerline. The VC furnace itself serves as a part of the exhaust stack to help create draft to increase the chimney effect [47]. In cabin heaters, which are rectangular, there are one or more rows of burners located in the floor, often referred to as hearth burners (see Fig. 15.11). Figure 15.12 shows some common burner arrangements in cabin heaters. Burners may also be fired parallel to and against the wall (see Fig. 15.13). Burners may be at a low level firing parallel to the floor. In that configuration, they may be firing from two opposite sides toward a partial wall in the middle of the furnace that acts as a radiator to distribute the heat (see Figs. 15.14 and 15.15. Burners may be located on the wall firing radially along the wall (see Figs. 15.16 and 15.17); those are referred to as radiant wall burners. There are also combinations of the above in certain heater designs. For example, in ethylene production heaters, both floor-mounted vertically fired burners (see Fig. 15.18) and radiant wall burners are used in the same heater. A schematic of the combination of floor and wall-mounted burners is shown in Fig. 15.19.

Typical examples of process heaters are shown in Figs. 15.20 and 15.21. A cabin heater is shown in Fig. 15.22. Burners firing in a crude unit are shown in Fig. 15.23, and typical burner arrangements are shown in Fig. 15.24. Berman [48] noted the following categories of process heaters: column reboilers, fractionating-column feed preheaters, reactor-feed preheaters including reformers, heat supplied to heat-transfer media (e.g., a circulating fluid or molten salt), heat supplied to viscous fluids, and fired reactors including steam reformers and pyrolysis heaters. Six types of vertical-cylindrical fired heaters were given: all radiant, helical coil, crossflow with convection section, integral convection section, arbor or wicket type, and single-row/double-fired. Six basic designs were also given for horizontal-tube fired heaters: cabin,

Table 15.6 Major Refinery Process Requiring a Fired Heater

Process	Process description	Heaters used	Process heat requirements		Feedstock temperature outlet of heater (°F)
			kJ/liter	10³ Btu/bbl feed	
	Distillation				
Atmospheric	Separates light hydrocarbons from crude in a distillation column under atmospheric conditions	Preheater, reboiler	590	89	700
Vacuum	Separates heavy gas oils from atmospheric distillation bottoms under vacuum	Preheater, reboiler	418	63	750–830
	Thermal processes				
Thermal cracking	Thermal decomposition of large molecules into lighter, more valuable products	Fired reactor	4650	700	850–1000
Coking	Cracking reactions allowed to go to completion Lighter products and coke produced	Preheater	1520	230	900–975
Visbreaking	Mild cracking of residuals to improve their viscosity and produce lighter gas oils	Fired reactor	961	145	850–950
	Catalytic cracking				
Fluidized catalytic cracking	Cracking of heavy petroleum products. A catalyst is used to aid the reaction	Preheater	663	100	600–885
Catalytic hydrocracking	Cracking heavy feedstocks to produce lighter products in the presence of hydrogen and a catalyst	Preheater	1290	195	400–850

		Hydroprocessing			
Hydrodesulfurization	Remove contaminating metals, sulfur, and nitrogen from the feedstock Hydrogen is added and reacted over a catalyst	Preheater	431	65[a]	390–850
Hydrotreating	Less severe than hydrodesulfurization Removes metals, nitrogen, and sulfur from lighter feedstocks Hydrogen is added and reacted over a catalyst	Preheater	497	75[b]	600–800
		Hydroconversion			
Alkylation	Combination of two hydrocarbons to produce a higher molecular weight hydrocarbon Heater used on the fractionator	Reboiler	2500	377[c]	400
Catalytic reforming	Low-octane naphthas are converted to high-octane, aromatic naphthas Feedstock is contacted with hydrogen over a catalyst	Preheater	1790	270	850–1000

[a] Heavy gas oils and middle distillates.
[b] Light distillate.
[c] Btu/bbl of total alkylate.
Source: Ref. 42.

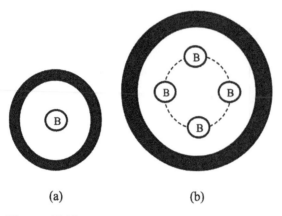

(a)   (b)

Figure 15.10 Plan view of burner (B) arrangement in the floor of vertical cylindrical furnaces: (a) small-diameter furnace with a single centered burner; (b) larger diameter furnace with four burners symmetrically arranged at a radius from the centerline. (From Ref. 46. Courtesy of CRC Press.)

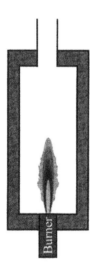

Figure 15.11 Elevation view of a hearth-fired burner configuration. (From Ref. 46. Courtesy of CRC Press.)

two-cell box, cabin with dividing bridgewall, end-fired box, end-fired box with side-mounted convection section, and horizontal-tube/single-row/double-fired.

A unique aspect of process heaters is that they are often natural draft. This means that no combustion air blower is used. The air is inspirated into the furnace by the suction created by the hot gases rising through the combustion chamber and exhausting to the atmosphere. Another unique aspect of these heaters is the wide range of fuels that are used, which are often by-products of the petroleum refining process. These fuels may contain significant amounts of hydrogen, which has a large impact on the burner design. It is also fairly common for multiple fuel compositions to be used, depending on the operating conditions of the plant at any given time. In addition to hydrocarbons ranging up to C_5, the gaseous fuels may also

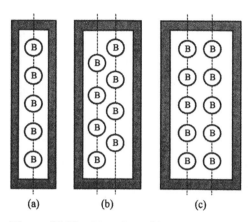

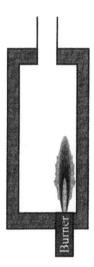

Figure 15.12 Plan view of burner (B) arrangement in the floor of cabin heaters: (a) single row of burners in a narrower heater; (b) two rows of staggered burners in a slightly wider heater; (c) two rows of parallel burners in an even wider heater. (From Ref. 46. Courtesy of CRC Press.)

Figure 15.13 Elevation view of a parallel, wall-fired burner configuration. (From Ref. 46. Courtesy of CRC Press.)

contain hydrogen and inerts (like CO_2 or N_2). The compositions can range from gases containing high levels of inerts to fuels containing high levels of H_2. The flame characteristics for fuels with high levels of inerts are very different from those for fuels with high levels of H_2. Figure 2.25 shows that the adiabatic flame temperature increases rapidly and nonlinearly as the hydrogen content in CH_4/H_2 flames increases. The figure also shows that the adiabatic flame temperature decreases rapidly and nonlinearly as the nitrogen content in CH_4/N_2 flames increases. Add to that the requirement for turndown conditions, and the burner design becomes very challenging to maintain stability, low emissions, and the desired heat-flux distribution over the range of conditions that are possible. Some plants use liquid fuels, like no. 2–no. 6 fuel oil, sometimes by themselves and sometimes in combination

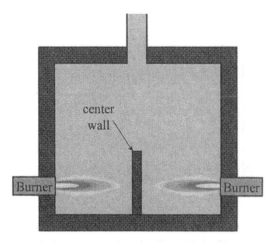

Figure 15.14   Schematic (elevation view) of center or target wall firing configuration. (From Ref. 31. Courtesy of John Zink Co. LLC.)

Figure 15.15   Horizontal floor-fired burners. (From Ref. 31. Courtesy of John Zink Co. LLC, Tulsa, OK.)

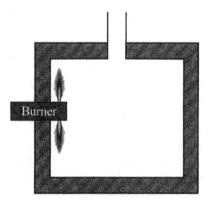

Figure 15.16   Elevation view of a perpendicular, wall-fired burner configuration. (From Ref. 46. Courtesy of CRC Press.)

Figure 15.17 Side view of wall-fired burner. (From Ref. 31. Courtesy of John Zink Co. LLC.)

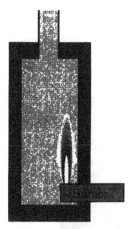

Figure 15.18 Schematic (elevation view) of a horizontally mounted, vertically-fired burner configuration. (From Ref. 31. Courtesy of John Zink Co. LLC.)

with gaseous fuels. So-called combination burners use both a liquid and a gaseous fuel, which are normally injected separately through each burner.

There are certain advantages to using cabin heaters compared to vertical cylindrical furnaces. Cabin heaters tend to have lower pressure drop through the tubes so less energy is required for pumping. They are more easily drainable, can fire from either the floor or the side, have all external fittings, and lower overall height. However, there are some disadvantages as well. The capital cost is slightly higher, they take up more space, the installation is longer and more costly, and the radiant heat flux may be less uniform.

There are also other types of control schemes used in process heaters besides natural draft. One type commonly used when the combustion system includes heat recovery is referred to as forced or mechanical draft where a blower is used to force combustion air through the system. There are some advantages to this type of control. One is that it is easy to modulate the blower to accurately account for changes in the system so that a more constant O_2 level can be maintained. Another

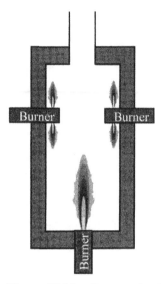

Figure 15.19 Cross-sectional elevation view of hearth burners and wall burners in a heater. (From Ref. 46. Courtesy of CRC Press.)

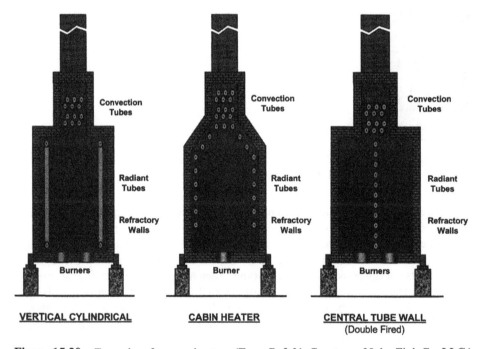

Figure 15.20 Examples of process heaters. (From Ref. 31. Courtesy of John Zink Co. LLC.)

advantage is that the system is not susceptible to environmental effects such as wind that can cause significant problems in natural-draft systems. The burners can usually be smaller because the blower can provide more air through a given cross-sectional area than can natural draft flow. The exhaust stack can also be shorter compared to

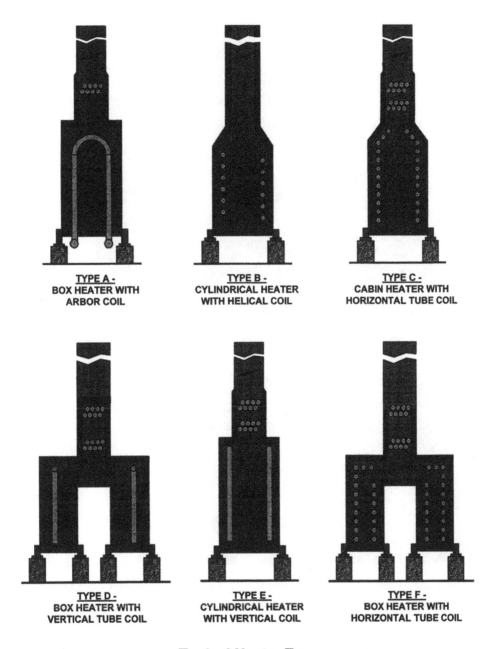

TYPE A -
BOX HEATER WITH
ARBOR COIL

TYPE B -
CYLINDRICAL HEATER
WITH HELICAL COIL

TYPE C -
CABIN HEATER WITH
HORIZONTAL TUBE COIL

TYPE D -
BOX HEATER WITH
VERTICAL TUBE COIL

TYPE E -
CYLINDRICAL HEATER
WITH VERTICAL COIL

TYPE F -
BOX HEATER WITH
HORIZONTAL TUBE COIL

Typical Heater Types

Figure 15.21   Typical heater types. (From Ref. 31. Courtesy of John Zink Co. LLC.)

natural draft where higher heights are required to achieve the necessary draft levels in the furnace. However, there are some disadvantages including higher capital costs for the blower and associated ductwork and possibly higher operating costs due to the electricity costs of the blower. These are normally more than offset by the fuel savings of using air preheating.

Figure 15.22 Cabin heater. (From Ref. 31. Courtesy of John Zink Company LLC.)

Figure 15.23 Crude unit burners. (From Ref. 31. Courtesy of John Zink Co. LLC.)

A related type of system is referred to as induced draft (ID) where an ID fan pulls the combustion gases through the heater. This is somewhere in between natural draft and forced draft operation. The ID fan pulls combustion air through the burners like natural draft operation, but more air can be pulled through because of the mechanical power of the fan, which is comparable to forced draft operation. An important advantage of the ID fan is the increased level of control compared to natural draft operation. Another advantage is that extensive ductwork is not required as is the case with forced draft operation where the combustion air from the blower must be ducted to all the burners. The ID fan also ensures that there will not be positive pressure inside the heater, which is a possibility with forced draft operation. However, the operating costs are higher than for natural draft and heat

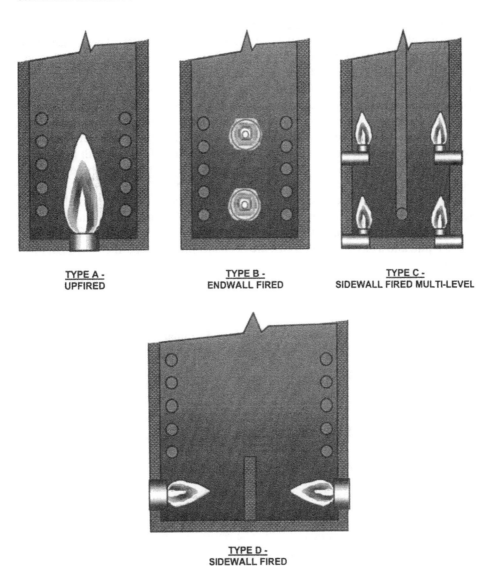

TYPE A -
UPFIRED

TYPE B -
ENDWALL FIRED

TYPE C -
SIDEWALL FIRED MULTI-LEVEL

TYPE D -
SIDEWALL FIRED

Typical Burner Arrangements (Elevation View)

Figure 15.24 Typical burner arrangements (elevation view). (From Ref. 31. Courtesy of John Zink Co. LLC.)

recovery by heating the incoming combustion air is not an option as it is with forced draft.

A hybrid control system is referred to as balanced draft, which uses both a combustion air blower and an ID fan. This is commonly used in systems with heat recovery for preheating the combustion air. The combustion air blower forces the incoming ambient combustion air through a heat exchanger that transfers energy from exiting warmer flue gases that are pulled through the exchanger by the ID fan.

Another unique aspect of these heaters is the wide range of fuels that are used, which are often by-products of the petroleum refining process. These fuels may contain significant amounts of hydrogen, which has a large impact on the burner design. It is also fairly common for multiple fuel compositions to be used, depending on the operating conditions of the plant at any given time. In addition to hydrocarbons ranging up to C_5, the gaseous fuels may also contain hydrogen and inerts (like CO_2 or N_2). The compositions can range from gases containing high levels of inerts to fuels containing high levels of H_2. The flame characteristics for fuels with high levels of inerts are very different from those for fuels with high levels of H_2. Add to that the requirement for turndown conditions, and it becomes very challenging to design burners that will maintain stability, low emissions, and the desired heat flux distribution over the range of conditions that are possible. Some plants use liquid fuels, like no. 2–no. 6 fuel oil, sometimes by themselves and sometimes in combination with gaseous fuels. So-called combination burners use both a liquid and a gaseous fuel, which are normally injected separately through each burner.

15.2.2 Reformers

As the name indicates, reformers are used to reformulate a material into another product. For example, a hydrogen reformer takes natural gas and reformulates it into hydrogen in a catalytic chemical process that involves a significant amount of heat. A sample set of reactions are given below for converting propane into hydrogen [49]:

$$C_3H_8 \rightarrow C_2H_4 + CH_4$$
$$C_2H_4 + 2H_2O \rightarrow 2CO + 4H_2$$
$$CH_4 + H_2O \rightarrow CO + 3H_2$$
$$CO + H_2O \rightarrow CO_2 + H_2$$

The reformer is a direct-fired combustor containing numerous tubes, filled with catalyst, inside the combustor [50]. The reformer is heated with burners, firing either vertically downward or upward, with the exhaust on the opposite end, depending on the specific design of the unit. Roof-mounted burners (see Fig. 12.14) are often used in ethylene production furnaces. The raw feed material flows through the catalyst in the tubes which, under the proper conditions, converts that material into the desired end product. The burners provide the heat needed for the highly endothermic chemical reactions. The fluid being reformulated typically flows through a reformer combustor containing many tubes (see Fig. 15.25). The side-fired reformer has multiple burners on the side of the furnace with a single row of tubes centrally located. The heat is transferred primarily by radiation from the hot refractory walls to the tubes. Top-fired reformers have multiple rows of tubes in the firebox. In that design, the heat is transferred primarily from radiation from the flame to the tubes. Figure 15.26 shows a downfired burner commonly used in top-fired reformers. In a design sometimes referred to as terrace firing, burners may be located in the side wall, but be firing up the wall, at a slight angle (see Fig. 15.27). Foster Wheeler uses terrace wall reformers in the production of hydrogen by steam

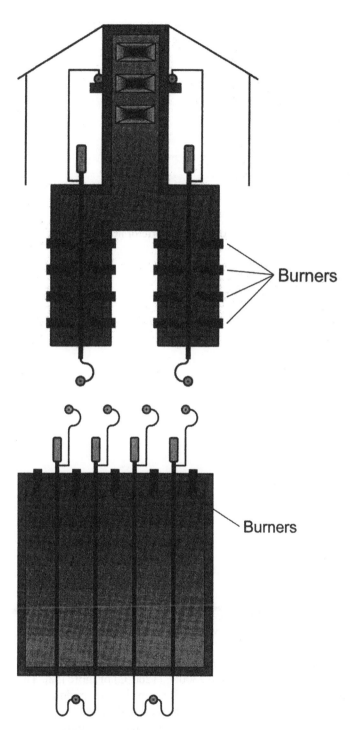

Figure 15.25 (Top) Side- and (bottom) top-fired reformers (elevation view). (From Ref. 31. Courtesy of John Zink Co. LLC.)

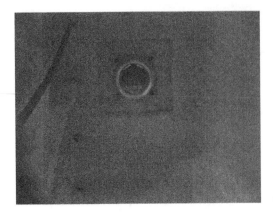

Figure 15.26 Downfired burner commonly used in top-fired reformers. (From Ref. 31. Courtesy of John Zink Co. LLC.)

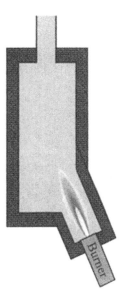

Figure 15.27 Elevation view of a terrace firing furnace. (From Ref. 31. Courtesy of John Zink Co. LLC.)

reformation of natural gas or light refinery gas [51]. Actual terrace wall process heaters typically have three to five terraces vertically with multiple burners on each terrace.

Ethylene cracking furnaces are among the more severe operating conditions for burners because of their higher temperature up to 2300°F (1260°C) or more [52]. The process tubes are usually suspended from top to bottom in the middle of a rectangular furnace volume in a single row. Burners are mounted on walls opposite the tubes where the heated walls radiate toward the tubes. Heat flux uniformity is especially important to minimize coke formation inside the tubes caused by hot spots from uneven heating. The burners typically fire against their adjacent walls for more uniform heat distribution so that the refractory can radiate to the tubes. These higher furnace temperatures cause higher NO_x emissions compared to other lower temperature applications such as refinery process heaters.

Funahashi et al. [53] discuss the use of selective catalytic reductions (SCRs) for reducing NO_x emissions from an ethylene cracking plant [53]. SCR was designed to reduce NO_x from 100 to 30 ppmvd (both at 6% O_2) from a furnace incorporating low NO_x burners and steam injection. The catalyst in SCR consisted of a honeycomb made of titanium and silicon (TiO_2–SiO_2) coated with vanadium oxide (V_2O_5) and tungsten oxide (WO_3). Ammonia was injected into the exhaust system near the catalyst where the temperature was in the range 290°–370°C (550°–700°F). The system successfully met the objective of reducing NO_x to below 30 ppmvd. Brundrett et al. [54] discussed the successful use of the Shell DeNO$_x$ process incorporating SCR in ethylene cracking furnaces. NO_x emissions were reduced by 90% and ammonia slip was less than 5 ppm. Suwa [55] describes the successful operation of SCR in an ethylene plant since 1985. Bussman et al. [56] describe the development of low-NO_x radiant wall burners for use in ethylene cracking furnaces that produce less than 20 ppm of NO_x using ultra-lean premix technology. The paper notes the historical progression of continually lower NO_x emissions from approximately 100 ppm in the 1970s to less than 20 ppm in the year 2000 (both corrected to 3% O_2). The lowest NO_x radiant wall burners used in reformers utilize ultra-lean premixing (see Chap. 6). These burners are natural draft and incorporate fuel staging. Cold-flow modeling and computational fluid dynamics [57] were utilized to optimize the design of this burner. Experimental results showed that NO_x increased with furnace temperature and decreased with fuel staging, as expected.

15.2.3 Flares

The API defines a flare as a "means to safely dispose of waste gases through the use of combustion" [58]. Flares (see Fig. 15.28) are used to combust unwanted hydrocarbon fuels, typically in the gaseous state. There are several conditions that may require flaring. The most common is in control of the process where gases or liquids are vented. Most refineries have a blowdown system used to flare unwanted gases and liquids, such as from process vents. When equipment is purged for maintenance or process modification, the purge gas containing hydrocarbon fluids is sent to the blowdown system. The blowdown gases are often combusted through a flare to ensure flammable gases are not vented into the atmosphere where they could later be ignited on contact with an ignition source. The U.S. EPA gives some guidelines for the use of flares as emission control devices [59]. One of the requirements is that flares should have no visible emissions such as smoke.

Another use of flares is in an upset condition where materials in the midst of processing need to be safely combusted to avoid a dangerous buildup and unsafe conditions during the restart of the process. Another common reason is excess by-product fuels that cannot be economically recovered during a transient condition such as a product change. Whatever the reason, flares must reliably combust fuels whenever they are called upon. One of the challenges for flares is maintaining a pilot flame to ignite the fuels, especially in very high wind conditions. Another challenge is an extremely wide turndown ratio, because of the wide variety of venting conditions. It is important to emphasize that flares are generally used as a safety device and,

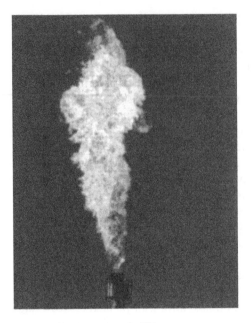

Figure 15.28 Steamizer™ steam-assisted smokeless flare. (From Ref. 63. Courtesy of John Zink Co. LLC.)

therefore, their design and operation are critical for a given plant. For this reason, they are often the first pieces of equipment started up for a plant and among the last to be shut down.

Flares are generally categorized by the location of the flare tip (ground or elevated) and by how the mixing is enhanced at the tip (e.g., steam-assisted, air-assisted, pressure-assisted, or nonassisted). The two types determined by elevation are: ground flares and elevated flares. Ground flares are designed to operate at or close to ground level. Some type of enclosure or screening is often required to shield the flaring from the surrounding area and personnel. Elevated flares operate with a flame at a height substantially above ground level with little or no attempt to hide the flame. Figure 15.29 shows a schematic of an elevated steam-assisted flare [78]. The flares are typically elevated to minimize the thermal radiation, smoke, and noise at ground level. The elevation also helps to disperse the products of combustion into the atmosphere so that they do not interfere with personnel, especially when strong winds and low fuel gas flow rates may deflect the flame from a flare towards equipment or personnel in the vicinity. Elevated flares may be literally hundreds of feet tall, depending on the design conditions. They are commonly used over ground flares because of their higher flow capacities [60]. Basic design guidelines for flares are given in API RP-521 [58].

The method of fuel-mixing enhancement is also used to categorize flares. In most cases, flares are designed to have diffusion flames [61], primarily for safety considerations to prevent flashbacks due to the wide range of fuel gas flow rates that may be flared. Therefore, as with most diffusion-style flames, mixing of the fuel and air is important for proper operation. The mixing in flares is commonly enhanced by the supplemental use of either steam or air. Steam injection increases the turbulence levels in the flame to increase mixing between the entrained ambient air and the

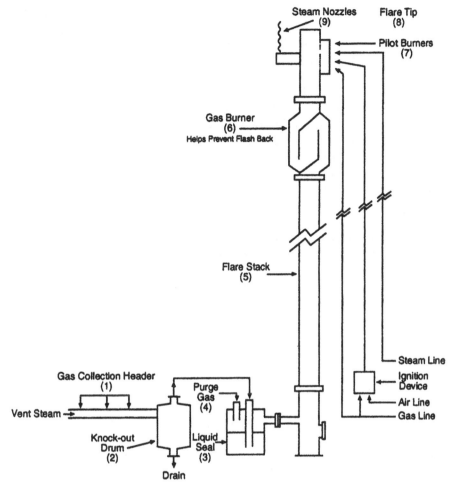

Figure 15.29 Schematic of an elevated steam-assisted flare. (From Ref. 78.)

injected fuel gas. In air-assisted flares, one or more blowers are used to force air through the flare to mix with the fuel gas being flared at the flare tip. The air not only increases mixing, but also enhances the combustion reaction because the air is supplied inside the flame, rather than relying on air entrainment from the ambient air surrounding the flare. Besides enhancing mixing, these fluids also help to increase the smokeless capacity of the flares so that more fuel can be combusted without generating smoke. The flare tip may also use the pressure of the fuel gas and the nozzle design to enhance mechanical mixing. This method can only be used when higher fuel gas pressures are available. Mechanical or pressure-assisted flares rely on the energy stored in the fuel gases in the form of pressure to improve the mixing of the fuel and ambient air, in conjunction with properly designed flare tip nozzles. Nonassisted flares may be used when the fuel gas being flared has a low carbon/hydrogen ratio, which burns readily without producing smoke (e.g., methane or natural gas).

The EPA prepared a report [62] on the flaring of very low heating value fuels containing hydrogen (12–22 vol%), inert gases (nitrogen, argon, carbon dioxide, and steam), oxygen (in some cases), and hazardous air pollutants (115 ppm to 5%) with heating values ranging from 59 to 120 Btu/scf. The heating values of these gases fell below the 40 CFR 60.18 and 40 CFR 63.11 guidelines. Supplemental hydrocarbon fuels would need to be added to the waste gases to bring them to the minimum heating values. The reported test results showed that the destruction efficiencies were all above the minimum of 98% as required by the regulations. Specifications were developed as a result of these tests for DuPont for maximum tip velocities for hydrogen-fueled flares. These specifications were limited to the conditions tested.

There are several environmental challenges concerning the use of flares. These include the radiation heat load and noise to the surrounding environment and the NO_x, CO, and particulate (smoke-producing) emissions to the atmosphere. Since NO_x and CO are rarely measured and reported in the literature at this time, radiation, noise, and smoke from flares will be considered here.

Schwartz et al. [63] provide a detailed discussion of all aspects of flares, including pollution emissions. Manning [64] describes a new type of mobile enclosed flare with high destruction efficiency for waste gases to minimize pollutant emissions, particularly SO_2, NO_x, and CO. Noise, thermal radiation, and smoke are also minimized. A key parameter that is controlled is the gas exit velocity. This is done through the use of a variable tip that adjusts according to the waste gas flow rate.

Flares are commonly of one of the following types or combinations thereof:

- Single point
- Multipoint
- Enclosed

Single-point flares are generally oriented upward with the discharge point at an elevated position relative to the surrounding equipment (see Fig. 15.30). Multipoint flares are designed to improve burning by distributing fuel gases to multiple burning locations that may be at or near grade level (see Fig. 15.31) or elevated

Figure 15.30 Typical elevated single-point flare. (From Ref. 63. Courtesy of John Zink Co. LLC.)

Figure 15.31 Typical grade-mounted multipoint flare system. (From Ref. 63. Courtesy of John Zink Co. LLC.)

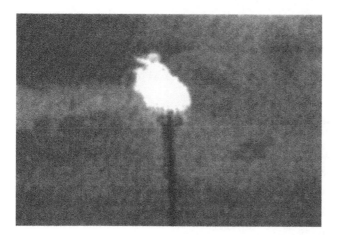

Figure 15.32 Typical elevated multipoint flare system. (From Ref. 63. Courtesy of John Zink Co. LLC.)

(see Fig. 15.32). Enclosed flares (see Fig. 15.33) are designed to conceal the flames from direct view, which can reduce noise and radiation levels to the surrounding area. A combination of techniques may be used, for example, when an enclosed flare is designed to handle most of the gases to be flared while an elevated flare may be added to handle additional gases that may be present only on rare occasions such as a major upset. Figure 15.34 shows an example of such a system consisting of an enclosed flare and an elevated single-point flare.

15.3 EMISSIONS

The Western States Petroleum Association (WSPA) and the American Petroleum Institute (API) worked with the California Air Resources Board (CARB) to develop air toxics emission factors for the petroleum industry [65]. Source data were provided

Figure 15.33 Typical installation of multiple enclosed flares. (From Ref. 63. Courtesy of John Zink Co. LLC.)

Figure 15.34 Combination of enclosed and elevated single-point flares. (From Ref. 63. Courtesy of John Zink Co. LLC.)

Table 15.7 Air Emissions from Typical Refinery Processes

Process	Air emissions (from combustion)
Crude oil desalting	Heater stack gas (CO, SO_x, NO_x, hydrocarbons, and particulates)
Atomspheric distillation	Heater stack gas (CO, SO_x, NO_x, hydrocarbons, and particulates)
Vacuum distillation	Heater stack gas (CO, SO_x, NO_x, hydrocarbons, and particulates)
Thermal cracking/ visbreaking	Heater stack gas (CO, SO_x, NO_x, hydrocarbons, and particulates)
Coking	Heater stack gas (CO, SO_x, NO_x, hydrocarbons, and particulates)
Catalytic cracking	Heater stack gas (CO, SO_x, NO_x, hydrocarbons, and particulates)
Catalytic hydrocracking	Heater stack gas (CO, SO_x, NO_x, hydrocarbons, and particulates)
Hydrotreating/ hydroproces	Heater stack gas (CO, SO_x, NO_x, hydrocarbons, and particulates)
Alkylation	Heater stack gas (CO, SO_x, NO_x, hydrocarbons, and particulates)
Isomerization	Heater stack gas (CO, SO_x, NO_x, hydrocarbons, and particulates)
Catalytic reforming	Heater stack gas (CO, SO_x, NO_x, hydrocarbons, and particulates)
Propane deasphalting	Heater stack gas (CO, SO_x, NO_x, hydrocarbons, and particulates)
Gas treatment and sulfur recovery	SO_x, NO_x, and H_2S from vent and tail gas emissions
Blowdown and flare	Combustion products (CO, SO_x, NO_x, and hydrocarbons) from flares

Source: Ref. 2.

in 18 groups. Some of those groups of relevance here include both refinery gas-fired and fuel oil-fired boilers and heaters fired on natural gas, refinery gas, oil, and a combination of natural gas and refinery oil. The U.S. Environmental Protection Agency has compiled an extensive list of emission factors for a wide range of industrial processes [66]. Chapter 1 of AP-42 concerns external combustion sources and focuses on the fuel type. Sections 1.3, 1.4, and 1.5 of AP-42 are on fuel oil combustion, natural gas combustion, and liquefied petroleum gas combustion, respectively. Chapter 5 of AP-42 is on the petroleum industry, where the reader is referred to Sections 1.3 and 1.4 for boilers and process heaters using fuel oil and natural gas, respectively. Chapter 6 of AP-42 is on the organic chemical process industry. Reis [67] has written a general book on environmental issues in petroleum engineering, including drilling and production operations.

Typical pollutants from refinery processes include NO_x, SO_x, CO, particulates, and unburned hydrocarbons (see Table 15.7) [2]. Figure 15.35 shows the dramatic reduction in pollutant emissions from U.S. refineries since 1970 [68]. The refinery processes include crude oil desalting, atmospheric distillation, vacuum distillation, thermal cracking/visbreaking, coking, catalytic cracking, catalytic hydrocracking, hydrotreating/hydroprocessing, alkylation, isomerization, catalytic reforming, propane deasphalting, gas treatment and sulfur recovery, and blowdown and flaring.

15.3.1 NO_x Emissions

Table 15.8 shows the uncontrolled NO_x emissions from typical process heaters. API 535 notes several parameters that impact NO_x emissions for burners in refinery

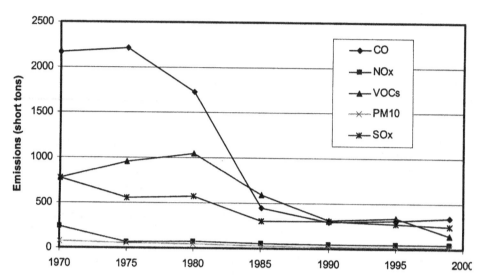

Figure 15.35 Pollution emissions from refineries and related processes in the United States since 1970. (From Ref. 68.)

Table 15.8 Uncontrolled NO_x Emission Factors for Typical Process Heaters

	Uncontrolled emission factor $(lb/10^6 Btu)$		
Model heater type	Thermal NO_x	Fuel NO_x	Total NO_x[a]
ND, natural gas fired[b]	0.098	N/A	0.098
MD, natural gas fired[b]	0.197	N/A	0.197
ND, distillate oil fired	0.140	0.060	0.200
ND, residual oil fired	0.140	0.280	0.420
MD, distillate oil fired	0.260	0.060	0.320
ND, residual oil fired	0.260	0.280	0.540
ND, pyrolysis, natural gas fired	0.104	N/A	0.104
ND, pyrolysis, high-hydrogen fuel gas fired[c]	0.140[d]	N/A	0.140

N/A = not applicable.
ND = natural draft.
MD = mechanical draft.
[a]Total NO_x = thermal NO_x + fuel NO_x.
[b]Heaters firing refinery fuel gas with up to 50 mol(%) hydrogen can have up to 20% higher NO_x emissions than similar heater firing gas.
[c]High-hydrogen fuel gas is fuel gas with 50 mol% or greater hydrogen content.
[d]Calculated assuming approximately 50 mol% hydrogen.
Source: Ref. 42.

service (see Chap. 6). NO_x increases with excess air in the combustion system. As an example, baseline emissions at 1% O_2 increase by a factor of approximately 1.8 at 8% O_2. NO_x increases with the combustion air temperature when forced-draft air preheat is used. As an example, baseline emissions at 125°F increase by a factor of two at air preheat levels of 650°F. NO_x increases with the furnace temperature. As an example, baseline emissions at 1300°F increase by more than a factor of two at a furnace temperature of 1900°F. NO_x also increases with the hydrogen content in the fuel [69]. As an example, baseline emissions at 0 vol% in the fuel increases by more than a factor of 1.5 at hydrogen contents above about 85% by volume. Fuel gases typically produce less NO_x than fuel oils.

Lin et al. [70] describe the application of selective noncatalytic reduction (SNCR) using urea in the hydrocarbon processing industry. They reported moderate to high NO_x reduction (58–70%) in a refinery process heater with minimum NH_3 slip under a wide range of temperature and load conditions.

McAdams et al. [71] have written a tutorial on evaluating NO_x control technologies, including cost effectiveness. They note that there is no single solution that fits every situation as the performance and costs of a given technology are dependent on the specific operating conditions and cost factors at the given plant. The following NO_x reduction technologies were considered in the tutorial: low NO_x burners, flue gas recirculation, low NO_x burners with flue gas recirculation, conventional burners with fuel dilution, selective catalytic reduction (SCR) with conventional burners, SCR with low NO_x burners, and advanced low NO_x burners. In some cases, the options may be limited by the regulations as only certain technologies are capable of achieving lower levels of NO_x. Easily retrofittable technologies like low NO_x burners are typically less expensive to purchase, install, and operate compared to other technologies like SCR, which requires significant initial, installation, and operating costs. Seebold et al. [72] discuss retrofitting lean premix burners (see Chap. 12) in an operating crude vacuum furnace where NO_x was reduced from 180 to 20 ppm. This eliminated the need for installing SCR. Bradford et al. [73] list some likely NO_x control strategies for fired heaters in the near future.

Baukal et al. [74] discuss the range of choices available to the petrochemical and refining industries for controlling NO_x emissions. These industries face a number of problems including increasing demand for their products, but with low margins and increasingly stringent environmental regulations. Choosing the proper technology for a given application needs to include capital and operating costs, required downtime, effects on operation and maintenance, safety, and the effects on the process (e.g., heater capacity and efficiency, tube fouling, or coking). While post-treatment control strategies can produce low NO_x emissions, the initial and ongoing costs can be significantly higher than for combustion modification techniques.

Laplante and Lindenhoff [75] discuss the use of SCRs in refineries to meet the increasingly more stringent NO_x regulations. Reductions of up to 95% are possible. According to the authors, this may be the only currently available technology to achieve single-digit NO_x performance. Examples of typical applications for this technology in refineries include hydrogen steam reformers, residual oil-fired boilers, cogeneration, and fluid catalytic crackers.

There are a number of challenges when choosing a NO_x control technology for process heaters. The fuel composition can vary widely as different fuels may be used during startup, one or more during normal operations depending on the product being made, and another during emergency conditions. The fuels are often by-products from a refining process and may contain hydrocarbons ranging from C_1 to C_4, hydrogen, and inert gases like N_2 and CO_2. Heaters are often overfired because of processing heavier feedstocks requiring more energy. Weather conditions can impact the operation of the natural-draft burners typically used in these heaters.

Table 15.9 Likely NO$_x$ Control Options in the Near Future for Fired Heaters

Unit	Options
All	• Replace boilers with cogeneration systems • Replace inefficient heaters with new more efficient heaters or with hot-oil systems • Provide new or upgrade existing convection sections to improve energy efficiency • Upgrade heater controls to reduce excess air requirements • Seal heater to minimize heat losses
Fired heaters (<40 MMBtu/h)	• First or second generation ULNBs • Possible dilution of fuel gas with steam or flue gas recirculation, for systems with high firebox temperatures or air preheat
Fired Heater (40–100 MMBtu/h)	• Latest generation of ULNB or SCR • ULNBs will require firebox sealing and may require the installation of tighter controls • SCR may not be economical for some cases if the area is congested or if tie-ins are expensive
Fired heaters (>100 MMBtu/h)	• Latest generation of ULNBs or SCR • Downfired reformers will require SCRs • ULNBs will require firebox sealing and will require the installation of tighter controls

Adapted from Ref. 73.
ULNB = ultra low NO$_x$ burner, SCR = selective catalytic reduction.

Refineries usually have very little available space to add large post-treatment systems like SCR systems. Because of the competitive nature of the industry, long downtimes cannot be tolerated so it is desirable to have control technologies that can be easily retrofitted, preferably without having to shut down the equipment for installation. Table 15.9 lists some of the likely NO$_x$ control options in the near future for chemical and petrochemical applications.

15.3.2 SO$_x$ Emissions

Nearly all of the sulfur in a fuel is converted into SO$_2$ in the exhaust gas, with a small amount of SO$_3$ generated. API 535 notes that between 94 and 98% of SO$_x$ emissions are SO$_2$ with the balance being SO$_3$ [16]. It is noted that low excess air operation reduces the conversion from SO$_2$ into SO$_3$. Liquid fuel oils like no. 6 oil that are used in fired heaters commonly contain some level of sulfur. Crane et al. [76] describe a method for estimating SO$_3$ emissions from fired heaters. The rate of conversion from SO$_2$ into SO$_3$ depends on the fuel sulfur content, excess oxygen in the combustion process, flue gas temperature, the catalytic effect of trace metals in the process, and the type of combustion unit. For example, the conversion rate of SO$_2$ into SO$_3$ decreases as the sulfur content in the fuel increases. Typical conversion rates are typically less than 3% based on EPA conversion tables for liquid fuels. Conversion rates may be higher for gaseous fuels that have relatively low sulfur contents, sometimes in the form of H$_2$S.

Table 15.10 Emission Factors for Flare Operations

Component	Emission factor (lb/10^6 Btu)
Total hydrocarbons[a]	0.14
Carbon monoxide	0.37
Nitrogen oxides	0.068
Soot[b]	0–274

Based on tests using crude propylene containing 80% propylene and 20% propane.
[a]Measured as methane equivalent.
[b]Soot in concentration values: nonsmoking flares, 0 micrograms per liter (µg/L); lightly smoking flares, 40 µg/L; average smoking, 177 µg/L; and heavily smoking flares, 274 µg/L.
Source: Ref. 77.

15.3.3 Other Pollutants

API 535 discusses CO emissions from refinery burners [16]. The point where the CO concentration in the exhaust gases approaches an asymptote is referred to as the CO breakpoint or CO breakthrough. Typical CO control ranges from 150 to 200 ppmv for best overall heater efficiency. The range could be lower, depending on local regulations. Fuel oils are more likely than fuel gases to produce unburned combustibles. Heavy fuel oils are likely to produce particulate emissions. Reducing excess air levels increases CO, combustibles, and particulates emissions. Increasing swirl tends to reduce CO and particulates emissions.

15.3.4 Flare Pollutants

There are some potential pollutants that may be generated during the flaring process. These include thermal-radiation heat loading of the surrounding area, noise caused by both the large gas flow rates through the flare and by the combustion reactions during flaring, and smoke and odor generated by the incomplete combustion of the flammables in the vent stream. The U.S. EPA has developed emission factors for industrial flares [77]. Besides noise and heat, flares also emit carbon particles (soot), CO and other unburned hydrocarbons, NO_x, SO_x (if the fuels contain any sulfur), and large quantities of visible light that can be especially noticeable at night. Flares do not typically lend themselves to conventional emission testing because of their configuration and open flames. Therefore, relatively little information is available on pollution emissions from flares. Some emission factors from the EPA for industrial flares are presented in Table 15.10.

15.3.4.1 Flare Radiation

Radiation is a primary concern when using flares to burn large quantities of hydrocarbon fuels. Guigard et al. [78] have written an extensive report on thermal radiation from flares. They note the importance of flare radiation in the design of flare stacks and siting of flares. The report contains an extensive literature search on the subject as well as a brief summary of important papers and articles. The report compares and summarizes the various techniques and methods that have been proposed for quantifying flare radiation. Instrumentation guidelines are presented for measuring flame radiation, gas temperature, gas exit velocity, fuel flow rate, gas composition, flare flame size, and ambient conditions.

Figure 15.36 shows the flame radiation from an offshore flare. Evans notes that one of the key concerns related to flares is the radiation heat load on both humans and equipment, which dictates the height and location of the flare [34]. Figure 15.37 shows a temperature profile, referred to as a thermogram, of the flame produced by a flare. As expected, the highest temperatures are close to the exit of the

Figure 15.36 Radiation from an offshore flare. (From Ref. 63. Courtesy of John Zink Co. LLC.)

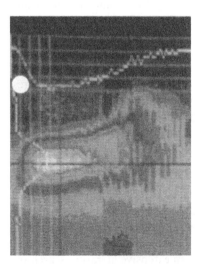

Figure 15.37 Thermogram of a flare flame. (From Ref. 63. Courtesy of John Zink Co. LLC.)

flare. Operating personnel need protection for radiant heat intensities exceeding 1500 Btu/hr-ft^2. Figure 15.38 shows that only a very short exposure time is needed before the threshold of pain is reached, depending on how high the thermal radiation levels are [79]. Equipment needs protection for radiant heat intensities exceeding 3000 Btu/hr-ft^2. Water injection has been shown to reduce flare radiation, which is particularly important in offshore flares where there are practical limits as to how far the flare can be located from personnel and other equipment [80]. Figure 15.39 shows an example of water being used to suppress thermal radiation from a flare.

A variety of factors affect the radiation of a flare. These include the fuel gas composition, flame type (especially the geometry), fuel/air mixing, soot and smoke

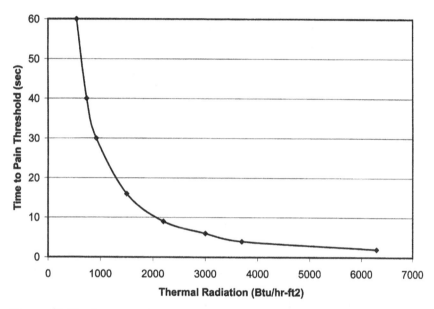

Figure 15.38 Exposure times necessary to reach the threshold of pain as a function of the thermal radiation level from a flare. (Adapted from Ref. 79.)

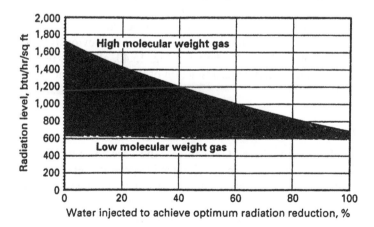

Figure 15.39 Radiation suppression using water injection. (From Ref. 80. Courtesy of PennWell.)

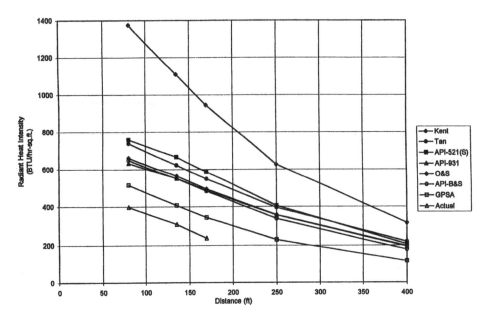

Figure 15.40 Predicted radiation intensity as a function of the radial distance from the base of a 150-ft flare stack. (From Ref. 82. Courtesy of John Zink Co. LLC.)

formation, fuel flow rate, flame temperature, and flare burner design [81]. Another important factor is atmospheric absorption of the radiation before it reaches a receiver such as personnel and equipment. A complicating environmental factor is the wind, which can deflect flare flames toward personnel and equipment. A variety of methods have been used to calculate the radiation from flares with varying degrees of accuracy. These methods are used to help determine the required height of the flare.

Schwartz and White [82] have done an extensive review of flare radiation prediction methods. Seven different methods were compared. Using an example flare problem, the authors showed that there is a wide variation in the predicted radiation, depending on the model. This significantly impacts the required flare stack height in order to achieve a given radiation level at grade level. The calculated stack heights for the example problem ranged from 190 to 660 ft (58–200 m). This obviously has a significant impact on the cost of the flare. As shown in Fig. 15.40, all of the models overestimated the actual measured radiation as a function of the radial distance from the base of a 150-ft tall flare stack.

15.3.4.2 Flare Noise

There are several sources of noise from flares. One source is the large flow rate of gases going through the flare tip. Sometimes flare tips are designed to produce very high fuel gas velocities, including up to sonic, in order to enhance turbulent mixing with air to achieve complete combustion. Another source of noise may be steam injection through the flare nozzle at very high velocities, again to enhance mixing with the fuel being flared. The steam is used to suppress smoke formation (discussed in the next section), but can be a significant source of noise. A third source of noise is the

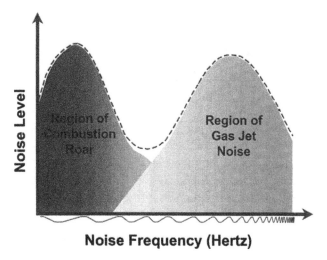

Figure 15.41 Flare noise characteristics. (From Ref. 83. Courtesy of John Zink Co. LLC.)

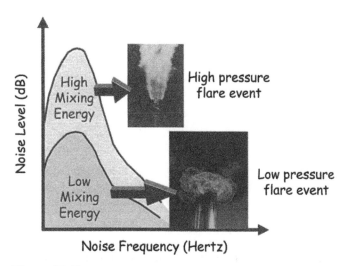

Figure 15.42 Comparison of noise signatures from high- and low-pressure flares. (Courtesy of John Zink Co. LLC.)

combustion process itself (see Chap. 10). Figure 15.41 shows two types of noise emitted from flares [83]. Figure 15.42 shows a relative comparison of noise emitted from high- and low-pressure flares. The high-pressure flare produces significantly more noise compared to the lower energy low-pressure flare. Other sources of flare noise that have received less attention include moisture condensation shock, seal drum sloshing, and low-flow instability [84]. Insulated grade-level flares have been found to produce substantially less noise than that from uninsulated elevated flares.

In addition to radiation suppression, water injection has also been shown to reduce flare noise [80]. Water injection reduced flare noise on a Caspian Sea platform by 10 dBA. Figure 15.43 shows an example of water being used for noise suppression on a flare. Another method used to reduce noise is by the use of mufflers on the flare.

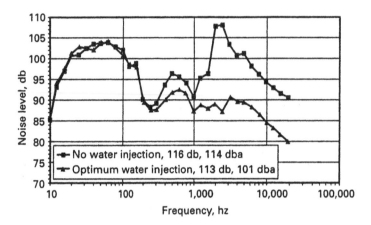

Figure 15.43 Noise suppression using water injection. (From Ref. 80. Courtesy of PennWell.)

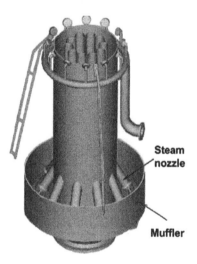

Figure 15.44 Muffler on a flare for noise suppression. (From Ref. 83. Courtesy of John Zink Co. LLC.)

Figure 15.44 shows a muffler on a flare for noise reduction. The design of the flare tip is critical to the amount of noise produced during flaring. Advances have been made in controlling the combustion-generated noise while maintaining proper mixing and combustion characteristics to prevent smoke formation.

15.3.4.3   Flare Smoke

Smoke is typically formed in industrial combustion processes by incomplete combustion of the hydrocarbon fuel, which generates carbon particles to form visible smoke. The incomplete combustion can result from several factors. An inadequate supply of air to combust fully the fuel is one source of incomplete combustion. This may result from too much fuel flow by exceeding the design fuel flow rates in the flare or it may result from an inadequate supply of air in air-assisted flares. This may also result from improper design of the flare tip properly to entrain enough air into the combustion region. Another source of incomplete combustion is improper mixing when there is enough air present to combust the fuel fully. This is generally the primary cause of smoke formation from flares [61]. This also is a function of the flare tip design to ensure adequate mixing and turbulence to combust the fuel. The number of nozzles in the flare tip, the flow characteristics such as swirl, the gas velocity, and the geometry of the tip all play important parts in ensuring adequate mixing. A third factor that can cause smoke formation is too low a temperature. While this is not usually a problem in typical industrial combustion processes, flames that are quenched quickly before fully combusting the fuel can form smoke particles. In flares, this could be caused by cold ambient temperatures, high winds, snow, rain, and other weather-related conditions that are not normally issues in typical industrial combustion applications using burners inside a combustor.

Visible smoke is undesirable in nearly all flare systems. A key specification is often the smokeless capacity, which is the maximum flow rate of waste gases that can be flared without producing visible smoke. The fuel composition plays an important role in designing a flare to minimize smoke formation. Figure 15.45 shows high levels of smoke produced by a flare designed in the 1950s, which is compared to the flare in Fig. 15.46 that produces no significant amount of smoke. The fuel gas being flared has an important impact on smoke formation. Another important factor is the state of the fuel, especially whether it is a liquid or a gas. Flares that have been designed to operate using gases will usually produce smoke if there is a significant amount of liquid hydrocarbons in the fuel [85]. This liquid is often eliminated by the use of liquid knock-out drums that separate the gas from the liquid so that only the gas goes to the flare. In addition, flaring liquids can generate a spray of burning chemicals that could reach ground level and cause a safety hazard.

A series of three photographs of the same flare shows the importance of fuel gas composition on smoke emissions. The gas being flared in Fig. 15.47 is natural gas where essentially no visible smoke is produced. Figure 15.48 shows the same flare with propane as the fuel and visible smoke can be seen, while that being flared

Figure 15.45 Typical flare performance in the 1950s. (From Ref. 63. Courtesy of John Zink Co. LLC.)

Figure 15.46   Major flaring event producing no significant smoke. (From Ref. 63. Courtesy of John Zink Co. LLC.)

Figure 15.47   Flaring of natural gas. (From Ref. 63. Courtesy of John Zink Co. LLC.)

in Fig. 15.49 is propylene, both at comparable firing rate. Propylene has a higher carbon-to-hydrogen weight ratio compared to propane and, therefore, a greater tendency to smoke.

One technique used to mitigate smoke formation is through the use of steam. Figure 15.50 shows how effectively steam is at minimizing smoke formation, which is very evident without the use of steam. Steam assisted are the predominant type of

Figure 15.48 Flaring of propane. (From Ref. 63. Courtesy of John Zink Co. LLC.)

Figure 15.49 Flaring of propylene. (From Ref. 63. Courtesy of John Zink Co. LLC.)

flare used in industry. Figure 15.51 shows the nonlinear relationship for the steam flow rate required for smokeless operation as a function of the molecular weight of the gaseous fuel being flared [58].

Similarly, air supplied through some type of blower system (see Fig. 15.52 [86]) is also effective for minimizing smoke formation, as shown in Fig. 15.53. Chaudhuri and Diefenderfer [87] discuss the choice between using air or steam to reduce smoke emissions. The majority of flares in industry use steam for smoke suppression. They have been in use for more than 50 years and have a long track record of successful operation. Steam flares can also be more easily expanded to meet increased smokeless capacity fuel flow rates. The cost of the steam can be an issue unless the plant already has a surplus where the cost is then minimal. Installation of steam

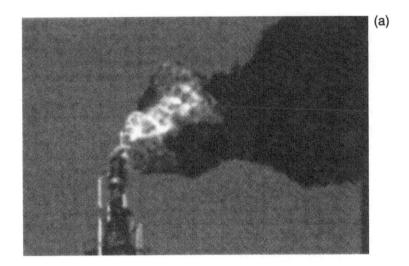

Figure 15.50 Flare (a) without and (b) with steam used for smoke suppression. (From Ref. 63. Courtesy of John Zink Co. LLC.)

flares is generally more costly than for air flares, which do not need steam piping and all the associated equipment to bring steam to the flare. The air flare has the blower mounted on or close to the flare, unlike the steam which normally must be brought to the flare from some distance away. An air flare may be preferable if there is not a cheap source of steam close to the flare. A potential problem of air flares is the loss of power, whereas a steam flare normally has enough residual steam in the system to run for some time after a power failure. Liquid water, sometimes seawater on an offshore platform, can be used to reduce smoke formation as shown in Fig. 15.54. As examples, API RP-521 suggests the following steam flow rates (lb/hr) to promote smokeless burning: 0.10–0.15, 0.25–0.30, 0.30–0.35, and 0.40–0.45 for ethane, propane, butane, and pentane plus, respectively [58]. The following empirical

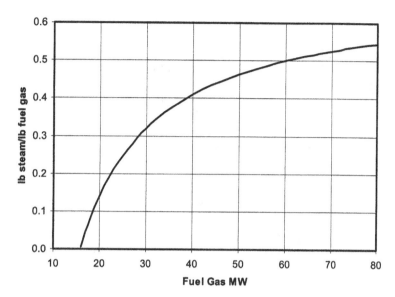

Figure 15.51 Steam flow required for smokeless operation as a function of the molecular weight (MW) of the fuel being flared. (From Ref. 58.)

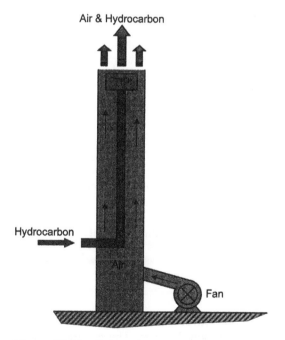

Figure 15.52 Schematic of an air-assisted flare. (From Ref. 86. Courtesy of John Zink Co. LLC.)

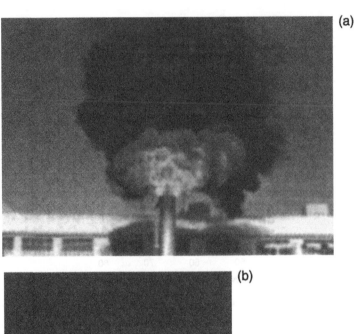

(a)

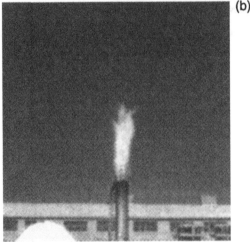

(b)

Figure 15.53   Flare (a) without and (b) with air supplied for smoke suppression. (From Ref. 63. Courtesy of John Zink Co. LLC.)

formula is suggested for determining the minimum steam flow rate to promote smokeless combustion:

$$W_{\text{steam}} = W_{\text{HC}}\left(0.68 - \frac{10.8}{M_{\text{HC}}}\right) \tag{15.1}$$

where W_{steam} = steam flow rate (lb/hr), W_{HC} = hydrocarbon fuel gas flow rate (lb/hr), and M_{HC} = molecular weight of the hydrocarbon fuel gas. The equation can be rewritten to show the ratio of steam to hydrocarbon rates:

$$\frac{W_{\text{steam}}}{W_{\text{HC}}} = 0.68 - \frac{10.8}{M_{\text{HC}}} \tag{15.2}$$

A graph of this equation is shown in Fig. 15.51.

Figure 15.54 Water-assisted flare on an offshore platform. (From Ref. 63. Courtesy of John Zink Co. LLC.)

Example 15.1

Given: Mixture of 50% methane and 50% propane by volume with a total
            flow rate of 100,000 lb/hr.
Find   Minimum steam flow rate for smokeless operation.
Solution: $W_{HC} = 100,000$ lb/hr
            $M_{CH4} = 16$, $M_{C3H8} = 44$, $M_{HC} = 0.5 M_{CH4} + 0.5 M_{C3H8} = 0.5(16) + 0.5(44) = 30$

$$W_{steam} = W_{HC}\left(0.68 - \frac{10.8}{M_{HC}}\right) = 100000\left(0.68 - \frac{10.8}{30}\right) = 32,000 \text{ lb/hr}$$

$W_{steam}/W_{HC} = 32,000/100,000 = 0.32$ lb steam/lb fuel gas

15.4 REGULATIONS

The Environmental Protection Agency (EPA) regulates emissions in the hydrocarbon and chemical processing industries (HPI and CPI, respectively) nation wide. For petroleum refineries, 40 CFR 60 (Subpart J) applies [88]. Guidelines for VOC emissions from synthetic organic chemical manufacturing for air oxidation unit processes (e.g., acetone and styrene), for distillation processes (e.g., butane, ethylene, and methanol), and for reactor processes (e.g., benzene, ethanol, and toluene) are given in Subparts III [89], NNN [90], and RRR [91], respectively. The VOC emission guidelines for the polymer manufacturing industry are given in Subpart DDD [92]. Some emission factors for various types of process heaters and flares are listed in Tables 15.3 and 15.10, respectively.

At the state level, additional agencies are free to adopt more stringent regulations. Examples are the California Air Resources Board (CARB) and the Texas National Resource Conservation Commission (TNRCC). TNRCC is proposing a very strict sub-10-ppm burner. For reference, traditional burners generate $\sim$100 ppm NO_x when firing gaseous fuels. Deason [93] notes that in the Houston–Galvaston (Texas) area (HGA), very strict regulations are being discussed for allowable emissions of VOCs and NO_x to reduce ground-level ozone levels dramatically. These much stricter limits would be imposed in 2007 and would likely require the addition of SCRs to many processes to achieve to these low NO_x levels.

Some states have even more local agencies such as California's SCAQMD regulating the greater Los Angeles area, or the Bay Area Air Quality Management District (BAAQMD) regulating the greater San Francisco area. Additionally, there are various voluntary standards recommended by various institutes. The general trend is towards more stringent regulation. The great number of governing bodies shows the general public support for stricter pollution control at all levels of government.

Fired process heaters are a significant source of noise emissions. American Petroleum Institute Recommended Practice 531 M gives information on measuring noise from process heaters [94]. This is used to ensure compliance with OSHA guidelines for maximum noise exposure limits for workers, depending on the length of the exposure (see Chap. 10). Suggested data sheets are given for reporting noise emissions.

15.5   EMISSION CONTROL STRATEGIES

Table 15.11 lists some common techniques for reducing NO_x in process heaters that include a variety of the strategies discussed next [95]. Table 15.12 provides a more detailed list of low-NO_x technologies for process heaters based on fuel and draft types.

15.5.1   Pretreatment

API 535 discusses low NO_x burners used in general refinery services [16]. Air staging, fuel staging, internal flue gas recirculation, and external flue gas recirculation are some common techniques used to reduce NO_x. API 560 notes that, while air preheat may increase the emissions of NO_x and possibly unburned combustibles, it may reduce particulate emissions [17]. More importantly, air preheating increases energy efficiency, which means that less fuel is combusted for a given unit of production (e.g., barrel of oil processed) and, therefore, all emissions are reduced per unit of production when less fuel is consumed. Noise emissions are higher with forced-draft systems that require fans and blowers so one option that may not be very attractive is to change from forced-draft or natural-draft systems.

Table 15.11 NO$_x$ Control Technologies for Process Heaters

Control technology	Controlled emissions	Percentage reduction
Low-NO$_x$ burners	0.1–0.3 lb/MMBtu	25–65
Staged air lances	Not available	35–51
Fiber burner	10–20 ppm	
Ammonia injection	Not available	43–70
Urea injection + low NO$_x$ burner	Not available	55–70
Selective catalytic reduction	20–40 ppm	65–90
Selective catalytic reduction + low-NO$_x$ burner	25–40 ppm	70–90

Note: uncontrolled emissions are in the range 0.1–0.53 lb/MMBtu.
Source: Ref. 95. (Courtesy of Gas Technology Institute.)

Anderson [96] notes the importance of damper control to minimize NO$_x$ emissions from fired process heaters. The dampers help ensure the proper draft and O$_2$ levels in the furnace. If the draft conditions are not optimized then tramp air may be pulled into the heater if the draft is too high or not enough air may be pulled in through the burner if the draft is too low. If the O$_2$ levels are too high, then NO$_x$ emissions often increase. If the O$_2$ levels in the exhaust gases are too low, then NO$_x$ emissions may decrease but CO emissions may dramatically increase. Too much excess O$_2$ in the furnace also reduces energy efficiency, which indirectly increases all pollutant emissions as more fuel must be burned for a given production rate. Because of constantly changing conditions such as production rates, fuel gas composition, and ambient conditions, dampers need to be constantly adjusted to maintain a given set of operating parameters. However, dampers are often manually controlled and set at a given point with little attention thereafter so that the heaters are often operating in suboptimal conditions.

Stansifer [97] notes the importance of the condition of the fuel on NO$_x$ emissions. In refineries and chemical plants, the fuel used in the burners often contains by-products from the chemical manufacturing process. In addition to fuels like methane, propane, and hydrogen, these fuels may also contain moisture and heavy hydrocarbons (sometimes tar-like) that can cause carbon buildup on the burner injector nozzles. This buildup is referred to as coking and can significantly degrade the performance of the burners, including increasing NO$_x$ emissions, as nozzles get plugged off and flame patterns become irregular. Fuel gas treatment systems (e.g., filters, coalescers, separators) upstream of the burners to remove moisture and heavy hydrocarbons can greatly improve overall system performance and reduce burner maintenance by reducing or eliminating coking of the burner nozzles. Clean fuels are especially important with newer generations of low-NO$_x$ burners that are more sensitive to system changes by virtue of their designs that are needed to achieve low NO$_x$.

Another important consideration for the pollution emission performance of a given burner design is how well the burner is maintained so that it operates as designed. McAdams et al. [52] note the importance of burner maintenance for maintaining the emission performance of low-NO$_x$ burners. Some potential problems include fuel tip orifice plugging, tip coking, fuel line flow restrictions, air inlet blockage, air leakage into the furnace, tip overheating, and refractory damage.

Table 15.12 Reduction Efficiencies for NO_x Control Techniques

Draft and fuel type	Control technique	Total effective NO_x reduction (%)
ND, distillate	(ND) LNB	40
	(MD) LNB	43
	(ND) ULNB	76
	(MD) ULNB	74
	SNCR[a]	60
	(MD) SCR	75
	(MD) LNB + FGR	43
	(ND) LNB + SNCR	76
	(MD) LNB + SNCR	77
	(MD) LNB + SCR	86
ND, residual	(ND) LNB	27
	(MD) LNB	33
	(ND) ULNB	77
	(MD) ULNB	73
	SNCR	60
	(MD) SCR	75
	(MD) LNB + FGR	28
	(ND) LNB + SNCR	71
	(MD) LNB + SNCR	73
	(MD) LNB + SCR	83
MD, distillate	(MD) LNB	45
	(MD) ULNB	74
	(MD) SNCR	60
	(MD) SCR	75
	(MD) LNB + FGR	48
	(MD) LNB + SNCR	78
	(MD) LNB + SCR	92
MD, residual	(MD) LNB	37
	(MD) ULNB	73
	(MD) SNCR	60
	(MD) SCR	75
	(MD) LNB + FGR	34
	(MD) LNB + SNCR	75
	(MD) LNB + SCR	91

MD = mechanical draft, ND = natural draft, LNB = low NO_x burner,
ULNB = ultra-low NO_x burner, SNCR = Selective noncatalytic reduction,
SCR = Selective catalytic reduction, FGR = Flue gas recirculation.
[a]Reduction efficiencies for ND or MD SNCR are equal.
Source: Ref. 42.

These problems can be mitigated by proper burner specification, appropriate materials of construction selection, rigorous preventative maintenance practices, proper operating procedures, and fuel conditioning where needed. Burners must operate as designed in order to optimize performance and minimize pollution emissions.

Another form of "pretreatment" is to capture the waste gases that would normally be flared and to reuse them as a fuel source. There has been increased interest in recovering the gases that have commonly been flared, especially to reduce

Flare Gas Recovery Unit
Process Flow Diagram

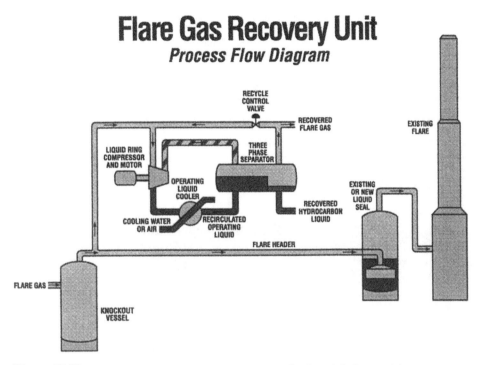

Figure 15.55   Flare gas recovery system. (Courtesy of John Zink Co. LLC.)

pollutant emissions where no product is being made from the combustion of the vented gases. While the process may only be practical and economic in some limited cases at present, flare gas recovery (see schematic in Figure 15.55) may be a viable option for some chemical plants. A flare gas recovery system (see Fig. 15.56) was installed at Lion Oil Company's refinery in El Dorado, AR [98]. The system is successfully being used to reduce plant emissions by capturing some of the vent gases that were previously flared and using those gases as a fuel.

15.5.2 Combustion Modifications

This is generally the least expensive option for reducing most pollutant emissions compared to pre- and post-treatment control strategies [38,42,43,71,74,99]. Numerous burner designs have been developed to reduce pollution emissions in chemical and petrochemical applications. Some of these designs are discussed in some detail in Chap. 12. Others are discussed in the chapters on specific pollutants. Table 15.13 shows some of the NO_x reductions possible in process heaters based on the burner type [100].

Kunz and coworkers discuss various techniques for reducing NO_x emissions from hydrogen/carbon monoxide reformers [101–103]. The common combustion modification techniques of lower air preheat, injecting steam or other inerts into the fuel, controlling O_2 to the lowest practical levels, and installing low NO_x burners are

Table 15.13 NO$_x$ Reduction for Low NO$_x$ Burner Types

Burner type	Typical No$_x$ reductions (%)
Staged-air burner	25–35
Staged-fuel burner	40–50
Low-excess-air burner	20–25
Burner with external FGR	50–60
Burner with internal FGR	40–50
Air or fuel–gas staging with internal FGR	55–75
Air or fuel–gas staging with external FGR	60–80

FGR = flue gas recirculation.
Adapted from Ref. 100.

Figure 15.56 Photograph of a flare gas recovery system. (Courtesy of John Zink Co. LLC.)

discussed. It is noted that CO_2-rich purge gas makes less NO$_x$ than natural gas and that measured NO$_x$ emissions are generally less than the values in AP-42 [66]. Experimental field data showed that NO$_x$ increases with hydrogen content in the fuel, which validates the theoretical and experimental results reported in Chap. 6. Experimental results also showed that steam injection of 0.3 lb steam/lb fuel reduced NO$_x$ by about 20%.

15.5.3 Post-Treatment

A variety of post-treatment techniques are used to control pollution emissions from fired heaters and furnaces. American Petroleum Institute Recommended Practice 536 discusses some of the more prominent techniques related to reducing NO$_x$ emissions in fired equipment used in refinery processes [104]. The two techniques considered in API 536 in some detail are selective catalytic reduction (SCR) and selective noncatalytic reduction (SNCR) where reagents like ammonia or urea are injected into the exhaust gases, within a specific temperature range, to convert the NO_x emissions into nitrogen and other constituents. Both of these are discussed in more detail in Chap. 6. API 536 has a number of useful data sheets for these control techniques to assist in developing appropriate specifications and the operating performance requirements for a given application.

Particulate emissions from petroleum processes are controlled by a variety of techniques [105]. These include cyclones, electrostatic precipitators, wet scrubbers, venturi scrubbers, and fabric filters. However, particulate emissions are generally low from these processes since the fuels in many cases are hydrocarbon gases that can be relatively easily fully combusted to prevent soot formation and unburned hydrocarbons that could make particulate emissions.

REFERENCES

1. MB Borup, EJ Middlebrooks. *Pollution Control for the Petrochemicals Industry*. Chelsea, MI: Lewis Publishers, 1987.
2. U.S. EPA. Profile of the Petroleum Refining Industry. Washington, DC: U.S. Environmental Protection Agency. Rep. EPA/310-R-95-013, 1995.
3. U.S. EPA. EPA Sector Notebook: Profile of the Petroleum Refining Industry. Rep. EPA/310-R-95-013. Washington, DC: U.S. Environmental Protection Agency, 1995.
4. U.S. EPA. EPA Sector Notebook: Profile of the Organic Chemical Industry. Rep. EPA/310-R-95-012. Washington, DC: U.S. Environmental Protection Agency, 1995.
5. DHF Liu. Pollution Prevention in Chemical Manufacturing. In DHF Liu, BG Lipták (eds.). *Environmental Engineers' Handbook*, 2nd edn. Boca Raton, FL: Lewis Publishers, 1997.
6. U.K. Environment Agency. Large-Volume Organic Chemicals. IPC Guidance Note S2 4.01. London: U.K. Environment Agency, 1999.
7. C Scott, S Mohan. Develop an effective startup, shutdown and malfunction plan. *Chem. Eng. Prog.*, Vol. 98, No. 8, pp. 50–52, 2002.
8. P Singh, K Seto. Analyzing APC performance. *Chem. Eng. Prog.*, Vol. 98, No. 8, pp. 60–66, 2002.
9. American Chemical Society. *Technology Vision 2020: The U.S. Chemical Industry*. Washington, DC: The American Chemical Society, 1996.
10. RO Anderson. *Fundamentals of the Petroleum Industry*. Norman, OK: University of Oklahoma Press, 1984.
11. WL Leffler. *Petroleum Refining for the Nontechnical Person*. Tulsa, OK: Pennwell Books, 1985.
12. JH Gary, GE Handwerk. *Petroleum Refining: Technology and Economics*, 3rd edn. New York: Marcel Dekker, 1994.
13. RA Meyers. *Handbook of Petroleum Refining Processes*, 2nd edn. New York: McGraw-Hill, 1997.
14. RD Reed. *Furnace Operations*, 3rd edn. Houston, TX: Gulf Publishing, 1981.
15. CE Baukal (ed.). *The John Zink Combustion Handbook*. Boca Raton, FL: CRC Press, 2001.
16. API. Publication 535: Burners for Fired Heaters in General Refinery Services. Washington, DC: American Publication Institute, 1995.
17. API. Standard 560: Fired Heaters for General Refinery Services, 2nd edn. Washington, DC: American Publication Institute, 1995.
18. U.K. Environment Agency. Petroleum Processes: Oil Refining and Associated Processes. Chief Inspector's Guidance Note S2 1.10. London: HMSO, Nov. 1995.

19. T Dark, J Ackland. Fuels. In CE Baukal (ed.). *The John Zink Combustion Handbook*. Boca Raton, FL: CRC Press, 2001.

20. AG Abilov, Z Zeybek, O Tuzunalp, Z Telatar. Fuzzy temperature control of industrial refineries furnaces through combined feedforward/feedback multivariable cascade systems. *Chem. Eng. Process.*, Vol. 41, pp. 87–98, 2002.

21. U.S. Dept. of Energy Office of Industrial Technology. Petroleum—Industry of the Future: Energy and Environmental Profile of the U.S. Petroleum Refining Industry, U.S. DOE, Washington, DC, Dec. 1998.

22. U.S. Dept. of Energy Office of Industrial Technology. Industrial Combustion Technology Roadmap, U.S. DOE, Washington, DC, Apr. 1999.

23. JA Stanislaw. Petroleum industry faces tectonic shifts changing global energy map. *Oil & Gas J.*, Vol. 97, No. 50, pp. 8–14, 1999.

24. American Petroleum Institute. Technology Vision 2020: A Report on Technology and the Future of the U.S. Petroleum Industry, API, Washington, DC, 1999.

25. U.S. Dept of Energy Office of Industrial Technology. Technology Roadmap for the Petroleum Industry, Draft. Washington, DC: U.S. Dept. of Energy, 2000.

26. U.S. Environmental Protection Agency. Profile of the Petroleum Refining Industry. Government Institutes, Rockville, MD (no date given).

27. U.S. Dept. of Energy Office of Industrial Technology. Petroleum—Industry of the Future: Energy and Environmental Profile of the U.S. Petroleum Refining Industry, U.S. DOE, Washington, DC, Dec. 1998.

28. SA Shareef, CL Anderson, LE Keller. Fired Heaters: Nitrogen Oxides Emissions and Controls, U.S. Environmental Protection Agency, Research Triangle Park, NC, EPA Contract No. 68-02-4286, June 1988.

29. Office of the Federal Register. 60.561 Definitions. U.S. Code of Federal Regulations Title 40, Part 60. Washington, DC: U.S. Government Printing Office, 2001.

30. DQ Kern. *Process Heat Transfer*. New York: McGraw-Hill, 1950.

31. CE Baukal. Introduction. In CE Baukal (ed.). *The John Zink Combustion Handbook*. Boca Raton, FL: CRC Press, 2001.

32. P Singh, M Henneke, JD Jayakaran, R Hayes, CE Baukal. Chap. 1: Heat Transfer. In CE Baukal (ed.). *The John Zink Combustion Handbook*. Boca Raton, FL: CRC Press, 2001.

33. NP Lieberman. *Troubleshooting Process Operations*. Tulsa, OK: PennWell Books, 1991.

34. FL Evans. *Equipment Design Handbook for Refineries and Chemical Plants*, Vol. 2, 2nd edn. Houston, TX: Gulf Publishing, 1980.

35. HS Bell, L Lowy. Equipment, In WF Bland, RL Davidson, (ed.), *Petroleum Processing Handbook*, New York: McGraw-Hill, Chap. 4, 1967.

36. A Garg. Optimize fired heater operations to save money. *Hydrocarbon Process.*, Vol. 76, No. 6, pp. 97–104, 1997.

37. WL Nelson. *Petroleum Refinery Engineering*, 2nd edn. New York: McGraw-Hill, 1941.

38. A Garg. Trimming NO_x from furnaces. *Chem. Eng.*, Vol. 99, No. 11, pp. 122–124, 1992.

39. A Garg, H Ghosh. Good heater specifications pay off. *Chem. Eng.*, Vol. 95, No. 10, pp. 77–80, 1988.

40. A Garg. How to boost the performance of fired heaters. *Chem. Eng.*, Vol. 96, No. 11, pp. 239–244, 1989.

41. A Garg. Better burner specifications. *Hydrocarbon Process.*, Vol. 68, No. 8, pp. 71–72, 1989.

42. EB Sanderford. Alternative Control Techniques Document—NO_x Emissions from Process Heaters. U.S. Environmental Protection Agency Rep. EPA-453/R-93-015, Feb. 1993.

43. U.S. EPA. Alternative Control Techniques Document—NO_x Emissions from Process Heaters (Revised). U.S. Environmental Protection Agency Rep. EPA-453/R-93-034, Sept. 1993.

44. LA Thrash. Annual refining survey. *Oil & Gas J.*, Vol. 89, No. 11, pp. 86–105, 1991.

45. HL Berman. Fired Heaters—II: Construction Materials, Mechanical Features, Performance Monitoring, In V. Cavaseno (ed.) *Process Heat Exchange*. New York: McGraw-Hill, pp. 293–302, 1979.

46. CE Baukal. *Heat Transfer in Industrial Combustion*. Boca Raton, FL: CRC Press, 2000.

47. AJ Johnson, GH Auth. *Fuels and Combustion Handbook*, 1st edn. New York: McGraw-Hill, 1951.

48. HL Berman. Fired Heaters—I: Finding the Basic Design for Your Application. In V Cavaseno (ed.), *Process Heat Exchange*. New York: McGraw-Hill, pp. 287–292, 1979.

49. H Futami, R Hashimoto, H Uchida. Development of new catalyst and heat-transfer design method for steam reformer. *J. Fuel Soc. Japan*, Vol. 68, No. 743, pp. 236–243, 1989.

50. H Gunardson. *Industrial Gases in Petrochemical Processing*. New York: Marcel Dekker, 1998.

51. JD Fleshman. FW Hydrogen Production. In RA Myers (ed.). *Handbook of Petroleum Refining Processes*, 2nd ed. New York: McGraw-Hill, Chap. 6.2, 1996.

52. J McAdams, J Karan, R Witte, M Claxton. Low NO_x Burner Maintenance in High Temperature Furnaces. *Proceedings of the 14th Ethylene Producers' Conference*, New York: American Institute of Chemical Engineers, Vol. 11, pp. 351–370, 2002.

53. K Funahashi, T Kobayakawa, K Ishii, H Hata. SCR $DeNO_x$ in New Maruzen Ethylene Plant. *Proceedings of the 13th Ethylene Producers' Conference*, New York: American Institute of Chemical Engineers, Vol. 10, pp. 741–755, 2001.

54. CP Brundrett, OL Maaskant, N Genty. Application and Operation of the Shell Low Temperature Technology on Ethylene Cracker Furnace. *Proceedings of the 13th Ethylene Producers' Conference*, New York: American Institute of Chemical Engineers, Vol. 10, pp. 756–765, 2001.

55. A Suwa. Operating Experiences of SCR $DeNO_x$ Unit in Idemitsu Ethylene Plant. *Proceedings of the 13th Ethylene Producers' Conference*, New York: American Institute of Chemical Engineers, Vol. 10, pp. 766–773, 2001.

56. W Bussman, R Poe, B Hayes, J McAdams, J Karan. Low NO_x Burner Technology for Ethylene Cracking Furnaces. *Proceedings of the 13th Ethylene Producers' Conference*, New York: American Institute of Chemical Engineers, Vol. 10, pp. 714–740, 2001.

57. CE Baukal (ed). *Computational Fluid Dynamics in Industrial Combustion*. Boca Raton, FL: CRC Press, 2001.

58. API. Guide for Pressure-Relieving and Depressuring Systems. Recommended Practic 521, 4th edn. Washington, DC: American Petroleum Institute, 1997.

59. Office of the Federal Register. 60.18 General Control Device Requirements. U.S. Code of Federal Regulations Title 40, Part 60. Washington, DC: U.S. Government Printing Office, 2001.

60. O Cunha-Leite. Design alternatives, components key to optimum flares. *Oil & Gas J.*, Vol. 90, No. 47, pp. 70–76, 1997.

61. LB Evans, WM Vatavuk, DK Stone, SK Lynch, RF Pandullo, W Koucky. Sec. 3.2: VOC Destruction Controls, Chap. 1: Flares. In DC Mussatti (ed.). *Air Pollution Control Cost Manual*, 6th edn. Rep. EPA/452/B-02-001. Washington, DC: U.S. Environmental Protection Agency, Jan. 2002.

62. U.S. EPA. Basis and Purpose Document on Specifications for Hydrogen-Fueled Flares. Rep. EPA-453/R-98-001. Washington, DC: U.S. Environmental Protection Agency, 1998.

63. R Schwartz, J White, W Bussman. Flares. Chap. 20 in the *John Zink Combustion Handbook*, ed. CE Baukal. Boca Raton, FL: CRC Press, 2001.

64. J Manning. Optimising [sic] waste gas emissions. *Hydrocarbon Eng.*, Vol. 7, No. 2, pp. 69–74, 2002.

65. D Hansell, G England. *Air Toxic Emission Factors for Combustion Sources Using Petroleum Based Fuels*, 3 Vols. Energy and Environmental Research Corp., Irvine, CA, 1998 (available at www.api.org/step/piep.htm).

66. U.S. EPA, AP-42: Compilation of Air Pollutant Emission Factors, 5th edn. U.S. Environmental Protection Agency, Jan. 1995.

67. JC Reis. *Environmental Control in Petroleum Engineering*. Houston, TX: Gulf Publishing, 1996.

68. J Elkins, N Frank, J Hemby, D Mintz, J Szykman, A Rush, T Fitz-Simons, T Rao, R Thompson, E Wildermann, G Lear. National Air Quality and Emissions Trends Report, 1999. Washington, DC: U.S. Environmental Protection Agency, Rep. EPA 454/R-01-004, 2001.

69. RR Hayes, CE Baukal, P Singh, D Wright. Fuel Composition Effects on NO_x. *Proceedings of the Air & Waste Managment Association's 94th Annual Conference & Exhibition*, Paper 434, Orlando, FL, 2001.

70. ML Lin, JR Comparato, WH Sun. Applications of Urea-Based Selective Noncatalytic Reduction in Hydrocarbon Processing Industry. In US Ozkan, SK Agarwal, G Marcelin. (eds.). *Reduction of Nitrogen Oxide Emissions*. Washington, DC: American Chemical Society, 1995.

71. JD McAdams, SD Reed, DC Itse. Minimize NO_x emissions cost-effectively. *Hydrocarbon Process.*, Vol. 80, No. 6, pp. 51–58, 2001.

72. JG Seebold, RG Miller, GW Spesert, DE Beckley, DJ Coutu, RT Waibel, D Venizelos, RR Hayes, W Bussman. Developing and Retrofitting Ultra Low NO_x Burners in a Refinery Furnace. *Proceedings of Joint International Combustion Symposium*, AFRC/JFRC/IEA 2001, Kauai, Hawaii, Sec. 2C, Paper 2, Sept. 2001.

73. M Bradford, R Grover, P Paul. Controlling NO_x emissions—Part 1. *CEP Magazine*, pp. 42–46, May, 2002.

74. CE Baukal, R Hayes, M Grant, P Singh, D Foote. NO_x Emissions Reduction Technologies in the Petrochemical and Refining Industries. Paper 42935. *Proceedings of the Air & Waste Management Association's 95th Annual Conf. & Exhibition*, Baltimore, MD, June 23–27, 2002.

75. MP Laplante, P Lindenhoff. How low can you go?—Catalytic NO_x reduction in refineries. *World Refining*, Vol. 13, No. 2, pp. 46–51, 2002.

76. ID Crane, RD Springer, JH Siegell. New method estimates sulfuric acid emissions from fired heaters. *Oil & Gas J.*, Vol. 100.40, pp. 78–80, 2002.

77. U.S. EPA. Compilation of Air Pollutant Emission Factors, Vol. I: Stationary Point and Area Sources, Sec. 13.5: Industrial Flares. Fifth Edition. Washington, DC: U.S. Environmental Protection Agency Rep. AP-42, 1995.

78. SE Guigard, WB Kindzierski, N Harper. Heat Radiation from Flares. Report prepared for Science and Technology Branch, Alberta Environment, ISBN 0-7785-1188-X. Edmonton, Canada, May 2000.

79. JKJ Buettner. *Heat Transfer and Safe Exposure Time for Man in Extreme Thermal Environment*, Paper No. 57-SA-20. New York: American Society of Mechanical Engineers, 1957.

80. K Leary, D Knott, R Thompson. Water-injected flare tips reduce radiated heat, noise. *Oil & Gas J.*, Vol. 100.18, pp. 76–83, 2002.

81. RE Schwartz, JW White. Predict radiation from flares. *Chem. Eng. Prog.*, Vol. 93, No. 7, pp. 42–49, 1997.

82. RE Schwartz, JW White. Flare Radiation Prediction: A Critical Review. Presented at the *30th Annual Loss Prevention Symposium of the American Institute of Chemical Engineers*, Feb. 28, 1996.

83. W Bussman, JD Jayakaran. Noise. Chap. 7 in CE Baukal (ed.). *The John Zink Combustion Handbook*. Boca Raton, FL: CRC Press, 2001.

84. JG Seebold. Flare noise: causes and cures. *Hydrocarbon Process*. pp. 143–147, Oct. 1972.

85. B Duck. Reducing Emissions. *Hydrocarbon Eng.*, Vol. 7, No. 9, pp. 109–112, 2002.

86. LD Berg, W Bussman, J Hong, M Henneke, I-P Chung, JD Smith. Modeling of Combustion Systems. Chap. 8 in CE Baukal (ed.). *The John Zink Combustion Handbook*. Boca Raton, FL: CRC Press, 2001.

87. M Chaudhuri, JJ Diefenderfer. Achieving smokeless flaring—air or steam assist? *Chem. Eng. Prog.*, Vol. 91, No. 6, pp. 40–43, 1995.

88. Office of the Federal Register. Subpart J: Standards of Performance for Petroleum Refineries. U.S. Code of Federal Regulations Title 40, Part 60. Washington, DC: U.S. Government Printing Office, 2001.

89. Office of the Federal Register. Subpart III: Standards of Performance for Volatile Organic Compound (VOC) Emissions From the Synthetic Organic Chemical Manufacturing Industry (SOCMI) Air Oxidation Unit Processes. U.S. Code of Federal Regulations Title 40, Part 60. Washington, DC: U.S. Government Printing Office, 2001.

90. Office of the Federal Register. Subpart NNN: Standards of Performance for Volatile Organic Compound (VOC) Emissions from Synthetic Organic Chemical Manufacturing (SOCMI) Distillation Operations. U.S. Code of Federal Regulations Title 40, Part 60. Washington, DC: U.S. Government Printing Office, 2001.

91. Office of the Federal Register. Subpart RRR: Standards of Performance for Volatile Organic Compound (VOC) Emissions from Synthetic Organic Chemical Manufacturing (SOCMI) Reactor Processes. U.S. Code of Federal Regulations Title 40, Part 60. Washington, DC: U.S. Government Printing Office, 2001.

92. Office of the Federal Register. Subpart DDD: Standards of Performance for Volatile Organic Compound (VOC) Emissions from the Polymer Manufacturing Industry. U.S. Code of Federal Regulations Title 40, Part 60. Washington, DC: U.S. Government Printing Office, 2001.

93. D Deason. Overview of NO_x Reduction Issues of Concern to Ethylene Producers. *Proceedings of the 13th Ethylene Producers' Conference*, New York: American Institute of Chemical Engineers, Vol. 10, pp. 714–740, 2001.

94. API. Measurement of Noise from Fired Process Heaters. Recommended Practice 531M. Washington, DC: American Petroleum Institute, reaff. Aug. 1995.

95. J Bluestein. NO_x Controls for Gas-Fired Industrial Boilers and Combustion Equipment: A Survey of Current Practices. Rep. GRI-92/0374. Chicago, IL: Gas Research Institute, 1992.

96. D Anderson. Damper drive performance helps meet NO_x-reduction requirements. *World Refining*, Vol. 12, No. 7, pp. 48–51, 2002.

97. MW Stansifer. Fuel Gas Clean Up for Low NO_x Burners. *Proceedings of the 14th Ethylene Producers' Conference*, New York: American Institute of Chemical Engineers, Vol. 11, pp. 166–172, 2002.

98. PW Fisher, D Brennan. Minimize flaring with flare gas recovery. *Hydrocarbon Process.*, Vol. 81, No. 6, pp. 83–85, 2002.

99. HL Shelton. Find the right low-NO_x solution. *Environ. Eng. World*, Vol. 2, No. 6, pp. 24–27, 1996.

100. A Garg. Specify better low-NO_x burners for furnaces. *Chem. Eng. Prog.*, Vol. 90, No. 1, pp. 46–49, 1994.

101. RG Kunz, DD Smith, NM Patel, GP Thompson, GS Patrick. Control NO_x from furnaces. *Hydrocarbon Process.*, Vol. 71, No. 8, pp. 57–62, 1996.

102. RG Kunz, DD Smith, EM Adamo. Predict NO_x from gas-fired furnaces. *Hydrocarbon Process.*, Vol. 75, No. 11, pp. 65–79, 1996.

103. RG Kunz, BR Keck, JM Repasky. Mitigate NO_x by steam injection. *Hydrocarbon Process.*, Vol. 77, No. 2, pp. 79–84, 1998.

104. API. Post-Combustion NO_x Control for Fired Equipment in General Refinery Services. Recommended Practice 536. Washington, DC: American Petroleum Institute, Mar. 1998.

105. U.S. EPA. Control Techniques for Particulate Emissions from Stationary Sources—Vol. 2, Rep. EPA-450/3-81-005b. Washington, DC: U.S. Environmental Protection Agency, Sep. 1982.

16

Waste Incineration

16.1 INTRODUCTION

Some of the material in this chapter has been partially adapted from previous publications [1–3]. Incinerators are designed to burn, and in many cases destroy, waste materials that may sometimes be contaminated with hazardous substances. The very word "waste" often means different things to different people. For example, newspapers thrown into the garbage of the average citizen can be recycled and made into more paper or can be burned in an incinerator to generate power and steam. To the paper recycler, the newspapers are feed materials and to the incinerator the newspapers are a fuel source.

Incineration can be a controversial topic, particularly among environmentalists. While it is preferable to eliminate waste materials or recycle the waste back into the process, this is not always technically or economically feasible. While landfilling is an option, it is generally less preferable to destroying the waste through burning, which dramatically reduces the amount of waste (e.g., in the form of ash) that needs to be landfilled. Today's incinerators are well managed and highly regulated to make them an acceptable alternative to dealing with most types of wastes. Incineration may be chosen to reduce dramatically the volume of a waste stream, destroy the toxics in so-called hazardous waste, and to recover energy in the forms of electricity and steam. Motyka and Mascarenhas [4] argue that incineration is more effective for disposing of waste natural gas products than flaring, in part because of more effective mixing of the waste gases with air for more complete combustion.

The waste materials burned in an incinerator usually have some heating value. For example, wood products such as paper and cardboard liberate significant amounts of heat when burned. However, the average heating value of most waste materials is not sufficient to make the process self-sustaining. Therefore, nearly all incineration processes require a substantial amount of auxiliary heat, which is commonly generated by the combustion of hydrocarbon fuels such as natural gas or oil. An example of the important processes in a rotary kiln incinerator is shown in Fig. 16.1 [5].

Waste incineration is different from other industrial combustion processes because of the nature of the materials processed in an incinerator that can vary widely, not only from one incinerator to the next, but also from one day to the next for a given incinerator. In some locations, waste materials are separated for

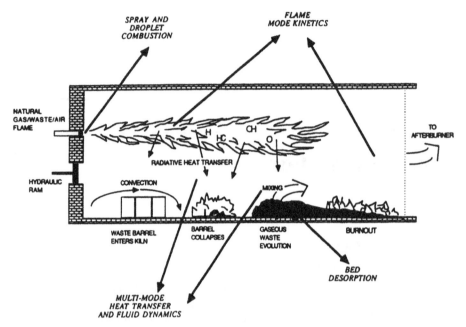

Figure 16.1 Fundamental processes in rotary kiln incinerators. (From Ref. 5. Courtesy of CRC Press.)

recycling. For example, plastics, aluminum cans, and glass products may be selectively removed from waste materials so that they can be recycled. This is very beneficial to the incineration process because the aluminum and glass products are an additional heat load and not an additional source of fuel. Plastics can burn but they can also produce some toxic pollutants like dioxins and furans. A study by Radian Corporation [6] for the U.S. EPA modeled the projected fatality rate due to cancer caused by dioxin and furan releases from municipal waste incinerators.

Waste incineration is a very complex combustion process that includes many processes: pyrolytic decomposition; surface and gas combustion; conductive, convective, and radiative heat transfer though heterogeneous media; and gas flow through randomly packed beds of material whose size, shape, and orientation are continually changing [7]. The physics, chemistry, and dynamic nature of the incineration process make it more difficult to control emissions compared to other steady-state industrial combustion applications.

Another problem with the waste feed to an incinerator concerns the weather. If it has recently rained and the waste materials are wet, this can add a huge additional heat load to the incinerator so the feed rate must be reduced or the firing rate increased to compensate. Incinerators can have a much more dynamic control scheme than other combustion processes as a result of the waste feed variability.

Unlike other industrial processes, the end product from the incinerator is another waste that must be disposed of in a landfill for the solids or further treated for disposal such as for sludges produced from the exhaust gas stream wet-scrubbing systems [8]. One objective of the incinerator is to reduce greatly the size of the incoming waste. Rather than burying the raw waste materials in a landfill, the materials are incinerated, which typically generates ash that is only a fraction of the size of the original waste.

Incineration is a common method for treating waste materials. An ASME Committee on Industrial and Municipal Wastes published a general guide to combustion in waste incineration [9]. Seeker [10] reviewed the use of combustion to dispose of waste materials and offered the following advantages compared to other methods of disposal:

- Destruction of hazardous constituents such as organics and pathogens in the waste.
- Reduction of the volume and mass of the waste.
- Potential for energy recovery by burning the wastes having heating value.
- Making the waste unrecognizable from its original form (often a requirement for medical waste).

The objective of most waste incineration processes is to destroy or burn any of the organic materials in the waste, leaving only an inert residue. The waste combustion system consists of the following components: (1) waste preparation and feeding, (2) combustion equipment, (3) heat recovery equipment, (4) air pollution control equipment, and (5) ash/solid residue stabilization and disposal.

Brunner [11] has written an extensive book on all aspects of waste incineration that includes two chapters on air pollution control equipment. The following benefits were offered for waste incineration:

- Volume and weight reduction to a fraction of its original size.
- Waste reduction is immediate—no long-term residence in a landfill is needed.
- Incineration can be on site without the need to transport long distances.
- Air discharges to the atmosphere can be monitored and controlled.
- The ash residue is sterile.
- Even hazardous wastes can often be essentially completely destroyed.
- A relatively small land area is needed for the incinerator compared to landfills.
- Energy can usually be economically recovered from the incineration process.

Some of the disadvantages include:

- High capital cost.
- Skilled operators are required.
- Not all materials are incinerable.
- Supplemental fuel may be required to achieve required destruction efficiencies.

Wendt [12] has also given a review of combustion in waste incineration. The paper gives a good overview of the technology with a specific emphasis on the pollutant emissions. Lighty and Veranth [13] have reviewed the effects of recent advances in combustion technology and their impact on practical incineration systems with specific emphasis on pollutant minimization, process upsets, and system simulations. The topic of pollutant minimization focused on dioxin and furan formation, criteria pollutants, and the impact of solid wastes from incineration systems on the environment.

Historically, air/fuel combustion has been used in waste processing to provide heat for thermal destruction of solid, liquid, and gaseous waste streams. Examples include medical and municipal waste (solid), spent solvents (liquid), and off-gas or vent streams (gaseous). Oxygen-based combustion systems are becoming more common in waste processing applications [14]. When traditional air/fuel combustion systems have been modified for oxygen-enhanced combustion (OEC), many benefits can be demonstrated. Typical improvements include higher destruction and removal efficiencies of the waste, increased thermal efficiency, increased processing rates, lower NO_x and particulates emissions, and less downtime for maintenance. OEC has been used in several different types of incineration applications to overcome thermal limitations in the process [15].

In many cases, the volatilization characteristics of waste materials are unpredictable so that incinerators are generally run with large amounts of excess air in order to ensure complete destruction of the volatiles [16].

Bartone [17] has identified four critical factors that determine the effectiveness of a waste incineration process:

1. Sufficient free oxygen available in the combustion zone to react with the combustible waste materials.
2. Adequate mixing of the waste with the oxygen usually due to turbulent flow conditions.
3. Sufficiently high gas temperatures to initiate and sustain the chemical reactions to destroy the waste completely.
4. Adequate residence time in the incinerator to ensure proper mixing and reaction time to destroy the waste materials fully.

16.2 TERMINOLOGY

Hesketh [18] defines a number of terms important in incineration:

Afterburner gas retention time	time that volatile matter is exposed to turbulent mixing, elevated temperature, and excess air for final combustion
Boiler	heat-transfer portion of incinerator where water is turned to steam
Bottom ash	residue from the furnace
Burn out	amount of combustible material in fuel converted into CO_2 and H_2O
Burn rate	total quantity of combined carbon and hydrogen that is converted into CO_2 and H_2O vapor (usually expressed in lb/hr)
Charge rate	quantity of waste material loaded into incinerator, but not necessarily burned (usually expressed in lb/hr)
Controlled air	controlling air flow to achieve desired rate of combustion when there are two or more combustion chambers

Endothermic	chemical reactions that absorb heat from the surroundings
Excess air	controlled burning at greater than stoichiometric air requirements
Exothermic	chemical reactions that liberate heat to the surroundings
Fixed carbon	nonvolatile portion of waste, which must be burned at a higher temperature
Grate	section of furnace used to support waste material
Grate heat release rate	measure of incinerator capacity (usually expressed as Btu/hr-ft^2)
Heat release	total energy released from combustion (usually expressed as Btu/hr)
Heating value	net energy available from chemical combustion (usually expressed in Btu/lb)
Moisture	water that must be evaporated from the waste material
Particulate emissions	fine solid and/or liquid matter suspended in the exhaust gas stream
Pathogenic waste material	waste capable of causing disease
Pathological waste material	waste from the study and treatment of diseases
Primary combustion chamber	combustion zone where solids are fed so they can heat, devolatize, and ignite, often operated in under reducing conditions
Proximate analysis	determination of the volatile matter, fixed carbon, moisture, and ash (non-combustible material) in the waste
Pyrolysis	chemical destruction of organic materials in presence of heat and absence of oxygen
Refuse-derived fuel (RDF)	processed municipal solid waste with heating value in the range 5000–5500 Btu/lb
Secondary combustion chamber	(sometimes called an afterburner) final combustion zone is a two-stage incinerator that is designed to combust completely any remaining combustibles; operated under excess air conditions
Starved air	controlled burning at greater than stoichiometric air requirements, usually in the primary chamber
Stoichiometric	theoretical air required for complete combustion of all combustibles to CO_2, H_2O vapor, and other oxides.
Stuff and burn	where charging rate of burnable material is greater than the actual burn rate
Tipping fee	cost charged to dispose of waste at incinerator or other disposal area

Tipping floor	area where solid waste is delivered at incinerator facility
Trammel	perforated, rotating horizontal cylinder used to separate solid waste and to remove cans from bottom ash
Volatile matter	part of the waste which can be liberated with application of heat only
Volumetric heat release rate	measure of incinerator capacity (usually expressed as Btu/hr-ft^3)
Waterwall facility	combustor with vertical water tubes to remove heat during burning

Destruction and removal efficiency (DRE) is defined as

$$\text{DRE} = \frac{w_{\text{in}} - w_{\text{out}}}{w_{\text{in}}} \tag{16.1}$$

where $w_{\text{in}} =$ inlet waste mass flow rate, and $w_{\text{out}} =$ outlet waste mass flow rate.

Example 16.1

Given: Incoming pollutant flow rate of 100 lb/hr, outgoing pollutant flow rate $= 0.05$ lb/hr.

Find: DRE.

Solution:

$$\text{DRE} = \frac{w_{\text{in}} - w_{\text{out}}}{w_{\text{in}}} = \frac{100 - 0.05}{100} = 99.95\%.$$

16.3 WASTE MATERIALS

The waste fed into incinerators is generally classified into seven categories as shown in Table 16.1 [18]. Theodore and Reynolds [19] recommend that the following data be collected to characterize properly the waste materials to be incinerated [19]:

- Concentrations of carbon, hydrogen, oxygen, nitrogen, sulfur, halogens, phosphorus, ash, metals, salts, and moisture in the waste.
- Trace metals.
- Ash composition.
- Special characteristics such as toxicity, corrosiveness, thermal stability, and chemical stability.

The U.S. EPA classifies solid waste into the following four categories: urban, industrial, mineral, and agricultural [20].

Brunner lists the following waste characteristics in determining the incinerability of a material [11]:

- Waste moisture content—more fuel is required for higher contents.
- Heating value—lower heating values require more supplemental fuel.

Table 16.1 Waste Classifications

Type	Name	Heating value/(Btu 1b)	Description
0	Trash	8500	Highly combustible waste such as paper, cardboard, and wood. Contains up to 10 wt% petrochemical waste, 10 wt% moisture, and 5 wt% noncombustible solids
1	Rubbish	6500	Mixture of combustible waste such as paper, cardboard, and floor sweepings from domestic, commercial, and industrial sources. Contains up to 20 wt% restaurant waste, up to 25 wt% moisture, and up to 10 wt% noncombustible solids, but not petrochemical wastes
2	Refuse	4300	Evenly distributed mixture of rubbish and garbage usually found in municipal waste. Contains up to 50% moisture and 7% noncombustible solids
3	Garbage	2500	Consists of animal and vegetable wastes from restaurants, cafeterias, hotels, hospitals, markets etc. Contains up to 70% moisture and 5% noncombustible solids
4	Human and animal remains	1000	Consists of carcasses, organs, and solid organic wastes from hospitals, laboratories, animal pounds, and similar sources. Contains up to 85% moisture and 5% non-combustable solids.
5	By-product waste	Must be calculated based on actual waste	Gaseous, liquid, or semiliquid material such as tar, paints, solvents, sludges, fumes, etc., from industrial operations
6	Solid by-product waste	Must be calculated based on actual waste	Contains material such as rubber, plastic, wood waste, etc., from industrial operations

Source: Ref. 18.

- Inorganic salts—these often become airborne inside the incinerator and collect on surfaces causing slags or cakes to build up and significantly increasing maintenance costs.
- High sulfur or halogen content—these can cause acids to attack the equipment causing maintenance and operational problems.
- Radioactive wastes—only incinerators specially designed to handle these wastes should be used.

Cozzani et al. [21] have shown the importance of the waste feed composition when combusting refuse derived fuel. They developed a model that included the chemical kinetics and heat transfer processes during the pyrolysis of the fuels in a fixed-bed reactor. Three different RDF compositions were tested with varying amounts of organic and aqueous tars. There was a good correlation between the experimental and modeling data.

It is also important to prepare the waste prior to incineration. This includes minimizing the amount of hazardous, flammable, explosive, toxic, and other related materials that enter the incinerator. It is desirable to minimize the moisture entering with the waste as this can be a significant added thermal load on the incinerator. This can be accomplished, for example, by covering the waste materials to keep them out of the rain. It is preferred to separate materials that can be recycled such as aluminum and glass that can be better used elsewhere and also further add to the thermal load on the incinerator.

The incoming waste can also be an important source of pollutant emissions exiting the exhaust stack. For example, plastics like polyvinyl chlorides (PVCs) often contain chlorine compounds that can produce dioxins when they are incinerated, which are toxic and must be removed. Some types of plastics are recyclable and again can be put to better use by removing them from the waste to be incinerated. Older thermometers and lead–acid batteries can lead to heavy metal emissions like mercury and lead. Table 16.2 lists examples of typical toxic substances that could produce heavy metals emissions among other things, and the recommended method for handling each product.

16.4 POLLUTION IN WASTE INCINERATION

Besides the typical combustion-generated pollutants like NO_x and SO_x, there are some pollutants unique to waste incineration that are not usually an issue in other types of industrial combustion processes. This is due to both the composition of the fuel sometimes combusted in waste incinerators and the material being processed in incinerators, either of which may contain hazardous chemicals. Therefore, the regulations governing waste incineration processes are generally much more comprehensive and restrictive because of the possibility of emitting toxic pollutants. Both the pollutants and the regulations are briefly considered in the next subsections.

16.4.1 Pollutants

Common pollutants from incineration applications include CO, NO_x, SO_x, and particulate matter. Particulate matter is particularly important in this application

Table 16.2 Potential Toxic Products in Waste Streams to be Incinerated

Product	Pollutant/medium	Proper management
Lead–acid batteries	Pb/to air, ash	Recycling
Mercury batteries	Hg/to air	Storage/R&D/WP
Silver oxide batteries	Ag/to ash	Storage/R&D/WP
Ni–Cd batteries	Ni, Cd/to air, ash	Storage/R&D
Other batteries	Zn, Mn/to air, ash	Storage/R&D
Fluorescent tubes	Hg/to air	Recycling
Mercury thermometers	Hg/to air	Recycling, WP
Mercury switches	Hg/to air	Recycling
Mercury thermostats	Hg/to air	Recycling
Inks containing metals	Various metals/air	WP
Plastic/metal pigments	Various metals/air	WP
Plastic w/stabilizers	Cd/air	WP
Leather clothing (Chrome-tanned)	Cr/air, ash?	WP
Bimetal cans	Zn, Pb?/air, ash	Recycling
Aluminum cans, foil	Al/ash	Recycling
Leaded glass	Pb/air, ash	?
Gypsum (wallboard)	Sulfur/to air	Recycling
PVC bottles	HCl/chl. orgs/air, ash	WP
PVC or PVDC wrap	HCl/chl. orgs/air, ash	WP
Bleached paper	HCl/chl. orgs/air, ash	WP
Yard waste	Nitrogen/to air	Composting
Food waste	N/to air	Composting

WP=waste prevention; can include legislation or other actions to limit or exclude pollutant in product, or to exclude or limit product from manufacture or disposal, or to promote substitute, less toxic products, product formulations, and manufacturing processes.
Source: Ref. 52. (Courtesy of Air and Waste management Association.)

Table 16.3 Particulate Emissions from Incinerators

Source	Number of sources tested	Uncontrolled emissions (lb/ton feed)	Controlled emissions (lb/ton feed)
Municipal incinerators	13	0.4–39	0.74–1.82
Pathological incinerators	37	1.04–32.6	0.10–34.2
Multiple-hearth sewage sludge	5	No data	0.28–4.0
Fluidized-bed burning sewage sludge	4	No data	0.70–5.6
Other incinerator designs burning sewage or industrial sludge	3	2.6–14.6	0.18–6.6
Trash incinerators	64	0.12–8.10	0.72–2.0

Adapted from Ref. 57.

compared to most other industrial combustion applications because of the nature of the material being processed where ash is normally formed and heavy metals may be present. Table 16.3 shows particulate emissions from various waste incinerators.

Some of the fuels combusted in incinerators and some of the materials processed in incinerators contain hazardous and toxic constituents. These pollutants are in addition to the "traditional" pollutants like NO_x produced in nearly all industrial combustion processes. Some of the nontraditional pollutants that may be present in incineration processes include: hydrogen chloride (HCl), chlorine (Cl_2), benzene (C_6H_6), arsenic (As), formaldehyde (HCHO), fluorine compounds, mercury compounds, heavy metal (e.g., lead) compounds, dioxins, and furans. In many cases, these compounds are already present in the waste fuel or waste being processed and are not formed as a result of the combustion process. Most hydrocarbon hazardous chemicals like benzene will be nearly or completely destroyed by combustion. Chlorine, fluorine, and heavy metals present in the fuel or waste being processed may be combined into other compounds during combustion, but are not destroyed and generally must be removed in some way from the stack gases, often by some type of scrubbing.

Dioxins and furans are of particular interest for incineration processes because they are carcinogenic [22]. Katami et al. [23] showed a clear correlation between the chloride content of incoming feed materials and the levels of dioxins generated in the exhaust gases. However, recent technology advances have significantly reduced the dioxin and furan content in the fly ash and slag from incinerators. Post-treatment equipment such as wet scrubbers has also significantly reduced dioxin and furan emissions from the exhaust stacks of incinerators. A (1987) inventory [24] of dioxins and furans showed that 70% came from municipal solid-waste incinerators, 22% from medical waste incinerators, and less than 1% from hazardous waste incinerators. In 1995, 57% came from municipal and medical waste incinerators. Other high emitters included secondary copper smelting (20%), fires (8%), and cement kilns burning hazardous wastes (6%).

Procaccini et al. [25] experimentally and computationally studied the formation of PICs (products of incomplete combustion) in waste incineration processes. PIC formation is strongly dependent on the mixing, the instantaneous reactant concentrations, and the temperature. The turbulent flow field controlled the rate of oxidation of the fuel and the levels of PIC formation. In many cases the waste-feed combustion content is continually varying, which means that the combustion air requirements are also varying. Significant excursions in the feed combustion heating value can cause the incinerator to become temporarily starved of air and lead to PICs. For example, if a large quantity of drums of waste oil are suddenly fed into the incinerator then there may not be adequate air to combust all of the oil fully. This is handled by a combination of combustion controls and waste feed handling. Sensors can determine when oxygen levels in an incinerator are getting too low and increase air or oxygen flow into the combustor. Properly blending the waste feed material can also mitigate excursions by blending high and low heating value waste materials to achieve a more consistent overall heating value. In addition to dioxins, VOCs and polycyclic aromatic hydrocarbons (PAHs) are also potential pollutants from incineration processes [26].

Sorum et al. [27] discuss the formation of NO from the combustion of volatiles from municipal solid wastes. This was studied numerically and experimentally with a pilot-scale fixed-bed reactor. One of the parameters studied was the oxygen content in the oxidizer, which was varied from 12 to 100% by volume. Another parameter studied was the conversion of fuel-bound nitrogen into NO for a variety of single-component wastes including newspaper, cardboard, glossy paper, low-density polyethylene (LDPE), and polyvinylchloride (PVC), as well as blends of these components. The conversion efficiency for paper and cardboard ranged from 26 to 99%. Efficiencies above 100% were calculated for LDPE and PVC, which suggested that NO was formed by thermal and/or prompt mechanisms in addition to fuel NO_x. NO increased with the oxygen content in the oxidizer.

The EPA [28] and Weitz et al. [29] discuss greenhouse gas emissions (primarily CO_2 and CH_4) from municipal solid-waste management processes in the United States, including recycling, composting, waste-to-energy combustion, and landfill gas recovery. Technology advancements and integrated pollution control strategies are expected to lead to significant reductions in greenhouse gas emissions from municipal solid-waste management processes.

16.4.2 Modeling

A number of models have been developed to understand better the complex processes occurring in waste incinerators, including pollution emissions. Frey et al. [30] numerically and experimentally studied a pilot waste incinerator. The primary pollutants studied were the products of incomplete combustion (PICs) including particulates. The model results were in good agreement with experimental measurements. The model can be used to optimize the design of the incinerator to minimize pollutant formation.

Shin et al. [31] have used CFD to model "good" performance in waste incinerators for use in minimizing pollutant emissions. Good performance was defined as the gases having a 2 sec residence time at 850°C (1560°F) and 6% O_2. The model was used to determine the residence time, mixing, and thermal decomposition of potential pollutants. The model was used to suggest improvements to an actual incinerator including reducing pollutant emissions. There was a dramatic reduction in dioxins/furans after the implementation of the changes.

Carlotti et al. [32] have developed a model called COGENE® to predict the performance in municipal waste incinerators. The modular approach simulates the various components in a system. It includes predictions of the major inputs, outputs, and operating conditions as well as particulate emissions.

16.4.3 Regulations

Figure 16.2 shows five of the six criteria pollutant emissions from incineration processes in the United States since 1970 [33]. The U.S. Environmental Protection Agency has compiled an extensive list of air pollutant emission factors for solid-waste disposal applications [20]. While these factors are generally not used for setting regulations, they may be used to estimate the emissions from different processes for the purposes of permitting, when actual emission data are not available. For example, when a new plant is being built there are no existing actual emission data available so the AP-42 factors may be useful as one means to estimate the emissions from the new plant. For municipal solid-waste incinerators, one of the units used is the mass of pollutant emitted per mass of waste combusted (e.g., pounds of particulate matter emitted per ton of trash combusted).

The destruction and removal efficiency (DRE) is typically regulated for hazardous pollutants as defined by the Resource Conservation and Recovery Act

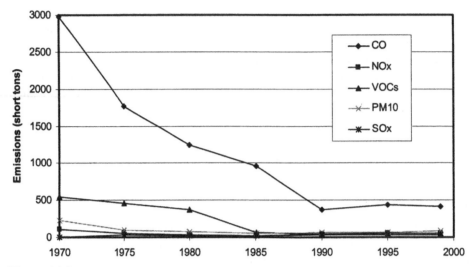

Figure 16.2 Emissions from incineration processes in the United States since 1970. (From Ref. 33.)

(RCRA). These pollutants include, for example, polychlorinated biphenyls (PCBs) and principal organic hazardous constituents (POHCs) that may be found, for example, in contaminated soil. The general strategy for achieving high waste destruction is often referred to as the "three Ts of incineration": time, temperature, and turbulence. Increasing any or all of these typically improves the DRE of the process. The three Ts are related to three characteristic times: a chemical time, a residence time, and a mixing time [34]. The chemical time can be estimated from

$$\tau_c = \frac{1}{k} \tag{16.2}$$

where k (sec^{-1}) is an overall reaction rate constant. The residence time can be estimated from

$$\tau_r = \frac{V}{Q} \tag{16.3}$$

where V is the volume (m^3 or ft^3) of the combustor and Q (m^3/sec or ft^3/sec) is the volumetric flow rate. The mixing time can be estimated from

$$\tau_m = \frac{L^2}{D_e} \tag{16.4}$$

where L (m or ft) is the length of the reaction zone and D_e (m^2/sec or ft^2/sec) is the effective turbulent diffusion coefficient.

Oxygen-enhanced combustion (OEC) can increase the DRE in an incineration process. This is accomplished by a combination of increased residence time within the incinerator and higher gas temperatures. These are two of the three Ts of incineration. The higher residence time is a result of the dramatically reduced flue gas volume that occurs when some or all of the combustion air, in an existing system, is replaced with oxygen. For a given PCC, the gas velocities through the system will be lower with OEC compared to air/fuel combustion. Therefore, the residence time will increase, which improves the mass transfer within the system. This increases the destruction efficiency of the hazardous pollutants in the process. The higher flame temperatures associated with OEC ensure that all hydrocarbons in the incinerator will be well above their ignition temperatures. Therefore, if there is sufficient oxygen for combustion, the hazardous pollutants should combust. The turbulence in the flame may be higher for OEC compared to conventional air/fuel combustion. However, the overall turbulence level in the combustion chamber will be reduced because of lower gas velocities. This must be offset by the increased residence time and higher gas temperatures in order to ensure adequate DRE. As will be shown, the DRE generally increases using OEC.

Table 16.4 shows U.S. EPA emission standards for new and existing municipal solid-waste incinerators for dioxins/furans, particulates, NO_x, and CO (note that there are standards for other pollutants like heavy metals, which are not considered here).

16.4.4 Pollution Control Strategies

The pollution control strategies are organized slightly differently here compared to other chapters, due to the unique aspects of incineration. The subsections below include waste pretreatment, thermal destruction, oxygen-enhanced combustion, combustion modification, process modification, and air pollution control equipment. Waste pretreatment and oxygen-enhanced combustion are pretreatment strategies. Air pollution control equipment is a post-treatment strategy. Thermal destruction does not strictly fit into any of the typical pollution control strategies as used here

Table 16.4 U.S. EPA Standards for Municipal-Waste Incinerators, Final Rule, December 1995

Pollutant	New MSW incinerators	Existing MSW incinerators
Dioxins/furans	13–7 ng/dscm total mass (this is equal to about 0.1 to 0.3 ng/dscm TEQ)	125–30 ng/dscm total mass depending on the unit (this is equal to about 0.3–3 ng/dscm TEQ)
Particulate matter	0.010 gr/dscf (24 mg/dscm)	0.012–0.030 gr/dscf (25–70 mg/dscm) depending on the unit
SO_2	30 ppmv or 80% reduction	31–80 ppmv or 75–50% reduction depending on the unit
NO_x	150 ppmv	200–250 ppmv depending on the unit
CO	50–150 ppmv depending on unit	50–200 ppmv depending on unit

Source: Adapted from Ref. 13.

because it is the function of the incinerator itself, rather than being an auxiliary process to control emissions.

16.4.4.1 Waste Pretreatment

Pretreating the incoming waste materials generally involves removing chemicals that may be hazardous or difficult to burn without generating pollutants that may be more difficult to remove prior to exhausting into the atmosphere. One example is removing automotive batteries that contain lead, which is not destroyed during incineration but can be carried with the exhaust products and would need to be removed prior to exiting the exhaust stack. In the case of preventing dioxin formation, it would be desirable to eliminate any sources of chlorine to prevent polychlorinated dibenzo-para-dioxin (PCDD) formation. It is also advisable to pretreat solid-waste materials by shredding or drying to ensure complete combustion to avoid PCDD formation.

16.4.4.2 Thermal Destruction

The primary technique used to control dioxin and furan emissions from combustion processes is to ensure adequate temperatures and residence times are achieved in the flame zone. Oxidation of dioxins and furans is essentially complete theoretically at temperatures above 500°C (900°F). In practice, it is generally agreed that combustion temperatures of 850°C (1600°F) and a gas residence time of 2 sec or 1000°C (1800°F) and a gas residence time of 1 sec are necessary for total destruction [22]. Ohta et al. [35] experimentally showed that gas temperatures above about 800°C (1500°F) produced far less dioxins/furans from the incineration of polyvinylidene chloride than those produced at temperatures below about 750°C (1400°F). Adequate excess oxygen and turbulent mixing are also required to ensure complete combustion. All of these are normally easily achieved in typical combustion processes. However, it is important that adequate oxygen is available for oxidation and there is proper mixing to ensure complete destruction. It is important to ensure that there is adequate temperature, residence time, mixing (turbulence), and O_2 to ensure dioxin and furan destruction. High temperatures, good mixing, and higher O_2 levels are good for dioxin and furan destruction but are not good for other pollutants such as NO_x. Therefore, a balance is needed to ensure proper destruction of dioxins and furans without adversely affecting other pollutants unless downstream cleanup equipment will be used.

Hatanaka et al. [36] experimentally studied the effect of gas temperature on dioxin/furan formation in a laboratory-scale fluidized-bed incinerator. The results showed that dioxin/furan formation increased with temperature in the primary combustion zone and decreased with temperature in the secondary combustion zone. The increase in formation with temperature in the primary zone was explained by the devolatilization of the solid surrogate waste. The results showed that high temperatures in the secondary combustion zone can ensure burnout of any dioxins coming from the primary combustion zone.

The formation of active carbon (e.g., from soot formation) should be avoided as this can act as a catalyst to enhance dioxin and furan formation. This means that the oxidation reactions in the incinerator must be controlled to ensure that adequate oxygen is available to prevent soot formation. While this is a relatively easy proposition for gaseous waste materials, it is more difficult for liquid and solid wastes, which are some of the primary materials to be incinerated.

Incinerators can also effectively be used to destroy dioxins and furans. Harjanto et al. [37] experimentally showed that destruction efficiencies of about 99% were achieved in a laboratory-scale waste incinerator for removing dioxins and furans from contaminated soil. Treatment of contaminated soil in an incinerator is referred to as thermal remediation.

16.4.4.3 Oxygen-Enhanced Combustion

Many of the benefits associated with OEC relate to increasing the partial pressure of O_2 in the incinerator. The combustion process becomes more reactive, which tends to increase the destruction efficiency of any hydrocarbons in the system. This lowers the pollutant emissions. A schematic of some of the benefits of using OEC is shown in Fig. 16.3. Figure 16.4 shows a schematic of oxygen lancing used to enhance the performance of an incinerator. Figure 16.5 shows the use of oxygen enrichment to enhance the performance of a municipal solid-waste incinerator, and Fig. 16.6 shows the use of oxygen enrichment in a sludge incinerator.

Higher DRE

The destruction and removal efficiency (DRE) is typically regulated for Body hazardous pollutants as defined by the Resource Conservation and Recovery Act (RCRA). These pollutants include, for example, polychlorinated biphenyls (PCBs) and principal organic hazardous constituents (POHCs) that may be found, for example, in contaminated soil. OEC can increase the DRE in an incineration process. This is accomplished by a combination of increased residence time within

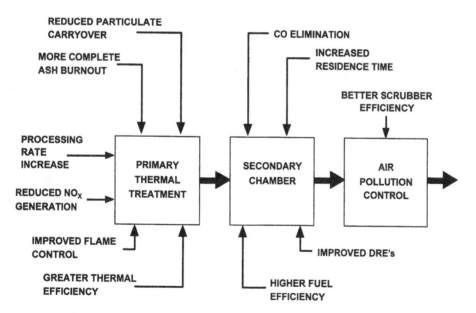

Figure 16.3 Benefits of using oxygen in waste incineration applications. (From Ref. 2. Courtesy of CRC Press.)

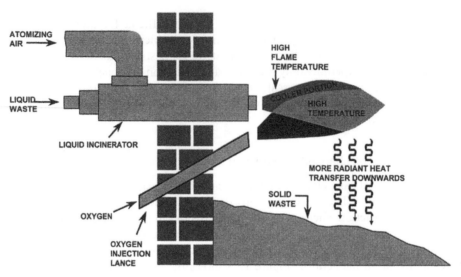

Figure 16.4   Oxygen lancing into an incinerator. (From Ref. 2. Courtesy of CRC Press.)

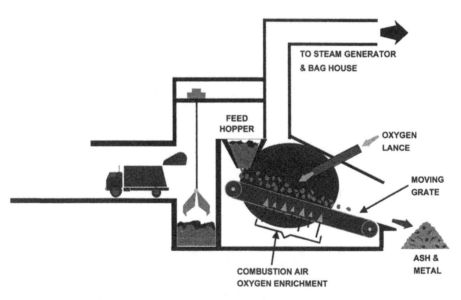

Figure 16.5   Oxygen enrichment of a municipal solid-waste incinerator. (From Ref. 2. Courtesy of CRC Press.)

the incinerator and higher gas temperatures. These are two of the "three Ts of incineration." The higher residence time is a result of the dramatically reduced flue gas volume that occurs when some or all of the combustion air, in an existing system, is replaced with oxygen. For a given PCC, the gas velocities through the system will be lower with OEC compared to air/fuel combustion. Therefore, the residence time will increase, which improves the mass transfer within the system. This increases the destruction efficiency of the hazardous pollutants in the process. The higher flame

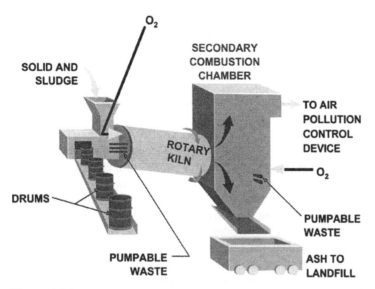

Figure 16.6 Oxygen enrichment of a sludge incinerator. (From Ref. 2. Courtesy of CRC Press.)

temperatures associated with OEC ensure that all hydrocarbons in the incinerator will be well above their ignition temperatures. Therefore, if there is sufficient oxygen for combustion, the hazardous pollutants should combust. As previously discussed, the turbulence in the flame may be higher for OEC compared to conventional air/fuel combustion. However, the overall turbulence level in the combustion chamber will be reduced because of lower gas velocities. This must be offset by the increased residence time and higher gas temperatures in order to ensure adequate DRE. As will be shown, the DRE generally increases using OEC.

Lower Combustibles

Combustible gases, like carbon monoxide and hydrogen, may be produced when fossil fuels are incompletely burned. These are sometimes referred to as products of incomplete combustion (PICs) or as unburned hydrocarbons (UHCs). There are regulatory limits on the amount of PICs that may be emitted into the atmosphere. High levels of PICs may occur when there is insufficient oxygen for complete combustion, generally referred to as fuel-rich combustion. High PIC levels may also occur when there is insufficient mixing between the fuel and the oxidizer, which may happen in a poorly mixed diffusion flame. In addition, high PICs may occur when there are transient increases in either the fuel flow rate or in the amount of hydrocarbons in the material being processed, or if the oxidizer flow is not properly adjusted and distributed. For example, in an incinerator the heating value of the material being incinerated may vary considerably with time.

Transient feeds of high heating value waste may lead to spikes or "puffs" of PICs in the exhaust gases. There are two common ways of handling this problem. One is to set the oxidizer flow rate high enough to handle the worst case. However, that means that there will be excess oxygen for the vast majority of the time. This lowers fuel efficiency and also produces higher NO_x emissions. Therefore, running with excess oxygen is not generally desirable. A better method for handling puffs is to have either a manual or automatic control system that can compensate for the changes. An example of a manual system would be where the material to be processed is analyzed before it enters the combustion system. The combustion system is then adjusted according to the incoming material. Another way is to mix the incoming materials in such a way as to keep the heat content nearly constant. This method is expensive and labor intensive. In an automatic control system, high levels of PICs, usually in the form of CO, are usually sensed in the primary combustion chamber. The PIC measurement system must respond quickly in order for this to be used in a real-time, feed-forward control system. Adjustments are then made in the secondary combustion chamber (SCC) to ensure complete combustion before the gases exit the exhaust system. These adjustments usually include increasing the oxidizer flow, somewhere downstream of the PCC. Oxygen has successfully been used to control these "puffs" in incinerators [38]. Oxygen is much more effective at controlling puffs than is air because it has much higher reactivity since it does not have the additional diluent nitrogen contained in air. The reduction in nitrogen in the combustion products also allows the system to react more quickly because there is less dilution so that adding oxygen to handle puffs produces a quicker response than adding air, which contains nearly 80% nitrogen. A coincidental benefit is that the gas flow through the combustion system is generally slower for an air/fuel incinerator that has been retrofitted with OEC compared to the original air/fuel system, which means that there is more time for the control system to react to the transient puffs.

Two examples of using O_2 to control CO spikes both concern fixed incinerators where the incoming waste feed would be automatically stopped, because of regulated limits, if high levels of CO were detected [15]. Figure 16.7 shows a schematic of a control scheme for controlling CO spikes. O_2 was added to the combustion air in the afterburner of a fixed-hearth incinerator to bring the total O_2 in the oxidizer up to a minimum of 27%. This nearly doubled the throughput by virtually eliminating the

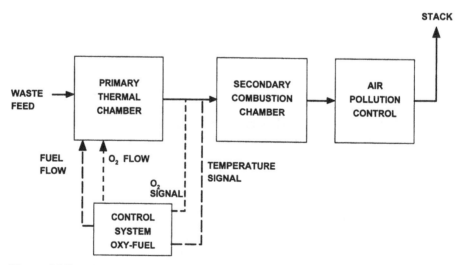

Figure 16.7 Control schematic for using oxygen-enhanced combustion to minimize CO spikes. (From Ref. 2. Courtesy of CRC Press.)

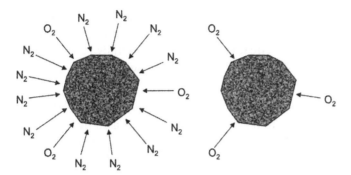

Figure 16.8 Oxygen-enhanced combustion overcomes mass-transfer limitations to increase solid waste and ash burnout: left-hand figure is for air/fuel combustion, right-hand figure is for oxy/fuel combustion.

CO spikes. In a fixed rotary kiln incinerator, pure O_2 was injected at the firing end of the kiln. This virtually eliminated the CO spikes, which increased the processing rates by 25–40%, depending on the waste composition.

Higher Ash Burnout

Because of higher temperatures and more chemical reactivity, OEC can increase the completeness of combustion of the solids being processed in an incinerator. This may have a number of important benefits. In certain geographic locations, regulations limit the amount of hydrocarbons that may be present in both the bottom ash and the fly ash. If the level is too high, then the ash may need to be either reprocessed or be classified as hazardous. In either case, this is usually costly. Oxygen has been used to solve this problem. Less ash is produced and the ash that is produced contains fewer hydrocarbons. Figure 16.8 shows how the removal of nitrogen from the combustion system improves the mass transfer, which leads to higher ash burnout.

A recently patented technology describes the use of oxygen to remove unburned carbon in fly ash [39]. The ash is fed through a heated stainless-steel chamber. Oxygen is used to accelerate the combustion of the carbon. The carbon content of the ash can be brought down below 0.7% so that the fly ash can be sold for use in making cement, instead of being sent to a landfill at a substantial cost.

Lower Particulate Emissions

Lower particulate emissions can result when an existing air/fuel incineration process is enhanced with oxygen. Because of the large reduction in flue gas volume, the gas velocity through the incinerator is reduced. This tends to reduce the amount of solid particles that are entrained by the gas flow through the incinerator. Therefore, there will usually be a substantial reduction in the particulate emissions from the incinerator. This benefit is particularly important where a high-ash waste is being processed. The quantity of fly ash that is produced is generally reduced. This decreases the load on the particulate removal system, which might consist of an electrostatic precipitator or a baghouse.

Increased Pollutant Removal Efficiency

OEC greatly reduces the flue gas volume in a combustion system by the removal of diluent N_2. This results in the concentration of the off-gases, where any pollutants are then contained in a smaller off-gas volume flow. This makes it easier to remove them with existing off-gas treatment equipment such as scrubbers, electrostatic precipitators, or baghouses. Also, the treatment equipment generally becomes more efficient at removing the pollutants because of the higher pollutant concentrations.

Reduction of Stack Opacity

OEC has been used to reduce the opacity in the exhaust gases of a waste chemical boiler that had no scrubbing system to remove PICs that were caused by inadequate residence time in the boiler [15]. O_2 was used to atomize the waste liquids. This not only reduced the opacity, but it also increased the overall thermal efficiency of the boiler.

Waste Processing Flexibility

OEC can give the incinerator operator more flexibility in the types of wastes that can be processed. One important example of the variation in waste composition is the moisture content. For example, in a municipal waste incinerator, the incoming moisture may vary dramatically, depending on the weather. If it has been warm and dry for several days, the trash will contain less moisture than if it has been raining for several days. Wet trash presents a much larger heat demand for the incinerator, compared to dry trash. Oxygen can be used to supplement an incinerator when more intense energy release is needed to vaporize large quantities of water contained in the waste. OEC is also effective for incinerating high-moisture (>20%), low heating value (500 Btu/lb or 1 MJ/kg) wastes, such as the contaminated soil at a Superfund site [40]. In a process that will be discussed in more detail later, OEC has been used to add the capability of sludge disposal at a municipal solid-waste incinerator facility [41].

Improved System Stability

OEC has been used in a commercial hazardous waste incinerator to improve the system stability [42]. The primary advantage of the OEC system was the reduction in excursions and system upsets that required operator intervention. With the old air/fuel system, the excursions were caused by the nonsteady-state nature of the containerized waste feed. That system was unable to compensate for the nonuniform release of hydrocarbons as the containers opened inside the incinerator. The waste feed would be cut off whenever these excursions led to low O_2 concentrations in the exhaust. These excursions were minimized by injecting O_2 into both the PCC and the SCC, through lances cooled by a water/glycol mixture. O_2 flow in the afterburner was controlled by an automatic feedback control system, set to a desired level of O_2 in the exhaust. Reducing the excursions caused the average solid-waste feed to increase by nearly 11%.

16.4.4.4   Combustion Modification

Rogaume et al. [43] experimentally studied waste incineration of plastics and wood in a pilot-scale incinerator to determine methods to reduce NO_x emissions. Careful control of air staging was shown to reduce NO_x emissions. The lowest emissions were found with approximately equal flow rates of air in the primary and secondary injectors, where each flow rate was between 80 and 110% of the air required for complete combustion.

16.4.4.5 Process Modification

Rogaume et al. [44] have shown experimentally that the air flow through the incinerator is very important in determining emission rate formation for both CO and NO_x. The objective of the study was to simulate the combustion of solid waste in a moving grate industrial incinerator. A pilot-scale fixed-bed counterflow reactor was used for the study. The results showed that the amount of primary air introduced with the fuel had a significant impact on the CO emissions. Substoichiometric air flows greatly enhanced CO production while superstoichiometric air flows had little effect on CO generation. Secondary air used downstream of the primary burning zone increased reactor temperatures and CO destruction. For the conditions studied, NO_x production was only influenced by the air flow in the secondary combustion region and increasing the total excess air flow increased NO_x (see Chap. 6).

16.4.4.6 Air Pollution Control Equipment

A common way to minimize emissions from incineration processes is to include some type of air pollution control equipment at the end of the process prior to emitting any exhaust gases into the atmosphere. Common equipment used in incinerators includes: electrostatic precipitators, fabric filters, spray dryers, dry sorbent injection systems, wet scrubbers, low NO_x combustion technologies, NO_x post-treatment controls such as selective catalytic reduction, and mercury control technologies. While incinerators can generate the typical pollutants similar to other industrial combustion processes (e.g., NO_x, SO_x, particulates), they may also emit some other pollutants that are generally much less common including heavy metals, dioxins, and furans. These pollutants are not due to the combustion process in the incinerator, but rather result from the waste materials being processed in the incinerator, which may include hazardous wastes. Most incineration facilities use some type of scrubbing system, an electrostatic precipitator (ESP), and/or a baghouse. Other treatment equipment may be required, depending on the incoming waste materials.

Fritsky et al. [45] described an air pollution control system on a medical-waste incineration plant designed to control both particulates and dioxins/furans in a baghouse. The conventional fiberglass filter bags were replaced with a special catalytic filter system designed to destroy dioxins/furans while simultaneously capturing particulates. The catalytic filter replaced a carbon-adsorption system. Measurements showed that the emissions of both particulates and dioxins/furans were well below regulatory limits with the new catalytic filter system.

Liljelind [46] experimentally showed in a laboratory-scale incinerator that a titanium/vanadium catalyst can effectively destroy (>99.9% destruction efficiency) dioxins/furans and also polycyclic aromatic hydrocarbons at temperatures at or above 150°C (300°F) and gas hour space velocities of 8000. This low-temperature effectiveness is important because of the possibility of forming dioxins/furans at low temperatures in ESPs. This catalyst could be located after the ESP to clean up any dioxins/furans that may have formed. Figure 16.9 shows a photograph of an ESP used in a hazardous waste incineration system.

Ruokojarvi et al. [47] demonstrated the use of urea as a dioxin/furan inhibitor in a pilot-scale waste incinerator. An aqueous solution of urea was injected into the flue

Figure 16.9   Electrostatic precipitator used at a hazardous waste incinerator. (Courtesy of Croll, Westfield, NJ.)

gas, which was at a temperature of about 730°C (1350°F). The results showed that proper urea concentrations and injection points can inhibit dioxin/furan formation.

Everaert and Baeyens [48] determined experimentally that temperatures in the electrostatic precipitator should be kept between 180° and 200°C (360°–400°F) to minimize the de novo synthesis of dioxins and to maximize dioxin adsorption on to fly ash where it can be more easily captured. They found a significant increase in dioxin/furan concentrations with temperature up to about 280°C (540°F).

Chang et al. [49] studied dioxin emissions from two municipal solid-waste incinerators in Taiwan. Incinerator A was equipped with an electrostatic precipitator followed by wet scrubbers. Incinerator B was equipped with cyclones, dry lime sorbent injection systems, and fabric filters. The dioxin concentrations were similar at the stack exits although incinerator B had 2.75 times more dioxins at the inlet of the air pollution control equipment. Incinerator A's removal efficiency suffered because dioxins were generated in the air pollution control equipment due to the temperature range. Activated carbon injection in incinerator B has significantly improved the removal efficiency.

16.5   INCINERATION APPLICATIONS

There are many types of incinerators including rotary kiln, fluidized bed, multiple hearth, and cement kiln cofiring to name a few. This section discusses some typical types of incineration processes including municipal solid-waste incineration, mobile incineration, transportable incineration, and fixed incineration. The U.S. EPA gives general emission guidelines for commercial and industrial solid-waste incinerators [50,51].

16.5.1 Municipal Waste Incineration

The U.S. EPA classifies refuse combustors for burning municipal solid waste (MSW) into three categories: mass burn, refuse-derived fuel (RDF), and modular combustors [20]. In mass burn, there is no preprocessing of the waste materials, which are placed on a moving grate that moves through the combustor. These are predominantly used for larger throughput processing rates. Combustion air is supplied above and below the grate. There are a variety of designs available for this type of system. Figure 16.10 shows a schematic of a typical mass burn combustor. RDF combustors burn preprocessed waste to ensure operational consistency. Figure 16.11 shows a schematic of a typical RDF combustor. Modular units generally burn waste that has not been preprocessed, but the combustors are typically shop-fabricated and are, therefore, of smaller size compared to mass burn units. Figure 16.12 shows a schematic of a modular MSW combustor.

Clarke [52] has written a good primer on MSW incineration. There has been a decided shift in preference away from landfilling and incineration with no energy recovery towards waste prevention, recycling, composting, and waste-to-energy (WTE) plants utilizing incineration for waste management. An example of a MSW incineration process is shown in Fig. 16.13. A photograph of an actual municipal waste incineration plant is shown in Fig. 16.14. Some of the important design factors for municipal incineration with energy recovery include:

1. Waste preparation and stoking (screening, processing, and feeding into the incinerator)
2. Waste incineration and emissions mitigation via efficient combustion
3. Energy extraction
4. Emissions mitigation via neutralization and/or capture in control devices
5. Ash handling

Table 16.5 shows a process schematic for MSW incineration systems incorporating energy recovery. Figure 16.15 shows a schematic of a MSW incinerator with energy recovery.

Rigo et al. [53] note that there are two predominant types of MSW incinerators: field erected modular mass-burn facilities that burn waste as it is received and those that burn refuse-derived fuel (RDF) where the waste has been processed first to improve its quality and uniformity for more consistent and predictable incineration operation.

Hamilton et al. [54] discuss the use of the Shell Denox selective catalytic reduction (SCR) system (see Chap. 6) to minimize NO_x emissions from the exhaust gases of waste incinerators. The unique aspect of the process is that it is effective even down to temperatures as low as 250°F (120°C), which is usually much lower than the preferred operating window for good removal efficiency. Field results at an incinerator in the Netherlands showed 90% removal efficiency with a system pressure drop of less than 2 in. of water column.

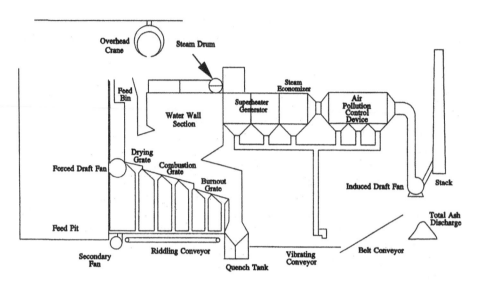

Figure 16.10 Schematic of a typical mass burn waterwall combustor. (From Ref. 20.)

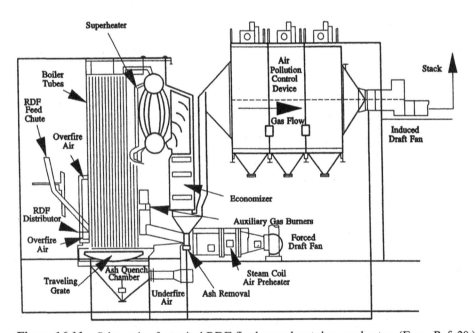

Figure 16.11 Schematic of a typical RDF-fired spreader stoker combustor. (From Ref. 20.)

Diestelkemper [55] has reviewed two different approaches to controlling emissions from WTE plants: one with an extensive process chain and one with a simple technology. Some of the factors that are important in determining the appropriate air pollution control equipment include:

- The waste composition impacts the concentrations of the individual pollutants.

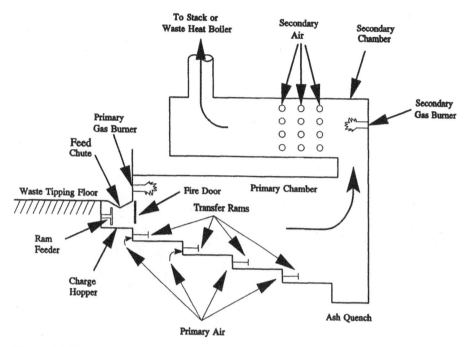

Figure 16.12   Schematic of a modular MSW combustor. (From Ref. 20.)

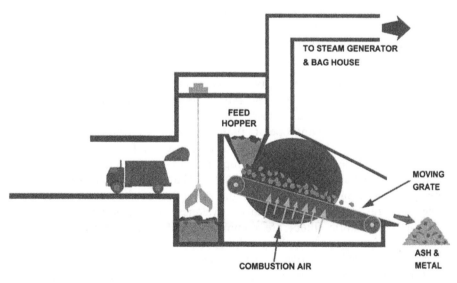

Figure 16.13   Municipal solid-waste incinerator. (From Ref. 3. Courtesy of CRC Press.)

- The properties of the flue gas (temperature, pressure, and moisture content) affect the choice of control equipment.
- The planned service life and capacity utilization impact the economics of the pollution control equipment.

Figure 16.14 Municipal waste incineration plant.

- The local conditions (e.g., landfill capacity in the vicinity, waste water permit requirements).
- Country-specific regulations.
- Space constraints.

Flue gas cleaning equipment at two different sites (one in Germany and one in The Netherlands) were compared. The more complicated cleaning system at the German plant, required in order to meet the stricter German regulations, was much more expensive to purchase and operate than the one in The Netherlands. The U.S. EPA has studied how to control particulate emissions from municipal waste incinerators [56]. Table 16.6 shows a comparison of uncontrolled and controlled emissions taken from a report issued in 1980 [57]. The table shows the wide range of potential particulate emissions, which is highly dependent on the type of material fed into the incinerators. Settling chambers, cyclones, wet scrubbers, electrostatic precipitators, and fabric filters are commonly used in incinerators to control particulate emissions (see Chap. 9 for a discussion of these techniques).

Liao et al. [58] investigated CO_2 removal from municipal solid-waste incinerators using chemical absorption [58]. Three different solvents were tested: NaOH, monoethanolamine (MEA), and ammonia (NH_3). The highest removal efficiency was 76% using NaOH under a specific set of operating conditions. The study investigated the effects of the solvent concentration, the liquid solvent-to-gas (L/G) ratios, CO_2 inlet concentration, and the gas inlet concentration. Removal efficiency increased with solvent concentration and L/G ratio, decreased with CO_2 inlet concentration, and was not affected by the gas inlet concentration for the range investigated of 30°–50°C (90°–120°F).

Table 16.5 Municipal Waste Process

Precombustion			Postcombustion	
Inputs	Preparation	Incineration	Emissions control	Outputs
MSW	Source separation	Single chamber	ESP and/or Baghouse	Emissions
Fossil fuels for maintaining	Or	or	Scrubber	Particulates
furnace temp.	Materials recovery at the			
	incinerator			
Water for scrubber		Multiple chamber	NO_x controls	Heavy metals
Reagents for air pollution	Tip floor screening to remove	or	Activated carbon	Acid gases
controls	toxics			
	Mixing	Fluidized	Continuous	NO_x
		bed	emission monitors	PICs
	Batch or continuous waste	or		
	feed			
	Drying	Rotary combustor	Ash management	Ash
		or	Containerized facility	Bottom ash
			and trucks	
		Refuse derived fuel	Ash landfill	Fly Ash
		Continuous process monitors	Ash reuse	
		Automatic combustion controls	Heat recovery	Energy
		Under fire and over fire air	Electricity generation	Steam
		injection		Electricity

Source: Ref. 52. (Courtesy of Air and Waste Management Association.)

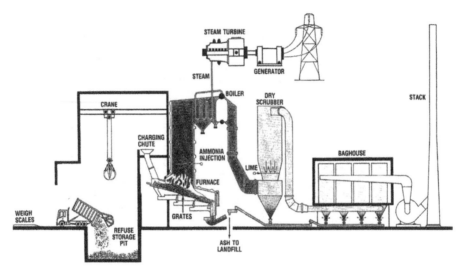

Figure 16.15   Municipal solid-waste incinerator incorporating energy recovery. (From Ref. 52. Courtesy of the Air & Waste Management Association.)

Table 16.6   Dioxin Emission Factors for Sludge Incinerators

Source category	2,3,7,8-TCDD		Total TCDD		Total PCDD	
	µg/Mg	lb/ton	µg/Mg	lb/ton	µg/Mg	lb/ton
Uncontrolled			6.3 E+01	1.3 E–07	2.7 E+00	5.4 E–09
Controlled						
Cyclone						
Cyclone/impingement						
Cyclone/venturi			1.4 E+00	2.8 E–09		
Cycone/venturi/ impingement	3.0 E–01	6.0 E–10				
Electrostatic precipitator (ESP)						
Fabric filter						
Impingement	5.0 E–01	1.0 E–09	2.8 E+01	5.6 E–08	3.7 E+00	7.4 E–09
Venturi						
Venturi/impingement/ afterburner	9.0 E–01	1.8 E–09				
Venturi/impingement	2.0 E+00	4.0 E–09				
Venturi/impingement/ wet ESP						
Venturi/wet ESP						

Source: Ref. 20.

The U.S. Environmental Protection Agency has developed extensive guidelines related to MSW incinerators. For large MSW incinerators where construction began after December 20, 1989 and on or before September 20, 1994, 40 CFR 60 Subpart Ea applies [59]. The standard defines MSW as "household, commercial/retail, and/or institutional waste." For large MSW incinerators constructed before September 20, 1994, 40 CFR 60 Subpart Cb applies [60]. Section 60.33b discusses emission guidelines for metals, acid gases, organics, and nitrogen oxides. For example, nitrogen oxide guidelines range from 205 to 250 ppmvd at 7% O_2 depending on the type of incinerator. For large MSW incinerators where construction started after September 20, 1994 or where modification or reconstruction started after June 19, 1996, 40 CFR 60 Subpart Eb applies [61]. Subpart AAAA applies to small MSW incinerators where construction started after August 30, 1999 or where modifications were started after June 6, 2001 [62]. That standard regulates the following pollution emissions: organics including dioxins/furans, cadmium, lead, mercury, opacity, particulates, hydrogen chloride, nitrogen oxides, hydrogen oxides, sulfur dioxide, carbon monoxide, and fugitive ash. Subpart BBBB applies to small MSW incinerators constructed on or before August 30, 1999 [63].

The performance of these incinerators can be improved by using OEC [64]. The economic incentives include: increased waste-processing capacity, greater thermal efficiency, increased production in a waste-to-energy facility, reducing the demand on the exhaust system, and a smaller air pollution control system. Increased capacity may be particularly important for many waste processors that are at their maximum capacity, since it is usually difficult to obtain permits to build new facilities. The environmental incentives include: improved ash burnout, lower hydrocarbon emissions, lower CO, greater flexibility and control, and the ability to burn low-heating-value wastes such as dewatered sludge. Another application of OEC to overcome thermal limitations is in thermal pyrolysis of municipal solid waste or refuse-derived fuel (RDF) [65]. The N_2 in air impedes the pyrolysis process, which is commonly used to recover chemicals and energy. Using oxygen enrichment, high-quality char or gas can be produced from the high ash and moisture content fuels. OEC can increase the heating value of the gas produced in gasification of municipal solid waste by enhancing the devolatilization and evolution of the gaseous products [65].

One of the earliest tests of oxygen enrichment in an MSW incinerator occurred at the Harrisburg, PA Waste-to-Energy Facility in 1987 [66]. The combustion air was enriched with 2% O_2 ($\Omega = 0.23$). The waste throughput, steam production, boiler efficiency, and sludge throughput all increased. The flame stability improved with OEC. OEC has been used in an MSW incinerator to overcome thermal limitations [15]. O_2 was injected, through a diffuser, into the air plenums beneath the waste bed. It was also injected, through a lance, directly on to the bed. This resulted in a 10% increase in the waste processing capacity, an increase in steam production, better overall boiler efficiency, and more complete burnout of the ash. Significant cost savings were realized in ash disposal due to its lower volume and increased density.

The U.S. Environmental Protection Agency (EPA) sponsored a demonstration program to investigate OEC in a pilot-scale incinerator [67]. With only 3% O_2 enrichment ($\Omega = 0.24$), the waste processing rate increased by 24%. OEC did not seem to have any effect on the metal content of the ash. There were some concerns about higher hydrocarbon emissions at the higher throughputs. Further research was recommended. One important commercial consideration was the impact this technology might have on permitting. New or amended permits normally require a lengthy and usually costly review process.

Niehoff et al. [68] discuss the use of oxygen lancing to enhance MSW incinerators. German regulations limit the carbon content in materials to be

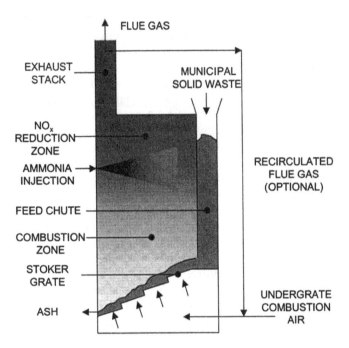

Figure 16.16 Ammonia injection into a municipal waste incinerator for NO_x control. (From Ref. 69. Courtesy of John Zink Co. LLC, Tulsa, OK.)

landfilled, including the ash by-products of waste incineration. Oxygen enrichment in field trials in an MSW incinerator showed the following results:

- Stabilization of the combustion process
- Improved quality of the bottom ash
- Increased waste throughput
- Increased steam production
- No negative effects on gaseous emissions such as CO and NO_x

Figure 16.16 shows a post-treatment technique for reducing NO_x emissions in municipal solid waste incinerators [69]. Éveraert et al. [70] discuss the use of activated carbon to adsorb dioxins and furans in municipal solid-waste incinerators. The carbon may be injected as particulates into the flue gas stream and then captured by fabric filtration.

16.5.2 Sludge Incinerators

The EPA classifies three main types of sludge incinerators: multiple hearth, fluidized bed, and electric infrared (not of interest here). The multiple-hearth incinerator is basically a vertical cylinder containing a series of horizontal refractory hearths, as shown in Fig. 16.17. Cooling air runs through a central pipe supporting the hearths. Sludge is fed into the top with ash exiting the bottom and exhaust gases out the top. Fluidized-bed sludge incinerators are also vertical

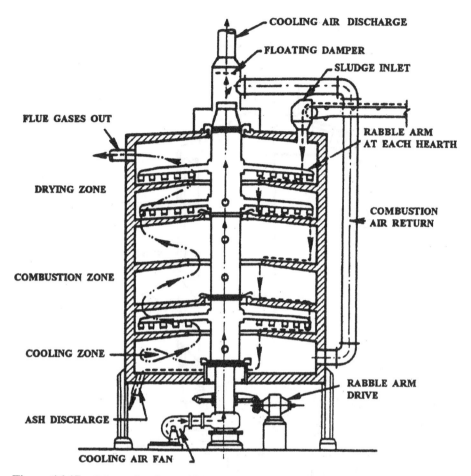

COOLING AIR DISCHARGE

FLOATING DAMPER

SLUDGE INLET

FLUE GASES OUT

RABBLE ARM
AT EACH HEARTH

DRYING ZONE

COMBUSTION
AIR RETURN

COMBUSTION ZONE

COOLING ZONE

RABBLE ARM
DRIVE

ASH DISCHARGE

COOLING AIR FAN

Figure 16.17 Schematic of a multiple-hearth sludge incinerator. (From Ref. 20.)

cylinders where waste is fed into the bottom and both ash and flue gases exit at the top, as shown in Fig. 16.18. Fluidizing air is injected through the floor to enhance heat transfer, mass transfer, and combustion processes.

A schematic of a sludge incineration process is shown in Fig. 16.19. The greatest challenge in sludge incineration is the large amount of energy required to evaporate the large quantity of water contained in the sludge. The heating value of the sludge is minimal. Therefore, large quantities of auxiliary fuel are required. Oxygen has been used to increase the capacity of a multiple-hearth sludge incinerator by 35–55% at a sewage treatment plant in Rochester, NY [71]. Oxygen was injected into the sludge-drying zone at a rate of 1 ton/hr through a series of lances. The amount of auxiliary natural gas fuel used to dry and burn the sludge was reduced by 57%. Emissions (per mass of dry sludge) of total hydrocarbons, NO_x, and CO were reduced by 58, 62, and 39%, respectively. Ruppert et al. [72] discuss the use of oxygen enrichment to enhance the incineration of sewage sludge in a rotary kiln.

Schifftner and Hesketh [73] describe the use of wet scrubbers to control particulate emissions from municipal sludge incinerators. One of the challenges is the

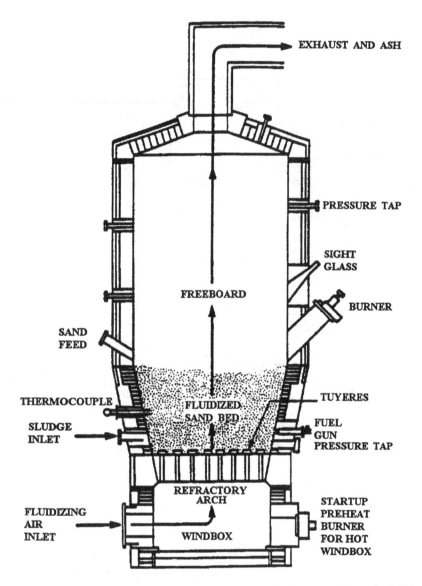

Figure 16.18 Schematic of a fluidized-bed sludge incinerator. (From Ref. 20.)

varying amount of moisture content in the sludge, which necessitates the use of adjustable venturi scrubbers.

An important concern from sludge incinerators is the emissions of dioxins and furans. Table 16.6 presents emission factors for these pollutants developed by the U.S. EPA for sludge incinerators equipped with a variety of air pollution control systems. These factors represent typical emissions from these processes. Actual emissions may vary significantly from these published values, which are often used as guidelines for permitting of new installations or for major modifications proposed for existing facilities.

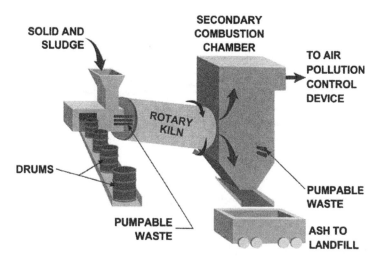

Figure 16.19 Sludge processing in a rotary kiln incinerator. (From Ref. 3. Courtesy of CRC Press.)

16.5.3 Mobile Incinerators

Mobile incinerators are commonly used to clean up contaminated soil and water at Superfund sites. The entire incineration system, including the PCC, the SCC, and the pollution control equipment, is small enough to be transported by road. It can be quickly set up and is usually preferred for smaller size cleanups. One of the first applications to use OEC in incineration was at the Superfund cleanup site at the Denney Farm in McDowell, MO, starting in 1987 [74]. The EPA mobile incinerator was used, along with an OEC incineration technology that was later awarded the prestigious Kirkpatrick Chemical Engineering Achievement Award for the results at this cleanup [75]. Dioxin-contaminated liquids and solids were successfully treated. The OEC system showed impressive performance compared to the original air/fuel system. The throughput was increased by 171%, the specific fuel consumption decreased by 61%, the residence time in the SCC increased by 21%, and CO spikes were reduced, while NO_x levels were unaffected.

A more recent example is a trial burn to destroy PCB-containing electrical transformers and related contaminated materials [76]. The waste material was fed into one end of a rotary kiln. A single oxy/fuel burner, located at the kiln entrance, fired co-current with the feed material. The ash was collected at the kiln exit. The combustion and process off-gases from the kiln were fed into the secondary chamber, which operated at a higher temperature than that of the primary chamber, to maximize destruction of any remaining combustible gases. A block diagram of the process is shown in Fig. 16.20. Before the contaminated soil was processed, the system was tested using surrogate wastes to ensure that the emission requirements could be met. Three different series of tests were conducted using various combinations of fuels and wastes. An oxy/propane burner in the PCC was used to incinerate PCB-contaminated soil for the series A and B tests. The only difference between A and B was that propane and oil with 1% PCB, respectively, were used as fuels in the SCC. In series C, the PCC was not operated, while the fuel for the SCC was oil with 42% PCB.

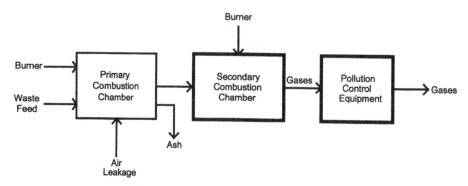

Figure 16.20 Mobile incineration system block diagram. (From Ref. 3. Courtesy of CRC Press.)

Vesta Technology, Ltd (Fort Lauderdale, FL) is a hazardous waste incineration company that provides on-site services throughout North America. Vesta's mobile systems use high-temperature rotary kiln incinerator technologies, coupled with innovative, proprietary designs for flue gas scrubbing. This proven technology destroys PCBs, dioxins, oil sludges, and other hazardous wastes with DREs that exceed Resource Conservation and Recovery Act (RCRA) and Toxic Substances Control Act (TSCA) standards. Vesta has used this technology with both air/fuel and oxy/fuel systems. The decision to try oxygen was mainly influenced by the requirement to reduce the flue gas volume within the system. The result was twofold. First, the lower flue gas velocity in the kiln resulted in less particulate carryover to the secondary combustion chamber (SCC) and the air pollution control system. Actual site operations showed that SCC cleanouts were reduced with the lower flue gas flow, while actually increasing soil processing rates. Second, the flue gas residence time in the SCC increased by 50% which was expected to give higher DREs. Representative data from two similar remediation projects using the same equipment showed that the instantaneous soil throughput rates increased by 50% using oxygen, compared to the air/fuel base case. Of greater importance, however, was the overall average hourly throughput, which increased by 150% due to the elimination of downtime for SCC particulate cleanout when using oxy/fuel. This rate enabled Vesta to complete the remediation project 28 days ahead of schedule [77]. The hourly NO_x emissions were reduced by 52%, while the pounds of NO_x per ton of soil processed were reduced by 66%.

16.5.4 Transportable Incinerators

Transportable incinerators are also commonly used at Superfund cleanup sites. They are larger in size than mobile incinerators and take longer to set up. Therefore, they are used at larger sites because of their increased processing capacity. The trend is to use these, instead of mobile incinerators [78]. OEC has been used to maximize the transportable incinerator throughput by reducing the gas volume and improving the heat-transfer performance. At the Bayou Bonfouca Superfund site in Saint Tammany Parish, LA, cost savings were estimated to be nearly $3 million using OEC instead of air/fuel.

The example given here shows how OEC reduced particulates in a transportable incinerator. Williams Environmental Services (WES), located in Auburn, AL, used an incineration system consisting of a co-current rotary kiln, a hot cyclone, a secondary combustion chamber, a quench tower, baghouses, an induced draft fan, an acid gas absorber, and an exhaust stack. Air/fuel and oxy/fuel were used in this incineration system at two Superfund sites: a bankrupt wood-treating operation in Prentiss, MO, and at the Bog Creek Farm site in Howell, NJ. At the Prentiss site, the kiln was equipped with two air/fuel burners while at the Bog Creek Farm site, a single oxy/fuel burner was installed. The Prentiss site contained 9200 tons (8300 metric tons) of creosote-containing soil. The Bog Creek Farm site contained 25,000 tons (23,000 metric tons) of soil that contained VOCs including benzene derivatives, chlorinated hydrocarbons, and semivolatile organics such as naphthalene and phthalates.

OEC was selected for the Bog Creek site, due to the particulate emissions criteria set by the New Jersey Department of Environmental Protection (NJDEP). The primary problem encountered during the startup at the Prentiss site was higher than expected fines carryover. About 50% of the ash output was from the air pollution control system. The Bog Creek site was located near the New Jersey coastline where the soil is naturally sandy. The existing particulate emission limit of 0.03 grain per dry standard cubic foot (gr/dscf) or 0.07 g per dry standard cubic meter (g/dscm) was reduced to a more rigorous standard of 0.015 gr/dscf (0.034 g/dscm) for this site, to prevent contaminated sand from entering the atmosphere. At a soil feed rate of 20 tons/hr (18 metric tons/hr), the equivalent Prentiss data indicated that the estimated combined emissions (soil and metals) would have to be reduced by 65% to fall below the newly prescribed limit. OEC technology was selected to meet the tougher particulate standard, without causing a delay in the schedule [79]. The Bog Creek site was the first Superfund site to use OEC incineration in the northeast region of the United States and the first North American site to use OEC in a commercial, transportable incinerator for the entire project. The processing rate increased using OEC. All emission and ash requirements were satisfied. The site was cleaned up 60 days ahead of schedule. In this case, there was no reduction in NO_x by using oxy/fuel, which was probably due to high air leakage into the kiln as noted by the increased oxygen in the flue gas.

16.5.5 Fixed Hazardous Waste Incinerators

Dioxin/furan emissions are a concern in hazardous waste incineration [80]. Emission data are available for incineration in cement kilns, lightweight aggregate kilns, commercial incinerators, on-site incinerators, and boilers. The amount of particulate matter in the system affects the levels of dioxin/furans, which form catalytically on solid surfaces. Devices such as electrostatic precipitators and fabric filters are often used to control particulate emissions and may lower dioxin/furan emissions indirectly by reducing the surface area for formation.

Oxygen-enhanced combustion has been used to reduce NO_x in a fixed-base resource recovery process [81,82]. Giant Resource Recovery (GRR) is a subsidiary of the Giant Group, Ltd, which is involved with cement manufacture and the use of waste materials as fuel and raw materials supplements. GRR processes creosote-contaminated soil through counter-current rotary kilns. By a patented process, the decontaminated soil is then used as a raw material for cement production, thus replacing a certain portion of the traditional feed material stream. The combustion products are ducted into the cement kilns. The kilns' processing rates were limited by two factors. Since the contaminated soil was high in moisture content and low in heating value, more heat transfer was required to increase the throughput while ensuring that the soil's creosote concentration did not exceed permitted levels at the kiln discharge. If the creosote concentration was too high, the soil had to be reprocessed before being sent to the cement kiln. Second, the flue gas volume needed to be minimized to prevent upsetting the cement kiln operation. These criteria were met by using an oxy/fuel burner. The kiln back-end temperature increased over $100°F$ ($38°C$) using an oxy/fuel burner. Over a range of several tests, NO_x emissions per ton of material processed were from 5 to 35% less using oxy/fuel. In this case, large amounts of air infiltration, evidenced by high O_2 concentrations in the flue gas, limited the NO_x reduction using O_2.

Oxygen has been injected into rotary kiln incinerators and the secondary combustion chamber to reduce CO emissions by more than 60% while increasing waste throughput by more than 15% at a German merchant incinerator [83]. In the process, $530\,ft^3$ ($15\,m^3$) of oxygen is injected for each 84 lb (38 kg) drum of waste material.

Acharya and Hay [84] describe a growing trend of converting transportable soil remediating incinerators (SRIs), designed for remediating soil at U.S. Superfund cleanup sites, into industrial waste incinerators. This trend was caused by a number of factors including permitting problems for SRIs, incorrect public perception that SRIs actually endanger the environment they are trying to clean up, and because of changing economics (price per ton of soil processed).

REFERENCES

1. CE Baukal. Oxygen-Enhanced Waste Incineration. In: RA Myers (ed.). *Encyclopedia of Environmental Analysis and Remediation*. New York: John Wiley, pp. 3283–3305, 1998.
2. CE Baukal. Waste Incineration. In: CE Baukal (ed.). *Oxygen-Enhanced Combustion*. Boca Raton, FL: CRC Press, pp. 237–259, 1998.
3. CE Baukal. Waste Incineration. In: CE Baukal (ed.). *Heat Transfer in Industrial Combustion*. Boca Raton, FL: CRC Press, pp. 419–431, 2000.
4. D Motyka. A Mascarenhas. Incineration innovation. *Hydrocarbon Eng.*, Vol. 7, No. 2, pp. 75–77, 2002.
5. AM Sterling, VA Cundy, TW Lester, AN Montestruc, JS Morse, C Leger, S Acharya. In Situ Sampling from an Industrial-Scale Rotary Kiln Incinerator. In R Clement, R Kagel (eds.). *Emissions from Combustion Processes: Origin, Measurement, Control*. Boca Raton, FL: CRC Press, pp. 319–335, 1990.
6. Radian Corporation (Research Triangle Park, NC). Municipal Waste Combustion Study: Assessment of Health Risks Associated with Municipal Waste Combustion Emissions. New York: Hemisphere, 1989.
7. RC Corey. 13 Incineration. In AC Stern (ed.). *Air Pollution*, 3rd ed., Vol. 4. New York: Academic Press, 1977.
8. AC Stern, RW Boubel, DB Turner, DL Fox. *Fundamentals of Air Pollution*, 2nd ed. Orlando, FL: Academic Press, 1984.
9. ASME Research Committee on Industrial and Municipal Wastes. *Combustion Fundamentals for Waste Incineration*, New York: American Society Mechanical Engineers, 1974.

10. WR Seeker. Waste Combustion. *Twenty-Third Symposium (International) on Combustion*. Pittsburgh, PA: The Combustion Institute, pp. 867–885, 1990.

11. CR Brunner. *Handbook of Incineration Systems*. New York: McGraw-Hill, 1991.

12. JOL Wendt. Combustion Science for Incineration Technology. *Twenty-Fifth Symposium (International) on Combustion*. Pittsburgh, PA: The Combustion Institute, pp. 277–289, 1994.

13. JS Lighty, JM Veranth. The Role of Research in Practical Incineration Systems—A Look at the Past and the Future. *Twenty-Seventh Symposium (International) on Combustion*. Pittsburgh, PA: The Combustion Institute, pp. 1255–1273, 1998.

14. D Fusaro. Incineration technology: still hot, getting hotter. *Chem. Process.*, Vol. 54, No. 6, pp. 26–32, 1991.

15. SD Reese. Diverse Experience Using Oxygen Systems in Waste Incineration. Presented at the *Fourth Annual National Symposium on Incineration of Industrial Wastes*. Feb. 28–Mar. 2, Houston, TX, 1990.

16. G Gitman, M Zwecker, F Kontz, T Wechsler. Oxygen Enhancement of Hazardous Waste Incineration with the PYRETRON Thermal Destruction System. In HM Freeman (ed.). *Thermal Processes*, Vol. 1, Lancaster, PA: Technomic, pp. 207–225, 1990.

17. RB Bartone. Incinerators. In L Theodore, J Reynolds (eds.). *Introduction to Hazardous Waste Incineration*. New York: John Wiley, 1987.

18. HE Hesketh. *Air Pollution Control: Traditional and Hazardous Pollutants*. Lancaster, PA: Technomic, 1991.

19. L Theodore, J Reynolds. *Introduction to Hazardous Waste Incineration*. New York: John Wiley, 1987.

20. U.S. EPA. Compilation of Air Pollutant Emission Factors. Vol. I: Stationary Point and Area Sources, 5th ed. Washington, DC: U.S. Environmental Protection Agency Rep. AP-42, 1995.

21. V Cozzani, C Nicolella, L Petarca, L Tognotti. Influence of the Feed Composition on the Product Yields in the Conventional Pyrolysis of Refuse Derived Fuels. In N Piccinini, R Delorenzo (eds.). *Chemical Industry and Environment II*, Vol. 2. Turin, Italy: Politecnico di Torino, 1996.

22. G McKay. Dioxin characterisation [sic], formation and minimisation [sic] during municipal solid waste (MSW) incineration: review. *Chem. Eng. J.*, Vol. 86, pp. 343–368, 2002.

23. T Katami, A Yasuhara, T Shibamoto. Formation of PCDDs, PCDFs, and coplanar PCBs from polyvinyl chloride during combustion in an incinerator. *Environ. Sci. Technol.*, Vol. 36, No. 6, pp. 1320–1324, 2002.

24. U.S. EPA. Inventory of Sources of Dioxin in the United States. Washington, DC: U.S. EPA Rep. EPA/600/P-98/002Aa, 1994.

25. C Procaccini, M Kraft, H Fey, H Bockhorn, JP Longwell, AF Sarofim, KA Smith. PIC Formation During the Combustion of Simple Hydrocarbons in Inhomogeneous Incineration Systems. *Twenty-Seventh Symposium (International) on Combustion*. Pittsburgh, PA: The Combustion Institute, pp. 1275–1281, 1998.

26. HK Chagger, JM Jones, M Pourkashanian, A Williams. The formation of VOC, PAH and dioxins during incineration. *Process Safety Environ. Protect.* Vol. 78, No. B1, pp. 53–59, 2000.

27. L Sorum, O Skreiberg, P Glarborg, A Jensen, K Dam-Johansen. Formation of NO from combustion of volatiles from municipal solid wastes. *Combust. Flame*, Vol. 123, pp. 195–212, 2001.

28. U.S. EPA. Solid Waste Management and Greenhouse Gases: A Life-Cycle Assessment of Emissions and Sinks. Rep. EPA-530-R-02-006. Washington, DC: Environmental Protection Agency, May 2002.

29. KA Weitz, SA Thorneloe, SR Nishtala, S Yarkosky, M Zannes. Impact of municipal solid waste management on greenhouse gas emissions in the United States. *J. Air Waste Mgmt. Assoc.*, Vol. 52, pp. 1000–1011, 2002.

30. HH Frey, B Peters, H Hunsinger, J Vehlow. Experimental and Numerical Evaluation of the Combustion Process in a Waste Incinerator. *Fifth International Conference on Technologies and Combustion for a Clean Environment*, Lisbon, Portugal, Vol. 1. The Combustion Institute—Portuguese Section, pp. 323–335, 1999.

31. D Shin, CK Ryu, S Choi. Computational Fluid Dynamics Evaluation of Good Combustion Performance in Waste Incinerators. *Fifth International Conference on Technologies and Combustion for a Clean Environment*, Lisbon, Portugal, Vol. 1. The Combustion Institute—Portuguese Section, pp. 353–357, 1999.

32. B Carlotti, P Martinetto, Y Denayrolles, G Pierotti. COGENE®: a Process Model for Heat and Power Plants—Application to the Household Waste Incinerators. *Fifth International Conference on Technologies and Combustion for a Clean Environment*, Lisbon, Portugal, Vol. 1. The Combustion Institute—Portuguese Section, pp. 359–364, 1999.

33. J Elkins, N Frank, J Hemby, D Mintz, J Szykman, A Rush, T Fitz-Simons, T Rao, R Thompson, E Wildermann, G Lear. National Air Quality and Emissions Trends Report, 1999. Washington, DC: U.S. Environmental Protection Agency. Rep. EPA 454/R-01-004, 2001.

34. CD Cooper. Vapor Incineration. In WT Davis (ed.). *Air Pollution Engineering Manual*, 2nd ed. New York: John Wiley, 2000.

35. M Ohta, S Oshima, T Iwasa, N Osawa, K Kumatoriya, A Yamazaki, T Takasuga, M Matsushita, N Umedzu. Formation of PCDDs and PCDFs during the combustion of polyvinylidene chloride. *Chemosphere*, Vol. 44, No. 6, pp. 1389–1394, 2001.

36. T Hatanaka, T Imagawa, A Kitajima, M Takeuchi. Effects of combustion temperature on PCDD/Fs formation in laboratory-scale fluidized-bed incineration. *Environ. Sci. Technol.*, Vol. 35, No. 24, pp. 4936–4940, 2001.

37. S Harjanto, E Kasai, T Terui, T Nakamura. Behavior of dioxin during thermal remediation in the zone combustion process. *Chemosphere*, Vol. 47, No. 7, pp. 687–693, 2002.

38. U.S. EPA. American Combustion Pyretron Destruction system—Applications Analysis Report. U.S. EPA Rep. EPA/540/A5-89/008, Office of Research and Development, Cincinnati, OH, June 1989.

39. MP Martinez. Apparatus and Process for Removing Unburned Carbon in Fly Ash. U.S. Patent 5 555 821, issued Sept. 17, 1996.

40. P Acharya, LL Schafer. Consider oxygen-based combustion for waste incineration. *Chem. Eng. Prog.*, Vol. 91, No. 3, p. 55, 1995.

41. GH Shahani, D Bucci, D DeVincentis, S Goff, MB Mucher. Intensify waste combustion with oxygen enrichment. *Chem. Eng.*, special supplement to Vol. 101, No. 2, p. 18, 1994.

42. SL Davidson, SR Fryer, M-D Ho. Optimization of Process Performance of a Commercial Hazardous Waste Incinerator Using Oxygen Enrichment. *Proceedings of the 1995 International Incineration Conference*, Bellevue, WA, p. 631, May 1995.

43. Y Rogaume, R Gadiou, D Schwartz, F Jabouille, M Auzanneau, J-C Goudeau. Reducing NO_x Emissions During Incineration. *Proceedings of 5th European Conference on Industrial Furnaces and Boilers*, Portugal, Vol. II, pp. 597–606, 2000.

44. T Rogaume, M Auzanneau, F Jabouille, JC Goudeau, JL Torero. The effects of different airflows on the formation of pollutants during waste incineration. *Fuel*, Vol. 81, pp. 2277–2288, 2002.

45. KJ Fritsky, JH Kumm, M Wilken. Combined PCDD/F destruction and particulate control in a baghouse: Experience with a catalytic filter system at a medical waste incineration plant. *J. Air and Waste Mgmt. Assoc.*, Vol. 51, No. 12, pp. 1642–1649, 2001.

46. P Liljelind, J Unsworth, O Maaskant, S Marklund. Removal of dioxins and related aromatic hydrocarbons from flue gas streams by adsorption and catalytic destruction. *Chemosphere*, Vol. 42, Nos. 5–7, pp. 615–623, 2001.

47. P Ruokojarvi, A Asikainen, J Ruuskanen, K Tuppurainen, C Mueller, P Kilpinen, N Yli-Keturi. Urea as a PCDD/F inhibitor in municipal waste incineration. *J. Air and Waste Mgmt. Assoc.*, Vol. 51, No. 3, pp. 422–431, 2001.

48. K Everaert, J Baeyens. The formation and emissions of dioxins in large scale thermal processes. *Chemosphere*, Vol. 46, No. 3, pp. 439–448, 2002.

49. MB Chang, JJ Lin, SH Chang. Characterization of dioxin emissions from two municipal solid waste incinerators in Taiwan. *Atmos. Environ.*, Vol. 36, No. 2, pp. 279–286, 2002.

50. Office of the Federal Register. Subpart DDDD: Emission Guidelines and Compliance Times for Commercial and Industrial Solid Waste Incineration Units That Commenced Construction On or Before November 30, 1999. U.S. Code of Federal Regulations Title 40, Part 60. Washington, DC: U.S. Government Printing Office, 2001.

51. Office of the Federal Register. Subpart CCCC: Standards of Performance for Commercial and Industrial Solid Waste Incineration Units for Which Construction is Commenced After November 30, 1999 or for Which Modification or Reconstruction is Commenced on or After June 1, 2001. U.S. Code of Federal Regulations Title 40, Part 60. Washington, DC: U.S. Government Printing Office, 2001.

52. MJ Clarke. Introduction to Municipal Solid Waste Incineration. *Proceedings of the Air and Waste Management Association's 95th Annual Conference and Exhibition*, Paper 45363. Baltimore, MD, June 23–27, 2002.

53. HG Rigo, AJ Chandler, WS Lanier. The Relationship Between Chlorine in Waste Streams and Dioxin Emissions from Waste Combustor Stacks. American Society of Mechanical Engineers' Rep. CRTD-Vol. 36, New York, 1995.

54. DM Hamilton, DM Clark, OL Maaskant, W Ylstra. A Novel Approach to the Removal of Nitrogen Oxides (NO$_x$) from Waste Incinerator Flue Gas Streams Using the Shell Denox System. *Proceedings of the 1996 International Incineration Conference*, Savannah, GA, May 1996.

55. J Diestelkemper. Control of Emissions from Waste to Energy Plants. In Controlling Industrial Emissions—Practical Experience. *IChemE Symposium Series 143*. Warwickshire, UK: Institution of Chemical Engineers, pp. 105–116, 1997.

56. U.S. EPA. Control Techniques for Particulate Emissions from Stationary Sources—Vol. 2, Rep. EPA-450/3-81-005b. Washington, DC: U.S. Environmental Protection Agency, Sept. 1982.

57. U.S. EPA. Source Category Survey: Industrial Incinerators. Rep. EPA 450/3/-80-013. Washington, DC: U.S. Environmental Protection Agency, May 1980.

58. K-J Liao, W-M Lee, S-H Yang, C-N Chen. Removal of CO$_2$ from Flue Gas of Municipal Solid Waste Incinerators. *Proceedings of the Air and Waste Management Association's 95th Annual Conference and Exhibition*, Paper 42863. Baltimore, MD, June 23–27, 2002.

59. Office of the Federal Register. Subpart Ea: Standards of Performance for Municipal Waste Combustors for Which Construction is Commenced After December 20, 1989 and on or Before September 20, 1994. U.S. Code of Federal Regulations Title 40, Part 60. Washington, DC: U.S. Government Printing Office, 2001.

60. Office of the Federal Register. Subpart Cb: Emissions Guidelines and Compliance Times for Large Municipal Waste Combustors That are Constructed on or Before September 20, 1994. U.S. Code of Federal Regulations Title 40, Part 60. Washington, DC: U.S. Government Printing Office, 2001.

61. Office of the Federal Register. Subpart Eb: Standards of Performance for Large Municipal Waste Combustors for Which Construction is commenced After September 20, 1994 or for Which Modification or Re-construction is Commenced

After June 19, 1996. U.S. Code of Federal Regulations Title 40, Part 60. Washington, DC: U.S. Government Printing Office, 2001.

62. Office of the Federal Register. Subpart AAAA: Standards of Performance for Small Municipal Waste Combustion Units for Which Construction is Commenced After August 30, 1999 or for Which Modification or Reconstruction is Commenced After June 6, 2001. U.S. Code of Federal Regulations Title 40, Part 60. Washington, DC: U.S. Government Printing Office, 2001.

63. Office of the Federal Register. Subpart BBBB: Emission Guidelines and Compliance Times for Small Municipal Waste Combustor Units Constructed on or Before August 30, 1999. U.S. Code of Federal Regulations Title 40, Part 60. Washington, DC: U.S. Government Printing Office, 2001.

64. GH Shahani, D Bucci, D DeVincentis, S Goff, MB Mucher. Intensify waste combustion with oxygen enrichment. *Chem. Eng.*, special supplement to Vol. 101, No. 2, pp. 18–24, 1994.

65. AK Gupta. Thermal destruction of solid wastes. *J. Energy Resour. Technol.*, Vol. 118, pp. 187–192, 1996.

66. WS Strauss, JA Lukens, FK Young, FB Bingham. Oxygen Enrichment of Combustion Air in a 360 TPD Mass Burn Refuse-Fired Waterwall Furnace. *Proceedings of 1988 National Waste Processing Conference*, 13th Bi-Annual Conference, Philadelphia, PA, pp. 315–320, May 1–4, 1988.

67. CSI Resource Systems and Solid Waste Association of North America. Evaluation of Oxygen-Enriched MSW/Sewage Sludge Co-Incineration Demonstration Program. U.S. Environmental Protection Agency Rep. EPA/600/R-94/145, Office of Research and Development, Cincinnati, OH, Sept. 1994.

68. T Niehoff, R Dudill, CE Baukal. Oxygen Lancing to Improve Municipal Solid Waste Incineration. *Proceedings of the International Conference on Incineration and Thermal Treatment Technologies*, Savannah, GA, pp. 527–530, 1996.

69. J Colannino. Experimental Design for Combustion Equipment. Chap. 13 in CE Baukal (ed.). *The John Zink Combustion Handbook*. Boca Raton, FL: CRC Press, 2001.

70. K Everaert, J Baeyens, J Degreve. Removal of PCDD/F from incinerator flue gases by entrained-phase adsorption. *J. Air and Waste Mgmt. Assoc.*, Vol. 52, pp. 1378–1388, 2002.

71. G Parkinson. Oxygen enrichment enhances sludge incineration. *Chem. Eng.*, Vol. 103, No. 12, p. 25, 1996.

72. H Ruppert, E Henrich, H Seifert. Incineration of Low-Calorific Wastes in Rotary Kilns. *Proceedings of 5th European Conference on Industrial Furnaces and Boilers*, Portugal, Vol. II, pp. 155–162, 2000.

73. KC Schifftner, HE Hesketh. *Wet Scrubbers*, 2nd ed. Lancaster, PA: Technomic, 1996.

74. M-D Ho, MG Ding. Field testing and computer modeling of an oxygen combustion system. *J. Air Pollut. and Waste Mgmt.*, Vol. 38, No. 9, pp. 1185–1191, 1988.

75. NP Chopey. The tops in chemical engineering achievement. *Chem. Eng.*, Vol. 96, No. 12, pp. 79–83, 1989.

76. CE Baukal, LL Schafer, EP Papadelis. PCB cleanup using an oxygen/fuel-fired mobile incinerator. *Environ. Prog.* Vol. 13, No. 3, pp. 188–191, 1994.

77. CR Griffith. PCB and PCP Destruction Using Oxygen in Mobile Incinerators. *Proceedings of the 1990 Incineration Conference*, San Diego, CA, May 4–18, 1990.

78. P Acharya, D Fogo, C McBride. Process challenges in rotary kiln-based incinerators in soil remediation projects. *Environ. Prog.* Vol. 15, No. 4, pp. 267–276, 1996.

79. FJ Romano, BM McLeod. The Use of Oxygen to Reduce Particulate Emissions Without Reducing Throughput. *Proceedings of 1990 Incineration Conference*, paper 3.3. San Diego, CA, May 14–18, pp. 589–596, 1990.

80. U.S. EPA. Combustion Emissions Technical Resource Document (CETRED). Rep. EPA/530-S-94-014. Washington, DC: U.S. Environmental Protection Agency, 1994.

81. FJ Romano, CE Baukal. How NO_x Emissions are Effected When Using Oxygen Enrichment. *Proceedings of the 1991 Incineration Conference*, Knoxville, TN, pp. 589–596, May 1991.

82. CE Baukal, FJ Romano. Reducing NO_x and particulate. *Pollut. Eng.*, Vol. 24, No. 15, pp. 76–79, 1992.

83. K Fouhy, G Ondrey. Incineration: turning up the heat on hazardous waste. *Chem. Eng.*, Vol. 101, No. 5, pp. 39–43, 1994.

84. P Acharya, GH Hay. Conversion of soil remediating incinerators into industrial waste incinerators. *Environ. Prog.*, Vol. 19, No. 3 pp. 207–217, 2000.

17

Other Industries

There are some lower temperature drying applications employing industrial combustion that are briefly considered in this chapter. These include the paper industry, printing and publishing, textile manufacturing, and food processing. Drying is defined as

"a process in which a wet solid is heated or contacted with a hot gas stream, causing some or all of the liquid wetting the solid to evaporate" [1].

Kudra and Mujumdar have written a new book [2] on advanced drying technologies that cover a wide range of industries, including those discussed in this chapter. Unfortunately, emissions are not included in that book. Some of the more advanced drying techniques include: impinging steam dryers, pulsed fluid beds, airless dryers (see Fig. 17.1), sonic dryers, plasma torch dryers, slush drying, and a variety of hybrid methods. The authors call for more R&D and innovation to advance the state-of-the-art in dryers, which have received much less attention than other types of heating equipment and has been mostly evolutionary up until fairly recently. Table 17.1 shows a classification of typical dryers.

17.1 PAPER INDUSTRY

The paper industry is composed of two primary sectors [3]:

- Pulp and paper mills, which produce mechanical, thermomechanical, and chemical pulps and process these pulps to form paper, paperboard, or building papers.
- Converting operations, which manufacture boxes, tablets, and other finished paper products.

The first sector involves production of paper products from raw wood while the second involves converting those initial products into more specialized end products. The pulp and paper industry produces commodity grades of wood pulp, primary paper products, and paper board products such as: printing and writing papers, sanitary tissue, industrial-type papers, container board, and boxboard [4]. Fig. 17.2 shows a schematic of an integrated paper mill. The only part of the mill that uses industrial combustion is the drying machine. Even there it is only supplemental to the steam-heated cylinders, which do the bulk of the dyring. Figure 17.3 shows an

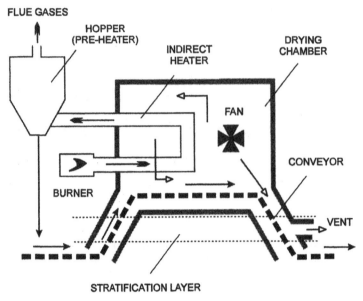

Figure 17.1 Continuous airless dryer. (From Ref. 2. Courtesy of Marcel Dekker.)

elevation view of a Fourdrinier paper machine commonly used to make paper. The steam-heated cylinders in the drying section can be seen in more detail. Figure 17.4 shows a schematic of the Kraft process for handling the pulp and bark used to make the paper, including the treatment of the chemicals in the various reactors, which are considered in more detail below.

Figure 17.5 shows pollution emissions from pulp and paper processes in the United States since 1970 [5]. The data show that CO emissions are the largest quantity of pollutants in this industry. The data also show that there has been a rise in emissions of VOCs since 1985. Other pollutants such as NO_x, PM_{10} and SO_x have been relatively flat since 1985. The Environmental Protection Agency (EPA) has established emission guidelines for Kraft pulp mills [6]. The combustion portion of the Kraft process includes the recovery furnace and the lime kiln as part of the recovery process shown in Fig. 17.6 [7] and the bark boiler to treat solid wastes from the process.

17.1.1 Black Liquor Recovery Boilers

A flow schematic of the Kraft process is shown in Fig. 17.7 [8]. It produces a strong, dark-colored fiber that is made from wood chips in either a batch or continuous digester, under pressure, in the presence of a cooking liquor [9]. The spent chemicals from the process are called black liquor, which is a highly viscous liquid waste containing inorganic cooking chemicals and organic materials such as lignin, aliphatic acids, and extractives. It is a by-product of the chemical pulping process. This black liquor is commonly concentrated and then burned in some type of recovery boiler to recover energy and chemicals. The molten inorganic process chemicals flow through the perforated floor of the boiler to water-cooled spouts and dissolving tanks for recovery in the recausticizing step. A significant pollutant from

Table 17.1 Classification of Dryers

Criterion	Types
Mode of operation	Batch
	Continuous[a]
Heat input type	Convection,[a] conduction, radiation, electromagnetic fields, combination of heat transfer modes
	Intermittent or continuous[a]
	Adiabatic or nonadiabatic
State of material in dryer	Stationary
	Moving, agitated, dispersed
Operating pressure	Vacuum[a]
	Atmospheric
Drying medium (convection)	Air[a]
	Superheated steam
	Flue gases
Drying temperature	Below boiling temperature[a]
	Above boiling temperature
	Below freezing point
Relative motion between drying medium and drying solids	Co-current
	Counter-current
	Mixed flow
Number of stages	Single[a]
	Multistage
Residence time	Short (< 1 min)
	Medium (1–60 min)
	Long (> 60 min)

[a]Most common in practice.
Source: Ref. 2. (Courtesy of Marcel Dekker, Inc.)

this process is particulates. Venturi scrubbers and electrostatic precipitators are commonly used to remove the particulates from recovery boilers. The particulates may also contain hazardous air pollutants (HAPs) [7]. Järvinen et al. [10] describe the development of a detailed computer model for simulating the combustion of black liquor. While the paper does not discuss pollution emissions, it does show the complexity of black liquor combustion

The primary air pollutants from Kraft recovery furnaces include fine particulates, sulfur oxides, and nitrogen oxides [4]. Proper process operation is used to control SO_x emissions [7]. Wallen et al. [11] experimentally demonstrated techniques for reducing NO_x emissions from recovery boilers. The NO_x emissions from recovery boilers are often lower than those from power boilers due to the differences in gas temperatures and fuels. Recovery boilers typically operate at lower temperatures and use fuels (wood chips and bark) that contain a significant amount of water, which lowers the gas temperatures in the combustor. Power boilers use fuels like natural gas that produce higher gas temperatures and therefore more thermal NO_x (see Chap. 6). Fuel NO_x is typically the dominant mechanism in recovery boilers, while thermal NO_x is typically dominant in power boilers. Approximately one-third of the nitrogen in the bark is converted into NO_x according to laboratory

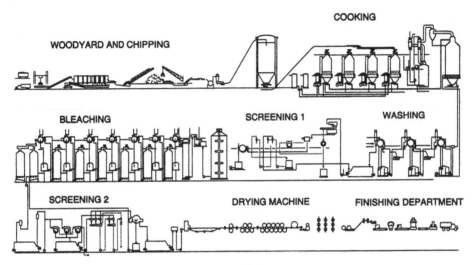

Figure 17.2   Integrated paper mill. (From Ref. 4.)

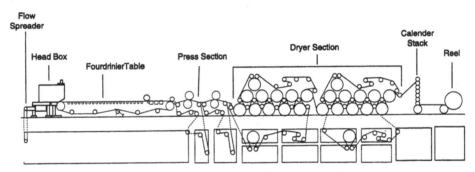

Figure 17.3   Fourdrinier paper machine. (From Ref. 4.)

experiments. Table 17.2 lists some of the factors that affect emissions from recovery furnaces.

The objective of the study by Wallen et al. [11] was to show that combustion modification techniques are preferred to post-treatment techniques like SCR or SNCR for controlling emissions from recovery boilers. Air staging in a recovery boiler was shown to reduce NO_x emissions by up to 50%. The more complete the burnout of the black liquor droplets in the furnace, the higher the NO_x emissions. Reducing excess O_2 reduced NO_x. This correlated with an increase in CO emissions as less air was available to combust fully the CO due to incomplete mixing. The results also showed that the location of the black liquor injection nozzles also influenced NO_x formation. When the burner was located below the liquor guns, NO_x emissions were reduced compared to the case with the burner located above the guns.

17.1.2   Lime Kiln

The concentrated waste black liquor is sprayed into a recovery furnace where the organic products are combusted. The inorganic compounds, mostly the cooking

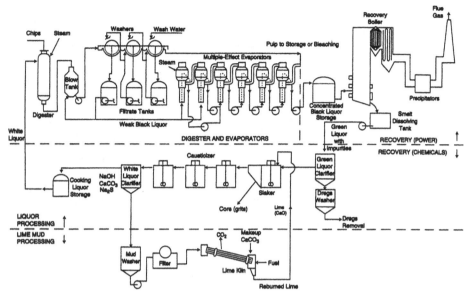

Figure 17.4 Kraft process flow diagram. (From Ref. 4.)

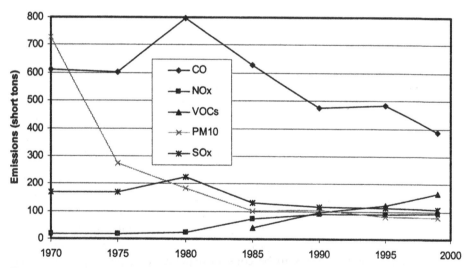

Figure 17.5 Emissions from pulp and paper processes in the United States since 1970 (From Ref. 5.)

chemicals, fall to the bottom of the furnace where chemical reactions occur in a reducing atmosphere. The chemicals are then removed from the furnace as a molten smelt containing mostly sodium sulfide and sodium carbonate. The smelt is dissolved in water in a tank and then treated with slaked lime in a causticizer to produce so-called "white liquor." The sludge resulting form the causticizer is burned or calcined to lime in a lime kiln (see Fig. 17.8). Particulate emissions are the primary pollutants from these lime kilns. The particulates contain sodium salts, calcium carbonate,

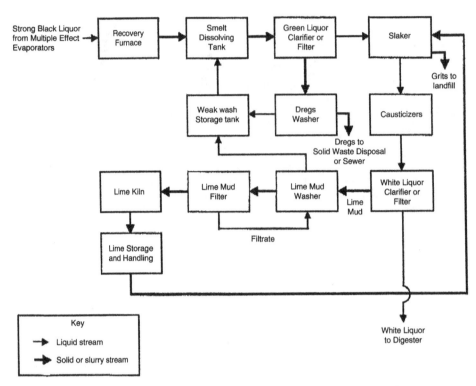

Figure 17.6 Flow diagram of the Kraft chemical recovery area. (From Ref. 7.)

calcium sulfate, calcium oxide, and insoluble ash. Cyclones and venturi scrubbers are commonly used to remove these particulates. Electrostatic precipitators may also be used. Fabric filters are not typically used because of the high moisture content of the particulates, which can cause caking and clogging of the filters [12].

The primary air pollutants from Kraft recovery furnaces include fine and coarse particulates [4]. SO_x emissions are generally not significant and are controlled by proper process operation [7]. Table 17.3 presents some of the factors that affect emissions from lime kilns including the kiln rotation rate, the oxygen level in the kiln, and the mud sodium content. Miner and Upton [13] discuss methods to calculate the CO_2 emissions produced by Kraft mill lime kilns. Paper mills have significantly increased fuel efficiencies over the past few decades, which indirectly reduces CO_2 emissions since less fuel needs to be consumed per unit of paper produced. The complication in estimating CO_2 emissions from Kraft mills is that much of the fuel is biomass—by-products of the waste from converting trees and recycled paper into paper products. These by-products may include wood chips, bark, pulp fibers, scrap paper products, pulping liquors, and the like. There are three sources of CO_2 from Kraft lime kilns: CO_2 produced from the combustion of fossil fuels, CO_2 released from the combustion of biomass fuels, and CO_2 released from $CaCO_3$ in the calcining process. The carbon in the last source originates from the wood chips and is therefore from a biomass source, which is not included in emissions inventories according to the U.S. EPA and the Intergovernmental Panel on Climate Change (IPCC). The fossil fuel is needed to provide the energy for the chemical reaction to

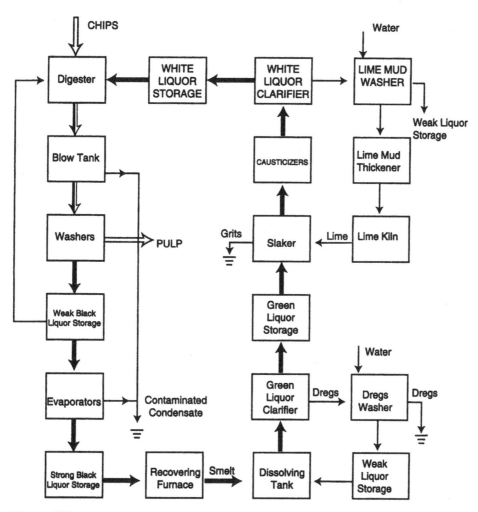

Figure 17.7 Typical Kraft pulping process. (From Ref. 8.)

produce calcium oxide (lime) from calcium carbonate in the lime kiln. The CO_2 emissions from a Kraft lime kiln can then be estimated using standard emission factors for the appropriate fuel being combusted in the kiln.

17.1.3 Bark Boilers

Bark boilers, sometimes referred to as hogged-fuel boilers, are used both to destroy waste products and produce steam for use in the plant. Most of the waste products in paper mills contain significant heating value and can be combusted to generate energy, rather than being disposed of. The primary pollutants from these boilers include particulates, NO_x, and SO_x [7]. Fuel sulfur levels are generally kept low to minimize SO_x emissions. Low-NO_x burners are commonly used to minimize NO_x emissions.

A variety of techniques are used to remove particulates from the products of combustion from bark boilers including gravity settling, cyclones, scrubbers,

Table 17.2 Recovery Furnace Operating and Maintenance Practices Affecting Uncontrolled Emissions

Operating parameter	Emission concern	Operation & maintenance/assessment technique
Fixing rate	Higher-than design firing rate (flue gas volume) leading to: • increased uncontrolled PM emission rate and concentration • nature of particulates altered • increased TRS emission rate • decreased ESP efficiency	Establish baseline comparison of boiler firing rate and (1) grain loading air volume and (2) temperature at the ESP. There monitor parameters would be expected to increase with increased firing rate
Black liquor heating value and solids content	Increased black liquor heating value/solids content leading to increased PM emission rate, especially for heating value increases	Difficult to control/evaluate due to significant daily variations. Ensure inlet grain loading remains within allowable variation for specific ESP.
Total combustion air (excess air) (include primary and secondary air)	Insufficient total combustion air leading to "black out" (incomplete combustion) Total combustion air greater than 125% of calculated theoretical (stoichiometric) air leading to: • increased PM emission ratio • increased flue gas volume to ESP • increased SO_3 formulation, causing particulates to become sticky and to build-up on ESP collection plates—reduces ESP power input and efficiency Primary air exceeding 45% of total air volume leading to: • sharp increase in PM emission rate • increased TRS emission rate	Check total amount of combustion air—the amount needed for complete combustion is normally between 110 and 125% of theoretical air Graph (using DCS if possible) the relationships between percent excess/primary air and • particulates loading to ESP • visible emissions observed from ESP • air volumes to ESP • flue gas temperature to ESP Also, check electrical data–possible indicators of buildup on ESP collection plates include high secondary voltage (> 50 kv) and low secondary current (<100 mA) in inlet fields
Char bed temperature	Increased cher bed temperature leading to • increased PM emission rate • increased flue gas volume to ESP	Assure proper combustion air and firing rate operation using techniques outlined above

Source: Ref. 7.

TRS = total sulfur; DCS = distributed control system.

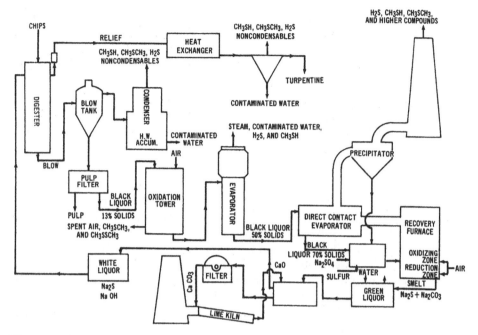

Figure 17.8 Kraft pulping and recovery process diagram. (From Ref. 8.)

Table 17.3 Lime Kiln Operating and Maintenance Practices Affecting Uncontrolled Emissions.

Operating parameter	Emission concern	Operation & maintenance/ assessment technique
Kiln rotation ratio	Increases above normal operating ranges can increase emissions	Compare rate to normal baseline rates using process monitor
O_2 level	Increases above normal operating O_2 levels exiting the kiln can increase emissions	Compare O_2 levels to normal baseline levels using O_2 process monitor, if available
Mud sodium content	Increased sodium in lime mud because of mud washing problems can lead to increased H_2S emissions and fine particulates	Check sodium content of lime mud entering kiln. Generally, should be in 0.5–1% range; 2–2.5% indicates likely problem

Source: Ref. 7.

electrostatic precipitators, and fabric filters. Schifftner and Hesketh [14] describe the use of wet scrubbers to control particulate emissions from bark boilers. They list some important considerations for this application:

1. Is the wood cut in sandy locations? Is it dragged on the ground after cutting? Is it hardwood or softwood? Does the mill stack pulpwood or furnish as logs or does it chip and then store?

2. Are ends, butts, and wood waste also burned? What percentage of the total feed does this represent?
3. If a multitube collector is used, is the char reinjected? If it is, the percentage of fines to the scrubber will likely increase, requiring higher pressure drops.
4. Four contributing factors combine to produce the net outlet emission:
   a. Percentage of reinjection
   b. Sand content of fuel
   c. Type and quantity of auxiliary fuel
   d. Fuel moisture

5. Is the mechanical collector functioning properly?

Yuan et al. [15] modeled NO_x emissions from bark boilers. A code developed at the University of British Columbia was used to simulate the combustion of wood chips. The numerical results were compared against experimental data. The NO_x predictions were generally higher than the measurements except for one case where they were lower. Because the temperatures were relatively low, the contribution from fuel NO_x (from the nitrogen in the bark) was dominant compared to thermal NO_x when the boiler was fired primarily on wood chips. NO_x increased significantly with fuel nitrogen content. Thermal NO_x was the dominant mechanism when the boiler was fired primarily on natural gas. The use of overfire air improved the burnout of the bark but did not reduce NO_x formation as is usually the case. Reducing excess air did reduce NO_x emissions as expected.

17.1.4 Paper Dryers

In many drying processes, moisture is removed from paper webs often traveling at high speeds. Radiant heating (see Fig. 17.9) is often used to supplement steam-heated

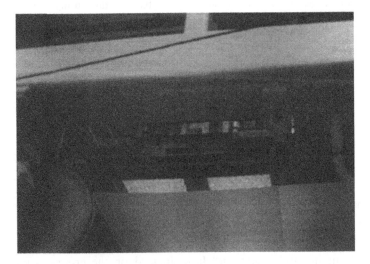

Figure 17.9 Infrared (IR) burner heating a continuously moving paper web. (From Ref. 16. Courtesy of Marsden, Inc., Pennsauken, NJ.)

Figure 17.10 Supplemental IR burners for paper drying located over steam cylinders. (Courtesy of Marsden, Inc., Pennsauken, NJ.)

cylinder drying or high-velocity hot-air dryers [16]. The radiant heaters are either electric or fired with a fuel gas such as natural gas. These supplemental heaters may be located before the steam-heated cylinders, in between cylinders, over cylinders (see Fig. 17.10), and after the cylinders. The heaters may also be partitioned across the machine to vary the drying capacity. The moisture content of the paper web often varies across the machine direction. Some streaks may be significantly wetter than others and therefore require more drying energy. A fictitious example is shown in Fig. 17.11 where the peak is near 14% while the minimum is at 10%. The paper is normally specified to have a maximum moisture content. Assume for the sake of argument that it is 8%. This means that the entire paper must be dried so that the wettest streaks are at or below the maximum allowable moisture content. If the heating is uniform across the web, then enough energy must be supplied to reduce the moisture streak of nearly 15% down to 8%. This means that the section of the web at 10% will be dried well below 8%. However, this is detrimental for several reasons. Since paper is sold by weight, removing more moisture than necessary in dryer sections of the paper reduces profitability. More fuel than necessary is used to dry some sections well below the maximum allowable moisture limit. The paper quality also suffers when some sections of the paper are overdried, which causes handling problems for machines like copiers.

So-called moisture profiling is where the radiant output of the burners varies across the machine direction (see Fig. 17.12) to match the moisture levels. This fixes the problems of overdrying some sections that occurs with a uniform radiant heating level. Not only is this much more fuel efficient, but it also indirectly reduces pollutant emissions because less fuel needs to be burned for a given production rate.

Another example of predrying is in the paper industry where infrared (IR) burners are installed after the coating machine and are used to set the coatings on the paper prior to the paper contacting a steam cylinder drum dryer that is used to complete the drying [17]. The IR predryer is primarily used to increase productivity

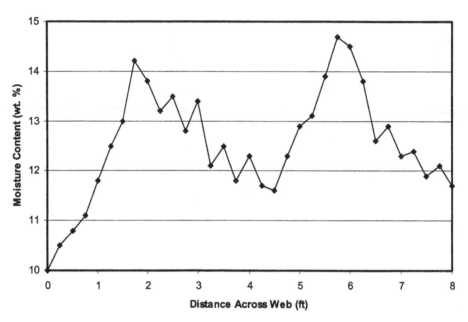

Figure 17.11 Moisture profile across a paper web prior to drying.

Figure 17.12 IR burners designed for moisture profiling. (Courtesy of Marsden, Inc., Pennsauken, NJ.)

and improve the paper coating quality. The productivity is increased because of the added heat. The quality is improved because the IR energy does not disturb the coating as convection or conduction heat-transfer methods could, which lets the coating set on the paper prior to contact with the steam cylinder, which relies on thermal conduction heat transfer.

Pettersson and Stenström [18] compared the use of gas-fired and electric IR burners used to set the coating before the paper reached the next cylinder in a paper

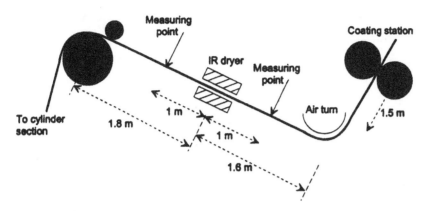

Figure 17.13 Schematic of IR heaters used to set the coating on paper, which is traveling from left to right in the figure. (From Ref. 18. Courtesy of Gas Technology Institute, Chicago, IL.)

line. Figure 17.13 shows the IR heaters between the coating station and the next steel cylinder in the paper machine. IR burners are preferred in this application, instead of convective or conductive dryers, because they are noncontact and have high power densities (10–40 kW/m^2 or 3000–13,000 Btu/hr-ft^2). The thermal efficiencies were calculated as 30 and 40%, respectively, for the gas-fired (propane) and electric IR burners. However, the burners were tested on two different machines and there was some uncertainty in the measurements.

The most common way for drying paper traveling at high velocities is by contact with steam heated drums or cans, usually referred to as steam cylinders. The paper wraps around the drums in a serpentine fashion to maximize the contact area with the drum. In this type of dryer, the primary method of drying is by thermal conduction [19]. One problem with this technique is that as the paper dries, the thermal conductivity goes down, which makes it harder to conduct the heat into the paper. This is known as the "falling rate period" where the downstream steam cylinders are much less effective at removing moisture than the upstream cylinders.

Infrared burners (see Fig. 17.14) are often used to supplement these dryers because the IR radiation can penetrate into the paper better when it is dry because there is less water to absorb the radiation. In a survey of paper makers by the Gas Research Institute (Chicago, IL), respondents believed that the best place to install IR burners on a paper drying line is in the preheat zone [20]. Other locations identified in the survey included in the forming section, above and below the steam cylinders in the constant rate zones, and above and below the steam cylinders in the falling rate zone.

One reason for the popularity of steam cylinder dryers is that there is usually plenty of steam available in paper mills because much of the waste bark and liquor from the trees used to make the paper is burned in hog fuel boilers. Another reason is that the cylinders help to guide and transport the paper. One disadvantage includes the large thermal inertia of the stainless-steel cylinders, which causes longer startup times and a reduced ability to change the drying rate quickly. An important disadvantage is the reduced drying effectiveness as the moisture content of the paper decreases because of the reduction in the thermal conductivity. There is a potential reduction in paper quality due to contact with the steel cylinders. These dryers do not typically have the capability to vary the drying capacity across the width of the

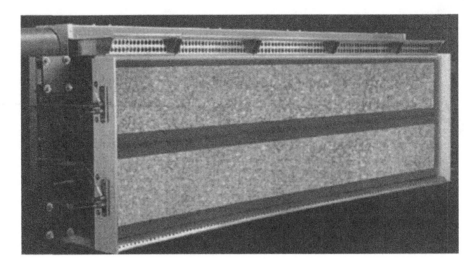

Figure 17.14 Example of a flat-panel gas-fired IR burner. (From Ref. 16. Courtesy of Marsden, Inc., Pennsauken, NJ.)

paper. Also, for thicker materials, the drying rate is reduced because the evolving water vapor is trapped between the cylinder and the paper as it is unable to exit the side of the paper in contact with the cylinder.

Another type of dryer uses very high velocity hot-air impingement on both sides of a moving web. The web "floats" through the nozzles, which is where this type of dryer gets its name—*floater dryer*. The primary mode of heat transfer for this dryer is convection. This technique combines heat and mass transfer in the same apparatus as the hot air both heats the web and carries away the moisture that evolves from it. This type of dryer has several potential advantages over other types of dryers. No additional systems are needed to remove the volatiles vaporizing from the material being dried. There is no direct contact with the product that could reduce the quality. It is possible to segment this type of dryer to vary the moisture removal rate across the width, although the reaction time is slow compared to that of IR dryers. There are also potential disadvantages. The air nozzles can become plugged because they are typically fairly small to achieve the high gas velocities. In drying materials like papers and textiles, this method also relies ultimately on conduction for the energy to reach the core of the product whose thermal conductivity decreases as the moisture content decreases. Another version of an air dryer is where heated air is blown only on to one side of a paper web traveling through a dryer where a coating is to be dried, as shown in Fig. 17.15 [21].

Hannum et al. [22] discuss the development of a high-intensity lean premix combustion system with low-NO_x emissions for use in drying applications such as tissue and plasterboard drying. The burner was used to heat air for use in drying in a loop drying system. Measured NO_x emissions were below 10 ppmvd (3% O_2).

In most paper drying applications, the only pollutant of significant concern is NO_x. The gaseous fuels used, such as natural gas, contain little or no sulfur to produce SO_x. These clean fuels and moist webs generate little if any particulates. There are no VOCs in the process. There is very little noise produced. While there are

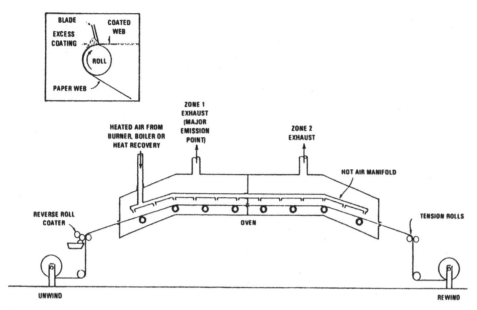

Figure 17.15   Schematic of a paper-coating dryer. (From Ref. 21.)

significant levels of thermal radiation, this is directed at the paper and is not usually a problem for personnel operating the equipment. Even NO_x emissions are typically low because of the relatively low operating temperatures.

17.2   PRINTING AND PUBLISHING

Web offset lithography is used to produce about 75% of books and pamphlets as well as an increasing number of newspapers [23]. Dryers and ovens are sometimes used to dry ink in the printing and publishing industry [24]. One of the major environmental concerns related to the printing and publishing industry is VOC emissions. VOCs are used in ink to promote fast drying. This ink is commonly referred to as "heatset" ink. In many applications, no ovens or dryers are required to dry the ink during the production process because of the rapid vaporization of the VOC solvents in the ink compared to the slower speed of the production process. However, high-speed paper webs are used in some printing applications where dryers are required to set the ink before it contacts downstream rollers that would cause the ink to smear if it were not set. Fig. 17.16 shows a schematic of a web-fed high-speed rotary flexographic press with a printing and drying section used to dry the ink on the paper [25].

These printing presses often use multiple colors, which means that each color must be individually set before the next color can be applied, or otherwise the colors will smear. The heater to set the ink may be as simple as an electric IR or UV burner. Figure 17.16 shows a heater between each color application of the roller ink system. In some cases, natural-gas-fired burners may be used because of their lower operating costs compared to using electricity. Whichever technology is used, the burners must either be capable of very rapid cool down or there must be some type of shielding from

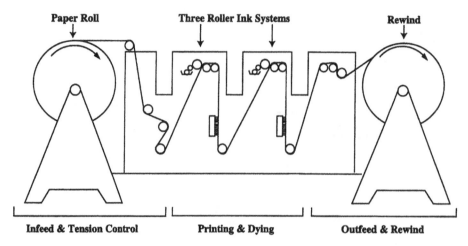

Figure 17.16 Schematic of a web-fed rotary flexographic press. (From Ref. 25.)

the web in the event of a sudden line stoppage in order to prevent the paper from catching on fire.

However, because of the increased concern about VOC emissions, inks are being made with less VOCs and with more aqueous-based solvents that require heat to set. The heaters may be gas-fired IR burners that produce NO_x emissions, although the temperatures are often low enough that these emissions are generally low. For larger and higher speed presses, gas-fired ovens may be used where air is heated and blown at high velocity on to the paper both to dry the paper and remove the vaporized solvents. These ovens are often designed to float the paper by having air nozzles on both sides of the paper. The floater dryer temperature is in the range 400°–500°F (200°–290°C). One or more burners are fired in the dryer to maintain a given air temperature. An afterburner may also be included in the drying system if there are significant VOC emissions in the recycled air containing ink solvents.

Besides VOCs, there are no other significant pollutants produced by the dryer. The fuel is typically natural gas, which contains little or no sulfur to produce SO_x. There are no significant sources for particulate emissions. Noise may be an issue for hot-air dryers because of the fans and high gas velocities. Radiation could be a consideration if radiant burners are used to set the ink.

17.3 TEXTILE MANUFACTURING

The textile manufacturing industry consists of the following segments [26]:

- Broad woven fabric mills and wool mills, including dyeing and finishing
- Knitting mills and knit goods finishing
- Other dyeing and finishing textile mills
- Floor covering mills, including dyeing and finishing.

The U.S. EPA ranks textile processing as number 44 on its prioritized list of 59 major source categories where the lower the number the higher the priority [27].

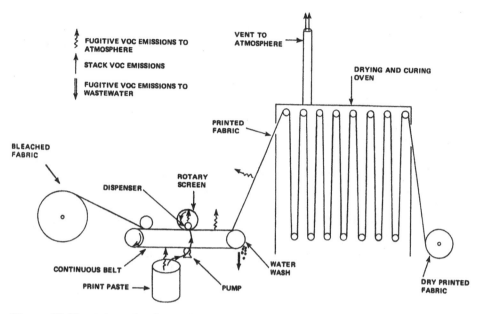

Figure 17.17 Schematic of a textile printing process. (From Ref. 30.)

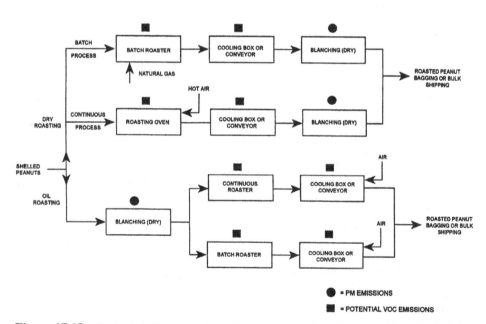

Figure 17.18 Typical shelled peanut roasting processing flow diagram. (From Ref. 32.)

Predryers are used in some applications prior to the final drying of the product. An example of this type of application is the use of IR burners to set the dyes in the dyeing of fabrics in textile manufacturing [28]. After the dyes are applied to the fabric, they must be set prior to contact with the dryer, otherwise the dyes will migrate to drier areas of the fabric, which reduces the quality of the textile. The IR burners in the predryer are used to set the dyes rapidly without the need to contact the material, as would be the case with, for example, a drum dryer [29]. Dryers and ovens are also used to cure inks used in textile printing, as shown in Fig. 17.17 [30].

17.4 FOOD PROCESSING

Cane sugar processing involves burning the fibrous residue, called bagasse, remaining after sugar extraction. This is usually burned in a boiler. This combustion process is the main source of air emissions in sugar cane production other than open field burning, which is completely uncontrolled and unmeasured [31]. The most common pollutant emissions include particulates, NO_x, unburned combustibles, and carbon dioxide.

Dryers are used in roasting seeds, almonds, and peanuts. Common pollutant emissions may include particulates and VOCs. Figure 17.18 shows a schematic of a typical peanut roasting process [32]. The primary air pollution emissions from fish processing are particulate emissions from the dryers [33].

REFERENCES

1. RM Fedler, RW Rousseau. *Elementary Principles of Chemical Process*, 3 edn. New York: John Wiley, 2000.
2. T Kudra, AS Mujumdar. *Advanced Drying Technologies*. New York: Marcel Dekker, 2002.
3. U.S. EPA. Paper Industry. Rep. EPA/530-SW-90-0270. Washington, DC: U.S. Environmental Protection Agency, 1990.
4. U.S. EPA. Profile of the Pulp and Paper Industry. Rep. EPA/310-R-95-015. Washington, DC: U.S. Environmental Protection Agency, 1995.
5. J Elkins, N Frank, J Hemby, D Mintz, J Szykman, A Rush, T Fitz-Simons, T Rao, R Thompson, E Wildermann, G Lear. National Air Quality and Emissions Trends Report, 1999. Rep. EPA 454/R-01-004. Washington, DC: U.S. Environmental Protection Agency, 2001.
6. Office of the Federal Register. Subpart BB: Standards of Performance for Kraft Pulp Mills. U.S. Code of Federal Regulations Title 40, Part 60. Washington, DC: U.S. Government Printing Office, 2001.
7. U.S. EPA. Kraft Pulp Mill Compliance Assessment Guide. Rep. EPA 310-8-99-001. Washington, DC: Environmental Protection Agency, May 1999.
8. U.S. EPA. Compilation of Air Pollutant Emission Factors, Vol. I: Stationary Point and Area Sources, Section 10.2: Chemical Wood Pulping, 5 edn. Washington, DC: U.S. Environmental Protection Agency Rep. AP-42, 1995.
9. ER Hendrickson. 16. The Forest Products Industry. In AC Stern (ed.). *Air Pollution*, 3 edn., Vol. 4. New York: Academic Press, 1977.
10. M Järvinen, R. Zevenhoven, E Vakkilainen. Implementation of a detailed physical black liquor combustion model into furnace calculations. *International Flame Research Foundation Journal*, Article No. 200206, June 2002.
11. J Wallen, T Ruohola, K Janka. NO_x Reduction in Recovery Boilers by Optimized Furnace Conditions—Field Studies. *Proceedings of 5th European Conference on Industrial Furnaces and Boilers*, Lisbon, Portugal, Vol. 1, pp. 669–678, 2000.
12. U.S. EPA. Control Techniques for Particulate Emissions from Stationary Sources—Vol. 2. Rep. EPA-450/3-81-005b. Washington, DC: U.S. Environmental Protection Agency, Sept. 1982.

13. R Miner, B Upton. Methods for estimating greenhouse gas emissions from lime kilns at Kraft pulp mills. *Energy*, Vol. 27, pp. 729–738, 2002.

14. KC Schifftner, HE Hesketh. *Wet Scrubbers*, 2 edn. Lancaster, PA: Technomic, 1996.

15. J Yuan, Z Xiao, I Garshore, M Salcudean. NO_x Emission Modeling and Control in Bark Boilers. *Proceedings of 5th European Conference on Industrial Furnaces and Boilers*, Lisbon, Portugal, Vol. 1, pp. 659–667, 2000.

16. CE Baukal. *Heat Transfer in Industrial Combustion*. Boca Raton, FL: CRC Press, 2000.

17. P Mattsson, J Perkonen, A Riikonen. Infrared Drying of Coated Paper. *Proceedings of 1989 International Gas Research Conference*, edn. TL Cramer. Govt. Institutes, Rockville, MD, pp. 1308–1316, 1989.

18. M Pettersson, S Stenström. Evaluation of Gas-Fired and Electrically Heated Industrial Infrared Paper Dryers. *Proceedings of 1998 International Gas Research Conference*, Vol. V: *Industrial Utilization and Power Generation*, ed. D Dolenc. Govt. Institutes, Rockville, MD, pp. 100–112, 1998.

19. T Berntsson, P-A Franck, A Åsblad. *Learning from Experiences with Process Heating in the Low and Medium Temperature Ranges*. CADDET Energy Efficiency, Sittard, The Netherlands, 1997.

20. CE Bean, JM Cocagne. Assessment of Gas-Fired Infrared Heaters in the Paper Industry. Gas Research Institute Rep. GRI-96/0087, Chicago, IL, 1996.

21. U.S. EPA. Compilation of Air Pollutant Emission Factors, Vol. I: Stationary Point and Area Sources, Sect. 4.2.2.6: Paper Coating, 5th edn. Washington, DC: U.S. Environmental Protection Agency Rep. AP-42, 1995.

22. M Hannum, T Robertson, J Winter, B Schmotzer, T Neville. High Intensity Lean Premix Combustion Systems for Drying Applications. *Proceedings of 5th European Conference on Industrial Furnaces and Boilers*, Lisbon, Portugal, Vol. 1, pp. 455–464, 2000.

23. U.S. EPA. Compilation of Air Pollutant Emission Factors, Vol. I: Stationary Point and Area Sources, Sect. 4.9.1: General Graphic Printing, 5th edn. Washington, DC: U.S. Environmental Protection Agency Rep. AP-42, 1995.

24. U.S. EPA. Profile of the Printing and Publishing Industry. Rep. EPA/310-R-95-014, 1995.Washington, DC: U.S. Environmental Protection Agency.

25. U.S. EPA. Printing Industry and Use Cluster Profile. Rep. EPA/744-R94-003. Washington, DC: U.S. Environmental Protection Agency, 1994.

26. U.S. EPA. Textile Manufacturing. Rep. EPA/530-SW-90-027e. Washington, DC: U.S. Environmental Protection Agency, 1990.

27. Office of the Federal Register. 60.16 Priority List. U.S. Code of Federal Regulations Title 40, Part 60. Washington, DC: U.S. Government Printing Office, 2001.

28. Anon. Ceramic tile burner improves performance of infrared predryer. *Process Heat.* Vol. 4, No. 7, pp. 43–45, 1997.

29. TM Smith, CE Baukal. Space-age refractory fibers improve gas-fired infrared generators for heat processing textile webs. *J. Coated Fabrics*, Vol. 12, pp. 160–173, 1983.

30. U.S. EPA. Compilation of Air Pollutant Emission Factors, Vol. I: Stationary Point and Area Sources, Sect. 4.11: Textile Fabric Printing, 5th edn. Washington, DC: U.S. Environmental Protection Agency Rep. AP-42, 1995.

31. U.S. EPA. Compilation of Air Pollutant Emission Factors, Vol. I: Stationary Point and Area Sources, Sect. 9.10.1.1: Cane Sugar Processing, 5th edn. Washington, DC: U.S. Environmental Protection Agency Rep. AP-42, 1995.

32. U.S. EPA. Compilation of Air Pollutant Emission Factors, Vol. I: Stationary Point and Area Sources, Sect. 9.10.2.2: Peanut Roasting, 5th edn. Washington, DC: U.S. Environmental Protection Agency Rep. AP-42, 1995.

33. U.S. EPA. Compilation of Air Pollutant Emission Factors, Vol. I: Stationary Point and Area Sources, Sect. 9.13.1: Fish Processing. 5th edn. Washington, DC: U.S. Environmental Protection Agency Rep. AP-42, 1995.

18

Future

18.1 INTRODUCTION

The U.S. Department of Energy (DOE) has been working on visions and roadmaps for the future of a number of industry areas including aluminum, metal casting, petroleum refining, glass, chemicals, and steel [1]. One of the technologies that crosscuts all of these industries is combustion. As a result, the DOE has also developed a vision and roadmap for industrial combustion. The industrial combustion vision foresees future furnace and process heating systems producing uniform, high-quality end products with minimal environmental impact [2]. Future burners will be compliant with emissions regulations.

There are many barriers to future combustion improvement including [3]:

- Most industrial firms are risk-averse
- Financial constraints
- Lack of performance standards
- Gap between industry and academia
- Inadequate technologies for measuring key combustion parameters
- Slow adoption of new technologies by industry

These barriers are balanced against ever more stringent emission regulations where firms are often forced to make changes to meet those regulations. Therefore, technological advances will continue to change industrial combustion processes. This chapter discusses some of those current and potential developments, including regulatory changes. It is by no means comprehensive nor exhaustive, but merely a sampling.

18.2 PROCESS PARAMETERS

A number of important process parameters may be affected by future changes and new developments. These parameters include fuels, oxidizers, burners, sensors and analyzers, feed materials, and energy sources. These are briefly discussed next.

18.2.1 Fuels

Concerning pollution emissions, there is a constant struggle between using cheaper but more-polluting fuels and using more expensive but less-polluting fuels. Technologies that can minimize pollution while using less expensive fuels will be in great demand, assuming that the overall process economics are still less expensive than using more expensive, cleaner fuels. There may also be increased interest in treating fuels prior to their use. One example is removing sulfur from liquid oil fuels prior to using them to minimize SO_x emissions.

Some consider hydrogen to be the fuel of the future because the main product is water [2]:

$$2H_2 + (O_2 + 3.76N_2) \rightarrow 2H_2O + 3.76N_2 \tag{18.1}$$

The only pollutant likely to be generated from the combustion of hydrogen is NO_x. It may be possible to reduce or eliminate even the NO_x emissions with advances in technology. One way to eliminate the possibility of NO_x formation is to use pure O_2:

$$2H_2 + O_2 \rightarrow 2H_2O \tag{18.2}$$

There is an energy cost (and possible pollutant emissions) to produce both the hydrogen and the oxygen, which must be taken into account. However, the pollution occurs at the power plant generating the electricity used in the production of hydrogen and oxygen. There is an advantage to that because it is usually easier and cheaper to handle the pollution emissions from a large source (i.e., the power plant) compared to many smaller sources (industrial combustion applications). Most industrial combustion processes rely to some degree on the radiation heat transfer from the flame to the load. This will be an issue with hydrogen combustion which produces very little flame radiation, which means that combustors may need to be redesigned for all convection heating. Hydrogen combustion deserves strong consideration as a potential solution to the pollution problem.

Another possible future trend regarding fuels is further use of nonstandard fuels. One example is the use of methane collected from landfills. Another example is further use of off-gases from production processes such as refinery gases generated during the processing of crude oil. It is likely that the widespread flaring of these gases (see Chap. 15) will be regulated. Therefore, it may become more cost-effective to collect those gases and sell them, rather than flaring them.

18.2.2 Oxidizers

Most combustion processes use air as the oxidant. In many cases, these processes can be enhanced by using an oxidant that contains a higher proportion of oxygen than that in air (approximately 21% O_2 by volume). This is known as *oxygen-enhanced combustion* or OEC [4]. There has been a continuing trend in a variety of industries for increased use of oxygen, instead of air, for combustion processes. Many of these applications have been discussed throughout this book.

New developments have made OEC technology more amenable to a wide range of applications. In the past, the benefits of using oxygen could not always offset the added costs. New oxygen generation technologies, such as pressure and vacuum swing adsorption, have substantially reduced the cost of separating oxygen from air. This has increased the number of applications where using oxygen to enhance performance is cost justified. Another important development is the increased emphasis on the environment. In many cases, OEC can substantially reduce pollutant emissions. This has also increased the number of cost-effective applications. The Gas Research Institute in Chicago, IL [5] and the U.S. Dept. of Energy [6] sponsored independent studies which predicted that OEC will be a critical combustion technology in the future.

Historically, air/fuel combustion has been the conventional technology used in nearly all industrial heating processes. Oxygen-enhanced combustion systems are becoming more common in a variety of industries. When traditional air/fuel combustion systems have been modified for OEC, many benefits have been demonstrated. Typical improvements include increased thermal efficiency, increased processing rates, reduced flue gas volumes, and reduced pollutant emissions.

Many industrial heating processes may be enhanced by replacing some or all of the air with high-purity oxygen [5,7]. Typical applications include metal heating and melting, glass melting, and calcining. In a report done for the Gas Research Institute [8], the following applications were identified as possible candidates for OEC:

- Processes that have high flue gas temperatures, typically in excess of 2000°F (1100°C).
- Processes that have low thermal efficiencies, typically due to heat-transfer limitations.
- Processes that have throughput limitations, which could benefit from additional heat transfer without adversely affecting product quality.
- Processes that have dirty flue gases, high NO_x emissions, or flue gas volume limitations.

When air is used as the oxidizer, only the O_2 is needed in the combustion process. By eliminating N_2 from the oxidizer, many benefits may be realized, particularly significant reductions in pollutants like NO_x and particulates.

18.2.3 Burners

Burners are the heart of the combustion system. Tomorrow's burners will produce very low NO_x, CO, and unburned hydrocarbon emissions while achieving high thermal efficiencies [2]. Ideally, the burners alone will be able to meet regulatory emission requirements without the need for post-treatment equipment. More advanced controls will be used to automatically adjust burners to changes in operating conditions while still meeting regulatory emission limits. These "smart" burners should be able to produce low emissions with fuel blends and liquid fuels [3]. The increased use of computational fluid dynamics should lead to more advanced burner designs [9].

18.2.4 Combustors

Combustors of the future will produce high-quality end products with minimal environmental impact [2]. Future combustors should be fully automated and adaptable to changing conditions while still producing high-quality products and low pollutant emissions. There is likely to be increased flexibility in some more advanced combustors to process a wider range of materials, including the ability to switch quickly to other products. Related to minimizing pollutant formation, new combustors will be more energy efficient, which indirectly reduces emissions as less fuel needs to be combusted.

18.2.5 Sensors and Analyzers

Advanced fast-reacting flame scanning may be necessary to adjust rapidly to changing operating conditions while minimizing emissions [3]. Robust emission sensors will be required to monitor quickly and reliably the effluents from combustion systems. Ruão et al. [10] describe a new technique for making in-flame NO measurements using a spectral ultraviolet/visible imaging device. The device was tested in a small-scale laboratory furnace using a propane- or ethylene-fired swirl burner. Besides NO_x, the in-flame measurements also included O_2, CO_2, CO, and unburned hydrocarbons. While the system needs to be tested on a larger scale and developed for commercial use, it could provide valuable data for automated, nonintrusive, and real-time control of combustion systems to minimize pollutant emissions.

Significant research has been done on using laser diode absorption sensors to make in-situ species and temperature measurements. Webber et al. [11] demonstrated their use in making CO, CO_2, and H_2O measurements in air–ethylene combustion. The accuracy was very good compared to that of laser measurements. Ebert et al. [12] measured H_2O, O_2, CH_4, CO_2, and temperature using laser diodes in a full-scale gas-fired power plant.

Zhuiykov [13] describes the development of a dual SO_x/O_2 in-situ sensor based on zirconia solid electrolyte and a composition of metal sulfates. The technology was demonstrated over a wide temperature range of 650°–1100°C (1200°–2000°F) with typical response times at 700°C (1300°F) in the range of 45–80 sec. While this is too slow for real-time control, it would still be useful for monitoring and for more course system adjustments to minimize SO_x emissions. Fleckl et al. [14] describe the use of Fourier transform infrared (FTIR) absorption spectroscopy as an in-situ diagnostic tool for measuring gas compositions in a traveling grate furnace. Lacas et al. [15] describe similar measurements made in a fluidized-bed waste incinerator. Using FTIR in situ could provide real-time data for use in controlling emissions.

18.2.6 Feed Materials

There are likely to be some changes in the feed materials processed in industrial combustion systems, in many cases to reduce pollution emissions. Some examples will illustrate this trend. In high-temperature glass processes, one of the feed materials referred to as niter contains nitrogen that contributes to NO_x emissions. Less niter and replacements for niter will be used to reduce NO_x. Incoming scrap metal will continue to be pretreated to reduce pollution. This includes removing materials that are not needed in the production process and that increase emissions. An example is plastics, which could produce dioxins and furans. The oil-based solvents in printing inks that cause VOC emissions are being replaced by aqueous solvents that are environmentally benign. The use of feed material preparation is likely to increase. An example is the shredding of incoming scrap metal to be charged into a melting furnace. More material can be charged and less energy is required per unit of charge to process shredded scrap. The increased energy efficiency reduces the pollution emissions per unit of production.

18.2.7   Energy Sources

Energy sources refer to the use of alternative energy sources to supplement industrial combustion and to the use of other sources of energy in the combustor. The former includes the use of, for example, electricity to replace some of the fossil-fuel combustion. This is commonplace in certain applications such as glass melting where electrodes under the molten glass assist in the heating. The use of electricity reduces the emissions at the plant site and may be an economical alternative for pollution control. There is an added benefit if some or all of the electricity has been generated by alternative sources such as solar and wind.

There may be some applications where other energy sources can be used in the combustor. One example discussed in Chap. 13 is the use of aluminum scrap contaminated with oil where the oil is a supplemental energy source. Another example in Chap. 16 is the processing of contaminated soils in an incinerator where there is some heating value in the contaminants. Waste oils are sometimes fired in an incinerator to dispose of them. A better use of the heating value would be to fire the waste oils in an industrial combustion application, making a usable product such as metal or glass.

18.3   REGULATIONS

Industrial combustion end-users see their most critical target for the year 2020 as a 90% reduction in total air emissions (including air toxics, but excluding CO_2) from combustion systems compared to uncontrolled 1990 levels [2]. It is noted that this is a particularly aggressive target as it is on an absolute basis and includes any new market growth. CO_2 emissions would be reduced to levels agreed upon by the international community by increasing energy efficiency, using fuels with higher hydrogen-to-carbon ratios, carbon sequestration, and the increased use of biomass. It is also noted that it is very difficult to predict the future of many factors including environmental regulations, global competitiveness, and fuel availability and pricing.

Davis et al. [16] note that future regulations will continue to emphasize reducing the emissions from a wide range of compounds including [16]:

1. Volatile organic compounds (VOCs) and other ozone precursors (CO and NO_x).
2. Hazardous air pollutants, including carcinogenic organic emissions and heavy metals emissions.
3. Acid rain precursors, including SO_x and NO_x.

Particulate emissions will continue to be reduced including reducing both the size and concentration limits.

Delaney et al. [17] discuss the importance of a greenhouse gas management strategy to ensure that companies are prepared for future regulations that may involve minimizing the emission of gases like CO_2 that are not currently regulated, for example, in the United States. It becomes more complicated for global companies that operate in some companies that already have CO_2 emissions restriction, but also operate in other companies that do not. Without advance preparation, it will be difficult and costly for companies to catch up and meet new regulations.

Zaborowsky [18] notes the importance of getting a head start to plan for impending changes in environmental regulations. There are many reasons for this. The first is that it takes a significant amount of time to determine what the best options are for a given situation. There is often a long lead time for the large and often sophisticated air pollution control equipment used by industrial combustion processes. Another reason is that in some cases, this may include an emissions trading program where companies can buy and sell pollution credits. There are numerous factors to consider including capital costs, operating costs, timing, impact on existing operations, and training of operating personnel to name a few. A company's strategy may change with time. For example, a company may buy credits from other companies in early years when those credits are often less expensive. The credits become more expensive in succeeding years when regulatory limits are phased and get more stringent with time. The first year may require only a 10% reduction in emissions but the fifth year may require a 50% reduction. There will be fewer excess emission credits available to sell in the fifth year, which increases the cost of the credits. The company may deliberately plan to install pollution control equipment after the first year but before the fifth year due to economics. In fact, the company may plan to reduce its emissions below the required levels in order to generate pollution credits that it can then sell to other companies.

18.4   TECHNOLOGY

18.4.1   Modeling

The U.S. Department of Energy has identified the following industrial combustion needs related to CFD modeling [3]:

- Computational tools that contain validated, high-fidelity models
- Reliable, efficient model of turbulent, reacting flow
- Common method for measuring furnace efficiency
- State-of-the-art combustion laboratories to validate CFD models.

Awais et al. [19] have shown that neural networks can be used to predict NO_x emissions. The authors believe neural networks have the potential to span the gap between numerical predictions and experimental measurements. This may be an alternative tool for studying combustion processes.

18.4.2   Combustion Modification

The U.S. EPA has sponsored much research on reducing NO_x [20] and SO_x [21] emissions, which has led to a number of combustion modification technologies that have been applied in industry. One example of a new yet-to-be commercialized research and development project is a demonstration of the use of pulse combustion in waste incineration to reduce pollutant emissions [22]. The technology was applied in the Environmental Protection Agency's Incineration Research Facility rotary kiln system. The demonstration showed that pulsed combustion

- Reduced NO_x, CO, and soot emissions
- Increased principal organic hazardous constituents (POHC) and destruction and removal efficiency (DRE)
- Decreased combustion air requirements
- Decreased auxiliary fuel requirements.
- Increased incinerator capacity

compared to nonpulsating combustion. Dioxin and furan emissions were similar for both pulsating and nonpulsating combustion.

Joshi et al. [23] have patented a process to minimize NO_x emissions in regenerative glass-melting furnaces. The process involves firing an oxidizing oxy/fuel burner at an angle to and underneath of an air/fuel burner operated under reducing conditions (see Fig. 18.1). The flame from the oxy/fuel burner should just intersect the end of the air/fuel burner. NO_x reductions from 30 to 70% are claimed.

Li and Garg [24] describe a process to reduce NO_x emissions in multizone metal reheating furnaces, as shown in Fig. 18.2. The oxidant/fuel stoichiometry is varied in different zones so that some zones are fuel rich and other zones are fuel lean, which are both less favorable for NO_x formation compared to near-stoichiometric operation. The overall stoichiometry of the system is near-stoichiometric. The stoichiometry in each zone is controlled by adjusting the oxidant and fuel flow rates to the burners in that zone, so no furnace modifications are required.

Pieper [25] has patented a process to reduce NO_x and CO in the exhaust gases from a glass manufacturing process [25]. Oxygen and fuel are introduced under the flames under slightly substoichiometric conditions. Secondary fuel and air are introduced downstream of the flames for reburning and postoxidation, respectively.

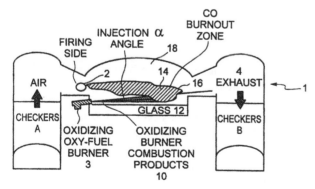

Figure 18.1 Process for reducing NO_x emissions from a regenerative glass furnace (From Ref. 23.)

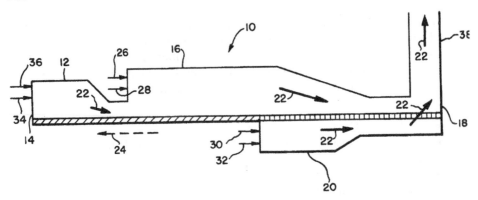

Figure 18.2 Schematic of a process for reducing NO$_x$ emissions in multizone metal reheating furnaces. (From Ref. 24.)

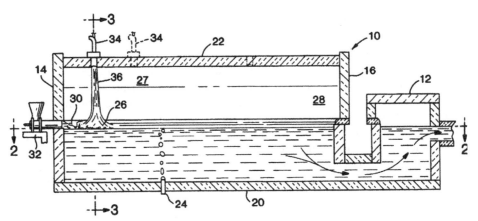

Figure 18.3 Method for improving the heat transfer and thermal efficiency of a glass-melting furnace. (From Ref. 27.)

This technology uses oxidant and fuel staging in the glass furnace to minimize pollution emissions.

18.4.3 Process Modifications

One alternative for incinerating municipal solid waste has been proposed, which would produce methane or methanol from a biomass process [26]. Figure 11.12 shows a schematic of the proposed process. This not only promises to reduce pollution emissions associated with combustion such as NO$_x$ and SO$_x$, but it also produces a valuable fuel that can be sold and used rather than being combusted.

Simpson et al. [27] describe a method for improving the heat transfer and thermal efficiency in a glass-melting furnace by using oxy–fuel burners mounted in the roof that directly impinge on the melt surface (see Fig. 18.3). By improving the thermal efficiency of the melting process, all pollutant emissions are reduced because less fuel is consumed for a given production rate. The proposed technique is a significant departure from current practices, but may become a viable option because of the increasing interest in reducing emissions and improving fuel efficiency.

18.4.4 Post-Treatment Equipment

In most cases, it is preferable to prevent pollutant formation rather than clean up emissions after they have already been formed. Some pollution control technologies (e.g., scrubbers) may generate another, although more benign, pollutant themselves (sludge in some cases) that must also be treated.

As with nearly all technologies, there is always a demand for cheaper, more efficient, smaller, easier-to-operate, lower maintenance, more durable, more flexible, retrofittable, more robust, etc., equipment. Because environmental regulations will only get stricter, the demand for pollution control equipment should only continue to rise. More advanced technologies will be developed for controlling emissions.

Altwicker [28] lists a number of emerging high-temperature technologies for thermally destroying a variety of pollutants. Molten salt oxidation can be used to destroy combustion liquid and solid wastes and can neutralize acidic species such as HCl and SO_2. Molten-metal reactors with temperatures up to 3000°F can be used to destroy wastes with low oxygen contents and produce off-gases composed primarily of CO and H_2 (due to the high-temperature dissociation of the hydrocarbons). These off-gases can be easily combusted to generate heat either to maintain the temperature of the molten metal or for use elsewhere in the plant. Molten-metal baths usually have very high waste destruction efficiencies and can even be used to treat wastes containing toxic metals. Molten-glass baths also have high destruction efficiencies and are especially effective for treating wastes containing some type of ash product that can be absorbed into the glass. The resulting glass is very stable and inert so it is highly resistant to leaching to make it amenable for landfilling. Plasmas can have very high destruction efficiencies due to the very high temperatures (5000°F or more). High combustion efficiencies make these systems especially portable. Corona discharges can be used to destroy NO_x, SO_x, and VOCs. Mok and Nam (29) modeled the use of a pulsed corona discharge to remove NO_x and SO_x from a gas stream [29]. Further research is needed for these technologies to make them commercially viable.

18.4.4.1 NO_x Control

One example currently under development, but not widely commercialized, is the use of aerobic biofiltration to convert NO into nitrite and nitrate and anaerobic biofiltration to convert NO into nitrogen [30]. Removal efficiencies of 70 and 90% have been demonstrated for the aerobic and anaerobic processes, respectively. Biofiltration has been used commercially for some time to control VOCs, but has only recently been applied to NO_x.

Leipertz et al. [31] discuss the use of UV emission tomography to measure NO_x emissions in real time for use in a closed-loop industrial combustion emission control system. The system measures both temperature and species including particulates. Because of the nonlinear nature of combustion processes, a control strategy using neural networks and fuzzy logic was recommended. The system needs further

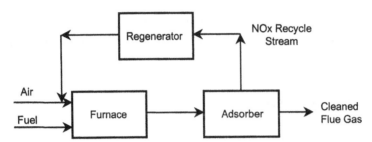

Figure 18.4 NO$_x$ recycle process. (From Ref. 32. Courtesy of the Combustion Institute, Pittsburgh, PA.)

development and more favorable economics before it is truly ready for the rugged demands of industrial combustion applications.

Zhou et al. [32] describe a technique referred to as the NO$_x$ recycle process (see Fig. 18.4) for reducing NO$_x$ emissions. Regenerable sorbent is used to adsorb NO$_x$ from the flue gas stream coming from the combustor. This is followed by desorption to produce a highly concentrated NO$_x$ stream containing both NO and NO$_2$. That stream is then sent back to the combustor where the NO$_x$ is reduced in the stream. Reduction efficiencies of 50–90% were experimentally demonstrated. The best reduction efficiencies were shown when the recycle stream was injected into the primary air inlet duct.

Zimmermann et al. [33] discuss the use of laser-induced resonance-enhanced multiphoton ionization (REMPI)/mass spectrometry for on-line control of emission from combustion processes. One of the advantages of the technique is the very fast response rate (up to 50 Hz) compared to most conventional gas-sampling systems that require seconds for in-situ measurements and sometimes minutes for extractive sampling systems. At this time the equipment is too expensive and requires highly trained technicians so it is not feasible for commercial applications.

Sung and Law [34] showed computationally that NO emissions from the combustion of pure O$_2$ with a fuel containing methane and nitrogen ranging from 0 to 1 wt% of the fuel, may potentially be reduced by increasing the velocities in the flame. The increased velocities increase the strain rate in the flame. While the flame temperatures are not significantly impacted by changing the strain rate, the residence time in the flame region is shortened, which can reduce NO$_x$ emissions. There is no known study that demonstrates this theory experimentally.

Lim et al. [35] describe the use of photocatalytic reduction of NO in flue gases in a fluidized bed containing titanium dioxide particles. The conversion efficiency depended on the UV intensity, the irradiation time, the TiO$_2$-to-silica gel ratio, the superficial gas velocity, the initial NO concentration, and the reaction temperature. Barranco et al. [36] experimentally demonstrated the feasibility of using palladium/zirconia catalysts as a future method for decomposing NO.

Cheng et al. [37] discuss the development of low-swirl burners for low NO$_x$ (<15 ppm) and CO (<10 ppm) emission furnaces and boilers [37]. The key aspect of the burner is a low-swirl vane for the combustion air that does not require flow recirculation to anchor the flame. The burner is capable of stabilizing very lean premixed flames close to the flammability limit. The experimental results showed that NO$_x$ was not dependent on heat input rate but was dependent on equivalence ratio.

Stern [38] has patented a post-treatment process for reducing NO_x emissions in a CO_2-rich stream by reacting the stream with ammonia in a reactor containing a reduction catalyst. Gas recycling is used to control the gas temperature and to reduce the CO_2 content in the stream before introduction into the reactor. Any SO_x in the stream should be removed prior to treatment in the reactor. The preferred catalyst is a zeolites/copper catalyst.

Yang [39] has patented a process using a corona discharge in conjunction with chemical scrubbers to minimize NO_x emissions in an exhaust gas stream. Figure 18.5 shows a schematic of the process, which is fairly complicated. It may take some time to develop a commercially viable system for use in industrial combustion processes.

Frohlich et al. [40] have patented a process for reducing NO_x emissions in exhaust gas streams with particular focus on the production of cement clinker. Figure 18.6 shows a schematic of the process. The exhaust gas containing suspended particles enters a cyclone where the particles drop out the bottom. The cyclone also increases the gas residence time in the NO_x reduction system. The exhaust gas is in the temperature range $450°-800°C$ ($840°-1500°F$) with some excess O_2. Heat is transferred in a heat exchanger, which substantially lowers the gas temperature and thus the NO_x emissions given adequate residence time and the above conditions.

Catton et al. [41] describe the development of carbon foam packings with a biofilm to reduce NO_x emissions. While the process physics are still being studied to understand the removal mechanisms, removal efficiencies up to 98% were demonstrated with sufficient residence times. Min et al. [42] showed a maximum NO removal rate of 74% for a hollow-fiber membrane bioreactor (HFMB) consisting of a bundle of microporous hydrophobic hollow-fiber membranes and a bioreactor containing nitrifying organisms and a nutrient supply. NO diffuses through the

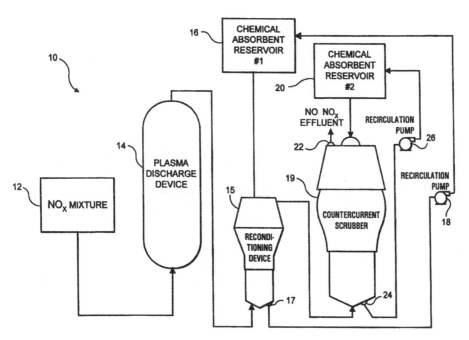

Figure 18.5 Post-treatment system for reducing NO_x emission. (From Ref. 39.)

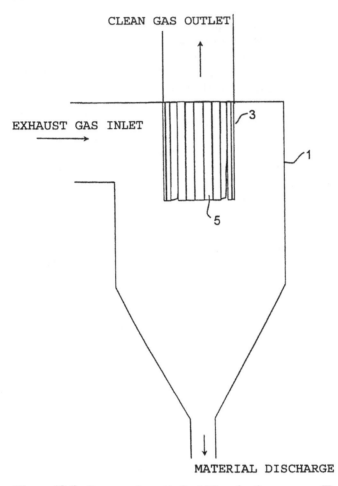

Figure 18.6   Process schematic for NO_x reduction process. (From Ref. 40.)

membrane pores and partitions into a nitrifying biofilm where it is oxidized to nitrate (NO_3) (see Fig. 18.7), which can be removed in a subsequent wastewater treatment step. The removal efficiency was not sensitive to the range of gas temperatures from 20° to 55°C (70°–130°F). van der Maas et al. [43] describe a BioDeNO$_x$ process for the biological removal of NO_x from flue gases, consisting of a two-step process. NO is absorbed into a chelate solution in the first stage and the chelate solution is regenerated by a biological denitrification process in the second stage (see Fig. 18.8) [44]. Hu et al. [45] describe a pulsed corona discharge for destroying N_2O. The study showed that argon was more effective than nitrogen as a background gas, which makes it less commercially attractive as combustion exhaust gases typically contain large fractions of nitrogen and only trace amounts of argon. Wang et al. [46] discuss NO adsorption and desorption on alumina-supported palladium. Zbicinski [47] discusses the use of pulse combustion in drying processes that offers the potential to increase the efficiency of drying processes and to reduce NO_x emissions.

Nam and Kim [48] describe the use a pulsed corona discharge for the simultaneous removal of both NO_x and SO_x specifically for an iron ore sintering

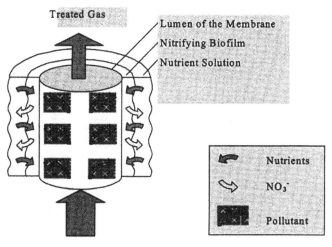

Figure 18.7 Hollow-fiber membrane bioreactor for NO_x removal. (From Ref. 42. Courtesy of Air & Waste Management Association, Pittsburgh, PA.)

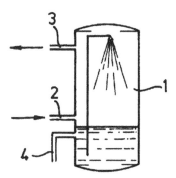

Figure 18.8 BioDeNO$_x$ process. (From Ref. 44.)

plant. However, the technology should be more generally applicable to a wide range of industrial processes where the flue gases contain NO_x and SO_x. Removal efficiencies of 90 and 65% were demonstrated for SO_2 and NO_x, respectively. Figure 18.9 shows a schematic of the process.

Ji et al. [49] have patented a vanadium-based catalyst for removing NO_x from flue gases, which also exhibits high durability against SO_2 poisoning. The inventors suggest a new way to prepare the catalyst, which has been used for some time to remove NO_x but which has had problems with plugging and poisoning. Barium oxide or calcium oxide are added as an active ingredient with vanadium pentoxide. The mixture is dried and calcined on the catalyst support surface, which gives it unique properties compared to conventional vanadium pentoxide catalysts.

18.4.4.2 SO$_x$ Control

Philip and Deshusses [50] discuss the use of a two-stage process consisting of a biotrickling filter followed by a biological post-treatment unit that has up to 100%

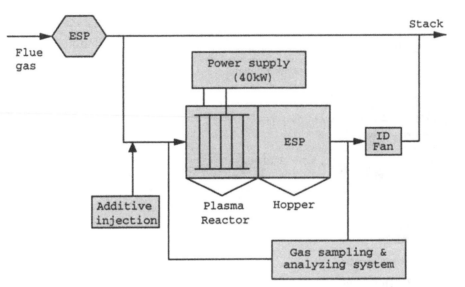

Figure 18.9 Schematic of a pilot-scale pulsed corona discharge process. (From Ref. 48. Courtesy of Air & Waste Management Association, Pittsburgh, PA.)

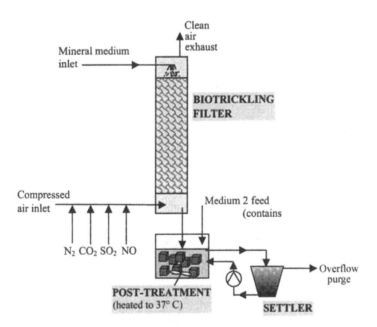

Figure 18.10 Biotrickling filter and post-treatment system for SO_2 removal. (From Ref. 50. Courtesy of Air & Waste Management Association, Pittsburgh, PA.)

removal efficiencies for SO_2 in concentrations from 300 to 1000 ppmv. In the first stage, SO_2 is converted into a liquid effluent containing sulfite and sulfate, which can be converted into elemental sulfur in the second stage. A schematic of the experimental setup is shown in Fig. 18.10.

Meikap et al. [51] discuss a novel modified multistage bubble column scrubber for flue gas desulfurization. Experimental results showed nearly 100% removal efficiency for SO_2 from an air–SO_2 mixture without the need for additives or pretreatment.

18.4.4.3 Soot

Martin and Ezekoye [52] describe the use of acoustics to control soot formation in flames. Applying a 940 Hz acoustic source to an acetylene diffusion flame reduced soot formation by nearly three orders of magnitude. The flame characteristics were also changed with the flame enlongating and brightening. However, applying a 3100 Hz acoustic source to the flame actually increased soot formation by almost a factor of four. This is a promising technique for controlling soot emissions from flames.

18.4.4.4 VOC Control

Ozone and membrane processes are available for controlling VOC emissions, but are not yet commercially viable [53]. Serageldin [54] notes that many of these devices are not widely used yet in the United States and discusses some of the programs the U.S. Environmental Protection Agency has sponsored to increase their use.

There are also numerous possible combinations of technologies that may be more optimal than any single technology, depending on the application. One example is referred to as "hybrid catalytic control" that combines fuel staging and a catalytic reaction [55]. Popov et al. [56] discuss a process called the BIOREACTOR that uses a trickling bed biofilter with VOC removal efficiencies of 80 to >99% for VOC concentrations of up to 1500 mg/m^3 and contact times of 2 to 20 sec. Figure 18.11 shows a schematic of the process. Commercial units have been successfully demonstrated in Russia. Kim et al. [57] describe a novel rotating biofilter for VOC removal that promises better distribution of moisture, oxygen, and nutrients on a large surface area of biofilm inside a rotating foam media. Figure 18.12 shows a schematic of the system which has VOC removal efficiencies of up to 99%.

Berry [58] describes the use of ordinary dirt as the ultimate catalyst for VOC catalytic thermal oxidation, as a type of biofilter. Three criteria are set forth as the requirements for the ultimate catalyst: easily available or renewable, minimal cost, and functions at atmospheric temperature. It is readily seen that dirt meets the first two criteria, but it is relatively unknown that dirt satisfies the third criterion. This is compared to the precious metal catalysts like platinum and palladium that are currently used, which are very expensive and not readily available.

18.4.4.5 CO$_2$ Control

Kumar et al. [59] describe the use of a new absorption liquid based on amino acid salts for removal of CO_2 from dilute gas streams in membrane gas–liquid contactors. The process was studied numerically and experimentally using a single tube made from relatively inexpensive polyolefin membrane.

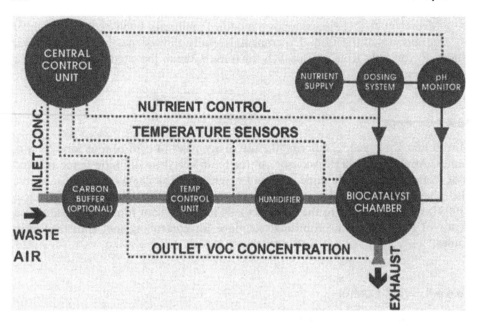

Figure 18.11   BIOREACTOR flow diagram for VOC removal. (From Ref. 56. Courtesy of the Air & Waste Management Association, Pittsburgh, PA.)

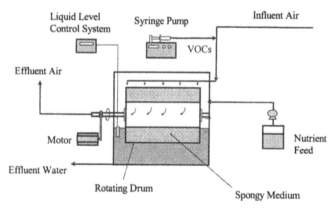

Figure 18.12   Rotating drum biofilter system schematic. (From Ref. 57. Courtesy of Air & Waste Management Association, Pittsburgh, PA.)

Tan et al. [60] describe experiments conducted with an oxidizer consisting of O_2 and recycled flue gas that they termed O_2/CO_2 recycle combustion. While the process is intended for large natural gas-fired power plants, it could be technically amenable to industrial combustion processes. However, the economics are not expected to be as favorable as for a large plant. The experiments showed that flames similar to those using air as the oxidant can be achieved. An added benefit is that NO_x could be eliminated in a tight system due to the absence of nitrogen.

18.4.4.6 Dioxins/Furans

Samaras et al. [61] experimentally demonstrated dioxin/furan prevention through the use of inorganic sulfur and nitrogen solid compounds in the fuel (refuse-derived fuel in this case). Reduction efficiencies of up to 98% were measured. This technology needs to be demonstrated on a full-scale combustion system. Alternative schemes may be needed for systems using gaseous or liquid fuels.

18.4.4.7 Multiple Pollutants

The U.S. EPA has published information on the use of advanced nonphotochemical oxidation processes for reducing pollutant emissions such as NO_x, SO_x, VOCs, and metals from both pilot-scale and industrial sources [62]. Two of these advanced techniques include electron-beam radiation and corona discharge. Further work is required for these to be commercially viable technologies for industrial combustion applications.

Sinha [63] has patented a process for removing particulates, hazardous substances, NO_x, and SO_x, in part by injecting a salt into the exhaust gas stream [63]. The hazardous substances could include arsenic, ammonia, ammonium sulfite, mercury, and the like. The salt comes from the group consisting of sodium nitrate, sodium nitrite, ammonium nitrate, lithium nitrate, barium nitrate, cerium nitrate, and mixtures of these. The salt is added upstream of some type of post-treatment system such as an electrostatic precipitator, cyclone, or baghouse.

Kurihara et al. [64] have patented a process for removing dust, dioxins, and NO_x from a combustion exhaust gas stream. The exhaust gas temperature is first reduced to $180°–230°C$ ($360°–450°F$) in a temperature-lowering unit. Next, dust is removed in a dry collector. Dioxins and NO_x are then removed in a catalytic reduction unit with a vanadium oxide-based catalyst. An air preheater can be incorporated into the system to reduce the exiting exhaust gas temperatures and to gain some thermal efficiency.

Penetrante et al. [65] have patented a two-stage noncatalytic process for removing NO_x and particulates from exhaust gas streams. A plasma is used in the first stage to convert NO into NO_2 in the presence of O_2 and hydrocarbons. The second stage converts the NO_2 and carbon soot particles into N_2 and CO_2. While this was originally developed for treating the exhaust from an engine, it should be adaptable for industrial combustion processes.

Das et al. [66] describe the use of a $Na–\gamma$-Al_2O_3 sorbent to adsorb SO_2, O_2, and NO simultaneously. Liu et al. [67] demonstrated experimentally the use of a spray dryer and a fabric filter along with some additives to control acid gas and polycyclic aromatic hydrocarbon (PAH) pollutant emissions. The inorganic additives to the spray dryer, such as SiO_2, $CaCl_2$, and $NaHCO_3$, increase the removal efficiency of SO_2 and PAHs. Additives to the feedstock, including polyvinyl chloride (PVC) and NaCl, also increase the removal efficiency of SO_2 and PAHs in the spray dryer.

Gerasimov [68] found that the use of electron-beam gas cleaning to remove SO_x and NO_x also had the beneficial effect of removing dioxins. This was shown kinetically to be caused by the degradation of dioxin molecules by OH radicals formed under the ionizing radiation. Schoubye et al. [69] describe the SNOX process used to remove NO_x and SO_x emissions produced by high-sulfur fuels. The process removes 95–99% of SO_x and 90–95% of NO_x emissions. While it is designed primarily for large sources such as power plants, it has successfully been applied in a refinery. Because of the economies of scale, the process is best applied to treating larger flue gas volumes. This could be applied in a plant where multiple sources may be combined and treated. The process does not generate any secondary pollutants such as wastewater, slurries, or solids.

Enviroscrub Technologies (Minneapolis, MN) has developed a new method for simultaneously removing NO_x and SO_x emissions called the Pahlman Process^{TM} [70]. It is designed for large sources that could be used at industrial plants combining multiple exhaust gas streams. This dry process is claimed to have operating costs lower than those of an SCR (see Chap. 6) and does not use any ammonia. Non-contaminated gypsum is produced, which can be sold along with nitrates and sulfates that can be recovered and sold. Removal efficiencies are up to 99% for both NO_x and SO_x are reported. The process has been successfully demonstrated at several sites.

18.4.4.8 Other

In some cases, waste materials to be incinerated contain inorganic salts such as NaCl, Na_2SO_4, CaCl, or KCl. Most of these wastes are salt-contaminated liquids that often contain water and may also contain organics, which makes them a candidate for thermal oxidation (see Chap. 11). The added heat load of the water often requires higher temperatures to ensure high destruction efficiency. The high temperatures and alkaline wastes are detrimental to typical refractory materials that would be used in a thermal oxidizer. This problem has led to the development of molten-salt incineration systems, as shown in Fig. 18.13 [71]. The incinerator is usually vertical to minimize buildup of any molten salts on the refractory-lined walls, which would reduce their life. The water-based wastes are injected into the thermal oxidizer downstream of the burner, which is usually fired with auxiliary fuel for the incinerator to reach the desired operating temperatures.

18.4.5 Controls

Diez de Ulzurran et al. [72] describe a sophisticated control system consisting of an array of sensors monitoring air flow, O_2 and CO concentrations in situ, water flow, and temperatures in a boiler for use in maximizing thermal efficiency and minimizing pollutant emissions. Further work was recommended to make the system robust and reliable enough for commercial use. Michel et al. [73] experimentally studied the use of existing flame detectors to produce so-called "flame signatures" to be used for controlling combustion systems including ensuring flame stability and minimizing CO and NO_x formation. The technique is potentially low cost and simply applied as most industrial combustion systems already have some type of flame detection system in place.

Homma and coworkers [74,75] proposed a combustion optimization control system using genetic algorithms to reduce NO_2 emissions from combustion processes. The optimization scheme is based on genetic evolution in biology. The system has the potential for controlling transient as well as slowly evolving "steady-state" systems that change over time. The numerical results were promising.

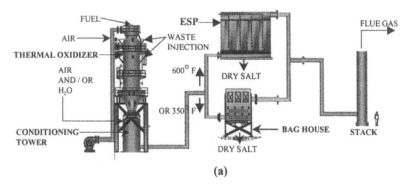

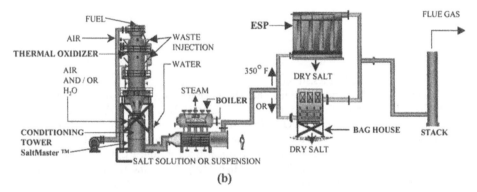

Figure 18.13 Molten salt thermal oxidation systems: (a) downfired salts—condition to low temperature; (b) downfired salts—condition to recover steam. (From Ref. 71. Courtesy of John Zink Co. LLC, Tulsa, OK.)

Further work is needed to apply this technology in large-scale industrial combustion processes.

Beisswenger et al. [76] have patented a feedback control process for minimizing NO_x emissions from glass furnaces. The O_2 and NO_x in the exhaust stream are fed back to the control system that adjusts which burners are on and which are off to minimize NO_x formation. The system works in conjunction with the ammonia injection in the post-treatment removal system.

18.5 RESEARCH NEEDS

The U.S. DOE has identified a number of burner R&D needs related to emissions including [3]:

- New combustion control algorithm for "smart" burners incorporating emissions numbers.
- Advanced fast-reacting flame scanning for low-NO_x burners.
- Equipment to measure low emissions economically and accurately.
- Ultra-low-emission liquid fuel burners.
- Low-emission burners that can operate on a wide range of fuel blends.
- Better understanding of how catalysts can be used to reduce NO_x.

18.6 CONCLUDING REMARKS

It is impossible to predict the future accurately, but it appears clear that regulatory limits will continue to get stricter. Therefore, whenever possible, it is advisable to choose pollution control strategies with maximum flexibility to meet increasingly more stringent regulations. It is also advisable to continue reviewing process requirements to see if modifications can be made to reduce emissions.

Technologies that may have previously been unfeasible or uneconomic may become feasible and economic in the future due to advances in technology and changes in regulations. Air pollution control is a dynamic field that is expected to remain so for the foreseeable future due to the importance of protecting people and the environment while maintaining and improving existing standards of living.

REFERENCES

1. www.oit.doe.gov/industries.shtml
2. U.S. Department of Energy (DOE). Industrial Combustion Vision: A Vision by and for the Industrial Combustion Community. Washington, DC: U.S. DOE, 1998.
3. U.S. Department of Energy (DOE). Industrial Combustion Roadmap: A Roadmap by and for the Industrial Combustion Community. Washington, DC: U.S. DOE, 1999.
4. CE Baukal (ed.). *Oxygen-Enhanced Combustion*. Boca Raton, FL: CRC Press, 1998.
5. SJ Williams, LA Cuervo, MA Chapman. High-Temperature Industrial Process Heating: Oxygen-Gas Combustion and Plasma Heating Systems, Gas Research Institute Rep. GRI-89/0256, Chicago, IL, July 1989.
6. AS Chace, HR Hazard, A Levy, AC Thekdi, EW Ungar. Combustion Research Opportunities for Industrial Applications—Phase II, U.S. Dept. of Energy Rep. DOE/ID-10204-2, Washington, DC, 1989.
7. CE Baukal, PB Eleazer, LK Farmer. Basis for enhancing combustion by oxygen enrichment. *Indust. Heat.*, Vol. LIX, No. 2, pp. 22–24, 1992.
8. KR Benedek, RP Wilson. The Competitive Position of Natural-Gas in Oxy–Fuel Burner Applications. Gas Research Institute Rep. no. GRI-96-0350, Chicago, IL, Sept. 1996.
9. CE Baukal (ed.). *Computational Fluid Dynamics in Industrial Combustion*. Boca Raton, FL: CRC Press, 2001.
10. M Ruão, M Costa, MG Carvalho. A NO_x diagnostic system based on a spectral ultraviolet/visible imaging device. *Fuel*, Vol. 78, No. 11, pp. 1283–1292, 1999.
11. ME Webber, J Wang, ST Sanders, DS Baer, RK Hanson. In Situ Combustion Measurements of CO, CO_2, H_2O and Temperature Using Diode Laser Absorption Sensors. *Proceedings of the Combustion Institute*, Pittsburgh, PA, Vol. 28, pp. 407–413, 2000.
12. V Ebert, T Fernholz, C Giesemann, H Pitz, H Teichert, J Wolfrum, H Jaritz. Simultaneous Diode-Laser-Based In Situ Detection of Multiple Species and Temperature in a Gas-Fired Power Plant. *Proceedings of the Combustion Institute*, Pittsburgh, PA, Vol. 28, pp. 423–430, 2000.
13. S Zhuiykov. Development of dual sulfur oxides and oxygen solid state sensor for "in situ" measurements. *Fuel*, Vol. 79, No. 10, pp. 1255–1265, 2000.
14. T Fleckl, H Jager, I Obernberger. Combustion diagnostics at a biomass-fired grate furnace using FT-IR absorption spectroscopy for hot gas measurements. *Proceedings of 5th European Conference on Industrial Furnaces and Boilers*, Lisbon, Portugal, Vol. II, pp. 55–65, 2000.

15. F Lacas, K Zahringer, D Veynante, O Gicquel. In Situ Measurements in a Real Scale Fluidized Bed Waste Incinerator. *Proceedings of 5th European Conference on Industrial Furnaces and Boilers*, Lisbon, Portugal, Vol. II, pp. 127–136, 2000.

16. WT Davis, AJ Buonicore, L Theodore. Air Pollution Control Engineering. In WT Davis (ed.). *Air Pollution Engineering Manual*. New York: John Wiley, 2000.

17. BT Delaney, J Wintergreen, L Jacobson. The Business Case for Greenhouse Gas Management. Paper 42589. *Proceedings of the Air & Waste Management Association's 95th Annual Conference & Exhibition*, Baltimore, MD, June 23–27, 2002.

18. P Zaborowsky. Getting a head start. *Environ. Protect.*, Vol. 13, No. 4, pp. 26–53, 2002.

19. MM Awais, S Godoy, FC Lockwood. The Development of a Neural Network Model for NO_x Predictions based on a Flat Flame Turbulent Jet Apparatus. *Fifth International Conference on Technologies and Combustion for a Clean Environment*, Lisbon, Portugal, Vol. 2. The Combustion Institute—Portuguese Section, pp. 825–837, 1999.

20. U.S. EPA. Combustion Modification Control of Nitrogen Oxides. Rep. EPA/600/F-95/012. Washington, DC: U.S. Environmental Protection Agency, 1995.

21. U.S. EPA. Flue Gas Desulfurization Technologies for Control of Sulfur Oxides: Research, Development, and Demonstration. Rep. EPA/600/F-95/013. Washington, DC: U.S. Environmental Protection Agency, 1995.

22. U.S. EPA. SITE Program Evaluation of the Sonotech Pulse Combustion Burner Technology. Rep. EPA/600/SR-97/061. Cincinnati, OH: U.S. Environmental Protection Agency, 1997.

23. ML Joshi, BJ Jurcik, J-F Simon. Oxidizing Oxygen–Fuel Burner Firing for Reducing NO_x Emissions from High Temperature Furnaces. U.S. Patent 6 171 100 B1, Jan. 9, 2001.

24. X Li, D Garg. Method of Reducing NO_x Emissions from Multi-Zone Reheat Furnaces. U.S. Patent 6 290 492 B1, issued Sept. 18, 2001.

25. H Pieper. Method and Apparatus for Melting Glass in U-Flame and Cross-Fired Tank Furnaces with a Reduction of the NOx and CO Content of the Waste Gases. U.S. Patent 6 289 694 B1, issued Sept. 18, 2001.

26. RA Geyer (ed.). *A Global Warming Forum: Scientific, Economic, and Legal Overview*. Boca Raton, FL: CRC Press, 1993.

27. NG Simpson, GF Prusia, SM Carney, AP Richardson, JR LeBlanc. Method of Boosting a Glass Furnace Using a Roof Mounted Oxygen-Fuel Burner. U.S. Patent 6 422 041 B1, issued July 23, 2002.

28. ER Altwicker. 5.21 Gaseous Emission Control: Thermal Destruction. In DHF Liu, BG Lipták (eds.). *Environmental Engineers' Handbook*, 2nd edn. Boca Raton, FL: Lewis Publishers, 1997.

29. YS Mok, I-S Nam. Modeling of pulsed corona discharge process for the removal of nitric oxide and sulfur dioxide. *Chem. Eng. J.*, Vol. 85, pp. 87–97, 2002.

30. JS Devinny, MA Deshusses, TS Webster. *Biofiltration for Air Pollution Control*. Boca Raton, FL: Lewis Publishers, 1999.

31. A Leipertz, R Obertacke, F Wintrich. Industrial Combustion Control Using UV Emission Tomography. *Twenty-Sixth Symposium (International) on Combustion*. Pittsburgh, PA: The Combustion Institute, pp. 2869–2875, 1996.

32. CQ Zhou, LG Neal, R Bolli, J Haslbeck, A Chang. Control of NO_x Emissions by NO_x Recycle Approach. *Twenty-Sixth Symposium (International) on Combustion*. Pittsburgh, PA: The Combustion Institute, pp. 2091–2097, 1996.

33. R Zimmermann, D Lenoir, A Kettrup, H Nagel, U Boesl. On-Line Emission Control of Combustion Processes by Laser-Induced Resonance-Enhanced Multi-Photon Ionization/Mass Spectrometry. *Twenty-Sixth Symposium (International) on Combustion*. Pittsburgh, PA: The Combustion Institute, pp. 2859–2868, 1996.

34. CJ Sung, CK Law. Dominant Chemistry and Physical Factors Affecting NO Formation and Control in Oxy–Fuel Burning. *Twenty-Seventh Symposium (International) on Combustion*. Pittsburgh, PA: The Combustion Institute, pp. 1411–1418, 1998.

35. TH Lim, SM Jeong, SD Kim. Photo-Catalytic Reduction of NO in Flue Gas over TiO_2 Particles. *Fifth International Conference on Technologies and Combustion for a Clean Environment*, Lisbon, Portugal, Vol. 1. The Combustion Institute—Portuguese Section, pp. 103–107, 1999.

36. M Barranco, A Guerrero-Ruiz, I Rodriguez-Ramos. Future DeNOx Catalysts for Turbines Without Modifying the Combustion Process. *Fifth International Conference on Technologies and Combustion for a Clean Environment*, Lisbon, Portugal, Vol. 1. The Combustion Institute—Portuguese Section, pp. 109–113, 1999.

37. RK Cheng, DT Yegian, MM Miyasato, GS Samuelsen, CE Benson, R Pellizzari, P Loftus. Scaling and Development of Low-Swirl Burners for Low Emission Furnaces and Boilers. *Proceedings of the Combustion Institute*, Pittsburgh, PA, Vol. 28, pp. 423–430, 2000.

38. SS Stern. Removal of Nitrogen Oxide from Gas Streams. U.S. Patent 6 136 283, issued Oct. 24, 2000.

39. C-L Yang. Corona-Induced Chemical Scrubber for the Control of NO_x Emissions. U.S. Patent 6 193 934 B1, issued Feb. 27, 2001.

40. A Frohlich, MR Monnig, S Happrich. Method for the Reduction of NO_x Emissions. U.S. Patent 6 296 820 B1, issued Oct. 2, 2001.

41. K Catton, L Hershman, DPY Chang, ED Schroeder, J Chen. Aerobic Removal of NO on Carbon Foam Packings. Paper 43550. *Proceedings of the Air & Waste Management Association's 95th Annual Conference & Exhibition*, Baltimore, MD, June 23–27, 2002.

42. K-N Min, SJ Ergas, JM Harrison. Hollow Fiber Membrane Bioreactor for Nitric Oxide Removal. Paper 42715. *Proceedings of the Air & Waste Management Association's 95th Annual Conference & Exhibition*, Baltimore, MD, June 23–27, 2002.

43. P van der Maas, S Weelink, L Hulshoff, B Klapwijk, P Lens. $BioDeNO_x$: Fe–EDTA as Electron Mediator Between Denitrification and Iron Reduction. Paper 55555. *Proceedings of the Air & Waste Management Association's 95th Annual Conference & Exhibition*, Baltimore, MD, June 23–27, 2002.

44. CJN Buisman, H Dijkman, PL Vergraak, AJ Den Hartog. Process for Purifying Flue Gas Containing Nitrogen Oxides. U.S. Patent 5 891 408, 1999.

45. X Hu, J Nicholas, J-J Zhang, TM Linjewile, P de Filippis, PK Agarwal. The destruction of N_2O in a pulsed corona discharge reactor. *Fuel*, Vol. 81, pp. 1259–1268, 2002.

46. C-B Wang, T-F Yeh, H-K Lin. Nitric oxide adsorption and desorption on alumina supported palladium. *J. Hazard. Mater.*, Vol. B92, pp. 241–251, 2002.

47. I Zbicinksi. Equipment, technology, perspectives and modeling of pulse combustion drying. *Chem. Eng. J.*, Vol. 86, pp. 33–46, 2002.

48. CM Nam, KT Kim. Pulsed Corona Discharge Process for Simultaneous Removal of SO_2/NO_x from a Sintering Plant of Steel Works. *Proceedings of the Air & Waste Management Association's 95th Conference & Exhibition*, Paper 42070. Baltimore, MD, June 23–27, 2002.

49. PS Ji, HM Eum, JB Lee, DH Kim, IY Lee, IS Nam, SW Ham, ST Choo. V_2O_5-Based Catalyst for Removing NO_x from Flue Gas and Preparing Method Therefor. U.S. Patent 6 380 128 B1, issued Apr. 30, 2002.

50. L Philip, MA Deshusses. Complete NO_x and SO_2 Treatment in a Biotrickling Filter—Bioreactor System. Paper 43571. *Proceedings of the Air & Waste Management Association's 95th Annual Conference & Exhibition*, Baltimore, MD, June 23–27, 2002.

51. BC Meikap, G Kundu, MN Biswas. Modeling of a novel multi-stage bubble column scrubber for flue gas desulfurization. *Chem. Eng. J.*, Vol. 86, pp. 331–342, 2002.

52. KM Martin, OA Ezekoye. Acoustic Control of Sooting Flames. *Fifth International Conference on Technologies and Combustion for a Clean Environment*, Lisbon, Portugal, Vol. 2, The Combustion Institute—Portuguese Section, pp. 1113–1120, 1999.

53. TT Shen, CE Schmidt, TR Card. *Assessment and Control of VOC Emissions from Waste Treatment and Disposal Facilities*. New York: Van Nostrand Reinhold, 1993.

54. M Serageldin. Air Bio-Reactor Systems: different designs and operational aspects. Paper 42952. *Proceedings of the Air & Waste Management Association's 95th Annual Conference & Exhibition*, Baltimore, MD, June 23–27, 2002.

55. CJ Pereira, MD Amiridis. NO_x Control from Stationary Sources. In US Ozkan, SK Agarwal, G Marcelin (eds.). *Reduction of Nitrogen Oxide Emissions*. Washington, DC: American Chemical Society, 1995.

56. VO Popov, AM Bezborodov, P Cross, W Jackson. Design, Construction and Long-Term Performance of Novel Type of Industrial Biotrickling Filters for VOC Control. Paper 42602. *Proceedings of the Air & Waste Management Association's 95th Annual Conference & Exhibition*, Baltimore, MD, June 23–27, 2002.

57. B Kim, X Zhu, M Suidan, C Yang. An Innovative Biofilter for Treating VOCs in Air Emissions. Paper 43187. *Proceedings of the Air & Waste Management Association's 95th Annual Conference & Exhibition*, Baltimore, MD, June 23–27, 2002.

58. J Berry. Dirt, The Ultimate Catalyst. *Proceedings of the Air & Waste Management Association's 95th Conference & Exhibition*, Paper 43184. Baltimore, MD, June 23–27, 2002.

59. PS Kumar, JA Hogendoorn, PHM Feron, GF Versteeg. New absorption liquid for the removal of CO_2 from dilute gas streams using membrane contactors. *Chem. Eng. Sci.*, Vol. 57, pp. 1639–1651, 2002.

60. Y Tan, MA Douglas, KV Thambimuthu. CO_2 capture using oxygen enhanced combustion strategies for natural gas power plants. *Fuel*, Vol. 81, pp. 1007–1016, 2002.

61. P Samaras, M Blumenstock, D Lenoir, KW Schramm, A Kettrup. PCDD/F prevention by novel inhibitors: Addition of inorganic S- and N-compounds in the fuel before combustion. *Environ. Sci. Tecnol.*, Vol. 34, No. 24, pp. 5092–5096, 2000.

62. U.S. EPA. Handbook on Advanced Nonphotochemical Oxidation Processes. Rep. EPA/ 625/R-01/004. Washington, DC: U.S. EPA, 2001.

63. RK Sinha. Flue Gas Conditioning for the Removal of Particulates, Hazardous Substances, NO_x, and SO_x. U.S. Patent 6 001 152, issued Dec. 14, 1999.

64. K Kurihara, M Kaneko, K Tsukamoto, K Ohya, H Kawaguchi, T Msuyama, K Shiraga, K Kadowaki, K Kiyono, K Ida, Y Taki, K Tanaka. Method and Apparatus for Treating Combustion Exhaust Gases. U.S. Patent 6 027 697, issued Feb. 22, 2000.

65. BM Penetrante, GE Vogtlin, BT Merritt, RM Brusasco. Plasma Regenerated Particulate Trap and NO_x Reduction System. U.S. Patent 6 038 854, issued Mar. 21, 2000.

66. AK Das, GB Marin, D Constales, GS Yablonsky. Effect of surface nonuniformity on the kinetics of simultaneous adsorption of SO_2–NO_x over Na–γ-Al_2O_3 sorbent: a coverage-dependent stoichiometry. *Chem. Eng. Sci.*, Vol. 57, pp. 1909–1922, 2002.

67. Z-S Liu, M-Y Wey, C-L Lin. Simultaneous control of acid gases and PAHs using a spray dryer combined with a fabric filter using different additives. *J. Hazard. Mater.*, Vol. B91, pp. 129–141, 2002.

68. GY Gerasimov. Degradation of dioxins in electron-beam gas cleaning of sulfur and nitrogen oxides. *High Energy Chem.*, Vol. 35, No. 6, pp. 389–393, 2001.

69. P Schoubye, S Enevoldsen, R Ricci. SNOX process for power plants using high-sulfur fuels. *World Refining*, Vol. 12, No. 8, pp. 24–29, 2002.

70. www.enviroscrub.com

71. P Melton, K Graham. Thermal Oxidizers. In CE Baukal (ed.). *The John Zink Combustion Handbook*. Boca Raton, FL: CRC Press, 2001.

72. JA Diez de Ulzurran, AM Gutierrez, K Mayora, JM Mitxelena, JM Chavarri, I Ocana. Sensor System for a Safe and Efficient Gas Combustion Control. *Fifth International Conference on Technologies and Combustion for a Clean Environment*, Lisbon, Portugal, Vol. 2. The Combustion Institute—Portuguese Section, pp. 1245–1250, 1999.

73. JB Michel, O Chetelat, N Weber, O Sari. Flame Signature as a Low-Cost Flame Control Method. *Fifth International Conference on Technologies and Combustion for a Clean Environment*, Lisbon, Portugal, Vol. 2. The Combustion Institute—Portuguese Section, pp. 1251–1253, 1999.

74. R Homma, J-Y Chen, RW Dibble. Combustion Optimization by Genetic Algorithms: Recution of NO_2 emission via Optimal Post-flame Process. *Fifth International Conference on Technologies and Combustion for a Clean Environment*, Lisbon, Portugal, Vol. 2. The Combustion Institute—Portuguese Section, pp. 1255–1263, 1999.

75. R Homma, J-Y Chen. Combustion Process Optimization by Genetic Algorithms: Reduction of NO_2 Emissions via Optimal Postflame Process. *Proceedings of the Combustion Institute*, Pittsburgh, PA, Vol. 28, pp. 2483–2489, 2000.

76. H Beisswenger, K Hasselwander, H Herden, G Mayer-Schwinning, G Samant, P Ludwig. Process for the Regulating or Controlling the NO_x Content of Exhaust Gases Given Off During the Operating of Glass Furnaces with Several Burners Run Alternately. U.S. Patent 6 237 368 B1, issued May 29, 2001.

Appendix A

Reference Sources for Further Information

The list below is not intended to be exhaustive as there are other journals that may occasionally have papers on industrial combustion, but the ones listed here are more likely to have such papers.

Air Pollution Journals and Magazines

Advances in Air Pollution
EM
Environment
Environment International
Environmental Engineer
Environmental Engineering Science
Environmental Pollution
Environmental Progress
Environmental Science & Technology
Environmental Technology
Industry & Environment
International Journal of Environment & Pollution
Journal of Air & Waste Management Association
Journal of Institute of Environmental Sciences & Technology
Noise Control Engineering Journal
Pollution Engineering
Pollution Equipment News
Progress in Environmental Science

Combustion Journals

Combustion and Flame
Combustion Science and Technology
Combustion Theory & Modelling
Fuel (formerly Journal of the Institute of Fuel)

Journal of the Institute of Energy
Progress in Energy and Combustion Science

Industrial Trade Journals and Magazines

33 Metal Producing
American Ceramic Society Bulletin
American Glass Review
Cement & Concrete
Cement Industry
Ceramic Age
Ceramic Industry
Ceramics Glass & Pottery
Chemical Engineering Journal
Chemical Processing
Concrete
Glass
Glass and Ceramics
Glass Digest
Glass Industry
Glass Packaging
Glass Technology
Heat Treating
Hydrocarbon Engineering
Hydrocarbon Processing
Industrial Heating
Iron and Steel Engineer
Iron and Steelmaker
Journal of Pulp & Paper Science
Light Metal Age
Modern Casting
New Steel (Iron Age)
Oil & Gas Journal
Process Engineering
Process Heating
Refractories
Refractories and Industrial Ceramics
Technical Association of Pulp & Paper Industry TAPPI Journal
World Refining

Industry Trade Associations

Aluminum Association (AA), The—Washington, DC
Aluminum Recycling Association (ARA)—Washington, DC
American Ceramic Society (ACerS)—Westerville, OH
American Concrete Institute (ACI)—Detroit, MI
American Forest and Paper Association (AF&PA)—Washington, DC
American Foundrymen's Society, Inc. (AFS)—Des Plaines, IL

American Iron and Steel Institute (AISI)—Washington, DC
American Petroleum Institute (API)—Washington, DC
American Zinc Association (AZA)—Washington, DC
ASM International (ASM)—Materials Park, OH
Association of Iron and Steel Engineers (AISE)—Pittsburgh, PA
Brick Institute of America (BIA)—Reston, VA
Casting Industry Suppliers Association (CISA)—Des Plaines, IL
Chemical Manufacturer's Association (CMA)—Washington, DC
Ferroalloys Association (TFA), The—Washington, DC
Flexographic Technical Association (FTA)—Ronkonkoma, NY
Forging Industry Association (FIA)—Cleveland, OH
Institute of Clean Air Companies (ICAC)—Washington, DC
International Copper Association (ICA)—New York
International Iron and Steel Institute—Brussels, Belgium
Investment Casting Institute—Dallas, TX
Iron and Steel Society (ISS)—Warrendale, PA
Lead Industries Association (LIA)—New York
Manufacturers of Emissions Controls Association (MECA)—Washington, DC
Mining Metals and Materials Society (TMS), The—Warrendale, PA
National Glass Association (NGA)—McLean, VA
National Petroleum Refiners Association (NPRA)—Washington, DC
Non-Ferrous Founders' Society—Des Plaines, IL
North American Die Casting Association (NADCA)—Rosemont, IL
Paper Industry Management Association (PIMA)—Arlington Heights, IL
Society for Mining, Metallurgy, and Exploration, Inc. (SME, Inc.)—Littleton, CO
Specialty Steel Industry of North America (SSINA)—Washington, DC
Steel Founders' Society of America (SFSA)—Des Plaines, IL
Steel Manufacturers Association (SMA)—Washington, DC
Technical Association of the Pulp and Paper Industry (TAPPI)—Atlanta, GA

Web Sites

Combustion Institute: www.combustioninstitute.org
Gas Technology Institute: www.gastechnology.org
International Flame Research Foundation: www.ifrf.net
South Coast Air Quality Management District (SCAQMD): www.aqmd.gov
Texas Commission on Environmental Quality: www.tceq.state.tx.us
United Nations Environment Programme: www.unep.or.jp/webm
RACT/MACT/LAER Clearinghouse (U.S. EPA): cfpub1.epa.gov/rblc/htm/bl02.cfm
U.S. Department of Energy: www.doe.gov
U.S. Environmental Protection Agency: www.epa.gov

Miscellaneous Sources

American Flame Research Committee conference proceedings
American Society of Mechanical Engineers combustion conference proceedings
(New York)

Gas Research Institute reports (Chicago, IL)—now Gas Technology Institute
International Flame Research Foundation reports (IJmuiden, The Netherlands)
International (Symposia) on Combustion proceedings (The Combustion Institute,
   Pittsburgh, PA)
U.S. Dept. of Energy reports (Washington, DC)
U.S. Patent & Trademark Office (Washington, DC)

Appendix B

Common Conversions

1 Btu	252.0 cal
	1055 J
1 Btu/ft^3	0.00890 cal/cm^3
	0.0373 MJ/m^3
1 Btu/hr	0.0003931 hp
	0.2520 kcal/hr
	0.2931 W
1 million Btu/hr	0.293 MW
1 Btu/hr-ft^2	0.003153 kW/m^2
1 Btu/hr-ft-°F	1.730 W/m-K
1 Btu/hr-ft^2-°F	5.67 W/m^2-K
1 Btu/lb	0.5556 cal/g
	2326 J/kg
1 Btu/lb-°F	1 cal/g-°C
	4187 J/kg-K
1 cal	0.003968 Btu
	4.187 J
1 cal/cm^2-sec	3.687 Btu/ft^2-sec
	41.87 kW/m^2
1 cal/cm-sec-°C	241.9 Btu/ft-hr-°F
	418.7 W/m-K
1 cal/g	1.80 Btu/lb
	4187 J/kg
1 cal/g-°C	1 Btu/lb-°F
	4187 J/kg-K
1 centipoise	2.421 lbm/hr-ft
1 cm^2/sec	100 centistokes
	3.874 ft^2/hr
1 ft	0.3048 m
1 ft^2/s	0.0929 m^2/sec
1 g/cm^3	1000 kg/m^3
	62.43 lb/ft^3
	0.03613 lb/in.^{3}

1 hp	33,000 ft-lb/min
	550 ft-lb/sec
	641.4 kcal/hr
	745.7 W
1 in.	2.540 cm
	25.40 mm
1 J	0.000948 Btu
	0.239 cal
	1 W/sec
1 kcal	3.968 Btu
	1000 cal
	4187 J
1 kcal/hr	3.968 Btu/hr
	1.162 J/sec
1 kcal/m^3	0.1124 Btu/ft^3
	4187 J/m^3
1 kg	2.205 lb
1 kg/hr-m	0.00278 g/sec-cm
	0.672 lb/hr-ft
1 kg/m^3	0.06243 lb/ft^3
1 kW	3413 Btu/hr
	1.341 hp
	660.6 kcal/hr
1 kW/m^2	317.2 Btu/hr-ft^2
1 kW/m^2-°C	176.2 Btu/hr-ft^2-°F
1 lb	0.4536 kg
1 lb/ft^3	0.0160 g/cm^3
	16.02 kg/m^3
1 lbm/hr-ft	0.413 centipoise
1 m	3.281 ft
1 mm	0.03937 in.
1 m^2/s	10.76 ft^2/sec
1 m-ton	1000 kg
	2205 lb
1 MW	3,413,000 Btu/hr
	1000 kW
1 therm	100,000 Btu
1 W	1 J/sec
1 W/m-K	0.5778 Btu/ft-hr-°F

Temperature Conversions

$°C = 5/9 \ (°F - 32)$ $°F = 9/5(°C) + 32$

$K = °C + 273.15$ $°R = °F + 459.67$

Appendix C

Methods of Expressing Mixture Ratios for CH_4, C_3H_8, and H_2

Table C.1a   Methods of Expressing Fuel-Rich Combustion Mixtures for CH_4

Oxidizer		S	S_1	S_2	ϕ	λ Act.		XA	XO_2
Vol% O_2 in oxid.	Vol% N_2 in oxid.	Stoich. vol. O_2/fuel	Actual vol. O_2/fuel	Actual vol. oxid./fuel	Theor. stoich./ act. stoich.	stoich./ theor. stoich.	Vol% fuel in mix	Vol% air above stoich.	Vol% O_2 above stoich.
0.21	0.79	2.00	0.500	2.381	4.000	0.250	0.2958	−75	−15.75
0.21	0.79	2.00	1.000	4.762	2.000	0.500	0.1736	−50	−10.50
0.21	0.79	2.00	1.053	5.012	1.900	0.526	0.1663	−47	−9.95
0.21	0.79	2.00	1.100	5.238	1.818	0.550	0.1603	−45	−9.45
0.21	0.79	2.00	1.111	5.290	1.800	0.556	0.1590	−44	−9.33
0.21	0.79	2.00	1.177	5.602	1.700	0.588	0.1515	−41	−8.65
0.21	0.79	2.00	1.200	5.714	1.667	0.600	0.1489	−40	−8.40
0.21	0.79	2.00	1.250	5.952	1.600	0.625	0.1438	−38	−7.88
0.21	0.79	2.00	1.300	6.190	1.538	0.650	0.1391	−35	−7.35
0.21	0.79	2.00	1.333	6.349	1.500	0.667	0.1361	−33	−7.00
0.21	0.79	2.00	1.400	6.667	1.429	0.700	0.1304	−30	−6.30
0.21	0.79	2.00	1.429	6.805	1.400	0.715	0.1281	−29	−6.00
0.21	0.79	2.00	1.500	7.143	1.333	0.750	0.1228	−25	−5.25
0.21	0.79	2.00	1.538	7.324	1.300	0.769	0.1201	−23	−4.85
0.21	0.79	2.00	1.600	7.619	1.250	0.800	0.1160	−20	−4.20
0.21	0.79	2.00	1.667	7.937	1.200	0.833	0.1119	−17	−3.50
0.21	0.79	2.00	1.700	8.095	1.176	0.850	0.1099	−15	−3.15
0.21	0.79	2.00	1.800	8.571	1.111	0.900	0.1045	−10	−2.10
0.21	0.79	2.00	1.818	8.657	1.100	0.909	0.1036	−9	−1.91
0.21	0.79	2.00	1.889	8.995	1.059	0.945	0.1000	−6	−1.17
0.21	0.79	2.00	1.900	9.048	1.053	0.950	0.0995	−5	−1.05

Table C.1b Methods of Expressing Stoichiometric and Fuel-Lean Combustion Mixtures for CH_4

Oxidizer		S Stoich. vol. O_2/fuel	S_1 Actual vol. O_2/fuel	S_2 Actual vol. oxid./fuel	ϕ Theor. stoich./act. stoich.	λ Act. stoich./theor. stoich.	Vol% fuel in mix	XA Vol% air above stoich.	XO_2 Vol% O_2 above stoich.
Vol% O_2 in oxid.	Vol% N_2 in oxid.								
0.21	0.79	2.00	2.000	9.524	1.000	1.000	0.0950	0	0.00
0.21	0.79	2.00	2.095	9.976	0.955	1.048	0.0911	5	1.00
0.21	0.79	2.00	2.100	10.000	0.952	1.050	0.0909	5	1.05
0.21	0.79	2.00	2.124	10.114	0.942	1.062	0.0900	6	1.30
0.21	0.79	2.00	2.190	10.429	0.913	1.095	0.0875	10	2.00
0.21	0.79	2.00	2.200	10.476	0.909	1.100	0.0871	10	2.10
0.21	0.79	2.00	2.221	10.576	0.900	1.111	0.0864	11	2.32
0.21	0.79	2.00	2.286	10.886	0.875	1.143	0.0841	14	3.00
0.21	0.79	2.00	2.300	10.952	0.870	1.150	0.0837	15	3.15
0.21	0.79	2.00	2.400	11.429	0.833	1.200	0.0805	20	4.20
0.21	0.79	2.00	2.500	11.905	0.800	1.250	0.0775	25	5.25
0.21	0.79	2.00	2.600	12.381	0.769	1.300	0.0747	30	6.30
0.21	0.79	2.00	2.667	12.698	0.750	1.333	0.0730	33	7.00
0.21	0.79	2.00	2.700	12.857	0.741	1.350	0.0722	35	7.35
0.21	0.79	2.00	2.800	13.333	0.714	1.400	0.0698	40	8.40
0.21	0.79	2.00	2.859	13.614	0.700	1.430	0.0684	43	9.02
0.21	0.79	2.00	2.900	13.810	0.690	1.450	0.0675	45	9.45
0.21	0.79	2.00	3.000	14.286	0.667	1.500	0.0654	50	10.50
0.21	0.79	2.00	3.500	16.667	0.571	1.750	0.0566	75	15.75
0.21	0.79	2.00	4.000	19.048	0.500	2.000	0.0499	100	21.00
0.30	0.70	2.00	2.000	6.667	1.000	1.000	0.1304	0	0.00
0.40	0.60	2.00	2.000	5.000	1.000	1.000	0.1667	0	0.00
0.50	0.50	2.00	2.000	4.000	1.000	1.000	0.2000	0	0.00
0.60	0.40	2.00	2.000	3.333	1.000	1.000	0.2308	0	0.00
0.70	0.30	2.00	2.000	2.857	1.000	1.000	0.2593	0	0.00
0.80	0.20	2.00	2.000	2.500	1.000	1.000	0.2857	0	0.00
0.90	0.10	2.00	2.000	2.222	1.000	1.000	0.3103	0	0.00
1.00	0.00	2.00	2.000	2.000	1.000	1.000	0.3333	0	0.00

Table C.2a Methods of Expressing Fuel-Rich Combustion Mixtures for C_3H_8

Oxidizer		S Stoich. vol. O_2/fuel	S_1 Actual vol. O_2/fuel	S_2 Actual vol. oxid./fuel	φ Ther. stoich./act. stoich.	λ Act. stoich./theor. stoich.	Vol% fuel in mix	XA Vol% air above stoich.	XO_2 Vol% O_2 above stoich.
Vol% O_2 in oxid.	Vol% N_2 in oxid.								
0.21	0.79	5.00	1.250	5.952	4.000	0.250	0.1438	−75	−15.75
0.21	0.79	5.00	2.500	11.905	2.000	0.500	0.0775	−50	−10.50
0.21	0.79	5.00	2.631	12.530	1.900	0.526	0.0739	−47	−9.95
0.21	0.79	5.00	2.750	13.095	1.818	0.550	0.0709	−45	−9.45
0.21	0.79	5.00	2.778	13.226	1.800	0.556	0.0703	−44	−9.33
0.21	0.79	5.00	2.941	14.006	1.700	0.588	0.0666	−41	−8.65
0.21	0.79	5.00	3.000	14.286	1.667	0.600	0.0654	−40	−8.40
0.21	0.79	5.00	3.125	14.881	1.600	0.625	0.0630	−38	−7.88
0.21	0.79	5.00	3.250	15.476	1.538	0.650	0.0607	−35	−7.35
0.21	0.79	5.00	3.333	15.873	1.500	0.667	0.0593	−33	−7.00
0.21	0.79	5.00	3.500	16.667	1.429	0.700	0.0566	−30	−6.30
0.21	0.79	5.00	3.573	17.012	1.400	0.715	0.0555	−29	−6.00
0.21	0.79	5.00	3.750	17.857	1.333	0.750	0.0530	−25	−5.25
0.21	0.79	5.00	3.845	18.310	1.300	0.769	0.0518	−23	−4.85
0.21	0.79	5.00	4.000	19.048	1.250	0.800	0.0499	−20	−4.20
0.21	0.79	5.00	4.167	19.842	1.200	0.833	0.0480	−17	−3.50
0.21	0.79	5.00	4.250	20.238	1.176	0.850	0.0471	−15	−3.15
0.21	0.79	5.00	4.500	21.429	1.111	0.900	0.0446	−10	−2.10
0.21	0.79	5.00	4.545	21.643	1.100	0.909	0.0442	−9	−1.91
0.21	0.79	5.00	4.723	22.488	1.059	0.945	0.0426	−6	−1.17
0.21	0.79	5.00	4.750	22.619	1.053	0.950	0.0423	−5	−1.05

Table C.2b Methods of Expressing Stoichiometric and Fuel-Lean Combustion Mixtures for C_3H_8

Oxidizer		S Stoich. vol. O_2/fuel	S_1 Actual vol. O_2/fuel	S_2 Actual vol. oxid./fuel	ϕ Theor. stoich./act. stoich.	λ Act. stoich./theor. stoich.	Vol% fuel in mix	XA Vol% air above stoich.	XO_2 Vol% O_2 above stoich.
Vol% O_2 in oxid.	Vol% N_2 in oxid.								
0.21	0.79	5.00	5.000	23.810	1.000	1.000	0.0403	0	0.00
0.21	0.79	5.00	5.238	24.940	0.955	1.048	0.0385	5	1.00
0.21	0.79	5.00	5.250	25.000	0.952	1.050	0.0385	5	1.05
0.21	0.79	5.00	5.310	25.286	0.942	1.062	0.0380	6	1.30
0.21	0.79	5.00	5.475	26.071	0.913	1.095	0.0369	10	2.00
0.21	0.79	5.00	5.500	26.190	0.909	1.100	0.0368	10	2.10
0.21	0.79	5.00	5.553	26.440	0.900	1.111	0.0364	11	2.32
0.21	0.79	5.00	5.715	27.214	0.875	1.143	0.0354	14	3.00
0.21	0.79	5.00	5.750	27.381	0.870	1.150	0.0352	15	3.15
0.21	0.79	5.00	6.000	28.571	0.833	1.200	0.0338	20	4.20
0.21	0.79	5.00	6.250	29.762	0.800	1.250	0.0325	25	5.25
0.21	0.79	5.00	6.500	30.952	0.769	1.300	0.0313	30	6.30
0.21	0.79	5.00	6.667	31.746	0.750	1.333	0.0305	33	7.00
0.21	0.79	5.00	6.750	32.143	0.741	1.350	0.0302	35	7.35
0.21	0.79	5.00	7.000	33.333	0.714	1.400	0.0291	40	8.40
0.21	0.79	5.00	7.148	34.036	0.700	1.430	0.0285	43	9.02
0.21	0.79	5.00	7.250	34.524	0.690	1.450	0.0282	45	9.45
0.21	0.79	5.00	7.500	35.714	0.667	1.500	0.0272	50	10.50
0.21	0.79	5.00	8.750	41.667	0.571	1.750	0.0234	75	15.75
0.21	0.79	5.00	10.000	47.619	0.500	2.000	0.0206	100	21.00
0.30	0.70	5.00	5.000	16.667	1.000	1.000	0.0566	0	0.00
0.40	0.60	5.00	5.000	12.500	1.000	1.000	0.0741	0	0.00
0.50	0.50	5.00	5.000	10.000	1.000	1.000	0.0909	0	0.00
0.60	0.40	5.00	5.000	8.333	1.000	1.000	0.1071	0	0.00
0.70	0.30	5.00	5.000	7.143	1.000	1.000	0.1228	0	0.00
0.80	0.20	5.00	5.000	6.250	1.000	1.000	0.1379	0	0.00
0.90	0.10	5.00	5.000	5.556	1.000	1.000	0.1525	0	0.00
1.00	0.00	5.00	5.000	5.000	1.000	1.000	0.1667	0	0.00

Table C.3a Methods of Expressing Fuel-Rich Combustion Mixtures for H_2

Oxidizer									
Vol% O_2 in oxid.	Vol% N_2 in oxid.	S Stoich. vol. O_2/fuel	S_1 Actual vol. O_2/fuel	S_2 Actual vol. oxid./fuel	ϕ Theor. stoich./act. stoich.	λ Act. stoich./theor. stoich.	Vol% fuel in mix	XA Vol% air above stoich.	XO_2 Vol% O_2 above stoich.
0.21	0.79	0.50	0.125	0.595	4.000	0.250	0.6269	−75	−15.75
0.21	0.79	0.50	0.250	1.190	2.000	0.500	0.4565	−50	−10.50
0.21	0.79	0.50	0.263	1.253	1.900	0.526	0.4439	−47	−9.95
0.21	0.79	0.50	0.275	1.310	1.818	0.550	0.4330	−45	−9.45
0.21	0.79	0.50	0.278	1.323	1.800	0.556	0.4305	−44	−9.33
0.21	0.79	0.50	0.294	1.401	1.700	0.588	0.4166	−41	−8.65
0.21	0.79	0.50	0.300	1.429	1.667	0.600	0.4118	−40	−8.40
0.21	0.79	0.50	0.313	1.488	1.600	0.625	0.4019	−38	−7.88
0.21	0.79	0.50	0.325	1.548	1.538	0.650	0.3925	−35	−7.35
0.21	0.79	0.50	0.333	1.587	1.500	0.667	0.3865	−33	−7.00
0.21	0.79	0.50	0.350	1.667	1.429	0.700	0.3750	−30	−6.30
0.21	0.79	0.50	0.357	1.701	1.400	0.715	0.3702	−29	−6.00
0.21	0.79	0.50	0.375	1.786	1.333	0.750	0.3590	−25	−5.25
0.21	0.79	0.50	0.385	1.831	1.300	0.769	0.3532	−23	−4.85
0.21	0.79	0.50	0.400	1.905	1.250	0.800	0.3443	−20	−4.20
0.21	0.79	0.50	0.417	1.984	1.200	0.833	0.3351	−17	−3.50
0.21	0.79	0.50	0.425	2.024	1.176	0.850	0.3307	−15	−3.15
0.21	0.79	0.50	0.450	2.143	1.111	0.900	0.3182	−10	−2.10
0.21	0.79	0.50	0.455	2.164	1.100	0.909	0.3160	−9	−1.91
0.21	0.79	0.50	0.472	2.249	1.059	0.945	0.3078	−6	−1.17
0.21	0.79	0.50	0.475	2.262	1.053	0.950	0.3066	−5	−1.05

Table C.3b Methods of Expressing Stoichiometric and Fuel Lean Combustion Mixtures for H_2

Oxidizer		S Stoich. vol. O_2/fuel	S_1 Actual vol. O_2/fuel	S_2 Actual vol. oxid./fuel	φ Theor. stoich./act. stoich.	λ Act. stoich./ theor. stoich.	Vol% fuel in mix	XA Vol% air above stoich.	XO_2 Vol% O_2 above stoich.
Vol% O_2 in oxid.	Vol% N_2 in oxid.								
0.21	0.79	0.50	0.500	2.381	1.000	1.000	0.2958	0	0.00
0.21	0.79	0.50	0.524	2.494	0.955	1.048	0.2862	5	1.00
0.21	0.79	0.50	0.525	2.500	0.952	1.050	0.2857	5	1.05
0.21	0.79	0.50	0.531	2.529	0.942	1.062	0.2834	6	1.30
0.21	0.79	0.50	0.548	2.607	0.913	1.095	0.2772	10	2.00
0.21	0.79	0.50	0.550	2.619	0.909	1.100	0.2763	10	2.10
0.21	0.79	0.50	0.555	2.644	0.900	1.111	0.2744	11	2.32
0.21	0.79	0.50	0.572	2.721	0.875	1.143	0.2687	14	3.00
0.21	0.79	0.50	0.575	2.738	0.870	1.150	0.2675	15	3.15
0.21	0.79	0.50	0.600	2.857	0.833	1.200	0.2593	20	4.20
0.21	0.79	0.50	0.625	2.976	0.800	1.250	0.2515	25	5.25
0.21	0.79	0.50	0.650	3.095	0.769	1.300	0.2442	30	6.30
0.21	0.79	0.50	0.667	3.175	0.750	1.333	0.2395	33	7.00
0.21	0.79	0.50	0.675	3.214	0.741	1.350	0.2373	35	7.35
0.21	0.79	0.50	0.700	3.333	0.714	1.400	0.2308	40	8.40
0.21	0.79	0.50	0.715	3.404	0.700	1.430	0.2271	43	9.02
0.21	0.79	0.50	0.725	3.452	0.690	1.450	0.2246	45	9.45
0.21	0.79	0.50	0.750	3.571	0.667	1.500	0.2188	50	10.50
0.21	0.79	0.50	0.875	4.167	0.571	1.750	0.1935	75	15.75
0.21	0.79	0.50	1.000	4.762	0.500	2.000	0.1736	100	21.00
0.30	0.70	0.50	0.500	1.667	1.000	1.000	0.3750	0	0.00
0.40	0.60	0.50	0.500	1.250	1.000	1.000	0.4444	0	0.00
0.50	0.50	0.50	0.500	1.000	1.000	1.000	0.5000	0	0.00
0.60	0.40	0.50	0.500	0.833	1.000	1.000	0.5455	0	0.00
0.70	0.30	0.50	0.500	0.714	1.000	1.000	0.5833	0	0.00
0.80	0.20	0.50	0.500	0.625	1.000	1.000	0.6154	0	0.00
0.90	0.10	0.50	0.500	0.556	1.000	1.000	0.6429	0	0.00
1.00	0.00	0.50	0.500	0.500	1.000	1.000	0.6667	0	0.00

Appendix D

Flame Properties for CH_4, C_3H_8, and H_2, and CH_4/H_2 Blends

Tables for various flame properties are given on the following pages.

Table D.1a H₂ (298 K) Adiabatically and Stoichiometrically (φ = 1.00) Combusted with a Variable Composition Oxidizer (298 K), Metric Units

O_2	N_2	Flame temp. (K)	H	HO_2	H_2	H_2O (g)	NO	N_2	O	OH	O_2	Enthalpy, H (kJ/kg)	Density (kg/m³)	Viscosity (kg/m-sec)	Equil. spec. heat (kJ/kg-K)	Equil. therm. cond. (W/m-K)	Equil. Pr	Le
0.21	0.79	2382	0.00179	0.00000	0.01526	0.32366	0.00264	0.64440	0.00054	0.00696	0.00475	2.09E−04	0.124	7.53E-05	2.72	0.383	0.536	1.28
0.30	0.70	2648	0.00933	0.00001	0.04074	0.38912	0.00607	0.51684	0.00321	0.02217	0.01251	0.00E+00	0.104	8.19E-05	4.72	0.851	0.454	1.45
0.40	0.60	2799	0.02092	0.00002	0.06662	0.43867	0.00851	0.39742	0.00770	0.03962	0.02052	4.19E−05	0.091	8.60E-05	7.11	1.420	0.431	1.49
0.50	0.50	2889	0.03269	0.00003	0.08793	0.47585	0.00960	0.29903	0.01253	0.05505	0.02729	4.19E−05	0.082	8.88E-05	9.47	1.962	0.429	1.49
0.60	0.40	2950	0.04357	0.00003	0.10540	0.50534	0.00973	0.21737	0.01717	0.06834	0.03303	8.37E−05	0.075	9.08E-05	11.75	2.459	0.434	1.47
0.70	0.30	2994	0.05333	0.00004	0.11988	0.52948	0.00911	0.14886	0.02148	0.07983	0.03798	1.26E−04	0.070	9.25E-05	13.94	2.913	0.442	1.45
0.80	0.20	3028	0.06203	0.00005	0.13199	0.54964	0.00782	0.09078	0.02545	0.08987	0.04236	1.67E−04	0.066	9.38E-05	16.04	3.328	0.452	1.43
0.90	0.10	3055	0.06980	0.00005	0.14217	0.56673	0.00569	0.04122	0.02915	0.09880	0.04637	0.00E+00	0.062	9.50E-05	18.06	3.709	0.462	1.41
1.00	0.00	3079	0.07695	0.00006	0.15027	0.58120	0.00000	0.00000	0.03302	0.10758	0.05092	0.00E+00	0.059	9.60E-05	20.01	4.0670	0.472	1.40

Pr = Prandtl number, Le = Lewis number.

Table D.1b H$_2$ (77°F) Adiabatically and Stoichiometrically ($\phi = 1.00$) Combusted with a Variable Composition Oxidizer (77°F), English Units

O$_2$	N$_2$	Flame temp. (°F)	H	HO$_2$	H$_2$	H$_2$O (g)	NO	N$_2$	O	OH	O$_2$	Enthalpy, H (Btu/lb)	Density (lb/ft^3)	Viscosity (lbm/ ft-hr)	Equil. spec. heat (Btu/ lb-°F)	Equil. therm. cond. (Btu/ ft-hr-°F)	Equil. Pr	Le
0.21	0.79	3827	0.00179	0.00000	0.01526	0.32366	0.00264	0.64440	0.00054	0.00696	0.00475	0	0.00775	5.06E-05	0.650	0.221	0.536	1.28
0.30	0.70	4307	0.00933	0.00001	0.04074	0.38912	0.00607	0.51684	0.00321	0.02217	0.01251	0	0.00649	5.50E-05	1.127	0.492	0.454	1.45
0.40	0.60	4579	0.02092	0.00002	0.06662	0.43867	0.00851	0.39742	0.00770	0.03962	0.02052	0	0.00568	5.78E-05	1.699	0.821	0.431	1.49
0.50	0.50	4741	0.03269	0.00003	0.08793	0.47585	0.00960	0.29903	0.01253	0.05505	0.02729	0	0.00512	5.97E-05	2.262	1.134	0.429	1.49
0.60	0.40	4850	0.04357	0.00003	0.10540	0.50534	0.00973	0.21737	0.01717	0.06834	0.03303	0	0.00470	6.11E-05	2.808	1.421	0.434	1.47
0.70	0.30	4929	0.05333	0.00004	0.11988	0.52948	0.00911	0.14886	0.02148	0.07983	0.03798	0	0.00437	6.22E-05	3.328	1.683	0.442	1.45
0.80	0.20	4990	0.06203	0.00005	0.13199	0.54964	0.00782	0.09078	0.02545	0.08987	0.04236	0	0.00410	6.31E-05	3.830	1.923	0.452	1.43
0.90	0.10	5039	0.06980	0.00005	0.14217	0.56673	0.00569	0.04122	0.02915	0.09880	0.04637	0	0.00387	6.39E-05	4.312	2.143	0.462	1.41
1.00	0.00	5082	0.07695	0.00006	0.15027	0.58120	0.00000	0.00000	0.03302	0.10758	0.05092	0	0.00367	6.45E-05	4.780	2.350	0.472	1.40

Table D.2a H_2 (298 K) Adiabatically and Stoichiometrically ($\phi = 1.00$) Combusted with Air (Variable Temp.), Metric Units

Air temp. (K)	Flame temp. (K)	H	HO_2	H_2	H_2O (g)	NO	N_2	O	OH	O_2	Enthalpy, H (kJ/kg)	Density (kg/m³)	Viscosity (kg/m-s)	Equil. spec. heat (kJ/kg-K)	Equil. therm. cond. (W/m-K)	Equil. Pr	Le
298	2382	0.00179	0.00000	0.01526	0.32366	0.00264	0.64440	0.00054	0.00696	0.00475	2.09E−04	0.124	7.53E-05	2.72	0.383	0.536	1.28
366	2406	0.00209	0.00000	0.01657	0.32145	0.00287	0.64356	0.00064	0.00768	0.00513	6.69E+01	0.123	7.58E-05	2.81	0.404	0.527	1.30
478	2444	0.00267	0.00000	0.01882	0.31760	0.00327	0.64207	0.00083	0.00895	0.00579	1.78E+02	0.121	7.66E-05	2.96	0.442	0.513	1.32
589	2482	0.00335	0.00000	0.02116	0.31350	0.00369	0.64047	0.00106	0.01030	0.00647	2.91E+02	0.119	7.74E-05	3.11	0.482	0.500	1.35
700	2518	0.00415	0.00000	0.02361	0.30913	0.00413	0.63874	0.00133	0.01174	0.00717	4.06E+02	0.117	7.82E-05	3.28	0.526	0.488	1.38
811	2553	0.00507	0.00000	0.02615	0.30448	0.00460	0.63687	0.00165	0.01328	0.00790	5.24E+02	0.115	7.89E-05	3.46	0.573	0.476	1.41
922	2587	0.00613	0.00000	0.02877	0.29958	0.00508	0.63487	0.00202	0.01490	0.00864	6.46E+02	0.113	7.96E-05	3.64	0.623	0.465	1.43
1033	2620	0.00733	0.00001	0.03145	0.29446	0.00557	0.63274	0.00245	0.01659	0.00939	7.69E+02	0.111	8.03E-05	3.83	0.677	0.455	1.46
1144	2652	0.00867	0.00001	0.03416	0.28915	0.00608	0.63050	0.00294	0.01833	0.01015	8.95E+02	0.109	8.09E-05	4.03	0.733	0.445	1.48
1255	2683	0.01015	0.00001	0.03688	0.28369	0.00659	0.62816	0.00348	0.02011	0.01091	1.02E+03	0.108	8.15E-05	4.24	0.791	0.437	1.50
1366	2712	0.01177	0.00001	0.03960	0.27811	0.00711	0.62573	0.00408	0.02193	0.01166	1.15E+03	0.106	8.21E-05	4.45	0.852	0.429	1.52

Table D.2b H_2 (77°F) Adiabatically and Stoichiometrically ($\phi = 1.00$) Combusted with Air (Variable Temp.), English Units

Air temp. (°F)	Flame temp. (°F)	H	HO_2	H_2	H_2O (g)	NO	N_2	O	OH	O_2	Enthalpy, H (Btu/lb)	Density (lb/ft³)	Viscosity (lbm/ft-hr)	Equil. spec. heat (Btu/lb-F)	Equil. therm. cond. (Btu/ft-hr-F)	Equil. Pr	Le
77	3827	0.00179	0.00000	0.01526	0.32366	0.00264	0.64440	0.00054	0.00696	0.00475	0	0.00775	5.06E-05	0.650	0.221	0.536	1.28
200	3871	0.00209	0.00000	0.01657	0.32145	0.00287	0.64356	0.00064	0.00768	0.00513	29	0.00767	5.10E-05	0.671	0.233	0.527	1.30
400	3940	0.00267	0.00000	0.01882	0.31760	0.00327	0.64207	0.00083	0.00895	0.00579	77	0.00753	5.15E-05	0.706	0.255	0.513	1.32
600	4007	0.00335	0.00000	0.02116	0.31350	0.00369	0.64047	0.00106	0.01030	0.00647	125	0.00740	5.21E-05	0.744	0.279	0.500	1.35
800	4072	0.00415	0.00000	0.02361	0.30913	0.00413	0.63874	0.00133	0.01174	0.00717	175	0.00728	5.26E-05	0.783	0.304	0.488	1.38
1000	4135	0.00507	0.00000	0.02615	0.30448	0.00460	0.63687	0.00165	0.01328	0.00790	225	0.00716	5.31E-05	0.825	0.331	0.476	1.41
1200	4197	0.00613	0.00000	0.02877	0.29958	0.00508	0.63487	0.00202	0.01490	0.00864	278	0.00705	5.35E-05	0.869	0.360	0.465	1.43
1400	4256	0.00733	0.00001	0.03145	0.29446	0.00557	0.63274	0.00245	0.01659	0.00939	331	0.00694	5.40E-05	0.915	0.391	0.455	1.46
1600	4314	0.00867	0.00001	0.03416	0.28915	0.00608	0.63050	0.00294	0.01833	0.01015	385	0.00683	5.44E-05	0.963	0.423	0.445	1.48
1800	4369	0.01015	0.00001	0.03688	0.28369	0.00659	0.62816	0.00348	0.02011	0.01091	439	0.00673	5.48E-05	1.012	0.457	0.437	1.50
2000	4422	0.01177	0.00001	0.03960	0.27811	0.00711	0.62573	0.00408	0.02193	0.01166	495	0.00664	5.52E-05	1.062	0.492	0.429	1.52

Table D.3a H$_2$ (Variable Temp.) Adiabatically and Stoichiometrically ($\phi = 1.00$) Combusted with Air (298 K), Metric Units

Fuel temp. (K)	Flame temp. (K)	H	H$_2$	H$_2$O (g)	NO	N$_2$	O	OH	O$_2$	Enthalpy, H (kJ/kg)	Density (kg/m^3)	Viscosity (kg/m-sec)	Equil. spec. heat (kJ/kg-K)	Equil. therm. cond. (W/m-K)	Equil. Pr	Le
298	2382	0.00179	0.01526	0.32366	0.00264	0.64440	0.00054	0.00696	0.00475	2.09E − 04	0.124	7.53E-05	2.72	0.383	0.536	1.28
311	2384	0.00181	0.01537	0.32348	0.00266	0.64434	0.00055	0.00702	0.00478	5.29E + 00	0.124	7.53E-05	2.73	0.384	0.535	1.28
366	2392	0.00191	0.01580	0.32275	0.00274	0.64406	0.00058	0.00726	0.00491	2.79E + 01	0.124	7.55E-05	2.76	0.391	0.532	1.28
422	2400	0.00202	0.01626	0.32198	0.00282	0.64376	0.00061	0.00751	0.00504	5.10E + 01	0.123	7.57E-05	2.79	0.399	0.529	1.29
478	2408	0.00213	0.01672	0.32121	0.00290	0.64347	0.00065	0.00776	0.00517	7.41E + 01	0.123	7.59E-05	2.82	0.406	0.526	1.30
533	2416	0.00224	0.01717	0.32043	0.00298	0.64317	0.00069	0.00802	0.00531	9.69E + 01	0.122	7.60E-05	2.85	0.414	0.523	1.30
589	2425	0.00236	0.01764	0.31963	0.00306	0.64286	0.00072	0.00828	0.00544	1.20E + 02	0.122	7.62E-05	2.88	0.422	0.520	1.31
644	2432	0.00248	0.01810	0.31884	0.00314	0.64256	0.00076	0.00854	0.00558	1.43E + 02	0.121	7.64E-05	2.91	0.429	0.517	1.31
700	2440	0.00260	0.01857	0.31802	0.00323	0.64224	0.00081	0.00881	0.00572	1.66E + 02	0.121	7.65E-05	2.94	0.437	0.514	1.32
755	2448	0.00273	0.01904	0.31721	0.00331	0.64192	0.00085	0.00907	0.00585	1.89E + 02	0.120	7.67E-05	2.97	0.445	0.512	1.33
811	2456	0.00287	0.01953	0.31637	0.00340	0.64160	0.00089	0.00935	0.00599	2.13E + 02	0.120	7.69E-05	3.00	0.454	0.509	1.33

Table D.3b H$_2$ (Variable Temp.) Adiabatically and Stoichiometrically ($\phi = 1.00$) Combusted with Air (77°F), English Units

Fuel temp. (K)	Flame temp. (K)	H	H$_2$	H$_2$O (g)	NO	N$_2$	O	OH	O$_2$	Enthalpy, H (Btu/lb)	Density (lb/ft^3)	Viscosity (lbm/ft-hr)	Equil. spec. heat (Btu/lb-°F)	Equil. therm. cond. (Btu/ft-hr-F)	Equil. Pr	Le
77	3827	0.00179	0.01526	0.32366	0.00264	0.64440	0.00054	0.00696	0.00475	0	0.00775	5.06E-05	0.650	0.221	0.536	1.28
100	3831	0.00181	0.01537	0.32348	0.00266	0.64434	0.00055	0.00702	0.00478	2	0.00775	5.07E-05	0.652	0.222	0.535	1.28
200	3846	0.00191	0.01580	0.32275	0.00274	0.64406	0.00058	0.00726	0.00491	12	0.00772	5.08E-05	0.659	0.226	0.532	1.28
300	3861	0.00202	0.01626	0.32198	0.00282	0.64376	0.00061	0.00751	0.00504	22	0.00769	5.09E-05	0.666	0.230	0.529	1.29
400	3875	0.00213	0.01672	0.32121	0.00290	0.64347	0.00065	0.00776	0.00517	32	0.00766	5.10E-05	0.673	0.235	0.526	1.30
500	3890	0.00224	0.01717	0.32043	0.00298	0.64317	0.00069	0.00802	0.00531	42	0.00763	5.11E-05	0.680	0.239	0.523	1.30
600	3904	0.00236	0.01764	0.31963	0.00306	0.64286	0.00072	0.00828	0.00544	52	0.00760	5.13E-05	0.687	0.244	0.520	1.31
700	3919	0.00248	0.01810	0.31884	0.00314	0.64256	0.00076	0.00854	0.00558	61	0.00757	5.14E-05	0.695	0.248	0.517	1.31
800	3933	0.00260	0.01857	0.31802	0.00323	0.64224	0.00081	0.00881	0.00572	71	0.00755	5.15E-05	0.702	0.253	0.514	1.32
900	3947	0.00273	0.01904	0.31721	0.00331	0.64192	0.00085	0.00907	0.00585	81	0.00752	5.16E-05	0.710	0.257	0.512	1.33
1000	3961	0.00287	0.01953	0.31637	0.00340	0.64160	0.00089	0.00935	0.00599	91	0.00749	5.17E-05	0.717	0.262	0.509	1.33

Table D.4a H₂ (298 K) Stoichiometrically ($\phi = 1.00$) Combusted with Air (298 K), Metric Units

Flame temp. (K)	H	H₂	H₂O (g)	NO	N₂	O	OH	O₂	Enthalpy, H (kJ/kg)	Density (kg/m³)	Viscosity (kg/m-sec)	Equil. spec. heat (kJ/kg-K)	Equil. therm. cond. (W/m-K)	Equil. Pr	Le	Avail. heat (%)
366	0.00000	0.00000	0.34711	0.00000	0.65289	0.00000	0.00000	0.00000	−3.34E+03	0.817	1.78E-05	1.26	0.027	0.825	1.00	82.2
478	0.00000	0.00000	0.34711	0.00000	0.65289	0.00000	0.00000	0.00000	−3.20E+03	0.626	2.23E-05	1.28	0.035	0.808	1.00	78.7
589	0.00000	0.00000	0.34711	0.00000	0.65289	0.00000	0.00000	0.00000	−3.05E+03	0.508	2.65E-05	1.31	0.044	0.795	1.00	75.2
700	0.00000	0.00000	0.34711	0.00000	0.65289	0.00000	0.00000	0.00000	−2.90E+03	0.427	3.06E-05	1.35	0.053	0.785	1.00	71.5
811	0.00000	0.00000	0.34711	0.00000	0.65289	0.00000	0.00000	0.00000	−2.75E+03	0.369	3.44E-05	1.39	0.061	0.777	1.00	67.8
922	0.00000	0.00000	0.34711	0.00000	0.65289	0.00000	0.00000	0.00000	−2.60E+03	0.324	3.80E-05	1.43	0.071	0.769	1.00	63.9
1033	0.00000	0.00000	0.34711	0.00000	0.65289	0.00000	0.00000	0.00000	−2.43E+03	0.290	4.15E-05	1.46	0.080	0.760	1.00	60.0
1144	0.00000	0.00000	0.34711	0.00000	0.65289	0.00000	0.00000	0.00000	−2.27E+03	0.261	4.48E-05	1.50	0.089	0.755	1.00	55.9
1255	0.00000	0.00001	0.34711	0.00000	0.65289	0.00000	0.00000	0.00000	−2.10E+03	0.238	4.80E-05	1.53	0.098	0.750	1.00	51.8
1366	0.00000	0.00003	0.34710	0.00000	0.65288	0.00000	0.00000	0.00001	−1.93E+03	0.219	5.10E-05	1.56	0.107	0.744	1.00	47.5
1478	0.00000	0.00008	0.34700	0.00001	0.65286	0.00000	0.00001	0.00003	−1.75E+03	0.202	5.40E-05	1.59	0.117	0.737	1.00	43.2
1589	0.00000	0.00022	0.34682	0.00004	0.65279	0.00000	0.00004	0.00008	−1.57E+03	0.188	5.69E-05	1.63	0.128	0.728	1.01	38.7
1700	0.00000	0.00051	0.34644	0.00008	0.65266	0.00000	0.00012	0.00018	−1.39E+03	0.176	5.97E-05	1.68	0.140	0.715	1.02	34.2
1811	0.00001	0.00106	0.34568	0.00018	0.65240	0.00001	0.00029	0.00037	−1.20E+03	0.165	6.24E-05	1.74	0.155	0.701	1.03	29.5
1922	0.00004	0.00203	0.34433	0.00034	0.65193	0.00001	0.00062	0.00069	−1.00E+03	0.155	6.50E-05	1.82	0.175	0.678	1.05	24.7
2033	0.00012	0.00361	0.34204	0.00061	0.65114	0.00003	0.00125	0.00120	−7.93E+02	0.147	6.76E-05	1.94	0.201	0.652	1.09	19.5
2144	0.00031	0.00605	0.33841	0.00103	0.64985	0.00008	0.00230	0.00197	−5.68E+02	0.139	7.01E-05	2.12	0.240	0.619	1.13	14.0
2255	0.00074	0.00958	0.33292	0.00164	0.64786	0.00021	0.00399	0.00305	−3.20E+02	0.132	7.26E-05	2.35	0.293	0.581	1.19	7.9
2382	0.00179	0.01526	0.32366	0.00264	0.64440	0.00054	0.00696	0.00475	−2.09E−04	0.124	7.53E-05	2.72	0.383	0.536	1.28	0.0

Table D.4b H₂ (Variable Temp.) Adiabatically and Stoichiometrically ($\phi = 1.00$) Combusted with Air (77°F), English Units

Flame temp. (°F)	H	H₂	H₂O (g)	NO	N₂	O	OH	O₂	Enthalpy, H (Btu/lb)	Density (lb/ft³)	Viscosity (lbm/ft-hr)	Equil. spec. heat (Btu/lb-°F)	Equil. therm. cond. (Btu/ft-hr-°F)	Equil. Pr	Le	Avail. heat (%)
200	0.00000	0.00000	0.34711	0.00000	0.65289	0.00000	0.00000	0.00000	−1435	0.05102	1.19E-05	0.300	0.016	0.825	1.00	82.2
400	0.00000	0.00000	0.34711	0.00000	0.65289	0.00000	0.00000	0.00000	−1374	0.03906	1.50E-05	0.306	0.020	0.808	1.00	78.7
600	0.00000	0.00000	0.34711	0.00000	0.65289	0.00000	0.00000	0.00000	−1312	0.03170	1.79E-05	0.313	0.025	0.795	1.00	75.2
800	0.00000	0.00000	0.34711	0.00000	0.65289	0.00000	0.00000	0.00000	−1248	0.02668	2.06E-05	0.322	0.030	0.785	1.00	71.5
1000	0.00000	0.00000	0.34711	0.00000	0.65289	0.00000	0.00000	0.00000	−1183	0.02302	2.31E-05	0.331	0.036	0.777	1.00	67.8
1200	0.00000	0.00000	0.34711	0.00000	0.65289	0.00000	0.00000	0.00000	−1116	0.02025	2.56E-05	0.341	0.041	0.769	1.00	63.9
1400	0.00000	0.00000	0.34711	0.00000	0.65289	0.00000	0.00000	0.00000	−1047	0.01808	2.79E-05	0.350	0.046	0.760	1.00	60.0
1600	0.00000	0.00001	0.34711	0.00000	0.65289	0.00000	0.00000	0.00000	−976	0.01632	3.01E-05	0.358	0.051	0.755	1.00	55.9
1800	0.00000	0.00003	0.34710	0.00000	0.65289	0.00000	0.00000	0.00000	−904	0.01488	3.23E-05	0.366	0.057	0.750	1.00	51.8
2000	0.00000	0.00008	0.34707	0.00000	0.65288	0.00000	0.00000	0.00001	−830	0.01367	3.43E-05	0.373	0.062	0.744	1.00	47.5
2200	0.00000	0.00022	0.34700	0.00001	0.65286	0.00000	0.00001	0.00003	−753	0.01263	3.63E-05	0.381	0.068	0.737	1.00	43.2
2400	0.00000	0.00051	0.34682	0.00004	0.65279	0.00000	0.00004	0.00008	−676	0.01175	3.83E-05	0.390	0.074	0.728	1.01	38.7
2600	0.00000	0.00106	0.34644	0.00008	0.65266	0.00000	0.00012	0.00018	−597	0.01098	4.01E-05	0.401	0.081	0.715	1.02	34.2
2800	0.00001	0.00203	0.34568	0.00018	0.65240	0.00000	0.00029	0.00037	−516	0.01030	4.20E-05	0.415	0.089	0.701	1.03	29.5
3000	0.00004	0.00361	0.34433	0.00034	0.65193	0.00001	0.00062	0.00069	−431	0.00970	4.37E-05	0.436	0.101	0.678	1.05	24.7
3200	0.00012	0.00605	0.34204	0.00061	0.65114	0.00003	0.00125	0.00120	−341	0.00916	4.55E-05	0.464	0.116	0.652	1.09	19.5
3400	0.00031	0.00958	0.33841	0.00103	0.64985	0.00008	0.00230	0.00197	−244	0.00868	4.72E-05	0.506	0.139	0.619	1.13	14.0
3600	0.00074	0.01200	0.33292	0.00164	0.64786	0.00021	0.00399	0.00305	−138	0.00823	4.88E-05	0.561	0.170	0.581	1.19	7.9
3827	0.00179	0.01526	0.32366	0.00264	0.64440	0.00054	0.00696	0.00475	0	0.00775	5.06E-05	0.650	0.221	0.536	1.28	0.0

Table D.5a H$_2$ (298 K) Adiabatically Combusted with Air (298 K) at Various Equivalence Ratios, Metric Units

Equiv. ratio	Flame temp. (K)	H	H$_2$	H$_2$O (g)	NH$_3$	NO	N$_2$	O	OH	O$_2$	Enthalpy, H (kJ/kg)	Density (kg/m^3)	Viscosity (kg/m-sec)	Equil. spec. heat (kJ/kg-K)	Equil. therm. cond. (W/m-K)	Equil. Pr	Le
4.00	1559	0.00002	0.51009	0.17004	0.00001	0.00000	0.31983	0.00000	0.00000	0.00000	4.19E−05	0.120	5.27E-05	2.75	0.285	0.509	1.00
2.00	2060	0.00122	0.25696	0.25735	0.00000	0.00001	0.48433	0.00000	0.00014	0.00000	8.37E−05	0.111	6.71E-05	2.23	0.283	0.529	1.09
1.33	2314	0.00332	0.10285	0.30829	0.00000	0.00021	0.58357	0.00004	0.00167	0.00005	2.51E−04	0.117	7.35E-05	2.22	0.312	0.523	1.22
1.00	2382	0.00179	0.01526	0.32366	0.00000	0.00264	0.64440	0.00054	0.00696	0.00475	2.09E−04	0.124	7.53E-05	2.72	0.383	0.536	1.28
0.80	2168	0.00018	0.00144	0.28314	0.00000	0.00454	0.67259	0.00040	0.00471	0.03300	1.67E−04	0.142	7.08E-05	1.92	0.223	0.610	1.15
0.67	1957	0.00002	0.00021	0.24439	0.00000	0.00358	0.69070	0.00012	0.00192	0.05907	0.00E+00	0.161	6.61E-05	1.66	0.158	0.694	1.03
0.57	1782	0.00000	0.00003	0.21370	0.00000	0.00242	0.70405	0.00003	0.00072	0.07903	4.19E−05	0.179	6.22E-05	1.53	0.133	0.718	1.00
0.50	1641	0.00000	0.00001	0.18989	0.00000	0.00158	0.71409	0.00001	0.00028	0.09415	0.00E+00	0.197	5.89E-05	1.46	0.119	0.723	1.00

Table D.5b H$_2$ (77°F) Combusted with air (77°F) at Various Equivalence Ratios, English Units

Equiv. ratio	Flame temp. (°F)	H	H$_2$	H$_2$O (g)	NH$_3$	NO	N$_2$	O	OH	O$_2$	Enthalpy, H (Btu/lb)	Density (lb/ft^3)	Viscosity (lbm/ft-hr)	Equil. spec. heat (Btu/lb-F)	Equil. therm. cond. (Btu/ft-hr-°F)	Equil. Pr	Le
4.00	2347	0.00002	0.51009	0.17004	0.00001	0.00000	0.31983	0.00000	0.00000	0.00000	0	0.00750	3.54E-05	0.657	0.165	0.509	1.00
2.00	3249	0.00122	0.25696	0.25735	0.00000	0.00001	0.48433	0.00000	0.00014	0.00000	0	0.00692	4.51E-05	0.532	0.163	0.529	1.09
1.33	3706	0.00332	0.10285	0.30829	0.00000	0.00021	0.58357	0.00004	0.00167	0.00005	0	0.00728	4.95E-05	0.530	0.180	0.523	1.22
1.00	3827	0.00179	0.01526	0.32366	0.00000	0.00264	0.64440	0.00054	0.00696	0.00475	0	0.00775	5.06E-05	0.650	0.221	0.536	1.28
0.80	3442	0.00018	0.00144	0.28314	0.00000	0.00454	0.67259	0.00040	0.00471	0.03300	0	0.00885	4.76E-05	0.459	0.129	0.610	1.15
0.67	3063	0.00002	0.00021	0.24439	0.00000	0.00358	0.69070	0.00012	0.00192	0.05907	0	0.01002	4.45E-05	0.396	0.091	0.694	1.03
0.57	2748	0.00000	0.00003	0.21370	0.00000	0.00242	0.70405	0.00003	0.00072	0.07903	0	0.01118	4.18E-05	0.366	0.077	0.718	1.00
0.50	2494	0.00000	0.00001	0.18989	0.00000	0.00158	0.71409	0.00001	0.00028	0.09415	0	0.01228	3.96E-05	0.350	0.069	0.723	1.00

Table D.6a H$_2$ (298 K) Stoichiometrically ($\phi = 1.00$) Combusted with Air (Variable Temp.) with an Exhaust Temp. of 1367 K, Metric Units

Oxid. temp. (K)	H$_2$	H$_2$O (g)	N$_2$	O$_2$	Enthalpy, H (kJ/kg)	Density (kg/m^3)	Viscosity (kg/m-sec)	Equil. spec. heat (kJ/kg-K)	Equil. therm. cond. (W/m-K)	Equil. Pr	Le	Avail. heat (%)
298	0.00003	0.34707	0.65288	0.00001	−1.93E + 03	0.219	5.11E−05	1.56	0.107	0.744	1.00	47.5
366	0.00003	0.34707	0.65288	0.00001	−1.93E + 03	0.219	5.11E−05	1.56	0.107	0.744	1.00	49.2
478	0.00003	0.34707	0.65288	0.00001	−1.93E + 03	0.219	5.11E−05	1.56	0.107	0.744	1.00	51.9
589	0.00003	0.34707	0.65288	0.00001	−1.93E + 03	0.219	5.11E−05	1.56	0.107	0.744	1.00	54.7
700	0.00003	0.34707	0.65288	0.00001	−1.93E + 03	0.219	5.11E−05	1.56	0.107	0.744	1.00	57.5
811	0.00003	0.34707	0.65288	0.00001	−1.93E + 03	0.219	5.11E−05	1.56	0.107	0.744	1.00	60.4
922	0.00003	0.34707	0.65288	0.00001	−1.93E + 03	0.219	5.11E−05	1.56	0.107	0.744	1.00	63.4
1033	0.00003	0.34707	0.65288	0.00001	−1.93E + 03	0.219	5.11E−05	1.56	0.107	0.744	1.00	66.4
1144	0.00003	0.34707	0.65288	0.00001	−1.93E + 03	0.219	5.11E−05	1.56	0.107	0.744	1.00	69.5
1255	0.00003	0.34707	0.65288	0.00001	−1.93E + 03	0.219	5.11E−05	1.56	0.107	0.744	1.00	72.7
1366	0.00003	0.34707	0.65288	0.00001	−1.93E + 03	0.219	5.11E−05	1.56	0.107	0.744	1.00	75.9

Table D.6b H$_2$ (77°F) Stoichiometrically (ϕ = 1.00) Combusted with Air (Variable Temp.) with an Exhaust Temp. of 2000°F, English Units

Oxid. temp. (°F)	H$_2$	H$_2$O (g)	N$_2$	O$_2$	Enthalpy, H (Btu/lb)	Density (lb/ft³)	Viscosity (lbm/ft-hr)	Equil. spec. heat (Btu/lb-°F)	Equil. therm. cond. (Btu/ft-hr-°F)	Equil. Pr	Le	Avail. heat (%)
77	0.00003	0.34707	0.65288	0.00001	−829	0.01366	3.43E-05	0.373	0.062	0.744	1.00	47.5
200	0.00003	0.34707	0.65288	0.00001	−829	0.01366	3.43E-05	0.373	0.062	0.744	1.00	49.2
400	0.00003	0.34707	0.65288	0.00001	−829	0.01366	3.43E-05	0.373	0.062	0.744	1.00	51.9
600	0.00003	0.34707	0.65288	0.00001	−829	0.01366	3.43E-05	0.373	0.062	0.744	1.00	54.7
800	0.00003	0.34707	0.65288	0.00001	−829	0.01366	3.43E-05	0.373	0.062	0.744	1.00	57.5
1000	0.00003	0.34707	0.65288	0.00001	−829	0.01366	3.43E-05	0.373	0.062	0.744	1.00	60.4
1200	0.00003	0.34707	0.65288	0.00001	−829	0.01366	3.43E-05	0.373	0.062	0.744	1.00	63.4
1400	0.00003	0.34707	0.65288	0.00001	−829	0.01366	3.43E-05	0.373	0.062	0.744	1.00	66.4
1600	0.00003	0.34707	0.65288	0.00001	−829	0.01366	3.43E-05	0.373	0.062	0.744	1.00	69.5
1800	0.00003	0.34707	0.65288	0.00001	−829	0.01366	3.43E-05	0.373	0.062	0.744	1.00	72.7
2000	0.00003	0.34707	0.65288	0.00001	−829	0.01366	3.43E-05	0.373	0.062	0.744	1.00	75.9

Table D.7a CH₄ (298 K) Adiabatically and Stoichiometrically ($\phi = 1.00$) Combusted with a Variable Composition Oxidizer (298 K), Metric Units

O₂	N₂	Flame temp. (K)	CO	CO₂	H	HO₂	H₂	H₂O (g)	NO	N₂	O	OH	O₂	Enthalpy, H (kJ/kg)	Density (kg/m³)	Viscosity (kg/m-sec)	Equil. spec. heat (kJ/kg-K)	Equil. therm cond. (W/m-K)	Equil Pr	Le
0.21	0.79	2225	0.00893	0.08539	0.00039	0.00000	0.00361	0.18338	0.00197	0.70866	0.00021	0.00291	0.00455	-2.58E+02	0.150	7.12E-05	2.19	0.248	0.630	1.12
0.30	0.70	2525	0.03158	0.09513	0.00312	0.00001	0.01249	0.23302	0.00594	0.58834	0.00204	0.01269	0.01565	-3.56E+02	0.129	7.82E-05	3.63	0.533	0.533	1.29
0.40	0.60	2703	0.05780	0.09934	0.00884	0.00001	0.02389	0.27268	0.00949	0.46666	0.00616	0.02653	0.02859	-4.57E+02	0.116	8.27E-05	5.27	0.909	0.479	1.39
0.50	0.50	2813	0.08072	0.10220	0.01587	0.00002	0.03474	0.30299	0.01155	0.36006	0.01138	0.04032	0.04013	-5.51E+02	0.108	8.57E-05	6.82	1.286	0.454	1.44
0.60	0.40	2889	0.10031	0.10472	0.02321	0.00003	0.04449	0.32738	0.01231	0.26721	0.01697	0.05312	0.05025	-6.38E+02	0.102	8.80E-05	8.28	1.646	0.443	1.46
0.70	0.30	2944	0.11712	0.10704	0.03039	0.00004	0.05311	0.34761	0.01197	0.18615	0.02255	0.06479	0.05922	-7.20E+02	0.097	8.99E-05	9.65	1.983	0.437	1.46
0.80	0.20	2988	0.13168	0.10920	0.03721	0.00005	0.06070	0.36473	0.01059	0.11514	0.02798	0.07540	0.06732	-7.97E+02	0.092	9.14E-05	10.95	2.297	0.436	1.45
0.90	0.10	3023	0.14438	0.11120	0.04363	0.00006	0.06736	0.37937	0.00789	0.05285	0.03325	0.08515	0.07487	-8.69E+02	0.089	9.27E-05	12.18	2.589	0.436	1.44
1.00	0.00	3054	0.15534	0.11310	0.04983	0.00007	0.07296	0.39154	0.00000	0.00000	0.03893	0.09486	0.08338	-9.36E+02	0.086	9.39E-05	13.35	2.868	0.437	1.44

Table D.7b CH$_4$ (77°F) Adiabatically and Stoichiometrically ($\phi = 1.00$) Combusted with a Variable Composition Oxidizer (77°F), English Units

O$_2$	N$_2$	Flame temp. (°F)	CO	CO$_2$	H	HO$_2$	H$_2$	H$_2$O (g)	NO	N$_2$	O	OH	O$_2$	Enthalpy H (Btu/lb)	Density (lb/ft^3)	Viscosity (lbm/ft-hr)	Equil. spec. heat (Btu/ lb-°F)	Equil. therm. cond. (Btu/ ft-hr-°F)	Equil Pr	Le
0.21	0.79	3545	0.00893	0.08539	0.00039	0.00000	0.00361	0.18338	0.00197	0.70866	0.00021	0.00291	0.00455	−111	0.00938	4.79E-05	0.524	0.143	0.536	1.28
0.30	0.70	4086	0.03158	0.09513	0.00312	0.00001	0.01249	0.23302	0.00594	0.58834	0.00204	0.01269	0.01565	−153	0.00805	5.26E-05	0.867	0.308	0.454	1.45
0.40	0.60	4406	0.05780	0.09934	0.00884	0.00001	0.02389	0.27268	0.00949	0.46666	0.00616	0.02653	0.02859	−196	0.00726	5.56E-05	1.258	0.525	0.431	1.49
0.50	0.50	4604	0.08072	0.10220	0.01587	0.00002	0.03474	0.30299	0.01155	0.36006	0.01138	0.04032	0.04013	−237	0.00673	5.76E-05	1.628	0.743	0.429	1.49
0.60	0.40	4740	0.10031	0.10472	0.02321	0.00003	0.04449	0.32738	0.01231	0.26721	0.01697	0.05312	0.05025	−274	0.00634	5.92E-05	1.977	0.951	0.434	1.47
0.70	0.30	4840	0.11712	0.10704	0.03039	0.00004	0.05311	0.34761	0.01197	0.18615	0.02255	0.06479	0.05922	−310	0.00603	6.04E-05	2.305	1.146	0.442	1.45
0.80	0.20	4918	0.13168	0.10920	0.03721	0.00005	0.06070	0.36473	0.01059	0.11514	0.02798	0.07540	0.06732	−343	0.00577	6.15E-05	2.615	1.327	0.452	1.43
0.90	0.10	4981	0.14438	0.11120	0.04363	0.00006	0.06736	0.37937	0.00789	0.05285	0.03325	0.08515	0.07487	−373	0.00555	6.24E-05	2.908	1.496	0.462	1.41
1.00	0.00	5037	0.15534	0.11310	0.04983	0.00007	0.07296	0.39154	0.00000	0.00000	0.03893	0.09486	0.08338	−403	0.00535	6.32E-05	3.189	1.657	0.472	1.40

Table D.8a   CH$_4$ (298 K) Adiabatically and Stoichiometrically ($\phi = 1.00$) Combusted with Air (Variable Temp.), Metric Units

Air temp. (K)	Flame temp. (K)	CO	CO$_2$	H	HO$_2$	H$_2$	H$_2$O (g)	NO	N$_2$	O	OH	O$_2$	Enthalpy, H (kJ/kg)	Density (kg/m^3)	Viscosity (kg/m-sec)	Equil. spec. heat (kJ/kg-K)	Equil. therm. cond. (W/m-K)	Equil. Pr	Le
298	2225	0.00893	0.08539	0.00039	0.00000	0.00361	0.18338	0.00197	0.70866	0.00021	0.00291	0.00455	−2.58E+02	0.150	7.12E-05	2.19	0.248	0.630	1.12
366	2254	0.01008	0.08415	0.00048	0.00000	0.00404	0.18249	0.00222	0.70782	0.00027	0.00335	0.00510	−1.93E+02	0.148	7.19E-05	2.27	0.262	0.622	1.14
478	2301	0.01212	0.08193	0.00067	0.00000	0.00482	0.18088	0.00266	0.70631	0.00039	0.00415	0.00607	−8.43E+01	0.145	7.28E-05	2.39	0.286	0.608	1.16
589	2345	0.01431	0.07955	0.00091	0.00000	0.00566	0.17908	0.00315	0.70464	0.00054	0.00505	0.00711	2.52E+01	0.142	7.38E-05	2.52	0.313	0.595	1.18
700	2389	0.01665	0.07700	0.00121	0.00000	0.00659	0.17709	0.00367	0.70279	0.00073	0.00605	0.00821	1.37E+02	0.139	7.47E-05	2.66	0.342	0.580	1.21
811	2431	0.01913	0.07429	0.00159	0.00000	0.00759	0.17488	0.00424	0.70078	0.00098	0.00715	0.00936	2.52E+02	0.136	7.56E-05	2.80	0.374	0.566	1.23
922	2471	0.02172	0.07146	0.00205	0.00000	0.00867	0.17248	0.00484	0.69860	0.00128	0.00836	0.01055	3.70E+02	0.134	7.64E-05	2.95	0.408	0.552	1.26
1033	2511	0.02437	0.06853	0.00260	0.00000	0.00981	0.16987	0.00547	0.69627	0.00165	0.00966	0.01177	4.90E+02	0.131	7.72E-05	3.10	0.445	0.538	1.29
1144	2550	0.02706	0.06556	0.00325	0.00000	0.01102	0.16708	0.00612	0.69380	0.00208	0.01103	0.01299	6.12E+02	0.129	7.80E-05	3.26	0.484	0.524	1.32
1255	2587	0.02975	0.06257	0.00400	0.00000	0.01228	0.16411	0.00679	0.69120	0.00259	0.01248	0.01420	7.36E+02	0.126	7.87E-05	3.41	0.526	0.511	1.34
1366	2623	0.03242	0.05958	0.00487	0.00001	0.01360	0.16098	0.00747	0.68849	0.00318	0.01399	0.01540	8.62E+02	0.124	7.95E-05	3.57	0.569	0.498	1.37

Table D.8b CH$_4$ (77°F) Adiabatically and Stoichiometrically ($\phi = 1.00$) Combusted with Air (Variable Temp.), English Units

Air temp. (°F)	Flame temp. (°F)	CO	CO$_2$	H	HO$_2$	H$_2$	H$_2$O (g)	NO	N$_2$	O	OH	O$_2$	Enthalpy, H (Btu/lb)	Density (lb/ft^3)	Viscosity (lbm/ft-hr)	Equil. spec. heat (Btu/lb-°F)	Equil. therm. cond. (Btu/ft-hr-°F)	Equil. Pr	Le
77	3545	0.00893	0.08539	0.00039	0.00000	0.00361	0.18338	0.00197	0.70866	0.00021	0.00291	0.00455	−111	0.00938	4.79E-05	0.524	0.143	0.630	1.12
200	3598	0.01008	0.08415	0.00048	0.00000	0.00404	0.18249	0.00222	0.70782	0.00027	0.00335	0.00510	−83	0.00925	4.83E-05	0.541	0.151	0.622	1.14
400	3682	0.01212	0.08193	0.00067	0.00000	0.00482	0.18088	0.00266	0.70631	0.00039	0.00415	0.00607	−36	0.00904	4.90E-05	0.571	0.165	0.608	1.16
600	3762	0.01431	0.07955	0.00091	0.00000	0.00566	0.17908	0.00315	0.70464	0.00054	0.00505	0.00711	11	0.00886	4.96E-05	0.603	0.181	0.595	1.18
800	3840	0.01665	0.07700	0.00121	0.00000	0.00659	0.17709	0.00367	0.70279	0.00073	0.00605	0.00821	59	0.00868	5.02E-05	0.636	0.198	0.580	1.21
1000	3915	0.01913	0.07429	0.00159	0.00000	-0.00759	0.17488	0.00424	0.70078	0.00098	0.00715	0.00936	108	0.00850	5.08E-05	0.670	0.216	0.566	1.23
1200	3989	0.02172	0.07146	0.00205	0.00000	0.00867	0.17248	0.00484	0.69860	0.00128	0.00836	0.01055	159	0.00834	5.14E-05	0.705	0.236	0.552	1.26
1400	4060	0.02437	0.06853	0.00260	0.00000	0.00981	0.16987	0.00547	0.69627	0.00165	0.00966	0.01177	211	0.00819	5.19E-05	0.741	0.257	0.538	1.29
1600	4129	0.02706	0.06556	0.00325	0.00000	0.01102	0.16708	0.00612	0.69380	0.00208	0.01103	0.01299	263	0.00804	5.25E-05	0.778	0.280	0.524	1.32
1800	4196	0.02975	0.06257	0.00400	0.00000	0.01228	0.16411	0.00679	0.69120	0.00259	0.01248	0.01420	317	0.00790	5.30E-05	0.815	0.304	0.511	1.34
2000	4261	0.03242	0.05958	0.00487	0.00001	0.01360	0.16098	0.00747	0.68849	0.00318	0.01399	0.01540	371	0.00776	5.34E-05	0.852	0.329	0.498	1.37

Table D.9a CH$_4$ (Variable Temp.) Adiabatically and Stoichiometrically (ϕ = 1.00) Combusted with Air (298 K), Metric Units

Fuel temp. (K)	Flame temp. (K)	CO	CO$_2$	H	H$_2$	H$_2$O (g)	NO	N$_2$	O	OH	O$_2$	Enthalpy, H (kJ/kg)	Density (kg/m^3)	Viscosity (kg/m-sec)	Equil. spec. heat (kJ/ kg-K)	Equil. therm. cond. (W/ m-K)	Equil. Pr	Le
298	2225	0.00893	0.08539	0.00039	0.00361	0.18338	0.00197	0.70866	0.00021	0.00291	0.00455	$-2.58E+02$	0.150	7.12E-05	2.19	0.248	0.630	1.12
311	2226	0.00896	0.08536	0.00039	0.00362	0.18336	0.00197	0.70864	0.00021	0.00292	0.00456	$-2.56E+02$	0.150	7.13E-05	2.19	0.248	0.630	1.12
366	2229	0.00908	0.08523	0.00040	0.00366	0.18327	0.00200	0.70856	0.00022	0.00297	0.00462	$-2.49E+02$	0.150	7.13E-05	2.20	0.250	0.629	1.13
422	2233	0.00922	0.08508	0.00041	0.00374	0.18316	0.00203	0.70846	0.00023	0.00302	0.00468	$-2.41E+02$	0.150	7.14E-05	2.21	0.251	0.628	1.13
478	2236	0.00936	0.08492	0.00042	0.00377	0.18305	0.00206	0.70835	0.00023	0.00308	0.00476	$-2.33E+02$	0.149	7.15E-05	2.22	0.253	0.627	1.13
533	2240	0.00952	0.08475	0.00043	0.00383	0.18293	0.00209	0.70823	0.00024	0.00313	0.00483	$-2.24E+02$	0.149	7.16E-05	2.23	0.255	0.626	1.13
589	2245	0.00969	0.08457	0.00045	0.00389	0.18280	0.00213	0.70811	0.00025	0.00320	0.00491	$-2.14E+02$	0.149	7.17E-05	2.24	0.257	0.625	1.13
644	2249	0.00987	0.08437	0.00046	0.00396	0.18266	0.00217	0.70798	0.00026	0.00327	0.00500	$-2.04E+02$	0.149	7.18E-05	2.25	0.259	0.624	1.13
700	2254	0.01007	0.08416	0.00048	0.00404	0.18250	0.00221	0.70783	0.00027	0.00334	0.00509	$-1.93E+02$	0.148	7.19E-05	2.26	0.262	0.622	1.14
755	2259	0.01027	0.08394	0.00050	0.00412	0.18234	0.00226	0.70768	0.00028	0.00342	0.00519	$-1.82E+02$	0.148	7.20E-05	2.28	0.264	0.621	1.14
811	2264	0.01050	0.08369	0.00052	0.00420	0.18217	0.00231	0.70752	0.00029	0.00351	0.00530	$-1.70E+02$	0.147	7.21E-05	2.29	0.267	0.619	1.14

Table D.9b CH$_4$ (Variable Temp.) Adiabatically and Stoichiometrically ($\phi = 1.00$) Combusted with Air (77°F), English Units

Fuel temp. (°F)	Flame temp. (°F)	CO	CO$_2$	H	H$_2$	H$_2$O (g)	NO	N$_2$	O	OH	O$_2$	Enthalpy, H (Btu/lb)	Density (lb/ft^3)	Viscosity (lbm/ft-hr)	Equil. spec. heat (Btu/lb-°F)	Equil. therm. cond. (Btu/ft-hr-°F)	Equil. Pr	Le
77	2225	0.00893	0.08539	0.00039	0.00361	0.18338	0.00197	0.70866	0.00021	0.00291	0.00455	−111	0.00938	4.79E-05	0.524	0.143	0.630	1.12
100	2226	0.00896	0.08536	0.00039	0.00362	0.18336	0.00197	0.70864	0.00021	0.00292	0.00456	−110	0.00938	4.79E-05	0.524	0.143	0.630	1.12
200	2229	0.00908	0.08523	0.00040	0.00366	0.18327	0.00200	0.70856	0.00022	0.00297	0.00462	−107	0.00936	4.80E-05	0.526	0.144	0.629	1.13
300	2233	0.00922	0.08508	0.00041	0.00371	0.18316	0.00203	0.70846	0.00023	0.00302	0.00468	−104	0.00935	4.80E-05	0.528	0.145	0.628	1.13
400	2236	0.00936	0.08492	0.00042	0.00377	0.18305	0.00206	0.70835	0.00023	0.00308	0.00476	−100	0.00933	4.81E-05	0.530	0.146	0.627	1.13
500	2240	0.00952	0.08475	0.00043	0.00383	0.18293	0.00209	0.70823	0.00024	0.00313	0.00483	−96	0.00931	4.81E-05	0.533	0.147	0.626	1.13
600	2245	0.00969	0.08457	0.00045	0.00389	0.18280	0.00213	0.70811	0.00025	0.00320	0.00491	−92	0.00929	4.82E-05	0.535	0.149	0.625	1.13
700	2249	0.00987	0.08437	0.00046	0.00396	0.18266	0.00217	0.70798	0.00026	0.00327	0.00500	−88	0.00927	4.83E-05	0.538	0.150	0.624	1.13
800	2254	0.01007	0.08416	0.00048	0.00404	0.18250	0.00221	0.70783	0.00027	0.00334	0.00509	−83	0.00925	4.83E-05	0.541	0.151	0.622	1.14
900	2259	0.01027	0.08394	0.00050	0.00412	0.18234	0.00226	0.70768	0.00028	0.00342	0.00519	−78	0.00923	4.84E-05	0.544	0.153	0.621	1.14
1000	2264	0.01050	0.08369	0.00052	0.00420	0.18217	0.00231	0.70752	0.00029	0.00351	0.00530	−73	0.00920	4.85E-05	0.547	0.154	0.619	1.14

Table D.10a CH_4 (298 K) Stoichiometrically ($\phi = 1.00$) Combusted with Air (298 K), Metric Units

Gas temp. (K)	CO	CO₂	H	H₂	H₂O (g)	NO	N₂	O	OH	O₂	Enthalpy, H (kJ/kg)	Density (kg/m³)	Viscosity (kg/m-sec)	Equil. spec. heat (kJ/kg-K)	Equil. therm. cond. (W/m-K)	Equil. Pr	Le	Avail. heat (%)
366	0.00000	0.09502	0.00000	0.00000	0.19004	0.00000	0.71493	0.00000	0.00000	0.00000	−2.94E+03	0.920	1.89E-05	1.13	0.027	0.782	1.00	87.6
478	0.00000	0.09502	0.00000	0.00000	0.19004	0.00000	0.71493	0.00000	0.00000	0.00000	−2.81E+03	0.705	2.35E-05	1.15	0.035	0.772	1.00	83.5
589	0.00000	0.09502	0.00000	0.00000	0.19005	0.00000	0.71493	0.00000	0.00000	0.00000	−2.68E+03	0.572	2.77E-05	1.19	0.043	0.766	1.00	79.2
700	0.00000	0.09502	0.00000	0.00000	0.19005	0.00000	0.71493	0.00000	0.00000	0.00000	−2.55E+03	0.481	3.16E-05	1.22	0.051	0.761	1.00	74.8
811	0.00000	0.09502	0.00000	0.00000	0.19005	0.00000	0.71493	0.00000	0.00000	0.00000	−2.41E+03	0.415	3.53E-05	1.26	0.059	0.756	1.00	70.3
922	0.00000	0.09502	0.00000	0.00000	0.19005	0.00000	0.71493	0.00000	0.00000	0.00000	−2.27E+03	0.365	3.87E-05	1.29	0.067	0.751	1.00	65.7
1033	0.00000	0.09502	0.00000	0.00000	0.19005	0.00000	0.71493	0.00000	0.00000	0.00000	−2.13E+03	0.326	4.21E-05	1.33	0.075	0.747	1.00	61.0
1144	0.00000	0.09502	0.00000	0.00000	0.19005	0.00000	0.71493	0.00000	0.00000	0.00000	−1.98E+03	0.294	4.52E-05	1.35	0.082	0.744	1.00	56.1
1255	0.00000	0.09502	0.00000	0.00000	0.19005	0.00000	0.71493	0.00000	0.00000	0.00000	−1.82E+03	0.268	4.83E-05	1.38	0.090	0.741	1.00	51.1
1366	0.00001	0.09501	0.00000	0.00000	0.19003	0.00000	0.71492	0.00000	0.00000	0.00001	−1.67E+03	0.247	5.12E-05	1.40	0.097	0.738	1.00	46.1
1478	0.00005	0.09496	0.00000	0.00004	0.18999	0.00002	0.71489	0.00000	0.00001	0.00004	−1.51E+03	0.228	5.40E-05	1.43	0.106	0.732	1.00	40.9
1589	0.00016	0.09485	0.00000	0.00011	0.18989	0.00004	0.71481	0.00000	0.00003	0.00010	−1.35E+03	0.212	5.68E-05	1.46	0.114	0.728	1.00	35.7
1700	0.00041	0.09458	0.00000	0.00024	0.18969	0.00010	0.71463	0.00000	0.00010	0.00025	−1.19E+03	0.198	5.94E-05	1.51	0.125	0.719	1.01	30.3
1811	0.00092	0.09403	0.00001	0.00048	0.18929	0.00022	0.71427	0.00000	0.00023	0.00054	−1.01E+03	0.186	6.20E-05	1.58	0.138	0.710	1.02	24.7
1922	0.00190	0.09298	0.00003	0.00090	0.18858	0.00044	0.71360	0.00001	0.00051	0.00105	−8.34E+02	0.175	6.46E-05	1.67	0.155	0.698	1.03	18.8
2033	0.00357	0.09118	0.00008	0.00157	0.18736	0.00081	0.71245	0.00004	0.00103	0.00191	−6.40E+02	0.165	6.71E-05	1.82	0.181	0.675	1.06	12.5
2144	0.00622	0.08832	0.00021	0.00259	0.18542	0.00138	0.71060	0.00011	0.00193	0.00323	−4.28E+02	0.156	6.95E-05	2.01	0.215	0.651	1.09	5.6
2255	0.00893	0.08539	0.00039	0.00361	0.18338	0.00197	0.70866	0.00021	0.00291	0.00455	−2.58E+02	0.150	7.12E-05	2.19	0.248	0.630	1.12	0.0

Table D.10b CH$_4$ (Variable Temp.) Adiabatically and Stoichiometrically ($\phi = 1.00$) Combusted with Air (77°F), English Units

Flame temp. (°F)	CO	CO$_2$	H	H$_2$	H$_2$O (g)	NO	N$_2$	O	OH	O$_2$	Enthalpy, H (Btu/lb)	Density (lb/ft^3)	Viscosity (lbm/ft-hr)	Equil. spec. heat (Btu/ lb-°F)	Equil. therm. cond. (Btu/ ft-hr-°F)	Equil. Pr	Le	Avail. heat (%)
200	0.00000	0.09502	0.00000	0.00000	0.19004	0.00000	0.71493	0.00000	0.00000	0.00000	−1265	0.05744	1.27E-05	0.269	0.016	0.782	1.00	87.6
400	0.00000	0.09502	0.00000	0.00000	0.19004	0.00000	0.71493	0.00000	0.00000	0.00000	−1210	0.04398	1.58E-05	0.276	0.020	0.772	1.00	83.5
600	0.00000	0.09502	0.00000	0.00000	0.19005	0.00000	0.71493	0.00000	0.00000	0.00000	−1154	0.03569	1.86E-05	0.284	0.025	0.766	1.00	79.2
800	0.00000	0.09502	0.00000	0.00000	0.19005	0.00000	0.71493	0.00000	0.00000	0.00000	−1096	0.03003	2.12E-05	0.292	0.029	0.761	1.00	74.8
1000	0.00000	0.09502	0.00000	0.00000	0.19005	0.00000	0.71493	0.00000	0.00000	0.00000	−1037	0.02592	2.37E-05	0.301	0.034	0.756	1.00	70.3
1200	0.00000	0.09502	0.00000	0.00000	0.19005	0.00000	0.71493	0.00000	0.00000	0.00000	−976	0.02280	2.61E-05	0.309	0.039	0.751	1.00	65.7
1400	0.00000	0.09502	0.00000	0.00000	0.19005	0.00000	0.71493	0.00000	0.00000	0.00000	−914	0.02035	2.83E-05	0.317	0.043	0.747	1.00	61.0
1600	0.00000	0.09502	0.00000	0.00000	0.19005	0.00000	0.71493	0.00000	0.00000	0.00000	−850	0.01838	3.04E-05	0.323	0.048	0.744	1.00	56.1
1800	0.00000	0.09502	0.00000	0.00000	0.19005	0.00000	0.71493	0.00000	0.00000	0.00000	−784	0.01675	3.25E-05	0.329	0.052	0.741	1.00	51.1
2000	0.00001	0.09501	0.00000	0.00001	0.19003	0.00000	0.71492	0.00000	0.00000	0.00001	−718	0.01539	3.44E-05	0.335	0.056	0.738	1.00	46.1
2200	0.00005	0.09496	0.00000	0.00004	0.18999	0.00002	0.71489	0.00000	0.00001	0.00004	−650	0.01422	3.63E-05	0.342	0.061	0.732	1.00	40.9
2400	0.00016	0.09485	0.00000	0.00011	0.18989	0.00004	0.71481	0.00000	0.00003	0.00010	−580	0.01323	3.82E-05	0.349	0.066	0.728	1.00	35.7
2600	0.00041	0.09458	0.00000	0.00024	0.18969	0.00010	0.71463	0.00000	0.00010	0.00025	−510	0.01236	4.00E-05	0.361	0.072	0.719	1.01	30.3
2800	0.00092	0.09403	0.00001	0.00048	0.18929	0.00022	0.71427	0.00000	0.00023	0.00054	−436	0.01160	4.17E-05	0.376	0.080	0.710	1.02	24.7
3000	0.00190	0.09298	0.00003	0.00090	0.18858	0.00044	0.71360	0.00001	0.00051	0.00105	−358	0.01092	4.34E-05	0.399	0.089	0.698	1.03	18.8
3200	0.00357	0.09118	0.00008	0.00157	0.18736	0.00081	0.71245	0.00004	0.00103	0.00191	−275	0.01031	4.51E-05	0.434	0.104	0.675	1.06	12.5
3400	0.00622	0.08832	0.00021	0.00259	0.18542	0.00138	0.71060	0.00011	0.00193	0.00323	−184	0.00976	4.67E-05	0.481	0.124	0.651	1.09	5.6
3600	0.00893	0.08539	0.00039	0.00361	0.18338	0.00197	0.70866	0.00021	0.00291	0.00455	−111	0.00938	4.79E-05	0.524	0.143	0.630	1.12	0.0

Table D.11a CH_4 (298 K) Adiabatically Combusted with Air (298 K) at Various Equivalence Ratios, Metric Units

Equiv. ratio	Flame temp. (K)	C(gr)	CH_4	CO	CO_2	H	H_2	H_2O (g)	NH_3	NO	N_2	O	OH	O_2	Enthalpy, H (kJ/kg)	Density (kg/m³)	Viscosity (kg/m-sec)	Equil. spec. heat (kJ/kg-K)	Equil. therm. cond. (W/m-K)	Equil. Pr	Le
4.00	944	0.0483	0.02281	0.11559	0.02752	0.00000	0.33907	0.04363	0.00012	0.00000	0.40293	0.00000	0.00000	0.00000	−8.84E+02	0.246	3.85E-05	6.71	0.401	0.644	0.80
2.00	1563	0.0000	0.00000	0.11939	0.02850	0.00002	0.17637	0.11939	0.00000	0.00000	0.55633	0.00000	0.00000	0.00000	−4.88E+02	0.177	5.55E-05	1.66	0.152	0.603	0.99
1.33	2030	0.0000	0.00000	0.06524	0.05041	0.00044	0.05021	0.18077	0.00000	0.00002	0.65273	0.00000	0.00017	0.00000	−3.37E+02	0.154	6.68E-05	1.61	0.168	0.640	1.06
1.00	2225	0.0000	0.00000	0.00893	0.08539	0.00039	0.00361	0.18338	0.00000	0.00197	0.70866	0.00021	0.00291	0.00455	−2.58E+02	0.150	7.12E-05	2.19	0.248	0.630	1.12
0.80	1995	0.0000	0.00000	0.00050	0.07692	0.00002	0.00022	0.15380	0.00000	0.00323	0.72654	0.00012	0.00164	0.03699	−2.08E+02	0.170	6.65E-05	1.59	0.151	0.698	1.03
0.67	1781	0.0000	0.00000	0.00004	0.06540	0.00000	0.00002	0.13059	0.00000	0.00223	0.73705	0.00003	0.00053	0.06411	−1.75E+02	0.192	6.18E-05	1.44	0.123	0.723	1.00
0.57	1612	0.0000	0.00000	0.00000	0.05656	0.00000	0.00000	0.11304	0.00000	0.00135	0.74460	0.00000	0.00017	0.08427	−1.51E+02	0.213	5.79E-05	1.38	0.110	0.726	1.00
0.50	1478	0.0000	0.00000	0.00000	0.04988	0.00000	0.00000	0.09973	0.00000	0.00080	0.75018	0.00000	0.00005	0.09934	−1.33E+02	0.233	5.48E-05	1.33	0.100	0.729	1.00

Table D.11b CH_4 (77°F) Combusted with Air (77°F) at Various Equivalence Ratios, English Units

Equiv. ratio	Flame temp. (°F)	C(gr)	CH_4	CO	CO_2	H	H_2	H_2O (g)	NH_3	NO	N_2	O	OH	O_2	Enthalpy, H (Btu/lb)	Density (lb/ft³)	Viscosity (lbm/ft-hr)	Equil. spec. heat (Btu/lb-°F)	Equil. therm. cond. (Btu/ft-hr-°F)	Equil. Pr	Le
4.00	1240	0.0483	0.02281	0.11559	0.02752	0.00000	0.33907	0.04363	0.00012	0.00000	0.40293	0.00000	0.00000	0.00000	−380	0.01537	2.59E-05	1.602	0.232	0.644	0.80
2.00	2353	0.0000	0.00000	0.11939	0.02850	0.00002	0.17637	0.11939	0.00000	0.00000	0.55633	0.00000	0.00000	0.00000	−210	0.01105	3.73E-05	0.395	0.088	0.603	0.99
1.33	3195	0.0000	0.00000	0.06524	0.05041	0.00044	0.05021	0.18077	0.00000	0.00002	0.65273	0.00000	0.00017	0.00000	−145	0.00963	4.49E-05	0.385	0.097	0.640	1.06
1.00	3545	0.0000	0.00000	0.00893	0.08539	0.00039	0.00361	0.18338	0.00000	0.00197	0.70866	0.00021	0.00291	0.00455	−111	0.00938	4.79E-05	0.524	0.143	0.630	1.12
0.80	3132	0.0000	0.00000	0.00050	0.07692	0.00002	0.00022	0.15380	0.00000	0.00323	0.72654	0.00012	0.00164	0.03699	−90	0.01061	4.47E-05	0.379	0.087	0.698	1.03
0.67	2747	0.0000	0.00000	0.00004	0.06540	0.00000	0.00002	0.13059	0.00000	0.00223	0.73705	0.00003	0.00053	0.06411	−75	0.01196	4.16E-05	0.344	0.071	0.723	1.00
0.57	2441	0.0000	0.00000	0.00000	0.05656	0.00000	0.00000	0.11304	0.00000	0.00135	0.74460	0.00000	0.00017	0.08427	−65	0.01328	3.90E-05	0.330	0.064	0.726	1.00
0.50	2201	0.0000	0.00000	0.00000	0.04988	0.00000	0.00000	0.09973	0.00000	0.00080	0.75018	0.00000	0.00005	0.09934	−57	0.01452	3.68E-05	0.319	0.058	0.729	1.00

Table D.12a CH_4 (298 K) Stoichiometrically ($\phi = 1.00$) Combusted with Air (Variable Temp.) with an Exhaust Temp. of 1367 K, Metric Units

Oxid. temp. (K)	CO	CO_2	H_2	H_2O (g)	N_2	O_2	Enthalpy, H (kJ/kg)	Density (kg/m³)	Viscosity (kg/m-sec)	Equil. spec. heat (kJ/kg-K)	Equil. therm. cond. (W/m-K)	Equil. Pr	Le	Avail. heat (%)
298	0.00001	0.09501	0.00001	0.19003	0.71492	0.00001	$-1.67E+03$	0.246	5.12E-05	1.40	0.097	0.737	1.00	46.1
366	0.00001	0.09501	0.00001	0.19003	0.71492	0.00001	$-1.67E+03$	0.246	5.12E-05	1.40	0.097	0.737	1.00	48.2
478	0.00001	0.09501	0.00001	0.19003	0.71492	0.00001	$-1.67E+03$	0.246	5.12E-05	1.40	0.097	0.737	1.00	51.7
589	0.00001	0.09501	0.00001	0.19003	0.71492	0.00001	$-1.67E+03$	0.246	5.12E-05	1.40	0.097	0.737	1.00	55.3
700	0.00001	0.09501	0.00001	0.19003	0.71492	0.00001	$-1.67E+03$	0.246	5.12E-05	1.40	0.097	0.737	1.00	59.0
811	0.00001	0.09501	0.00001	0.19003	0.71492	0.00001	$-1.67E+03$	0.246	5.12E-05	1.40	0.097	0.737	1.00	62.7
922	0.00001	0.09501	0.00001	0.19003	0.71492	0.00001	$-1.67E+03$	0.246	5.12E-05	1.40	0.097	0.737	1.00	66.6
1033	0.00001	0.09501	0.00001	0.19003	0.71492	0.00001	$-1.67E+03$	0.246	5.12E-05	1.40	0.097	0.737	1.00	70.5
1144	0.00001	0.09501	0.00001	0.19003	0.71492	0.00001	$-1.67E+03$	0.246	5.12E-05	1.40	0.097	0.737	1.00	74.5
1255	0.00001	0.09501	0.00001	0.19003	0.71492	0.00001	$-1.67E+03$	0.246	5.12E-05	1.40	0.097	0.737	1.00	78.5
1366	0.00001	0.09501	0.00001	0.19003	0.71492	0.00001	$-1.67E+03$	0.246	5.12E-05	1.40	0.097	0.737	1.00	82.6

Table D.12b CH$_4$ (77°F) Stoichiometrically ($\phi = 1.00$) Combusted with Air (Variable Temp.) with an Exhaust Temp. of 2000°F, English Units

Oxid. temp. (°F)	CO	CO$_2$	H$_2$	H$_2$O (g)	N$_2$	O$_2$	Enthalpy, H (Btu/lb)	Density (lb/ft^3)	Viscosity (lbm/ft-hr)	Equil. spec. heat (Btu/lb-°F)	Equil. therm. cond. (Btu/ ft-hr-°F)	Equil. Pr	Le	Avail. heat (%)
77	0.00001	0.09501	0.00001	0.19003	0.71492	0.00001	−718	0.01538	3.44E-05	0.335	0.056	0.737	1.00	46.1
200	0.00001	0.09501	0.00001	0.19003	0.71492	0.00001	−718	0.01538	3.44E-05	0.335	0.056	0.737	1.00	48.2
400	0.00001	0.09501	0.00001	0.19003	0.71492	0.00001	−718	0.01538	3.44E-05	0.335	0.056	0.737	1.00	51.7
600	0.00001	0.09501	0.00001	0.19003	0.71492	0.00001	−718	0.01538	3.44E-05	0.335	0.056	0.737	1.00	55.3
800	0.00001	0.09501	0.00001	0.19003	0.71492	0.00001	−718	0.01538	3.44E-05	0.335	0.056	0.737	1.00	59.0
1000	0.00001	0.09501	0.00001	0.19003	0.71492	0.00001	−718	0.01538	3.44E-05	0.335	0.056	0.737	1.00	62.7
1200	0.00001	0.09501	0.00001	0.19003	0.71492	0.00001	−718	0.01538	3.44E-05	0.335	0.056	0.737	1.00	66.6
1400	0.00001	0.09501	0.00001	0.19003	0.71492	0.00001	−718	0.01538	3.44E-05	0.335	0.056	0.737	1.00	70.5
1600	0.00001	0.09501	0.00001	0.19003	0.71492	0.00001	−718	0.01538	3.44E-05	0.335	0.056	0.737	1.00	74.5
1800	0.00001	0.09501	0.00001	0.19003	0.71492	0.00001	−718	0.01538	3.44E-05	0.335	0.056	0.737	1.00	78.5
2000	0.00001	0.09501	0.00001	0.19003	0.71492	0.00001	−718	0.01538	3.44E-05	0.335	0.056	0.737	1.00	82.6

Table D.13a C_3H_8 (298 K) Adiabatically and Stoichiometrically ($\phi = 1.00$) Combusted with a Variable Composition Oxidizer (298 K), Metric Units

O_2	N_2	Flame temp. (K)	CO	CO_2	H	HO_2	H_2	H_2O (g)	NO	N_2	O	OH	O_2	Enthalpy, H (kJ/kg)	Density (kg/m³)	Viscosity (kg/m-sec)	Equil. spec. heat (kJ/kg-K)	Equil. therm cond. (W/m-K)	Equil. Pr	Le
0.21	0.79	2266	0.01248	0.10269	0.00046	0.00000	0.00331	0.14838	0.00245	0.72084	0.00031	0.00327	0.00581	−1.42E+02	0.151	7.20E-05	2.27	0.259	0.631	1.13
0.30	0.70	2562	0.04149	0.11386	0.00340	0.00001	0.01089	0.18788	0.00700	0.60063	0.00266	0.01332	0.01887	−1.96E+02	0.131	7.87E-05	3.76	0.550	0.538	1.29
0.40	0.60	2739	0.07453	0.11889	0.00932	0.00002	0.02044	0.21919	0.01097	0.47806	0.00773	0.02718	0.03366	−2.51E+02	0.119	8.31E-05	5.37	0.920	0.484	1.39
0.50	0.50	2849	0.10359	0.12235	0.01654	0.00003	0.02954	0.24298	0.01326	0.36994	0.01409	0.04090	0.04677	−3.02E+02	0.111	8.61E-05	6.86	1.288	0.458	1.44
0.60	0.40	2925	0.12862	0.12538	0.02410	0.00004	0.03774	0.26203	0.01410	0.27517	0.02089	0.05366	0.05827	−3.49E+02	0.105	8.83E-05	8.25	1.637	0.445	1.45
0.70	0.30	2982	0.15027	0.12816	0.03152	0.00005	0.04501	0.27779	0.01370	0.19203	0.02770	0.06531	0.06846	−3.93E+02	0.100	9.01E-05	9.56	1.962	0.439	1.46
0.80	0.20	3027	0.16913	0.13074	0.03860	0.00006	0.05144	0.29108	0.01211	0.11888	0.03435	0.07594	0.07768	−4.35E+02	0.096	9.17E-05	10.78	2.265	0.436	1.45
0.90	0.10	3063	0.18565	0.13312	0.04529	0.00006	0.05709	0.30240	0.00901	0.05452	0.04083	0.08571	0.08629	−4.73E+02	0.093	9.30E-05	11.94	2.548	0.436	1.44
1.00	0.00	3095	0.19999	0.13537	0.05178	0.00007	0.06184	0.31164	0.00000	0.00000	0.04783	0.09546	0.09602	−5.09E+02	0.090	9.42E-05	13.05	2.818	0.436	1.43

Table D.13b C_3H_8 (77°F) Adiabatically and Stoichiometrically ($\phi = 1.00$) Combusted with a Variable Composition Oxidizer (77°F), English Units

O_2	N_2	Flame temp. (°F)	CO	CO_2	H	HO_2	H_2	H_2O (g)	NO	N_2	O	OH	O_2	Enthalpy, H (Btu/lb)	Density (lb/ft³)	Viscosity (lbm/ ft-hr)	Equil. spec. heat (Btu/ lb-°F)	Equil. therm cond. (Btu/ft-hr -°F)	Equil Pr	Le
0.21	0.79	3620	0.01248	0.10269	0.00046	0.00000	0.00331	0.14838	0.00245	0.72084	0.00031	0.00327	0.00581	−61	0.00942	4.84E-05	0.543	0.150	0.631	1.13
0.30	0.70	4151	0.04149	0.11386	0.00340	0.00001	0.01089	0.18788	0.00700	0.60063	0.00266	0.01332	0.01887	−84	0.00817	5.29E-05	0.897	0.318	0.538	1.29
0.40	0.60	4470	0.07453	0.11889	0.00932	0.00002	0.02044	0.21919	0.01097	0.47806	0.00773	0.02718	0.03366	−108	0.00742	5.59E-05	1.281	0.532	0.484	1.39
0.50	0.50	4668	0.10359	0.12235	0.01654	0.00003	0.02954	0.24298	0.01326	0.36994	0.01409	0.04090	0.04677	−130	0.00692	5.79E-05	1.639	0.744	0.458	1.44
0.60	0.40	4806	0.12862	0.12538	0.02410	0.00004	0.03774	0.26203	0.01410	0.27517	0.02089	0.05366	0.05827	−150	0.00655	5.94E-05	1.971	0.946	0.445	1.45
0.70	0.30	4908	0.15027	0.12816	0.03152	0.00005	0.04501	0.27779	0.01370	0.19203	0.02770	0.06531	0.06846	−169	0.00625	6.06E-05	2.283	1.134	0.439	1.46
0.80	0.20	4988	0.16913	0.13074	0.03860	0.00006	0.05144	0.29108	0.01211	0.11888	0.03435	0.07594	0.07768	−187	0.00601	6.17E-05	2.575	1.309	0.436	1.45
0.90	0.10	5053	0.18565	0.13312	0.04529	0.00006	0.05709	0.30240	0.00901	0.05452	0.04083	0.08571	0.08629	−203	0.00580	6.26E-05	2.852	1.472	0.436	1.44
1.00	0.00	5111	0.19999	0.13537	0.05178	0.00007	0.06184	0.31164	0.00000	0.00000	0.04783	0.09546	0.09602	−219	0.00561	6.34E-05	3.116	1.628	0.436	1.43

Table D.14a C₃H₈ (298 K) Adiabatically and Stoichiometrically (φ = 1.00) Combusted with Air (Variable Temp.), Metric Units

Air temp. (K)	Flame temp. (K)	CO	CO₂	H	HO₂	H₂	H₂O (g)	NO	N₂	O	OH	O₂	Enthalpy, H (kJ/kg)	Density (kg/m³)	Viscosity (kg/m-sec)	Equil. spec. heat (kJ/kg-K)	Equil. therm. cond. (W/m-K)	Equil. Pr	Le
298	2266	0.01248	0.10269	0.00046	0.00000	0.00331	0.14838	0.00245	0.72084	0.00031	0.00327	0.00581	−1.42E+02	0.151	7.20E-05	2.27	0.259	0.631	1.13
366	2294	0.01394	0.10110	0.00057	0.00000	0.00368	0.14756	0.00274	0.71987	0.00038	0.00372	0.00646	−7.75E+01	0.149	7.25E-05	2.35	0.273	0.624	1.14
478	2339	0.01649	0.09831	0.00077	0.00000	0.00432	0.14609	0.00324	0.71814	0.00054	0.00453	0.00758	3.03E+01	0.146	7.35E-05	2.48	0.298	0.611	1.16
589	2382	0.01920	0.09534	0.00103	0.00000	0.00502	0.14447	0.00378	0.71625	0.00073	0.00542	0.00875	1.39E+02	0.143	7.44E-05	2.61	0.325	0.598	1.18
700	2423	0.02207	0.09219	0.00134	0.00000	0.00578	0.14269	0.00436	0.71420	0.00097	0.00641	0.00998	2.51E+02	0.140	7.52E-05	2.75	0.354	0.585	1.20
811	2464	0.02508	0.08888	0.00173	0.00000	0.00659	0.14074	0.00498	0.71198	0.00128	0.00749	0.01126	3.65E+02	0.137	7.61E-05	2.89	0.385	0.572	1.23
922	2503	0.02820	0.08543	0.00219	0.00000	0.00746	0.13862	0.00563	0.70960	0.00164	0.00865	0.01257	4.82E+02	0.135	7.69E-05	3.04	0.418	0.558	1.25
1033	2542	0.03139	0.08189	0.00274	0.00000	0.00838	0.13634	0.00631	0.70708	0.00208	0.00990	0.01390	6.02E+02	0.132	7.76E-05	3.18	0.453	0.545	1.28
1144	2579	0.03462	0.07829	0.00338	0.00000	0.00935	0.13390	0.00701	0.70442	0.00259	0.01121	0.01522	7.23E+02	0.130	7.84E-05	3.33	0.491	0.532	1.31
1255	2615	0.03784	0.07468	0.00413	0.00001	0.01036	0.13132	0.00772	0.70165	0.00318	0.01259	0.01652	8.46E+02	0.128	7.91E-05	3.48	0.530	0.519	1.33
1366	2650	0.04104	0.07109	0.00498	0.00001	0.01140	0.12860	0.00845	0.69877	0.00386	0.01401	0.01780	9.71E+02	0.126	7.98E-05	3.62	0.571	0.507	1.36

Table D.14b C_3H_8 (77°F) Adiabatically and Stoichiometrically ($\phi = 1.00$) Combusted with Air (Variable Temp.), English Units

Air temp. (°F)	Flame temp. (°F)	CO	CO_2	H	HO_2	H_2	H_2O (g)	NO	N_2	O	OH	O_2	Enthalpy, H (Btu/lb)	Density (lb/ft³)	Viscosity (lbm/ft-hr)	Equil. spec. heat (Btu/ lb-F)	Equil. therm. cond. (Btu/ ft-hr-F)	Equil. Pr	Le
77	3620	0.01248	0.10269	0.00046	0.00000	0.00331	0.14838	0.00245	0.72084	0.00031	0.00327	0.00581	−61	0.00942	4.84E-05	0.543	0.150	0.631	1.13
200	3670	0.01394	0.10110	0.00057	0.00000	0.00368	0.14756	0.00274	0.71987	0.00038	0.00372	0.00646	−33	0.00930	4.88E-05	0.561	0.158	0.624	1.14
400	3750	0.01649	0.09831	0.00077	0.00000	0.00432	0.14609	0.00324	0.71814	0.00054	0.00453	0.00758	13	0.00910	4.94E-05	0.592	0.172	0.611	1.16
600	3827	0.01920	0.09534	0.00103	0.00000	0.00502	0.14447	0.00378	0.71625	0.00073	0.00542	0.00875	60	0.00892	5.00E-05	0.624	0.188	0.598	1.18
800	3902	0.02207	0.09219	0.00134	0.00000	0.00578	0.14269	0.00436	0.71420	0.00097	0.00641	0.00998	108	0.00874	5.06E-05	0.657	0.204	0.585	1.20
1000	3975	0.02508	0.08888	0.00173	0.00000	0.00659	0.14074	0.00498	0.71198	0.00128	0.00749	0.01126	157	0.00857	5.12E-05	0.691	0.222	0.572	1.23
1200	4046	0.02820	0.08543	0.00219	0.00000	0.00746	0.13862	0.00563	0.70960	0.00164	0.00865	0.01257	207	0.00841	5.17E-05	0.725	0.242	0.558	1.25
1400	4115	0.03139	0.08189	0.00274	0.00000	0.00838	0.13634	0.00631	0.70708	0.00208	0.00990	0.01390	259	0.00826	5.22E-05	0.760	0.262	0.545	1.28
1600	4182	0.03462	0.07829	0.00338	0.00000	0.00935	0.13390	0.00701	0.70442	0.00259	0.01121	0.01522	311	0.00812	5.27E-05	0.795	0.284	0.532	1.31
1800	4248	0.03784	0.07468	0.00413	0.00001	0.01036	0.13132	0.00772	0.70165	0.00318	0.01259	0.01652	364	0.00798	5.32E-05	0.830	0.306	0.519	1.33
2000	4311	0.04104	0.07109	0.00498	0.00001	0.01140	0.12860	0.00845	0.69877	0.00386	0.01401	0.01780	418	0.00784	5.37E-05	0.865	0.330	0.507	1.36

Table D.15a C$_3$H$_8$ (Variable Temp.) Adiabatically and Stoichiometrically ($\phi = 1.00$) Combusted with Air (298 K), Metric Units

Fuel temp. (K)	Flame temp. (K)	CO	CO$_2$	H	H$_2$	H$_2$O (g)	NO	N$_2$	O	OH	O$_2$	Enthalpy, H (kJ/kg)	Density (kg/m^3)	Viscosity (kg/m-sec)	Equil. spec. heat (kJ/kg-K)	Equil. therm. cond. (W/m-K)	Equil. Pr	Le
298	2266	0.01248	0.10269	0.00046	0.00331	0.14838	0.00245	0.72084	0.00031	0.00327	0.00581	−1.42E + 02	0.151	7.20E-05	2.27	0.259	0.631	1.13
311	2267	0.01251	0.10266	0.00047	0.00332	0.14836	0.00246	0.72082	0.00031	0.00328	0.00582	−1.41E + 02	0.151	7.20E-05	2.27	0.259	0.631	1.13
366	2270	0.01264	0.10251	0.00047	0.00335	0.14829	0.00248	0.72073	0.00032	0.00332	0.00588	−1.35E + 02	0.151	7.20E-05	2.28	0.261	0.631	1.13
422	2273	0.01280	0.10233	0.00049	0.00339	0.14820	0.00251	0.72063	0.00032	0.00337	0.00596	−1.28E + 02	0.150	7.21E-05	2.29	0.262	0.630	1.13
478	2276	0.01298	0.10214	0.00050	0.00344	0.14810	0.00255	0.72051	0.00033	0.00342	0.00603	−1.20E + 02	0.150	7.22E-05	2.30	0.264	0.629	1.13
533	2280	0.01317	0.10193	0.00051	0.00348	0.14799	0.00259	0.72038	0.00034	0.00348	0.00612	−1.11E + 02	0.150	7.22E-05	2.31	0.266	0.628	1.13
589	2284	0.01338	0.10170	0.00053	0.00354	0.14787	0.00263	0.72024	0.00035	0.00355	0.00621	−1.02E + 02	0.150	7.23E-05	2.32	0.268	0.627	1.13
644	2288	0.01361	0.10146	0.00054	0.00359	0.14774	0.00267	0.72009	0.00037	0.00362	0.00631	−9.18E + 01	0.149	7.24E-05	2.33	0.270	0.626	1.14
700	2293	0.01385	0.10119	0.00056	0.00365	0.14761	0.00272	0.71993	0.00038	0.00369	0.00642	−8.12E + 01	0.149	7.25E-05	2.35	0.272	0.624	1.14
755	2297	0.01410	0.10091	0.00058	0.00372	0.14746	0.00277	0.71976	0.00039	0.00377	0.00653	−7.02E + 01	0.149	7.26E-05	2.36	0.275	0.623	1.14
811	2302	0.01438	0.10062	0.00060	0.00379	0.14731	0.00282	0.71957	0.00041	0.00385	0.00665	−5.84E + 01	0.148	7.27E-05	2.37	0.278	0.622	1.14

Table D.15b C$_3$H$_8$ (Variable Temp.) Adiabatically and Stoichiometrically ($\phi = 1.00$) Combusted with Air (77°F), English Units

Fuel temp. (°F)	Flame temp. (°F)	CO	CO$_2$	H	H$_2$	H$_2$O (g)	NO	N$_2$	O	OH	O$_2$	Enthalpy, H (Btu/lb)	Density (lb/ft^3)	Viscosity (lbm/ft-hr)	Equil. spec. heat (Btu/lb-°F)	Equil. therm. cond. (Btu/ft-hr-°F)	Equil. Pr	Le
77	3620	0.01248	0.10269	0.00046	0.00331	0.14838	0.00245	0.72084	0.00031	0.00327	0.00581	−61	0.00942	4.84E-05	0.543	0.150	0.631	1.13
100	3621	0.01251	0.10266	0.00047	0.00332	0.14836	0.00246	0.72082	0.00031	0.00328	0.00582	−61	0.00942	4.84E-05	0.543	0.150	0.631	1.13
200	3626	0.01264	0.10251	0.00047	0.00335	0.14829	0.00248	0.72073	0.00032	0.00332	0.00588	−58	0.00941	4.84E-05	0.545	0.151	0.631	1.13
300	3631	0.01280	0.10233	0.00049	0.00339	0.14820	0.00251	0.72063	0.00032	0.00337	0.00596	−55	0.00939	4.85E-05	0.547	0.152	0.630	1.13
400	3637	0.01298	0.10214	0.00050	0.00344	0.14810	0.00255	0.72051	0.00033	0.00342	0.00603	−51	0.00938	4.85E-05	0.549	0.152	0.629	1.13
500	3644	0.01317	0.10193	0.00051	0.00348	0.14799	0.00259	0.72038	0.00034	0.00348	0.00612	−48	0.00936	4.86E-05	0.552	0.154	0.628	1.13
600	3652	0.01338	0.10170	0.00053	0.00354	0.14787	0.00263	0.72024	0.00035	0.00355	0.00621	−44	0.00934	4.86E-05	0.554	0.155	0.627	1.13
700	3659	0.01361	0.10146	0.00054	0.00359	0.14774	0.00267	0.72009	0.00037	0.00362	0.00631	−39	0.00932	4.87E-05	0.557	0.156	0.626	1.13
800	3667	0.01385	0.10119	0.00056	0.00365	0.14761	0.00272	0.71993	0.00038	0.00369	0.00642	−35	0.00930	4.88E-05	0.560	0.157	0.624	1.14
900	3676	0.01410	0.10091	0.00058	0.00372	0.14746	0.00277	0.71976	0.00039	0.00377	0.00653	−30	0.00928	4.88E-05	0.563	0.159	0.623	1.14
1000	3685	0.01438	0.10062	0.00060	0.00379	0.14731	0.00282	0.71957	0.00041	0.00385	0.00665	−25	0.00926	4.89E-05	0.567	0.160	0.622	1.14

Table D.16a C_3H_8 (298 K) Stoichiometrically ($\phi = 1.00$) Combusted with Air (298 K), Metric Units

Flame temp. (K)	CO	CO$_2$	H	H$_2$	H$_2$O (g)	NO	N$_2$	O	OH	O$_2$	Enthalpy, H (kJ/kg)	Density (kg/m^3)	Viscosity (kg/m-sec)	Equil. spec. heat (kJ/kg-K)	Equil. therm. cond. (W/m-K)	Equil. Pr	Le	Avail. heat (%)
366	0.00000	0.11624	0.00000	0.00000	0.15498	0.00000	0.72878	0.00000	0.00000	0.00000	−2.87E+03	0.943	1.92E-05	1.10	0.027	0.772	1.00	89.6
478	0.00000	0.11624	0.00000	0.00000	0.15498	0.00000	0.72878	0.00000	0.00000	0.00000	−2.74E+03	0.722	2.38E-05	1.13	0.035	0.764	1.00	85.5
589	0.00000	0.11624	0.00000	0.00000	0.15498	0.00000	0.72878	0.00000	0.00000	0.00000	−2.61E+03	0.586	2.79E-05	1.16	0.043	0.760	1.00	81.3
700	0.00000	0.11624	0.00000	0.00000	0.15498	0.00000	0.72878	0.00000	0.00000	0.00000	−2.48E+03	0.493	3.18E-05	1.20	0.051	0.756	1.00	76.9
811	0.00000	0.11624	0.00000	0.00000	0.15498	0.00000	0.72878	0.00000	0.00000	0.00000	−2.35E+03	0.426	3.54E-05	1.24	0.058	0.752	1.00	72.5
922	0.00000	0.11624	0.00000	0.00000	0.15498	0.00000	0.72878	0.00000	0.00000	0.00000	−2.21E+03	0.374	3.89E-05	1.27	0.066	0.748	1.00	67.9
1033	0.00000	0.11624	0.00000	0.00000	0.15498	0.00000	0.72878	0.00000	0.00000	0.00000	−2.07E+03	0.334	4.22E-05	1.30	0.074	0.744	1.00	63.2
1144	0.00000	0.11624	0.00000	0.00000	0.15498	0.00000	0.72878	0.00000	0.00000	0.00000	−1.92E+03	0.302	4.53E-05	1.32	0.081	0.742	1.00	58.4
1255	0.00000	0.11624	0.00000	0.00000	0.15498	0.00000	0.72878	0.00000	0.00000	0.00000	−1.77E+03	0.275	4.83E-05	1.35	0.088	0.739	1.00	53.6
1366	0.00002	0.11622	0.00000	0.00001	0.15497	0.00000	0.72877	0.00000	0.00000	0.00001	−1.62E+03	0.253	5.12E-05	1.37	0.095	0.737	1.00	48.6
1478	0.00006	0.11617	0.00000	0.00003	0.15493	0.00002	0.72874	0.00000	0.00001	0.00004	−1.46E+03	0.234	5.40E-05	1.39	0.102	0.734	1.00	43.5
1589	0.00019	0.11603	0.00000	0.00009	0.15486	0.00004	0.72865	0.00000	0.00003	0.00011	−1.31E+03	0.217	5.67E-05	1.43	0.111	0.729	1.00	38.3
1700	0.00048	0.11571	0.00000	0.00019	0.15469	0.00011	0.72847	0.00000	0.00009	0.00026	−1.15E+03	0.203	5.94E-05	1.48	0.122	0.722	1.01	33.0
1811	0.00110	0.11505	0.00001	0.00038	0.15436	0.00023	0.72808	0.00000	0.00021	0.00057	−9.78E+02	0.190	6.20E-05	1.54	0.134	0.713	1.02	27.5
1922	0.00224	0.11381	0.00002	0.00071	0.15377	0.00046	0.72737	0.00001	0.00047	0.00113	−8.01E+02	0.179	6.45E-05	1.64	0.150	0.704	1.03	21.7
2033	0.00421	0.11167	0.00007	0.00123	0.15277	0.00084	0.72615	0.00004	0.00095	0.00206	−6.11E+02	0.169	6.69E-05	1.79	0.175	0.682	1.05	15.4
2144	0.00733	0.10829	0.00018	0.00203	0.15115	0.00145	0.72418	0.00011	0.00178	0.00350	−4.02E+02	0.160	6.93E-05	1.99	0.208	0.661	1.08	8.5
2266	0.01248	0.10269	0.00046	0.00331	0.14838	0.00245	0.72084	0.00031	0.00327	0.00581	−1.42E+02	0.151	7.20E-05	2.27	0.259	0.631	1.13	0.0

Table D.16b C$_3$H$_8$ (Variable Temp.) Adiabatically and Stoichiometrically ($\phi = 1.00$) Combusted with Air (77°F), English Units

Flame temp. (°F)	CO	CO$_2$	H	H$_2$	H$_2$O (g)	NO	N$_2$	O	OH	O$_2$	Enthalpy, H (Btu/lb)	Density (lb/ft^3)	Viscosity (lbm/ft-hr)	Equil. spec. heat (Btu/lb-°F)	Equil. therm. cond. (Btu/ft-hr-°F)	Equil. Pr	Le	Avail. heat (%)
200	0.00000	0.11624	0.00000	0.00000	0.15498	0.00000	0.72878	0.00000	0.00000	0.00000	−1232	0.05888	1.29E-05	0.263	0.016	0.772	1.00	89.6
400	0.00000	0.11624	0.00000	0.00000	0.15498	0.00000	0.72878	0.00000	0.00000	0.00000	−1178	0.04508	1.60E-05	0.270	0.020	0.764	1.00	85.5
600	0.00000	0.11624	0.00000	0.00000	0.15498	0.00000	0.72878	0.00000	0.00000	0.00000	−1124	0.03659	1.88E-05	0.278	0.025	0.760	1.00	81.3
800	0.00000	0.11624	0.00000	0.00000	0.15498	0.00000	0.72878	0.00000	0.00000	0.00000	−1067	0.03078	2.14E-05	0.287	0.029	0.756	1.00	76.9
1000	0.00000	0.11624	0.00000	0.00000	0.15498	0.00000	0.72878	0.00000	0.00000	0.00000	−1009	0.02657	2.38E-05	0.295	0.034	0.752	1.00	72.5
1200	0.00000	0.11624	0.00000	0.00000	0.15498	0.00000	0.72878	0.00000	0.00000	0.00000	−949	0.02337	2.61E-05	0.303	0.038	0.748	1.00	67.9
1400	0.00000	0.11624	0.00000	0.00000	0.15498	0.00000	0.72878	0.00000	0.00000	0.00000	−888	0.02086	2.83E-05	0.310	0.042	0.744	1.00	63.2
1600	0.00000	0.11624	0.00000	0.00000	0.15498	0.00000	0.72878	0.00000	0.00000	0.00000	−825	0.01884	3.05E-05	0.316	0.047	0.742	1.00	58.4
1800	0.00000	0.11624	0.00000	0.00000	0.15498	0.00000	0.72878	0.00000	0.00000	0.00000	−761	0.01717	3.25E-05	0.322	0.051	0.739	1.00	53.6
2000	0.00002	0.11622	0.00000	0.00001	0.15497	0.00000	0.72877	0.00000	0.00000	0.00001	−696	0.01577	3.44E-05	0.327	0.055	0.737	1.00	48.6
2200	0.00006	0.11617	0.00000	0.00003	0.15493	0.00002	0.72874	0.00000	0.00001	0.00004	−630	0.01458	3.63E-05	0.332	0.059	0.734	1.00	43.5
2400	0.00019	0.11603	0.00000	0.00009	0.15486	0.00004	0.72865	0.00000	0.00003	0.00011	−562	0.01356	3.82E-05	0.341	0.064	0.729	1.00	38.3
2600	0.00048	0.11571	0.00000	0.00019	0.15469	0.00011	0.72847	0.00000	0.00009	0.00026	−493	0.01267	3.99E-05	0.353	0.070	0.722	1.01	33.0
2800	0.00110	0.11505	0.00001	0.00038	0.15436	0.00023	0.72808	0.00000	0.00021	0.00057	−420	0.01189	4.17E-05	0.369	0.077	0.713	1.02	27.5
3000	0.00224	0.11381	0.00002	0.00071	0.15377	0.00046	0.72737	0.00001	0.00047	0.00113	−345	0.01119	4.34E-05	0.392	0.087	0.704	1.03	21.7
3200	0.00421	0.11167	0.00007	0.00123	0.15277	0.00084	0.72615	0.00004	0.00095	0.00206	−263	0.01057	4.50E-05	0.427	0.101	0.682	1.05	15.4
3400	0.00733	0.10829	0.00018	0.00203	0.15115	0.00145	0.72418	0.00011	0.00178	0.00350	−173	0.01000	4.66E-05	0.474	0.120	0.661	1.08	8.5
3620	0.01248	0.10269	0.00046	0.00331	0.14838	0.00245	0.72084	0.00031	0.00327	0.00581	−61	0.00942	4.84E-05	0.543	0.150	0.631	1.13	0.0

Table D.17a C_3H_8 (298 K) Adiabatically Combusted with Air (298 K) at Various Equivalence Ratios, Metric Units

Equiv. ratio	Flame temp. (K)	C(gr)	CH_4	CO	CO_2	H	H_2	H_2O (g)	NH_3	NO	N_2	O	OH	O_2	Enthalpy, H (kJ/kg)	Density (kg/m³)	Viscosity (kg/m-sec)	Equil. spec. heat (kJ/kg-K)	Equil. therm. cond. (W/m-K)	Equil. Pr	Le
4.00	1004	0.0701	0.00975	0.16583	0.01569	0.00000	0.30829	0.02061	0.00007	0.00000	0.40966	0.00000	0.00000	0.00000	-4.82E+02	0.245	4.06E-05	4.18	0.269	0.631	0.82
2.00	1633	0.0000	0.00000	0.15310	0.02977	0.00003	0.15167	0.09215	0.00000	0.00000	0.57329	0.00000	0.00000	0.00000	-2.68E+02	0.176	5.72E-05	1.59	0.148	0.613	1.00
1.33	2098	0.0000	0.00000	0.07936	0.06268	0.00060	0.03881	0.15013	0.00000	0.00005	0.66807	0.00000	0.00028	0.00001	-1.86E+02	0.154	6.81E-05	1.58	0.169	0.635	1.08
1.00	2266	0.0000	0.00000	0.01248	0.10269	0.00046	0.00331	0.14838	0.00000	0.00245	0.72084	0.00031	0.00327	0.00581	-1.42E+02	0.151	7.20E-05	2.27	0.259	0.631	1.13
0.80	2042	0.0000	0.00000	0.00089	0.09346	0.00003	0.00025	0.12460	0.00000	0.00371	0.73755	0.00018	0.00185	0.03748	-1.15E+02	0.169	6.74E-05	1.60	0.154	0.697	1.03
0.67	1821	0.0000	0.00000	0.00008	0.07949	0.00000	0.00003	0.10575	0.00000	0.00258	0.74663	0.00004	0.00061	0.06479	-9.68E+01	0.191	6.26E-05	1.44	0.125	0.720	1.01
0.57	1646	0.0000	0.00000	0.00001	0.06864	0.00000	0.00000	0.09143	0.00000	0.00157	0.75302	0.00001	0.00019	0.08513	-8.33E+01	0.211	5.87E-05	1.37	0.110	0.727	1.00
0.50	1508	0.0000	0.00000	0.00000	0.06046	0.00000	0.00000	0.08058	0.00000	0.00093	0.75768	0.00000	0.00006	0.10028	-7.33E+01	0.231	5.55E-05	1.32	0.101	0.729	1.00

Table D.17b C_3H_8 (77°F) Combusted with Air (77°F) at Various Equivalence Ratios, English Units

Equiv. ratio	Flame temp. (°F)	C(gr)	CH_4	CO	CO_2	H	H_2	H_2O (g)	NH_3	NO	N_2	O	OH	O_2	Enthalpy, H (Btu/lb)	Density (lb/ft³)	Viscosity (lbm/ft-hr)	Equil. spec. heat (Btu/lb-°F)	Equil. therm. cond. (Btu/ft-hr-°F)	Equil. Pr	Le
4.00	1348	0.0701	0.00975	0.16583	0.01569	0.00000	0.30829	0.02061	0.00007	0.00000	0.40966	0.00000	0.00000	0.00000	−207	0.01532	2.73E-05	0.997	0.155	0.631	0.82
2.00	2479	0.0000	0.00000	0.15310	0.02977	0.00003	0.15167	0.09215	0.00000	0.00000	0.57329	0.00000	0.00000	0.00000	−115	0.01101	3.84E-05	0.379	0.086	0.613	1.00
1.33	3317	0.0000	0.00000	0.07936	0.06268	0.00060	0.03881	0.15013	0.00000	0.00005	0.66807	0.00000	0.00028	0.00001	−80	0.00960	4.58E-05	0.377	0.098	0.635	1.08
1.00	3620	0.0000	0.00000	0.01248	0.10269	0.00046	0.00331	0.14838	0.00000	0.00245	0.72084	0.00031	0.00327	0.00581	−61	0.00942	4.84E-05	0.543	0.150	0.631	1.13
0.80	3215	0.0000	0.00000	0.00089	0.09346	0.00003	0.00025	0.12460	0.00000	0.00371	0.73755	0.00018	0.00185	0.03748	−49	0.01058	4.53E-05	0.382	0.089	0.697	1.03
0.67	2818	0.0000	0.00000	0.00008	0.07949	0.00000	0.00003	0.10575	0.00000	0.00258	0.74663	0.00004	0.00061	0.06479	−42	0.01190	4.21E-05	0.343	0.072	0.720	1.01
0.57	2502	0.0000	0.00000	0.00001	0.06864	0.00000	0.00000	0.09143	0.00000	0.00157	0.75302	0.00001	0.00019	0.08513	−36	0.01319	3.95E-05	0.327	0.064	0.727	1.00
0.50	2254	0.0000	0.00000	0.00000	0.06046	0.00000	0.00000	0.08058	0.00000	0.00093	0.75768	0.00000	0.00006	0.10028	−32	0.01442	3.73E-05	0.316	0.058	0.729	1.00

Table D.18a C$_3$H$_8$ (298 K) Stoichiometrically ($\phi = 1.00$) Combusted with Air (Variable Temp.) with an Exhaust Temp. of 1367 K, Metric Units

Oxid. temp. (K)	CO	CO$_2$	H$_2$	H$_2$O (g)	N$_2$	O$_2$	Enthalpy, H (kJ/kg)	Density (kg/m^3)	Viscosity (kg/m-sec)	Equil. spec. heat (kJ/kg-K)	Equil. therm. cond. (W/m-K)	Equil. Pr	Le	Avail. heat (%)
298	0.00002	0.11622	0.00001	0.15497	0.72877	0.00001	$-1.62E+03$	0.253	5.12E-05	1.37	0.095	0.737	1.00	48.6
366	0.00002	0.11622	0.00001	0.15497	0.72877	0.00001	$-1.62E+03$	0.253	5.12E-05	1.37	0.095	0.737	1.00	50.7
478	0.00002	0.11622	0.00001	0.15497	0.72877	0.00001	$-1.62E+03$	0.253	5.12E-05	1.37	0.095	0.737	1.00	54.2
589	0.00002	0.11622	0.00001	0.15497	0.72877	0.00001	$-1.62E+03$	0.253	5.12E-05	1.37	0.095	0.737	1.00	57.8
700	0.00002	0.11622	0.00001	0.15497	0.72877	0.00001	$-1.62E+03$	0.253	5.12E-05	1.37	0.095	0.737	1.00	61.5
811	0.00002	0.11622	0.00001	0.15497	0.72877	0.00001	$-1.62E+03$	0.253	5.12E-05	1.37	0.095	0.737	1.00	65.2
922	0.00002	0.11622	0.00001	0.15497	0.72877	0.00001	$-1.62E+03$	0.253	5.12E-05	1.37	0.095	0.737	1.00	69.1
1033	0.00002	0.11622	0.00001	0.15497	0.72877	0.00001	$-1.62E+03$	0.253	5.12E-05	1.37	0.095	0.737	1.00	73.0
1144	0.00002	0.11622	0.00001	0.15497	0.72877	0.00001	$-1.62E+03$	0.253	5.12E-05	1.37	0.095	0.737	1.00	77.0
1255	0.00002	0.11622	0.00001	0.15497	0.72877	0.00001	$-1.62E+03$	0.253	5.12E-05	1.37	0.095	0.737	1.00	81.1
1366	0.00002	0.11622	0.00001	0.15497	0.72877	0.00001	$-1.62E+03$	0.253	5.12E-05	1.37	0.095	0.737	1.00	85.2

Table D.18b C_3H_8 (77°F) Stoichiometrically ($\phi = 1.00$) Combusted with Air (Variable Temp.) with an Exhaust Temp. of 2000°F, English Units

Oxid. temp. (°F)	CO	CO_2	H_2	H_2O (g)	N_2	O_2	Enthalpy, H (Btu/lb)	Density (lb/ft³)	Viscosity (lbm/ft-hr)	Equil. spec. heat (Btu/lb-°F)	Equil. therm. cond. (Btu/ft-hr-°F)	Equil. Pr	Le	Avail. heat (%)
77	0.00002	0.11622	0.00001	0.15497	0.72877	0.00001	−696	0.01577	3.44E-05	0.327	0.055	0.737	1.00	48.6
200	0.00002	0.11622	0.00001	0.15497	0.72877	0.00001	−696	0.01577	3.44E-05	0.327	0.055	0.737	1.00	50.7
400	0.00002	0.11622	0.00001	0.15497	0.72877	0.00001	−696	0.01577	3.44E-05	0.327	0.055	0.737	1.00	54.2
600	0.00002	0.11622	0.00001	0.15497	0.72877	0.00001	−696	0.01577	3.44E-05	0.327	0.055	0.737	1.00	57.8
800	0.00002	0.11622	0.00001	0.15497	0.72877	0.00001	−696	0.01577	3.44E-05	0.327	0.055	0.737	1.00	61.5
1000	0.00002	0.11622	0.00001	0.15497	0.72877	0.00001	−696	0.01577	3.44E-05	0.327	0.055	0.737	1.00	65.2
1200	0.00002	0.11622	0.00001	0.15497	0.72877	0.00001	−696	0.01577	3.44E-05	0.327	0.055	0.737	1.00	69.1
1400	0.00002	0.11622	0.00001	0.15497	0.72877	0.00001	−696	0.01577	3.44E-05	0.327	0.055	0.737	1.00	73.0
1600	0.00002	0.11622	0.00001	0.15497	0.72877	0.00001	−696	0.01577	3.44E-05	0.327	0.055	0.737	1.00	77.0
1800	0.00002	0.11622	0.00001	0.15497	0.72877	0.00001	−696	0.01577	3.44E-05	0.327	0.055	0.737	1.00	81.1
2000	0.00002	0.11622	0.00001	0.15497	0.72877	0.00001	−696	0.01577	3.44E-05	0.327	0.055	0.737	1.00	85.2

Table D.19a $H_2 + CH_4$ (298 K) Stoichiometrically ($\phi = 1.00$) Combusted with Air (298 K), Metric Units

H_2	CH_4	Exhaust temp. (K)	CO	CO_2	H	H_2	H_2O (g)	NO	N_2	O	OH	O_2	Enthalpy, H (kJ/kg)	Density (kg/m^3)	Viscosity (kg/m-sec)	Equil. spec. heat (kJ/kg-K)	Equil. therm. cond. (W/m-K)	Equil. Pr	Le
0.0	1.0	2225	0.00893	0.08539	0.00039	0.00361	0.18338	0.00197	0.70866	0.00021	0.00291	0.00455	-2.58E+02	0.150	7.12E-05	2.19	0.248	0.630	1.12
0.1	0.9	2230	0.00888	0.08265	0.00041	0.00378	0.18773	0.00199	0.70676	0.00022	0.00300	0.00458	-2.51E+02	0.149	7.14E-05	2.21	0.251	0.627	1.13
0.2	0.8	2235	0.00881	0.07945	0.00043	0.00400	0.19281	0.00202	0.70453	0.00023	0.00311	0.00461	-2.43E+02	0.148	7.15E-05	2.22	0.255	0.624	1.13
0.3	0.7	2241	0.00872	0.07567	0.00046	0.00427	0.19881	0.00205	0.70189	0.00024	0.00325	0.00465	-2.33E+02	0.147	7.17E-05	2.24	0.259	0.620	1.14
0.4	0.6	2249	0.00858	0.07114	0.00050	0.00461	0.20602	0.00209	0.69871	0.00025	0.00341	0.00469	-2.22E+02	0.146	7.19E-05	2.27	0.265	0.616	1.14
0.5	0.5	2258	0.00837	0.06562	0.00055	0.00505	0.21483	0.00214	0.69480	0.00027	0.00362	0.00474	-2.07E+02	0.144	7.21E-05	2.30	0.272	0.610	1.15
0.6	0.4	2270	0.00805	0.05875	0.00062	0.00567	0.22585	0.00219	0.68990	0.00029	0.00390	0.00479	-1.89E+02	0.142	7.24E-05	2.34	0.281	0.603	1.16
0.7	0.3	2285	0.00751	0.04997	0.00072	0.00655	0.24001	0.00227	0.68354	0.00032	0.00427	0.00485	-1.64E+02	0.140	7.28E-05	2.39	0.293	0.593	1.18
0.8	0.2	2306	0.00654	0.03839	0.00088	0.00792	0.25886	0.00236	0.67500	0.00036	0.00479	0.00489	-1.31E+02	0.136	7.33E-05	2.45	0.310	0.580	1.20
0.9	0.1	2336	0.00461	0.02255	0.00117	0.01030	0.28507	0.00249	0.66289	0.00042	0.00560	0.00489	-8.08E+01	0.131	7.41E-05	2.55	0.337	0.562	1.23
1.0	0.0	2382	0.00000	0.00000	0.00179	0.01526	0.32366	0.00264	0.64440	0.00054	0.00696	0.00475	2.09E-04	0.124	7.53E-05	2.72	0.383	0.536	1.28

Table D.19b $H_2 + CH_4$ (298 K) Stoichiometrically ($\phi = 1.00$) Combusted with Air (77°F), English Units

H_2	CH_4	Product temp. (°F)	CO	CO_2	H	H_2	H_2O (g)	NO	N_2	O	OH	O_2	Enthalpy, H (Btu/lb)	Density (lb/ft³)	Viscosity (lbm/ft-hr)	Equil. spec. heat (Btu/lb-°F)	Equil. therm. cond. (Btu/ft-hr-°F)	Equil. Pr	Le
0.0	1.0	3545	0.00893	0.08539	0.00039	0.00361	0.18338	0.00197	0.70866	0.00021	0.00291	0.00455	-111	0.00938	4.79E-05	0.524	0.143	0.630	1.12
0.1	0.9	3554	0.00888	0.08265	0.00041	0.00378	0.18773	0.00199	0.70676	0.00022	0.00300	0.00458	-108	0.00933	4.80E-05	0.527	0.145	0.627	1.13
0.2	0.8	3563	0.00881	0.07945	0.00043	0.00400	0.19281	0.00202	0.70453	0.00023	0.00311	0.00461	-104	0.00927	4.81E-05	0.531	0.147	0.624	1.13
0.3	0.7	3574	0.00872	0.07567	0.00046	0.00427	0.19881	0.00205	0.70189	0.00024	0.00325	0.00465	-100	0.00920	4.82E-05	0.536	0.150	0.620	1.14
0.4	0.6	3588	0.00858	0.07114	0.00050	0.00461	0.20602	0.00209	0.69871	0.00025	0.00341	0.00469	-95	0.00911	4.83E-05	0.542	0.153	0.616	1.14
0.5	0.5	3605	0.00837	0.06562	0.00055	0.00505	0.21483	0.00214	0.69480	0.00027	0.00362	0.00474	-89	0.00901	4.85E-05	0.549	0.157	0.610	1.15
0.6	0.4	3626	0.00805	0.05875	0.00062	0.00567	0.22585	0.00219	0.68990	0.00029	0.00390	0.00479	-81	0.00888	4.87E-05	0.558	0.162	0.603	1.16
0.7	0.3	3654	0.00751	0.04997	0.00072	0.00655	0.24001	0.00227	0.68354	0.00032	0.00427	0.00485	-71	0.00872	4.90E-05	0.570	0.169	0.593	1.18
0.8	0.2	3691	0.00654	0.03839	0.00088	0.00792	0.25886	0.00236	0.67500	0.00036	0.00479	0.00489	-56	0.00850	4.93E-05	0.586	0.179	0.580	1.20
0.9	0.1	3744	0.00461	0.02255	0.00117	0.01030	0.28507	0.00249	0.66289	0.00042	0.00560	0.00489	-35	0.00820	4.98E-05	0.610	0.195	0.562	1.23
1.0	0.0	3827	0.00000	0.00000	0.00179	0.01526	0.32366	0.00264	0.64440	0.00054	0.00696	0.00475	0	0.00775	5.06E-05	0.650	0.221	0.536	1.28

Table D.20a H$_2$ + CH$_4$ (298 K) Stoichiometrically ($\phi = 1.00$) Combusted with Air (533 K), Metric Units

H$_2$	CH$_4$	Exhaust temp. (K)	CO	CO$_2$	H	H$_2$	H$_2$O (g)	NO	N$_2$	O	OH	O$_2$	Enthalpy, H (kJ/kg)	Density (kg/m^3)	Viscosity (kg/m-sec)	Equil. spec. heat (kJ/kg-K)	Equil. therm. cond. (W/m-K)	Equil. Pr	Le
0.0	1.0	2323	0.01319	0.08077	0.00078	0.00523	0.18001	0.00290	0.70550	0.00046	0.00458	0.00658	−3.02E+01	0.143	7.33E−05	2.46	0.299	0.601	1.17
0.1	0.9	2327	0.01306	0.07812	0.00082	0.00547	0.18425	0.00292	0.70359	0.00047	0.00471	0.00659	−2.32E+01	0.143	7.34E−05	2.47	0.303	0.599	1.17
0.2	0.8	2332	0.01289	0.07503	0.00086	0.00576	0.18921	0.00295	0.70136	0.00048	0.00485	0.00661	−1.50E+01	0.142	7.35E−05	2.49	0.308	0.595	1.18
0.3	0.7	2337	0.01268	0.07138	0.00091	0.00612	0.19506	0.00298	0.69872	0.00050	0.00503	0.00663	−5.26E+00	0.141	7.37E−05	2.51	0.313	0.591	1.19
0.4	0.6	2344	0.01238	0.06703	0.00098	0.00657	0.20208	0.00302	0.69554	0.00052	0.00525	0.00664	6.66E+00	0.140	7.39E−05	2.54	0.320	0.586	1.19
0.5	0.5	2352	0.01198	0.06172	0.00106	0.00716	0.21066	0.00306	0.69164	0.00054	0.00552	0.00665	2.15E+01	0.138	7.41E−05	2.57	0.328	0.580	1.20
0.6	0.4	2363	0.01139	0.05514	0.00118	0.00797	0.22137	0.00311	0.68673	0.00057	0.00587	0.00666	4.04E+01	0.136	7.44E−05	2.61	0.339	0.572	1.22
0.7	0.3	2377	0.01048	0.04677	0.00135	0.00912	0.23511	0.00318	0.68039	0.00062	0.00634	0.00665	6.54E+01	0.134	7.47E−05	2.66	0.354	0.562	1.23
0.8	0.2	2395	0.00897	0.03579	0.00161	0.01088	0.25335	0.00326	0.67185	0.00068	0.00700	0.00660	1.00E+02	0.131	7.52E−05	2.73	0.375	0.549	1.26
0.9	0.1	2422	0.00616	0.02089	0.00206	0.01389	0.27863	0.00336	0.65976	0.00078	0.00799	0.00647	1.51E+02	0.126	7.59E−05	2.84	0.406	0.531	1.29
1.0	0.0	2463	0.00000	0.00000	0.00299	0.01997	0.31560	0.00348	0.64130	0.00094	0.00960	0.00612	2.34E+02	0.120	7.70E−05	3.03	0.461	0.507	1.34

Table D.20b H$_2$ + CH$_4$ (298 K) Stoichiometrically (ϕ = 1.00) Combusted with Air (500°F), English Units

H$_2$	CH$_4$	Exhaust temp. (°F)	CO	CO$_2$	H	H$_2$	H$_2$O (g)	NO	N$_2$	O	OH	O$_2$	Enthalpy, H (Btu/lb)	Density (lb/ft^3)	Viscosity (lbm/ft-hr)	Equil. spec. heat (Btu/lb-°F)	Equil. therm. cond. (Btu/ft-hr-°F)	Equil. Pr	Le
0.0	1.0	3722	0.01319	0.08077	0.00078	0.00523	0.18001	0.00290	0.70550	0.00046	0.00458	0.00658	-13	0.00895	4.93E-05	0.587	0.173	0.601	1.17
0.1	0.9	3729	0.01306	0.07812	0.00082	0.00547	0.18425	0.00292	0.70359	0.00047	0.00471	0.00659	-10	0.00890	4.94E-05	0.590	0.175	0.599	1.17
0.2	0.8	3737	0.01289	0.07503	0.00086	0.00576	0.18921	0.00295	0.70136	0.00048	0.00485	0.00661	-6	0.00885	4.95E-05	0.594	0.178	0.595	1.18
0.3	0.7	3747	0.01268	0.07138	0.00091	0.00612	0.19506	0.00298	0.69872	0.00050	0.00503	0.00663	-2	0.00879	4.96E-05	0.599	0.181	0.591	1.19
0.4	0.6	3760	0.01238	0.06703	0.00098	0.00657	0.20208	0.00302	0.69554	0.00052	0.00525	0.00664	3	0.00871	4.97E-05	0.605	0.185	0.586	1.19
0.5	0.5	3774	0.01198	0.06172	0.00106	0.00716	0.21066	0.00306	0.69164	0.00054	0.00552	0.00665	9	0.00862	4.98E-05	0.613	0.190	0.580	1.20
0.6	0.4	3793	0.01139	0.05514	0.00118	0.00797	0.22137	0.00311	0.68673	0.00057	0.00587	0.00666	17	0.00850	5.00E-05	0.623	0.196	0.572	1.22
0.7	0.3	3818	0.01048	0.04677	0.00135	0.00912	0.23511	0.00318	0.68039	0.00062	0.00634	0.00665	28	0.00835	5.03E-05	0.635	0.204	0.562	1.23
0.8	0.2	3851	0.00897	0.03579	0.00161	0.01088	0.25335	0.00326	0.67185	0.00068	0.00700	0.00660	43	0.00816	5.06E-05	0.653	0.216	0.549	1.26
0.9	0.1	3899	0.00616	0.02089	0.00206	0.01389	0.27863	0.00336	0.65976	0.00078	0.00799	0.00647	65	0.00788	5.11E-05	0.679	0.235	0.531	1.29
1.0	0.0	3974	0.00000	0.00000	0.00299	0.01997	0.31560	0.00348	0.64130	0.00094	0.00960	0.00612	101	0.00747	5.18E-05	0.724	0.267	0.507	1.34

Table D.21a $H_2 + CH_4$ (298 K) Stoichiometrically ($\phi = 1.00$) Combusted with Air (811 K), Metric Units

H_2	CH_4	Exhaust temp. (K)	CO	CO_2	H	H_2	H_2O (g)	NO	N_2	O	OH	O_2	Enthalpy, H (kJ/kg)	Density (kg/m³)	Viscosity (kg/m-sec)	Equil. spec. heat (kJ/kg-K)	Equil. therm. cond. (W/m-K)	Equil. Pr	Le
0.0	1.0	2431	0.01913	0.07429	0.00159	0.00759	0.17489	0.00424	0.70078	0.00098	0.00715	0.00936	2.52E+02	0.136	7.56E-05	2.80	0.374	0.566	1.23
0.1	0.9	2434	0.01887	0.07179	0.00165	0.00791	0.17898	0.00426	0.69888	0.00100	0.00731	0.00934	2.59E+02	0.136	7.57E-05	2.82	0.379	0.563	1.24
0.2	0.8	2438	0.01854	0.06888	0.00172	0.00830	0.18376	0.00428	0.69666	0.00102	0.00751	0.00932	2.68E+02	0.135	7.58E-05	2.84	0.384	0.559	1.24
0.3	0.7	2443	0.01812	0.06545	0.00181	0.00878	0.18941	0.00430	0.69403	0.00104	0.00774	0.00930	2.78E+02	0.134	7.59E-05	2.86	0.391	0.555	1.25
0.4	0.6	2449	0.01759	0.06137	0.00192	0.00939	0.19618	0.00433	0.69087	0.00107	0.00802	0.00926	2.90E+02	0.133	7.61E-05	2.89	0.399	0.550	1.26
0.5	0.5	2456	0.01687	0.05641	0.00207	0.01018	0.20443	0.00436	0.68698	0.00111	0.00837	0.00921	3.06E+02	0.131	7.63E-05	2.92	0.410	0.544	1.27
0.6	0.4	2465	0.01587	0.05028	0.00227	0.01124	0.21473	0.00440	0.68210	0.00116	0.00882	0.00914	3.25E+02	0.130	7.65E-05	2.96	0.423	0.536	1.29
0.7	0.3	2477	0.01441	0.04251	0.00255	0.01274	0.22792	0.00444	0.67579	0.00122	0.00941	0.00902	3.51E+02	0.128	7.68E-05	3.02	0.441	0.526	1.30
0.8	0.2	2493	0.01210	0.03239	0.00297	0.01501	0.24537	0.00450	0.66729	0.00131	0.01022	0.00883	3.86E+02	0.125	7.73E-05	3.10	0.467	0.513	1.33
0.9	0.1	2517	0.00809	0.01880	0.00368	0.01879	0.26946	0.00455	0.65526	0.00144	0.01141	0.00851	4.39E+02	0.121	7.79E-05	3.23	0.506	0.497	1.36
1.0	0.0	2553	0.00000	0.00000	0.00507	0.02615	0.30448	0.00460	0.63687	0.00165	0.01328	0.00790	5.24E+02	0.115	7.89E-05	3.45	0.573	0.476	1.41

Table D.21b $H_2 + CH_4$ (298 K) Stoichiometrically ($\phi = 1.00$) Combusted with Air (1000°F), English Units

H_2	CH_4	Exhaust temp. (°F)	CO	CO_2	H	H_2	H_2O (g)	NO	N_2	O	OH	O_2	Enthalpy, H (Btu/lb)	Density (lb/ft³)	Viscosity (lbm/ft-hr)	Equil. spec. heat (Btu/lb-°F)	Equil. therm. cond. (Btu/ft-hr-°F)	Equil. Pr	Le
0.0	1.0	3915	0.01913	0.07429	0.00159	0.00759	0.17489	0.00424	0.70078	0.00098	0.00715	0.00936	108	0.00850	5.08E-05	0.670	0.216	0.566	1.23
0.1	0.9	3922	0.01887	0.07179	0.00165	0.00791	0.17898	0.00426	0.69888	0.00100	0.00731	0.00934	112	0.00846	5.09E-05	0.674	0.219	0.563	1.24
0.2	0.8	3929	0.01854	0.06888	0.00172	0.00830	0.18376	0.00428	0.69666	0.00102	0.00751	0.00932	115	0.00841	5.10E-05	0.678	0.222	0.559	1.24
0.3	0.7	3938	0.01812	0.06545	0.00181	0.00878	0.18941	0.00430	0.69403	0.00104	0.00774	0.00930	120	0.00836	5.10E-05	0.683	0.226	0.555	1.25
0.4	0.6	3948	0.01759	0.06137	0.00192	0.00939	0.19618	0.00433	0.69087	0.00107	0.00802	0.00926	125	0.00829	5.11E-05	0.690	0.231	0.550	1.26
0.5	0.5	3961	0.01687	0.05641	0.00207	0.01018	0.20443	0.00436	0.68698	0.00111	0.00837	0.00921	131	0.00821	5.13E-05	0.698	0.237	0.544	1.27
0.6	0.4	3978	0.01587	0.05028	0.00227	0.01124	0.21473	0.00440	0.68210	0.00116	0.00882	0.00914	140	0.00810	5.15E-05	0.708	0.245	0.536	1.29
0.7	0.3	3999	0.01441	0.04251	0.00255	0.01274	0.22792	0.00444	0.67579	0.00122	0.00941	0.00902	151	0.00797	5.17E-05	0.722	0.255	0.526	1.30
0.8	0.2	4028	0.01210	0.03239	0.00297	0.01501	0.24537	0.00450	0.66729	0.00131	0.01022	0.00883	166	0.00779	5.20E-05	0.741	0.270	0.513	1.33
0.9	0.1	4070	0.00809	0.01880	0.00368	0.01879	0.26946	0.00455	0.65526	0.00144	0.01141	0.00851	189	0.00754	5.24E-05	0.771	0.293	0.497	1.36
1.0	0.0	4135	0.00000	0.00000	0.00507	0.02615	0.30448	0.00460	0.63687	0.00165	0.01328	0.00790	225	0.00716	5.30E-05	0.825	0.331	0.476	1.41

Table D.22a H₂ + CH₄ (298 K) Stoichiometrically ($\phi = 1.00$) Combusted with Air (1089 K), Metric Units

H₂	CH₄	Exhaust temp. (K)	CO	CO₂	H	H₂	H₂O (g)	NO	N₂	O	OH	O₂	Enthalpy, H (kJ/kg)	Density (kg/m³)	Viscosity (kg/m-sec)	Equil. spec. heat (kJ/kg-K)	Equil. therm. cond. (W/m-K)	Equil. Pr	Le
0.0	1.0	2531	0.02572	0.06705	0.00291	0.01041	0.16849	0.00579	0.69505	0.00186	0.01034	0.01238	5.51E+02	0.130	7.76E-05	3.18	0.465	0.531	1.30
0.1	0.9	2534	0.02528	0.06474	0.00301	0.01083	0.17242	0.00580	0.69317	0.00188	0.01054	0.01233	5.59E+02	0.129	7.77E-05	3.20	0.470	0.528	1.31
0.2	0.8	2537	0.02474	0.06206	0.00312	0.01134	0.17701	0.00581	0.69097	0.00191	0.01078	0.01226	5.67E+02	0.129	7.78E-05	3.22	0.477	0.524	1.31
0.3	0.7	2541	0.02408	0.05891	0.00327	0.01196	0.18241	0.00582	0.68836	0.00194	0.01107	0.01218	5.78E+02	0.128	7.79E-05	3.24	0.486	0.520	1.32
0.4	0.6	2546	0.02323	0.05516	0.00344	0.01273	0.18889	0.00583	0.68522	0.00198	0.01141	0.01208	5.90E+02	0.127	7.81E-05	3.27	0.496	0.515	1.33
0.5	0.5	2553	0.02213	0.05063	0.00367	0.01373	0.19679	0.00585	0.68137	0.00203	0.01184	0.01195	6.06E+02	0.126	7.82E-05	3.31	0.509	0.509	1.34
0.6	0.4	2561	0.02065	0.04503	0.00398	0.01507	0.20663	0.00586	0.67653	0.00209	0.01237	0.01178	6.26E+02	0.124	7.85E-05	3.36	0.525	0.501	1.36
0.7	0.3	2571	0.01853	0.03798	0.00441	0.01694	0.21920	0.00587	0.67027	0.00217	0.01308	0.01154	6.53E+02	0.122	7.88E-05	3.42	0.547	0.492	1.38
0.8	0.2	2585	0.01533	0.02885	0.00504	0.01973	0.23581	0.00588	0.66185	0.00228	0.01403	0.01119	6.90E+02	0.119	7.92E-05	3.52	0.579	0.481	1.40
0.9	0.1	2605	0.01002	0.01668	0.00607	0.02429	0.25866	0.00588	0.64990	0.00245	0.01538	0.01066	7.44E+02	0.116	7.97E-05	3.66	0.626	0.467	1.43
1.0	0.0	2636	0.00000	0.00000	0.00799	0.03281	0.29182	0.00583	0.63163	0.00269	0.01746	0.00977	8.32E+02	0.110	8.06E-05	3.93	0.704	0.450	1.47

Table D.22b $H_2 + CH_4$ (298 K) Stoichiometrically ($\phi = 1.00$) Combusted with Air (1500°F), English Units

H_2	CH_4	Exhaust temp. (°F)	CO	CO_2	H	H_2	H_2O (g)	NO	N_2	O	OH	O_2	Enthalpy, H (Btu/lb)	Density (lb/ft³)	Viscosity (lbm/ft-hr)	Equil. spec. heat (Btu/lb-°F)	Equil. therm. cond. (Btu/ft-hr-°F)	Equil. Pr	Le
0.0	1.0	4095	0.02572	0.06705	0.00291	0.01041	0.16849	0.00579	0.69505	0.00186	0.01034	0.01238	237	0.00811	5.22E-05	0.759	0.268	0.531	1.30
0.1	0.9	4101	0.02528	0.06474	0.00301	0.01083	0.17242	0.00580	0.69317	0.00188	0.01054	0.01233	240	0.00807	5.23E-05	0.763	0.272	0.528	1.31
0.2	0.8	4107	0.02474	0.06206	0.00312	0.01134	0.17701	0.00581	0.69097	0.00191	0.01078	0.01226	244	0.00803	5.23E-05	0.768	0.276	0.524	1.31
0.3	0.7	4114	0.02408	0.05891	0.00327	0.01196	0.18241	0.00582	0.68836	0.00194	0.01107	0.01218	248	0.00798	5.24E-05	0.774	0.281	0.520	1.32
0.4	0.6	4124	0.02323	0.05516	0.00344	0.01273	0.18889	0.00583	0.68522	0.00198	0.01141	0.01208	254	0.00792	5.25E-05	0.781	0.286	0.515	1.33
0.5	0.5	4135	0.02213	0.05063	0.00367	0.01373	0.19679	0.00585	0.68137	0.00203	0.01184	0.01195	261	0.00784	5.26E-05	0.790	0.294	0.509	1.34
0.6	0.4	4149	0.02065	0.04503	0.00398	0.01507	0.20663	0.00586	0.67653	0.00209	0.01237	0.01178	269	0.00775	5.28E-05	0.802	0.303	0.501	1.36
0.7	0.3	4168	0.01853	0.03798	0.00441	0.01694	0.21920	0.00587	0.67027	0.00217	0.01308	0.01154	281	0.00762	5.30E-05	0.817	0.316	0.492	1.38
0.8	0.2	4193	0.01533	0.02885	0.00504	0.01973	0.23581	0.00588	0.66185	0.00228	0.01403	0.01119	296	0.00746	5.32E-05	0.840	0.334	0.481	1.40
0.9	0.1	4230	0.01002	0.01668	0.00607	0.02429	0.25866	0.00588	0.64990	0.00245	0.01538	0.01066	320	0.00723	5.36E-05	0.875	0.362	0.467	1.43
1.0	0.0	4285	0.00000	0.00000	0.00799	0.03281	0.29182	0.00583	0.63163	0.00269	0.01746	0.00977	358	0.00688	5.42E-05	0.939	0.407	0.450	1.47

Appendix E

Fluid Dynamics Equations

E.1 FLUID DYNAMICS EQUATIONS IN CYLINDRICAL COORDINATES

The unsteady equations of motion for an incompressible Newtonian fluid with constant viscosity in cylindrical coordinates (r, θ, z) are given as follows:

$$
\frac{\partial v_r}{\partial \tau} + v_r \frac{\partial v_r}{\partial r} + \frac{v_\theta}{r} \frac{\partial v_r}{\partial \theta} - \frac{v_\theta^2}{r} + w \frac{\partial v_r}{\partial z}
$$
$$
= f_r - \frac{1}{\rho} \frac{\partial p}{\partial r} + \nu \left\{ \frac{\partial}{\partial r} \left[\frac{1}{r} \frac{\partial (r v_r)}{\partial r} \right] + \frac{1}{r^2} \frac{\partial^2 v_r}{\partial \theta^2} - \frac{2}{r^2} \frac{\partial v_\theta}{\partial \theta} + \frac{\partial^2 v_r}{\partial z^2} \right\} \tag{E.1}
$$

$$
\frac{\partial v_\theta}{\partial \tau} + v_r \frac{\partial v_\theta}{\partial r} + \frac{v_\theta}{r} \frac{\partial v_\theta}{\partial \theta} + \frac{v_r v_\theta}{r} + w \frac{\partial v_\theta}{\partial z}
$$
$$
= f_\theta - \frac{1}{\rho r} \frac{\partial p}{\partial \theta} + \nu \left\{ \frac{\partial}{\partial r} \left[\frac{1}{r} \frac{\partial (r v_\theta)}{\partial r} \right] + \frac{1}{r^2} \frac{\partial^2 v_\theta}{\partial \theta^2} + \frac{2}{r^2} \frac{\partial v_r}{\partial \theta} + \frac{\partial^2 v_\theta}{\partial z^2} \right\} \tag{E.2}
$$

$$
\frac{\partial w}{\partial \tau} + v_r \frac{\partial w}{\partial r} + \frac{v_\theta}{r} \frac{\partial w}{\partial \theta} + w \frac{\partial w}{\partial z}
$$
$$
= f_z - \frac{1}{\rho} \frac{\partial p}{\partial z} + \nu \left\{ \frac{1}{r} \frac{\partial}{\partial r} \left(r \frac{\partial w}{\partial r} \right) + \frac{1}{r^2} \frac{\partial^2 w}{\partial \theta^2} + \frac{\partial^2 w}{\partial z^2} \right\} \tag{E.3}
$$

where f_i is some type of body force such as buoyancy. The energy equation for an incompressible fluid can be written as

$$
\rho c_p \left(\frac{\partial t}{\partial \tau} + v_r \frac{\partial t}{\partial r} + \frac{v_\theta}{r} \frac{\partial t}{\partial \theta} + w \frac{\partial t}{\partial z} \right)
$$
$$
= \frac{1}{r} \frac{\partial}{\partial r} \left(r k \frac{\partial t}{\partial r} \right) + \frac{1}{r^2} \frac{\partial}{\partial \theta} \left(k \frac{\partial t}{\partial \theta} \right) + \frac{\partial}{\partial z} \left(k \frac{\partial t}{\partial z} \right) + \dot{q} + \Phi \tag{E.4}
$$

where

$$
\Phi = 2\mu \left\{ \begin{array}{l} \left(\frac{\partial v_r}{\partial r} \right)^2 + \left[\frac{1}{r} \left(\frac{\partial v_\theta}{\partial \theta} + v_r \right) \right]^2 + \left(\frac{\partial w}{\partial z} \right)^2 + \frac{1}{2} \left(\frac{\partial v_\theta}{\partial z} + \frac{1}{r} \frac{\partial w}{\partial \theta} \right)^2 \\ + \frac{1}{2} \left(\frac{\partial w}{\partial r} + \frac{\partial v_r}{\partial z} \right)^2 + \frac{1}{2} \left[\frac{1}{r} \frac{\partial v_r}{\partial \theta} + r \frac{\partial}{\partial r} \left(\frac{v_\theta}{r} \right) \right]^2 \end{array} \right\} \tag{E.5}
$$

E.2   FLUID DYNAMICS EQUATIONS IN SPHERICAL COORDINATES

The unsteady equations of motion for an incompressible Newtonian fluid with constant viscosity in spherical coordinates (r, θ, ϕ) are given as follows:

$$
\frac{\partial v_r}{\partial \tau} + v_r \frac{\partial v_r}{\partial r} + \frac{v_\theta}{r} \frac{\partial v_r}{\partial \theta} + \frac{v_\phi}{r \sin \theta} \frac{\partial v_r}{\partial \phi} - \frac{v_\theta^2 + v_\phi^2}{r}
$$
$$
= f_r - \frac{1}{\rho} \frac{\partial p}{\partial r} + \nu \left\{ \nabla^2 v_r - \frac{2 v_r}{r^2} - \frac{2}{r^2} \frac{\partial v_\theta}{\partial \theta} - \frac{2}{r^2} v_\theta \cot \theta - \frac{2}{r^2 \sin \theta} \frac{\partial v_\phi}{\partial \phi} \right\} \tag{E.6}
$$

$$
\frac{\partial v_\theta}{\partial \tau} + v_r \frac{\partial v_\theta}{\partial r} + \frac{v_\theta}{r} \frac{\partial v_\theta}{\partial \theta} + \frac{v_\phi}{r \sin \theta} \frac{\partial v_\theta}{\partial \phi} + \frac{v_r v_\theta}{r} - \frac{v_\phi^2 \cot \theta}{r}
$$
$$
= f_\theta - \frac{1}{\rho r} \frac{\partial p}{\partial \theta} + \nu \left\{ \nabla^2 v_\theta + \frac{2}{r^2} \frac{\partial v_r}{\partial \theta} - \frac{v_\theta}{r^2 \sin^2 \theta} - \frac{2 \cos \theta}{r^2 \sin^2 \theta} \frac{\partial v_\phi}{\partial \phi} \right\} \tag{E.7}
$$

$$
\frac{\partial v_\phi}{\partial \tau} + v_r \frac{\partial v_\phi}{\partial r} + \frac{v_\theta}{r} \frac{\partial v_\phi}{\partial \theta} + \frac{v_\phi}{r \sin \theta} \frac{\partial v_\phi}{\partial \phi} + \frac{v_r v_\phi}{r} + \frac{v_\theta v_\phi \cot \theta}{r}
$$
$$
= f_\phi - \frac{1}{\rho r \sin \theta} \frac{\partial p}{\partial \phi} + \nu \left\{ \nabla^2 v_\phi - \frac{v_\phi}{r^2 \sin^2 \theta} + \frac{2}{r^2 \sin \theta} \frac{\partial v_r}{\partial \phi} + \frac{2 \cos \theta}{r^2 \sin^2 \theta} \frac{\partial v_\theta}{\partial \phi} \right\} \tag{E.8}
$$

where f_i is some type of body force such as buoyancy, and

$$
\nabla^2 \equiv \frac{1}{r^2} \frac{\partial}{\partial r} \left(r^2 \frac{\partial}{\partial r} \right) + \frac{1}{r^2 \sin \theta} \frac{\partial}{\partial \theta} \left(\sin \theta \frac{\partial}{\partial \theta} \right) + \frac{1}{r^2 \sin^2 \theta} \frac{\partial^2}{\partial \phi^2}
$$

The energy equation for an incompressible fluid can be written as

$$
\rho c_p \left(\frac{\partial t}{\partial \tau} + v_r \frac{\partial t}{\partial r} + \frac{v_\theta}{r} \frac{\partial t}{\partial \theta} + \frac{v_\phi}{r \sin \theta} \frac{\partial t}{\partial \phi} \right)
$$
$$
= \frac{1}{r^2} \frac{\partial}{\partial r} \left(r^2 k \frac{\partial t}{\partial r} \right) + \frac{1}{r^2 \sin \theta} \frac{\partial}{\partial \theta} \left(k \sin \theta \frac{\partial t}{\partial \theta} \right) + \frac{1}{r^2 \sin^2 \theta} \frac{\partial}{\partial \phi} \left(k \frac{\partial t}{\partial \phi} \right) + \dot{q} + \Phi \tag{E.9}
$$

where

$$
\Phi = 2\mu \left\{ \begin{array}{l} \left(\dfrac{\partial v_r}{\partial r} \right)^2 + \left[\dfrac{1}{r} \left(\dfrac{\partial v_\theta}{\partial \theta} + v_r \right) \right]^2 \\[2mm] + \dfrac{1}{r^2} \left(\dfrac{1}{\sin \theta} \dfrac{\partial v_\phi}{\partial \phi} + v_r + v_\theta \cot \theta \right)^2 + \dfrac{1}{2} \left[r \dfrac{\partial}{\partial r} \left(\dfrac{v_\theta}{r} \right) + \dfrac{1}{r} \dfrac{\partial v_r}{\partial \theta} \right]^2 \\[2mm] + \dfrac{1}{2} \left[\dfrac{1}{r \sin \theta} \dfrac{\partial v_r}{\partial \phi} + r \dfrac{\partial}{\partial r} \left(\dfrac{v_\phi}{r} \right) \right]^2 + \dfrac{1}{2} \left[\dfrac{\sin \theta}{r} \dfrac{\partial}{\partial \theta} \left(\dfrac{v_\phi}{\sin \theta} \right) + \dfrac{1}{r \sin \theta} \dfrac{\partial v_\theta}{\partial \phi} \right]^2 \end{array} \right\}
$$
$$
\tag{E.10}
$$

Appendix F

Gas Properties

Table F.1a Properties of Air (English Units)

T (°F)	ρ (lb/ft^3)	c_p (Btu/lb-°F)	μ (lb$_m$/hr-ft)	k (Btu/hr-ft-°F)	Pr
−9	0.08707	0.2403	0.0386	0.0129	0.720
81	0.07251	0.2405	0.0447	0.0152	0.707
171	0.06212	0.2410	0.0504	0.0173	0.700
261	0.05438	0.2422	0.0557	0.0195	0.690
351	0.04832	0.2439	0.0607	0.0216	0.686
441	0.04348	0.2460	0.0654	0.0235	0.684
531	0.03951	0.2484	0.0698	0.0254	0.683
621	0.03623	0.2510	0.0740	0.0271	0.685
711	0.03344	0.2539	0.0781	0.0287	0.690
801	0.03106	0.2567	0.0820	0.0303	0.695
891	0.02899	0.2596	0.0859	0.0317	0.702
981	0.02718	0.2625	0.0895	0.0331	0.709
1071	0.02558	0.2651	0.0931	0.0344	0.716
1161	0.02415	0.2677	0.0964	0.0358	0.720
1251	0.02289	0.2701	0.0996	0.0372	0.723
1341	0.02174	0.2725	0.1028	0.0385	0.726
1521	0.01977	0.2768	0.1087	0.0413	0.728
1701	0.01812	0.2806	0.1145	0.0441	0.728
1881	0.01673	0.2840	0.1201	0.0474	0.719
2061	0.01553	0.2883	0.1283	0.0526	0.703
2241	0.01450	0.2938	0.1349	0.0578	0.685

Source: Table A.8 on pp. A-32–A-33 of *The CRC Handbook of Mechanical Engineering*, ed. F Kreith. Boca Raton, FL: CRC Press, 1998.

Table F.1b   Properties of Air (Metric Units)

T (K)	ρ (kg/m^3)	c_p [kJ/(kg·K)]	$\mu \times (10^7)$ [(N·s/m^2)]	$\lambda \times (10^3)$ [(W/(m·K)]	Pr
100	3.5562	1.032	71.1	9.34	0.786
150	2.3364	1.012	103.4	13.8	0.758
200	1.7458	1.007	132.5	18.1	0.737
250	1.3947	1.006	159.6	22.3	0.720
300	1.1614	1.007	184.6	26.3	0.707
350	0.9950	1.009	208.2	30.0	0.700
400	0.8711	1.014	230.1	33.8	0.690
450	0.7740	1.021	250.7	37.3	0.686
500	0.6964	1.030	270.1	40.7	0.684
550	0.6329	1.040	288.4	43.9	0.683
600	0.5804	1.051	305.8	46.9	0.685
650	0.5356	1.063	322.5	49.7	0.690
700	0.4975	1.075	338.8	52.4	0.695
750	0.4643	1.087	354.6	54.9	0.702
800	0.4354	1.099	369.8	57.3	0.709
850	0.4097	1.110	384.3	59.6	0.716
900	0.3868	1.121	398.1	62.0	0.720
950	0.3666	1.131	411.3	64.3	0.723
1000	0.3482	1.141	424.4	66.7	0.726
1100	0.3166	1.159	449.0	71.5	0.728
1200	0.2902	1.175	473.0	76.3	0.728
1300	0.2679	1.189	496.0	82	0.719
1400	0.2488	1.207	530	91	0.703
1500	0.2322	1.230	557	100	0.685

Source: From GF Hewitt, GL Shires, TR Bott, eds *Process Heat Transfer*. Boca Raton, FL: CRC Press, 1994.

Table F.2 Gas Properties

	Acetylene (Ethyne), C_2H_2	Air (mixture)	Butadiene, C_4H_6	n-Butane, C_4H_{10}
CHEMICAL AND PHYSICAL PROPERTIES				
Molecular weight	26.04	28.966	54.09	58.12
Specific gravity, air $= 1$	0.90	1.00	1.87	2.07
Specific volume, ft^3/lb	14.9	13.5	7.1	6.5
Specific volume, m^3/kg	0.93	0.842	0.44	0.405
Density of liquid (at atm bp), Ib/ft^3	43.0	54.6		37.5
Density of liquid (at atm bp), kg/m^3	693	879		604
Vapor pressure at 25 deg C, psia				35.4
Vapor pressure at 25 deg C, MN/m^2				0.0244
Viscosity (abs), Ibm/ft sec	6.72×10^{-6}	12.1×10^{-6}		4.8×10^{-6}
Viscosity (abs), centipoises[a]	0.01	0.018		0.007
Sound velocity in gas, m/sec	343	346	226	216
THERMAL AND THERMO-DYNAMIC PROPERTIES				
Specific heat, c_p, Btu/lb $\cdot$ deg F or cal/g $\cdot$ deg C	0.40	0.2403	0.341	0.39
Specific heat, c_p, J/kg $\cdot$ K	1,674	1,005.	1,427.	1,675
Specific heat ratio, c_p/c_v	1.25	1.40	1.12	1.096
Gas constant R, ft-lb/lb $\cdot$ deg F	59.3	53.3	28.55	26.56
Gas constant R, J/kg $\cdot$ deg C	319	286.8	154	143
Thermal conductivity, Btu/hr $\cdot$ ft $\cdot$ deg F	0.014	0.0151		0.01
Thermal conductivity, W/m $\cdot$ deg C	0.024	0.026		0.017
Boiling point (sat 14.7 psia), deg F	-103	-320	24.1	31.2
Boiling point (sat 760 mm), deg C	-75	-195	-4.5	-0.4
Latent beat of evap (at bp), Btu/lb	264	88.2		165.6
Latent heat of evap (at bp), J/kg	614,000	205,000		386,000
Freezing (melting) point, deg F (1 atm)	-116	-357.2	-164	-217
Freezing (melting) point, deg C (1 atm)	-82.2	-216.2	-109	-138
Latent heat of fusion, Btu/lb	23	10.0		19.2
Latent heat of fusion, J/kg	53,500	23,200		44,700
Critical temperature, deg F	97.1	-220.5		306
Critical temperature, deg C	36.2	-140.3	171	152
Critical pressure, psia	907	550	652	550
Critical pressure, MN/m^2	6.25	38		3.8
Critical volume, ft^3/lb		0.050		0.070
Critical volume, m^3/kg		0.003		0.0043
Flammable (yes or no)	Yes	No	Yes	Yes
Heat of combustion, Btu/ft^3	1,450	—	2,950	3,300
Heat of combustion, Btu/lb	21,600	—	20,900	21,400
Heat of combustion, kJ/kg	50,200	—	48,600	49,700

[a]For N$\cdot$ sec/m^2 divide by 1000.

Source: From F. Kreith ed. *The CRC Handbook of Mechanical Engineering.* Boca Ration, FL: CRC Press, 1998.

(Continued)

Table F.2   Continued

	Isobutane (2-Methyl-propane), C_4H_{10}	1-Butene (Butylene), C_4H_8	Carbon monoxide, CO	Ethane, C_2H_6
CHEMICAL AND PHYSICAL PROPERTIES				
Molecular weight	58.12	56.108	28.011	30.070
Specific gravity, air = 1	2.07	1.94	0.967	1.04
Specific volume, ft^3/lb	6.5	6.7	14.0	13.025
Specific volume, m^3/kg	0.418	0.42	0.874	0.815
Density of liquid (at atm bp), Ib/ft^3	37.2			28
Density of liquid (at atm bp), kg/m^3	599			449
Vapor pressure at 25 deg C, psia	50.4			
Vapor pressure at 25 deg C, MN/m^2	0.347			
Viscosity (abs), Ibm/ft sec			12.1×10^{-6}	64×10^{-6}
Viscosity (abs), centipoises[a]			0.018	0.095
Sound velocity in gas, m/sec	216	222	352	316
THERMAL AND THERMO-DYNAMIC PROPERTIES				
Specific heat, c_p, Btu/lb·deg F or cal/g·deg C	0.39	0.36	0.25	0.41
Specific heat, c_p, J/kg·K	1,630	1,505	1,046	1,715
Specific heat ratio, c_p/c_v	1.10	1.112	1.40	1.20
Gas constant R, ft-lb/lb·deg F	26.56	27.52	55.2	51.4
Gas constant R, J/kg·deg C	143	148	297	276
Thermal conductivity, Btu/hr·ft·deg F	0.01		0.014	0.010
Thermal conductivity, W/m·deg C	0.0017		0.024	0.017
Boiling point (sat 14.7 psia), deg F	10.8	20.6	−312.7	−127
Boiling point (sat 760 mm), deg C	−11.8	−6.3	−191.5	−88.3
Latent beat of evap (at bp), Btu/lb	157.5	167.9	92.8	210
Latent heat of evap (at bp), J/kg	366,000	391,000	216,000	488,000
Freezing (melting) point, deg F (1 atm)	−229	−301.6	−337	−278
Freezing (melting) point, deg C (1 atm)	−145	−185.3	−205	−172.2
Latent heat of fusion, Btu/lb		16.4	12.8	41
Latent heat of fusion, J/kg		38,100		95,300
Critical temperature, deg F	273	291	−220	90.1
Critical temperature, deg C	134	144	−140	32.2
Critical pressure, psia	537	621	507	709
Critical pressure, MN/m^2	3.7	4.28	3.49	4.89
Critical volume, ft^3/lb		0.068	0.053	0.076
Critical volume, m^3/kg		0.0042	0.0033	0.0047
Flammable (yes or no)	Yes	Yes	Yes	Yes
Heat of combustion, Btu/ft^3	3,300	3,150	310	
Heat of combustion, Btu/lb	21,400	21,000	4,340	22,300
Heat of combustion, kJ/kg	49,700	48,800	10,100	51,800

(Continued)

Table F.2 Continued

	Ethylene (Ethene), C_2H_4	Hydrogen, H_2	Hydrogen sulfide, H_2S	Methane CH_4
CHEMICAL AND PHYSICAL PROPERTIES				
Molecular weight	28.054	2.016	34.076	16.044
Specific gravity, air = 1	0.969	0.070	1.18	0.554
Specific volume, ft^3/lb	13.9	194	11.5	24.2
Specific volume, m^3/kg	0.87	12.1	0.0930	1.51
Density of liquid (at atm bp), Ib/ft^3	35.5	4.43	62	26.3
Density of liquid (at atm bp), kg/m^3	569	71.0	993	421
Vapor pressure at 25 deg C, psia				
Vapor pressure at 25 deg C, MN/m^2				
Viscosity (abs), Ibm/ft sec	6.72×10^{-6}	6.05×10^{-6}	8.74×10^{-6}	7.39×10^{-6}
Viscosity (abs), centipoises[a]	0.010	0.009	0.013	0.011
Sound velocity in gas, m/sec	331	1,315	302	446
THERMAL AND THERMO-DYNAMIC PROPERTIES				
Specific heat, c_p, Btu/lb · deg F or cal/g · deg C	0.37	3.42	0.23	0.054
Specific heat, c_p, J/kg · K	1,548	14,310	962	2,260
Specific heat ratio, c_p/c_v	1.24	1.405	1.33	1.31
Gas constant R, ft-lb/lb · deg F	55.1	767	45.3	96
Gas constant R, J/kg · deg C	296	4,126	244	518
Thermal conductivity, Btu/hr · ft · deg F	0.010	0.105	0.008	0.02
Thermal conductivity, W/m · deg C	0.017	0.0182	0.014	0.035
Boiling point (sat 14.7 psia), deg F	−155	−423	−76	−259
Boiling point (sat 760 mm), deg C	−103.8	20.4 K	−60	−434.2
Latent beat of evap (at bp), Btu/lb	208	192	234	219.2
Latent heat of evap (at bp), J/kg	484,000	447,000	544,000	510,000
Freezing (melting) point, deg F (1 atm)	−272	−434.6	−119.2	−296.6
Freezing (melting) point, deg C (1 atm)	−169	−259.1	−84	−182.6
Latent heat of fusion, Btu/lb	51.5	25.0	30.2	14
Latent heat of fusion, J/kg	120,000	58,000	70,200	32,600
Critical temperature, deg F	49	−399.8	213	−116
Critical temperature, deg C	9.5	−240.0	100.4	−82.3
Critical pressure, psia	741	189	1,309	673
Critical pressure, MN/m^2	5.11	1.30	9.02	4.64
Critical volume, ft^3/lb	0.073	0.53	0.046	0.099
Critical volume, m^3/kg	0.0046	0.033	0.0029	0.0062
Flammable (yes or no)	Yes	Yes	Yes	Yes
Heat of combustion, Btu/ft^3	1,480	320	700	985
Heat of combustion, Btu/lb	20,600	62,050	8,000	22,900
Heat of combustion, kJ/kg	47,800	144,000	18,600	—

(Continued)

Table F.2 Continued

	Nitric oxide, NO	Nitrogen, N$_2$	Nitrous oxide, N$_2$O	Oxygen O$_2$
CHEMICAL AND PHYSICAL PROPERTIES				
Molecular weight	30.006	28.0134	44.012	31.9988
Specific gravity, air = 1	1.04	0.967	1.52	1.105
Specific volume, ft^3/lb	13.05	13.98	8.90	12.24
Specific volume, m^3/kg	0.814	0.872	0.555	0.764
Density of liquid (at atm bp), Ib/ft^3		50.46	76.6	71.27
Density of liquid (at atm bp), kg/m^3		808.4	1,227	1,142
Vapor pressure at 25 deg C, psia				
Vapor pressure at 25 deg C, MN/m^2				
Viscosity (abs), Ibm/ft sec	12.8×10^{-6}	12.1×10^{-6}	10.1×10^{-6}	13.4×10^{-6}
Viscosity (abs), centipoises[a]	0.019	0.018	0.015	0.020
Sound velocity in gas, m/sec	341	353	268	329
THERMAL AND THERMO-DYNAMIC PROPERTIES				
Specific heat, c_p, Btu/lb · deg F or cal/g · deg C	0.235	0.249	0.21	0.220
Specific heat, c_p, J/kg · K	983	1,040	879	920
Specific heat ratio, c_p/c_p	1.40	1.40	1.31	1.40
Gas constant R, ft-lb/lb · deg F	51.5	55.2	35.1	48.3
Gas constant R, J/kg · deg C	277	297	189	260
Thermal conductivity, Btu/hr · ft · deg F	0.015	0.015	0.010	0.015
Thermal conductivity, W/m · deg C	0.026	0.026	0.017	0.026
Boiling point (sat 14.7 psia), deg F	−240	−320.4	−127.3	−297.3
Boiling point (sat 760 mm), deg C	−151.5	−195.8	−88.5	−182.97
Latent beat of evap (at bp), Btu/lb		85.5	161.8	91.7
Latent heat of evap (at bp), J/kg		199,000	376,000	213,000
Freezing (melting) point, deg F (1 atm)	−258	−346	−131.5	−361.1
Freezing (melting) point, deg C (1 atm)	−161	−210	−90.8	−218.4
Latent heat of fusion, Btu/lb	32.9	11.1	63.9	5.9
Latent heat of fusion, J/kg	76,500	25,800	149,000	13,700
Critical temperature, deg F	−136	−232.6	97.7	−181.5
Critical temperature, deg C	−93.3	−147	36.5	−118.6
Critical pressure, psia	945	493	1,052	726
Critical pressure, MN/m^2	6.52	3.40	7.25	5.01
Critical volume, ft^3/lb	0.0332	0.051	0.036	0.040
Critical volume, m^3/kg	0.00207	0.00318	0.0022	0.0025
Flammable (yes or no)	No	No	No	No
Heat of combustion, Btu/ft^3	—	—	—	—
Heat of combustion, Btu/lb	—	—	—	—
Heat of combustion, kJ/kg	–	—	—	—

(Continued)

Table F.2 Continued

	Ozone, O_3	Propane, C_3H_8	Propylene (Propene), C_3H_6	Sulfur dioxide, SO_2
CHEMICAL AND PHYSICAL PROPERTIES				
Molecular weight	47.998	44.097	42.08	64.06
Specific gravity, air = 1	1.66	1.52	1.45	2.21
Specific volume, ft^3/lb	8.16	8.84	9.3	6.11
Specific volume, m^3/kg	0.509	0.552	0.58	
Density of liquid (at atm bp), Ib/ft^3		36.2	37.5	42.8
Density of liquid (at atm bp), kg/m^3		580	601.1	585
Vapor pressure at 25 deg C, psia		135.7	166.4	56.6
Vapor pressure at 25 deg C, MN/m^2		0.936	1.147	0.390
Viscosity (abs), Ibm/ft sec	8.74×10^{-6}	53.8×10^{-6}	57.1×10^{-6}	8.74×10^{-6}
Viscosity (abs), centipoises[a]	0.013	0.080	0.085	0.013
Sound velocity in gas, m/sec		253	261	220
THERMAL AND THERMO-DYNAMIC PROPERTIES				
Specific heat, c_p, Btu/lb $\cdot$ deg F or cal/g $\cdot$ deg C	0.196	0.39	0.36	0.11
Specific heat, c_p, J/kg $\cdot$ K	820	1,630	1,506	460
Specific heat ratio, c_p/c_v		1.2	1.16	1.29
Gas constant R, ft-lb/lb $\cdot$ deg F	32.2	35.0	36.7	24.1
Gas constant R, J/kg $\cdot$ deg C	173	188	197	130
Thermal conductivity, Btu/hr $\cdot$ ft $\cdot$ deg F	0.019	0.010	0.010	0.006
Thermal conductivity, W/m $\cdot$ deg C	0.033	0.017	0.017	0.010
Boiling point (sat 14.7 psia), deg F	-170	-44	-54	14.0
Boiling point (sat 760 mm), deg C	-112	-42.2	-48.3	-10
Latent heat of evap (at bp), Btu/lb		184	188.2	155.5
Latent heat of evap (at bp), J/kg		428,000	438,000	362,000
Freezing (melting) point, deg F (1 atm)	-315.5	-309.8	-301	-104
Freezing (melting) point, deg C (1 atm)	-193	-189.9	-185	-75.5
Latent heat of fusion, Btu/lb	97.2	19.1		58.0
Latent heat of fusion, J/kg	226,000	44,400		135,000
Critical temperature, deg F	16	205	197	315.5
Critical temperature, deg C	-9	96	91.7	157.6
Critical pressure, psia	800	618	668	1,141
Critical pressure, MN/m^2	5.52	4.26	4.61	7.87
Critical volume, ft^3/lb	0.0298	0.073	0.069	0.03
Critical volume, m^3/kg	0.00186	0.0045	0.0043	0.0019
Flammable (yes or no)	No	Yes	Yes	No
Heat of combustion, Btu/ft^3	—	2,450	2,310	—
Heat of combustion, Btu/lb	—	21,660	21,500	—
Heat of combustion, kJ/kg	—	50,340	50,000	—

(Continued)

Table F.2 Continued

	cis-2-Butene, C₄H₈	*trans*-2-Butene C₄H₄	Isobutene, C₄H₈	Carbon dioxide, CO₂
CHEMICAL AND PHYSICAL PROPERTIES				
Molecular weight	56.108	56.108	56.108	44.01
Specific gravity, air = 1	1.94	1.94	1.94	1.52
Specific volume, ft³/lb	6.7	6.7	6.7	8.8
Specific volume, m³/kg	0.42	0.42	0.42	0.55
Density of liquid (at atm bp), Ib/ft³				—
Density of liquid (at atm bp), kg/m³				—
Vapor pressure at 25 deg C, psia				931
Vapor pressure at 25 deg C, MN/m²				6.42
Viscosity (abs), Ibm/ft sec				9.4×10^{-6}
Viscosity (abs), centipoises[a]				0.014
Sound velocity in gas, m/sec	223	221	221	270
THERMAL AND THERMO-DYNAMIC PROPERTIES				
Specific heat, c_p, Btu/lb · deg F or cal/g · deg C	0.327	0.365	0.37	0.205
Specific heat, c_p, J/kg · K	1,368	1,527	1,548	876
Specific heat ratio, c_p/c_v	1.121	1.107	1.10	1.30
Gas constant R, ft-lb/lb · deg F				35.1
Gas constant R, J/kg · deg C				189
Thermal conductivity, Btu/hr · ft · deg F				0.01
Thermal conductivity, W/m · deg C				0.017
Boiling point (sat 14.7 psia), deg F	38.6	33.6	19.2	-109.4^{b}
Boiling point (sat 760 mm), deg C	3.7	0.9	−7.1	−78.5
Latent heat of evap (at bp), Btu/lb	178.9	174.4	169	246
Latent heat of evap (at bp), J/kg	416,000	406,000	393,000	572,000
Freezing (melting) point, deg F (1 atm)	−218	−158		
Freezing (melting) point, deg C (1 atm)	−138.9	−105.5		
Latent heat of fusion, Btu/lb	31.2	41.6	25.3	—
Latent heat of fusion, J/kg	72,600	96,800	58,800	—
Critical temperature, deg F				88
Critical temperature, deg C	160	155		31
Critical pressure, psia	595	610		1,072
Critical pressure, MN/m²	4.10	4.20		7.4
Critical volume, ft³/lb				
Critical volume, m³/kg				
Flammable (yes or no)	Yes	Yes	Yes	No
Heat of combustion, Btu/ft³	3,150	3,150	3,150	—
Heat of combustion, Btu/lb	21,000	21,000	21,000	—
Heat of combustion, kJ/kg	48,800	48,00	48,800	—

[b]Sublimes.

Table F.3a Requirements for fuel oils (per ASTM D 396)

Classification	No. 1 Distillate	No. 2 Distillate	No. 4 Distillate (Heavy)	No. 6 Residual
Density (kg/m³) @ 60°F (15°C), max	850	876	—	—
Viscosity @ 104°F (40°C) mm/s²				
min	1.3	1.9	> 5.5	—
max	2.1	3.4	24	—
Viscosity @ 212°F (100°C) mm/s²				
min	—	—	—	—
max	—	—	—	—
Flash point °F (°C), min	100 (38)	100 (38)	131 (55)	140 (60)
Pour point °F (°C), max	−0.4 (−18)	21 (−6)	21 (−6)	—
Ash, % mass, max	—	—	0.01	—
Sulfur, % mass, max	0.5	0.5	—	—
Water sediment, % vol., max	0.05	0.05	0.5	2.0
Distillation temperature °F (°C)				
0% volume recovered, max	419 (215)	—	—	—
90% volume recovered, min	—	540 (282)	—	—
90% volume recovered, max	550 (228)	640 (338)	—	—

Source: CE Baukal, ed. *The John Zink Combustion Handbook.* Boca Raton, FL: CRC Press, 2001.

Table F.3b Typical analysis of different fuel oils

	No. 1 Fuel oil	No. 2 Fuel oil	No. 4 Fuel oil	No. 6 Fuel oil (sour)
Ash %	< 0.01	< 0.01	0.02	0.05
Hydrogen %	13.6	13.6	11.7	11.2
Nitrogen %	0.003	0.007	0.24	0.37
Sulfur %	0.09	0.1	1.35	2.1
Carbon %	86.4	86.6	86.5	85.7
Heat of combustion (HHV), Btu/lb	20,187	19,639	19,382	18,343
Specific gravity 60/60°F	0.825	0.84	0.898	0.97
Density (lb/U.S.gal	6.877	6.96	7.488	8.08

Source: CE Baukal, ed. *The John Zink Combustion Handbook.* Boca Raton, FL: CRC Press, 2001.

Appendix G

EPA Sample Methods

Method	Title
1	Sample and velocity traverses for stationary sources
1A	Sample and velocity traverses for stationary sources with small stacks or ducts
2	Determination of stack gas velocity and volumetric flow rate (Type S pitot tube)
2A	Direct measurement of gas volume through pipes and small ducts
2B	Determination of exhaust gas volume flow rate from gasoline vapor incinerators
2C	Determination of gas velocity and volumetric flow rate in small stacks or ducts (standard pitot tube)
2D	Measurement of gas volume flow rate in small pipes and ducts
2E	Determination of landfill gas production flow rate
2F	Determination of stack gas velocity and volumetric flow rate with three-dimensional probes
2G	Determination of stack gas velocity and volumetric flow rate with two-dimensional probes
2H	Determination of stack gas velocity taking into account velocity decay near the stack wall
3	Gas analysis for the determination of dry molecular weight
3A	Determination of oxygen and carbon dioxide concentrations in emissions from stationary sources (instrumental analyzer procedure)
3B	Gas analysis for the determination of emission rate correction factor or excess air
3C	Determination of carbon dioxide, methane, nitrogen, and oxygen from stationary sources
4	Determination of moisture content in stack gases
5	Determination of particulate matter emissions from stationary sources
5A	Determination of particulate matter emissions from asphalt processing and asphalt roofing industry

12	Determination of inorganic lead emissions from stationary sources
13A	Determination of total fluoride emissions from stationary sources—SPADNS zirconium lake method
13B	Determination of total fluoride emissions from stationary sources—specific ion electrode method
14	Determination of fluoride emissions from potroom roof monitors for primary aluminum plants
14A	Determination of total fluoride emissions from selected sources at primary aluminum production facilities
15	Determination of hydrogen sulfide, carbonyl sulfide, and carbon disulfide emissions from stationary sources
15A	Determination of total reduced sulfur emissions from sulfur recovery plants in petroleum refineries
16	Semicontinuous determination of sulfur emissions from stationary sources
16A	Determination of total reduced sulfur emissions from stationary sources (impinger technique)
17	Determination of particulate emissions from stationary sources (in-stack filtration method)
18	Measurement of gaseous organic compounds by gas chromatography
19	Determination of sulfur dioxide removal efficiency and particulate, sulfur dioxide, and nitrogen oxides emission rates
20	Determination of nitrogen oxides, sulfur dioxide, and diluent emissions from stationary gas turbines
21	Determination of volatile organic compound leaks
22	Visual determination of fugitive emissions from material sources and smoke emissions from flares
23	Determination of polychlorinated dibenzo-p-dioxins and polychlorinated dibenzofurans from stationary sources
24	Determination of volatile matter content, water content, density, volume solids, and weight solids of surface coatings
24A	Determination of volatile matter content and density of printing inks and related coatings
25	Determination of total gaseous nonmethane organic emissions as carbon
25A	Determination of total gaseous organic concentration using a flame ionization analyzer
25B	Determination of total gaseous organic concentration using a nondispersive infrared analyzer
25C	Determination of nonmethane organic compounds (NMOC) in MSW landfill gases
25D	Determination of volatile organic concentration of waste samples
25E	Determination of vapor phase organic concentration in waste samples
26	Determination of hydrogen chloride emissions from stationary sources

Appendix H

EPA Performance Specifications

Glossary

Absorption	transfer of molecules from the bulk of the gas to a liquid surface followed by diffusion to the bulk liquid
adsorption	attraction of molecules from a gas to a solid surface
BACT	best available control technology
BART	best available retrofit technology
CAA	Clean Air Act
CAAA	Clean Air Act Amendments (1990)
CEMS	continuous emission monitoring system
CFR	Code of Federal Regulations
CO	carbon monoxide
dBA	decibels, A-weighted (noise)
DOE	U.S. Department of Energy
EPA	U.S. Environmental Protection Agency (Washington, DC)
FlGR	flue gas recirculation
FuGR	furnace gas recirculation
HAP	hazardous air pollutant
LAER	lowest achievable emission rate
MACT	maximum achievable control technology
NAAQS	National Ambient Air Quality Standards
NO_x	oxides of nitrogen (NO, NO_2, N_2O, etc.)
NSPS	New Source Performance Standards (CAA)
opacity	measure of the fraction of light attenuated by suspended particulates
PIC	product of incomplete combustion
PM	particulate matter
PM_{10}	particulate matter $10\,\mu m$ in diameter or smaller
$PM_{2.5}$	particulate matter $2.5\,\mu m$ in diameter or smaller
POM	polycyclic organic matter
ppm	parts per million
ppmv	parts per million, volume basis
ppmvd	parts per million, volume dry basis
ppmvw	parts per million, volume wet basis

RACT	reasonably available control technology
SCR	selective catalytic reduction
smoke	small gasborne particulates (primarily consisting of carbon) resulting from incomplete combustion
SNCR	selective noncatalytic reduction
SO_x	oxides of sulfur (primarily SO_2 and SO_3)
UHC	unburned hydrocarbon
VOC	volatile organic compound

Index